THE OXFORD AUTHORS

General Editor: Frank Kermode

JOHN DRYDEN (1631–1700) developed late as a poet. His poems written after the Restoration of Charles II in 1660 won him the office of Poet Laureate (1668) and for the period 1660 to 1685 he may be said to have dominated the literary scene, especially with his satires *MacFlecknoe* (probably written 1676), *Absalom and Achitophel* (1681), and *The Medal* (1682). He turned briefly to religious engagement with *Religio Laici* (1682) and *The Hind and the Panther* (1687). On the accession of William III in 1688, Dryden, as a convert to Roman Catholicism, could no longer comment publicly upon public affairs, and turned increasingly towards translation as a means of expression and sometimes covert commentary on politics and society. In his later years he formed an association with Jacob Tonson which was to initiate the future model for relationships between writers and publishers. The products of this relationship culminated in the complete translation of Virgil (1697) and *Fables Ancient and Modern*, published in 1700, the year of Dryden's death.

Throughout his life Dryden also wrote plays of disputable value, and, in a large body of prose commentary, he began the practice of literary criticism in England.

KEITH WALKER is Senior Lecturer in English Language and Literature at University College London. He has written on eighteenth-century literature, and has edited the poems of John Wilmot Earl of Rochester.

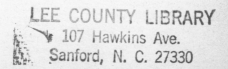

THE OXFORD AUTHORS

JOHN DRYDEN

EDITED BY
KEITH WALKER

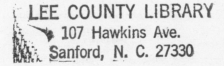
Oxford　New York
OXFORD UNIVERSITY PRESS
1987

Oxford University Press, Walton Street, Oxford OX2 6DP

Oxford New York Toronto
Delhi Bombay Calcutta Madras Karachi
Petaling Jaya Singapore Hong Kong Tokyo
Nairobi Dar es Salaam Cape Town
Melbourne Auckland

and associated companies in
Beirut Berlin Ibadan Nicosia

Oxford is a trade mark of Oxford University Press

Introduction, edited text, and editorial matter © Keith Walker 1987

British Library Cataloguing in Publication Data

Dryden, John, 1631–1700
John Dryden.—(The Oxford authors)
I. Title II. Walker, Keith, 1936–
821'.4 PR3410

ISBN 0–19–254192–7
ISBN 0–19–281402–8 Pbk

Library of Congress Cataloging in Publication Data

Dryden, John, 1631–1700.
John Dryden.
(The Oxford authors)
Bibliography: p. Includes index.
I. Walker, Keith, 1936– .
II. Title. III. Series.
PR3412.W35 1986 821'.4 85-32000

ISBN 0–19–254192–7
ISBN 0–19–281402–8 (pbk.)

Set by Eta Services Ltd.
Printed in Great Britain by
Richard Clay Ltd.
Bungay, Suffolk

CONTENTS

Italics indicate a prose work

viii   CONTENTS

INTRODUCTION

DRYDEN'S poetry is straightforward, bold, and energetic. He was in the public eye for some forty years, holding positions at court for a long period of that time. He was indisputably perceived as the leading writer of his day. He excelled in all the types of writing practised at the time. He wrote more, and in more genres than anyone. He accumulated to himself (it is an odd distinction) a huge mass of attacks, ranging from the reasoned to the scabrous. Dryden explained his attitudes and intentions in a large number of prologues, epilogues, prefaces, defences, and vindications—thereby quite casually producing the first body of what we now call 'criticism' in English. And yet his life and character remain something of a mystery. The manuscript of John Aubrey's 'Brief Life' of Dryden is headed 'John Dryden, esq., Poet Laureate. He will write it for me himself'. The rest of the page remains blank.

1631–1659

Dryden was born in Northamptonshire at Aldwincle where his maternal grandfather was rector. His family was puritan. His cousin, Sir Gilbert Pickering, was to be one of the judges at Charles I's trial, and high in Cromwell's favour. Gilbert was a 'committee-man' of the Commonwealth, responsible for ejecting ministers and sequestrating estates on behalf of the government.

Dryden tells us that as a child he read Spenser and Sylvester's Du Bartas and an English translation of the Greek historian of Rome, Polybius. That apart, we know nothing of his early education until he went up to Westminster School as a King's Scholar, perhaps in 1645. Here he received a thorough education in the classics from Dr Richard Busby. Dryden later wrote in the argument to his translation of the Third Satire of Persius, 'I translated this satire when I was a King's Scholar at Westminster school for a Thursday night's exercise and believe that it, and many other of my exercises of this nature in English verse are still in the hands of my learned master, the Reverend Dr Busby.' Before Dryden left Westminster he published his first poem, an elegy for Lord Hastings, among thirty-three similar poems, some by fellow schoolboys.

In 1650 Dryden went up to Cambridge, not to his father's college which was the puritan Emmanuel, but to Trinity. Trinity College was at the time at the forefront of the study of the 'new science', which was not formally part of the University curriculum (that was dominated by Aristotle), and his manifest interest in the natural sciences may have been stimulated while he was there.

He took his BA in 1654. A fellow Trinity-man was to remember later:

Dryden . . . was reckoned a man of good parts and learning while in college; he had . . . read over and very well understood all the Greek and Latin poets. He stayed to take his bachelor's degree; but his head was too roving and active or what else you'll call it, to confine himself to a college life, and so he left it and went to London into gayer company and set up for a poet.[1]

His father's death in 1654 left him a modest, but never adequate, income. He was to live in London until his death, returning to Northampton for vacations from time to time. Almost uniquely among the English poets, he never travelled outside England.

In London Dryden worked as a secretary for the protectorate. He appeared alongside Marvell and Milton at Cromwell's funeral in 1658 and the next year published *Heroic Stanzas* on the death of Cromwell, which came out alongside elegies by Edmund Waller and Thomas Sprat, later historian of the Royal Society. *Heroic Stanzas* was the first work of Dryden's maturity as a poet. It was sometimes represented as containing praise for regicide and Dryden was attacked for the poem at the restoration, but he never disowned it, and his publisher issued a reprint about 1692.

1660–1678

At the restoration of the monarchy Dryden, like any civil servant today when a new government comes to power, transferred his allegiance to the new regime. He welcomed the new government in two panegyrics to the king, and one each to the Lord Chancellor and to Sir Robert Howard, a staunch royalist whose sister, Lady Elizabeth, he was to marry in 1663. She bore him three sons, and outlived him.

The poems to the king and others suggest that Dryden was looking for a patron. Although he could hardly have realized the fact at the time, he had already found a patron of sorts in his publisher Henry Herringman for whom he worked from before 1660 until 1679. Dryden's career spans a period when poets were ceasing to write for

[1] *Notes and Queries*, 4th series, 10 (1872), p. 370.

aristocratic patrons, and finding their living in writing for publishers and thus ultimately for the reading public.

In 1662 Dryden was elected an early fellow of the Royal Society and busied himself with plans to found a British Academy after the French model. The plans came to nothing.

With the opening of the theatres after the puritan ban there came the possibility of a new kind of employment. Dryden's first play, the comedy *The Wild Gallant*, failed in 1663, and he did not write another comedy for nearly six years, but his tragicomedy *The Rival Ladies* had better success the following year.

Dryden established himself as the leading dramatist of the period with *The Indian Queen* written in collaboration with Sir Robert Howard (1664) and its sequel *The Indian Emperor* (1665) which was written partly to recover some of the money spent on scenery and costumes, as the prologue engagingly admits. *The Indian Queen* and *The Indian Emperor* were heroic tragedies of the kind Dryden was later to be particularly associated with. The line was continued with *Tyrannic Love* (1669) and *The Conquest of Granada* in two parts (1671, 1672), which pushed the heroic play as far into heroic absurdity as it was capable of going. *Aureng-Zebe* followed in 1675, Dryden's last play in rhymed couplets for some time.

There were two London theatrical companies. At first, Dryden worked for the King's Company of which he was a shareholder and for which he engaged himself to write three plays a year, a promise he proved unable to keep. He fell out with the King's Company in 1678 and moved to the better managed Duke's Company.

During the 1660s and the 1670s we may suppose Dryden to have made the bulk of his earned income from the theatre (a playwright received the profits from the third night of a play's run), but he found the stage uncongenial. He wrote many years later:

Having been longer acquainted with the stage than any poet now living, and having observed how difficult it was to please; that the humours of comedy were almost spent; that love and honour (the mistaken topics of tragedy) were quite worn out; that the theatres could not support their charges; that the audience forsook them; that young men without learning set up for judges, and that they talked loudest who understood the least; all these discouragements had not only weaned me from the stage, but had also given me a loathing of it.[2]

---

[2] Preface to *Don Sebastian* (1690), in '*Of Dramatic Poesy' and Other Critical Essays*, ed. George Watson (London, 1962), vol. ii. p. 45.

We should, however, note the range and experimental variety of Dryden's output for the stage. Between 1663 and his death he wrote, collaborated in, or adapted nearly thirty plays, including comedies, tragedies, political plays, operas, and a masque.

In 1671 *The Rehearsal* by the Duke of Buckingham and others was acted. It satirized the heroic plays, and seems to have had a vogue alongside the plays which it mocked. Dryden appears as 'Mr Bays', a nickname which stuck. Dryden was to have a terrible revenge ten years later, portraying Buckingham as 'Zimri' in *Absalom and Achitophel*.

In 1665 the Great Plague closed the theatres. Dryden retreated to Charlton in Wiltshire, the seat of his father-in-law, the Earl of Berkshire, where he had leisure to read and reflect. He wrote *An Essay of Dramatic Poesy* here, and in 1666 he made a bold bid for poetic fame in a 'modern' epic (that is, without mythology and legend), *Annus Mirabilis*. The publication of Milton's *Paradise Lost* two months after *Annus Mirabilis* we may suppose effectively pre-empted the possibility of Dryden's ever fulfilling his often proclaimed intention of writing an epic on classical lines. Still, he was almost the only poet of his time to have taken proper note of *Paradise Lost*. It is true that he turned it into a rhymed 'opera' (really a dramatic poem), but the presence of *Paradise Lost* broods over and enriches *MacFlecknoe* and *Absalom and Achitophel*.

In 1668 Dryden was rewarded with the office of Poet Laureate, and two years later with the additional office of Historiographer Royal. These posts brought a welcome, if irregularly paid, addition to his income.

1678–1688

Probably in 1676, spurred on by some wounding reference to him in the preface to Thomas Shadwell's play *The Virtuoso*, Dryden began a period of heroic satires with *MacFlecknoe* (which was not immediately published). This line of satire (as we can conveniently call it, though the poems are very different from one another) continued with *Absalom and Achitophel* (1681) and *The Medal* (1682). *MacFlecknoe*'s concerns may loosely be described as the politics of the theatre. The other two satires are very much concerned with the politics of the nation in the troubled years 1679–83, years of 'the popish plot', attempts to prevent the succession to the throne of the king's brother James on the grounds that he was a Roman Catholic, and the births of the Tory and Whig parties. Dryden's position was one of support for legitimacy, the king, and the duke. He was to remain recklessly loyal to James until his death.

Possibly out of this loyalty to James, Dryden converted to Roman Catholicism some time in the middle 1680s. Certainly his poems record a mind seriously occupied with questions of religion, faith, and authority. He wrote *Religio Laici* (1682) from the position of a member of the Church of England; *The Hind and the Panther* (1687) celebrates his conversion to the Church of Rome.

1688–1700

James was deposed in 1688; catholics had been a legally disadvantaged minority even under the catholic James. Under the fiercely protestant William of Orange Dryden could not keep his public offices, but was forced to turn away from public concerns and involvement in public affairs. The despised Thomas Shadwell succeeded him in the office of Poet Laureate. Despite a law which forbade catholics to live in London, Dryden lived on, apparently tolerated. He wrote the occasional play and poem, but the bulk of his writing consisted in translating, an occupation he found more congenial than the writing of plays. It has recently been suggested that Dryden, aided by Jacob Tonson, his publisher from 1679, acted as the Poundian figure of his age, 'invigorating the talents of others', a group of young translators, among them Addison and Congreve, whose work was to bear fruit in Samuel Garth's great composite translation of Ovid's *Metamorphoses* in 1717, in which Dryden's translations figure largely.

Dryden's pension was £1,075 in arrears when Charles II died in 1685. It was not possible for him to write poems which commented on public affairs. A few poems were commissioned, among them *A Song for St Cecilia's Day*, 1687, and *Eleonora* (1692), but Dryden had to turn to translating to support his family. His translations form the vast bulk of his poems, totalling nearly 40,000 lines. In the last two decades of the seventeenth century, Dryden translated Virgil complete; from Ovid, two whole books of the *Metamorphoses*, and parts of six others, along with Book I of *The Art of Love*, three *Epistles* and three books of the *Amores*; substantial passages of Lucretius; Persius complete; five Satires of Juvenal; four poems of Horace and four of Theocritus; and the first book of the *Iliad*. Besides, there are translations of three tales from Boccaccio, rehandlings of four tales from Chaucer, and a mass of prose translations, mainly from French, but including parts of Polybius, Plutarch, and Tacitus.

The summit of Dryden's career as a translator was *The Works of Virgil* (1697), which was published by subscription. This was not a new method, for John Ogilby's *Virgil* (1654) had employed it; but it had never before achieved so much success. Dryden wrote to William Walsh in 1693:

I propose to do it by subscription, having a hundred and two brass cuts [plates], with the coats of arms of the subscriber to each cut; and every subscriber to pay five guineas: half in hand; besides another inferior subscription of two guineas for the rest whose names are only written in a catalogue printed with the book.[3]

The publication of the translation of Virgil was a national event, and may have consoled Dryden for the epic he had always wished to write. It brought him £1,400 and showed the means by which Pope, a boy of just nine at the time, would achieve financial independence as a poet.

Dryden's last volume, *Fables*, is more relaxed. It seems a miscellany of translations with a few original poems. But the plan of the volume we might think based on that of Ovid's *Metamorphoses*, that collection of interlocked stories in which narratives melt into one another. For all its seeming casual diversity, *Fables* forms a unity, beautifully bringing together the concerns of Dryden's last years. Dryden wrote to his kinswoman Elizabeth Stewart:

In the meantime, betwixt my intervals of physic and other remedies which I am using for my gravel [a condition characterized by difficulty in passing urine], I am still drudging on: always a poet, and never a good one. I pass my time sometimes with Ovid, and sometimes with our old English poet Chaucer; translating such stories as best please my fancy; and intend besides them to add somewhat of my own . . .[4]

Dryden died in 1700 two months after *Fables* was published.

---

[3] *The Letters of John Dryden*, ed. Charles E. Ward (Chapel Hill, NC, 1942), p. 64.
[4] *Letters*, p. 109.

ACKNOWLEDGEMENTS

DRYDEN's editors have served him well. I am indebted to most of them, in particular James Kinsley, George Watson, and the editors of the Californian Dryden.

Friends and colleagues have given me ready help in the preparation of this book. For information, comments, and other assistance I would like to give my thanks to Elsie Duncan-Jones, David Edmonds, Hilary Feldman, Alan Gaylord, Paul Hammond, Brian Oatley, Kenneth Palmer, Oonagh Sayce, James A. Winn, David Wykes, and especially Henry Woudhuysen.

CHRONOLOGY

1631   Dryden born (9 August OS) at Aldwincle, North-amptonshire.

1645?   Enters Westminster School as King's Scholar.

1649   First poem published, *Upon the Death of Lord Hastings.*

1650   Admitted as pensioner at Trinity College, Cambridge.

c.1657 In the service of the Protectorate.

1659   *Heroic Stanzas* on the death of Cromwell.

1660   Charles II restored. *Astraea Redux.* Poem to Sir Robert Howard.

1662   Elected to the Royal Society. Poem to Dr Charleton.

1663   Marries Lady Elizabeth Howard, sister of Sir Robert Howard. First play, *The Wild Gallant*, acted.

1664   *The Rival Ladies.*

1666   At Charlton where he writes *An Essay of Dramatic Poesy* and *Annus Mirabilis.* Son Charles born.

1667   *Annus Mirabilis* published.

1668   *An Essay of Dramatic Poesy* published. Son John born. Dryden appointed Poet Laureate. Shareholder in King's Company.

1669   Dryden appointed Historiographer Royal. Son Erasmus-Henry born.

1670   *The Conquest of Granada, Part One* performed.

1671   *The Conquest of Granada, Part Two* performed. Dryden satirized in *The Rehearsal* by Buckingham and others. *Marriage à la Mode* performed (published 1673).

1675   *Aureng-Zebe* performed (published 1676).

1676   Writes *MacFlecknoe.*

1677   *All for Love* (based on *Antony and Cleopatra*) performed (published 1678).

1679   Dryden breaks with his publisher Henry Herringman. Begins association with Jacob Tonson. Beaten up by hired thugs near Covent Garden.

1680   Dryden edits *Ovid's Epistles* for Tonson. This volume contains Dryden's first verse translations. *The Spanish Friar* performed (published 1681).

1681   *Absalom and Achitophel.*

1682   *The Medal* and *Religio Laici.*

1684   Dryden edits *Miscellany Poems* for Tonson, the first of a series
        (1684, 1685, 1693, 1694, and 1704).

1685   James II king. Dryden converts to Roman Catholicism about this
        time.

1687   *The Hind and the Panther. A Song for St Cecilia's Day.*

1688   James II deposed. Tonson publishes first collected edition of
        Dryden's poems.

1689   Dryden loses his public posts.

1691   *King Arthur* by Dryden and Purcell.

1692   *The Satires of Juvenal and Persius.*

1693   Dryden writes his last play, *Love Triumphant.*

1697   *The Works of Virgil. Alexander's Feast.*

1698   Dryden attacked in Jeremy Collier's *Short View of the Immorality
        and Profaneness of the English Stage.*

1700   *Fables.* Dryden dies 1 May.

NOTE ON THE TEXT

DRYDEN had close, even intimate, relations with both his publishers Henry Herringman and Jacob Tonson, and we may suppose him to have had some control over the publication of his works. Virtually none of his manuscripts has survived. The poems and prose works in this edition are normally given whole, and are based on the earliest published versions. (A few trifling exceptions are indicated in the notes.) The ordering of selections is based on the date of publication except when we know this to have been separated considerably from the date of composition.

Dryden almost invariably elides the final letter of 'the' and 'to' when an initial vowel follows in the next word. In modernizing, I have spelt such words out in full.

The degree sign (°) indicates a note at the end of the book. General headnotes, giving details of publication etc., are not cued.

To John Hoddesdon on his Divine Epigrams

Thou hast inspired me with thy soul, and I,
Who ne'er before could ken of poetry,
Am grown so good proficient, I can lend
A line in commendation of my friend.
Yet 'tis but of the second hand; if ought
There be in this, 'tis from thy fancy brought.
Good thief, who dar'st, Prometheus-like, aspire
And fill thy poems with celestial fire,
Enlivened by these sparks divine, their rays
Add a bright lustre to thy crown of bays.   10
Young eaglet, who thy nest thus soon forsook,
So lofty and divine a course hast took,
As all admire, before the down begin
To peep, as yet, upon thy smoother chin;
And making heaven thy aim, hast had the grace
To look the sun of righteousness i' the face.°
What may we hope, if thou goest on thus fast?
Scriptures at first, enthusiasms at last!
Thou hast commenced betimes a saint: go on,
Mingling diviner streams with Helicon,°   20
That they who view what epigrams here be,
May learn to make like, in just praise of thee.
Reader, I've done, nor longer will withhold
Thy greedy eyes; looking on this pure gold,
Thou'lt know adulterate copper; which, like this,
Will only serve to be a foil to his.

Heroic Stanzas Consecrated to the Glorious Memory of his most Serene and Renowned Highness Oliver, late Lord Protector of this Commonwealth. Written after the Celebration of his Funeral

And now 'tis time; for their officious haste°
Who would before have borne him to the sky,
Like eager Romans, ere all rites were past,
Did let too soon the sacred eagle fly.°

Though our best notes are treason to his fame
Joined with the loud applause of public voice,
Since heaven, what praise we offer to his name,
Hath rendered too authentic by its choice:

Though in his praise no arts can liberal be,°
Since they whose muses have the highest flown   10
Add not to his immortal memory,
But do an act of friendship to their own:

Yet 'tis our duty and our interest too
Such monuments as we can build to raise,
Lest all the world prevent what we should do
And claim a title in him by their praise.

How shall I then begin or where conclude
To draw a fame so truly circular?
For in a round what order can be showed,
Where all the parts so equal-perfect are?   20

His grandeur he derived from heaven alone,
For he was great, ere fortune made him so;
And wars, like mists that rise against the sun,
Made him but greater seem, not greater grow.

No borrowed bays his temples did adorn,
But to our crown he did fresh jewels bring;
Nor was his virtue poisoned, soon as born,
With the too early thoughts of being king.

Fortune, that easy mistress of the young,
But to her ancient servants coy and hard,   30
Him at that age her favourites ranked among
When she her best-loved Pompey did discard.°

He, private, marked the faults of others' sway°
And set as sea-marks for himself to shun;
Not like rash monarchs, who their youth betray
By acts their age too late would wish undone.

And yet dominion was not his design;
We owe that blessing not to him but heaven,
Which to fair acts unsought rewards did join,
Rewards that less to him than us were given.   40

Our former chiefs, like sticklers of the war,
First sought to inflame the parties, then to poise.
The quarrel loved, but did the cause abhor,
And did not strike to hurt, but make a noise.

War, our consumption, was their gainful trade;
We inward bled, whilst they prolonged our pain:
He fought to end our fighting, and essayed
To stanch the blood by breathing of the vein.

Swift and resistless through the land he passed,
Like that bold Greek who did the East subdue,°   50
And made to battles such heroic haste
As if on wings of victory he flew.

He fought, secure of fortune as of fame,
Till by new maps the island might be shown,°
Of conquests, which he strewed where'er he came,
Thick as the galaxy with stars is sown.

His palms, though under weights they did not stand,°
Still thrived; no winter could his laurels fade:
Heaven in his portrait showed a workman's hand
And drew it perfect, yet without a shade.   60

Peace was the prize of all his toils and care,
Which war had banished and did now restore:
Bologna's walls thus mounted in the air°
To seat themselves more surely than before.

Her safety rescued Ireland to him owes;°
And treacherous Scotland to no interest true,°
Yet blessed that fate which did his arms dispose
Her land to civilize as to subdue.

Nor was he like those stars which only shine
When to pale mariners they storms portend;   70
He had his calmer influence, and his mien
Did love and majesty together blend.

'Tis true his countenance did imprint an awe
And naturally all souls to his did bow,
As wands of divination downward draw
And point to beds where sovereign gold doth grow.

When, past all offerings to Feretrian Jove,°
He Mars deposed and arms to gowns made yield,
Successful counsels did him soon approve
As fit for close intrigues as open field.   80

To suppliant Holland he vouchsafed a peace,
Our once bold rival in the British main,
Now tamely glad her unjust claim to cease
And buy our friendship with her idol, gain.

Fame of the asserted sea, through Europe blown,
Made France and Spain ambitious of his love;
Each knew that side must conquer he would own
And for him fiercely as for empire strove.

No sooner was the Frenchman's cause embraced
Than the light Monsieur the grave Don outweighed: 90
His fortune turned the scale where it was cast,
Though Indian mines were in the other laid.

When absent, yet we conquered in his right:
For though some meaner artist's skill were shown
In mingling colours or in placing light,
Yet still the fair designment was his own.

For from all tempers he could service draw;
The worth of each with its alloy he knew;
And, as the confident of nature, saw
How she complexions did divide and brew: 100

Or he their single virtues did survey
By intuition in his own large breast,
Where all the rich ideas of them lay
That were the rule and measure to the rest.

When such heroic virtue heaven sets out,°
The stars, like commons, sullenly obey,
Because it drains them, when it comes about,
And therefore is a tax they seldom pay.

From this high spring our foreign conquests flow
Which yet more glorious triumphs do portend,   110
Since their commencement to his arms they owe,
If springs as high as fountains may ascend.

He made us freemen of the continent°
Whom nature did like captives treat before,
To nobler preys the English lion sent,
And taught him first in Belgian walks to roar.

That old unquestioned pirate of the land,
Proud Rome, with dread the fate of Dunkirk heard,
And trembling wished behind more Alps to stand,
Although an Alexander were her guard.°   120

By his command we boldly crossed the line,°
And bravely fought where southern stars arise;
We traced the far-fetched gold unto the mine,
And that which bribed our fathers made our prize.

Such was our prince; yet owned a soul above
The highest acts it could produce to show:
Thus poor mechanic arts in public move,
Whilst the deep secrets beyond practice go.

Nor died he when his ebbing fame went less,
But when fresh laurels courted him to live;   130
He seemed but to prevent some new success,
As if above what triumphs earth could give.

His latest victories still thickest came,
As near the centre motion does increase;
Till he, pressed down by his own weighty name,
Did, like the vestal, under spoils decease.°

But first the ocean as a tribute sent
That giant-prince of all her watery herd;°
And the isle, when her protecting genius went,
Upon his obsequies loud sighs conferred.   140

No civil broils have since his death arose,
But faction now by habit does obey;
And wars have that respect for his repose
As winds for halcyons when they breed at sea.

His ashes in a peaceful urn shall rest;
His name a great example stands to show
How strangely high endeavours may be blessed
Where piety and valour jointly go.

To my Honoured Friend
Sir Robert Howard, on his Excellent Poems

As there is music uninformed by art
In those wild notes, which with a merry heart
The birds in unfrequented shades express,
Who, better taught at home, yet please us less:
So in your verse, a native sweetness dwells,
Which shames composure, and its art excels.
Singing no more can your soft numbers grace
Than paint adds charms unto a beauteous face.
Yet as when mighty rivers gently creep
Their even calmness does suppose them deep,   10
Such is your muse: no metaphor swelled high
With dangerous boldness lifts her to the sky:
Those mounting fancies, when they fall again,
Show sand and dirt at bottom do remain.
So firm a strength, and yet withal so sweet,
Did never but in Samson's riddle meet.°
'Tis strange each line so great a weight should bear,

And yet no sign of toil, no sweat appear.
Either your art hides art, as stoics feign
Then least to feel, when most they suffer pain;   20
And we, dull souls, admire, but cannot see
What hidden springs within the engine be:
Or 'tis some happiness that still pursues
Each act and motion of your graceful muse,
Or is it fortune's work, that in your head
The curious net that is for fancies spread°
Lets through its meshes every meaner thought,
While rich ideas there are only caught?
Sure that's not all; this is a piece too fair
To be the child of chance, and not of care.   30
No atoms casually together hurled
Could e'er produce so beautiful a world.
Nor dare I such a doctrine here admit
As would destroy the providence of wit.
'Tis your strong genius then which does not feel
Those weights would make a weaker spirit reel.
To carry weight, and run so lightly too,
Is what alone your Pegasus can do.
Great Hercules himself could ne'er do more
Than not to feel those heavens and gods he bore.   40
Your easier odes, which for delight were penned,
Yet our instruction make their second end;
We're both enriched and pleased, like them that woo
At once a beauty and a fortune too.
Of moral knowledge poesy was queen,
And still she might, had wanton wits not been,
Who, like ill guardians, lived themselves at large,
And, not content with that, debauched their charge.
Like some brave captain, your successful pen
Restores the exiled to her crown again;   50
And gives us hope that having seen the days
When nothing flourished but fanatic bays,
All will at length in this opinion rest:
'A sober prince's government is best.'
This is not all; your art the way has found
To make improvement of the richest ground,
That soil which those immortal laurels bore,
That once the sacred Maro's temples wore.°

Elissa's griefs are so expressed by you,°
They are too eloquent to have been true.   60
Had she so spoke, Aeneas had obeyed
What Dido rather than what Jove had said.
If funeral rites can give a ghost repose,
Your muse so justly has discharged those,
Elissa's shade may now its wandering cease,
And claim a title to the fields of peace.
But if Aeneas be obliged, no less
Your kindness great Achilles doth confess,
Who, dressed by Statius in too bold a look,
Did ill become those virgin's robes he took.   70
To understand how much we owe to you
We must your numbers with your author's view:
Then we shall see his work was lamely rough,
Each figure stiff, as if designed in buff;
His colours laid so thick on every place,
As only showed the paint, but hid the face.
But as in a perspective we beauties see,°
Which in the glass, not in the picture, be;
So here our sight obligingly mistakes
That wealth which his your bounty only makes.   80
Thus vulgar dishes are by cooks disguised,
More for their dressing than their substance prized.
Your curious notes so search into that age,°
When all was fable but the sacred page,
That, since in that dark night we needs must stray,
We are at least misled in pleasant way.
But what we most admire, your verse no less
The prophet than the poet doth confess.
Ere our weak eyes discerned the doubtful streak
Of light, you saw great Charles's morning break.   90
So skilful seamen ken the land from far,
Which shows like mists to the dull passenger.
To Charles your muse first pays her duteous love,°
As still the ancients did begin from Jove;
With Monck you end, whose name preserved shall be,
As Rome recorded Rufus' memory;°
Who thought it greater honour to obey
His country's interest, than the world to sway.
But to write worthy things of worthy men

Is the peculiar talent of your pen: 100
Yet let me take your mantle up, and I
Will venture in your right to prophesy:

'This work, by merit first of fame secure,
Is likewise happy in its geniture:
For since 'tis born when Charles ascends the throne,
It shares at once his fortune and its own.'

Astraea Redux

A POEM

ON THE HAPPY RESTORATION AND RETURN OF HIS SACRED MAJESTY CHARLES THE SECOND.

Iam redit et virgo, redeunt Saturnia regna.°   Virgil

Now with a general peace the world was blessed,
While ours, a world divided from the rest,
A dreadful quiet felt, and worser far
Than arms, a sullen interval of war.
Thus when black clouds draw down the labouring skies,
Ere yet abroad the winged thunder flies,
A horrid stillness first invades the ear
And in that silence we the tempest fear.
The ambitious Swede, like restless billows tossed,°
On this hand gaining what on that he lost, 10
Though in his life he blood and ruin breathed,
To his now guideless kingdom peace bequeathed;
And heaven that seemed regardless of our fate,°
For France and Spain did miracles create,
Such mortal quarrels to compose in peace
As nature bred and interest did increase.
We sighed to hear the fair Iberian bride
Must grow a lily to the lily's side;
While our cross stars denied us Charles's bed,
Whom our first flames and virgin love did wed. 20
For his long absence church and state did groan;
Madness the pulpit, faction seized the throne:

Experienced age in deep despair was lost,
To see the rebel thrive, the loyal crossed:
Youth, that with joys had unacquainted been,
Envied grey hairs, that once good days had seen:
We thought our sires, not with their own content,
Had ere we came to age our portion spent.
Nor could our nobles hope their bold attempt
Who ruined crowns, would coronets exempt: 30
For when, by their designing leaders taught
To strike at power, which for themselves they sought,
The vulgar, gulled into rebellion, armed,
Their blood to action by the prize was warmed.
The sacred purple, then, and scarlet gown,°
Like sanguine dye to elephants, was shown.°
Thus when the bold Typhoeus scaled the sky°
And forced great Jove from his own heaven to fly,
(What king, what crown, from treason's reach is free,
If Jove and heaven can violated be?)   40
The lesser gods that shared his prosperous state
All suffered in the exiled thunderer's fate.
The rabble now such freedom did enjoy,
As winds at sea that use it to destroy:
Blind as the Cyclops, and as wild as he,°
They owned a lawless savage liberty,
Like that our painted ancestors so prized,
Ere empire's arts their breasts had civilized.
How great were then our Charles's woes, who thus
Was forced to suffer for himself and us!   50
He, tossed by fate, and hurried up and down,
Heir to his father's sorrows with his crown
Could taste no sweets of youth's desired age,
But found his life too true a pilgrimage.
Unconquered yet in that forlorn estate,
His manly courage overcame his fate.
His wounds he took like Romans on his breast,
Which by his virtue were with laurels dressed.
As souls reach heaven, while yet in bodies pent,
So did he live above his banishment.   60
That sun which we beheld with cozened eyes
Within the water, moved along the skies.
How easy 'tis when destiny proves kind

With full-spread sails to run before the wind!
But those that 'gainst stiff gales laveering go,
Must be at once resolved, and skilful too.
He would not like soft Otho hope prevent,°
But stayed, and suffered fortune to repent.
These virtues Galba in a stranger sought,
And Piso to adopted empire brought.   70
How shall I then my doubtful thoughts express,
That must his sufferings both regret and bless?
For when his early valour heaven had crossed,
And all at Worcester but the honour lost,°
Forced into exile from his rightful throne,
He made all countries where he came his own;
And viewing monarchs' secret arts of sway,
A royal factor for their kingdoms lay.
Thus banished David spent abroad his time,°
When to be God's anointed was his crime;   80
And when restored made his proud neighbours rue
Those choice remarks he from his travels drew.
Nor is he only by afflictions shown
To conquer others' realms, but rule his own;
Recovering hardly what he lost before,
His right endears it much, his purchase more.
Inured to suffer ere he came to reign,
No rash procedure will his actions stain.
To business ripened by digestive thought,
His future rule is into method brought,   90
As they who first proportion understand,
With easy practice reach a master's hand.
Well might the ancient poets then confer
On night the honoured name of counsellor;
Since, struck with rays of prosperous fortune blind,
We light alone in dark afflictions find.
In such adversities to sceptres trained,
The name of Great his famous grandsire gained;°
Who yet a king alone in name and right,
With hunger, cold, and angry Jove did fight;   100
Shocked by a covenanting League's vast powers,°
As holy and as catholic as ours:
Till fortune's fruitless spite had made it known
Her blows not shook but riveted his throne.

Some lazy ages lost in sleep and ease
No action leave to busy chronicles:
Such whose supine felicity but makes
In story chasms, in epochés mistakes;°
O'er whom time gently shakes his wings of down
Till with his silent sickle they are mown.   110
Such is not Charles's too too active age,
Which, governed by the wild distempered rage
Of some black star infecting all the skies,
Made him at his own cost like Adam wise.
Tremble, ye nations, who secure before,
Laughed at those arms that 'gainst ourselves we bore;
Roused by the lash of his own stubborn tail,
Our lion now will foreign foes assail.
With alga who the sacred altar strows?
To all the sea-gods Charles an offering owes: 120
A bull to thee Portunus shall be slain,
A lamb to you the tempests of the main:
For those loud storms that did against him roar
Have cast his shipwrecked vessel on the shore.
Yet as wise artists mix their colours so
That by degrees they from each other go,
Black steals unheeded from the neighbouring white
Without offending the well-cozened sight:
So on us stole our blessed change; while we
The effect did feel but scarce the manner see.   130
Frosts that constrain the ground and birth deny
To flowers that in its womb expecting lie,
Do seldom their usurping power withdraw,
But raging floods pursue their hasty thaw;
Our thaw was mild, the cold not chased away,
But lost in kindly heat of lengthened day.
Heaven would no bargain for its blessings drive,
But what we could not pay for, freely give.
The prince of peace would, like himself, confer
A gift unhoped, without the price of war: 140
Yet as he knew his blessing's worth, took care,
That we should know it by repeated prayer;
Which stormed the skies and ravished Charles from thence,
As heaven itself is took by violence.
Booth's forward valour only served to show°

He durst that duty pay, we all did owe:
The attempt was fair; but heaven's prefixed hour
Not come: so, like the watchful traveller
That by the moon's mistaken light did rise,
Lay down again and closed his weary eyes.   150
'Twas Monck whom Providence designed to loose°
Those real bonds false freedom did impose.
The blessed saints that watched this turning scene
Did from their stars with joyful wonder lean,
To see small clews draw vastest weights along,
Not in their bulk but in their order strong.
Thus pencils can by one slight touch restore
Smiles to that changed face that wept before.
With ease such fond chimeras we pursue,
As fancy frames for fancy to subdue;   160
But when ourselves to action we betake,
It shuns the mint like gold that chemists make.
How hard was then his task, at once to be
What in the body natural we see
Man's architect distinctly did ordain
The charge of muscles, nerves, and of the brain,
Through viewless conduits spirits to dispense,
The springs of motion from the seat of sense.
'Twas not the hasty product of a day,
But the well-ripened fruit of wise delay.   170
He, like a patient angler, ere he struck,
Would let them play a while upon the hook.
Our healthful food the stomach labours thus,
At first embracing what it straight doth crush.
Wise leeches will not vain receipts obtrude,
While growing pains pronounce the humours crude:
Deaf to complaints, they wait upon the ill,
Till some safe crisis authorize their skill.
Nor could his acts too close a visor wear,
To scape their eyes whom guilt had taught to fear,   180
And guard with caution that polluted nest
Whence Legion twice before was dispossessed:°
Once sacred house, which when they entered in,
They thought the place could sanctify a sin;
Like those that vainly hoped kind heaven would wink
While to excess on martyrs' tombs they drink.

And as devouter Turks first warn their souls
To part before they taste forbidden bowls,°
So these when their black crimes they went about
First timely charmed their useless conscience out.   190
Religion's name against itself was made;
The shadow served the substance to invade:
Like zealous missions they did care pretend
Of souls in show but made the gold their end.
The incensed powers beheld with scorn from high
A heaven so far distant from the sky,
Which durst with horses' hoofs that beat the ground
And martial brass belie the thunder's sound.
'Twas hence at length just vengeance thought it fit
To speed their ruin by their impious wit.   200
Thus Sforza, cursed with a too fertile brain,°
Lost by his wiles the power his wit did gain.
Henceforth their fougue must spend at lesser rate
Than in its flames to wrap a nation's fate.
Suffered to live, they are like helots set,°
A virtuous shame within us to beget.
For by example most we sinned before,
And glass-like clearness mixed with frailty bore.
But since reformed by what we did amiss
We by our sufferings learn to prize our bliss: 210
Like early lovers whose unpractised hearts
Were long the May-game of malicious arts,
When once they find their jealousies were vain,
With double heat renew their fires again.
'Twas this produced the joy that hurried o'er
Such swarms of English to the neighbouring shore,°
To fetch that prize, by which Batavia made°
So rich amends for our impoverished trade.
Oh had you seen from Scheveline's barren shore,°
(Crowded with troops and barren now no more,)   220
Afflicted Holland to his farewell bring
True sorrow, Holland to regret a king,
While waiting him his royal fleet did ride,
And willing winds to their lowered sails denied.
The wavering streamers, flags, and standard out,
The merry seamen's rude but cheerful shout,
And last the cannon's voice that shook the skies,

And as it fares in sudden ecstasies
At once bereft us both of ears and eyes.
The Naseby, now no longer England's shame,°   230
But better to be lost in Charles's name,
(Like some unequal bride in nobler sheets)
Receives her lord; the joyful London meets
The princely York, himself alone a freight;
The Swiftsure groans beneath great Gloucester's weight:°
Secure as when the halcyon breeds, with these
He that was born to drown might cross the seas.
Heaven could not own a providence, and take
The wealth three nations ventured at a stake.
The same indulgence Charles's voyage blessed   240
Which in his right had miracles confessed.
The winds that never moderation knew
Afraid to blow too much, too faintly blew;
Or out of breath with joy could not enlarge
Their straightened lungs, or conscious of their charge.
The British Amphitrite, smooth and clear,°
In richer azure never did appear;
Proud her returning prince to entertain
With the submitted fasces of the main.°

    And welcome now, great monarch, to your own!   250
Behold the approaching cliffs of Albion.
It is no longer motion cheats your view;
As you meet it, the land approacheth you.
The land returns, and in the white it wears
The marks of penitence and sorrow bears.
But you, whose goodness your descent doth show,
Your heavenly parentage and earthly too,
By that same mildness which your father's crown
Before did ravish, shall secure your own.
Not tied to rules of policy, you find   260
Revenge less sweet than a forgiving mind.
Thus when the almighty would to Moses give°
A sight of all he could behold and live,
A voice before his entry did proclaim
Long-suffering, goodness, mercy, in his name.
Your power to justice doth submit your cause,
Your goodness only is above the laws;

Whose rigid letter, while pronounced by you,°
Is softer made. So winds that tempests brew,
When through Arabian groves they take their flight,   270
Made wanton with rich odours, lose their spite.
And as those lees that trouble it refine
The agitated soul of generous wine;
So tears of joy, for your returning spilt,
Work out and expiate our former guilt.
Methinks I see those crowds on Dover's strand
Who in their haste to welcome you to land
Choked up the beach with their still growing store,
And made a wilder torrent on the shore:
While spurred with eager thoughts of past delight,   280
Those who had seen you court a second sight;
Preventing still your steps and making haste
To meet you often wheresoe'er you passed.
How shall I speak of that triumphant day,
When you renewed the expiring pomp of May!
(A month that owns an interest in your name:
You and the flowers are its peculiar claim.)
That star, that at your birth shone out so bright,°
It stained the duller sun's meridian light,
Did once again its potent fires renew,   290
Guiding our eyes to find and worship you.

    And now time's whiter series is begun,°
Which in soft centuries shall smoothly run:
Those clouds that overcast your morn shall fly,
Dispelled to farthest corners of the sky.
Our nation, with united interest blest,
Not now content to poise, shall sway the rest.
Abroad your empire shall no limits know,
But like the sea in boundless circles flow;
Your much-loved fleet shall with a wide command   300
Besiege the petty monarchs of the land;
And as old time his offspring swallowed down,
Our ocean in its depths all seas shall drown.
Their wealthy trade from pirates' rapine free,
Our merchants shall no more adventurers be;
Nor in the farthest east those dangers fear,
Which humble Holland must dissemble here.

Spain to your gift alone her Indies owes;
For what the powerful takes not, he bestows:
And France that did an exile's presence fear,   310
May justly apprehend you still too near.
At home the hateful names of parties cease
And factious souls are wearied into peace.
The discontented now are only they
Whose crimes before did your just cause betray:
Of those your edicts some reclaim from sins,
But most your life and blest example wins.
Oh happy prince whom heaven hath taught the way
By paying vows to have more vows to pay!
Oh happy age! Oh times like those alone   320
By fate reserved for great Augustus' throne!
When the joint growth of arms and arts foreshow
The world a monarch, and that monarch you.

To His Sacred Majesty, a Panegyric
on His Coronation

In that wild deluge where the world was drowned,°
When life and sin one common tomb had found,
The first small prospect of a rising hill
With various notes of joy the ark did fill:
Yet when that flood in its own depths was drowned,
It left behind it false and slippery ground;
And the more solemn pomp was still deferred,
Till new-born nature in fresh looks appeared.
Thus, royal sir, to see you landed here,
Was cause enough of triumph for a year: 10
Nor would your care those glorious joys repeat,
Till they at once might be secure and great;
Till your kind beams, by their continued stay,
Had warmed the ground, and called the damps away.
Such vapours, while your powerful influence dries,
Then soonest vanish when they highest rise.
Had greater haste these sacred rites prepared,
Some guilty months had in your triumphs shared;°
But this untainted year is all your own,
Your glories may without our crimes be shown.   20

We had not yet exhausted all our store,
When you refreshed our joys by adding more:
As heaven, of old, dispensed celestial dew,°
You gave us manna, and still give us new.
    Now our sad ruins are removed from sight,
The season too comes fraught with new delight:
Time seems not now beneath his years to stoop,
Nor do his wings with sickly feathers droop:
Soft western winds waft o'er the gaudy spring,
And opened scenes of flowers and blossoms bring   30
To grace this happy day, while you appear
Not king of us alone but of the year.
All eyes you draw, and with the eyes the heart;
Of your own pomp yourself the greatest part:
Loud shouts the nation's happiness proclaim
And heaven this day is feasted with your name.
Your cavalcade the fair spectators view
From their high standings, yet look up to you.
From your brave train each singles out a prey,
And longs to date a conquest from your day.   40
Now charged with blessings while you seek repose,
Officious slumbers haste your eyes to close;
And glorious dreams stand ready to restore
The pleasing shapes of all you saw before.
Next to the sacred temple you are led,
Where waits a crown for your more sacred head.
How justly from the church that crown is due,
Preserved from ruin and restored by you!
The grateful choir their harmony employ
Not to make greater but more solemn joy.   50
Wrapped soft and warm your name is sent on high,
As flames do on the wing of incense fly.
Music herself is lost; in vain she brings
Her choicest notes to praise the best of kings:
Her melting strains in you a tomb have found,
And lie like bees in their own sweetness drowned.
He that brought peace and discord could atone,
His name is music of itself alone.
Now while the sacred oil anoints your head,
And fragrant scents, begun from you, are spread   60
Through the large dome, the people's joyful sound

Sent back, is still preserved in hallowed ground;
Which in one blessing mixed descends on you,
As heightened spirits fall in richer dew.
Not that our wishes do increase your store;
Full of yourself you can admit no more.
We add not to your glory, but employ
Our time like angels in expressing joy.
Nor is it duty, or our hopes alone,
Create that joy, but full fruition: 70
We know those blessings which we must possess
And judge of future by past happiness.
No promise can oblige a prince so much
Still to be good, as long to have been such.
A noble emulation heats your breast,
And your own fame now robs you of your rest.
Good actions still must be maintained with good,
As bodies nourished with resembling food.
You have already quenched sedition's brand;
And zeal, which burnt it, only warms the land.   80
The jealous sects that dare not trust their cause
So far from their own will as to the laws,
You for their umpire and their synod take,
And their appeal alone to Caesar make.
Kind heaven so rare a temper did provide,
That guilt repenting might in it confide.
Among our crimes oblivion may be set,
But 'tis our king's perfection to forget.
Virtues unknown to these rough northern climes
From milder heavens you bring, without their crimes.   90
Your calmness does no after-storms provide,
Nor seeming patience mortal anger hide.
When empire first from families did spring,
Then every father governed as a king;
But you, that are a sovereign prince, allay
Imperial power with your paternal sway.
From those great cares when ease your soul unbends,
Your pleasures are designed to noble ends;
Born to command the mistress of the seas,
Your thoughts themselves in that blue empire please.   100
Hither in summer evenings you repair
To take the fraischeur of the purer air:

Undaunted here you ride, when winter raves,
With Caesar's heart that rose above the waves.°
More I could sing, but fear my numbers stays;
No loyal subject dares that courage praise.
In stately frigates most delight you find,
Where well-drawn battles fire your martial mind.
What to your cares we owe is learnt from hence,
When e'en your pleasures serve for our defence.   110
Beyond your court flows in the admitted tide,
Where in new depths the wondering fishes glide:
Here in a royal bed the waters sleep;
When tired at sea, within this bay they creep.
Here the mistrustful fowl no harm suspects,
So safe are all things which our king protects.
From your loved Thames a blessing yet is due,
Second alone to that it brought in you;
A queen from whose chaste womb, ordained by fate,
The souls of kings unborn for bodies wait.   120
It was your love before made discord cease:
Your love is destined to your country's peace.
Both Indies, rivals in your bed, provide
With gold or jewels to adorn your bride.
This to a mighty king presents rich ore,
While that with incense does a god implore.
Two kingdoms wait your doom; and, as you choose,°
This must receive a crown, or that must lose.
Thus from your Royal Oak, like Jove's of old,°
Are answers sought and destinies foretold: 130
Propitious oracles are begged with vows,
And crowns that grow upon the sacred boughs.
Your subjects, while you weigh the nations' fate,
Suspend to both their doubtful love or hate:
Choose only, sir, that so they may possess
With their own peace their children's happiness.

To my Honoured Friend Dr Charleton,

ON HIS LEARNED AND USEFUL WORKS, AND MORE PARTI-
CULARLY THIS OF STONEHENGE, BY HIM RESTORED TO THE
TRUE FOUNDERS

The longest tyranny that ever swayed
Was that wherein our ancestors betrayed
Their free-born reason to the Stagirite,°
And made his torch their universal light.
So truth, while only one supplied the state,
Grew scarce and dear, and yet sophisticate;
Until 'twas bought, like emp'ric wares or charms,
Hard words sealed up with Aristotle's arms.
Columbus was the first that shook his throne
And found a temperate in a torrid zone: 10
The feverish air fanned by a cooling breeze,
The fruitful vales set round with shady trees;
And guiltless men, who danced away their time,
Fresh as their groves, and happy as their clime.
Had we still paid that homage to a name
Which only God and nature justly claim,
The western seas had been our utmost bound,
Where poets still might dream the sun was drowned:
And all the stars that shine in southern skies
Had been admired by none but savage eyes   20
    Among the asserters of free reason's claim,
The English are not the least in worth or fame.
The world to Bacon does not only owe
Its present knowledge, but its future too.
Gilbert shall live, till loadstones cease to draw,°
Or British fleets the boundless ocean awe;
And noble Boyle, not less in nature seen,°
Than his great brother, read in states and men.°
The circling streams, once thought but pools, of blood
(Whether life's fuel or the body's food),   30
From dark oblivion Harvey's name shall save;°
While Ent keeps all the honour that he gave.°
Nor are *you*, learned friend, the least renowned;
Whose fame, not circumscribed with English ground,

Flies like the nimble journeys of the light,
And is, like that, unspent too in its flight.
Whatever truths have been, by art or chance,
Redeemed from error, or from ignorance,
Thin in their authors, like rich veins of ore,
Your works unite, and still discover more.   40
Such is the healing virtue of your pen,
To perfect cures on books, as well as men.
Nor is this work the least: you well may give
To men new vigour, who make stones to live.
Through you, the Danes, their short dominion lost,
A longer conquest than the Saxons boast.
Stonehenge, once thought a temple, you have found
A throne where kings, our earthly gods, were crowned;
Where by their wondering subjects they were seen,
Joyed with their stature, and their princely mien.   50
Our sovereign here above the rest might stand,
And here be chose again to rule the land.
    These ruins sheltered once his sacred head,
Then when from Worcester's fatal field he fled,°
Watched by the genius of this royal place,
And mighty visions of the Danish race.
His refuge then was for a temple shown:
But, he restored, 'tis now become a throne.

Prologue to The Rival Ladies

'Tis much desired, you judges of the town
Would pass a vote to put all prologues down;
For who can show me, since they first were writ,
They e'er converted one hard-hearted wit?
Yet the world's mended well; in former days
Good prologues were as scarce as now good plays.
For the reforming poets of our age,
In this first charge, spend their poetic rage:
Expect no more when once the prologue's done;
The wit is ended ere the play's begun.   10
You now have habits, dances, scenes, and rhymes,
High language often, ay, and sense sometimes.

As for a clear contrivance, doubt it not;
They blow out candles to give light to the plot.
And for surprise, two bloody-minded men
Fight till they die, then rise and dance again.
Such deep intrigues you're welcome to this day:
But blame yourselves, not him who writ the play.
Though his plot's dull as can be well desired,
Wit stiff as any you have e'er admired,   20
He's bound to please, not to write well; and knows
There is a mode in plays as well as clothes:
Therefore, kind judges—

A Second Prologue enters.

2.   Hold! would you admit
For judges all you see within the pit?
    1. Whom would he then except, or on what score?
    2. All who (like him) have writ ill plays before;
For they, like thieves condemned, are hangmen made,
To execute the members of their trade.
All that are writing now he would disown,
But then he must except—e'en all the town: 30
All choleric losing gamesters, who in spite
Will damn today, because they lost last night;
All servants, whom their mistress' scorn upbraids;
All maudlin lovers, and all slighted maids;
All who are out of humour, or severe;
All that want wit, or hope to find it here.

ANNUS MIRABILIS:
THE YEAR OF WONDERS, *1666.*

AN ACCOUNT OF THE ENSUING POEM,
IN A LETTER TO THE
HONOURABLE SIR ROBERT HOWARD

Sir,

I am so many ways obliged to you, and so little able to return your
favours, that, like those who owe too much, I can only live by getting
further into your debt. You have not only been careful of my fortune,
which was the effect of your nobleness, but you have been solicitous of

my reputation, which is that of your kindness. It is not long since I gave you the trouble of perusing a play° for me, and now, instead of an acknowledgment, I have given you a greater, in the correction of a poem. But since you are to bear this persecution, I will at least give you the encouragement of a martyr, you could never suffer in a nobler cause. For I have chosen the most heroic subject which any poet could desire: I have taken upon me to describe the motives, the beginning, progress and successes of a most just and necessary war; in it, the care, management and prudence of our king; the conduct and valour of a royal admiral,° and of two incomparable generals:° the invincible courage of our captains and seamen, and three glorious victories, the result of all. After this I have, in the Fire, the most deplorable, but withal the greatest argument that can be imagined: the destruction being so swift, so sudden, so vast and miserable, as nothing can parallel in story. The former part of this poem, relating to the war, is but a due expiation for my not serving my king and country in it. All gentlemen are almost obliged to it: and I know no reason we should give that advantage to the commonalty of England to be foremost in brave actions, which the noblesse of France would never suffer in their peasants. I should not have written this but to a person who has been ever forward to appear in all employments, whither his honour and generosity have called him. The latter part of my poem, which describes the Fire, I owe first to the piety and fatherly affection of our monarch to his suffering subjects; and, in the second place, to the courage, loyalty and magnanimity of the City: both which were so conspicuous, that I have wanted words to celebrate them as they deserve. I have called my poem *historical*, not *epic*, though both the actions and actors are as much heroic, as any poem can contain. But since the action is not properly one, nor that accomplished in the last successes, I have judged it too bold a title for a few stanzas, which are little more in number than a single Iliad,° or the longest of the Aeneids. For this reason (I mean not of length, but broken action, tied too severely to the laws of history) I am apt to agree with those who rank Lucan° rather among historians in verse, than epic poets: in whose room, if I am not deceived, Silius Italicus,° though a worse writer, may more justly be admitted. I have chosen to write my poem in quatrains or stanzas of four in alternate rhyme, because I have ever judged them more noble, and of greater dignity, both for the sound and number, than any other verse in use amongst us; in which I am sure I have your approbation. The learned languages have, certainly, a great advantage of us, in not being tied to the slavery of any rhyme;

and were less constrained in the quantity of every syllable, which they might vary with spondees or dactyls, besides so many other helps of grammatical figures, for the lengthening or abbreviation of them, than the modern are in the close of that one syllable, which often confines, and more often corrupts the sense of all the rest. But in this necessity of our rhymes, I have always found the couplet verse most easy (though not so proper for this occasion) for there the work is sooner at an end, every two lines concluding the labour of the poet: but in quatrains he is to carry it further on; and not only so, but to bear along in his head the troublesome sense of four lines together. For those who write correctly in this kind must needs acknowledge, that the last line of the stanza is to be considered in the composition of the first. Neither can we give ourselves the liberty of making any part of a verse for the sake of rhyme, or concluding with a word which is not current English, or using the variety of female rhymes, all which our fathers practised; and for the female rhymes, they are still in use amongst other nations: with the Italian in every line, with the Spaniard promis- cuously, with the French alternately, as those who have read the *Alaric*,° the *Pucelle*,° or any of their later poems, will agree with me. And besides this, they write in alexandrines, or verses of six feet, such as amongst us is the old translation of Homer, by Chapman,° all which, by lengthening of their chain, makes the sphere of their activity the larger. I have dwelt too long upon the choice of my stanza, which you may remember is much better defended in the preface to *Gondibert*,° and therefore I will hasten to acquaint you with my endeavours in the writing. In general I will only say I have never yet seen the description of any naval fight in the proper terms which are used at sea; and if there be any such in another language, as that of Lucan in the third of his *Pharsalia*, yet I could not prevail myself of it in the English; the terms of arts in every tongue bearing more of the idiom of it than any other words. We hear, indeed, among our poets, of the thundering of guns, the smoke, the disorder and the slaughter; but all these are common notions. And certainly as those who, in a logical dispute, keep in general terms, would hide a fallacy, so those who do it in any poetical description would veil their ignorance.

> descriptas servare vices operumque colores
> cur ego, si nequeo ignoroque, poeta salutor?°

[if in my ignorance I lack the skill to maintain a part, and to use rhetorical figures in my writing, why should I be hailed a poet?]

For my own part, if I had little knowledge of the sea, yet I have thought it no shame to learn: and if I have made some few mistakes, 'tis only, as you can bear me witness, because I have wanted opportunity to correct them, the whole poem being first written, and now sent you from a place, where I have not so much as the converse of any seaman. Yet, though the trouble I had in writing it was great, it was more than recompensed by the pleasure; I found myself so warm in celebrating the praises of military men, two such especially as the prince and general, that it is no wonder if they inspired me with thoughts above my ordinary level. And I am well satisfied, that as they are incomparably the best subject I have ever had, excepting only the royal family; so also, that this I have written of them is much better than what I have performed on any other. I have been forced to help out other arguments, but this has been bountiful to me; they have been low and barren of praise, and I have exalted them, and made them fruitful: but here—*omnia sponte sua reddit justissima tellus* [the earth with justice returns everything freely]. I have had a large, a fair and a pleasant field, so fertile, that, without my cultivating, it has given me two harvests in a summer, and in both oppressed the reaper. All other greatness in subjects is only counterfeit, it will not endure the test of danger; the greatness of arms is only real: other greatness burdens a nation with its weight, this supports it with its strength. And as it is the happiness of the age, so is it the peculiar goodness of the best of kings, that we may praise his subjects without offending him: doubtless it proceeds from a just confidence of his own virtue, which the lustre of no other can be so great as to darken in him: for the good or the valiant are never safely praised under a bad or a degenerate prince. But to return from this digression to a further account of my poem. I must crave leave to tell you, that as I have endeavoured to adorn it with noble thoughts, so much more to express those thoughts with elocution. The composition of all poems is, or ought to be, of wit, and wit in the poet, or wit writing, (if you will give me leave to use a school distinction) is no other than the faculty of imagination in the writer, which, like a nimble spaniel, beats over and ranges through the field of memory, till it springs the quarry it hunted after; or, without metaphor, which searches over all the memory for the species or ideas of those things which it designs to represent. Wit written, is that which is well defined, the happy result of thought, or product of that imagination. But to proceed from wit in the general notion of it, to the proper wit of a heroic or historical poem, I judge it chiefly to consist in the delightful imaging of persons, actions, passions, or things. 'Tis not the jerk or sting of an epigram, nor the seeming contradiction of a poor antithesis, (the delight of an ill-judging audience in a play of rhyme) nor the jingle of a more poor paranomasia: neither is it so much the morality of a grave sentence, affected by Lucan, but more sparingly used by Virgil; but it is some lively and apt description, dressed in such colours of speech, that it sets before your eyes the absent object, as perfectly and more delightfully than nature. So then, the first happiness of the poet's imagination is properly invention, or finding of the thought; the second is fancy, or the variation, driving or moulding of that thought, as the judgement represents it proper to the subject; the third is elocution, or the art of clothing and adorning that thought so found and varied, in apt, significant and sounding words: the quickness of the imagination is seen in the invention, the fertility in the fancy, and the accuracy in the expression. For the two first of these Ovid is famous amongst the poets, for the latter Virgil. Ovid images more often the movements and affections of the mind, either combating between two contrary passions, or extremely discomposed by one: his words therefore are the least part of his care, for he pictures nature in disorder, with which the study and choice of words is inconsistent. This is the proper wit of dialogue or discourse, and, consequently, of the drama, where all that is said is to be supposed the effect of sudden thought; which, though it excludes not the quickness of wit in repartees, yet admits not a too curious election of words, too frequent allusions, or use of tropes, or, in fine, anything that shows remoteness of thought, or labour in the writer. On the other side, Virgil speaks not so often to us in the person of another, like Ovid, but in his own; he relates almost all things as from himself, and thereby gains more liberty than the other to express his thoughts with all the graces of elocution, to write more figuratively, and to confess as well the labour as the force of his imagination. Though he describes his Dido° well and naturally, in the violence of her passions, yet he must yield in that to the Myrrha, the Biblis, the Althaea, of Ovid;° for, as great an admirer of him as I am, I must acknowledge that if I see not more of their souls than I see of Dido's, at least I have a greater concernment for them: and that convinces me that Ovid has touched those tender strokes more delicately than Virgil could. But when action or persons are to be described, when any such image is to be set before us, how bold, how masterly are the strokes of Virgil! We see the objects he represents us with in their native figures, in their proper motions; but we so see them, as our own eyes could never have beheld them so beautiful in themselves. We see the soul of the poet, like that universal one of which he speaks, informing and moving through all his pictures, *totamque infusa per artus mens agitat molem, et magno se corpore miscet* [and mind, pervading everything, moves the mass and mingles with its mighty bulk];° we behold him embellishing his images, as he makes Venus breathing beauty upon her son Aeneas,

> —lumenque juventae
> purpureum, et laetos oculis afflarat honores:
> quale manus addunt ebori decus, aut ubi flavo
> argentum, Pariusve lapis circundatur auro.°

[and the radiant light of youth, and his eyes shone bright, like the beauty which the hand gives to ivory, or when silver or Parian marble is set in gold.]

See his tempest, his funeral sports, his combat of Turnus and Aeneas, and in his *Georgics*, which I esteem the divinest part of all his writings, the plague, the country, the battle of bulls, the labour of the bees, and those many other excellent images of nature, most of which are neither great in themselves, nor have any natural ornament to bear them up: but the words wherewith he describes them are so excellent, that it might be well applied to him which was said by Ovid, *materiam superabat opus* [the workmanship is better than the material];° the very sound of his words has often somewhat that is connatural to the subject, and while we read him, we sit, as in a play, beholding the scenes of what he represents. To perform this, he made frequent use of tropes, which you know change the nature of a known word, by applying it to some other signification; and this is it which Horace means in his *Epistle to the Pisos*,

> dixeris egregie notum si callida verbum
> reddiderit junctura novum—°

[you will write well if a skilful setting gives new force to a well-known word]

But I am sensible I have presumed too far, to entertain you with a rude discourse of that art, which you both know so well, and put into practice with so much happiness. Yet before I leave Virgil, I must own the vanity to tell you, and by you the world, that he has been my master in this poem: I have followed him everywhere, I know not with what success, but I am sure with diligence enough: my images are many of them copied from him, and the rest are imitations of him. My expressions also are as near as the idioms of the two languages would admit of in translation. And this, Sir, I have done with that boldness, for which I will stand accountable to any of our little critics, who, perhaps, are not better acquainted with him than I am. Upon your first perusal of this poem, you have taken notice of some words which I have innovated (if it be too bold for me to say refined) upon his Latin;

which, as I offer not to introduce into English prose, so I hope they are neither improper, nor altogether unelegant in verse; and, in this, Horace will again defend me.

> et nova, fictaque nuper habebunt verba fidem, si
> Graeco fonte cadant, parce detorta—°

[new and newly coined words will be accepted if they come from a Greek source only slightly altered]

The inference is exceeding plain; for if a Roman poet might have liberty to coin a word, supposing only that it was derived from the Greek, was put into a Latin termination, and that he used this liberty but seldom, and with modesty: how much more justly may I challenge that privilege to do it with the same prerequisites, from the best and most judicious of Latin writers? In some places, where either the fancy, or the words, were his, or any others', I have noted it in the margin, that I might not seem a plagiary: in others I have neglected it, to avoid as well the tediousness, as the affectation of doing it too often. Such descriptions or images, well wrought, which I promise not for mine, are, as I have said, the adequate delight of heroic poesy, for they beget admiration, which is its proper object; as the images of the burlesque, which is contrary to this, by the same reason beget laughter; for the one shows nature beautified, as in the picture of a fair woman, which we all admire; the other shows her deformed, as in that of a lazar, or of a fool with distorted face and antic gestures, at which we cannot forbear to laugh, because it is a deviation from nature. But though the same images serve equally for the epic poesy, and for the historic and panegyric, which are branches of it, yet a several sort of sculpture is to be used in them: if some of them are to be like those of Juvenal,° *stantes in curribus aemiliani* [Aemiliani standing in their chariots], heroes drawn in their triumphal chariots, and in their full proportion; others are to be like that of Virgil,° *spirantia mollius aera* [give gentler lines to breathing statues of bronze]: there is somewhat more of softness and tenderness to be shown in them. You will soon find I write not this without concern. Some who have seen a paper of verses which I wrote last year to her Highness the Duchess, have accused them of that only thing I could defend in them; they have said I did *humi serpere* [crawl on the earth],° that I wanted not only height of fancy, but dignity of words to set it off; I might well answer with that of Horace,° *nunc non erat his locus* [this was not the place for such things], I knew I addressed them to a lady, and accordingly I affected the softness of expression, and the smoothness of measure, rather than

the height of thought; and in what I did endeavour, it is no vanity to
say I have succeeded. I detest arrogance, but there is some difference
betwixt that and a just defence. But I will not further bribe your candour, or the reader's. I leave them to speak for me, and, if they can, to
make out that character, not pretending to a greater, which I have
given them.

Verses to her Highness the Duchess, on the memorable victory gained
by the Duke against the Hollanders, June the 3, 1665, and on her
journey afterwards into the North

MADAM.

> When, for our sakes, your hero you resigned
> To swelling seas, and every faithless wind;
> When you released his courage, and set free
> A valour fatal to the enemy,
> You lodged your country's cares within your breast;
> (The mansion where soft love should only rest:)
> And ere our foes abroad were overcome,
> The noblest conquest you had gained at home.
> Ah, what concerns did both your souls divide!
> Your honour gave us what your love denied: 10
> And 'twas for him much easier to subdue
> Those foes he fought with, than to part from you.
> That glorious day, which two such navies saw,
> As each, unmatched, might to the world give law,
> Neptune, yet doubtful whom he should obey,
> Held to them both the trident of the sea:
> The winds were hushed, the waves in ranks were cast,
> As awfully as when God's people passed:°
> Those, yet uncertain on whose sails to blow,
> These, where the wealth of nations ought to flow.   20
> Then with the Duke your highness ruled the day:
> While all the brave did his command obey,
> The fair and pious under you did pray.
> How powerful are chaste vows! the wind and tide
> You bribed to combat on the English side.
> Thus to your much loved lord you did convey
> An unknown succour, sent the nearest way.
> New vigour to his wearied arms you brought;
> (So Moses was upheld while Israel fought.)°

While, from afar, we heard the cannon play,   30
Like distant thunder on a shiny day,
For absent friends we were ashamed to fear,
When we considered what you ventured there.
Ships, men and arms our country might restore,
But such a leader could supply no more.
With generous thoughts of conquest he did burn,
Yet fought not more to vanquish than return.
Fortune and victory he did pursue,
To bring them, as his slaves, to wait on you.
Thus beauty ravished the rewards of fame,   40
And the fair triumphed when the brave o'ercame.
Then, as you meant to spread another way
By land your conquests far as his by sea,
Leaving our southern clime, you marched along
The stubborn north, ten thousand Cupids strong.
Like commons the nobility resort
In crowding heaps, to fill your moving court:
To welcome your approach the vulgar run,
Like some new envoy from the distant sun.
And country beauties by their lovers go,   50
Blessing themselves, and wondering at the show.
So when the new-born Phoenix first is seen,
Her feathered subjects all adore their queen.
And while she makes her progress through the east,
From every grove her numerous train's increased:
Each poet of the air her glory sings,
And round him the pleased audience clap their wings.

And now, sir, 'tis time I should relieve you from the tedious length
of this account. You have better and more profitable employment for
your hours, and I wrong the public to detain you longer. In
conclusion, I must leave my poem to you with all its faults, which I
hope to find fewer in the printing by your emendations. I know you
are not of the number of those, of whom the younger Pliny° speaks,
Nec sunt parum multi qui carpere amicos suos judicium vocant [there is no
lack of people who think they show good judgment by criticizing their
friends]; I am rather too secure of you on that side. Your candour in
pardoning my errors may make you more remiss in correcting them; if
you will not withal consider that they come into the world with your
approbation, and through your hands. I beg from you the greatest

favour you can confer upon an absent person, since I repose upon your management what is dearest to me, my fame and reputation; and therefore I hope it will stir you up to make my poem fairer by many of your blots. If not, you know the story of the gamester who married the rich man's daughter, and when her father denied the portion, christened all the children by his surname, that if, in conclusion, they must beg, they should do so by one name, as well as by the other. But since the reproach of my faults will light on you, 'tis but reason I should do you that justice to the readers, to let them know that if there be anything tolerable in this poem, they owe the argument to your choice, the writing to your encouragement, the correction to your judgement, and the care of it to your friendship, to which he must ever acknowledge himself to owe all things, who is,

> SIR,
> The most obedient and most faithful of
> your servants,
>
> JOHN DRYDEN

From Charlton in Wiltshire,
November 10, 1666.

In thriving arts long time had Holland grown,
   Crouching at home, and cruel when abroad;
Scarce leaving us the means to claim our own.
   Our king they courted, and our merchants awed.

Trade, which like blood should circularly flow,
   Stopped in their channels, found its freedom lost:
Thither the wealth of all the world did go,
   And seemed but shipwrecked on so base a coast.°

For them alone the heavens had kindly heat,
   In eastern quarries ripening precious dew;°   10
For them the Idumaean balm did sweat,°
   And in hot Ceylon spicy forests grew.

The sun but seemed the labourer of their year;
   Each waxing moon supplied her watery store°
To swell those tides, which from the Line did bear
   Their brim-full vessels to the Belgian shore.

Thus mighty in her ships stood Carthage long,
   And swept the riches of the world from far;
Yet stooped to Rome, less wealthy, but more strong:
   And this may prove our second Punic war.°   20

What peace can be, where both to one pretend,
   But they more diligent, and we more strong?
Or if a peace, it soon must have an end,
   For they would grow too powerful, were it long.

Behold two nations then, engaged so far
   That each seven years the fit must shake each land:
Where France will side to weaken us by war,
   Who only can his vast designs withstand.

See how he feeds the Iberian with delays,°
   To render us his timely friendship vain;   30
And while his secret soul on Flanders preys,
   He rocks the cradle of the babe of Spain.°

Such deep designs of empire does he lay
   O'er them whose cause he seems to take in hand;
And, prudently, would make them lords at sea,
   To whom with ease he can give laws by land.

This saw our king; and long within his breast
   His pensive counsels balanced to and fro;
He grieved the land he freed should be oppressed,
   And he less for it than usurpers do.°   40

His generous mind the fair ideas drew
   Of fame and honour, which in dangers lay;
Where wealth, like fruit on precipices, grew,
   Not to be gathered but by birds of prey.

The loss and gain each fatally were great;
   And still his subjects called aloud for war:
But peaceful kings o'er martial people set,
   Each other's poise and counterbalance are.

He, first, surveyed the charge with careful eyes,
   Which none but mighty monarchs could maintain;   50
Yet judged, like vapours that from limbecs rise,
   It would in richer showers descend again.

At length resolved to assert the watery ball,
   He in himself did whole armados bring;
Him aged seamen might their master call,
   And choose for general, were he not their king.

It seems as every ship their sovereign knows,
   His awful summons they so soon obey;
So hear the scaly herd when Proteus blows,°
   And so to pasture follow through the sea.   60

To see this fleet upon the ocean move
   Angels drew wide the curtains of the skies;
And heaven, as if there wanted lights above,
   For tapers made two glaring comets rise;

Whether they unctuous exhalations are,
   Fired by the sun, or seeming so alone;
Or each some more remote and slippery star,
   Which loses footing when to mortals shown;

Or one that bright companion of the sun,
   Whose glorious aspect sealed our new-born king;   70
And now, a round of greater years begun,
   New influence from his walks of light did bring.

Victorious York did first, with famed success,
   To his known valour make the Dutch give place:
Thus heaven our monarch's fortune did confess,
   Beginning conquest from his royal race.

But since it was decreed, auspicious king,
   In Britain's right that thou shouldst wed the main,
Heaven, as a gage, would cast some precious thing,
   And therefore doomed that Lawson should be slain.°   80

Lawson amongst the foremost met his fate,
   Whom sea-green Sirens from the rocks lament:
Thus as an offering for the Grecian state,
   He first was killed who first to battle went.

Their chief blown up, in air, not waves, expired,°
   To which his pride presumed to give the law:
The Dutch confessed heaven present and retired,
   And all was Britain the wide ocean saw.

To nearest ports their shattered ships repair,
   Where by our dreadful cannon they lay awed: 90
So reverently men quit the open air
   When thunder speaks the angry gods abroad.

The attempt at Bergen

And now approached their fleet from India, fraught
   With all the riches of the rising sun,
And precious sand from southern climates brought,°
   (The fatal regions where the war begun.)

Like hunted castors conscious of their store.°
   Their way-laid wealth to Norway's coasts they bring:
There first the north's cold bosom spices bore,
   And winter brooded on the eastern spring.   100

By the rich scent we found our perfumed prey,
   Which, flanked with rocks, did close in covert lie;
And round about their murdering cannon lay,
   At once to threaten and invite the eye.

Fiercer than cannon, and than rocks more hard,
   The English undertake the unequal war:
Seven ships alone, by which the port is barred,
   Besiege the Indies and all Denmark dare.

These fight like husbands, but like lovers those;
   These fain would keep, and those more fain enjoy;   110
And to such height their frantic passion grows,
   That what both love both hazard to destroy.

Amidst whole heaps of spices lights a ball,
    And now their odours armed against them fly:
Some preciously by shattered porcelain fall,
    And some by aromatic splinters die.

And though by tempests of the prize bereft,
    In heaven's inclemency some ease we find:
Our foes we vanquished by our valour left,
    And only yielded to the seas and wind.   120

Nor wholly lost we so deserved a prey,
    For storms, repenting, part of it restored;
Which, as a tribute from the Baltic sea,
    The British ocean sent her mighty lord.

Go, mortals, now, and vex yourselves in vain
    For wealth, which so uncertainly must come;
When what was brought so far and with such pain
    Was only kept to lose it nearer home.

The son who, twice three months on the ocean tossed,
    Prepared to tell what he had passed before,   130
Now sees in English ships the Holland coast,
    And parents' arms in vain stretched from the shore.

This careful husband had been long away
    Whom his chaste wife and little children mourn,
Who on their fingers learned to tell the day
    On which their father promised to return.

Such are the proud designs of human kind,
    And so we suffer shipwreck everywhere!°
Alas, what port can such a pilot find
    Who in the night of fate must blindly steer!   140

The undistinguished seeds of good and ill
    Heaven in his bosom from our knowledge hides,
And draws them in contempt of human skill,
    Which oft, for friends, mistaken foes provides.

Let Munster's prelate ever be accurst,°
    In whom we seek the German faith in vain:°
Alas, that he should teach the English first
    That fraud and avarice in the Church could reign!

Happy who never trust a stranger's will
    Whose friendship's in his interest understood!   150
Since money given but tempts him to be ill,
    When power is too remote to make him good.

War declared by France

Till now, alone the mighty nations strove:
    The rest at gaze without the lists did stand;
And threatening France, placed like a painted Jove,
    Kept idle thunder in his lifted hand.

That eunuch guardian of rich Holland's trade,
    Who envies us what he wants power to enjoy,
Whose noiseful valour does no foe invade
    And weak assistance will his friends destroy;   160

Offended that we fought without his leave,
    He takes this time his secret hate to show;
Which Charles does with a mind so calm receive
    As one that neither seeks nor shuns his foe.

With France, to aid the Dutch, the Danes unite;
    France as their tyrant, Denmark as their slave;
But when with one three nations join to fight,
    They silently confess that one more brave.

Lewis had chased the English from his shore,
    But Charles the French as subjects does invite;   170
Would heaven for each some Solomon restore,
    Who by their mercy may decide their right!

Were subjects so but only by their choice,
    And not from birth did forced dominion take,
Our prince alone would have the public voice,
    And all his neighbours' realms would deserts make.

He without fear a dangerous war pursues,
　　Which without rashness he began before:
As honour made him first the danger choose,
　　So still he makes it good on virtue's score.   180

The doubled charge his subjects' love supplies,°
　　Who in that bounty to themselves are kind:
So glad Egyptians see their Nilus rise,
　　And in his plenty their abundance find.

Prince Rupert and Duke
Albemarle sent to sea

With equal power he does two chiefs create,
　　Two such, as each seemed worthiest when alone:
Each able to sustain a nation's fate,
　　Since both had found a greater in their own.

　　　　　　　　　　　　　　　　　　　　　　　　　190

Both great in courage, conduct, and in fame,
　　Yet neither envious of the other's praise;
Their duty, faith, and interest too the same,
　　Like mighty partners, equally they raise.

The prince long time had courted fortune's love,
　　But once possessed did absolutely reign:
Thus with their Amazons the heroes strove,
　　And conquered first those beauties they would gain.

The duke beheld, like Scipio, with disdain°
　　That Carthage which he ruined rise once more,
And shook aloft the fasces of the main°   200
　　To fright those slaves with what they felt before.

Together to the watery camp they haste,
　　Whom matrons passing to their children show;
Infants' first vows for them to heaven are cast,
　　And future people bless them as they go.°

With them no riotous pomp nor Asian train
　　To infect a navy with their gaudy fears:
To make slow fights and victories but vain;
　　But war severely, like itself, appears.

Diffusive of themselves, where'er they pass,
   They make that warmth in others they expect: 210
Their valour works like bodies on a glass,
   And does its image on their men project.

Duke of Albemarle's battle, first day

Our fleet divides, and straight the Dutch appear,
   In number and a famed commander bold:°
The narrow seas can scarce their navy bear,
   Or crowded vessels can their soldiers hold.

The duke, less numerous, but in courage more,
   On wings of all the winds to combat flies;
His murdering guns a loud defiance roar
   And bloody crosses on his flag-staffs rise.°   220

Both furl their sails and strip them for the fight;
   Their folded sheets dismiss the useless air;
The Elean plains could boast no nobler sight,°
   When struggling champions did their bodies bare.

Borne each by other in a distant line,
   The sea-built forts in dreadful order move:
So vast the noise, as if not fleets did join,
   But lands unfixed and floating nations strove.°

Now passed, on either side they nimbly tack;
   Both strive to intercept and guide the wind: 230
And in its eye more closely they come back
   To finish all the deaths they left behind.

On high-raised decks the haughty Belgians ride,
   Beneath whose shade our humble frigates go;
Such port the elephant bears, and so defied°
   By the rhinoceros, her unequal foe.

And as the build, so different is the fight;
   Their mounting shot is on our sails designed:
Deep in their hulls our deadly bullets light
   And through the yielding planks a passage find.   240

Our dreaded admiral from far they threat,°
   Whose battered rigging their whole war receives.
All bare, like some old oak which tempests beat,
   He stands, and sees below his scattered leaves.

Heroes of old when wounded shelter sought;
   But he, who meets all danger with disdain,
E'en in their face his ship to anchor brought
   And steeple-high stood propped upon the main.

At this excess of courage, all amazed,
   The foremost of his foes a while withdraw.   250
With such respect in entered Rome they gazed
   Who on high chairs the godlike fathers saw.°

And now, as where Patroclus' body lay,°
   Here Trojan chiefs advanced, and there the Greek,
Ours o'er the duke their pious wings display,
   And theirs the noblest spoils of Britain seek.

Meantime, his busy mariners he hastes
   His shattered sails with rigging to restore;
And willing pines ascend his broken masts,
   Whose lofty heads rise higher than before.   260

Straight to the Dutch he turns his dreadful prow,
   More fierce the important quarrel to decide:
Like swans, in long array his vessels show,
   Whose crests, advancing, do the waves divide.

They charge, recharge, and all along the sea
   They drive, and squander the huge Belgian fleet.
Berkeley alone, not making equal way,°
   Did a like fate with lost Creusa meet.°

The night comes on, we eager to pursue
   The combat still, and they ashamed to leave: 270
Till the last streaks of dying day withdrew,
   And doubtful moonlight did our rage deceive.

In the English fleet each ship resounds with joy
   And loud applause of their great leader's fame:
In fiery dreams the Dutch they still destroy,
   And, slumbering, smile at the imagined flame.

Not so the Holland fleet, who, tired and done,
   Stretched on their decks like weary oxen lie:
Faint sweats all down their mighty members run,
   (Vast bulks, which little souls but ill supply.)   280

In dreams they fearful precipices tread,
   Or, shipwrecked, labour to some distant shore,
Or in dark churches walk among the dead:
   They wake with horror, and dare sleep no more.

Second day's battle

The morn they look on with unwilling eyes,
   Till from their maintop joyful news they hear
Of ships, which by their mould bring new supplies,
   And in their colours Belgian lions bear.

Our watchful general had discerned from far
   This mighty succour which made glad the foe.   290
He sighed, but, like a father of the war,
   His face spake hope, while deep his sorrows flow.°

His wounded men he first sends off to shore,
   (Never, till now, unwilling to obey:)
They not their wounds but want of strength deplore,
   And think them happy who with him can stay.

Then to the rest, 'Rejoice,' said he, 'today
   In you the fortune of Great Britain lies:
Among so brave a people you are they
   Whom heaven has chose to fight for such a prize.   300

'If number English courages could quell,
   We should at first have shunned, not met our foes,
Whose numerous sails the fearful only tell:
   Courage from hearts and not from numbers grows.'

He said, nor needed more to say: with haste
   To their known stations cheerfully they go;
And all at once, disdaining to be last,
   Solicit every gale to meet the foe.

Nor did the encouraged Belgians long delay,
   But bold in others, not themselves, they stood: 310
So thick, our navy scarce could sheer their way,
   But seemed to wander in a moving wood.

Our little fleet was now engaged so far
   That like the swordfish in the whale they fought.
The combat only seemed a civil war,
   Till through their bowels we our passage wrought.

Never had valour, no, not ours before,
   Done aught like this upon the land or main;
Where not to be o'ercome was to do more
   Than all the conquests former kings did gain.   320

The mighty ghosts of our great Harrys rose,
   And armed Edwards looked with anxious eyes,
To see this fleet among unequal foes,
   By which fate promised them their Charles should rise.

Meantime the Belgians tack upon our rear,
   And raking chase-guns through our sterns they send:°
Close by, their fire-ships like jackals appear,
   Who on their lions for the prey attend.

Silent in smoke of cannons they come on:
   (Such vapours once did fiery Cacus hide:)°   330
In these the height of pleased revenge is shown,
   Who burn contented by another's side.

Sometimes from fighting squadrons of each fleet,
   (Deceived themselves or to preserve some friend,)
Two grappling Etnas on the ocean meet,
   And English fires with Belgian flames contend.

Now at each tack our little fleet grows less;
   And, like maimed fowl, swim lagging on the main.
Their greater loss their numbers scarce confess
   While they lose cheaper than the English gain.   340

Have you not seen when, whistled from the fist,
   Some falcon stoops at what her eye designed,
And, with her eagerness the quarry missed,
   Straight flies at check, and clips it down the wind;°

The dastard crow, that to the wood made wing
   And sees the groves no shelter can afford,
With her loud caws her craven kind does bring,
   Who, safe in numbers, cuff the noble bird?

Among the Dutch thus Albemarle did fare:
   He could not conquer and disdained to fly: 350
Past hope of safety, 'twas his latest care,
   Like falling Caesar, decently to die.

Yet pity did his manly spirit move,
   To see those perish who so well had fought;
And, generously, with his despair he strove,
   Resolved to live till he their safety wrought.

Let other muses write his prosperous fate,
   Of conquered nations tell, and kings restored:
But mine shall sing of his eclipsed estate,
   Which, like the sun's, more wonders does afford.   360

He drew his mighty frigates all before,
   On which the foe his fruitless force employs;
His weak ones deep into the rear he bore,
   Remote from guns as sick men are from noise.

His fiery cannon did their passage guide,
   And following smoke obscured them from the foe.
Thus Israel safe from the Egyptian's pride°
   By flaming pillars and by clouds did go.

Elsewhere the Belgian force we did defeat,
   But here our courages did theirs subdue: 370
So Xenophon once led that famed retreat°
   Which first the Asian empire overthrew.

The foe approached; and one, for his bold sin,
   Was sunk, (as he that touched the Ark was slain;)°
The wild waves mastered him and sucked him in,
   And smiling eddies dimpled on the main.

This seen, the rest at awful distance stood;
   As if they had been there as servants set,
To stay, or to go on, as he thought good,
   And not pursue, but wait on his retreat.   380

So Libyan huntsmen on some sandy plain,
   From shady coverts roused, the lion chase:
The kingly beast roars out with loud disdain,
   And slowly moves, unknowing to give place.°

But if some one approach to dare his force,
   He swings his tail and swiftly turns him round,
With one paw seizes on his trembling horse,
   And with the other tears him to the ground.

Amidst these toils succeeds the balmy night;
   Now hissing waters the quenched guns restore;   390
And weary waves, withdrawing from the fight,°
   Lie lulled and panting on the silent shore.

The moon shone clear on the becalmed flood,
   Where, while her beams like glittering silver play,
Upon the deck our careful general stood,
   And deeply mused on the succeeding day.°

'That happy sun', said he, 'will rise again,
   Who twice victorious did our navy see:
And I alone must view him rise in vain,
   Without one ray of all his star for me.   400

'Yet, like an English general will I die,
    And all the ocean make my spacious grave;
Women and cowards on the land may lie,
    The sea's a tomb that's proper for the brave.'

Restless he passed the remnants of the night,
    Till the fresh air proclaimed the morning nigh,
And burning ships, the martyrs of the fight,
    With paler fires beheld the eastern sky.

Third day

But now, his stores of ammunition spent,
    His naked valour is his only guard: 410
Rare thunders are from his dumb cannon sent,
    And solitary guns are scarcely heard.

Thus far had Fortune power, here forced to stay,
    Nor longer durst with virtue be at strife:
This as a ransom Albemarle did pay
    For all the glories of so great a life.

For now brave Rupert's navy did appear,
    Whose waving streamers from afar he knows,
As in his fate something divine there were,
    Who dead and buried the third day arose.°   420

The anxious prince had heard the cannon long,
    And from that length of time dire omens drew
Of English overmatched, and Dutch too strong
    Who never fought three days but to pursue.

Then, as an eagle, who with pious care
    Was beating widely on the wing for prey,
To her now silent eyrie does repair,
    And finds her callow infants forced away;

Stung with her love she stoops upon the plain,
    The broken air loud whistling as she flies;   430
She stops and listens, and shoots forth again,
    And guides her pinions by her young ones' cries:

With such kind passion hastes the prince to fight,
  And spreads his flying canvas to the sound:
Him whom no danger, were he there, could fright,
  Now absent, every little noise can wound.

As, in a drought, the thirsty creatures cry,
  And gape upon the gathered clouds for rain,
And first the martlet meets it in the sky,
  And with wet wings joys all the feathered train;   440

With such glad hearts did our despairing men
  Salute the appearance of the prince's fleet:
And each ambitiously would claim the ken
  That with first eyes did distant safety meet.

The Dutch, who came like greedy hinds before
  To reap the harvest their ripe ears did yield,
Now look like those, when rolling thunders roar,
  And sheets of lightning blast the standing field.

Full in the prince's passage, hills of sand
  And dangerous flats in secret ambush lay,   450
Where the false tides skim o'er the covered land,
  And seamen with dissembled depths betray.

The wily Dutch, who, like fallen angels, feared
  This new Messiah's coming, there did wait,
And round the verge their braving vessels steered
  To tempt his courage with so fair a bait.

But he, unmoved, contemns their idle threat,
  Secure of fame whene'er he please to fight:
His cold experience tempers all his heat,
  And inbred worth does boasting valour slight.   460

Heroic virtue did his actions guide,
  And he the substance, not the appearance, chose:
To rescue one such friend he took more pride
  Than to destroy whole thousands of such foes.

But, when approached, in strict embraces bound,
  Rupert and Albemarle together grow:
He joys to have his friend in safety found,
  Which he to none but to that friend would owe.

The cheerful soldiers, with new stores supplied,
  Now long to execute their spleenful will;   470
And in revenge for those three days they tried,
  Wish one like Joshua's, when the sun stood still.°

Fourth day's battle

Thus reinforced, against the adverse fleet,
  Still doubling ours, brave Rupert leads the way.
With the first blushes of the morn they meet,
  And bring night back upon the new-born day.

His presence soon blows up the kindling fight,
  And his loud guns speak thick like angry men:
It seemed as slaughter had been breathed all night,
  And Death new pointed his dull dart again.   480

The Dutch too well his mighty conduct knew,
  And matchless courage, since the former fight:
Whose navy like a stiff stretched cord did show,
  Till he bore in and bent them into flight.

The wind he shares, while half their fleet offends
  His open side, and high above him shows;
Upon the rest at pleasure he descends,
  And, doubly harmed, he double harms bestows.

Behind, the general mends his weary pace,
  And sullenly to his revenge he sails: 490
So glides some trodden serpent on the grass,°
  And long behind his wounded volume trails.

The increasing sound is borne to either shore
  And for their stakes the throwing nations fear,°
Their passions double with the cannons' roar,
  And with warm wishes each man combats there.

Plied thick and close as when the fight begun,
　　Their huge unwieldy navy wastes away:
So sicken waning moons too near the sun,
　　And blunt their crescents on the edge of day.　　　　500

And now, reduced on equal terms to fight,
　　Their ships like wasted patrimonies show,
Where the thin scattering trees admit the light
　　And shun each other's shadows as they grow.

The warlike prince had severed from the rest
　　Two giant ships, the pride of all the main;
Which, with his one, so vigorously he pressed,
　　And flew so home they could not rise again.

Already battered, by his lee they lay;
　　In vain upon the passing winds they call:　　　　510
The passing winds through their torn canvas play,
　　And flagging sails on heartless sailors fall.

Their opened sides receive a gloomy light,
　　Dreadful as day let in to shades below;
Without, grim Death rides barefaced in their sight,
　　And urges entering billows as they flow.

When one dire shot, the last they could supply,
　　Close by the board the prince's main-mast bore:
All three now helpless by each other lie,
　　And this offends not and those fear no more.　　　　520

So have I seen some fearful hare maintain
　　A course, till tired before the dog she lay;
Who, stretched behind her, pants upon the plain,
　　Past power to kill as she to get away.

With his lolled tongue he faintly licks his prey,
　　His warm breath blows her flix up as she lies;
She, trembling, creeps upon the ground away,
　　And looks back to him with beseeching eyes.

The prince unjustly does his stars accuse,
   Which hindered him to push his fortune on;   530
For what they to his courage did refuse
   By mortal valour never must be done.

This lucky hour the wise Batavian takes
   And warns his tattered fleet to follow home:
Proud to have so got off with equal stakes,
   Where 'twas a triumph not to be o'ercome.°

The general's force, as kept alive by fight,
   Now, not opposed, no longer can pursue:
Lasting till heaven had done his courage right,
   When he had conquered he his weakness knew.   540

He casts a frown on the departing foe,
   And sighs to see him quit the watery field:
His stern fixed eyes no satisfaction show
   For all the glories which the fight did yield.

Though, as when fiends did miracles avow,°
   He stands confessed e'en by the boastful Dutch;
He only does his conquest disavow,
   And thinks too little what they found too much.

Returned, he with the fleet resolved to stay;
   No tender thoughts of home his heart divide;   550
Domestic joys and cares he puts away,
   For realms are households which the great must guide.

As those who unripe veins in mines explore°
   On the rich bed again the warm turf lay,
Till time digests the yet imperfect ore,
   And know it will be gold another day;

So looks our monarch on this early fight,
   The essay and rudiments of great success,
Which all-maturing time must bring to light,
   While he, like heaven, does each day's labour bless.   560

Heaven ended not the first or second day,
 Yet each was perfect to the work designed:
God and kings work, when they their work survey,
 And passive aptness in all subjects find.

His Majesty repairs the fleet

In burdened vessels, first, with speedy care,
 His plenteous stores do seasoned timber send:
Thither the brawny carpenters repair,
 , And as the surgeons of maimed ships attend.

With cord and canvas from rich Hamburg sent
 His navy's moulted wings he imps once more;   570
Tall Norway fir their masts in battle spent,
 And English oak sprung leaks and planks restore.

All hands employed, the royal work grows warm:°
 Like labouring bees on a long summer's day,
Some sound the trumpet for the rest to swarm,
 And some on bells of tasted lilies play;

With gluey wax some new foundation lay
 Of virgin-combs, which from the roof are hung;
Some armed within doors upon duty stay,
 Or tend the sick, or educate the young.   580

So here, some pick out bullets from the sides,
 Some drive old oakum through each seam and rift:°
Their left hand does the caulking-iron guide,
 The rattling mallet with the right they lift.

With boiling pitch another near at hand,
 From friendly Sweden brought, the seams instops,
Which well paid o'er, the salt sea waves withstand,
 And shakes them from the rising beak in drops.

Some the galled ropes with dauby marling bind,
 Or cerecloth masts with strong tarpaulin coats:° 590
To try new shrouds one mounts into the wind,
 And one, below, their ease or stiffness notes.

Our careful monarch stands in person by,
   His new-cast cannons' firmness to explore:
The strength of big-corned powder loves to try,°
   And ball and cartridge sorts for every bore.

Each day brings fresh supplies of arms and men,
   And ships which all last winter were abroad;
And such as fitted since the fight had been,
   Or new from stocks were fallen into the road.   600

Loyal London described

The goodly London in her gallant trim,°
   (The phoenix-daughter of the vanished old,)
Like a rich bride does to the ocean swim,
   And on her shadow rides in floating gold.

Her flag aloft, spread ruffling to the wind,
   And sanguine streamers seem the flood to fire:
The weaver, charmed with what his loom designed,
   Goes on to sea and knows not to retire.

With roomy decks, her guns of mighty strength,
   (Whose low-laid mouths each mounting billow laves,) 610
Deep in her draught, and warlike in her length,
   She seems a sea-wasp flying on the waves.

This martial present, piously designed,
   The loyal City give their best-loved king:
And, with a bounty ample as the wind,
   Built, fitted, and maintained to aid him bring.

Digression concerning shipping and navigation

By viewing nature, nature's handmaid, art,
   Makes mighty things from small beginnings grow:
Thus fishes first to shipping did impart
   Their tail the rudder, and their head the prow.   620

Some log, perhaps, upon the waters swam,
  A useless drift, which, rudely cut within,
And hollowed, first a floating trough became,
  And cross some rivulet passage did begin.

In shipping such as this the Irish kern,
  And untaught Indian, on the stream did glide:
Ere sharp-keeled boats to stem the flood did learn,
  Or fin-like oars did spread from either side.

Add but a sail, and Saturn so appeared,°
  When from lost empire he to exile went,   630
And with the golden age to Tiber steered,
  Where coin and first commerce he did invent.

Rude as their ships was navigation then,
  No useful compass or meridian known:
Coasting, they kept the land within their ken,
  And knew no north but when the pole-star shone.

Of all who since have used the open sea,
  Than the bold English none more fame have won:
Beyond the year, and out of heaven's highway,°
  They make discoveries where they see no sun.   640

But what so long in vain, and yet unknown,
  By poor mankind's benighted wit is sought,
Shall in this age to Britain first be shown,
  And hence be to admiring nations taught.

The ebbs of tides and their mysterious flow
  We as art's elements shall understand,
And as by line upon the ocean go,
  Whose paths shall be familiar as the land.

Instructed ships shall sail to quick commerce,°
  By which remotest regions are allied;   650
Which makes one city of the universe,
  Where some may gain and all may be supplied.

Then, we upon our globe's last verge shall go°
   And view the ocean leaning on the sky:
From thence our rolling neighbours we shall know,°
   And on the lunar world securely pry.

Apostrophe to the Royal Society°

This I foretell, from your auspicious care,
   Who great in search of God and nature grow;
Who best your wise creator's praise declare,
   Since best to praise his works is best to know.   660

O, truly royal! who behold the law
   And rule of beings in your maker's mind,
And thence, like limbecs, rich ideas draw
   To fit the levelled use of human kind.

But first the toils of war we must endure
   And from the injurious Dutch redeem the seas.
War makes the valiant of his right secure,
   And gives up fraud to be chastised with ease.

Already were the Belgians on our coast,
   Whose fleet more mighty every day became   670
By late success, which they did falsely boast,
   And now by first appearing seemed to claim.

Designing, subtle, diligent, and close,
   They know to manage war with wise delay:
Yet all those arts their vanity did cross,
   And by their pride their prudence did betray.

Nor stayed the English long; but, well supplied,
   Appear as numerous as the insulting foe.
The combat now by courage must be tried
   And the success the braver nation show.   680

There was the Plymouth squadron new come in,
   Which in the Straits last winter was abroad:°
Which twice on Biscay's working bay had been,°
   And on the midland sea the French had awed.

Old expert Allin, loyal all along,°
    Famed for his action on the Smyrna fleet;
And Holmes, whose name shall live in epic song,°
    While music numbers, or while verse has feet;

Holmes, the Achates of the generals' fight,
    Who first bewitched our eyes with Guinea gold,   690
As once old Cato in the Romans' sight
    The tempting fruits of Afric did unfold.°

With him went Spragge, as bountiful as brave,°
    Whom his high courage to command had brought;
Harman, who did the twice-fired Harry save°
    And in his burning ship undaunted fought.

Young Holles, on a muse by Mars begot,°
    Born, Caesar-like, to write and act great deeds:
Impatient to revenge his fatal shot,
    His right hand doubly to his left succeeds.   700

Thousands were there in darker fame that dwell,
    Whose deeds some nobler poem shall adorn;
And though to me unknown, they, sure, fought well
    Whom Rupert led, and who were British born.

Of every size a hundred fighting sail,
    So vast the navy now at anchor rides,
That underneath it the pressed waters fail,
    And with its weight it shoulders off the tides.

Now, anchors weighed, the seamen shout so shrill
    That heaven and earth and the wide ocean rings: 710
A breeze from westward waits their sails to fill
    And rests, in those high beds, his downy wings.

The wary Dutch this gathering storm foresaw,
    And durst not bide it on the English coast:
Behind their treacherous shallows they withdraw,
    And there lay snares to catch the British host.

So the false spider, when her nets are spread,
  Deep ambushed in her silent den does lie,
And feels far off the trembling of her thread,
  Whose filmy cord should bind the struggling fly;   720

Then, if at last she finds him fast beset,
  She issues forth and runs along her loom:
She joys to touch the captive in her net,
  And drags the little wretch in triumph home,

The Belgians hoped that with disordered haste
  Our deep-cut keels upon the sands might run;
Or, if with caution leisurely were passed,
  Their numerous gross might charge us one by one.°

But, with a fore-wind pushing them above,
  And swelling tide that heaved them from below,   730
O'er the blind flats our warlike squadrons move°
  And with spread sails to welcome battle go.

It seemed as there the British Neptune stood.
  With all his host of waters at command,
Beneath them to submit the officious flood,
  And with his trident shoved them off the sand.°

To the pale foes they suddenly draw near,
  And summon them to unexpected fight:
They start like murderers when ghosts appear,
  And draw their curtains in the dead of night.   740

Second battle

Now van to van the foremost squadrons meet,
  The midmost battles hasting up behind,°
Who view, far off, the storm of falling sleet,
  And hear their thunder rattling in the wind.

At length the adverse admirals appear;
  (The two bold champions of each country's right:)
Their eyes describe the lists as they come near,
  And draw the lines of death before they fight.

The distance judged for shot of every size,
    The linstocks touch, the ponderous ball expires: 750
The vigorous seaman every porthole plies,
    And adds his heart to every gun he fires.

Fierce was the fight on the proud Belgians' side
    For honour, which they seldom sought before;
But now they by their own vain boasts were tied,
    And forced at least in show to prize it more.

But sharp remembrance on the English part,
    And shame of being matched by such a foe,
Rouse conscious virtue up in every heart,
    And seeming to be stronger makes them so.°   760

Nor long the Belgians could that fleet sustain,
    Which did two generals' fates, and Caesar's bear:°
Each several ship a victory did gain,
    As Rupert or as Albemarle were there.

Their battered admiral too soon withdrew,
    Unthanked by ours for his unfinished fight:
But he the minds of his Dutch masters knew,
    Who called that providence which we called flight.

Never did men more joyfully obey,
    Or sooner understand the sign to fly: 770
With such alacrity they bore away,
    As if to praise them all the states stood by.

O famous leader of the Belgian fleet,°
    Thy monument inscribed such praise shall wear
As Varro, timely flying, once did meet,°
    Because he did not of his Rome despair.

Behold that navy, which a while before
    Provoked the tardy English to the fight,
Now draw their beaten vessels close to shore,
    As larks lie dared to shun the hobby's flight.   780

Whoe'er would English monuments survey
  In other records may our courage know:
But let them hide the story of this day,
  Whose fame was blemished by too base a foe.

Or if too busily they will enquire
  Into a victory which we disdain,
Then let them know the Belgians did retire
  Before the patron saint of injured Spain.°

Repenting England, this revengeful day,
  To Philip's manes did an offering bring:°   790
England, which first, by leading them astray,
  Hatched up rebellion to destroy her king.

Our fathers bent their baneful industry
  To check a monarchy that slowly grew;
But did not France or Holland's fate foresee,
  Whose rising power to swift dominion flew.

In fortune's empire blindly thus we go,
  And wander after pathless destiny;
Whose dark resorts since prudence cannot know,
  In vain it would provide for what shall be.   800

But whate'er English to the blessed shall go,
  And the fourth Harry or first Orange meet;
Find him disowning of a Bourbon foe,°
  And him detesting a Batavian fleet.

Now on their coasts our conquering navy rides,
  Waylays their merchants, and their land besets;
Each day new wealth without their care provides;
  They lie asleep with prizes in their nets.

So, close behind some promontory lie
  The huge leviathans to attend their prey;   810
And give no chase, but swallow in the fry,
  Which through their gaping jaws mistake the way.

Burning of the fleet in the Vlie by Sir Robert Holmes

Nor was this all: in ports and roads remote,
   Destructive fires among whole fleets we send;
Triumphant flames upon the water float,
   And out-bound ships at home their voyage end.

Those various squadrons, variously designed,
   Each vessel freighted with a several load,
Each squadron waiting for a several wind,
   All find but one, to burn them in the road.   820

Some bound for Guinea, golden sand to find,
   Bore all the gauds the simple natives wear;
Some, for the pride of Turkish courts designed,
   For folded turbans finest holland bear.

Some English wool, vexed in a Belgian loom,°
   And into cloth of spungy softness made,
Did into France or colder Denmark doom,
   To ruin with worse ware our staple trade.

Our greedy seamen rummage every hold,°
   Smile on the booty of each wealthier chest,   830
And, as the priests who with their gods make bold,
   Take what they like and sacrifice the rest.

Transitum to the Fire of London°

But ah! how unsincere are all our joys!
   Which sent from heaven, like lightning make no stay:
Their palling taste the journey's length destroys,
   Or grief, sent post, o'ertakes them on the way.

Swelled with our late successes on the foe,
   Which France and Holland wanted power to cross,
We urge an unseen fate to lay us low,
   And feed their envious eyes with English loss.   840

Each element his dread command obeys,
    Who makes or ruins with a smile or frown;
Who, as by one he did our nation raise,
    So now he with another pulls us down.

Yet, London, empress of the northern clime,
    By a high fate thou greatly didst expire;
Great as the world's, which at the death of time°
    Must fall, and rise a nobler frame by fire.

As when some dire usurper heaven provides
    To scourge his country with a lawless sway, 850
His birth perhaps some petty village hides,
    And sets his cradle out of fortune's way;

Till, fully ripe, his swelling fate breaks out,
    And hurries him to mighty mischiefs on;
His prince, surprised at first, no ill could doubt,
    And wants the power to meet it when 'tis known:

Such was the rise of this prodigious fire,
    Which, in mean buildings first obscurely bred,
From thence did soon to open streets aspire,
    And straight to palaces and temples spread. 860

The diligence of trades, and noiseful gain,
    And luxury, more late, asleep were laid;
All was the night's, and in her silent reign
    No sound the rest of nature did invade.

In this deep quiet, from what source unknown,
    Those seeds of fire their fatal birth disclose:
And first few scattering sparks about were blown,
    Big with the flames that to our ruin rose.

Then, in some close-pent room it crept along,
    And, smouldering as it went, in silence fed: 870
Till the infant monster, with devouring strong,
    Walked boldly upright with exalted head.

Now, like some rich or mighty murderer,
    Too great for prison, which he breaks with gold,
Who fresher for new mischiefs does appear,
    And dares the world to tax him with the old;

So scapes the insulting fire his narrow gaol,
    And makes small outlets into open air:
There the fierce winds his tender force assail,
    And beat him downward to his first repair.   880

The winds, like crafty courtesans, withheld°
    His flames from burning but to blow them more:
And, every fresh attempt, he is repelled
    With faint denials, weaker than before.

And now, no longer letted of his prey,
    He leaps up at it with enraged desire,
O'erlooks the neighbours with a wide survey,
    And nods at every house his threatening fire.

The ghosts of traitors from the bridge descend,
    With bold fanatic spectres to rejoice;   890
About the fire into a dance they bend,
    And sing their sabbath notes with feeble voice.

Our guardian angel saw them where he sate
    Above the palace of our slumbering king:
He sighed, abandoning his charge to fate,
    And, drooping, oft looked back upon the wing.

At length the crackling noise and dreadful blaze
    Called up some waking lover to the sight;
And long it was ere he the rest could raise,
    Whose heavy eyelids yet were full of night.   900

The next to danger, hot pursued by fate,
    Half-clothed, half-naked, hastily retire;
And frighted mothers strike their breasts, too late,
    For helpless infants left amidst the fire.

Their cries soon waken all the dwellers near;
    Now murmuring noises rise in every street;
The more remote run stumbling with their fear,
    And in the dark men jostle as they meet.

So weary bees in little cells repose;
    But if night-robbers lift the well-stored hive,   910
A humming through their waxen city grows,
    And out upon each other's wings they drive.

Now streets grow thronged and busy as by day;
    Some run for buckets to the hallowed choir;°
Some cut the pipes, and some the engines play,
    And some more bold mount ladders to the fire.

In vain; for from the east a Belgian wind
    His hostile breath through the dry rafters sent;
The flames impelled soon left their foes behind,
    And forward with a wanton fury went.   920

A quay of fire ran all along the shore,
    And lightened all the river with a blaze;°
The wakened tides began again to roar,
    And wondering fish in shining waters gaze.

Old Father Thames raised up his reverend head,
    But feared the fate of Simois would return;°
Deep in his ooze he sought his sedgy bed,
    And shrank his waters back into his urn.

The fire meantime walks in a broader gross;
    To either hand his wings he opens wide;   930
He wades the streets, and straight he reaches cross
    And plays his longing flames on the other side.

At first they warm, then scorch, and then they take;
    Now with long necks from side to side they feed;
At length, grown strong, their mother-fire forsake,
    And a new colony of flames succeed.

To every nobler portion of the town
   The curling billows roll their restless tide;
In parties now they straggle up and down,
   As armies, unopposed, for prey divide.   940

One mighty squadron, with a side-wind sped,
   Through narrow lanes his cumbered fire does haste,
By powerful charms of gold and silver led
   The Lombard bankers and the Change to waste.°

Another backward to the Tower would go,
   And slowly eats his way against the wind;
But the main body of the marching foe
   Against the imperial palace is designed.

Now day appears, and with the day the king,
   Whose early care had robbed him of his rest: 950
Far off the cracks of falling houses ring,
   And shrieks of subjects pierce his tender breast.

Near as he draws, thick harbingers of smoke
   With gloomy pillars cover all the place;
Whose little intervals of night are broke
   By sparks that drive against his sacred face.

More than his guards his sorrows made him known,
   And pious tears which down his cheeks did shower:
The wretched in his grief forgot their own;
   (So much the pity of a king has power.)   960

He wept the flames of what he loved so well,
   And what so well had merited his love;
For never prince in grace did more excel,
   Or royal city more in duty strove.

Nor with an idle care did he behold:
   (Subjects may grieve, but monarchs must redress;)
He cheers the fearful, and commends the bold,
   And makes despairers hope for good success.

Himself directs what first is to be done,
  And orders all the succours which they bring.   970
The helpful and the good about him run,
  And form an army worthy such a king.

He sees the dire contagion spread so fast
  That, where it seizes, all relief is vain;
And therefore must unwillingly lay waste
  That country which would, else, the foe maintain.

The powder blows up all before the fire:
  The amazed flames stand gathered on a heap,
And from the precipice's brink retire,
  Afraid to venture on so large a leap.   980

Thus fighting fires a while themselves consume,
  But straight, like Turks, forced on to win or die,
They first lay tender bridges of their fume,
  And o'er the breach in unctuous vapours fly.

Part stays for passage till a gust of wind
  Ships o'er their forces in a shining sheet;
Part, creeping under ground, their journey blind,
  And, climbing from below, their fellows meet.

Thus, to some desert plain, or old wood-side,
  Dire night-hags come from far to dance their round;   990
And o'er broad rivers on their fiends they ride,
  Or sweep in clouds above the blasted ground.

No help avails: for, Hydra-like, the fire
  Lifts up his hundred heads to aim his way;
And scarce the wealthy can one half retire
  Before he rushes in to share the prey.

The rich grow suppliant, and the poor grow proud;
  Those offer mighty gain and these ask more:
So void of pity is the ignoble crowd,
  When others' ruin may increase their store.   1000

As those who live by shores with joy behold
   Some wealthy vessel split or stranded nigh;
And, from the rocks, leap down for shipwrecked gold,
   And seek the tempest which the others fly:

So these but wait the owners' last despair,
   And what's permitted to the flames invade:°
E'en from their jaws they hungry morsels tear,
   And on their backs the spoils of Vulcan lade.

The days were all in this lost labour spent:
   And when the weary king gave place to night,   1010
His beams he to his royal brother lent,
   And so shone still in his reflective light.

Night came, but without darkness or repose,
   A dismal picture of the general doom;
Where souls distracted when the trumpet blows,
   And half unready with their bodies come.

Those who have homes, when home they do repair,
   To a last lodging call their wandering friends.
Their short uneasy sleeps are broke with care,
   To look how near their own destruction tends.   1020

Those who have none sit round where once it was,
   And with full eyes each wonted room require;
Haunting the yet warm ashes of the place,
   As murdered men walk where they did expire.

Some stir up coals and watch the vestal fire,
   Others in vain from sight of ruin run;
And, while through burning labyrinths they retire,
   With loathing eyes repeat what they would shun.

The most in fields like herded beasts lie down,
   To dews obnoxious on the grassy floor: 1030
And while their babes in sleep their sorrows drown,
   Sad parents watch the remnants of their store.

While by the motion of the flames they guess
   What streets are burning now, and what are near;
An infant, waking, to the paps would press
   And meets, instead of milk, a falling tear.

No thought can ease them but their sovereign's care,
   Whose praise the afflicted as their comfort sing:
E'en those whom want might drive to just despair,
   Think life a blessing under such a king.   1040

Meantime he sadly suffers in their grief,
   Outweeps a hermit, and outprays a saint:
All the long night he studies their relief,
   How they may be supplied, and he may want.

King's prayer

'O God,' said he, 'thou patron of my days,
   Guide of my youth in exile and distress!
Who me unfriended broughtst by wondrous ways,
   The kingdom of my fathers to possess:

'Be thou my judge, with what unwearied care
   I since have laboured for my people's good;   1050
To bind the bruises of a civil war,
   And stop the issues of their wasting blood.

'Thou, who hast taught me to forgive the ill,
   And recompense, as friends, the good misled;
If mercy be a precept of thy will,
   Return that mercy on thy servant's head.

'Or, if my heedless youth has stepped astray,
   Too soon forgetful of thy gracious hand;
On me alone thy just displeasure lay,
   But take thy judgments from this mourning land.   1060

'We all have sinned, and thou hast laid us low,
   As humble earth from whence at first we came:
Like flying shades before the clouds we show,
   And shrink like parchment in consuming flame.

'O let it be enough what thou hast done,
   When spotted deaths ran armed through every street,°
With poisoned darts, which not the good could shun,
   The speedy could outfly, or valiant meet.

'The living few, and frequent funerals then,°
   Proclaimed thy wrath on this forsaken place;   1070
And now those few, who are returned again,
   Thy searching judgments to their dwellings trace.

'O pass not, lord, an absolute decree,
   Or bind thy sentence unconditional;
But in thy sentence our remorse foresee,
   And, in that foresight, this thy doom recall.

'Thy threatenings, lord, as thine thou mayst revoke;
   But if immutable and fixed they stand,
Continue still thyself to give the stroke,
   And let not foreign foes oppress thy land.'   1080

The eternal heard, and from the heavenly choir
   Chose out the cherub with the flaming sword;
And bade him swiftly drive the approaching fire
   From where our naval magazines were stored.

The blessed minister his wings displayed,
   And like a shooting star he cleft the night;
He charged the flames, and those that disobeyed
   He lashed to duty with his sword of light.

The fugitive flames, chastised, went forth to prey
   On pious structures, by our fathers reared;   1090
By which to heaven they did affect the way,
   Ere faith in churchmen without works was heard.

The wanting orphans saw, with watery eyes,°
   Their founders' charity in dust laid low;
And sent to God their ever-answered cries,
   (For he protects the poor who made them so.)

Nor could thy fabric, Paul's, defend thee long,
  Though thou wert sacred to thy maker's praise;
Though made immortal by a poet's song,°
  And poet's songs the Theban walls could raise.°   1100

The daring flames peeped in and saw from far
  The awful beauties of the sacred choir;
But, since it was profaned by civil war,
  Heaven thought it fit to have it purged by fire.

Now down the narrow streets it swiftly came,
  And, widely opening, did on both sides prey.
This benefit we sadly owe the flame,
  If only ruin must enlarge our way.

And now four days the sun had seen our woes,
  Four nights the moon beheld the incessant fire: 1110
It seemed as if the stars more sickly rose,
  And farther from the feverish north retire.

In the empyrean heaven (the blessed abode)
  The thrones and the dominions prostrate lie,
Not daring to behold their angry God;
  And a hushed silence damps the tuneful sky.

At length the almighty cast a pitying eye,
  And mercy softly touched his melting breast:
He saw the town's one half in rubbish lie,
  And eager flames give on to storm the rest.   1120

A hollow crystal pyramid he takes,
  In firmamental waters dipped above;
Of it a broad extinguisher he makes,
  And hoods the flames that to their quarry strove.

The vanquished fires withdraw from every place,
  Or, full with feeding, sink into a sleep:
Each household genius shows again his face,
  And from the hearths the little Lares creep.

Our king this more than natural change beholds,
    With sober joy his heart and eyes abound;   1130
To the all-good his lifted hands he folds,
    And thanks him low on his redeemed ground.

As, when sharp frosts had long constrained the earth,
    A kindly thaw unlocks it with mild rain,
And first the tender blade peeps up to birth,
    And straight the green fields laugh with promised grain:

By such degrees the spreading gladness grew
    In every heart, which fear had froze before;
The standing streets with so much joy they view,
    That with less grief the perished they deplore.   1140

The father of the people opened wide
    His stores, and all the poor with plenty fed:
Thus God's anointed God's own place supplied,
    And filled the empty with his daily bread.

This royal bounty brought its own reward,
    And in their minds so deep did print the sense,
That, if their ruins sadly they regard,
    'Tis but with fear the sight might drive him thence.

City's request to the king not to leave them

But so may he live long, that town to sway,
    Which by his auspice they will nobler make,   1150
As he will hatch their ashes by his stay,
    And not their humble ruins now forsake.

They have not lost their loyalty by fire;
    Nor is their courage or their wealth so low,
That from his wars they poorly would retire,
    Or beg the pity of a vanquished foe.

Not with more constancy the Jews of old,°
    By Cyrus from rewarded exile sent,
Their royal city did in dust behold,
    Or with more vigour to rebuild it went.   1160

The utmost malice of their stars is past,
  And two dire comets which have scourged the town,
In their own plague and fire have breathed their last,
  Or, dimly, in their sinking sockets frown.

Now frequent trines the happier lights among,°
  And high-raised Jove from his dark prison freed,
(Those weights took off that on his planet hung,)
  Will gloriously the new-laid work succeed.

Methinks already, from this chemic flame,°
  I see a city of more precious mould;   1170
Rich as the town which gives the Indies name,°
  With silver paved, and all divine with gold.

Already, labouring with a mighty fate,
  She shakes the rubbish from her mounting brow,
And seems to have renewed her charter's date,
  Which heaven will to the death of time allow.

More great than human, now, and more august,°
  New deified she from her fires does rise:
Her widening streets on new foundations trust,
  And, opening, into larger parts she flies.   1180

Before, she like some shepherdess did show,
  Who sat to bathe her by a river's side;
Not answering to her fame, but rude and low,
  Nor taught the beauteous arts of modern pride.

Now, like a maiden queen, she will behold,
  From her high turrets, hourly suitors come;
The East with incense, and the West with gold,
  Will stand, like suppliants, to receive her doom.

The silver Thames, her own domestic flood,
  Shall bear her vessels, like a sweeping train,   1190
And often wind (as of his mistress proud)
  With longing eyes to meet her face again.

The wealthy Tagus, and the wealthier Rhine,°
    The glory of their towns no more shall boast;
And Seine, that would with Belgian rivers join,
    Shall find her lustre stained and traffic lost.

The venturous merchant, who designed more far,
    And touches on our hospitable shore,
Charmed with the splendour of this northern star,
    Shall here unlade him and depart no more.   1200

Our powerful navy shall no longer meet,
    The wealth of France or Holland to invade:
The beauty of this town, without a fleet,
    From all the world shall vindicate her trade.

And, while this famed emporium we prepare,
    The British ocean shall such triumphs boast,
That those who now disdain our trade to share
    Shall rob like pirates on our wealthy coast.

Already we have conquered half the war,
    And the less dangerous part is left behind;°   1210
Our trouble now is but to make them dare,
    And not so great to vanquish as to find.

Thus to the Eastern wealth through storms we go;
    But now, the Cape once doubled, fear no more:
A constant trade-wind will securely blow,
    And gently lay us on the spicy shore.

AN ESSAY OF DRAMATIC POESY

To
The Right Honourable Charles
Lord Buckhurst°

My Lord,
As I was lately reviewing my loose papers, amongst the rest I found
this essay, the writing of which,—in this rude and indigested manner
wherein your lordship now sees it,—served as an amusement to me in
the country, when the violence of the last plague° had driven me from

the town. Seeing then our theatres shut up, I was engaged in these kind of thoughts with the same delight with which men think upon their absent mistresses. I confess I find many things in this discourse which I do not now approve—my judgment being a little altered since the writing of it—but whether for the better or the worse, I know not. Neither indeed is it much material in an essay where all I have said is problematical.

For the way of writing plays in verse, which I have seemed to favour, I have since that time laid the practice of it aside till I have more leisure, because I find it troublesome and slow. But I am no way altered from my opinion of it,—at least with any reasons which have opposed it. For your lordship may easily observe that none are very violent against it but those who either have not attempted it, or who have succeeded ill in their attempt.

'Tis enough for me to have your lordship's example for my excuse in that little which I have done in it; and I am sure my adversaries can bring no such arguments against verse as the fourth act of *Pompey*° will furnish me with in its defence. Yet, my lord, you must suffer me a little to complain of you that you too soon withdraw from us a contentment, of which we expected the continuance, because you gave it us so early. 'Tis a revolt without occasion from your party, where your merits had already raised you to the highest commands, and where you have not the excuse of other men—that you have been ill used, and therefore laid down arms. I know no other quarrel you can have to verse than that which Spurina° had to his beauty, when he tore and mangled the features of his face only because they pleased too well the lookers-on. It was an honour which seemed to wait for you, to lead out a new colony of writers from the mother nation: and upon the first spreading of your ensigns, there had been many in a readiness to have followed so fortunate a leader—if not all, yet the better part of writers.

> pars indocili melior grege; mollis et exspes
> inominata perprimat cubilia.°

[the part better than the ignorant mob; the weak and faint-hearted may remain on their unlucky beds]

I am almost of opinion that we should force you to accept of the command, as sometimes the praetorian bands have compelled their captains to receive the empire. The court, which is the best and surest judge of writing, has generally allowed of verse, and in the town it has found favourers of wit and quality. As for your own particular, my lord, you have yet youth and time enough to give part of it to the divertisement of the public, before you enter into the serious and more unpleasant business of the world. That which the French poet said of the temple of love, may be as well applied to the temple of the muses. The words, as near as I can remember them, were these:

> Le jeune homme a mauvaise grâce,
> N'ayant pas adoré dans le temple d'Amour,
> Il faut qu'il entre: et pour le sage,
> Si ce n'est pas son vrai séjour,
> C'est un gîte sur son passage.°

[the youth is in ill-humour for not having worshipped in the temple of love. He must go in. And for the wise man, if it is not a permanent abode, it is at least a halt on his way]

I leave the words to work their effect upon your lordship in their own language (because no other can so well express the nobleness of the thought); and wish you may be soon called to bear a part in the affairs of the nation, where I know the world expects you, and wonders why you have been so long forgotten—there being no person amongst our young nobility, on whom the eyes of all men are so much bent. But in the meantime your lordship may imitate the course of nature, who gives us the flower before the fruit, that I may speak to you in the language of the muses, which I have taken from an excellent poem to the king:

> As nature, when she fruit designs, thinks fit
> By beauteous blossoms to proceed to it;
> And while she does accomplish all the spring,
> Birds to her secret operations sing.°

I confess I have no greater reason in addressing this essay to your lordship than that it might awaken in you the desire of writing something (in whatever kind it be) which might be an honour to our age and country. And methinks it might have the same effect upon you which Homer° tells us the fight of the Greeks and Trojans before the fleet had on the spirit of Achilles, who, though he had resolved not to engage, yet found a martial warmth to steal upon him at the sight of blows, the sound of trumpets, and the cries of fighting men.

For my own part, if in treating of this subject I sometimes dissent from the opinion of better wits, I declare it is not so much to combat their opinions as to defend mine own,° which were first made public.° Sometimes, like a scholar in a fencing-school, I put forth myself, and show my own ill play, on purpose to be better taught. Sometimes I stand desperately to my arms,—like the foot when deserted by their horse, not in hope to overcome, but only to yield on more honourable terms.

And yet, my lord, this war of opinions, you well know, has fallen out among the writers of all ages, and sometimes betwixt friends. Only it has been prosecuted by some, like pedants, with violence of words, and managed by others, like gentlemen, with candour and civility. Even Tully had a controversy with his dear Atticus, and in one of his dialogues° makes him sustain the part of an enemy in philosophy, who in his letters is his confident of state and made privy to the most weighty affairs of the Roman senate. And the same respect which was paid by Tully to Atticus we find returned to him afterwards by Caesar on a like occasion, who, answering his book in praise of Cato, made it not so much his business to condemn Cato as to praise Cicero.°

But that I may decline some part of the encounter with my adversaries, whom I am neither willing to combat nor well able to resist, I will give your lordship the relation of a dispute betwixt some of our wits upon this subject, in which they did not only speak of plays in verse, but mingled in the freedom of discourse some things of the ancient, many of the modern ways of writing—comparing those with these, and the wits of our nation with those of others. 'Tis true, they differed in their opinions—as 'tis probable they would: neither do I take upon me to reconcile, but to relate them; and that as Tacitus professes of himself, *sine studio partium aut ira°* without passion or interest, leaving your lordship to decide it in favour of which part you shall judge most reasonable, and withal to pardon the many errors of

Your lordship's most obedient humble servant,
JOHN DRYDEN.

To
The Reader

The drift of the ensuing discourse was chiefly to vindicate the honour of our English writers from the censure of those who unjustly prefer the French before them. This I intimate, lest any should think me so exceeding vain as to teach others an art which they understand much better than myself. But if this incorrect essay, written in the country without the help of books or advice of friends, shall find any acceptance in the world, I promise to myself a better success of the second part, wherein the virtues and faults of the English poets who have written either in this, the epic, or the lyric way, will be more fully treated of, and their several styles impartially imitated.

An
ESSAY OF DRAMATIC POESY

It was that memorable day° in the first summer of the late war° when our navy engaged the Dutch—a day wherein the two most mighty and best appointed fleets which any age had ever seen disputed the command of the greater half of the globe, the commerce of nations, and the riches of the universe.

While these vast floating bodies, on either side, moved against each other in parallel lines, and our countrymen, under the happy conduct of his royal highness,° went breaking by little and little into the line of the enemies, the noise of the cannon from both navies reached our ears about the city, so that all men being alarmed with it, and in a dreadful suspense of the event which we knew was then deciding, everyone went following the sound as his fancy led him; and, leaving the town almost empty, some took towards the park,° some cross the river, others down it—all seeking the noise in the depth of silence.

Amongst the rest, it was the fortune of Eugenius, Crites, Lisideius, and Neander° to be in company together—three of them persons whom their wit and quality have made known to all the town, and whom I have chosen to hide under these borrowed names, that they may not suffer by so ill a relation as I am going to make of their discourse.

Taking then a barge which a servant of Lisideius had provided for them, they made haste to shoot the bridge, and left behind them that great fall of waters which hindered them from hearing what they desired: after which, having disengaged themselves from many vessels which rode at anchor in the Thames, and almost blocked up the passage towards Greenwich, they ordered the watermen to let fall their oars more gently; and then—everyone favouring his own curiosity with a strict silence, it was not long ere they perceived the air break about them like the noise of distant thunder, or of swallows in a chimney—those little undulations of sound, though almost vanishing before they reached them, yet still seeming to retain somewhat of their first horror which they had betwixt the fleets.

After they had attentively listened till such time as the sound by little and little went from them, Eugenius, lifting up his head, and taking notice of it, was the first who congratulated to the rest that happy omen of our nation's victory, adding that we had but this to desire in confirmation of it, that we might hear no more of that noise which was now leaving the English coast.

When the rest had concurred in the same opinion, Crites—a person of a sharp judgment, and somewhat too delicate a taste in wit, which the world has mistaken in him for ill-nature—said, smiling to us, that if the concernment of this battle had not been so exceeding great, he could scarce have wished the victory at the price he knew he must pay for it, in being subject to the reading and hearing of so many ill verses as he was sure would be made upon it,—adding that no argument could scape some of those eternal rhymers, who watch a battle with more diligence than the ravens and birds of prey, and the worst of them surest to be first in upon the quarry, while the better able either out of modesty writ not at all, or set that due value upon their poems, as to let them be often called for and long expected.

'There are some of those impertinent people you speak of', answered Lisideius, 'who to my knowledge are already so provided either way, that they can produce not only a panegyric upon the victory, but, if need be, a funeral elegy on the duke—and after they have crowned his valour with many laurels, at last deplore the odds under which he fell, concluding that his courage deserved a better destiny.'

All the company smiled at the conceit of Lisideius, but Crites, more eager than before, began to make particular exceptions against some writers, and said the public magistrate ought to send betimes to forbid them, and that it concerned the peace and quiet of all honest people, that ill poets should be as well silenced as seditious preachers.

'In my opinion', replied Eugenius, 'you pursue your point too far; for, as to my own particular, I am so great a lover of poesy that I could wish them all rewarded who attempt but to do well—at least, I would not have them worse used than Sylla the dictator did one of their brethren heretofore, *Quem in concione vidimus* (says Tully speaking of him) *cum ei libellum malus poeta de populo subjecisset, quod epigramma in eum fecisset tantummodo alternis versibus longiusculis, statim, ex iis rebus quas tunc vendebat jubere ei praemium tribui, sub ea conditione ne quid postea scriberet* [we have seen him at a public meeting, when a bad poet from the crowd handed up an epigram on him, improvised in bad verses, immediately order him to be paid out of the proceeds of the sale on condition that he should never write again].'°

'I could wish with all my heart', replied Crites, 'that many whom we know were as bountifully thanked upon the same condition, that they would never trouble us again. For, amongst others, I have a mortal apprehension of two poets,° whom this victory with the help of both her wings will never be able to escape.'

''Tis easy to guess whom you intend', said Lisideius, 'and without naming them, I ask you if one of them does not perpetually pay us with clenches upon words, and a certain clownish kind of raillery:—if now and then he does not offer at a catachresis° or Clevelandism, wresting and torturing a word into another meaning:—in fine, if he be not one of those whom the French would call *un mauvais buffon* [a poor clown],—one who is so much a well-willer to the satire, that he spares no man, and, though he cannot strike a blow to hurt any, yet ought to be punished for the malice of the action—as our witches are justly hanged because they think themselves so, and suffer deservedly for believing they did mischief, because they meant it.'

'You have described him', said Crites, 'so exactly that I am afraid to come after you with my other extremity of poetry. He is one of those who, having had some advantage of education and converse, knows better than the other what a poet should be, but puts it into practice more unluckily than any man. His style and matter are everywhere alike. He is the most calm, peaceable writer you ever read. He never disquiets your passions with the least concernment, but still leaves you in as even a temper as he found you. He is a very leveller° in poetry. He creeps along with ten little words in every line, and helps out his numbers with *for to* and *unto*, and all the pretty expletives he can find, till he drags them to the end of another line, while the sense is left tired half-way behind it. He doubly starves all his verses,—first, for want of thought, and then of expression. His poetry neither has wit in it, nor seems to have it. Like him in Martial,

> pauper videri Cinna vult, et est pauper.°

[Cinna wished to seem poor and thereby is poor]

He affects plainness, to cover his want of imagination. When he writes the serious way, the highest flight of his fancy is some miserable antithesis or seeming contradiction, and in the comic he is still reaching at some thin conceit, the ghost of a jest, and that too flies before him, never to be caught. These swallows which we see before us on the Thames are the just resemblance of his wit. You may observe how near the water they stoop, how many proffers they make to dip, and yet how seldom they touch it; and when they do, 'tis but the surface. They skim over it but to catch a gnat, and then mount into the air and leave it.'

(Well, gentlemen,' said Eugenius, 'you may speak your pleasure of these authors. But though I and some few more about the town may give you a peaceable hearing, yet assure yourselves there are

multitudes who would think you malicious and them injured—
especially him whom you first described. He is the very Withers° of
the city. They have bought more editions of his works than would
serve to lay under all their pies at the lord mayor's Christmas. When
his famous poem first came out in the year 1660, I have seen them
reading it in the midst of Change-time.° Nay, so vehement they were
at it, that they lost their bargain by the candles' ends.° But what will
you say if he has been received amongst the great ones? I can assure
you he is this day the envy of a great person who is lord in the art of
quibbling, and who does not take it well that any man should intrude
so far into his province.'

'All I would wish', replied Crites, 'is that they who love his writings
may still admire him and his fellow-poet. *Qui Bavium non odit etc.* [let
him who does not hate Bavius (love your songs, Maevius)],° is curse
sufficient.'

'And further', added Lisideius, 'I believe there is no man who
writes well but would think himself very hardly dealt with if their
admirers should praise anything of his: *nam quos contemnimus, eorum
quoque laudes contemnimus* [we despise those who praise what we
despise].'°

'There are so few who write well in this age', said Crites, 'that
methinks any praises should be welcome. They neither rise to the
dignity of the last age, nor to any of the ancients, and we may cry out
of the writers of this time, with more reason than Petronius of his, *pace
vestra liceat dixisse, primi omnium eloquentiam perdidistis* [if you will
allow me, I must tell you that you teachers have killed real eloquence].°
You have debauched the true old poetry so far that nature, which is
the soul of it, is not in any of your writings.'

'If your quarrel', said Eugenius, 'to those who now write be
grounded only upon your reverence to antiquity, there is no man more
ready to adore those great Greeks and Romans than I am. But, on the
other side, I cannot think so contemptibly of the age I live in, or so
dishonourably of my own country, as not to judge we equal the
ancients in most kinds of poesy, and in some surpass them. Neither
know I any reason why I may not be as zealous for the reputation of
our age as we find the ancients themselves in reference to those who
lived before them. For you hear your Horace saying,

> indignor quidquam reprehendi, non quia crasse
> compositum illepideve putetur, sed quia nuper.°

[I am angry when any work is censured, not because it is thought coarse or
inelegant in style, but because it is modern]

And after,

> si meliora dies, ut vina, poemata reddit,
> scire velim pretium chartis quotus arroget annus?°

[if poems are like wine which is improved by time, I should like to know
which year gives fresh value to literature]

'But I see I am engaging in a wide dispute, where the arguments are
not like to reach close on either side. For poesy is of so large extent,
and so many both of the ancients and moderns have done well in all
kinds of it, that, in citing one against the other, we shall take up more
time this evening than each man's occasions will allow him. Therefore
I would ask Crites to what part of poesy he would confine his
arguments, and whether he would defend the general cause of the
ancients against the moderns, or oppose any age of the moderns
against this of ours?'

Crites, a little while considering upon this demand, told Eugenius
he approved his propositions, and, if he pleased, he would limit their
dispute to dramatic poesy, in which he thought it not difficult to
prove, either that the ancients were superior to the moderns, or the
last age to this of ours.

Eugenius was somewhat surprised when he heard Crites make
choice of that subject. 'For ought I see', said he, 'I have undertaken a
harder province than I imagined. For though I never judged the
plays of the Greek or Roman poets comparable to ours, yet, on the
other side, those we now see acted come short of many which were
written in the last age. But my comfort is, if we are overcome, it will be
only by our own countrymen, and if we yield to them in this one part
of poesy, we more surpass them in all the other. For in the epic or lyric
way it will be hard for them to show us one such amongst them as we
have many now living, or who lately were so. They can produce
nothing so courtly writ, or which expresses so much the conversation
of a gentleman, as Sir John Suckling,° nothing so even, sweet, and
flowing, as Mr Waller,° nothing so majestic, so correct, as Sir John
Denham,° nothing so elevated, so copious, and full of spirit, as Mr
Cowley.° As for the Italian, French, and Spanish plays, I can make it
evident that those who now write surpass them, and that the drama is
wholly ours.'

All of them were thus far of Eugenius's opinion that the sweetness
of English verse was never understood or practised by our fathers—
even Crites himself did not much oppose it and everyone was willing
to acknowledge how much our poesy is improved by the happiness of

some writers yet living, who first taught us to mould our thoughts into easy and significant words, to retrench the superfluities of expression, and to make our rhyme so properly a part of the verse, that it should never mislead the sense, but itself be led and governed by it.

Eugenius was going to continue this discourse, when Lisideius told him it was necessary, before they proceeded further, to take a standing measure of their controversy. For how was it possible to be decided who writ the best plays, before we know what a play should be? But, this once agreed on by both parties, each might have recourse to it, either to prove his own advantages, or discover the failings of his adversary.

He had no sooner said this but all desired the favour of him to give the definition of a play, and they were the more importunate because neither Aristotle, nor Horace, nor any other who writ of that subject, had ever done it.

Lisideius, after some modest denials, at last confessed he had a rude notion of it, indeed, rather a description than a definition, but which served to guide him in his private thoughts when he was to make a judgment of what others writ, that he conceived a play ought to be,

A just and lively image of human nature, representing its passions and humours, and the changes of fortune to which it is subject, for the delight and instruction of mankind.

This definition (though Crites raised a logical objection against it, that it was only *a genere et fine*,° and so not altogether perfect) was yet well received by the rest: and, after they had given order to the watermen to turn their barge and row softly, that they might take the cool of the evening in their return, Crites, being desired by the company to begin, spoke on behalf of the ancients in this manner:

'If confidence presage a victory, Eugenius—in his own opinion—has already triumphed over the ancients. Nothing seems more easy to him than to overcome those whom it is our greatest praise to have imitated well. For we do not only build upon their foundation, but by their models. Dramatic poesy had time enough, reckoning from Thespis (who first invented it) to Aristophanes, to be born, to grow up, and to flourish in maturity. It has been observed of arts and sciences, that in one and the same century they have arrived to a great perfection, and no wonder, since every age has a kind of universal genius, which inclines those that live in it to some particular studies. The work then, being pushed on by many hands, must of necessity go forward.

'Is it not evident, in these last hundred years (when the study of philosophy has been the business of all the virtuosi in Christendom), that almost a new nature has been revealed to us—that more errors of the school° have been detected, more useful experiments in philosophy have been made, more noble secrets in optics, medicine, anatomy, astronomy, discovered, than in all those credulous and doting ages from Aristotle to us? So true it is that nothing spreads more fast than science, when rightly and generally cultivated.

'Add to this the more than common emulation that was in those times of writing well, which, though it be found in all ages and all persons that pretend to the same reputation, yet poesy, being then in more esteem than now it is, had greater honours decreed to the professors of it, and consequently the rivalship was more high between them. They had judges ordained to decide their merit, and prizes to reward it, and historians have been diligent to record of Aeschylus, Euripides, Sophocles, Lycophron,° and the rest of them, both who they were that vanquished in these wars of the theatre, and how often they were crowned, while the Asian kings and Grecian commonwealths scarce afforded them a nobler subject than the unmanly luxuries of a debauched court, or giddy intrigues of a factious city. *Alit aemulatio ingenia* (says Paterculus), *et nunc invidia, nunc admiratio incitationem accendit.*° "Emulation is the spur of wit; and sometimes envy, sometimes admiration, quickens our endeavours."

'But now, since the rewards of honour are taken away, that virtuous emulation is turned into direct malice, yet so slothful, that it contents itself to condemn and cry down others without attempting to do better. 'Tis a reputation too unprofitable to take the necessary pains for it. Yet wishing they had it is incitement enough to hinder others from it. And this, in short, Eugenius, is the reason why you have now so few good poets, and so many severe judges.

'Certainly, to imitate the ancients well, much labour and long study is required, which pains I have already shown, our poets would want encouragement to take, if yet they had ability to go through with it. Those ancients have been faithful imitators and wise observers of that nature which is so torn and ill-represented in our plays. They have handed down to us a perfect resemblance of her, which we, like ill copiers, neglecting to look on, have rendered monstrous and disfigured.

'But, that you may know how much you are indebted to those your masters, and be ashamed to have so ill requited them, I must remember you that all the rules by which we practise the drama at this day—either such as relate to the justness and symmetry of the plot, or the episodical ornaments, such as descriptions, narrations, and other beauties, which are not essential to the play—were delivered to us from the observations that Aristotle made of those poets which either lived before him or were his contemporaries. We have added nothing of our own, except we have the confidence to say our wit is better, which none boast of in our age but such as understand not theirs. Of that book which Aristotle has left us, περὶ τῆς Ποιητικῆς,° Horace's *Art of Poetry* is an excellent comment, and, I believe, restores to us that second book of his concerning comedy, which is wanting in him.

'Out of these two has been extracted° the famous rules which the French call *Des Trois Unités*, or The Three Unities, which ought to be observed in every regular play, namely, of time, place, and action.

'The unity of time they comprehend in twenty-four hours, the compass of a natural day, or as near it as can be contrived; and the reason of it is obvious to everyone, that the time of the feigned action or fable of the play should be proportioned as near as can be to the duration of that time in which it is represented. Since therefore all plays are acted on the theatre in a space of time much within the compass of twenty-four hours, that play is to be thought the nearest imitation of nature whose plot or action is confined within that time; and, by the same rule which concludes this general proportion of time, it follows that all the parts of it are to be equally subdivided—as, namely, that one act take not up the supposed time of half a day, which is out of proportion to the rest, since the other four are then to be straitened within the compass of the remaining half. For it is unnatural that one act, which being spoke or written is not longer than the rest, should be supposed longer by the audience: 'Tis therefore the poet's duty to take care that no act should be imagined to exceed the time in which it is represented on the stage, and that the intervals and inequalities of time be supposed to fall out between the acts.

'This rule of time, how well it has been observed by the ancients, most of their plays will witness. You see them in their tragedies (wherein to follow this rule is certainly most difficult), from the very beginning of their plays, falling close into that part of the story which they intend for the action or principal object of it, leaving the former part to be delivered by narration: so that they set the audience, as it were, at the post where the race is to be concluded: and, saving them the tedious expectation of seeing the poet set out and ride the beginning of the course, you behold him not till he is in sight of the goal, and just upon you.

'For the second unity, which is that of place, the ancients meant by it that the scene ought to be continued through the play in the same place where it was laid in the beginning: for the stage, on which it is represented, being but one and the same place, it is unnatural to conceive it many, and those far distant from one another. I will not deny but by the variation of painted scenes the fancy (which in these cases will contribute to its own deceit) may sometimes imagine it several places, with some appearance of probability. Yet it still carries the greater likelihood of truth, if those places be supposed so near each other, as in the same town or city, which may all be comprehended under the larger denomination of one place: for a greater distance will bear no proportion to the shortness of time which is allotted in the acting to pass from one of them to another.

'For the observation of this, next to the ancients, the French are to be most commended. They tie themselves so strictly to the unity of place that you never see in any of their plays a scene changed in the middle of an act. If the act begins in a garden, a street, or chamber, 'tis ended in the same place; and that you may know it to be the same, the stage is so supplied with persons that it is never empty all the time, he that enters the second has business with him who was on before, and before the second quits the stage, a third appears who has business with him.

'This Corneille calls la liaison des scènes, "the continuity or joining of the scenes", and 'tis a good mark of a well-contrived play when all the persons are known to each other, and every one of them has some affairs with all the rest.

'As for the third unity, which is that of action, the ancients meant no other by it than what the logicians do by their *finis*, the end or scope of any action, that which is the first in intention and last in execution. Now the poet is to aim at one great and complete action, to the carrying on of which all things in his play, even the very obstacles, are to be subservient; and the reason of this is as evident as any of the former.

'For two actions, equally laboured and driven on by the writer, would destroy the unity of the poem. It would be no longer one play, but two. Not but that there may be many actions in a play, as Ben Jonson has observed in his *Discoveries*,° but they must be all subservient to the great one, which our language happily expresses in the name of underplots,—such as in Terence's *Eunuch* is the difference and reconcilement of Thais and Phaedria, which is not the chief business of the play, but promotes the marriage of Chaerea and

Chremes's sister, principally intended by the poet. There ought to be but one action, says Corneille,°—that is one complete action, which leaves the mind of the audience in a full repose: but this cannot be brought to pass but by many other imperfect ones which conduce to it, and hold the audience in a delightful suspense of what will be.

'If by these rules (to omit many others drawn from the precepts and practice of the ancients) we should judge our modern plays, 'tis probable that few of them would endure the trial. That which should be the business of a day, takes up in some of them an age. Instead of one action, they are the epitomes of a man's life. And for one spot of ground (which the stage should represent) we are sometimes in more countries than the map can show us.

'But if we will allow the ancients to have contrived well, we must acknowledge them to have writ better. Questionless we are deprived of a great stock of wit in the loss of Menander° amongst the Greek poets, and of Caecilius, Affranius, and Varius among the Romans. We may guess of Menander's excellence by the plays of Terence, who translated some of his (and yet wanted so much of him that he was called by C. Caesar the half-Menander),° and of Varius by the testimonies of Horace, Martial, and Velleius Paterculus.° 'Tis probable that these, could they be recovered, would decide the controversy. But so long as Aristophanes in the old comedy and Plautus in the new are extant, while the tragedies of Euripides, Sophocles, and Seneca are to be had, I can never see one of those plays which are now written, but it increases my admiration of the ancients.

'And yet I must acknowledge further that, to admire them as we ought, we should understand them better than we do. Doubtless many things appear flat to us, whose wit depended upon some custom or story which never came to our knowledge, or perhaps upon some criticism in their language, which being so long dead, and only remaining in their books, 'tis not possible they should make us know it perfectly. To read Macrobius° explaining the propriety and elegancy of many words in Virgil, which I had before passed over without consideration as common things, is enough to assure me that I ought to think the same of Terence, and that in the purity of his style (which Tully so much valued that he ever carried his works about him) there is yet left in him great room for admiration, if I knew but where to place it.

'In the meantime I must desire you to take notice that the greatest man of the last age, Ben Jonson, was willing to give place to them in all things. He was not only a professed imitator of Horace, but a learned plagiary of all the others. You track him everywhere in their snow. If Horace, Lucan, Petronius Arbiter, Seneca, and Juvenal had their own from him, there are few serious thoughts which are new in him. You will pardon me, therefore, if I presume he loved their fashion when he wore their clothes. But since I have otherwise a great veneration for him, and you, Eugenius, prefer him above all other poets, I will use no further argument to you than his example. I will produce Father Ben to you dressed in all the ornaments and colours of the ancients. You will need no other guide to our party if you follow him; and—whether you consider the bad plays of our age, or regard the good ones of the last,—both the best and worst of the modern poets will equally instruct you to esteem the ancients.'

Crites had no sooner left speaking but Eugenius, who waited with some impatience for it, thus began:

'I have observed in your speech that the former part of it is convincing as to what the moderns have profited by the rules of the ancients; but in the latter you are careful to conceal how much they have excelled them. We own all the helps we have from them, and want neither veneration nor gratitude while we acknowledge that to overcome them we must make use of all the advantages we have received from them. But to these assistances we have joined our own industry. For, had we sat down with a dull imitation of them, we might then have lost somewhat of the old perfection, but never acquired any that was new. We draw not therefore after their lines, but those of nature, and having the life before us, besides the experience of all they knew, it is no wonder if we hit some airs and features which they have missed.

'I deny not what you urge of arts and sciences, that they have flourished in some ages more than others. But your instance in philosophy makes for me. For if natural causes be more known now that in the time of Aristotle, because more studied, it follows that poesy and other arts may, with the same pains, arrive still nearer to perfection. And, that granted, it will rest for you to prove that they wrought more perfect images of human life than we,—which seeing in your discourse you have avoided to make good, it shall now be my task to show you some part of their defects, and some few excellences of the moderns. And I think there is none among us can imagine I do it enviously, or with purpose to detract from them. For what interest of fame or profit can the living lose by the reputation of the dead? On the other side, it is a great truth which Velleius Paterculus affirms,—*audita visis libentius laudamus; et praesentia invidia, praeterita admiratione prosequimur; et his nos obrui, illis instrui credimus* [we tend to praise what we have heard more than what we have seen; we view the present with envy, the past with admiration, and believe that the present eclipses us, while the past teaches].° That praise or censure is certainly the most sincere which unbribed posterity shall give us.

'Be pleased then, in the first place, to take notice that the Greek poesy, which Crites has affirmed to have arrived to perfection in the reign of the old comedy, was so far from it, that the distinction of it into acts was not known to them; or if it were, it is yet so darkly delivered to us that we cannot make it out.

'All we know of it is from the singing of their chorus,—and that too is so uncertain that in some of their plays we have reason to conjecture they sang more than five times. Aristotle° indeed divides the integral parts of a play into four:

'First, the *protasis*, or entrance, which gives light only to the characters of the persons, and proceeds very little into any part of the action.

'Secondly, the *epitasis*, or working up of the plot, where the play grows warmer. The design or action of it is drawing on, and you see something promising that it will come to pass.

'Thirdly, the *catastasis*, or counter-turn, which destroys that expectation, embroils the action in new difficulties, and leaves you far distant from that hope in which it found you,—as you may have observed in a violent stream resisted by a narrow passage, it runs round to an eddy, and carries back the waters with more swiftness than it brought them on.

'Lastly, the *catastrophe*, which the Grecians called λύσις,° the French *le dénouement*, and we the discovery or unravelling of the plot. There you see all things settling again upon their first foundations, and, the obstacles which hindered the design or action of the play once removed, it ends with that resemblance of truth and nature, that the audience are satisfied with the conduct of it.

'Thus this great man delivered to us the image of a play. And I must confess it is so lively, that from thence much light has been derived to the forming it more perfectly into acts and scenes. But what poet first limited to five the number of the acts, I know not, only we see it so firmly established in the time of Horace, that he gives it for a rule in comedy,

> neu brevior quinto, neu sit productior actu.°

[let the acts be neither fewer than five, nor more]

So that you see the Grecians cannot be said to have consummated this art, writing rather by entrances than by acts, and having rather a general indigested notion of a play, than knowing how and where to bestow the particular graces of it.

'But since the Spaniards at this day allow but three acts, which they call *jornadas*, to a play, and the Italians in many of theirs follow them, when I condemn the ancients, I declare it is not altogether because they have not five acts to every play, but because they have not confined themselves to one certain number. 'Tis building a house without a model. And when they succeeded in such undertakings, they ought to have sacrificed to fortune, not to the muses.

'Next, for the plot (which Aristotle called τὸ μῦθος, and often τῶν πραγμάτων σύνθεσις [the arrangement of the incidents], and from him the Romans *fabula*). It has already been judiciously observed by a late writer° that in their tragedies it was only some tale derived from Thebes or Troy,—or at least something that happened in those two ages,—which was worn so threadbare by the pens of all the epic poets, and even by tradition itself of the talkative Greeklings (as Ben Jonson° calls them), that before it came upon the stage, it was already known to all the audience. And the people, so soon as ever they heard the name of Oedipus, knew as well as the poet that he had killed his father by a mistake, and committed incest with his mother, before the play; that they were now to hear of a great plague, an oracle, and the ghost of Laius. So that they sat with a yawning kind of expectation till he was to come with his eyes pulled out, and speak a hundred or two of verses in a tragic tone in complaint of his misfortunes. But one Oedipus, Hercules, or Medea had been tolerable. Poor people, they scaped not so good cheap!° They had still the *chapon bouilli* [boiled capon] set before them, till their appetites were cloyed with the same dish, and, the novelty being gone, the pleasure vanished. So that one main end of dramatic poesy in its definition, which was to cause delight, was of consequence destroyed.

'In their comedies, the Romans generally borrowed their plots from the Greek poets. And theirs were commonly a little girl stolen or wandered from her parents, brought back unknown to the same city, there got with child by some lewd young fellow, who, by the help of his servants, cheats his father; and when her time comes to cry *Juno Lucina fer opem* [goddess of childbirth help me],° one or other sees a little box or cabinet which was carried away with her, and so discovers her to her friends, if some god do not prevent it, by coming down in a machine and take the thanks of it to himself.

'By the plot you may guess much of the characters of the persons. An old father that would willingly before he dies see his son well married, his debauched son, kind in his nature to his wench, but miserably in want of money; a servant or slave, who has so much wit to strike in with him, and help to dupe his father; a braggadocio captain, a parasite, and a lady of pleasure.

'As for the poor honest maid, whom all the story is built upon, and who ought to be one of the principal actors in the play, she is commonly a mute in it. She has the breeding of the old Elizabeth way, for maids to be seen and not to be heard; and it is enough you know she is willing to be married when the fifth act requires it.

'These are plots built after the Italian mode of houses; you see through them all at once. The characters are indeed the imitations of nature, but so narrow as if they had imitated only an eye or a hand, and did not dare to venture on the lines of a face or the proportion of a body.

'But in how strait a compass soever they have bounded their plots and characters, we will pass it by if they have regularly pursued them, and perfectly observed those three unities of time, place, and action,—the knowledge of which, you say, is derived to us from them.

'But in the first place give me leave to tell you that the unity of place, however it might be practised by them, was never any of their rules. We neither find it in Aristotle, Horace, or any who have written of it, till in our age, the French poets first made it a precept of the stage.°

'The unity of time even Terence himself, who was the best and most regular of them, has neglected. His *Heautontimorumenos*, or *Self-punisher*, takes up visibly two days. Therefore, says Scaliger,° the two first acts concluding the first day were acted overnight, the three last on the ensuing day. And Euripides, in tying himself to one day, has committed an absurdity never to be forgiven him. For in one of his tragedies he has made Theseus go from Athens to Thebes, which was about forty English miles, under the walls of it to give battle, and appear victorious in the next act, and yet, from the time of his departure to the return of the Nuntius [messenger] who gives the relation of his victory, Aethra and the Chorus have but thirty-six verses, which is not for every mile a verse.

'The like error is as evident in Terence's *Eunuch*, when Laches, the old man, enters in a mistake the house of Thais, where betwixt his exit and the entrance of Pythias, who comes to give an ample

relation of the garboils he has raised within, Parmeno, who was left
upon the stage, has not above five lines to speak.

C'est bien employer un temps si court,

[he makes good use of so short a time]

says the French poet° who furnished me with one of the observations.
And almost all their tragedies will afford us examples of the like
nature.

''Tis true, they have kept the continuity, or, as you called it, *liaison
des scènes* [joining of scenes], somewhat better. Two do not perpetually
come in together, talk, and go out together, and other two succeed
them, and do the same throughout the act, which the English call by
the name of single scenes. But the reason is because they have seldom
above two or three scenes, properly so called, in every act. For it is to
be accounted a new scene, not every time the stage is empty, but every
person who enters, though to others, makes it so, because he
introduces a new business. Now the plots of their plays being narrow,
and the persons few, one of their acts was written in a less compass
than one of our well-wrought scenes. And yet they are often deficient
even in this.

'To go no further than Terence, you find in the *Eunuch* Antipho
entering single in the midst of the third act, after Chremes and Pythias
were gone off. In the same play you have likewise Dorias beginning
the fourth act alone, and, after she has made a relation of what was
done at the soldier's entertainment (which by the way was very
inartificial to do, because she was presumed to speak directly to the
audience, and to acquaint them with what was necessary to be known,
but yet should have been so contrived by the poet as to have been told
by persons of the drama to one another, and so by them to have come
to the knowledge of the people), she quits the stage, and Phaedria
enters next, alone likewise. He also gives you an account of himself,
and of his returning from the country, in monologue, to which
unnatural way of narration Terence is subject in all his plays. In his
Adelphi, or *Brothers*, Syrus and Demea enter after the scene was
broken by the departure of Sostrata, Geta, and Canthara. And indeed
you can scarce look into any of his comedies where you will not
presently discover the same interruption.

'But as they have failed both in laying of their plots, and managing
of them, swerving from the rules of their own art by misrepresenting
nature to us, in which they have ill satisfied one intention of a play,
which was delight, so in the instructive part they have erred worse.

Instead of punishing vice and rewarding virtue, they have often shown a prosperous wickedness and an unhappy piety. They have set before us a bloody image of revenge in Medea,° and given her dragons to convey her safe from punishment, a Priam and Astyanax murdered, and Cassandra ravished, and the lust and murder ending in the victory of him who acted them. In short, there is no indecorum in any of our modern plays, which, if I would excuse, I could not shadow with some authority from the ancients.

'And one further note of them let me leave you. Tragedies and comedies were not writ then as they are now, promiscuously, by the same person, but he who found his genius bending to the one, never attempted the other way. This is so plain that I need not instance to you that Aristophanes, Plautus, Terence, never any of them writ a tragedy. Aeschylus, Euripides, Sophocles, and Seneca never meddled with comedy. The sock and buskin° were not worn by the same poet. Having then so much care to excel in one kind, very little is to be pardoned them if they miscarried in it.

'And this would lead me to the consideration of their wit, had not Crites given me sufficient warning not to be too bold in my judgment of it, because the languages being dead, and many of the customs and little accidents on which it depended lost to us, we are not competent judges of it. But though I grant that here and there we may miss the application of a proverb or a custom, yet a thing well said will be wit in all languages, and though it may lose something in the translation, yet, to him who reads it in the original, 'tis still the same. He has an idea of its excellency, though it cannot pass from his mind into any other expression or words than those in which he finds it.

'When Phaedria in the *Eunuch* had a command from his mistress to be absent two days, and encouraging himself to go through with it, said,

> tandem ego non illa caream, si opus sit, vel totum triduum?

[but can't I manage without her for three days together?]

Parmeno, to mock the softness of his master, lifting up his hands and eyes, cried out, as it were in admiration,

> hui! universum triduum!°

[what? three whole days!]

the elegancy of which 'universum', though it cannot be rendered in our language, yet leaves an impression of the wit upon our souls. But this happens seldom in him, in Plautus oftener, who is infinitely too

bold in his metaphors and coining words, out of which many times his
wit is nothing,—which questionless was one reason why Horace falls
upon him so severely in those verses:

> sed proavi nostri Plautinos et numeros et
> laudavere sales, nimium patienter utrumque,
> ne dicam stolide.°

[yet our forefathers praised both Plautus' verses, and his wit, being over-
tolerant, not to say stupid]

For Horace himself was cautious to obtrude a new word upon his
readers, and makes custom and common use the best measure of
receiving it into our writings.

> multa renascentur quae nunc cecidere, cadentque
> quae nunc sunt in honore vocabula, si volet usus,
> quem penes arbitrium est, et jus, et norma loquendi.°

[many terms now obsolete shall be born again, and those now in favour shall
fall, if usage so decrees, in whose hands lies the judgment, the law, and the
rule of speech]

  'The not observing this rule is that which the world has blamed in
our satirist Cleveland. To express a thing hard and unnaturally is his
new way of elocution. 'Tis true, no poet but may sometimes use a
catachresis. Virgil does it,

> mistaque ridenti colocasia fundet acantho,°

[the Egyptian bean mixed with the joyful acanthus will grow abundantly]

in his eclogue of Pollio, and in his seventh *Aeneid*,

> mirantur et undae,
> miratur nemus, insuetum fulgentia longe
> scuta virum fluvio, pictasque innare carinas,°

[the waves and the woods marvel, startled by the flashing shields of the
warriors and the painted ships]

and Ovid once so modestly that he asks leave to do it,

> si verbo audacia detur,
> haud metuam summi dixisse Palatia caeli,°

[if I may speak boldly, I would not be afraid to call this place the palace of
heaven itself]

calling the court of Jupiter by the name of Augustus' palace—though
in another place he is more bold, where he says,

> et longas visent Capitolia pompas.°

[and capitols witness long processions]

But to do this always, and never be able to write a line without it, though it may be admired by some few pedants, will not pass upon those who know that wit is best conveyed to us in the most easy language, and is most to be admired when a great thought comes dressed in words so commonly received that it is understood by the meanest apprehensions, as the best meat is the most easily digested.

'But we cannot read a verse of Cleveland's without making a face at it, as if every word were a pill to swallow. He gives us many times a hard nut to break our teeth, without a kernel for our pains. So that there is this difference betwixt his *Satires* and Doctor Donne's,—that the one gives us deep thoughts in common language, though rough cadence, the other gives us common thoughts in abstruse words. 'Tis true, in some places his wit is independent of his words, as in that of the rebel Scot:

> Had Cain been Scot, God would have changed his doom;
> Not forced him wander, but confined him home.°

"Si sic omnia dixisset [if only he had always spoken like this]!"° This is wit in all languages. 'Tis like Mercury, never to be lost or killed. And so that other,

> For beauty, like white powder, makes no noise,
> And yet the silent hypocrite destroys.°

You see the last line is highly metaphorical, but it is so soft and gentle that it does not shock us as we read it.

'But to return from whence I have digressed, to the consideration of the ancients' writing and their wit, of which by this time you will grant us in some measure to be fit judges. Though I see many excellent thoughts in Seneca, yet he of them who had a genius most proper for the stage was Ovid. He had a way of writing so fit to stir up a pleasing admiration and concernment, which are the objects of a tragedy, and to show the various movements of a soul combating betwixt two different passions, that, had he lived in our age, or in his own could have writ with our advantages, no man but must have yielded to him; and therefore I am confident the *Medea* is none of his. For though I esteem it for the gravity and sententiousness of it, which he himself concludes to be suitable to a tragedy,

> omne genus scripti gravitate tragoedia vincit,°

[tragedy surpasses all other kinds of writing]

yet it moves not my soul enough to judge that he, who in the epic way wrote things so near the drama, as the story of Myrrha, of Caunus and

Biblis,° and the rest, should stir up no more concernment where he most endeavoured it.

'The masterpiece of Seneca I hold to be that scene in the *Troades* where Ulysses is seeking for Astynax to kill him. There you see the tenderness of a mother so represented in Andromache, that it raises compassion to a high degree in the reader, and bears the nearest resemblance of anything in their tragedies to the excellent scenes of passion in Shakespeare or in Fletcher—for love-scenes you will find few among them. Their tragic poets dealt not with that soft passion, but with lust, cruelty, revenge, ambition, and those bloody actions they produced, which were more capable of raising horror than compassion in an audience, leaving love untouched, whose gentleness would have tempered them, which is the most frequent of all the passions, and which, being the private concernment of every person, is soothed by viewing its own image in a public entertainment.

'Among their comedies we find a scene or two of tenderness, and that, where you would least expect it,—in Plautus. But to speak generally, their lovers say little when they see each other, but *anima mea, vita mea*; ζωὴ καὶ ψυχή [my life and soul; life and soul],° as the women in Juvenal's time used to cry out in the fury of their kindness. Then indeed to speak sense were an offence. Any sudden gust of passion (as an ecstasy of love in an unexpected meeting) cannot better be expressed than in a word and a sigh, breaking one another. Nature is dumb on such occasions, and to make her speak would be to represent her unlike herself. But there are a thousand other concernments of lovers, as jealousies, complaints, contrivances, and the like, where not to open their minds at large to each other, were to be wanting to their own love, and to the expectation of the audience, who watch the movements of their minds as much as the changes of their fortunes. For the imaging of the first is properly the work of a poet; the latter he borrows from the historian.'

Eugenius was proceeding in that part of his discourse when Crites interrupted him.

'I see', said he, 'Eugenius and I are never likely to have this question decided betwixt us. For he maintains the moderns have acquired a new perfection in writing; I can only grant they have altered the mode of it. Homer described his heroes men of great appetites, lovers of beef broiled upon the coals, and good fellows— contrary to the practice of the French romances, whose heroes neither eat, nor drink, nor sleep for love. Virgil makes Aeneas a bold avower of his own virtues, sum pius Aeneas, fama super aethera notus;°

[I am the faithful Aeneas, renowned throughout the world]

which in the civility of our poets is the character of a fanfaron° or Hector. For with us the knight takes occasion to walk out, or sleep, to avoid the vanity of telling his own story, which the trusty squire is ever to perform for him.

'So in their love scenes, of which Eugenius spoke last, the ancients were more hearty, we more talkative. They writ love as it was then the mode to make it, and I will grant thus much to Eugenius, that perhaps one of their poets, had he lived in our age,—

si foret hoc nostrum fato delapsus in aevum°

[if, by fate, he had been born in this age of ours]

(as Horace says of Lucilius),—he had altered many things,—not that they were not as natural before, but that he might accommodate himself to the age he lived in. Yet in the meantime we are not to conclude anything rashly against those great men, but preserve to them the dignity of masters, and give that honour to their memories (*quos Libitina sacravit* [whom the funeral goddess has hallowed]),° part of which we expect may be paid to us in future times.'

This moderation of Crites, as it was pleasing to all the company, so it put an end to that dispute which Eugenius, who seemed to have the better of the argument, would urge no further. But Lisideius, after he had acknowledged himself of Eugenius's opinion concerning the ancients, yet told him he had forborne till his discourse were ended to ask him why he preferred the English plays above those of other nations, and whether we ought not to submit our stage to the exactness of our next neighbours.

'Though', said Eugenius, 'I am at all times ready to defend the honour of my country against the French, and to maintain we are as well able to vanquish them with our pens as our ancestors have been with their swords, yet, if you please', added he, looking upon Neander, 'I will commit this cause to my friend's management. His opinion of our plays is the same with mine, and besides, there is no reason that Crites and I, who have now left the stage, should re-enter so suddenly upon it—which is against the laws of comedy.'

'If the question had been stated', replied Lisideius, 'who had writ best—the French or English—forty years ago, I should have been of your opinion, and adjudged the honour to our own nation. But since that time', said he, turning towards Neander, 'we have been so long together bad Englishmen that we had not leisure to be good poets. Beaumont, Fletcher, and Jonson (who were only capable of bringing us to that degree of perfection which we have) were just then leaving the world°—as if in an age of so much horror wit and those milder studies of humanity had no further business among us. But the muses, who ever follow peace, went to plant in another country.

'It was then that the great Cardinal Richelieu began to take them into his protection, and that by his encouragement Corneille and some other Frenchmen reformed their theatre, which before was as much below ours as it now surpasses it and the rest of Europe. But because Crites in his discourse for the ancients has prevented me, by touching upon many rules of the stage which the moderns have borrowed from them, I shall only, in short, demand of you whether you are not convinced that of all nations the French have best observed them.

'In the unity of time you find them so scrupulous that it yet remains a dispute among their poets whether the artificial day of twelve hours, more or less, be not meant by Aristotle, rather than the natural one of twenty-four, and consequently, whether all plays ought not to be reduced into that compass. This I can testify, that in all their dramas writ within these last twenty years and upwards I have not observed any that have extended the time to thirty hours.

'In the unity of place they are full as scrupulous. For many of their critics limit it to that spot of ground where the play is supposed to begin. None of them exceed the compass of the same town or city.

'The unity of action in all their plays is yet more conspicuous. For they do not burden them with under-plots, as the English do—which is the reason why many scenes of our tragicomedies carry on a design that is nothing of kin to the main plot, and that we see two distinct webs in a play, like those in ill-wrought stuffs, and two actions—that is two plays—carried on together to the confounding of the audience, who, before they are warm in their concernments for one part, are diverted to another, and by that means espouse the interest of neither. From hence likewise it arises that one half of our actors are not known to the other. They keep their distances as if they were Montagues and Capulets, and seldom begin an acquaintance till the last scene of the fifth act, when they are all to meet upon the stage.

'There is no theatre in the world has anything so absurd as the English tragicomedy. 'Tis a drama of our own invention, and the fashion of it is enough to proclaim it so—here a course of mirth, there another of sadness and passion, a third of honour and fourth a duel. Thus in two hours and a half we run through all the fits of Bedlam.

The French affords you as much variety on the same day, but they do it not so unseasonably, or *mal à propos* [at the wrong time], as we. Our poets present you the play and the farce together, and our stages still retain somewhat of the original civility of the Red Bull:°

> atque ursum et pugiles media inter carmina poscunt.°

[and they call for a bear or for boxers in the middle of a play]

'The end of tragedies or serious plays, says Aristotle, is to beget admiration,° compassion, or concernment. But are not mirth and compassion things incompatible? And is it not evident that the poet must of necessity destroy the former by intermingling of the latter— that is, he must ruin the sole end and object of his tragedy to introduce somewhat that is forced into it, and is not of the body of it? Would you not think that physician mad who, having prescribed a purge, should immediately order you to take restringents upon it?

'But to leave our plays and return to theirs. I have noted one great advantage they have had in the plotting of their tragedies. That is, they are always grounded upon some known history, according to that of Horace,

> ex noto fictum carmen sequar,°

[my aim shall be to create from familiar matter]

and in that they have so imitated the ancients that they have surpassed them. For the ancients, as was observed before, took for the foundation of their plays some poetical fiction, such as under that consideration could move but little concernment in the audience, because they already knew the event of it. But the French goes further:

> atque ita mentitur, sic veris falsa remiscet,
> primo ne medium, medio ne discrepet imum.°

[and so does [Homer] invent, so closely does he mix fact and fiction, that the middle is not out of harmony with the beginning, nor the end with the middle]

He so interweaves truth with probable fiction that he puts a pleasing fallacy upon us, mends the intrigues of fate, and dispenses with the severity of history, to reward that virtue which has been rendered to us there unfortunate.

'Sometimes the story has left the success so doubtful that the writer is free, by the privilege of a poet, to take that which of two or more relations will best suit with his design—as, for example, the death of Cyrus, whom Justin and some others report to have perished in the Scythian

war, but Xenophon affirms to have died in his bed of extreme old age.°
Nay more, when the event is past dispute, even then we are willing to
be deceived, and the poet, if he contrives it with appearance of truth,
has all the audience of his party,—at least during the time his play is
acting. So naturally we are kind to virtue, when our own interest is not
in question, that we take it up as the general concernment of mankind.

'On the other side, if you consider the historical plays of
Shakespeare, they are rather so many chronicles of kings, or the
business many times of thirty or forty years cramped into a
representation of two hours and a half, which is not to imitate or paint
nature, but rather to draw her in miniature, to take her in little, to look
upon her through the wrong end of a perspective,° and receive her
images not only much less, but infinitely more imperfect than the life.
This, instead of making a play delightful, renders it ridiculous.

quodcunque ostendis mihi sic, incredulus odi.°

[whatever you show me in this manner I disbelieve and hate]

For the spirit of man cannot be satisfied but with truth, or at least
verisimility, and a poem is to contain, if not τὰ ἔτυμα [the truth], yet
ἐτύμοισιν ὁμοῖα [things resembling the truth],° as one of the Greek
poets has expressed it.

'Another thing in which the French differ from us and from the
Spaniards, is that they do not embarrass or cumber themselves with
too much plot. They only represent so much of a story as will
constitute one whole and great action sufficient for a play. We, who
undertake more, do but multiply adventures, which, not being
produced from one another as effects from causes, but barely
following, constitute many actions in the drama, and consequently
make it many plays.

'But by pursuing closely one argument which is not cloyed with
many turns the French have gained more liberty for verse, in which
they write. They have leisure to dwell upon a subject which deserves
it, and to represent the passions (which we have acknowledged to be
the poet's work) without being hurried from one thing to another,—as
we are in the plays of Calderon, which we have seen lately upon our
theatres, under the name of Spanish plots.

'I have taken notice but of one tragedy of ours whose plot has that
uniformity and unity of design in it which I have commended in the
French and that is *Rollo*,°—or rather under the name of Rollo, the
story of Bassianus and Geta in Herodian.° There indeed the plot is
neither large nor intricate, but just enough to fill the minds of the

audience, not to cloy them. Besides, you see it founded upon the truth of history, only the time of the action is not reduceable to the strictness of the rules, and you see in some places a little farce mingled, which is below the dignity of the other parts; and in this all our poets are extremely peccant. Even Ben Jonson himself, in *Sejanus* and *Catiline*, has given us this oleo° of a play, this unnatural mixture of comedy and tragedy, which to me sounds just as ridiculously as the history of David with the merry humours of Golias.° In *Sejanus*° you may take notice of the scene betwixt Livia and the physician, which is a pleasant satire upon the artificial helps of beauty. In *Catiline*,° you may see the parliament of women, the little envies of them to one another, and all that passes betwixt Curio and Fulvia—scenes admirable in their kind, but of an ill mingle with the rest.

'But I return again to the French writers, who, as I have said, do not burden themselves too much with plot, which has been reproached to them by an ingenious person° of our nation as a fault. For he says they commonly make but one person considerable in a play; they dwell upon him and his concernments, while the rest of the persons are only subservient to set him off.

'If he intends this by it, that there is one person in the play who is of greater dignity than the rest, he must tax not only theirs, but those of the ancients, and—which he would be loath to do—the best of ours. For 'tis impossible but that one person must be more conspicuous in it than any other; and consequently the greatest share in the action must devolve on him. We see it so in the management of all affairs,—even in the most equal aristocracy, the balance cannot be so justly poised but some one will be superior to the rest, either in parts, fortune, interest, or the consideration of some glorious exploit, which will reduce the greatest part of business into his hands.

'But if he would have us to imagine that in exalting of one character the rest of them are neglected, and that all of them have not some share or other in the action of the play, I desire him to produce any of Corneille's tragedies wherein every person (like so many servants in a well-governed family) has not some employment, and who is not necessary to the carrying on of the plot, or at least to your understanding it.

'There are indeed some protatic persons° in the ancients, whom they make use of in their plays, either to hear or give the relation. But the French avoid this with great address, making their narrations only to or by such who are some way interested° in the main design.

'And now I am speaking of relations, I cannot take a fitter

opportunity to add this in favour of the French, that they often use them with better judgment and more *à propos* [to the purpose] than the English do. Not that I commend narrations in general, but there are two sorts of them: one, of those things which are antecedent to the play, and are related to make the conduct of it more clear to us. But 'tis a fault to choose such subjects for the stage as will force us on that rock, because we see they are seldom listened to by the audience, and that is many times the ruin of the play, for, being once let pass without attention, the audience can never recover themselves to understand the plot—and indeed it is somewhat unreasonable that they should be put to so much trouble as that to comprehend what passes in their sight they must have recourse to what was done perhaps ten or twenty years ago.

'But there is another sort of relations, that is, of things happening in the action of the play, and supposed to be done behind the scenes;— and this is many times both convenient and beautiful, for by it the French avoid the tumult which we are subject to in England by representing duels, battles, and the like, which renders our stage too like the theatres where they fight prizes.° For what is more ridiculous than to represent an army with a drum and five men behind it, all which the hero of the other side is to drive in before him, or to see a duel fought, and one slain with two or three thrusts of the foils, which we know are so blunted that we might give a man an hour to kill another in good earnest with them?

'I have observed that in all our tragedies the audience cannot forbear laughing when the actors are to die. 'Tis the most comic part of the whole play. All *passions* may be lively represented on the stage, if to the well writing of them the actor supplies a good commanded voice, and limbs that move easily, and without stiffness. But there are many *actions* which can never be imitated to a just height. Dying especially is a thing which none but a Roman gladiator could naturally perform upon the stage when he did not imitate or represent, but naturally do it, and therefore it is better to omit the representation of it.

'The words of a good writer which describe it lively, will make a deeper impression of belief in us than all the actor can persuade us to when he seems to fall dead before us—as a poet, in the description of a beautiful garden or a meadow, will please our imagination more than the place itself can please our sight. When we see death represented, we are convinced it is but fiction, but when we hear it related, our eyes, the strongest witnesses, are wanting, which might have undeceived us, and we are all willing to favour the sleight when the poet does not too grossly impose upon us.

'They therefore who imagine these relations would make no concernment in the audience are deceived, by confounding them with the other, which are of things antecedent to the play. Those are made often in cold blood, as I may say, to the audience. But these are warmed with our concernments, which are before awakened in the play. What the philosophers say° of motion, that when it is once begun it continues of itself and will do so to eternity without some stop put to it, is clearly true on this occasion. The soul, being already moved with the characters and fortunes of those imaginary persons, continues going of its own accord, and we are no more weary to hear what becomes of them when they are not on the stage, than we are to listen to the news of an absent mistress.

'But it is objected that if one part of the play may be related, then why not all. I answer, some parts of the action are more fit to be represented, some to be related. Corneille° says judiciously that the poet is not obliged to expose to view all particular actions which conduce to the principal. He ought to select such of them to be seen which will appear with the greatest beauty, either by the magnificence of the show, or the vehemence of passions which they produce, or some other charm which they have in them, and let the rest arrive to the audience by narration.

''Tis a great mistake in us to believe the French present no part of the action upon the stage. Every alteration or crossing of a design, every new-sprung passion and turn of it is a part of the action, and much the noblest, except we conceive nothing to be action till they come to blows—as if the painting of the hero's mind were not more properly the poet's work than the strength of his body. Nor does this anything contradict the opinion of Horace, where he tells us,

> segnius irritant animos demissa per aurem,
> quam quae sunt oculis subjecta fidelibus.

[the mind is stirred less vividly by what we hear than by what we see, which we rely upon]

For he says immediately after,

> non tamen intus
> digna geri promes in scenam, multaque tolles
> ex oculis, quae mox narret facundia praesens.

[yet you do not bring on the stage actions that should be performed behind the scenes, and keep much from sight which an actor's ready tongue will warrant before us]

Among which many he recounts some,

> nec pueros coram populo Medea trucidet,
> aut in avem Progne mutetur, Cadmus in anguem, etc.°

[Medea must not slaughter her boys in front of the audience, nor Procne be turned into a bird, Cadmus into a snake]

That is, those actions which by reason of their cruelty will cause aversion in us, or by reason of their impossibility unbelief, ought either wholly to be avoided by a poet, or only delivered by narration. To which we may have leave to add such as to avoid tumult (as was before hinted), or to reduce the plot into a more reasonable compass of time, or for defect of beauty in them, are rather to be related than presented to the eye.

'Examples of all these kinds are frequent, not only among all the ancients, but in the best received of our English poets. We find Ben Jonson using them in his *Magnetic Lady*,° where one comes out from dinner and relates the quarrels and disorders of it to save the undecent appearing of them on the stage, and to abbreviate the story—and this in express imitation of Terence, who had done the same before him in his *Eunuch*,° where Pythias makes the like relation of what had happened within at the soldier's entertainment. The relations likewise of Sejanus's° death and the prodigies before it are remarkable; the one of which was hid from sight to avoid the horror and tumult of the representation, the other, to shun the introducing of things impossible to be believed. In that excellent play, the *King and no King*,° Fletcher goes yet further. For the whole unravelling of the plot is done by narration in the fifth act, after the manner of the ancients, and it moves great concernment in the audience, though it be only a relation of what was done many years before the play. I could multiply other instances, but these are sufficient to prove that there is no error in choosing a subject which requires this sort of narration. In the ill managing of them, there may.

'But I find I have been too long in this discourse, since the French have many other excellences not common to us, as that you never see any of their plays end with a conversion, or simple change of will, which is the ordinary way our poets use to end theirs.

'It shows little art in the conclusion of a dramatic poem, when they who have hindered the felicity during the four acts desist from it in the fifth without some powerful cause to take them off. And though I deny not but such reasons may be found, yet it is a path that is cautiously to be trod, and the poet is to be sure he convinces the audience that the motive is strong enough. As for example, the conversion of the usurer in the *Scornful Lady*° seems to me a little forced. For, being a usurer,—which implies a lover of money to the highest degree of covetousness (and such the poet has represented him),—the account he gives for the sudden change is that he has been duped by the wild young fellow, which in reason might render him more wary another time, and make him punish himself with harder fare and coarser clothes, to get it up again. But that he should look upon it as a judgment, and so repent, we may expect to hear of in a sermon, but I should never endure it in a play.

'I pass by this. Neither will I insist upon the care they take that no person after his first entrance shall ever appear, but the business which brings him upon the stage shall be evident, which, if observed, must needs render all the events of the play more natural. For there you see the probability of every accident in the cause that produced it, and that which appears chance in the play will seem so reasonable to you that you will there find it almost necessary.° So that in the exits of their actors you have a clear account of their purpose and design in the next entrance (though, if the scene be well wrought, the event will commonly deceive you), for there is nothing so absurd, says Corneille, as for an actor to leave the stage only because he has no more to say.

'I should now speak of the beauty of their rhyme, and the just reason I have to prefer that way of writing in tragedies before ours in blank verse. But, because it is partly received by us and therefore not altogether peculiar to them, I will say no more of it in relation to their plays. For our own, I doubt not but it will exceedingly beautify them, and I can see but one reason why it should not generally obtain,—that is, because our poets write so ill in it. This indeed may prove a more prevailing argument than all others which are used to destroy it, and therefore I am only troubled when great and judicious poets,° and those who are acknowledged such, have writ or spoke against it. As for others, they are to be answered by that one sentence of an ancient author:°

sed ut primo ad consequendos eos quos priores ducimus accendimur, ita ubi aut praeteriri aut aequari eos posse desperavimus, studium cum spe senescit: quod, scilicet, assequi non potest, sequi desinit; praeteritoque eo in quo eminere non possumus, aliquid in quo nitamur conquirimus.

[but as in the beginning we burn to surpass those whom we regard as leaders, so when we have despaired of being able either to surpass or even to equal them, our zeal wanes with our hopes; it stops following what it cannot overtake, and leaving aside that in which we cannot excel, we seek for some new object of our effort].'

Lisideius concluded in this manner, and Neander after a little pause thus answered him:

'I shall grant Lisideius, without much dispute, a great part of what he has urged against us. For I acknowledge the French contrive their plots more regularly, and observe the laws of comedy and decorum of the stage (to speak generally) with more exactness than the English. Further, I deny not but he has taxed us justly in some irregularities of ours which he has mentioned.

'Yet, after all, I am of opinion that neither our faults nor their virtues are considerable enough to place them above us, for, the lively imitation of nature being in the definition of a play, those which best fulfil that law ought to be esteemed superior to the others. 'Tis true those beauties of the French poesy are such as will raise perfection higher where it is, but are not sufficient to give it where it is not. They are indeed the beauties of a statue, but not of a man, because not animated with the soul of poesy, which is imitation of humour and passions: and this Lisideius himself, or any other, however biased to their party, cannot but acknowledge, if he will either compare the humours of our comedies, or the characters of our serious plays with theirs. He that will look upon theirs which have been written till these last ten years, or thereabouts, will find it an hard matter to pick out two or three passable humours amongst them.

'Corneille himself, their arch poet, what has he produced except the *Liar*?°—and you know how it was cried up in France. But when it came upon the English stage, though well translated, and that part of Dorant acted to so much advantage by Mr Hart, as I am confident it never received in its own country, the most favourable to it would not put in competition with many of Fletcher's or Ben Jonson's. In the rest of Corneille's comedies you have little humour. He tells you himself° his way is, first to show two lovers in good intelligence with each other, in the working up of the play to embroil them by some mistake, and in the latter end to clear it up.

'But of late years de Molière, the younger Corneille,° Quinault,° and some others have been imitating afar off the quick turns and graces of the English stage. They have mixed their serious plays with mirth, like our tragicomedies, since the death of Cardinal Richelieu,° which Lisideius and many others not observing have commended that in them for a virtue which they themselves no longer practise. Most of their new plays are, like some of ours, derived from the Spanish novels. There is scarce one of them without a veil, and a trusty Diego who drolls much after the rate of the *Adventures.*° But their humours—if I may grace them with that name—are so thin sown that never above one of them comes up in any play. I dare take upon me to find more variety of them in some one play of Ben Jonson's than in all theirs together, as he who has seen the *Alchemist*, the *Silent Woman*, or *Bartholomew Fair*, cannot but acknowledge with me.

'I grant the French have performed what was possible on the ground-work of the Spanish plays. What was pleasant before they have made regular. But there is not above one good play to be writ upon all those plots. They are too much alike to please often, which we need not the experience of our own stage to justify.

'As for their new way of mingling mirth with serious plot, I do not with Lisideius condemn the thing, though I cannot approve their manner of doing it. He tells us we cannot so speedily recollect ourselves after a scene of great passion and concernment as to pass to another of mirth and humour, and to enjoy it with any relish. But why should he imagine the soul of man more heavy than his senses? Does not the eye pass from an unpleasant object to a pleasant in a much shorter time than is required to this? And does not the unpleasantness of the first commend the beauty of the latter?

'The old rule of logic might have convinced him that contraries, when placed near, set off each other. A continued gravity keeps the spirit too much bent. We must refresh it sometimes, as we bait upon a journey° that we may go on with greater ease. A scene of mirth mixed with tragedy has the same effect upon us which our music has betwixt the acts, and that we find a relief to us from the best plots and language of the stage, if the discourses have been long.

'I must therefore have stronger arguments ere I am convinced that compassion and mirth in the same subject destroy each other, and in the meantime cannot but conclude to the honour of our nation, that we have invented, increased, and perfected a more pleasant way of writing for the stage than was ever known to the ancients or moderns of any nation, which is tragicomedy.

'And this leads me to wonder why Lisideius and many others should cry up the barrenness of the French plots above the variety and copiousness of the English. Their plots are single. They carry on one design, which is pushed forward by all the actors, every scene in the play contributing and moving towards it. Ours, besides the main design, have under-plots or by-concernments of less considerable persons and intrigues, which are carried on with the motion of the main plot,—just as they say the orb of the fixed stars, and those of the planets, though they have motions of their own, are whirled about by the motion of the *primum mobile*° in which they are contained.

'That similitude expresses much of the English stage. For if contrary motions may be found in nature to agree, if a planet can go east and west at the same time, one way by virtue of his own motion, the other by the force of the first mover, it will not be difficult to imagine how the under-plot, which is only different, not contrary to the great design, may naturally be conducted along with it.

'Crites° has already shown us from the confession of the French poets that the unity of action is sufficiently preserved if all the imperfect actions of the play are conducing to the main design. But, when those petty intrigues of a play are so ill ordered that they have no coherence with the other, I must grant that Lisideius has reason to tax that want of due connexion. For co-ordination° in a play is as dangerous and unnatural as in a state. In the meantime he must acknowledge our variety, if well ordered, will afford a greater pleasure to the audience.

'As for his other argument,—that by pursuing one single theme they gain an advantage to express and work up the passions,—I wish any example he could bring from them would make it good. For I confess their verses are to me the coldest I have ever read. Neither indeed is it possible for them, in the way they take, so to express passion as that the effects of it should appear in the concernment of an audience,—their speeches being so many declamations, which tire us with the length, so that instead of persuading us to grieve for their imaginary heroes, we are concerned for our own trouble,—as we are in the tedious visits of bad company:—we are in pain till they are gone.

'When the French stage came to be reformed by Cardinal Richelieu, those long harangues were introduced to comply with the gravity of a churchman. Look upon the *Cinna* and *Pompey*. They are not so properly to be called plays as long discourses of reason of state, and *Polyeucte*° in matters of religion is as solemn as the long stops upon our organs.

'Since that time it is grown into a custom, and their actors speak by the hour-glass, as our parsons do. Nay, they account it the grace of their parts, and think themselves disparaged by the poet if they may not twice or thrice in a play entertain the audience with a speech of an hundred or two hundred lines.

'I deny not but this may suit well enough with the French. For as we, who are a more sullen people, come to be diverted at our plays, they, who are of an airy and gay temper, come thither to make themselves more serious. And this I conceive to be one reason why comedy is more pleasing to us, and tragedies to them.

'But to speak generally, it cannot be denied that short speeches and replies are more apt to move the passions and beget concernment in us than the other. For it is unnatural for anyone in a gust of passion to speak long together, or for another in the same condition to suffer him without interruption. Grief and passion are like floods raised in little brooks by a sudden rain. They are quickly up, and if the concernment be poured unexpectedly in upon us, it overflows us. But a long sober shower gives them leisure to run out as they came in, without troubling the ordinary current. As for comedy, repartee is one of its chiefest graces. The greatest pleasure of the audience is a chase of wit, kept up on both sides, and swiftly managed. And this our forefathers, if not we, have had in Fletcher's plays to a much higher degree of perfection than the French poets can arrive at.

'There is another part of Lisideius's discourse, in which he has rather excused our neighbours than commended them,—that is, for aiming only to make one person considerable in their plays. 'Tis very true what he has urged, that one character in all plays, even without the poet's care, will have advantage of all the others, and that the design of the whole drama will chiefly depend on it. But this hinders not that there may be more shining characters in the play, many persons of a second magnitude,—nay, some so very near, so almost equal to the first that greatness may be opposed to greatness, and all the persons be made considerable, not only by their quality, but by their action.

'Tis evident that the more the persons are, the greater will be the variety of the plot. If then the parts are managed so regularly that the beauty of the whole be kept entire, and that the variety become not a perplexed and confused mass of accidents, you will find it infinitely pleasing to be led in a labyrinth of design, where you see some of your way before you, yet discern not the end till you arrive at it.

'And that all this is practicable, I can produce for examples many of our English plays, as *The Maid's Tragedy*,° *The Alchemist*,° *The Silent Woman*.° I was going to have named *The Fox*,° but that the unity of design seems not exactly observed in it. For there appear two actions in the play,—the first naturally ending with the fourth act, the second forced from it in the fifth, which yet is the less to be condemned in him, because the disguise of Volpone—though it suited not with his

character as a crafty or covetous person—agreed well enough with
that of a voluptuary, and by it the poet gained the end he aimed at, the
punishment of vice and the reward of virtue, which that disguise
produced. So that to judge equally of it, it was an excellent fifth act,
but not so naturally proceeding from the former.

'But to leave this, and pass to the latter part of Lisideius's discourse,
which concerns relations. I must acknowledge with him that the
French have reason when they hide that part of the action which
would occasion too much tumult upon the stage, and choose rather to
have it made known by narration to the audience.

'Further, I think it very convenient, for the reasons he has given,
that all incredible actions were removed. But, whether custom has so
insinuated itself into our countrymen, or nature has so formed them to
fierceness, I know not, but they will scarcely suffer combats or other
objects of horror to be taken from them. And indeed the indecency of
tumults is all which can be objected against fighting. For why may not
our imagination as well suffer itself to be deluded with the probability
of it as with any other thing in the play?

'For my part, I can with as great ease persuade myself that the
blows which are struck are given in good earnest, as I can that they
who strike them are kings or princes, or those persons which they
represent.

'For objects of incredibility, I would be satisfied from Lisideius
whether we have any so removed from all appearance of truth as are
those of Corneille's *Andromède*—a play which has been frequented the
most of any he has writ. If the Perseus, or the son of a heathen god,
the Pegasus, and the Monster, were not capable to choke a strong
belief, let him blame any representation of ours hereafter. Those
indeed were objects of delight, yet the reason is the same as to the
probability: for he makes it not a ballet or masque, but a play, which is
to resemble truth.

'But for death, that it ought not to be represented, I have, besides
the arguments alleged by Lisideius, the authority of Ben Jonson, who
has forborne it in his tragedies; for both the death of Sejanus and
Catiline are related—though in the latter I cannot but observe one
irregularity of that great poet. He has removed the scene in the same
act from Rome to Catiline's army, and from thence again to Rome,
and besides, has allowed a very inconsiderable time, after Catiline's
speech, for the striking of the battle and the return of Petreius, who is
to relate the event of it to the senate,—which I should not animadvert
upon him, who was otherwise a painful observer of τὸ πρέπον, or the

decorum of the stage, if he had not used extreme severity in his judgment° upon the incomparable Shakespeare for the same fault.

'To conclude on this subject of relations, if we are to be blamed for showing too much of the action, the French are as faulty for discovering too little of it. A mean betwixt both should be observed by every judicious writer, so as the audience may neither be left unsatisfied by not seeing what is beautiful, or shocked by beholding what is either incredible or undecent.

'I hope I have already proved in this discourse that though we are not altogether so punctual as the French in observing the laws of comedy, yet our errors are so few and little, and those things wherein we excel them so considerable, that we ought of right to be preferred before them.

'But what will Lisideius say if they themselves acknowledge they are too strictly tied up by those laws, for breaking which he has blamed the English? I will allege Corneille's words, as I find them in the end of his *Discourse of the Three Unities*:

Il est facile aux spéculatifs d'être sévères, etc.
''Tis easy for speculative persons to judge severely. But if they would produce to public view ten or twelve pieces of this nature, they would perhaps give more latitude to the rules than I have done, when by experience they had known how much we are bound up and constrained by them, and how many beauties of the stage they banished from it.'

'To illustrate a little what he has said:—by their servile observations of the unities of time and place, and integrity of scenes, they have brought upon themselves that dearth of plot and narrowness of imagination which may be observed in all their plays. How many beautiful accidents might naturally happen in two or three days, which cannot arrive with any probability in the compass of twenty-four hours?

'There is time to be allowed also for maturity of design, which, amongst great and prudent persons, such as are often represented in tragedy, cannot, with any likelihood of truth, be brought to pass at so short a warning.

'Further, by tying themselves strictly to the Unity of Place and unbroken scenes, they are forced many times to omit some beauties which cannot be shown where the act began, but might, if the scene were interrupted, and the stage cleared for the persons to enter in another place.

'And therefore the French poets are often forced upon absurdities.

For if the act begins in a chamber, all the persons in the play must have some business or other to come thither, or else they are not to be shown that act. And sometimes their characters are very unfitting to appear there:—as suppose it were the king's bedchamber, yet the meanest man in the tragedy must come and dispatch his business there, rather than in the lobby or courtyard (which is fitter for him), for fear the stage should be cleared and the scenes broken.

'Many times they fall by it into a greater inconvenience, for they keep their scenes unbroken, and yet change the place—as in one of their newest plays,° where the act begins in the street. There a gentleman is to meet his friend; he sees him with his man, coming out from his father's house; they talk together, and the first goes out; the second, who is a lover, has made an appointment with his mistress; she appears at the window, and then we are to imagine the scene lies under it. This gentleman is called away, and leaves his servant with his mistress; presently her father is heard from within; the young lady is afraid the serving-man should be discovered, and thrusts him in through a door which is supposed to be her closet. After this, the father enters to the daughter, and now the scene is in a house—for he is seeking from one room to another for this poor Philipin, or French Diego, who is heard from within, drolling and breaking many a miserable conceit upon his sad condition. In this ridiculous manner the play goes on, the stage being never empty all the while: so that the street, the window, the two houses, and the closet, are made to walk about, and the persons to stand still.

'Now what, I beseech you, is more easy than to write a regular French play, or more difficult than to write an irregular English one, like those of Fletcher or of Shakespeare?

'If they content themselves, as Corneille did, with some flat design, which, like an ill riddle, is found out ere it be half proposed, such plots we can make every way regular as easily as they. But whenever they endeavour to rise up to any quick turns and counter-turns of plot,—as some of them have attempted, since Corneille's plays have been less in vogue,—you see they write as irregularly as we, though they cover it more speciously.

'Hence the reason is perspicuous why no French plays, when translated, have, or ever can succeed upon the English stage. For, if you consider the plots, our own are fuller of variety, if the writing, ours are more quick and fuller of spirit: and therefore 'tis a strange mistake in those who decry the way of writing plays in verse, as if the English therein imitated the French.

'We have borrowed nothing from them. Our plots are weaved in English looms. We endeavour therein to follow the variety and greatness of characters which are derived to us from Shakespeare and Fletcher. The copiousness and well-knitting of the intrigues we have from Jonson, and for the verse itself we have English precedents of older date than any of Corneille's plays.

'Not to name our old comedies before Shakespeare, which were all° writ in verse of six feet, or Alexandrines, such as the French now use, I can show in Shakespeare many scenes of rhyme together, and the like in Ben Jonson's tragedies—in *Catiline* and *Sejanus* sometimes thirty or forty lines (I mean besides the chorus, or the monologues, which, by the way, showed Ben no enemy to this way of writing, especially if you look upon his *Sad Shepherd*, which goes sometimes upon rhyme, sometimes upon blank verse, like a horse who eases himself upon trot and amble). You find him likewise commending Fletcher's pastoral of *The Faithful Shepherdess*, which is for the most part rhyme, though not refined to that purity to which it hath since been brought. And these examples are enough to clear us from a servile imitation of the French.

'But to return whence I have digressed, I dare boldly affirm these two things of the English drama:—

'First, that we have many plays of ours as regular as any of theirs, and which, besides, have more variety of plot and characters;

'And secondly, that in most of the irregular plays of Shakespeare or Fletcher (for Ben Jonson's are for the most part regular), there is a more masculine fancy and greater spirit in all the writing than there is in any of the French.

'I could produce even in Shakespeare's and Fletcher's works some plays which are almost exactly formed:—as *The Merry Wives of Windsor* and *The Scornful Lady*. But because, generally speaking, Shakespeare, who writ first, did not perfectly observe the laws of comedy, and Fletcher, who came nearer to perfection, yet through carelessness made many faults, I will take the pattern of a perfect play from Ben Jonson, who was a careful and learned observer of the dramatic laws, and from all his comedies I shall select *The Silent Woman*, of which I will make a short examen,° according to those rules which the French observe.'

As Neander was beginning to examine *The Silent Woman*, Eugenius looking earnestly upon him, 'I beseech you, Neander,' said he, 'gratify the company, and me in particular, so far as, before you speak of the play, to give us a character of the author, and tell us frankly your opinion, whether you do not think all writers, both French and English, ought to give place to him?'

'I fear', replied Neander, 'that in obeying your commands I shall draw a little envy upon myself. Besides, in performing them, it will be first necessary to speak somewhat of Shakespeare and Fletcher, his rivals in poesy—and one of them, in my opinion, at least his equal, perhaps his superior.

'To begin then with Shakespeare. He was the man who, of all modern and perhaps ancient poets, had the largest and most comprehensive soul. All the images of nature were still present to him, and he drew them not laboriously, but luckily. When he describes anything, you more than see it, you feel it too. Those who accuse him to have wanted learning give him the greater commendation. He was naturally learned. He needed not the spectacles of books to read nature. He looked inwards, and found her there.

'I cannot say he is everywhere alike. Were he so, I should do him injury to compare him with the greatest of mankind. He is many times flat, insipid, his comic wit degenerating into clenches, his serious swelling into bombast. But he is always great when some great occasion is presented to him. No man can say he ever had a fit subject for his wit and did not then raise himself as high above the rest of poets,

> quantum lenta solent inter viburna cupressi.°

[as cypresses often do among the bending osiers]

'The consideration of this made Mr Hales of Eton° say that there was no subject of which any poet ever writ, but he would produce it much better treated of in Shakespeare. And however others are now generally preferred before him, yet the age wherein he lived, which had contemporaries with him Fletcher and Jonson, never equalled them to him in their esteem, and in the last king's court, when Ben's reputation was at highest, Sir John Suckling, and with him the greater part of the courtiers, set our Shakespeare far above him.

'Beaumont and Fletcher, of whom I am next to speak, had, with the advantage of Shakespeare's wit, which was their precedent, great natural gifts improved by study,—Beaumont especially being so accurate a judge of plays, that Ben Jonson, while he lived, submitted all his writings to his censure, and, 'tis thought, used his judgment in correcting, if not contriving, all his plots. What value he had for him appears by the verses° he writ to him, and therefore I need speak no further of it.

'The first play which brought Fletcher and him in esteem was their *Philaster*,° for before that they had written two or three very unsuccessfully,—as the like is reported of Ben Jonson before he wrote *Every Man in his Humour*.

'Their plots were generally more regular than Shakespeare's, especially those which were made before Beaumont's death, and they understood and imitated the conversation of gentlemen much better, whose wild debaucheries, and quickness of wit in repartees, no poet can ever paint as they have done. This humour, of which Ben Jonson derived from particular persons, they made it not their business to describe. They represented all the passions very lively, but above all, love.

'I am apt to believe the English language in them arrived to its highest perfection. What words have since been taken in are rather superfluous than necessary. Their plays are now the most pleasant and frequent entertainments of the stage, two of theirs being acted through the year for one of Shakespeare's or Jonson's. The reason is because there is a certain gaiety in their comedies, and pathos in their more serious plays, which suits generally with all men's humours. Shakespeare's language is likewise a little obsolete, and Ben Jonson's wit comes short of theirs.

'As for Jonson, to whose character I am now arrived, if we look upon him while he was himself (for his last plays were but his dotages), I think him the most learned and judicious writer which any theatre ever had. He was a most severe judge of himself as well as others. One cannot say he wanted wit, but rather that he was frugal of it. In his works you find little to retrench or alter. Wit and language, and humour also in some measure, we had before him. But something of art was wanting to the drama till he came. He managed his strength to more advantage than any who preceded him.

'You seldom find him making love in any of his scenes, or endeavouring to move the passions. His genius was too sullen and saturnine to do it gracefully, especially when he knew he came after those who had performed both to such a height. Humour was his proper sphere, and in that he delighted most to represent mechanic people.

'He was deeply conversant in the ancients, both Greek and Latin, and he borrowed boldly from them. There is scarce a poet or historian among the Roman authors of those times whom he has not translated in *Sejanus* and *Catiline*. But he has done his robberies so openly that one may see he fears not to be taxed by any law. He invades authors like a monarch, and what would be theft in other poets is only victory in him. With the spoils of these writers he so represents old Rome to us, in its rites, ceremonies, and customs, that if one of their own poets had written either of his tragedies, we had seen less of it than in him.

'If there was any fault in his language, 'twas that he weaved it too closely and laboriously, in his serious plays. Perhaps, too, he did a little too much romanize our tongue, leaving the words which he translated almost as much Latin as he found them, wherein, though he learnedly followed the idiom of their language, he did not enough comply with ours.

'If I would compare him with Shakespeare, I must acknowledge him the more correct poet, but Shakespeare the greater wit. Shakespeare was the Homer, or father of our dramatic poets. Jonson was the Virgil, the pattern of elaborate writing. I admire him, but I love Shakespeare.

'To conclude of him. As he has given us the most correct plays, so in the precepts which he has laid down in his *Discoveries*, we have as many and profitable rules for perfecting the stage as any wherewith the French can furnish us.

'Having thus spoken of the author, I proceed to the examination of his comedy, *The Silent Woman*.°

EXAMEN OF THE SILENT WOMAN

'To begin first with the length of the action:—it is so far from exceeding the compass of a natural day, that it takes not up an artificial one. 'Tis all included in the limits of three hours and a half, which is no more than is required for the presentment on the stage—a beauty perhaps not much observed; if it had, we should not have looked on the Spanish translation of *Five Hours* with so much wonder.

'The scene of it is laid in London. The latitude of place is almost as little as you can imagine, for it lies all within the compass of two houses, and after the first act in one. The continuity of scenes is observed more than in any of our plays, excepting his own *Fox* and *Alchemist*. They are not broken above twice or thrice at most in the whole comedy—and in the two best of Corneille's plays, the *Cid* and *Cinna*, they are interrupted once apiece.

'The action of the play is entirely one, the aim or end of which is the settling Morose's estate on Dauphine. The intrigue of it is the greatest and most noble of any pure unmixed comedy in any language. You see in it many persons of various characters and humours, and all delightful—as first, Morose, or an old man, to whom all noise but his own talking is offensive.

'Some who would be thought critics say this humour of his is forced. But to remove that objection, we may consider him first to be naturally of a delicate hearing, as many are to whom all sharp sounds are unpleasant, and secondly, we may attribute much of it to the peevishness of his age, or the wayward authority of an old man in his own house, where he may make himself obeyed—and this the poet seems to allude to in his name Morose. Besides this, I am assured from divers persons that Ben Jonson was actually acquainted with such a man, one altogether as ridiculous as he is here represented.

'Others say it is not enough to find one man of such a humour, it must be common to more, and the more common the more natural. To prove this, they instance in the best of comical characters, Falstaff. There are many men resembling him—old, fat, merry, cowardly, drunken, amorous, vain, and lying. But to convince these people, I need but tell them that humour is the ridiculous extravagance of conversation, wherein one man differs from all others. If then it be common, or communicated to many, how differs it from other men's? Or what indeed causes it to be ridiculous so much as the singularity of it? As for Falstaff, he is not properly one humour, but a miscellany of humours or images, drawn from so many several men. That wherein he is singular is his wit, or those things he says *praeter expectatum* [contrary to expectation], unexpected by the audience, his quick evasions, when you imagine him surprised, which, as they are extremely diverting of themselves, so receive a great addition from his person—for the very sight of such an unwieldy old debauched fellow is a comedy alone.

'And here, having a place so proper for it, I cannot but enlarge somewhat upon this subject of humour into which I am fallen. The ancients had little of it in their comedies. For the τὸ γελοῖον° [the ludicrous] of the old comedy, of which Aristophanes was chief, was not so much to imitate a man as to make the people laugh at some odd conceit, which had commonly somewhat of unnatural or obscene in it. Thus when you see Socrates brought upon the stage, you are not to imagine him made ridiculous by the imitation of his actions, but rather by making him perform something very unlike himself— something so childish and absurd as, by comparing it with the gravity of the true Socrates, makes a ridiculous object for the spectators.°

'In their new comedy which succeeded, the poets sought indeed to express the ἦθος [character], as in their tragedies the πάθος [emotion] of mankind. But this ἦθος contained only the general characters of men and manners,—as old men, lovers, serving-men, courtesans, parasites, and such other persons as we see in their comedies, all which they made alike—that is, one old man or father, one lover, one courtesan, so like another, as if the first of them had begot the rest of every sort:

> ex homine hunc natum dicas.°

[one would say this fellow was born of a human]

The same custom they observed likewise in their tragedies.

'As for the French, though they have the word *humeur* among them, yet they have small use of it in their comedies or farces, they being but ill imitations of the *ridiculum*, or that which stirred up laughter in the old comedy.

'But among the English 'tis otherwise, where by humour is meant some extravagant habit, passion, or affection, particular (as I said before) to some one person, by the oddness of which he is immediately distinguished from the rest of men, which, being lively and naturally represented, most frequently begets that malicious pleasure in the audience which is testified by laughter—as all things which are deviations from common customs are ever the aptest to produce it (though, by the way, this laughter is only accidental, as the person represented is fantastic or bizarre; but pleasure is essential to it, as the imitation of what is natural). The description of these humours, drawn from the knowledge and observation of particular persons, was the peculiar genius and talent of Ben Jonson, to whose play I now return.

'Besides Morose, there are at least nine or ten different characters and humours in *The Silent Woman*, all which persons have several concernments of their own, yet are all used by the poet to the conducting of the main design to perfection. I shall not waste time in commending the writing of this play, but I will give you my opinion that there is more wit and acuteness of fancy in it than in any of Ben Jonson's—besides that he has here described the conversation of gentlemen in the persons of True-Wit and his friends, with more gaiety, air, and freedom than in the rest of his comedies.

'For the contrivance of the plot, 'tis extreme elaborate, and yet withal easy. For the λύσις, or untying of it, 'tis so admirable that, when it is done, no one of the audience would think the poet could have missed it, and yet it was concealed so much before the last scene, that any other way would sooner have entered into your thoughts. But I dare not take upon me to commend the fabric of it, because it is altogether so full of art, that I must unravel every scene in it to commend it as I ought.

'And this excellent contrivance is still the more to be admired, because 'tis comedy where the persons are only of common rank, and their business private, not elevated by passions or high concernments as in serious plays. Here everyone is a proper judge of all he sees; nothing is represented but that with which he daily converses, so that by consequence all faults lie open to discovery, and few are pardonable. 'Tis this which Horace has judiciously observed:

> creditur, ex medio quia res arcessit, habere
> sudoris minimum; sed habet comoedia tanto
> plus oneris, quanto veniae minus.°

[it is thought that comedy, drawing its themes from daily life, involves less effort; but it calls for proportionately more as it is allowed less indulgence]

But our poet, who was not ignorant of these difficulties, had prevailed himself° of all advantages—as he who designs a large leap takes his rise from the highest ground.

'One of these advantages is that which Corneille has laid down as the greatest which can arrive to any poem, and which he himself could never compass above thrice in all his plays, viz. the making choice of some signal and long-expected day whereon the action of the play is to depend. This day was that designed by Dauphine for the settling of his uncle's estate upon him, which to compass, he contrives to marry him. That the marriage had been plotted by him long beforehand is made evident by what he tells True-Wit in the second act, that in one moment he had destroyed what he had been raising many months.

'There is another artifice of the poet, which I cannot here omit, because by the frequent practice of it in his comedies he has left it to us almost as a rule. That is, when he has any character or humour wherein he would show a *coup de maître*, or his highest skill, he recommends it to your observation by a pleasant description of it before the person first appears. Thus in *Bartholomew Fair* he gives you the pictures of Numps and Cokes, and in this those of Daw, Lafoole, Morose, and the Collegiate Ladies—all which you hear described before you see them. So that before they come upon the stage, you have a longing expectation of them, which prepares you to receive them favourably, and when they are there, even from their first appearance you are so far acquainted with them, that nothing of their humour is lost to you.

'I will observe yet one thing further of this admirable plot. The business of it rises on every act. The second is greater than the first, the third than the second, and so forward to the fifth. There too you see, till the very last scene, new difficulties arising to obstruct the action of the play; and when the audience is brought into despair that the business can naturally be effected, then, and not before, the discovery is made. But that the poet might entertain you with more variety all this while, he reserves some new characters to show you, which he opens not till the second and third act: in the second Morose, Daw, the Barber, and Otter: in the third the Collegiate Ladies:—all which he moves afterwards in by-walks, or under-plots, as diversions to the main design, lest it should grow tedious, though they are still naturally joined with it, and somewhere or other subservient to it. Thus, like a skilful chess-player, by little and little he draws out his men, and makes his pawns of use to his greater persons.

'If this comedy and some others of his were translated into French prose (which would now be no wonder to them, since Molière has lately given them plays out of verse, which have not displeased them), I believe the controversy would soon be decided betwixt the two nations, even making them the judges.

'But we need not call our heroes to our aid. Be it spoken to the honour of the English, our nation can never want in any age such who are able to dispute the empire of wit with any people in the universe. And though the fury of a civil war, and power for twenty years together abandoned to a barbarous race of men, enemies of all good learning, had buried the muses under the ruins of monarchy, yet, with the restoration of our happiness, we see revived poesy lifting up its head, and already shaking off the rubbish which lay so heavy on it. We have seen since his majesty's return many dramatic poems which yield not to those of any foreign nation, and which deserve all laurels but the English.

'I will set aside flattery and envy. It cannot be denied but we have had some little blemish either in the plot or writing of all those plays which have been made within these seven years (and perhaps there is no nation in the world so quick to discern them, or so difficult to pardon them, as ours); yet, if we can persuade ourselves to use the candour of that poet, who, though the most severe of critics, has left us this caution by which to moderate our censures,

> . . . ubi plura nitent in carmine, non ego paucis
> offendar maculis;°

[when the beauties of a poem are numerous, I am not one to be offended at a few faults] if, in consideration of their many and great beauties, we can wink at some slight and little imperfections, if we, I say, can be thus equal to ourselves, I ask no favour from the French.

'And if I do not venture upon any particular judgment of our late plays, 'tis out of the consideration which an ancient writer gives me: *vivorum, ut magna admiratio, ita censura difficilis:*° "betwixt the extremes of admiration and malice, 'tis hard to judge uprightly of the living". Only I think it may be permitted me to say that as it is no lessening to us to yield to some plays—and those not many—of our own nation in the last age, so can it be no addition to pronounce of our present poets that they have far surpassed all the ancients, and the modern writers of other countries.'

This, my lord, was the substance of what was then spoke on that occasion, and Lisideius, I think, was going to reply, when he was prevented thus by Crites: 'I am confident', said he, 'the most material things that can be said have been already urged on either side. If they have not, I must beg of Lisideius that he will defer his answer till another time. For I confess I have a joint quarrel to you both, because you have concluded, without any reason given for it, that rhyme is proper for the stage.

'I will not dispute how ancient it hath been among us to write this way; perhaps our ancestors knew no better till Shakespeare's time. I will grant it was not altogether left by him, and that Fletcher and Ben Jonson used it frequently in their pastorals, and sometimes in other plays. Further, I will not argue whether we received it originally from our own countrymen or from the French;—for that is an inquiry of as little benefit as theirs who, in the midst of the great plague, were not so solicitous to provide against it as to know whether we had it from the malignity of our own air or by transportation from Holland. I have therefore only to affirm that it is not allowable in serious plays. For comedies, I find you already concluding with me.

'To prove this, I might satisfy myself to tell you how much in vain it is for you to strive against the stream of the people's inclination, the greatest part of which are prepossessed so much with those excellent plays of Shakespeare, Fletcher, and Ben Jonson, which have been written out of rhyme, that, except you could bring them such as were written better in it,—and those too by persons of equal reputation with them,—it would be impossible for you to gain your cause with them, who will still be judges. This it is to which, in fine, all your reasons must submit. The unanimous consent of an audience is so powerful that even Julius Caesar (as Macrobius reports of him), when he was perpetual dictator, was not able to balance it on the other side, but when Laberius, a Roman knight, at his request contended in the mime with another poet, he was forced to cry out *etiam favente me victus es, Laberi* [you were beaten, Laberius, even though you had my support].°

'But I will not on this occasion take the advantage of the greater number, but only urge such reasons against rhyme as I find in the writings of those who have argued for the other way.

'First then, I am of opinion that rhyme is unnatural in a play, because dialogue there is presented as the effect of sudden thought. For a play is the imitation of nature, and since no man, without premeditation, speaks in rhyme, neither ought he to do it on the stage. This hinders not but the fancy may be there elevated to an higher pitch of thought than it is in ordinary discourse, for there is a probability that men of excellent and quick parts may speak noble things *ex tempore* [without premeditation]. But those thoughts are never fettered with the numbers or sound of verse without study, and therefore it cannot be but unnatural to present the most free way of speaking in that which is the most constrained.

'For this reason, says Aristotle,° 'tis best to write tragedy in that kind of verse which is the least such, or which is nearest prose. And this amongst the ancients was the iambic, and with us is blank verse, or the measure of verse kept exactly without rhyme. These numbers therefore are fittest for a play, the others for a paper of verses, or a poem—blank verse being as much below them as rhyme is improper for the drama. And if it be objected that neither are blank verses made *ex tempore* [without premeditation], yet, as nearest nature, they are still to be preferred.

'But there are two particular exceptions, which many besides myself have had to verse, by which it will appear yet more plainly how improper it is in plays. And the first of them is grounded upon that very reason for which some have commended rhyme. They say the quickness of repartees in argumentative scenes receives an ornament from verse. Now what is more unreasonable than to imagine that a man should not only light upon the wit, but the rhyme too, upon the sudden? This nicking° of him who spoke before both in sound and measure is so great a happiness that you must at least suppose the persons of your play to be born poets,

Arcades omnes,
et cantare pares et respondere parati,°

[Arcadians all ready both to sing and to make reply]

they must have arrived to the degree of *quicquid conabar dicere* [whatever I was trying to say],° to make verses almost whether they will or no. If they are anything below this, it will look rather like the design of two than the answer of one. It will appear that your actors hold intelligence together, that they perform their tricks like fortune-tellers, by confederacy. The hand of art will be too visible in it against that maxim of all professions, *ars est celare artem*:—that it is the greatest perfection of art to keep itself undiscovered.

'Nor will it serve you to object that, however you manage it, 'tis still known to be a play, and, consequently, the dialogue of two persons understood to be the labour of one poet. For a play is still an imitation of nature. We know we are to be deceived, and we desire to be so. But no man ever was deceived but with a probability of truth, for who will suffer a gross lie to be fastened on him? Thus we sufficiently understand that the scenes which represent cities and countries to us are not really such but only painted on boards and canvas. But shall that excuse the ill painture or designment of them? Nay, rather ought they not to be laboured with so much the more diligence and exactness to help the imagination, since the mind of man does naturally tend to, and seek after truth, and therefore the nearer anything comes to the imitation of it, the more it pleases?

'Thus, you see, your rhyme is uncapable of expressing the greatest thought naturally, and the lowest it cannot with any grace. For what is more unbefitting the majesty of verse than to call a servant or bid a door be shut in rhyme? And yet this miserable necessity you are forced upon. But verse, you say, circumscribes a quick and luxuriant fancy, which would extend itself too far on every subject, did not the labour which is required to well-turned and polished rhyme set bounds to it. Yet this argument, if granted, would only prove that we may write better in verse, but not more naturally. Neither is it able to evince that. For he who wants judgment to confine his fancy in blank verse, may want it as much in rhyme, and he who has it will avoid errors in both kinds.

'Latin verse was as great a confinement to the imagination of those poets as rhyme to ours: and yet you find Ovid saying too much on every subject. *Nescivit* (says Seneca°) *quod bene cessit relinquere* [he does not know when to leave well alone] of which he gives you one famous instance in his description of the deluge,

> omnia pontus erat, deerant quoque litora ponto.°

"Now all was sea, nor had that sea a shore."

Thus Ovid's fancy was not limited by verse, and Virgil needed not verse to have bounded his.

'In our own language we see Ben Jonson confining himself to what ought to be said, even in the liberty of blank verse. And yet Corneille, the most judicious of the French poets, is still varying the same sense a hundred ways, and dwelling eternally upon the same subject, though confined by rhyme.

'Some other exceptions I have to verse. But, since these I have named are for the most part already public, I conceive it reasonable they should first be answered.'

'It concerns me less than any', said Neander (seeing he had ended), 'to reply to this discourse; because when I should have proved that verse may be natural in plays, yet I should always be ready to confess that those which I have written in this kind come short of that perfection which is required. Yet since you are pleased I should undertake this province, I will do it,—though with all imaginable respect and deference, both to that person° from whom you have borrowed your strongest arguments, and to whose judgment, when I have said all, I finally submit.

'But before I proceed to answer your objections, I must first remember you that I exclude all comedy from my defence, and next, that I deny not but blank verse may be also used, and content myself only to assert that in serious plays where the subject and characters are great, and the plot unmixed with mirth, which might allay or divert these concernments which are produced, rhyme is there as natural and more effectual than blank verse.

'And now having laid down this as a foundation, to begin with Crites, I must crave leave to tell him that some of his arguments against rhyme reach no further than, from the faults or defects of ill rhyme, to conclude against the use of it in general. May not I conclude against blank verse by the same reason? If the words of some poets who write in it are either ill-chosen or ill-placed (which makes not only rhyme, but all kind of verse in any language unnatural), shall I, for their vicious affectation, condemn those excellent lines of Fletcher which are written in that kind? Is there anything in rhyme more constrained than this line in blank verse—

> I heaven invoke, and strong resistance make;°

where you see both the clauses are placed unnaturally,—that is, contrary to the common way of speaking,—and that without the excuse of a rhyme to cause it? Yet you would think me very ridiculous if I should accuse the stubbornness of blank verse for this, and not rather the stiffness of the poet. Therefore, Crites, you must either prove that words, though well chosen and duly placed, yet render not rhyme natural in itself, or that, however natural and easy the rhyme may be, yet it is not proper for a play.

'If you insist upon the former part, I would ask you what other conditions are required to make rhyme natural in itself, besides an election of apt words and a right disposing of them. For the due choice of your words expresses your sense naturally, and the due placing them adapts the rhyme to it.

'If you object that one verse may be made for the sake of another, though both the words and rhyme be apt, I answer it cannot possibly so fall out. For either there is a dependence of sense betwixt the first line and the second, or there is none. If there be that connection, then in the natural position of the words the latter line must of necessity flow from the former. If there be no dependence, yet still the due ordering of words makes the last line as natural in itself as the other,— so that the necessity of a rhyme never forces any but bad or lazy writers to say what they would not otherwise.

''Tis true, there is both care and art required to write in verse. A good poet never concludes upon the first line till he has sought out such a rhyme as may fit the sense, already prepared to heighten the second. Many times the close of the sense falls into the middle of the next verse, or farther off, and he may often prevail himself° of the same advantages in English which Virgil had in Latin,—he may break off in the hemistich and begin another line.

'Indeed, the not observing these two last things makes plays which are writ in verse so tedious. For though, most commonly, the sense is to be confined to the couplet, yet nothing that does *perpetuo tenore fluere*,° run in the same channel, can please always. 'Tis like the murmuring of a stream, which, not varying in the fall, causes at first attention, at last drowsiness. Variety of cadences is the best rule, the greatest help to the actors, and refreshment to the audience.

'If then verse may be made natural in itself, how becomes it unnatural in a play? You say the stage is the representation of nature, and no man in ordinary conversation speaks in rhyme. But you foresaw, when you said this, that it might be answered, neither does any man speak in blank verse, or in measure without rhyme. Therefore you concluded that which is nearest nature is still to be preferred. But you took no notice that rhyme might be made as natural as blank verse, by the well placing of the words, etc. All the difference between them, when they are both correct, is the sound in one which the other wants. And if so, the sweetness of it, and all the advantage resulting from it, which are handled in the Preface to *The Rival Ladies*,° will yet stand good. As for that place of Aristotle where he says plays should be writ in that kind of verse which is nearest prose, it makes little for you, blank verse being properly but measured prose.

'Now measure alone, in any modern language, does not constitute verse. Those of the ancients in Greek and Latin consisted in quantity of words and a determinate number of feet. But when, by the inundation of the Goths and Vandals into Italy, new languages were brought in, and barbarously mingled with the Latin, of which the Italian, Spanish, French, and ours (made out of them and the Teutonic) are dialects, a new way of poesy was practised—new, I say, in those countries, for in all probability it was that of the conquerors in their own nations.° This new way consisted in measure or number of feet, and rhyme, the sweetness of rhyme and observation of accent supplying the place of quantity in words, which could neither exactly be observed by those barbarians, who knew not the rules of it, neither was it suitable to their tongues, as it had been to the Greek and Latin.

'No man is tied in modern poesy to observe any further rule in the feet of his verse but that they be dissyllables—whether spondee, trochee, or iambic, it matters not—only he is obliged to rhyme. Neither do the Spanish, French, Italians, or Germans acknowledge at all, or very rarely, any such kind of poesy as blank verse amongst them. Therefore, at most, 'tis but a poetic prose, a *sermo pedestris* [prose discourse], and, as such, most fit for comedies, where I acknowledge rhyme to be improper.

'Further, as to that quotation of Aristotle, our couplet verses may be rendered as near prose as blank verse itself by using those advantages I lately named, as breaks in a hemistich, or running the sense into another line, thereby making art and order appear as loose and free as nature. Or, not tying ourselves to couplets strictly, we may use the benefit of the Pindaric way,° practised in *The Siege of Rhodes*,° where the numbers vary, and the rhyme is disposed carelessly, and far from often chiming.

'Neither is that other advantage of the ancients to be despised, of changing the kind of verse when they please with the change of the scene or some new entrance. For they confine not themselves always to iambics, but extend their liberty to all lyric numbers, and sometimes even to hexameter.

'But I need not go so far to prove that rhyme, as it succeeds to all other offices of Greek and Latin verse, so especially to this of plays, since the custom of all nations at this day confirms it. The French, Italian, and Spanish tragedies are generally writ in it, and sure the universal consent of the most civilized parts of the world ought in this, as it doth in other customs, include the rest.

'But perhaps you may tell me I have proposed such a way to make rhyme natural, and consequently proper to plays, as is impracticable, and that I shall scarce find six or eight lines together in any play, where the words are so placed and chosen as is required to make it natural. I answer, no poet need constrain himself at all times to it. It is enough he makes it his general rule. For I deny not but sometimes there may be a greatness in placing the words otherwise. And sometimes they may sound better. Sometimes also the variety itself is excuse enough. But if, for the most part, the words be placed as they are in the negligence of prose, it is sufficient to denominate the way practicable. For we esteem that to be such which in the trial oftener succeeds than misses. And thus far you may find the practice made good in many plays. Where you do not, remember still that if you cannot find six natural rhymes together, it will be as hard for you to produce as many lines in blank verse, even among the greatest of our poets, against which I cannot make some reasonable exception.

'And this, sir, calls to my remembrance the beginning of your discourse, where you told us we should never find the audience favourable to this kind of writing till we could produce as good plays in rhyme as Ben Jonson, Fletcher, and Shakespeare had writ out of it. But it is to raise envy to the living to compare them with the dead. They are honoured, and almost adored by us, as they deserve. Neither do I know any so presumptuous of themselves as to contend with them.

'Yet give me leave to say thus much, without injury to their ashes, that not only we shall never equal them, but they could never equal themselves were they to rise and write again. We acknowledge them our fathers in wit, but they have ruined their estates themselves before they came to their children's hands. There is scarce a humour, a character, or any kind of plot, which they have not blown upon. All comes sullied or wasted to us. And were they to entertain this age, they could not make so plenteous treatments out of such decayed fortunes. This, therefore, will be a good argument to us, either not to

write at all, or to attempt some other way. There is no bays to be
expected in their walks:

> tentanda via est qua me quoque possim tollere humo.°

[I must venture on a theme which will raise me too from the earth]

'This way of writing in verse they had only left free to us. Our age is
arrived to a perfection in it which they never knew, and which (if we
may guess by what of theirs we have seen in verse, as *The Faithful
Shepherdess* and *Sad Shepherd*)° 'tis probable they never could have
reached. For the genius of every age is different, and though ours excel
in this, I deny not but that to imitate nature in that perfection which
they did in prose is a greater commendation than to write in verse
exactly.

'As for what you have added, that the people are not generally
inclined to like this way,—if it were true, it would be no wonder that
betwixt the shaking off an old habit and the introducing of a new there
should be difficulty. Do we not see them stick to Hopkins and
Sternhold's psalms, and forsake those of David,—I mean Sandys's°
translation of them?

'If by the people you understand the multitude, the οἱ πολλοί, 'tis no
matter what they think. They are sometimes in the right, sometimes in
the wrong. Their judgment is a mere lottery.

> est ubi plebs recte putat, est ubi peccat.°

[sometimes the crowd thinks correctly, sometimes they err]

Horace says it of the vulgar, judging poesy. But if you mean the mixed
audience of the populace and the noblesse, I dare confidently affirm
that a great part of the latter sort are already favourable to verse, and
that no serious plays written since the king's return have been more
kindly received by them than *The Siege of Rhodes*, the *Mustapha*,° *The
Indian Queen*, and *Indian Emperor*.°

'But I come now to the inference of your first argument. You said
the dialogue of plays is presented as the effect of sudden thought, but
no man speaks suddenly or *ex tempore* in rhyme, and you inferred
from thence that rhyme, which you acknowledge to be proper to epic
poesy, cannot equally be proper to dramatic, unless we could suppose
all men born so much more than poets that verses should be made in
them, not by them.

'It has been formerly urged by you, and confessed by me, that,
since no man spoke any kind of verse *ex tempore*, that which was
nearest nature was to be preferred. I answer you, therefore, by

distinguishing betwixt what is nearest to the nature of comedy,—
which is the imitation of common persons and ordinary speaking,—
and what is nearest the nature of a serious play.

'This last is indeed the representation of nature, but 'tis nature
wrought up to an higher pitch. The plot, the characters, the wit, the
passions, the descriptions are all exalted above the level of common
converse, as high as the imagination of the poet can carry them, with
proportion to verisimility. Tragedy, we know, is wont to image to us
the minds and fortunes of noble persons, and to portray these exactly.
Heroic rhyme is nearest nature, as being the noblest kind of modern
verse.

> indignatur enim privatis et prope socco
> dignis carminibus narrari cena Thyestae,°

[for the feast of Thyestes scorns to be told in the language of daily life that is
appropriate to comedy]

says Horace. And in another place,

> effutire leves indigna tragoedia versus.°

[tragedy scorns to babble trivial verses]

Blank verse is acknowledged to be too low for a poem—nay more, for
a paper of verses; but if too low for an ordinary sonnet,° how much
more for tragedy, which is by Aristotle,° in the dispute betwixt the
epic poesy and the dramatic, for many reasons he there alleges, ranked
above it?

'But setting this defence aside, your argument is almost as strong
against the use of rhyme in poems as in plays. For the epic way is
everywhere interlaced with dialogue or discursive scenes. And
therefore you must either grant rhyme to be improper there,—which
is contrary to your assertion,—or admit it into plays by the same title
which you have given it to poems. For though tragedy be justly
preferred above the other, yet there is a great affinity between them, as
may easily be discovered in that definition of a play which Lisideius
gave us.

'The *genus* of them is the same:—a just and lively image of human
nature in its actions, passions, and traverses of fortune. So is the end,
namely for the delight and benefit of mankind. The characters and
persons are still the same, viz. the greatest of both sorts. Only the
manner of acquainting us with those actions, passions, and fortunes is
different. Tragedy performs it *viva voce* [orally], or by action in
dialogue, wherein it excels the epic poem, which does it chiefly by
narration, and therefore is not so lively an image of human nature.

However, the agreement betwixt them is such that if rhyme be proper for one, it must be for the other.

'Verse, 'tis true, is not the effect of sudden thought. But this hinders not that sudden thought may be represented in verse, since those thoughts are such as must be higher than nature can raise them without premeditation, especially to a continuance of them even out of verse. And consequently you cannot imagine them to have been sudden either in the poet or the actors. A play, as I have said, to be like nature, is to be set above it—as statues which are placed on high are made greater than the life that they may descend to the sight in their just proportion.

'Perhaps I have insisted too long upon this objection. But the clearing of it will make my stay shorter on the rest. You tell us, Crites, that rhyme appears most unnatural in repartees or short replies, when he who answers—it being presumed he knew not what the other would say—yet makes up that part of the verse which was left incomplete, and supplies both the sound and measure of it. This, you say, looks rather like the confederacy of two than the answer of one.

'This, I confess, is an objection which is in everyone's mouth who likes not rhyme. But suppose, I beseech you, the repartee were made only in blank verse, might not part of the same argument be turned against you? For the measure is as often supplied there as it is in rhyme, the latter half of the hemistich as commonly made up, or a second line subjoined as a reply to the former—which any one leaf in Jonson's plays will sufficiently clear to you.

'You will often find in the Greek tragedians, and in Seneca, that when a scene grows up into the warmth of repartees (which is the close fighting of it), the latter part of the trimeter is supplied by him who answers. And yet it was never observed as a fault in them by any of the ancient or modern critics. The case is the same in our verse as it was in theirs—rhyme to us being in lieu of quantity to them.

'But if no latitude is to be allowed a poet, you take from him not only his license of *quidlibet audendi* [hazarding anything],° but you tie him up in a straiter compass than you would a philosopher. This is indeed

musas colere severiores.°

[to cultivate the stricter muses]

You would have him follow nature, but he must follow her on foot. You have dismounted him from his Pegasus.

'But you tell us this supplying the last half of a verse, or adjoining a whole second to the former, looks more like the design of two than the answer of one. Suppose we acknowledge it. How comes this confederacy to be more displeasing to you than in a dance which is well contrived? You see there the united design of many persons to make up one figure. After they have separated themselves in many petty divisions, they rejoin one by one into a gross. The confederacy is plain amongst them, for chance could never produce anything so beautiful. And yet there is nothing in it that shocks your sight.

'I acknowledge the hand of art appears in repartee, as of necessity it must in all kind of verse. But there is also the quick and poignant brevity of it (which is a high imitation of nature in those sudden gusts of passion) to mingle with it. And this, joined with the cadency and sweetness of the rhyme, leaves nothing in the soul of the hearer to desire. 'Tis an art which appears. But it appears only like the shadowings of painture, which, being to cause the rounding of it, cannot be absent; but, while that is considered, they are lost. So, while we attend to the other beauties of the matter, the care and labour of the rhyme is carried from us, or at least drowned in its own sweetness, as bees are sometimes buried in their honey.

'When a poet has found the repartee, the last perfection he can add to it is to put it into verse. However good the thought may be, however apt the words in which 'tis couched, yet he finds himself at a little unrest while rhyme is wanting. He cannot leave it till that comes naturally, and then is at ease, and sits down contented.

'From replies, which are the most elevated thoughts of verse, you pass to the most mean ones—those which are common with the lowest of household conversation. In these, you say, the majesty of verse suffers. You instance in the calling of a servant or commanding a door to be shut in rhyme. This, Crites, is a good observation of yours, but no argument. For it proves no more but that such thoughts should be waived, as often as may be, by the address of the poet.

'But suppose they are necessary in the places where he uses them, yet there is no need to put them into rhyme. He may place them in the beginning of a verse, and break it off as unfit, when so debased, for any other use. Or, granting the worst,—that they require more room than the hemistich will allow,—yet still there is a choice to be made of the best words, and least vulgar—provided they be apt—to express such thoughts.

'Many have blamed rhyme in general for this fault, when the poet, with a little care, might have redressed it. But they do it with no more justice than if English poesy should be made ridiculous for the sake of the water poet's° rhymes. Our language is noble, full, and significant,

and I know not why he who is master of it may not clothe ordinary
things in it as decently as the Latin, if he use the same diligence in his
choice of words. *Delectus verborum origo est eloquentiae* [the choice of
words is the beginning of eloquence].° It was the saying of Julius
Caesar, one so curious in his, that none of them can be changed but for
a worse. One would think *unlock the door* was a thing as vulgar as could
be spoken; and yet Seneca could make it sound high and lofty in his
Latin:

> reserate clusos regii postes laris.°

[set wide the palace gates]

'But I turn from this exception, both because it happens not above
twice or thrice in any play that those vulgar thoughts are used, and
then too, were there no other apology to be made, yet the necessity of
them, which is alike in all kind of writing, may excuse them. Besides
that the great eagerness and precipitation with which they are spoken
makes us rather mind the substance than the dress—that for which
they are spoken rather than what is spoke. For they are always the
effect of some hasty concernment, and something of consequence
depends upon them.

'Thus, Crites, I have endeavoured to answer your objections. It
remains only that I should vindicate an argument for verse which you
have gone about to overthrow. It had formerly been said that the
easiness of blank verse renders the poet too luxuriant, but that the
labour of rhyme bounds and circumscribes an over-fruitful fancy—
the sense there being commonly confined to the couplet, and the
words so ordered that the rhyme naturally follows them, not they the
rhyme. To this you answered that it was no argument to the question
in hand, for the dispute was not which way a man may write best, but
which is most proper for the subject on which he writes.

'First, give me leave, sir, to remember you that the argument
against which you raised this objection was only secondary. It was
built upon this hypothesis, that to write in verse was proper for serious
plays. Which supposition being granted (as it was briefly made out in
that discourse, by showing how verse might be made natural), it
asserted that this way of writing was a help to the poet's judgment,
by putting bounds to a wild overflowing fancy. I think therefore it will
not be hard for me to make good what it was to prove.

'But you add that, were this let pass, yet he who wants judgment in
the liberty of his fancy may as well show the defect of it when he is
confined to verse. For he who has judgment will avoid errors, and he
who has it not will commit them in all kinds of writing.

'This argument, as you have taken it from a most acute person,° so, I confess, it carries much weight in it. But by using the word judgment here indefinitely, you seem to have put a fallacy upon us. I grant he who has judgment,—that is, so profound, so strong, so infallible a judgment that he needs no helps to keep it always poised and upright,—will commit no faults either in rhyme or out of it. And, on the other extreme, he who has a judgment so weak and crazed that no helps can correct or amend it, shall write scurvily out of rhyme, and worse in it. But the first of these judgments is nowhere to be found, and the latter is not fit to write at all.

'To speak therefore of judgment as it is in the best poets, they who have the greatest proportion of it want other helps than from it within—as for example, you would be loath to say that he who was endued with a sound judgment had no need of history, geography, or moral philosophy, to write correctly. Judgment is indeed the master-workman in a play. But he requires many subordinate hands, many tools to his assistance.

'And verse I affirm to be one of these. 'Tis a rule and line by which he keeps his building compact and even, which otherwise lawless imagination would raise either irregularly or loosely. At least, if the poet commits errors with this help, he would make greater and more without it. 'Tis, in short, a slow and painful, but the surest kind of working.

'Ovid, whom you accuse for luxuriancy in verse, had perhaps been further guilty of it had he writ in prose. And for your instance of Ben Jonson, who, you say, wrote exactly without the help of rhyme, you are to remember 'tis only an aid to a luxuriant fancy, which his was not. As he did not want imagination, so none ever said he had much to spare. Neither was verse then refined so much to be a help to that age as it is to ours.

'Thus then the second thoughts being usually the best, as receiving the maturest digestion from judgment, and the last and most mature product of those thoughts being artful and laboured verse, it may well be inferred that verse is a great help to a luxuriant fancy. And this is what that argument which you opposed was to evince.'

Neander was pursuing this discourse so eagerly that Eugenius had called to him twice or thrice ere he took notice that the barge stood still, and that they were at the foot of Somerset-Stairs,° where they had appointed it to land. The company were all sorry to separate so soon, though a great part of the evening was already spent, and stood awhile looking back upon the water, which the moonbeams played upon, and made it appear like floating quicksilver.

At last they went up through a crowd of French people who were merrily dancing in the open air, and nothing concerned for the noise of guns which had alarmed the town that afternoon. Walking thence together to the Piazze,° they parted there—Eugenius and Lisideius to some pleasant appointment they had made, and Crites and Neander to their several lodgings.

Prologues to Secret Love

PROLOGUE

He who writ this, not without pains and thought
From French and English theatres has brought
The exactest rules by which a play is wrought:

The unities of action, place, and time;
The scenes unbroken; and a mingled chime
Of Jonson's humour with Corneille's rhyme.

But while dead colours he with care did lay,°
He fears his wit or plot he did not weigh,
Which are the living beauties of a play.

Plays are like towns, which, howe'er fortified   10
By engineers, have still some weaker side
By the o'er-seen defendant unespied.

And with that art you make approaches now;
Such skilful fury in assaults you show,
That every poet without shame may bow.

Ours therefore humbly would attend your doom,
If, soldier-like, he may have terms to come
With flying colours and with beat of drum.

The prologue *goes out, and stays while a tune is played, after which he returns again.*

SECOND PROLOGUE

I had forgot one half, I do protest,
And now am sent again to speak the rest.   20

He bows to every great and noble wit;
But to the little Hectors of the pit
Our poet's sturdy, and will not submit.
He'll be beforehand with 'em, and not stay
To see each peevish critic stab his play:
Each puny censor, who, his skill to boast,
Is cheaply witty on the poet's cost.
No critic's verdict should of right stand good;
They are excepted all, as men of blood;
And the same law should shield him from their fury   30
Which has excluded butchers from a jury.
You'd all be wits—
But writing's tedious, and that way may fail;
The most compendious method is to rail;
Which you so like, you think yourselves ill used
When in smart prologues you are not abused.
A civil prologue is approved by no man;
You hate it as you do a civil woman:
Your fancy's palled, and liberally you pay
To have it quickened, ere you see a play;   40
Just as old sinners, worn from their delight,
Give money to be whipped to appetite.
But what a pox keep I so much ado
To save our poet? He is one of you;
A brother judgment, and, as I hear say,
A cursed critic as e'er damned a play.
Good savage gentlemen, your own kind spare;
He is, like you, a very wolf or bear.
Yet think not he'll your ancient rights invade,
Or stop the course of your free damning trade;   50
For he, he vows, at no friend's play can sit,
But he must needs find fault to show his wit.
Then, for his sake, ne'er stint your own delight;
Throw boldly, for he sets to all that write:°
With such he ventures on an even lay,
For they bring ready money into play.
Those who write not, and yet all writers nick,
Are bankrupt gamesters, for they damn on tick.

Prologue and Epilogue to
Sir Martin Mar-all

PROLOGUE

Fools, which each man meets in his dish each day,
Are yet the great regalios of a play;°
In which to poets you but just appear,
To prize that highest which costs them so dear.
Fops in the town more easily will pass;
One story makes a statutable ass:°
But such in plays must be much thicker sown,
Like yolks of eggs, a dozen beat to one.
Observing poets all their walks invade,
As men watch woodcocks gliding through a glade;°   10
And when they have enough for comedy,
They stow their several bodies in a pie.
The poet's but the cook to fashion it,
For, gallants, you yourselves have found the wit.
To bid you welcome would your bounty wrong;
None welcome those who bring their cheer along.

EPILOGUE

As country vicars, when the sermon's done,
Run huddling to the benediction;
Well knowing, though the better sort may stay,
The vulgar rout will run unblest away:
So we, when once our play is done, make haste
With a short epilogue to close your taste.
In thus withdrawing we seem mannerly,
But when the curtain's down we peep and see
A jury of the wits who still stay late,
And in their club decree the poor play's fate: 10
Their verdict back is to the boxes brought,
Thence all the town pronounces it their thought.
Thus, gallants, we like Lilly can foresee;°
But if you ask us what our doom will be,
We by tomorrow will our fortune cast,
As he tells all things when the year is past.

Prologue to The Wild Gallant *revived*

As some raw squire, by tender mother bred,
Till one-and-twenty keeps his maidenhead,
(Pleased with some sport, which he alone does find,
And thinks a secret to all humankind,)
Till mightily in love, yet half afraid,
He first attempts the gentle dairy-maid.
Succeeding there, and led by the renown
Of Whetstone's Park, he comes at length to town;°
Where entered, by some school-fellow or friend,
He grows to break glass-windows in the end: 10
His valour too, which with the watch began,
Proceeds to duel, and he kills his man.
By such degrees, while knowledge he did want,
Our unfledged author writ a *Wild Gallant*.
He thought him monstrous lewd (I'll lay my life)
Because suspected with his landlord's wife;
But, since his knowledge of the town began,
He thinks him now a very civil man;
And, much ashamed of what he was before,
Has fairly played him at three wenches more.   20
'Tis some amends his frailties to confess;
Pray pardon him his want of wickedness.
He's towardly, and will come on apace;
His frank confession shows he has some grace.
You balked him when he was a young beginner.°
And almost spoiled a very hopeful sinner;
But, if once more you slight his weak endeavour,
For aught I know, he may turn tail for ever.

Prologue and Epilogue to
The Tempest

PROLOGUE

As when a tree's cut down, the secret root
Lives under ground, and thence new branches shoot,
So, from old Shakespeare's honoured dust, this day
Springs up and buds a new reviving play.

Shakespeare, who (taught by none) did first impart
To Fletcher wit, to labouring Jonson art.
He monarch-like gave those his subjects law,
And is that nature which they paint and draw.
Fletcher reached that which on his heights did grow,
Whilst Jonson crept and gathered all below.   10
This did his love, and this his mirth digest:
One imitates him most, the other best.
If they have since out-writ all other men,
'Tis with the drops which fell from Shakespeare's pen.
The storm which vanished on the neighbouring shore,°
Was taught by Shakespeare's Tempest first to roar.
That innocence and beauty which did smile
In Fletcher, grew on this Enchanted Isle.°
But Shakespeare's magic could not copied be,
Within that circle none durst walk but he.   20
I must confess 'twas bold, nor would you now,
That liberty to vulgar wits allow,
Which works by magic supernatural things:
But Shakespeare's power is sacred as a king's.
Those legends from old priest-hood were received,
And he then writ, as people then believed.
But, if for Shakespeare we your grace implore,
We for our theatre shall want it more:
Who by our dearth of youths are forced to employ
One of our women to present a boy.   30
And that's a transformation you will say
Exceeding all the magic in the play.
Let none expect in the last act to find
Her sex transformed from man to woman-kind.
Whate'er she was before the play began,
All you shall see of her is perfect man.
Or if your fancy will be farther led,
To find her woman, it must be abed.

EPILOGUE

Gallants, by all good signs it does appear
That sixty-seven's a very damning year,
For knaves abroad, and for ill poets here.

Among the muses there's a general rot:°
The rhyming Mounsieur and the Spanish plot,
Defy or court, all's one, they go to pot.

The ghosts of poets walk within this place,
And haunt us actors wheresoe'er we pass,
In visions bloodier than King Richard's was.

For this poor wretch he has not much to say,   10
But quietly brings in his part o' th' play,
And begs the favour to be damned today.

He sends me only like a sheriff's man here,
To let you know the malefactor's near,
And that he means to die *en cavalier*.

For if you should be gracious to his pen,
The example will prove ill to other men,
And you'll be troubled with 'em all again.

Prologue and Epilogue to Tyrannic Love

PROLOGUE

Self-love (which never rightly understood)
Makes poets still conclude their plays are good:
And malice in all critics reigns so high,
That for small errors they whole plays decry;
So that to see this fondness, and that spite,
You'd think that none but madmen judge or write.
Therefore our poet, as he thinks not fit
To impose upon you what he writes for wit,
So hopes that leaving you your censures free,
You equal judges of the whole will be: 10
They judge but half who only faults will see.
Poets like lovers should be bold and dare,
They spoil their business with an over-care.
And he who servilely creeps after sense,
Is safe, but ne'er will reach an excellence.

Hence 'tis our poet in his conjuring,
Allowed his fancy the full scope and swing.
But when a tyrant for his theme he had,
He loosed the reins, and bid his muse run mad:
And though he stumbles in a full career;   20
Yet rashness is a better fault than fear.
He saw his way; but in so swift a pace,
To choose the ground might be to lose the race.
They then who of each trip the advantage take
Find but those faults which they want wit to make.

EPILOGUE

Spoken by Mrs Ellen,° when she was to
be carried off dead by the bearers

To the bearer

Hold! are you mad? you damned, confounded dog!
I am to rise and speak the epilogue.

To the audience

I come, kind gentlemen, strange news to tell ye,
I am the ghost of poor departed Nelly.
Sweet ladies, be not frighted, I'll be civil;
I'm what I was, a little harmless devil:
For after death, we sprites have just such natures
We had, for all the world, when human creatures;
And therefore I that was an actress here,
Play all my tricks in hell, a goblin there.   10
Gallants, look to't, you say there are no sprites;
But I'll come dance about your beds at nights.
And faith you'll be in a sweet kind of taking,°
When I suprise you between sleep and waking.
To tell you true, I walk because I die
Out of my calling in a tragedy.
O poet, damned dull poet, who could prove
So senseless, to make Nelly die for love!
Nay, what's yet worse, to kill me in the prime
Of Easter term, in tart and cheese-cake time!   20
I'll fit the fop; for I'll not one word say
To excuse his godly out-of-fashion play;

A play which if you dare but twice sit out,
You'll all be slandered, and be thought devout.
But farewell, gentlemen, make haste to me,
I'm sure ere long to have your company.
As for my epitaph when I am gone,
I'll trust no poet, but will write my own:

Here Nelly lies, who, though she lived a slattern,
Yet died a princess, acting in St Catharn.°   30

Prologue to The First Part of
The Conquest of Granada

Spoken by Mrs Ellen Gwyn in a broadbrimmed
hat and waistbelt

This jest was first of t'other house's making,°
And, five times tried, has never failed of taking;
For 'twere a shame a poet should be killed
Under the shelter of so broad a shield.
This is that hat, whose very sight did win ye
To laugh and clap as though the devil were in ye.
As then, for Nokes, so now I hope you'll be,°
So dull, to laugh, once more, for love of me.
'I'll write a play,' says one, 'for I have got
A broad-brimmed hat, and waist-belt towards a plot.'   10
Says t'other, 'I have one more large than that.'
Thus they out-write each other with a hat!
The brims still grew with every play they writ;
And grew so large, they covered all the wit.
Hat was the play; 'twas language, wit, and tale:
Like them that find meat, drink, and cloth in ale.
What dulness do these mongrel wits confess,
When all their hope is acting of a dress!
Thus two, the best comedians of the age°
Must be worn out, with being blocks of the stage:°   20
Like a young girl, who better things has known,
Beneath their poet's impotence they groan.
See now what charity it was to save!
They thought you liked what only you forgave;

And brought you more dull sense, dull sense much worse
Than brisk gay nonsense, and the heavier curse.
They bring old iron and glass upon the stage,
To barter with the Indians of our age.
Still they write on, and like great authors show;
But 'tis as rollers in wet gardens grow   30
Heavy with dirt, and gathering as they go.
May none, who have so little understood,
To like such trash, presume to praise what's good!
And may those drudges of the stage, whose fate
Is damned dull farce more dully to translate,
Fall under that excise the state thinks fit
To set on all French wares, whose worst is wit.
French farce, worn out at home, is sent abroad;
And, patched up here, is made our English mode.
Henceforth, let poets, ere allowed to write,   40
Be searched, like duellists, before they fight,
For wheel-broad hats, dull humour, all that chaff,
Which makes you mourn, and makes the vulgar laugh.
For these, in plays, are as unlawful arms,
As, in a combat, coats of mail and charms.

Epilogue to The Second Part of
The Conquest of Granada

They who have best succeeded on the stage,
Have still conform'd their genius to their age.
Thus Jonson did mechanic humour show,
When men were dull, and conversation low.
Then comedy was faultless, but 'twas coarse:
Cobb's tankard was a jest, and Otter's horse.°
And, as their comedy, their love was mean:
Except, by chance, in some one laboured scene,
Which must atone for an ill-written play.
They rose, but at their height could seldom stay.   10
Fame then was cheap, and the first comer sped;
And they have kept it since, by being dead.
But, were they now to write, when critics weigh
Each line, and every word, throughout a play,

None of them, no, not Jonson in his height
Could pass without allowing grains for weight.°
Think it not envy that these truths are told,
Our poet's not malicious, though he's bold.
'Tis not to brand 'em that their faults are shown,
But, by their errors, to excuse his own.   20
If love and honour now are higher raised,
'Tis not the poet but the age is praised.
Wit's now arrived to a more high degree;
Our native language more refined and free.
Our ladies and our men now speak more wit
In conversation than those poets writ.
Then one of these is, consequently, true;
That what this poet writes comes short of you,
And imitates you ill (which most he fears),
Or else his writing is not worse than theirs.   30
Yet, though you judge (as sure the critics will),
That some before him writ with greater skill,
In this one praise he has their fame surpassed,
To please an age more gallant than the last.

Prologue and Epilogue to Aureng-Zebe

PROLOGUE

Our author by experience finds it true,
'Tis much more hard to please himself than you:
And out of no feigned modesty, this day
Damns his laborious trifle of a play:
Not that it's worse than what before he writ,
But he has now another taste of wit;
And to confess a truth (though out of time)
Grows weary of his long-loved mistress, Rhyme.°
Passion's too fierce to be in fetters bound,
And Nature flies him like enchanted ground.   10
What verse can do, he has performed in this,
Which he presumes the most correct of his:
But spite of all his pride, a secret shame
Invades his breast at Shakespeare's sacred name:

Awed when he hears his godlike Romans rage,
He, in a just despair, would quit the stage.
And to an age less polished, more unskilled,
Does with disdain the foremost honours yield.
As with the greater dead he dares not strive,
He would not match his verse with those who live: 20
Let him retire, betwixt two ages cast,
The first of this, and hindmost of the last.
A losing gamester, let him sneak away;
He bears no ready money from the play.
The fate which governs poets thought it fit,
He should not raise his fortunes by his wit.
The clergy thrive, and the litigious bar;
Dull heroes fatten with the spoils of war:
All southern vices, heaven be praised, are here;
But wit's a luxury you think too dear.   30
When you to cultivate the plant are loath,
'Tis a shrewd sign 'twas never of your growth:
And wit in northern climates will not blow,
Except, like orange-trees, 'tis housed from snow.
There needs no care to put a playhouse down,
'Tis the most desert place of all the town.
We and our neighbours, to speak proudly, are°
Like monarchs ruined with expensive war.
While, like wise English, unconcerned, you sit,
And see us play the tragedy of wit.   40

EPILOGUE

A pretty task! and so I told the fool,
Who needs would undertake to please by rule:
He thought that if his characters were good,
The scenes entire, and freed from noise and blood,
The action great, yet circumscribed by time,
The words not forced, but sliding into rhyme,
The passions raised and calmed by just degrees,
As tides are swelled, and then retire to seas;
He thought, in hitting these, his business done,
Though he, perhaps, has failed in every one: 10
But, after all, a poet must confess,
His art's like physic, but a happy guess.

Your pleasure on your fancy must depend:
The lady's pleased just as she likes her friend.
No song! no dance! no show! he fears you'll say
You love all naked beauties but a play.
He much mistakes your methods to delight,
And like the French, abhors our target-fight:
But those damned dogs can ne'er be in the right.
True English hate your Monsieur's paltry arts,   20
For you are all silk-weavers in your hearts.°
Bold Britons, at a brave bear-garden fray
Are roused, and clattering sticks cry, 'Play, play, play!'
Meantime, your filthy foreigner will stare,
And mutter to himself, '*ha! gens barbares!*'°
And, gad, 'tis well he mutters; well for him;
Our butchers else would tear him limb from limb.
'Tis true, the time may come your sons may be
Infected with this French civility:
But this, in after-ages will be done: 30
Our poet writes a hundred years too soon.
This age comes on too slow, or he too fast:
And early springs are subject to a blast!
Who would excel, when few can make a test
Betwixt indifferent writing and the best?
For favours, cheap and common, who would strive
Which like abandoned prostitutes, you give?
Yet scattered here and there, I some behold
Who can discern the tinsel from the gold:
To these he writes; and if by them allowed,   40
'Tis their prerogative to rule the crowd.
For he more fears, like a presuming man,
Their votes who cannot judge, than theirs who can.

Epilogue to The Man of Mode

Most modern wits such monstrous fools have shown,
They seemed not of heaven's making, but their own.
Those nauseous harlequins in farce may pass;°
But there goes more to a substantial ass!
Something of man must be exposed to view,
That, gallants, they may more resemble you.

Sir Fopling is a fool so nicely writ,°
The ladies would mistake him for a wit;
And, when he sings, talks loud, and cocks, would cry,°
'Ay now, methinks, he's pretty company—   10
So brisk, so gay, so travelled, so refined,
As he took pains to graft upon his kind.'°
True fops help nature's work, and go to school,
To file and finish God almighty's fool.
Yet none Sir Fopling him, or him, can call;
He's knight of the shire, and represents ye all.°
From each he meets he culls whate'er he can;
Legion's his name, a people in a man.
His bulky folly gathers as it goes,
And, rolling o'er you, like a snowball grows.   20
His various modes from various fathers follow;
One taught the toss, and one the new French wallow;°
His sword-knot this, his cravat this designed;
And this the yard-long snake he twirls behind.°
From one the sacred periwig he gained,
Which wind ne'er blew, nor touch of hat profaned.
Another's diving bow he did adore,
Which with a shog casts all the hair before,°
Till he with full decorum brings it back,
And rises with a water spaniel shake.   30
As for his songs (the ladies' dear delight),
These sure he took from most of you who write.
Yet every man is safe from what he feared;
For no one fool is hunted from the herd.

MacFlecknoe

All human things are subject to decay,
And when fate summons, monarchs must obey.
This Flecknoe found, who, like Augustus, young°
Was called to empire, and had governed long;
In prose and verse, was owned, without dispute,
Through all the realms of Nonsense, absolute.
This aged prince, now flourishing in peace,
And blest with issue of a large increase,
Worn out with business, did at length debate
To settle the succession of the state;   10

And, pondering which of all his sons was fit
To reign, and wage immortal war with wit,
Cried: "'Tis resolved; for nature pleads, that he
Should only rule, who most resembles me.
Shadwell alone my perfect image bears,°
Mature in dullness from his tender years:
Shadwell alone, of all my sons, is he
Who stands confirmed in full stupidity.
The rest to some faint meaning make pretence,
But Shadwell never deviates into sense.   20
Some beams of wit on other souls may fall,
Strike through, and make a lucid interval;
But Shadwell's genuine night admits no ray;
His rising fogs prevail upon the day.
Besides, his goodly fabric fills the eye,°
And seems designed for thoughtless majesty;
Thoughtless as monarch oaks that shade the plain,
And, spread in solemn state, supinely reign.
Heywood and Shirley were but types of thee,°
Thou last great prophet of tautology.   30
Even I, a dunce of more renown than they,
Was sent before but to prepare thy way:°
And, coarsely clad in Norwich drugget, came°
To teach the nations in thy greater name.
My warbling lute, the lute I whilom strung,°
When to King John of Portugal I sung,°
Was but the prelude to that glorious day,
When thou on silver Thames didst cut thy way,
With well-timed oars before the royal barge,
Swelled with the pride of thy celestial charge;   40
And big with hymn, commander of a host,
The like was ne'er in Epsom blankets tossed.°
Methinks I see the new Arion sail,
The lute still trembling underneath thy nail.
At thy well-sharpened thumb from shore to shore
The treble squeaks for fear, the basses roar;
Echoes from Pissing Alley "Shadwell" call,°
And "Shadwell" they resound from Ashton Hall.°
About thy boat the little fishes throng,
As at the morning toast that floats along.°   50
Sometimes, as prince of thy harmonious band,
Thou wield'st thy papers in thy threshing hand.

St. André's feet ne'er kept more equal time,°
Not e'en the feet of thy own *Psyche's* rhyme;°
Though they in number as in sense excel:
So just, so like tautology, they fell,
That, pale with envy, Singleton forswore°
The lute and sword, which he in triumph bore,
And vowed he ne'er would act Villerius more.'°
Here stopped the good old sire, and wept for joy   60
In silent raptures of the hopeful boy.
All arguments, but most his plays, persuade,
That for anointed dullness he was made.

    Close to the walls which fair Augusta bind,°
(The fair Augusta much to fears inclined,)
An ancient fabric raised to inform the sight,
There stood of yore, and Barbican it hight:°
A watchtower once; but now, so fate ordains,
Of all the pile an empty name remains.
From its old ruins brothel-houses rise,   70
Scenes of lewd loves, and of polluted joys,
Where their vast courts the mother-strumpets keep,°
And, undisturbed by watch, in silence sleep.
Near these a nursery erects its head,
Where queens are formed, and future heroes bred;
Where unfledged actors learn to laugh and cry,°
Where infant punks their tender voices try,
And little Maximins the gods defy.°
Great Fletcher never treads in buskins here,°
Nor greater Jonson dares in socks appear;°   80
But gentle Simkin just reception finds°
Amidst this monument of vanished minds:
Pure clenches the suburban muse affords,
And Panton waging harmless war with words.°
Here Flecknoe, as a place to fame well known,
Ambitiously designed his Shadwell's throne;
For ancient Dekker prophesied long since,°
That in this pile should reign a mighty prince,
Born for a scourge of wit, and flail of sense;
To whom true dullness should some *Psyches* owe,   90
But worlds of *Misers* from his pen should flow;°
Humourists and *Hypocrites* it should produce,
Whole Raymond families, and tribes of Bruce.°

    Now Empress Fame had published the renown

Of Shadwell's coronation through the town.
Roused by report of Fame, the nations meet,
From near Bunhill, and distant Watling Street.°
No Persian carpets spread the imperial way,
But scattered limbs of mangled poets lay;
From dusty shops neglected authors come,   100
Martyrs of pies, and relics of the bum.°
Much Heywood, Shirley, Ogilby there lay,°
But loads of Shadwell almost choked the way.
Bilked stationers for yeomen stood prepared,°
And Herringman was captain of the guard.°
The hoary prince in majesty appeared,
High on a throne of his own labours reared.°
At his right hand our young Ascanius sate,
Rome's other hope, and pillar of the State.°
His brows thick fogs, instead of glories, grace,°   110
And lambent dullness played around his face.
As Hannibal did to the altars come,
Sworn by his sire a mortal foe to Rome;
So Shadwell swore, nor should his vow be vain,
That he till death true dullness would maintain;
And, in his father's right, and realm's defence,
Ne'er to have peace with wit, nor truce with sense.
The king himself the sacred unction made,
As king by office, and as priest by trade.
In his sinister hand, instead of ball,   120
He placed a mighty mug of potent ale;
Love's Kingdom to his right he did convey,°
At once his sceptre, and his rule of sway;
Whose righteous lore the prince had practised young,
And from whose loins recorded *Psyche* sprung.
His temples, last, with poppies were o'erspread,°
That nodding seemed to consecrate his head.
Just at that point of time, if fame not lie,
On his left hand twelve reverend owls did fly.°
So Romulus, 'tis sung, by Tiber's brook,   130
Presage of sway from twice six vultures took.°
The admiring throng loud acclamations make,
And omens of his future empire take.
The sire then shook the honours of his head,
And from his brows damps of oblivion shed
Full on the filial dullness: long he stood,

Repelling from his breast the raging god;
At length burst out in this prophetic mood:
    'Heavens bless my son, from Ireland let him reign
To far Barbadoes on the western main;   140
Of his dominion may no end be known,
And greater than his father's be his throne;
Beyond *Love's Kingdom* let him stretch his pen!'
He paused, and all the people cried, 'Amen.'
Then thus continued he: 'My son, advance
Still in new impudence, new ignorance.
Success let others teach, learn thou from me
Pangs without birth, and fruitless industry.
Let *Virtuosos* in five years be writ;
Yet not one thought accuse thy toil of wit.   150
Let gentle George in triumph tread the stage,°
Make Dorimant betray, and Loveit rage;°
Let Cully, Cockwood, Fopling, charm the pit,°
And in their folly show the writer's wit.
Yet still thy fools shall stand in thy defence,
And justify their author's want of sense.
Let 'em be all by thy own model made
Of dullness, and desire no foreign aid;
That they to future ages may be known,
Not copies drawn, but issue of thy own.   160
Nay, let thy men of wit too be the same,
All full of thee, and differing but in name.
But let no alien Sedley interpose,°
To lard with wit thy hungry *Epsom* prose.
And when false flowers of rhetoric thou wouldst cull,
Trust nature, do not labour to be dull;
But write thy best, and top; and, in each line,
Sir Formal's oratory will be thine:°
Sir Formal, though unsought, attends thy quill,
And does thy northern dedications fill.°   170
Nor let false friends seduce thy mind to fame,
By arrogating Jonson's hostile name.
Let father Flecknoe fire thy mind with praise,
And uncle Ogilby thy envy raise.
Thou art my blood, where Jonson has no part:
What share have we in nature, or in art?
Where did his wit on learning fix a brand,

And rail at arts he did not understand?°
Where made he love in Prince Nicander's vein,°
Or swept the dust in *Psyche's* humble strain?   180
Where sold he bargains, "whip-stitch, kiss my arse",°
Promised a play and dwindled to a farce?
When did his muse from Fletcher scenes purloin,
As thou whole Etherege dost transfuse to thine?
But so transfused, as oil on water's flow,
His always floats above, thine sinks below.
This is thy province, this thy wondrous way,
New humours to invent for each new play:°
This is that boasted bias of thy mind,°
By which one way, to dullness, 'tis inclined;   190
Which makes thy writings lean on one side still,
And, in all changes, that way bends thy will.
Nor let thy mountain-belly make pretence
Of likeness; thine's a tympany of sense.°
A tun of man in thy large bulk is writ,°
But sure thou'rt but a kilderkin of wit.°
Like mine, thy gentle numbers feebly creep;
Thy tragic muse gives smiles, thy comic sleep.
With whate'er gall thou settst thyself to write,
Thy inoffensive satires never bite.   200
In thy felonious heart though venom lies,
It does but touch thy Irish pen, and dies.
Thy genius calls thee not to purchase fame
In keen iambics, but mild anagram.°
Leave writing plays, and choose for thy command
Some peaceful province in acrostic land.°
There thou mayst wings display and altars raise,°
And torture one poor word ten thousand ways.
Or, if thou wouldst thy different talents suit,
Set thy own songs, and sing them to thy lute.'   210

   He said: but his last words were scarcely heard;
For Bruce and Longvil had a trap prepared,°
And down they sent the yet declaiming bard.
Sinking he left his drugget robe behind,
Borne upwards by a subterranean wind.
The mantle fell to the young prophet's part,°
With double portion of his father's art.

Heads of an Answer to Rymer

1. He who undertakes to answer this excellent critique of Mr. Rymer, in behalf of our English poets against the Greek, ought to do it in this manner.

2. Either by yielding to him the greatest part of what he contends for, which consists in this, that the μῦθος [plot], i.e. the design and conduct of it, is more conducing in the Greeks to those ends of tragedy which Aristotle and he propose, namely to cause terror and pity; yet the granting this does not set the Greeks above the English poets.

3. But the answerer ought to prove two things: first, that the fable is not the greatest masterpiece of a tragedy, though it be the foundation of it.

4. Secondly, that other ends as suitable to the nature of tragedy may be found in the English, which were not in the Greek.

5. Aristotle places the fable first;° not *quoad dignitatem, sed quoad fundamentum* [because of its dignity, but because it is basic]; for a fable, never so movingly contrived to those ends of his, pity and terror, will operate nothing on our affections, except the characters, manners, thoughts, and words are suitable.

6. So that it remains for Mr Rymer to prove that in all those, or the greatest part of them, we are inferior to Sophocles and Euripides; and this he has offered at in some measure but, I think, a little partially to the Ancients.

7. To make a true judgement in this competition, between the Greek poets and the English in tragedy, consider

    I. How Aristotle has defined a tragedy.
   II. What he assigns the end of it to be.
  III. What he thinks the beauties of it.
  IV. The means to attain the end proposed.

Compare the Greek and English tragic poets justly and without partiality, according to those rules.

8. Then, secondly, consider whether Aristotle has made a just definition of tragedy, of its parts, of its ends, of its beauties; and whether he, having not seen any others but those of Sophocles, Euripides, etc., had or truly could determine what all the excellencies of tragedy are, and wherein they consist.

9. Next show in what ancient tragedy was deficient: for example, in the narrowness of its plots, and fewness of persons, and try whether

that be not a fault in the Greek poets; and whether their excellency was so great when the variety was visibly so little; or whether what they did was not very easy to do.

10. Then make a judgement on what the English have added to their beauties: as, for example, not only more plot, but also new passions; as namely, that of love, scarce touched on by the Ancients, except in this one example of Phaedra, cited by Mr Rymer;° and in that how short they were of Fletcher.°

11. Prove also that love, being a heroic passion, is fit for tragedy, which cannot be denied, because of the example alleged of Phaedra; and how far Shakespeare has outdone them in friendship, etc.

12. To return to the beginning of this enquiry: consider if pity and terror be not enough for tragedy to move; and I believe, upon a true definition of tragedy, it will be found that its work extends farther, and that it is to reform manners by delightful representation of human life in great persons, by way of dialogue. If this be true, then not only pity and terror are to be moved as the only means to bring us to virtue, but generally love to virtue and hatred to vice; by showing the rewards of one, and punishments of the other; at least by rendering virtue always amiable, though it be shown unfortunate; and vice detestable, though it be shown triumphant.

13. If then the encouragement of virtue and discouragement of vice be the proper ends of poetry in tragedy: pity and terror, though good means, are not the only. For all the passions in their turns are to be set in a ferment: as joy, anger, love, fear are to be used as the poet's commonplaces; and a general concernment for the principal actors is to be raised, by making them appear such in their characters, their words and actions, as will interest the audience in their fortunes.

14. And if after all, in a large sense, pity comprehends this concernment for the good, and terror includes detestation for the bad, then let us consider whether the English have not answered this end of tragedy as well as the Ancients, or perhaps better.

15. And here Mr Rymer's objections against these plays are to be impartially weighed, that we may see whether they are of weight enough to turn the balance against our countrymen.

16. 'Tis evident those plays which he arraigns° have moved both those passions in a high degree upon the stage.

17. To give the glory of this away from the poet, and to place it upon the actors,° seems unjust.

18. One reason is, because whatever actors they have found, the event has been the same, that is, the same passions have been always

moved; which shows that there is something of force and merit in the plays themselves, conducing to the design of raising these two passions: and suppose them ever to have been excellently acted, yet action only adds grace, vigour, and more life upon the stage; but cannot give it wholly where it is not first. But secondly, I dare appeal to those who have never seen them acted, if they have not found those two passions moved within them; and if the general voice will carry it, Mr Rymer's prejudice will take off his single testimony.

19. This being matter of fact, is reasonably to be established by this appeal; as if one man says 'tis night, when the rest of the world conclude it to be day, there needs no further argument against him that it is so.

20. If he urge that the general taste is depraved, his arguments to prove this can at best but evince that our poets took not the best way to raise those passions; but experience proves against him that those means which they have used have been successful and have produced them.

21. And one reason of that success is, in my opinion, this, that Shakespeare and Fletcher have written to the genius of the age and nation in which they lived; for though nature, as he objects, is the same in all places,° and reason too the same, yet the climate, the age, the dispositions of the people to whom a poet writes, may be so different that what pleased the Greeks would not satisfy an English audience.

22. And if they proceeded upon a foundation of truer reason to please the Athenians than Shakespeare and Fletcher to please the English, it only shows that the Athenians were a more judicious people; but the poet's business is certainly to please the audience.

23. Whether our English audience have been pleased hitherto with acorns, as he calls it, or with bread,° is the next question; that is, whether the means which Shakespeare and Fletcher have used in their plays to raise those passions before named, be better applied to the ends by the Greek poets than by them; and perhaps we shall not grant him this wholly. Let it be yielded that a writer is not to run down with the stream, or to please the people by their own usual methods, but rather to reform their judgements: it still remains to prove that our theatre needs this total reformation.

24. The faults which he has found in their designs are rather wittily aggravated in many places than reasonably urged; and as much may be returned on the Greeks by one who were as witty as himself.

25. Secondly, they destroy not, if they are granted, the foundation

of the fabric, only take away from the beauty of the symmetry: for example, the faults in the character of the *King and No King* are not, as he makes them, such as render him detestable, but only imperfections which accompany human nature, and for the most part excused by the violence of his love; so that they destroy not our pity or concernment for him. This answer may be applied to most of his objections of that kind.

26. And Rollo° committing many murders, when he is answerable but for one, is too severely arraigned by him; for it adds to our horror and detestation of the criminal: and poetic justice° is not neglected neither, for we stab him in our minds for every offence which he commits; and the point which the poet is to gain on the audience is not so much in the death of an offender, as the raising a horror of his crimes.

27. That the criminal should neither be wholly guilty, nor wholly innocent, but so participating of both as to move both pity and terror, is certainly a good rule, but not perpetually to be observed; for that were to make all tragedies too much alike; which objection he foresaw, but has not fully answered.

28. To conclude, therefore: if the plays of the Ancients are more correctly plotted, ours are more beautifully written; and if we can raise passions as high on worse foundations, it shows our genius in tragedy is greater, for in all other parts of it the English have manifestly excelled them.

29. For the fable itself, 'tis in the English more adorned with episodes, and larger than in the Greek poets; consequently more diverting, for, if the action be but one, and that plain, without any counter-turn° of design or episode, i.e. under-plot, how can it be so pleasing as the English, which have both under-plot and a turned design,° which keeps the audience in expectation of the catastrophe? whereas in the Greek poets we see through the whole design at first.

30. For the characters, they are neither so many nor so various in Sophocles and Euripides as in Shakespeare and Fletcher; only they are more adapted to those ends of tragedy which Aristotle commends to us: pity and terror.

31. The manners flow from the characters, and consequently must partake of their advantages and disadvantages.

32. The thoughts and words, which are the fourth and fifth beauties of tragedy, are certainly more noble and more poetical in the English than in the Greek, which must be proved by comparing them somewhat more equitably than Mr Rymer has done.

33. After all, we need not yield that the English way is less conducing to move pity and terror, because they often show virtue oppressed and vice punished: where they do not both, or either, they are not to be defended.

34. That we may the less wonder why pity and terror are not now the only springs on which our tragedies move, and that Shakespeare may be more excused, Rapin confesses° that the French tragedies now all run on the *tendre* [tender]; and gives the reason, because love is the passion which most predominates in our souls, and that therefore the passions represented become insipid, unless they are conformable to the thoughts of the audience. But it is to be concluded that this passion works not now among the French so strongly as the other two did amongst the Ancients: amongst us, who have a stronger genius for writing, the operations from the writing are much stronger; for the raising of Shakespeare's passions are more from the excellency of the words and thoughts than the justness of the occasion; and if he has been able to pick single occasions, he has never founded the whole reasonably; yet by the genius of poetry, in writing he has succeeded.

35. The parts of a poem,° tragic or heroic, are:

   I. The fable itself.
  II. The order or manner of its contrivance in relation of the parts to the whole.
 III. The manners or decency of the characters in speaking or acting what is proper for them, and proper to be shown by the poet.
 IV. The thoughts which express the manners.
  V. The words which express those thoughts.

36. In the last of these Homer excels Virgil, Virgil all other ancient poets, and Shakespeare all modern poets.

37. For the second of these, the order: the meaning is that a fable ought to have a beginning, middle, and an end, all just and natural, so that that part which is the middle, could not naturally be the beginning or end, and so of the rest: all are depending on one another, like the links of a curious chain.

38. If terror and pity are only to be raised, certainly this author follows Aristotle's rules, and Sophocles's and Euripides's example; but joy may be raised too, and that doubly, either by seeing a wicked man punished, or a good man at last fortunate; or perhaps indignation, to see wickedness prosperous and goodness depressed: both these may be profitable to the end of tragedy, reformation of manners; but the

last improperly, only as it begets pity in the audience: though Aristotle,° I confess, places tragedies of this kind in the second form.

39. And, if we should grant that the Greeks performed this better, perhaps it may admit a dispute whether pity and terror are either the prime, or at least the only ends of tragedy.

40. 'Tis not enough that Aristotle has said so, for Aristotle drew his models of tragedy from Sophocles and Euripides; and if he had seen ours, might have changed his mind.

41. And chiefly we have to say (what I hinted on pity and terror in the last paragraph save one)° that the punishment of vice and reward of virtue are the most adequate ends of tragedy, because most conducing to good example of life. Now pity is not so easily raised for a criminal (as the ancient tragedy always represents its chief person such) as it is for an innocent man, and the suffering of innocence and punishment of the offender is of the nature of English tragedy: contrary, in the Greek, innocence is unhappy often, and the offender escapes.

42. Then, we are not touched with the sufferings of any sort of men so much as of lovers; and this was almost unknown to the Ancients; so that they neither administered poetical justice (of which Mr Rymer boasts) so well as we; neither knew they the best commonplace of pity, which is love.

43. He therefore unjustly blames us for not building upon what the Ancients left us, for it seems, upon consideration of the premisses, that we have wholly finished what they began.

44. My judgement on this piece is this; that it is extremely learned; but that the author of it is better read in the Greek than in the English poets; that all writers ought to study this critique as the best account I have ever seen of the Ancients; that the model of tragedy he has here given is excellent and extremely correct; but that it is not the only model of all tragedy, because it is too much circumscribed in plot, characters, etc.; and lastly, that we may be taught here justly to admire and imitate the Ancients, without giving them the preference with this author in prejudice to our own country.

45. Want of method in this excellent treatise makes the thoughts of the author sometimes obscure.

46. His meaning, that pity and terror are to be moved, is that they are to be moved as the means conducing to the ends of tragedy, which are pleasure and instruction.

47. And these two ends may be thus distinguished. The chief end of the poet is to please; for his immediate reputation depends on it.

48. The great end of the poem is to instruct, which is performed by

making pleasure the vehicle of that instruction; for poetry is an art,° and all arts are made to profit.

49. The pity which the poet is to labour for is for the criminal, not for those, or him, whom he has murdered, or who have been the occasion of the tragedy. The terror is likewise in the punishment of the same criminal who, if he be represented too great an offender, will not be pitied; if altogether innocent, his punishment will be unjust.°

50. Another obscurity is, where he says Sophocles perfected tragedy by introducing the third actor;° that is, he meant three kinds of action, one company singing or speaking, another playing on the music, a third dancing.

51. Rapin attributes more to the *dictio*, that is, to the words and discourses of a tragedy, than Aristotle has done, who places them in the last rank of beauties; perhaps only last in order, because they are the last product of the design, of the disposition or connection of its parts; of the characters, of the manners of those characters, and of the thoughts proceeding from those manners.

52. Rapin's words are remarkable:° ''Tis not the admirable intrigue, the surprising events, the extraordinary incidents that make the beauty of a tragedy; 'tis the discourses when they are natural and passionate.'

53. So are Shakespeare's.

Prologue to Oedipus

When Athens all the Grecian state did guide,
And Greece gave laws to all the world beside;
Then Sophocles with Socrates did sit,
Supreme in wisdom one, and one in wit:
And wit from wisdom differed not in those,
But as 'twas sung in verse, or said in prose.
Then, Oedipus, on crowded theatres,
Drew all admiring eyes and listening ears:
The pleased spectator shouted every line,
The noblest, manliest, and the best design!   10
And every critic of each learned age,
By this just model has reformed the stage.
Now, should it fail, (as heaven avert our fear!)
Damn it in silence, lest the world should hear.

For were it known this poem did not please,
You might set up for perfect savages:
Your neighbours would not look on you as men,
But think the nation all turned Picts again.
Faith, as you manage matters, 'tis not fit
You should suspect yourselves of too much wit: 20
Drive not the jest too far, but spare this piece;
And, for this once, be not more wise than Greece.
See twice! do not pell-mell to damning fall,
Like true-born Britons, who ne'er think at all:
Pray be advised; and though at Mons you won,°
On pointed cannon do not always run.
With some respect to ancient wit proceed;
You take the four first councils for your creed.°
But, when you lay tradition wholly by,
And on the private spirit alone rely,   30
You turn fanatics in your poetry.
If, notwithstanding all that we can say,
You needs will have your penn'orths of the play,
And come resolved to damn, because you pay,
Record it, in memorial of the fact,
The first play buried since the Woollen Act.°

Preface to Ovid's Epistles
Translated by Several Hands

The life of Ovid being already written in our language before the translation of his *Metamorphoses*, I will not presume so far upon myself, to think I can add anything to Mr. Sandys's undertaking.° The English reader may there be satisfied, that he flourished in the reign of Augustus Caesar, that he was extracted from an ancient family of Roman knights; that he was born to the inheritance of a splendid fortune, that he was designed to the study of the law; and had made considerable progress in it, before he quitted that profession for this of poetry, to which he was more naturally formed. The cause of his banishment is unknown; because he was himself unwilling further to provoke the emperor, by ascribing it to any other reason, than what was pretended by Augustus, which was the lasciviousness of his

Elegies, and his *Art of Love*. 'Tis true they are not to be excused in the severity of manners, as being able to corrupt a larger empire, if there were any, than that of Rome; yet this may be said in behalf of Ovid, that no man has ever treated the passion of love with so much delicacy of thought, and of expression, or searched into the nature of it more philosophically than he. And the emperor who condemned him, had as little reason as another man to punish that fault with so much severity, if at least he were the author of a certain epigram, which is ascribed to him,° relating to the cause of the first Civil War betwixt himself and Mark Anthony the triumvir, which is more fulsome° than any passage I have met with in our poet. To pass by the naked familiarity of his expressions to Horace, which are cited in that author's life,° I need only mention one notorious act of his in taking Livia to his bed, when she was not only married, but with child by her husband, then living. But deeds, it seems, may be justified by arbitrary power, when words are questioned in a poet. There is another guess of the grammarians, as far from truth as the first from reason; they will have him banished for some favours, which they say he received from Julia, the daughter of Augustus, whom they think he celebrates under the name of Corinna in his *Elegies*: but he who will observe the verses which are made to that mistress, may gather from the whole contexture of them, that Corinna was not a woman of the highest quality: if Julia were then married to Agrippa, why should out poet make his petition to Isis,° for her safe delivery, and afterwards condole her miscarriage; which for aught he knew might be by her own husband? Or indeed how durst he be so bold to make the least discovery of such a crime, which was no less than capital, especially committed against a person of Agrippa's rank? Or if it were before her marriage, he would surely have been more discreet than to have published an accident, which must have been fatal to them both. But what most confirms me against this opinion is, that Ovid himself complains° that the true person of Corinna was found out by the fame of his verses to her: which if it had been Julia, he durst not have owned; and besides, an immediate punishment must have followed. He seems himself more truly to have touched at the cause of his exile in those obscure verses,

> cur aliquid vidi, cur noxia lumina feci?°

[what did I see? Why were harmful rays made to shine?]

namely, that he had either seen, or was conscious to somewhat, which had procured him his disgrace. But neither am I satisfied that this was the incest of the emperor with his own daughter, for Augustus was of a nature too vindicative to have contented himself with so small a revenge, or so unsafe to himself, as that of simple banishment, and would certainly have secured his crimes from public notice by the death of him who was witness to them. Neither have histories given us any sight into such an action of this emperor: nor would he (the greatest politician of his time), in all probability, have managed his crimes with so little secrecy, as not to shun the observation of any man. It seems more probable that Ovid was either the confidant of some other passion, or that he had stumbled by some inadvertency upon the privacies of Livia, and seen her in a bath. For the words

nudam sine veste Dianam,°

[Diana naked and unclothed]

agree better with Livia, who had the fame of chastity, than with either of the Julias, who were both noted of incontinency. The first verses° which were made by him in his youth, and recited publicly according to the custom were, as he himself assures us, to Corinna. His banishment happened not till the age of fifty; from which it may be deduced, with probability enough, that the love of Corinna did not occasion it. Nay he tells us plainly, that his offence was that of error only, not of wickedness: and in the same paper of verses also, that the cause was notoriously known at Rome, though it be left so obscure to after ages.

But to leave conjectures on a subject so uncertain, and to write somewhat more authentic of this poet, that he frequented the court of Augustus, and was well received in it, is most undoubted. All his poems bear the character of a court, and appear to be written as the French call it *cavalièrement* [offhandedly]. Add to this, that the titles of many of his elegies, and more of his letters in his banishment, are addressed to persons well known to us, even at this distance, to have been considerable in that court.

Nor was his acquaintance less with the famous poets of his age, than with the noblemen and ladies; he tells you himself, in a particular account of his own life,° that Macer, Horace, Tibullus, Propertius, and many others of them were his familiar friends, and that some of them communicated their writings to him, but that he had only seen Virgil.

If the imitation of nature be the business of a poet, I know no author who can justly be compared with ours, especially in the description of the passions. And to prove this, I shall need no other judges than the generality of his readers: for all passions being inborn with us, we are almost equally judges when we are concerned in the representation of them. Now I will appeal to any man who has read this poet, whether he find not the natural emotion of the same passion in himself, which the poet describes in his feigned persons? His thoughts which are the pictures and results of those passions, are generally such as naturally arise from those disorderly motions of our spirits. Yet, not to speak too partially on his behalf, I will confess that the copiousness of his wit was such, that he often writ too pointedly for his subject, and made his persons speak more eloquently than the violence of their passion would admit: so that he is frequently witty out of season, leaving the imitation of nature, and the cooler dictates of his judgement, for the false applause of fancy. Yet he seems to have found out this imperfection in his riper age: for why else should he complain that his *Metamorphosis* was left unfinished? Nothing sure can be added to the wit of that poem, or of the rest, but many things ought to have been retrenched; which I suppose would have been the business of his age, if his misfortunes had not come too fast upon him. But take him uncorrected as he is transmitted to us, and it must be acknowledged in spite of his Dutch friends, the commentators, even of Julius Scaliger himself, that Seneca's censure will stand good against him; *nescivit quod bene cessit relinquere*:° he never knew how to give over, when he had done well: but continually varying the same sense a hundred ways, and taking up in another place what he had more than enough inculcated before, he sometimes cloys his readers instead of satisfying them, and gives occasion to his translators, who dare not cover him, to blush at the nakedness of their father. This then is the alloy of Ovid's writing, which is sufficiently recompensed by his other excellencies; nay this very fault is not without its beauties: for the most severe censor cannot but be pleased with the prodigality of his wit, though at the same time he could have wished that the master of it had been a better manager. Everything which he does becomes him, and if sometimes he appear too gay, yet there is a secret gracefulness of youth, which accompanies his writings, though the staidness and sobriety of age be wanting. In the most material part, which is the conduct, 'tis certain that he seldom has miscarried: for if his elegies be compared with those of Tibullus and Propertius his contemporaries, it will be found that those poets seldom designed before they writ. And though the language of Tibullus be more polished, and the learning of Propertius, especially in his fourth book, more set out to ostentation, yet their common practice, was to look no further before them than the next line: whence it will inevitably follow, that they can drive to no certain point, but ramble from one subject to another, and conclude with somewhat which is not of a piece with their beginning:

> purpureus late qui splendeat, unus et alter
> assuitur pannus:

[one to two purple patches are sewn on, to shine far and wide]

as Horace° says, though the verses are golden, they are but patched into the garment. But our poet has always the goal in his eye, which directs him in his race, some beautiful design, which he first establishes, and then contrives the means, which will naturally conduct it to his end. This will be evident to judicious readers in this work of his Epistles, of which somewhat, at least in general, will be expected.

The title of them in our late editions is *Epistolae Heroidum*, The Letters of the Heroines. But Heinsius° has judged more truly, that the inscription of our author was barely, *Epistles*, which he concludes from his cited verses, where Ovid asserts this work as his own invention, and not borrowed from the Greeks, whom (as the masters of their learning) the Romans usually did imitate. But it appears not from their writers, that any of the Grecians ever touched upon this way, which our poet therefore justly has vindicated to° himself. I quarrel not at the word *Heroidum*, because 'tis used by Ovid in his *Art of Love*:°

> Jupiter ad veteres supplex heroidas ibat.

[Jupiter went a supplicant to the heroines of old]

But sure he could not be guilty of such an oversight, to call his work by the name of heroines, when there are divers men or heroes, as namely Paris, Leander, and Acontius, joined in it. Except Sabinus, who writ some answers to Ovid's letters,

> (quam celer e toto rediit meus orbe Sabinus,)°

[how quickly my Sabinus has returned from the ends of the earth]

I remember not any of the Romans who have treated this subject, save only Propertius, and that but once, in his *Epistle of Arethusa to Lycotas*,° which is written so near the style of Ovid that it seems to be but an imitation, and therefore ought not to defraud our poet of the glory of his invention.

Concerning this work of the epistles, I shall content myself to observe these few particulars. First, that they are generally granted to be the most perfect piece of Ovid, and that the style of them is tenderly passionate and courtly; two properties well agreeing with the persons which were heroines, and lovers. Yet where the characters were lower, as in Oenone, and Hero, he has kept close to nature in drawing his images after a country life, though perhaps he has romanized his Grecian dames too much, and made them speak sometimes as if they had been born in the city of Rome, and under the Empire of Augustus. There seems to be no great variety in the particular subjects which he has chosen, most of the epistles being written from ladies who were forsaken by their lovers: which is the reason that many of the same thoughts come back upon us in divers letters. But of the general character of women which is modesty, he has taken a most becoming care; for his amorous expressions go no further than virtue may allow, and therefore may be read, as he intended them, by matrons without a blush.

Thus much concerning the poet: whom you find translated by divers hands, that you may at least have that variety in the English, which the subject denied to the author of the Latin. It remains that I should say somewhat of poetical translations in general, and give my opinion (with submission to better judgements) which way of version seems to me most proper.

All translation I suppose may be reduced to these three heads.

First, that of metaphrase, or turning an author word by word, and line by line, from one language into another. Thus, or near this manner, was Horace's *Art of Poetry*° translated by Ben Jonson. The second way is that of paraphrase, or translation with latitude, where the author is kept in view by the translator, so as never to be lost, but his words are not so strictly followed as his sense, and that too is admitted to be amplified, but not altered. Such is Mr Waller's translation of Virgil's fourth *Aeneid*.° The third way is that of imitation, where the translator (if now he has not lost that name) assumes the liberty not only to vary from the words and sense, but to forsake them both as he sees occasion: and taking only some general hints from the original, to run division on the groundwork,° as he pleases. Such is Mr Cowley's practice in turning two odes of Pindar, and one of Horace into English.°

Concerning the first of these methods, our master Horace has given us this caution,

> nec verbum verbo curabis reddere, fidus
> interpres—°

'Nor word for word too faithfully translate', as the Earl of Roscommon has excellently rendered it.° Too faithfully is indeed pedantically: 'tis a faith like that which proceeds from superstition, blind and zealous.

Take it in the expression of Sir John Denham, to Sir Richard
Fanshaw, on his version of the *Pastor Fido*.

> That servile path, thou nobly dost decline,
> Of tracing word by word and line by line
> A new and nobler way thou dost pursue,
> To make translations, and translators too:
> They but preserve the ashes, thou the flame,
> True to his sense, but truer to his fame.°

'Tis almost impossible to translate verbally, and well, at the same
time; for the Latin (a most severe and compendious language) often
expresses that in one word, which either the barbarity, or the
narrowness of modern tongues cannot supply in more. 'Tis frequent
also that the conceit is couched in some expression, which will be lost
in English:

> atque idem venti vela fidemque ferent.°

[While you, with loosened sails and vows prepare
To seek a land that flies the searcher's care]

What poet of our nation is so happy as to express this thought literally
in English, and to strike wit or almost sense out of it?

In short the verbal copier is encumbered with so many difficulties at
once, that he can never disentangle himself from all. He is to consider
at the same time the thought of his author, and his words, and to find
out the counterpart to each in another language: and besides this he is
to confine himself to the compass of numbers, and the slavery of
rhyme. 'Tis much like dancing on ropes with fettered legs. A man may
shun a fall by using caution, but the gracefulness of motion is not to be
expected, and when we have said the best of it, 'tis but a foolish task;
for no sober man would put himself into a danger for the applause of
escaping without breaking his neck. We see Ben Jonson could not
avoid obscurity in his literal translation of Horace, attempted in the
same compass of lines: nay Horace himself could scarce have done it to
a Greek poet:

> brevis esse laboro, obscurus fio.°

[I labour to be brief and become obscure]

Either perspicuity or gracefulness will frequently be wanting. Horace
has indeed avoided both these rocks in his translation of the three first
lines of Homer's *Odyssey*, which he has contracted into two:

> dic mihi Musa virum captae post tempora Trojae
> qui mores hominum multorum vidit et urbes.

> Muse, speak the man, who since the Siege of Troy,
> So many towns, such change of manners saw.
>
> Earl of Roscommon

But then the sufferings of Ulysses, which are a considerable part of that sentence are omitted:

$$\text{ὃς μάλα πολλὰ πλάγχθη.}°$$

[who was very much tossed on the sea]

The consideration of these difficulties, in a servile, literal translation, not long since made two of our famous wits, Sir John Denham, and Mr Cowley to contrive another way of turning authors into our tongue, called by the latter of them, imitation. As they were friends, I suppose they communicated their thoughts on this subject to each other, and therefore their reasons for it are little different: though the practice of one° is much more moderate. I take imitation of an author in their sense to be an endeavour of a later poet to write like one who has written before him on the same subject: that is, not to translate his words, or to be confined to his sense, but only to set him as a pattern, and to write as he supposes that author would have done had he lived in our age, and in our country. Yet I dare not say that either of them have carried this libertine way of rendering authors (as Mr Cowley calls it) so far as my definition reaches. For in the Pindaric Odes, the customs and ceremonies of ancient Greece are still preserved, but I know not what mischief may arise hereafter from the example of such an innovation, when writers of unequal parts to him shall imitate so bold an undertaking. To add and to diminish what we please, which is the way avowed by him, ought only to be granted to Mr Cowley, and that too only in his translation of Pindar, because he alone was able to make him amends, by giving him better of his own, whenever he refused his author's thoughts. Pindar is generally known to be a dark writer, to want connexion (I mean as to our understanding) to soar out of sight, and leave his reader at a gaze. So wild and ungovernable a poet cannot be translated literally; his genius is too strong to bear a chain, and Samson-like he shakes it off. A genius so elevated and unconfined as Mr Cowley's was but necessary to make Pindar speak English, and that was to be performed by no other way than imitation. But if Virgil or Ovid, or any regular intelligible authors be thus used, 'tis no longer to be called their work, when neither the thoughts nor words are drawn from the original: but instead of them there is something new produced, which is almost the creation of another hand. By this way 'tis true, somewhat that is excellent may be invented, perhaps more excellent than the first design, though Virgil must be still excepted when that perhaps takes place. Yet he who is inquisitive to know an author's thoughts will be disappointed in his expectation. And 'tis not always that a man will be contented to have a present made him, when he expects the payment of a debt. To state it fairly, imitation of an author is the most advantageous way for a translator to show himself, but the greatest wrong which can be done to the memory and reputation of the dead. Sir John Denham (who advised more liberty than he took himself,) gives this reason for his innovation, in his admirable preface before the translation of the second *Aeneid*:° 'Poetry is of so subtle a spirit, that in pouring out of one language into another, it will all evaporate; and if a new spirit be not added in the transfusion, there will remain nothing but a *caput mortuum*.' I confess this argument holds good against a literal translation, but who defends it? Imitation and verbal version are in my opinion the two extremes, which ought to be avoided: and therefore when I have proposed the mean betwixt them, it will be seen how far his argument will reach.

No man is capable of translating poetry, who besides a genius to that art, is not a master both of his author's language, and of his own. Nor must we understand the language only of the poet, but his particular turn of thoughts, and of expression, which are the characters that distinguish, and as it were individuate him from all other writers. When we are come thus far, 'tis time to look into ourselves, to conform our genius to his, to give his thought either the same turn if our tongue will bear it, or if not, to vary but the dress, not to alter or destroy the substance. The like care must be taken of the more outward ornaments, the words: when they appear (which is but seldom) literally graceful, it were an injury to the author that they should be changed. But since every language is so full of its own proprieties, that what is beautiful in one, is often barbarous, nay sometimes nonsense in another, it would be unreasonable to limit a translator to the narrow compass of his author's words: 'tis enough if he choose out some expression which does not vitiate the sense. I suppose he may stretch his chain to such a latitude, but by innovation of thoughts, methinks he breaks it. By this means the spirit of an author may be transfused, and yet not lost: and thus 'tis plain that the reason alleged by Sir John Denham, has no farther force than to expression: for thought, if it be translated truly, cannot be lost in another language, but the words that convey it to our apprehension (which are the image and ornament of that thought) may be so ill chosen as to make it appear in an unhandsome dress, and rob it of its native lustre. There is therefore a liberty to be allowed for the expression, neither is it necessary that words and lines should be confined to the measure of their original. The sense of an author, generally speaking, is to be sacred and inviolable. If the fancy of Ovid be luxuriant, 'tis his character to be so, and if I retrench it, he is no longer Ovid. It will be replied that he receives advantage by this lopping of his superfluous branches, but I rejoin that a translator has no such right. When a painter copies from the life, I suppose he has no privilege to alter features, and lineaments, under pretence that his picture will look better. Perhaps the face which he has drawn would be more exact, if the eyes, or nose were altered, but 'tis his business to make it resemble the original. In two cases only there may a seeming difficulty arise, that is, if the thought be notoriously trivial or dishonest; but the same answer will serve for both, that then they ought not to be translated.

> et quae
> desperes tractata nitescere posse, relinquas.°

[And abandon these things which you despair of treating effectively]

Thus I have ventured to give my opinion on this subject against the authority of two great men, but I hope without offence to either of their memories, for I both loved them living, and reverence them now they are dead. But if after what I have urged, it be thought by better judges that the praise of a translation consists in adding new beauties to the piece, thereby to recompense the loss which it sustains by change of language, I shall be willing to be taught better, and to recant. In the meantime it seems to me that the true reason why we have so few versions which are tolerable, is not from the too close pursuing of the author's sense: but because there are so few who have all the talents which are requisite for translation: and that there is so little praise and so small encouragement for so considerable a part of learning.

To apply in short, what has been said, to this present work, the reader will here find most of the translations, with some little latitude or variation from the author's sense. That of Oenone to Paris,° is in Mr Cowley's way of imitation only. I was desired to say that the author who is of the fair sex, understood not Latin. But if she does not, I am afraid she has given us occasion to be ashamed who do.

For my own part I am ready to acknowledge that I have transgressed the rules which I have given; and taken more liberty than a just translation will allow. But so many gentlemen whose wit and learning are well known, being joined in it, I doubt not but that their excellencies will make you ample satisfaction for my errors.

J. DRYDEN

Canace to Macareus

THE ARGUMENT

Macareus and Canace, son and daughter to Aeolus, god of the winds, loved each other incestuously: Canace was delivered of a son, and committed him to her nurse, to be secretly conveyed away. The infant crying out, by that means was discovered to Aeolus, who enraged at the wickedness of his children commanded the babe to be exposed to wild beasts on the mountains; and withal sent a sword to Canace with this message, that her crimes would instruct her how to use it. With this sword she slew herself; but before she died, she writ the following letter to her brother Macareus, who had taken sanctuary in the temple of Apollo.

If streaming blood my fatal letter stain,
Imagine, ere you read, the writer slain;
One hand the sword, and one the pen employs,
And in my lap the ready paper lies.
Think in this posture thou behold'st me write;
In this my cruel father would delight.
O! were he present, that his eyes and hands
Might see, and urge the death which he commands!
Than all his raging winds more dreadful, he,
Unmoved, without a tear, my wounds would see.   10
Jove justly placed him on a stormy throne,
His people's temper is so like his own.
The north and south, and each contending blast,
Are underneath his wide dominion cast:
Those he can rule; but his tempestuous mind
Is, like his airy kingdom, unconfined.
Ah! what avail my kindred gods above,
That in their number I can reckon Jove!
What help will all my heavenly friends afford,
When to my breast I lift the pointed sword?   20
That hour, which joined us, came before its time;
In death we had been one without a crime.
Why did thy flames beyond a brother's move?
Why loved I thee with more than sister's love?
For I loved too; and, knowing not my wound,
A secret pleasure in thy kisses found;
My cheeks no longer did their colour boast,
My food grew loathsome, and my strength I lost:

Still ere I spoke, a sigh would stop my tongue;
Short were my slumbers, and my nights were long.   30
I knew not from my love these griefs did grow,
Yet was, alas! the thing I did not know.
My wily nurse, by long experience, found,
And first discovered to my soul its wound.
''Tis love,' said she; and then my downcast eyes,
And guilty dumbness, witnessed my surprise.
Forced at the last my shameful pain I tell;
And oh, what followed, we both know too well!
   When half denying, more than half content,
Embraces warmed me to a full consent,   40
Then with tumultuous joys my heart did beat,
And guilt, that made them anxious, made them great.
But now my swelling womb heaved up my breast,
And rising weight my sinking limbs oppressed.
What herbs, what plants, did not my nurse produce,
To make abortion by their powerful juice!
What medicines tried we not, to thee unknown!
Our first crime common; this was mine alone.
But the strong child, secure in his dark cell,
With nature's vigour did our arts repel.   50
And now the pale-faced empress of the night
Nine times had filled her orb with borrowed light;
Not knowing 'twas my labour, I complain
Of sudden shootings and of grinding pain;
My throes came thicker, and my cries increased,
Which with her hand the conscious nurse suppressed.
To that unhappy fortune was I come,
Pain urged my clamours, but fear kept me dumb.
With inward struggling I restrained my cries,
And drunk the tears that trickled from my eyes.   60
Death was in sight, Lucina gave no aid,
And e'en my dying had my guilt betrayed.
Thou cam'st, and in thy countenance sat despair;
Rent were thy garments all, and torn thy hair;
Yet feigning comfort, which thou couldst not give,
Pressed in thy arms, and whispering me to live;
'For both our sakes,' saidst thou, 'preserve thy life;
Live, my dear sister, and my dearer wife.'
Raised by that name, with my last pangs I strove;

Such power have words, when spoke by those we love.   70
The babe, as if he heard what thou hadst sworn,
With hasty joy sprung forward to be born.
What helps it to have weathered out one storm!
Fear of our father does another form.
High in his hall, rocked in a chair of state,
The king with his tempestuous council sate;
Through this large room our only passage lay,
By which we could the new-born babe convey.
Swathed in her lap, the bold nurse bore him out
With olive branches covered round about,   80
And muttering prayers, as holy rites she meant,
Through the divided crowd unquestioned went.
Just at the door the unhappy infant cried;
The grandsire heard him, and the theft he spied.
Swift as a whirlwind to the nurse he flies,
And deafs his stormy subjects with his cries.
With one fierce puff he blows the leaves away,
Exposed the self-discovered infant lay.
The noise reached me, and my presaging mind
Too soon its own approaching woes divined.   90
Not ships at sea with winds are shaken more,
Nor seas themselves, when angry tempests roar,
Than I, when my loud father's voice I hear;
The bed beneath me trembled with my fear.
He rushed upon me, and divulged my stain;
Scarce from my murder could his hands refrain.
I only answered him with silent tears;
They flowed; my tongue was frozen up with fears.
His little grandchild he commands away,
To mountain wolves and every bird of prey.   100
The babe cried out, as if he understood,
And begged his pardon with what voice he could.
By what expressions can my grief be shown?
Yet you may guess my anguish by your own,
To see my bowels, and what yet was worse,
Your bowels too, condemned to such a curse!
Out went the king; my voice its freedom found,
My breasts I beat, my blubbered cheeks I wound.
And now appeared the messenger of death;
Sad were his looks and scarce he drew his breath   110

To say, 'Your father sends you' (with that word
His trembling hands presented me a sword)
'Your father sends you this; and lets you know,
That your own crimes the use of it will show.'
Too well I know the sense those words impart;
His present shall be treasured in my heart.
Are these the nuptial gifts a bride receives?
And this the fatal dower a father gives?
Thou god of marriage, shun thy own disgrace,
And take thy torch from this detested place!   120
Instead of that, let furies light their brands,
And fire my pile with their infernal hands!
With happier fortune may my sisters wed,
Warned by the dire example of the dead.
For thee, poor babe, what crime could they pretend?
How could thy infant innocence offend?
A guilt there was; but oh, that guilt was mine!
Thou sufferest for a sin that was not thine.
Thy mother's grief and crime! but just enjoyed,
Shown to my sight, and born to be destroyed!   130
Unhappy offspring of my teeming womb!
Dragged headlong from thy cradle to thy tomb!
Thy unoffending life I could not save,
Nor weeping could I follow to thy grave;
Nor on thy tomb could offer my shorn hair,
Nor show the grief which tender mothers bear.
Yet long thou shalt not from my arms be lost;
For soon I will o'ertake thy infant ghost.
But thou, my love, and now my love's despair,
Perform his funerals with paternal care;   140
His scattered limbs with my dead body burn,
And once more join us in the pious urn.
If on my wounded breast thou dropp'st a tear,
Think for whose sake my breast that wound did bear;
And faithfully my last desires fulfil,
As I perform my cruel father's will.

Dido to Aeneas

THE ARGUMENT

Aeneas, the son of Venus and Anchises, having, at the destruction of Troy, saved his gods, his father, and son Ascanius, from the fire, put to sea with twenty sail of ships; and, having been long tossed with tempests, was at last cast upon the shore of Libya, where Queen Dido (flying from the cruelty of Pygmalion, her brother, who had killed her husband Sychaeus) had lately built Carthage. She entertained Aeneas and his fleet with great civility, fell passionately in love with him, and in the end denied him not the last favours. But Mercury admonishing Aeneas to go in search of Italy (a kingdom promised him by the gods), he readily prepared to obey him. Dido soon perceived it, and, having in vain tried all other means to engage him to stay, at last, in despair, writes to him as follows.

So, on Maeander's banks, when death is nigh,
The mournful swan sings her own elegy.
Not that I hope (for, oh, that hope were vain!)
By words your lost affection to regain;
But having lost whate'er was worth my care,
Why should I fear to lose a dying prayer?
'Tis then resolved poor Dido must be left,
Of life, of honour, and of love bereft!
While you, with loosened sails and vows prepare
To seek a land that flies the searcher's care;   10
Nor can my rising towers your flight restrain,
Nor my new empire, offered you in vain.
Built walls you shun, unbuilt you seek; that land
Is yet to conquer, but you this command.
Suppose you landed where your wish designed,
Think what reception foreigners would find,
What people is so void of common sense,
To vote succession from a native prince?
Yet there new sceptres and new loves you seek,
New vows to plight, and plighted vows to break.   20
When will your towers the height of Carthage know?
Or when your eyes discern such crowds below?
If such a town and subjects you could see,
Still would you want a wife who loved liked me.
For, oh! I burn like fires with incense bright;
Not holy tapers flame with purer light.

Aeneas is my thoughts' perpetual theme,
Their daily longing, and their nightly dream.
Yet he ungrateful and obdurate still;
Fool that I am to place my heart so ill!   30
Myself I cannot to myself restore;
Still I complain, and still I love him more.
Have pity, Cupid, on my bleeding heart,
And pierce thy brother's with an equal dart.
I rave; nor canst thou Venus' offspring be,
Love's mother could not bear a son like thee.
From hardened oak, or from a rock's cold womb,
At least thou art from some fierce tigress come;
Or on rough seas, from their foundation torn,
Got by the winds, and in a tempest born: 40
Like that, which now thy trembling sailors fear;
Like that, whose rage should still detain thee here.
Behold how high the foamy billows ride!
The winds and waves are on the juster side.
To winter weather, and a stormy sea,
I'll owe what rather I would owe to thee.
Death thou deserv'st from heaven's avenging laws;
But I'm unwilling to become the cause.
To shun my love, if thou wilt seek thy fate,
'Tis a dear purchase, and a costly hate.   50
Stay but a little, till the tempest cease,
And the loud winds are lulled into a peace.
May all thy rage, like theirs, inconstant prove!
And so it will, if there be power in love.
Know'st thou not yet what dangers ships sustain?
So often wrecked, how darst thou tempt the main?
Which were it smooth, were every wave asleep,
Ten thousand forms of death are in the deep.
In that abyss the gods their vengeance store,
For broken vows of those who falsely swore;   60
There winged storms on sea-born Venus wait,
To vindicate the justice of her state.
Thus I to thee the means of safety show;
And, lost myself, would still preserve my foe.
False as thou art, I not thy death design;
O rather live, to be the cause of mine!
Should some avenging storm thy vessel tear,

(But heaven forbid my words should omen bear!)
Then in thy face thy perjured vows would fly,
And my wronged ghost be present to thy eye;   70
With threatening looks think thou behold'st me stare,
Gasping my mouth, and clotted all my hair.
Then, should forked lightning and red thunder fall,
What couldst thou say, but 'I deserved them all'?
Lest this should happen, make not haste away;
To shun the danger will be worth thy stay.
Have pity on thy son, if not on me;
My death alone is guilt enough for thee.
What has his youth, what have thy gods deserved,
To sink in seas, who were from fires preserved?   80
But neither gods nor parent didst thou bear;
Smooth stories all, to please a woman's ear,
False was the tale of thy romantic life.
Nor yet am I thy first-deluded wife;
Left to pursuing foes Creusa stayed,
By thee, base man, forsaken and betrayed.
This, when thou told'st me, struck my tender heart,
That such requital followed such desert.
Nor doubt I but the gods, for crimes like these,
Seven winters kept thee wandering on the seas.   90
Thy starved companions, cast ashore, I fed,
Thyself admitted to my crown and bed.
To harbour strangers, succour the distressed,
Was kind enough; but, oh, too kind the rest!
Cursed be the cave which first my ruin brought,
Where, from the storm, we common shelter sought!
A dreadful howling echoed round the place;
The mountain nymphs, thought I, my nuptials grace.
I thought so then, but now too late I know
The furies yelled my funerals from below.   100
O chastity and violated fame,
Exact your dues to my dead husband's name!
By death redeem my reputation lost,
And to his arms restore my guilty ghost!
Close by my palace, in a gloomy grove,
Is raised a chapel to my murdered love;
There, wreathed with boughs and wool, his statue stands,
The pious monument of artful hands.

Last night, methought, he called me from the dome,
And thrice, with hollow voice cried, 'Dido, come!'   110
She comes; thy wife thy lawful summons hears,
But comes more slowly, clogged with conscious fears.
Forgive the wrong I offered to thy bed;
Strong were his charms, who my weak faith misled.
His goddess mother, and his aged sire
Borne on his back, did to my fall conspire.
Oh, such he was, and is, that, were he true,
Without a blush I might his love pursue;
But cruel stars my birthday did attend,
And, as my fortune opened, it must end.   120
My plighted lord was at the altar slain,
Whose wealth was made my bloody brother's gain;
Friendless, and followed by the murderer's hate,
To foreign countries I removed my fate;
And here, a suppliant, from the natives' hands
I bought the ground on which my city stands,
With all the coast that stretches to the sea,
E'en to the friendly port that sheltered thee;
Then raised these walls, which mount into the air,
At once my neighbours' wonder, and their fear.   130
For now they arm; and round me leagues are made,
My scarce established empire to invade.
To man my new-built walls I must prepare,
A helpless woman, and unskilled in war
Yet thousand rivals to my love pretend,
And for my person would my crown defend;
Whose jarring votes in one complaint agree,
That each unjustly is disdained for thee.
To proud Iarbas give me up a prey,°
For that must follow, if thou goest away;   140
Or to my husband's murderer leave my life,
That to the husband he may add the wife.
Go then, since no complaints can move thy mind;
Go, perjured man, but leave thy gods behind.
Touch not those gods, by whom thou art forsworn,
Who will in impious hands no more be borne;
Thy sacrilegious worship they disdain,
And rather would the Grecian fires sustain.
Perhaps my greatest shame is still to come,

And part of thee lies hid within my womb;   150
The babe unborn must perish by thy hate,
And perish, guiltless, in his mother's fate.
Some god, thou sayest, thy voyage does command;
Would the same god had barred thee from my land!
The same, I doubt not, thy departure steers,
Who kept thee out at sea so many years,
Where thy long labours were a price so great,
As thou, to purchase Troy, wouldst not repeat.
But Tiber now thou seek'st, to be at best,
When there arrived, a poor precarious guest.   160
Yet it deludes thy search; perhaps it will
To thy old age lie undiscovered still.
A ready crown and wealth in dower I bring,
And, without conquering, here thou art a king.
Here thou to Carthage may'st transfer thy Troy;
Here young Ascanius may his arms employ;
And, while we live secure in soft repose,
Bring many laurels home from conquered foes.
   By Cupid's arrows, I adjure thee stay!°
By all the gods, companions of thy way!   170
So may thy Trojans, who are yet alive,
Live still, and with no future fortune strive;
So may thy youthful son old age attain,
And thy dead father's bones in peace remain;
As thou hast pity on unhappy me,
Who knows no crime, but too much love of thee.
I am not born from fierce Achilles' line,
Nor did my parents against Troy combine.
To be thy wife if I unworthy prove,
By some inferior name admit my love.   180
To be secured of still possessing thee,
What would I do, and what would I not be!
   Our Libyan coasts their certain seasons know,
When, free from tempests, passengers may go;
But now with northern blasts the billows roar,
And drive the floating seaweed to the shore.
Leave to my care the time to sail away;
When safe, I will not suffer thee to stay.
Thy weary men would be with ease content;
Their sails are tattered, and their masts are spent.   190

If by no merit I thy mind can move,
What thou deniest my merit, give my love.
Stay, till I learn my loss to undergo,
And give me time to struggle with my woe:
If not, know this, I will not suffer long;
My life's too loathsome, and my love too strong.
Death holds my pen, and dictates what I say,
While cross my lap thy Trojan sword I lay.
My tears flow down; the sharp edge cuts their flood,
And drinks my sorrows, that must drink my blood.   200
How well thy gift does with my fate agree!
My funeral pomp is cheaply made by thee.
To no new wounds my bosom I display;
The sword but enters where love made the way.
But thou, dear sister, and yet dearer friend,°
Shalt my cold ashes to their urn attend.
Sychaeus' wife let not the marble boast;
I lost that title, when my fame I lost.
This short inscription only let it bear;
'Unhappy Dido lies in quiet here.   210
The cause of death, and sword by which she died,
Aeneas gave; the rest her arm supplied.'

Prologue to The Spanish Friar

Now luck for us and a kind hearty pit;
For he who pleases never fails of wit.
Honour is yours:
And you like kings at City treats bestow it;
The writer kneels and is bid rise a poet.
But you are fickle sovereigns to our sorrow,
You dub today and hang a man tomorrow;
You cry the same sense up and down again,
Just like brass money once a year in Spain.°
Take you i' th' mood, whate'er base metal come,   10
You coin as fast as groats at Bromingam;°
Though 'tis no more like sense in ancient plays
Than Rome's religion like St Peter's days.
In short, so swift your judgements turn and wind,
You cast our fleetest wits a mile behind.

'Twere well your judgements but in plays did range,
But e'en your follies and debauches change
With such a whirl, the poets of your age
Are tired and cannot score 'em on the stage,
Unless each vice in short-hand they indite,   20
E'en as notched prentices whole sermons write.°
The heavy Hollanders no vices know
But what they used a hundred years ago;
Like honest plants, where they were stuck they grow;
They cheat, but still from cheating sires they come;
They drink, but they were christened first in mum.°
Their patrimonial sloth the Spaniards keep,
And Philip first taught Philip how to sleep.
The French and we still change, but here's the curse,
They change for better and we change for worse;   30
They take up our old trade of conquering,
And we are taking theirs, to dance and sing.
Our fathers did for change to France repair,
And they for change will try our English air.
As children when they throw one toy away,
Straight a more foolish gewgaw comes in play;
So we, grown penitent on serious thinking,
Leave whoring and devoutly fall to drinking.
Scouring the watch grows out of fashion wit;°
Now we set up for tilting in the pit,   40
Where 'tis agreed by bullies, chicken-hearted,
To fright the ladies first and then be parted.
A fair attempt has twice or thrice been made
To hire night murderers and make death a trade.
When murder's out, what vice can we advance?
Unless the new found poisoning trick of France;
And when their art of rats-bane we have got,
By way of thanks, we'll send 'em o'er our Plot.°

Epilogue to Mithridates

SPOKEN BY MR. GOODMAN

Pox on this playhouse; 'tis an old tired jade!
'Twill do no longer; we must force a trade.
What if we all turn witness of the Plot?°
That's overstocked; there's nothing to be got.

Shall we take orders? That will parts require,
Our colleges give no degrees for hire.
Would Salamanca was a little nigher!°
Will nothing do? Oh, now 'tis found, I hope!
Have not you seen the dancing of the rope?°
When Andrew's wit was clean run off the score°   10
And Jacob's capering tricks could do no more,°
A damsel does to the ladder's top advance
And with two heavy buckets drags a dance;
The yawning crown perked up to see the sight
And slavered at the mouth for vast delight.
Oh, friends, there's nothing to enchant the mind,
Nothing like that cleft sex to draw mankind:
The foundered horse that switching will not stir,
Trots to the mare afore without a spur.
Faith, I'll go scour the scene room and engage   20
Some toy within to save the falling stage.

                                                        [*Exit.*

[*Re-enters with Mrs. Cox*

Who have we here again? what numps i'th' stocks?°
Your most obedient slave, sweet Madam Cox.
You'd best be coy and blush for a pretence:
For shame, say something in your own defence!
MRS. COX. What shall I say? I have been hence so long
I've e'en almost forgot my mother tongue.
If I can act I wish I were ten fathom
Beneath—
MR. GOODMAN. Oh, Lord! Pray, no swearing madam!
MRS. COX. Why, sir, if I had sworn to save the nation   30
I could find out some mental reservation.
Well, in plain terms, gallants, without a sham,
Will you be pleased to take me as I am:
Quite out of countenance, with a downcast look,
Just like a truant that returns to book.
Yet I'm not old; but if I were, this place
Ne'er wanted art to piece a ruined face.
When graybeards governed I forsook the stage;
You know 'tis piteous work to act with age.
Though there's no sense amongst these beardless boys,   40
There's what we women love: that's mirth and noise.
These young beginners may grow up in time,
And the devil's in't if I'm past my prime.

ABSALOM AND ACHITOPHEL

A POEM

si propius stes
te capiet magis.°
[if you stand nearer it will attract you more]

TO THE READER

It is not my intention to make an apology for my poem: some will think it needs no excuse, and others will receive none. The design, I am sure, is honest; but he who draws his pen for one party must expect to make enemies of the other. For wit and fool are consequents of Whig and Tory; and every man is a knave or an ass to the contrary side. There's a treasury of merits in the fanatic church, as well as in the Papist; and a pennyworth to be had of saintship, honesty, and poetry, for the lewd, the factious, and the blockheads; but the longest chapter in Deuteronomy has not curses enough for an anti-Bromingham.° My comfort is, their manifest prejudice to my cause will render their judgment of less authority against me. Yet if a poem have a genius, it will force its own reception in the world; for there's a sweetness in good verse, which tickles even while it hurts, and no man can be heartily angry with him who pleases him against his will. The commendation of adversaries is the greatest triumph of a writer, because it never comes unless extorted. But I can be satisfied on more easy terms: if I happen to please the more moderate sort, I shall be sure of an honest party, and, in all probability, of the best judges; for the least concerned are commonly the least corrupt. And I confess, I have laid in for those by rebating the satire (where justice would allow it) from carrying too sharp an edge. They who can criticise so weakly as to imagine I have done my worst, may be convinced, at their own cost, that I can write severely with more ease than I can gently. I have but laughed at some men's follies, when I could have declaimed against their vices; and other men's virtues I have commended as freely as I have taxed their crimes. And now, if you are a malicious reader, I expect you should return upon me that I affect to be thought more impartial than I am. But if men are not to be judged by their professions, God forgive you commonwealth's-men° for professing so plausibly for the government. You cannot be so unconscionable as to charge me for not subscribing of my name; for that would reflect too grossly upon your own party, who never dare, though they have the advantage of a jury to secure them. If you like not my poem, the fault may, possibly, be in my writing (though it is hard for an author to judge against himself); but, more probably, it is in your morals, which cannot bear the truth of it. The violent, on both sides, will condemn the character of Absalom, as either too favourably or too hardly drawn. But they are not the violent whom I desire to please. The fault on the right hand is to extenuate, palliate, and indulge, and, to confess freely, I have endeavoured to commit it. Besides the respect which I owe his birth, I have a greater for his heroic virtues, and David himself could not be more tender of the young man's life than I would be of his reputation. But since the most excellent natures are always the most easy, and, as being such, are the soonest perverted by ill counsels, especially when baited with fame and glory; it is no more a wonder that he withstood not the temptations of Achitophel, than it was for Adam not to have resisted the two devils, the serpent and the woman. The conclusion of the story I purposely forbore to prosecute, because I could not obtain from myself to show Absalom unfortunate. The frame of it was cut out but for a picture to the waist, and if the draft be so far true, it is as much as I designed.

Were I the inventor, who am only the historian, I should certainly conclude the piece with the reconcilement of Absalom to David. And who knows but this may come to pass? Things were not brought to an extremity where I left the story; there seems yet to be room left for a composure; hereafter there may only be for pity. I have not so much as an uncharitable wish against Achitophel, but am content to be accused of a good-natured error, and to hope with Origen,° that the Devil himself may at last be saved. For which reason, in this poem, he is neither brought to set his house in order, nor to dispose of his person afterwards as he in wisdom shall think fit. God is infinitely merciful; and his vicegerent is only not so, because he is not infinite.

The true end of satire is the amendment of vices by correction. And he who writes honestly is no more an enemy to the offender, than the physician to the patient, when he prescribes harsh remedies to an inveterate disease; for those are only in order to prevent the surgeon's work of an *ense rescindendum* [something that must be cut off with the sword],° which I wish not to my very enemies. To conclude all: if the body politic have any analogy to the natural, in my weak judgment, an act of oblivion were as necessary in a hot, distempered state, as an opiate would be in a raging fever.

> In pious times, ere priestcraft did begin,
> Before polygamy was made a sin;

When man on many multiplied his kind,
Ere one to one was cursedly confined;
When nature prompted, and no law denied
Promiscuous use of concubine and bride;
Then Israel's monarch after heaven's own heart,°
His vigorous warmth did variously impart
To wives and slaves; and, wide as his command,
Scattered his maker's image through the land.   10
Michal, of royal blood, the crown did wear;°
A soil ungrateful to the tiller's care:
Not so the rest; for several mothers bore
To godlike David several sons before.°
But since like slaves his bed they did ascend,
No true succession could their seed attend.
Of all this numerous progeny was none
So beautiful, so brave, as Absalom:°
Whether, inspired by some diviner lust,
His father got him with a greater gust;   20
Or that his conscious destiny made way,
By manly beauty, to imperial sway.
Early in foreign fields he won renown,
With kings and states allied to Israel's crown:
In peace the thoughts of war he could remove,
And seemed as he were only born for love.
Whate'er he did was done with so much ease,
In him alone 'twas natural to please:
His motions all accompanied with grace;
And paradise was opened in his face.   30
With secret joy indulgent David viewed
His youthful image in his son renewed:
To all his wishes nothing he denied,
And made the charming Annabel his bride.°
What faults he had (for who from faults is free?)
His father could not, or he would not see.
Some warm excesses which the law forbore,
Were construed youth that purged by boiling o'er,
And Amnon's murder, by a specious name,°
Was called a just revenge for injured fame.   40
Thus praised and loved the noble youth remained,
While David, undisturbed, in Sion reigned.°
But life can never be sincerely blest;

Heaven punishes the bad, and proves the best.
The Jews, a headstrong, moody, murmuring race,°
As ever tried the extent and stretch of grace;
God's pampered people, whom, debauched with ease,
No king could govern, nor no God could please
(Gods they had tried of every shape and size,
That god-smiths could produce, or priests devise);   50
These Adam-wits, too fortunately free,
Began to dream they wanted liberty;
And when no rule, no precedent was found,
Of men by laws less circumscribed and bound,
They led their wild desires to woods and caves,
And thought that all but savages were slaves.
They who, when Saul was dead, without a blow,°
Made foolish Ishbosheth the crown forgo;°
Who banished David did from Hebron bring,°
And with a general shout proclaimed him king: 60
Those very Jews, who, at their very best,
Their humour more than loyalty expressed,
Now wondered why so long they had obeyed
An idol monarch, which their hands had made;
Thought they might ruin him they could create,
Or melt him to that golden calf, a state.
But these were random bolts; no formed design,
Nor interest made the factious crowd to join:
The sober part of Israel, free from stain,
Well knew the value of a peaceful reign,   70
And, looking backward with a wise affright,
Saw seams of wounds, dishonest to the sight:
In contemplation of whose ugly scars
They cursed the memory of civil wars.
The moderate sort of men, thus qualified,
Inclined the balance to the better side;
And David's mildness managed it so well,
The bad found no occasion to rebel.
But when to sin our biased nature leans,
The careful devil is still at hand with means;   80
And providently pimps for ill desires.
The Good Old Cause revived, a plot requires:°
Plots, true or false, are necessary things,
To raise up commonwealths, and ruin kings.

The inhabitants of old Jerusalem°
Were Jebusites, the town so called from them;°
And theirs the native right—
But when the chosen people grew more strong,°
The rightful cause at length became the wrong;
And every loss the men of Jebus bore,   90
They still were thought God's enemies the more.
Thus worn and weakened, well or ill content,
Submit they must to David's government:
Impoverished and deprived of all command,°
Their taxes doubled as they lost their land;
And what was harder yet to flesh and blood,
Their gods disgraced, and burnt like common wood.
This set the heathen priesthood in a flame;
For priests of all religions are the same:
Of whatsoe'er descent their godhead be,   100
Stock, stone, or other homely pedigree,
In his defence his servants are as bold,
As if he had been born of beaten gold.
The Jewish rabbins, though their enemies,
In this conclude them honest men and wise:
For 'twas their duty, all the learned think,
To espouse his cause, by whom they eat and drink.°
From hence began that Plot, the nation's curse,°
Bad in itself, but represented worse;
Raised in extremes, and in extremes decried;   110
With oaths affirmed, with dying vows denied.
Not weighed or winnowed by the multitude;
But swallowed in the mass, unchewed and crude.
Some truth there was, but dashed and brewed with lies,
To please the fools, and puzzle all the wise.
Succeeding times did equal folly call,
Believing nothing, or believing all.
The Egyptian rites the Jebusites embraced;°
Where gods were recommended by their taste.
Such savoury deities must needs be good,   120
As served at once for worship and for food.
By force they could not introduce these gods,
For ten to one in former days was odds;
So fraud was used (the sacrificer's trade):
Fools are more hard to conquer than persuade.

Their busy teachers mingled with the Jews,
And raked for converts even the court and stews:
Which Hebrew priests the more unkindly took,
Because the fleece accompanies the flock.°
Some thought they God's anointed meant to slay°   130
By guns, invented since full many a day:
Our author swears it not; but who can know
How far the Devil and Jebusites may go?
This Plot, which failed for want of common sense,
Had yet a deep and dangerous consequence:
For, as when raging fevers boil the blood,
The standing lake soon floats into a flood,
And every hostile humour, which before
Slept quiet in its channels, bubbles o'er;
So several factions from this first ferment   140
Work up to foam, and threat the government.
Some by their friends, more by themselves thought wise,
Opposed the power to which they could not rise.
Some had in courts been great, and thrown from thence,
Like fiends were hardened in impenitence.
Some, by their monarch's fatal mercy, grown
From pardoned rebels kinsmen to the throne,
Were raised in power and public office high;
Strong bands, if bands ungrateful men could tie.
   Of these the false Achitophel was first,°   150
A name to all succeeding ages cursed:
For close designs and crooked counsels fit;
Sagacious, bold, and turbulent of wit;
Restless, unfixed in principles and place;
In power unpleased, impatient of disgrace:
A fiery soul, which, working out its way,
Fretted the pigmy body to decay,
And o'er-informed the tenement of clay.
A daring pilot in extremity;
Pleased with the danger, when the waves went high,   160
He sought the storms; but, for a calm unfit,
Would steer too nigh the sands, to boast his wit.
Great wits are sure to madness near allied,
And thin partitions do their bounds divide;
Else why should he, with wealth and honour blest,
Refuse his age the needful hours of rest?

Punish a body which he could not please;
Bankrupt of life, yet prodigal of ease?
And all to leave what with his toil he won,
To that unfeathered two-legged thing, a son,   170
Got, while his soul did huddled notions try;
And born a shapeless lump, like anarchy.
In friendship false, implacable in hate;
Resolved to ruin or to rule the state.
To compass this the triple bond he broke,°
The pillars of the public safety shook;
And fitted Israel for a foreign yoke:
Then seized with fear, yet still affecting fame,
Usurped a patriot's all-atoning name.
So easy still it proves in factious times,   180
With public zeal to cancel private crimes.
How safe is treason, and how sacred ill,
Where none can sin against the people's will!
Where crowds can wink, and no offence be known,
Since in another's guilt they find their own!
Yet fame deserved no enemy can grudge;
The statesman we abhor, but praise the judge.
In Israel's courts ne'er sat an Abbethdin°
With more discerning eyes, or hands more clean;
Unbribed, unsought, the wretched to redress;   190
Swift of dispatch, and easy of access.
O! had he been content to serve the crown,
With virtues only proper to the gown;
Or had the rankness of the soil been freed
From cockle, that oppressed the noble seed;
David for him his tuneful harp had strung,
And heaven had wanted one immortal song.
But wild ambition loves to slide, not stand,
And fortune's ice prefers to virtue's land.
Achitophel, grown weary to possess°   200
A lawful fame, and lazy happiness,
Disdained the golden fruit to gather free,
And lent the crowd his arm to shake the tree.
Now, manifest of crimes contrived long since,
He stood at bold defiance with his prince;
Held up the buckler of the people's cause
Against the crown, and skulked behind the laws.

The wished occasion of the Plot he takes;
Some circumstances finds, but more he makes.
By buzzing emissaries fills the ears   210
Of listening crowds with jealousies and fears
Of arbitrary counsels brought to light,
And proves the king himself a Jebusite.
Weak arguments! which yet he knew full well
Were strong with people easy to rebel.
For, governed by the moon, the giddy Jews
Tread the same track when she the prime renews;°
And once in twenty years, their scribes record,
By natural instinct they change their lord.
Achitophel still wants a chief, and none   220
Was found so fit as warlike Absalom:
Not that he wished his greatness to create
(For politicians neither love nor hate),
But, for he knew his title not allowed,°
Would keep him still depending on the crowd:
That kingly power, thus ebbing out, might be
Drawn to the dregs of a democracy.
Him he attempts with studied arts to please,
And sheds his venom in such words as these:
    'Auspicious prince, at whose nativity   230
Some royal planet ruled the southern sky;
Thy longing country's darling and desire;
Their cloudy pillar and their guardian fire:°
Their second Moses, whose extended wand°
Divides the seas, and shows the promised land;
Whose dawning day in every distant age
Has exercised the sacred prophets' rage:
The people's prayer, the glad diviners' theme,
The young men's vision, and the old men's dream!°
Thee, saviour, thee, the nation's vows confess,   240
And, never satisfied with seeing, bless:
Swift unbespoken pomps thy steps proclaim,
And stammering babes are taught to lisp thy name.
How long wilt thou the general joy detain,
Starve and defraud the people of thy reign?
Content ingloriously to pass thy days
Like one of virtue's fools that feeds on praise,
Till thy fresh glories, which now shine so bright,

Grow stale and tarnish with our daily sight.
Believe me, royal youth, thy fruit must be   250
Or gathered ripe, or rot upon the tree.
Heaven has to all allotted, soon or late,
Some lucky revolution of their fate;
Whose motions if we watch and guide with skill
(For human good depends on human will),
Our fortune rolls as from a smooth descent,
And from the first impression takes the bent:
But, if unseized, she glides away like wind,
And leaves repenting folly far behind.
Now, now she meets you with a glorious prize,   260
And spreads her locks before her as she flies.
Had thus old David, from whose loins you spring,
Not dared, when fortune called him, to be king,
At Gath an exile he might still remain,°
And heaven's anointing oil had been in vain.
Let his successful youth your hopes engage,
But shun the example of declining age:
Behold him setting in his western skies,
The shadows lengthening as the vapours rise.
He is not now, as when on Jordan's sand°   270
The joyful people thronged to see him land,
Covering the beach, and blackening all the strand;
But, like the prince of angels, from his height°
Comes tumbling downward with diminished light,
Betrayed by one poor plot to public scorn
(Our only blessing since his cursed return);
Those heaps of people which one sheaf did bind,
Blown off and scattered by a puff of wind.
What strength can he to your designs oppose,
Naked of friends, and round beset with foes?   280
If Pharaoh's doubtful succour he should use,°
A foreign aid would more incense the Jews.
Proud Egypt would dissembled friendship bring,
Foment the war, but not support the king.
Nor would the royal party e'er unite
With Pharaoh's arms to assist the Jebusite;
Or if they should, their interest soon would break,
And with such odious aid make David weak.
All sorts of men by my successful arts,

Abhorring kings, estrange their altered hearts   290
From David's rule: and 'tis the general cry,
"Religion, commonwealth, and liberty."
If you, as champion of the public good,
Add to their arms a chief of royal blood,
What may not Israel hope, and what applause
Might such a general gain by such a cause?
Not barren praise alone, that gaudy flower
Fair only to the sight, but solid power;
And nobler is a limited command,
Given by love of all your native land,   300
Than a successive title, long and dark,
Drawn from the mouldy rolls of Noah's ark.'
    What cannot praise effect in mighty minds,
When flattery soothes, and when ambition blinds!
Desire of power, on earth a vicious weed,
Yet, sprung from high, is of celestial seed:
In God 'tis glory, and when men aspire,
'Tis but a spark too much of heavenly fire.
The ambitious youth, too covetous of fame,
Too full of angels' metal in his frame,°   310
Unwarily was led from virtue's ways,
Made drunk with honour, and debauched with praise.
Half loath, and half consenting to the ill
(For loyal blood within him struggled still),
He thus replied: 'And what pretence have I
To take up arms for public liberty?
My father governs with unquestioned right;
The faith's defender, and mankind's delight,
Good, gracious, just, observant of the laws:
And heaven by wonders has espoused his cause.°   320
Whom has he wronged in all his peaceful reign?
Who sues for justice to his throne in vain?
What millions has he pardoned of his foes,
Whom just revenge did to his wrath expose?
Mild, easy, humble, studious of our good;
Inclined to mercy, and averse from blood;
If mildness ill with stubborn Israel suit,
His crime is God's beloved attribute.
What could he gain, his people to betray,

Or change his right for arbitrary sway?   330
Let haughty Pharaoh curse with such a reign
His fruitful Nile, and yoke a servile train.
If David's rule Jerusalem displease,
The dog-star heats their brains to this disease.°
Why then should I, encouraging the bad,
Turn rebel and run popularly mad?
Were he a tyrant, who, by lawless might
Oppressed the Jews, and raised the Jebusite,
Well might I mourn; but nature's holy bands
Would curb my spirits and restrain my hands: 340
The people might assert their liberty;
But what was right in them were crime in me.
His favour leaves me nothing to require,
Prevents my wishes, and outruns desire.
What more can I expect while David lives?
All but his kingly diadem he gives:
And that'—But there he paused; then sighing, said—
'Is justly destined for a worthier head.
For when my father from his toils shall rest,
And late augment the number of the blessed,   350
His lawful issue shall the throne ascend,
Or the collateral line, where that shall end.
His brother, though oppressed with vulgar spite,°
Yet dauntless, and secure of native right,
Of every royal virtue stands possessed;
Still dear to all the bravest and the best.
His courage foes, his friends his truth proclaim;
His loyalty the king, the world his fame.
His mercy e'en the offending crowd will find;
For sure he comes of a forgiving kind.   360
Why should I then repine at heaven's decree,
Which gives me no pretence to royalty?
Yet O that fate, propitiously inclined,
Had raised my birth, or had debased my mind;
To my large soul not all her treasure lent,
And then betrayed it to a mean descent!
I find, I find my mounting spirits bold,
And David's part disdains my mother's mould.
Why am I scanted by a niggard birth?

My soul disclaims the kindred of her earth;   370
And, made for empire, whispers me within,
"Desire of greatness is a godlike sin." '
    Him staggering so when hell's dire agent found,
While fainting virtue scarce maintained her ground,
He pours fresh forces in, and thus replies:
    'The eternal God, supremely good and wise,
Imparts not these prodigious gifts in vain:
What wonders are reserved to bless your reign!
Against your will, your arguments have shown,
Such virtue's only given to guide a throne.   380
Not that your father's mildness I contemn;
But manly force becomes the diadem.
'Tis true he grants the people all they crave;
And more, perhaps, than subjects ought to have:
For lavish grants suppose a monarch tame,
And more his goodness than his wit proclaim.
But when should people strive their bonds to break,
If not when kings are negligent or weak?
Let him give on till he can give no more,
The thrifty Sanhedrin shall keep him poor;°   390
And every shekel which he can receive,
Shall cost a limb of his prerogative.
To ply him with new plots shall be my care;
Or plunge him deep in some expensive war;
Which when his treasure can no more supply,
He must, with the remains of kingship, buy.
His faithful friends, our jealousies and fears
Call Jebusites, and Pharaoh's pensioners;
Whom when our fury from his aid has torn,
He shall be naked left to public scorn.   400
The next successor, whom I fear and hate,
My arts have made obnoxious to the state;
Turned all his virtues to his overthrow,
And gained our elders to pronounce a foe.
His right, for sums of necessary gold,
Shall first be pawned, and afterwards be sold;
Till time shall ever-wanting David draw,
To pass your doubtful title into law:
If not, the people have a right supreme
To make their kings; for kings are made for them.   410

All empire is no more than power in trust,
Which, when resumed, can be no longer just.
Succession, for the general good designed,
In its own wrong a nation cannot bind;
If altering that the people can relieve,
Better one suffer than a nation grieve.
The Jews well know their power: ere Saul they chose,
God was their king, and God they durst depose.
Urge now your piety, your filial name,
A father's right, and fear of future fame,   420
The public good, that universal call,
To which e'en heaven submitted, answers all.
Nor let his love enchant your generous mind,
'Tis nature's trick to propagate her kind.
Our fond begetters, who would never die,
Love but themselves in their posterity.
Or let his kindness by the effects be tried,
Or let him lay his vain pretence aside.
God said he loved your father; could he bring
A better proof, than to anoint him king?   430
It surely showed he loved the shepherd well,
Who gave so fair a flock as Israel.
Would David have you thought his darling son?
What means he then, to alienate the crown?
The name of godly he may blush to bear:
'Tis after God's own heart to cheat his heir.
He to his brother gives supreme command,
To you a legacy of barren land:
Perhaps the old harp, on which he thrums his lays,
Or some dull Hebrew ballad in your praise.   440
Then the next heir, a prince severe and wise,
Already looks on you with jealous eyes;
Sees through the thin disguises of your arts,
And marks your progress in the people's hearts.
Though now his mighty soul its grief contains,
He meditates revenge who least complains;
And, like a lion, slumbering in the way,
Or sleep dissembling, while he waits his prey,
His fearless foes within his distance draws,
Constrains his roaring, and contracts his paws: 450
Till at the last, his time for fury found,

He shoots with sudden vengeance from the ground;
The prostrate vulgar passes o'er and spares,
But with a lordly rage his hunters tears.
Your case no tame expedients will afford:
Resolve on death, or conquest by the sword,
Which for no less a stake than life you draw;
And self-defence is nature's eldest law.
Leave the warm people no considering time;
For then rebellion may be thought a crime.   460
Prevail yourself of what occasion gives,
But try your title while your father lives;
And that your arms may have a fair pretence,
Proclaim you take them in the king's defence;
Whose sacred life each minute would expose
To plots, from seeming friends, and secret foes.
And who can sound the depth of David's soul?
Perhaps his fear his kindness may control.
He fears his brother, though he loves his son,
For plighted vows too late to be undone.   470
If so, by force he wishes to be gained;
Like women's lechery, to seem constrained.
Doubt not: but, when he most affects the frown,
Commit a pleasing rape upon the crown.
Secure his person to secure your cause:
They who possess the prince, possess the laws.'
   He said, and this advice above the rest,
With Absalom's mild nature suited best:
Unblamed of life (ambition set aside),
Not stained with cruelty, nor puffed with pride;   480
How happy had he been, if destiny
Had higher placed his birth, or not so high!
His kingly virtues might have claimed a throne,
And blessed all other countries but his own.
But charming greatness since so few refuse,
'Tis juster to lament him than accuse.
Strong were his hopes a rival to remove,
With blandishments to gain the public love;
To head the faction while their zeal was hot,
And popularly prosecute the Plot.   490
To further this, Achitophel unites
The malcontents of all the Israelites;

Whose differing parties he could wisely join,
For several ends, to serve the same design:
The best (and of the princes some were such)
Who thought the power of monarchy too much;
Mistaken men, and patriots in their hearts;
Not wicked, but seduced by impious arts.
By these the springs of property were bent,
And wound so high, they cracked the government.   500
The next for interest sought to embroil the state,
To sell their duty at a dearer rate;
And make their Jewish markets of the throne,
Pretending public good, to serve their own.
Others thought kings a useless heavy load,
Who cost too much, and did too little good.
These were for laying honest David by,
On principles of pure good husbandry.
With them joined all the haranguers of the throng,
That thought to get preferment by the tongue.   510
Who follow next, a double danger bring,
Not only hating David, but the king:
The Solymaean rout, well-versed of old°
In godly faction, and in treason bold;
Cowering and quaking at a conqueror's sword;
But lofty to a lawful prince restored;
Saw with disdain an ethnic plot begun,°
And scorned by Jebusites to be outdone.
Hot Levites headed these; who, pulled before°
From the ark, which in the Judges' days they bore,   520
Resumed their cant, and with a zealous cry
Pursued their old beloved theocracy:
Where Sanhedrin and priest enslaved the nation,
And justified their spoils by inspiration:
For who so fit for reign as Aaron's race,°
If once dominion they could found in grace.
These led the pack; though not of surest scent,
Yet deepest mouthed against the government.°
A numerous host of dreaming saints succeed,°
Of the true old enthusiastic breed: 530
'Gainst form and order they their power employ,
Nothing to build, and all things to destroy.
But far more numerous was the herd of such,

Who think too little, and who talk too much.
These, out of mere instinct, they knew not why,
Adored their fathers' God and property;
And, by the same blind benefit of fate,
The devil and the Jebusite did hate:
Born to be saved, e'en in their own despite,
Because they could not help believing right.   540
Such were the tools; but a whole Hydra more
Remains, of sprouting heads too long to score.
     Some of their chiefs were princes of the land:
In the first rank of these did Zimri stand;°
A man so various, that he seemed to be
Not one, but all mankind's epitome:
Stiff in opinions, always in the wrong;
Was everything by starts, and nothing long;
But, in the course of one revolving moon,
Was chemist, fiddler, statesman, and buffoon: 550
Then all for women, painting, rhyming, drinking,
Besides ten thousand freaks that died in thinking.
Blest madman, who could every hour employ,
With something new to wish, or to enjoy!
Railing and praising were his usual themes;
And both (to show his judgment) in extremes:
So over-violent, or over-civil,
That every man, with him, was God or devil.
In squandering wealth was his peculiar art:
Nothing went unrewarded but desert.   560
Beggared by fools, whom still he found too late,
He had his jest, and they had his estate.
He laughed himself from court; then sought relief
By forming parties, but could ne'er be chief;
For, spite of him, the weight of business fell
On Absalom and wise Achitophel:
Thus, wicked but in will, of means bereft,
He left not faction, but of that was left.
     Titles and names 'twere tedious to rehearse
Of lords, below the dignity of verse.   570
Wits, warriors, commonwealth's-men, were the best;
Kind husbands, and mere nobles, all the rest.
And therefore, in the name of dullness, be
The well-hung Balaam and cold Caleb, free;°

And canting Nadab let oblivion damn,°
Who made new porridge for the Paschal Lamb.°
Let friendship's holy band some names assure;
Some their own worth, and some let scorn secure.
Nor shall the rascal rabble here have place,
Whom kings no titles gave, and God no grace: 580
Not bull-faced Jonas, who could statutes draw°
To mean rebellion, and make treason law.
But he, though bad, is followed by a worse,
The wretch who heaven's anointed dared to curse:
Shimei, whose youth did early promise bring°
Of zeal to God and hatred to his king,
Did wisely from expensive sins refrain,
And never broke the sabbath, but for gain;
Nor ever was he known an oath to vent,
Or curse, unless against the government.   590
Thus heaping wealth, by the most ready way
Among the Jews, which was to cheat and pray,
The city, to reward his pious hate
Against his master, chose him magistrate.
His hand a vare of justice did uphold;
His neck was loaded with a chain of gold.
During his office, treason was no crime;
The sons of Belial had a glorious time;°
For Shimei, though not prodigal of pelf,
Yet loved his wicked neighbour as himself.   600
When two or three were gathered to declaim
Against the monarch of Jerusalem,
Shimei was always in the midst of them;
And if they cursed the king when he was by,
Would rather curse than break good company.
If any durst his factious friends accuse,
He packed a jury of dissenting Jews;
Whose fellow-feeling in the godly cause
Would free the suffering saint from human laws.
For laws are only made to punish those   610
Who serve the king, and to protect his foes.
If any leisure time he had from power
(Because 'tis sin to misemploy an hour),
His business was, by writing, to persuade
That kings were useless, and a clog to trade;

And, that his noble style he might refine,
No Rechabite more shunned the fumes of wine.°
Chaste were his cellars, and his shrieval board
The grossness of a city feast abhorred:
His cooks, with long disuse, their trade forgot;   620
Cool was his kitchen, though his brains were hot.
Such frugal virtue malice may accuse,
But sure 'twas necessary to the Jews;
For towns once burnt such magistrates require
As dare not tempt God's providence by fire.
With spiritual food he fed his servants well,
But free from flesh that made the Jews rebel;
And Moses' laws he held in more account,
For forty days of fasting in the mount.

　　To speak the rest, who better are forgot,   630
Would tire a well-breathed witness of the Plot.
Yet, Corah, thou shalt from oblivion pass:°
Erect thyself, thou monumental brass,°
High as the serpent of thy metal made,
While nations stand secure beneath thy shade.
What though his birth were base, yet comets rise
From earthy vapours, ere they shine in skies.
Prodigious actions may as well be done
By weaver's issue as by prince's son.
This arch-attestor for the public good   640
By that one deed ennobles all his blood
Who ever asked the witnesses' high race,
Whose oath with martyrdom did Stephen grace?°
Ours was a Levite, and as times went then,
His tribe were God almighty's gentlemen.°
Sunk were his eyes, his voice was harsh and loud,
Sure signs he neither choleric was nor proud:
His long chin proved his wit; his saintlike grace
A church vermilion, and a Moses' face.°
His memory, miraculously great,   650
Could plots exceeding man's belief, repeat;
Which therefore cannot be accounted lies,
For human wit could never such devise.
Some future truths are mingled in his book;
But where the witness failed, the prophet spoke:
Some things like visionary flights appear;

The spirit caught him up, the Lord knows where;
And gave him his rabbinical degree,
Unknown to foreign university.
His judgment yet his memory did excel;   660
Which pieced his wondrous evidence so well,
And suited to the temper of the times,
Then groaning under Jebusitic crimes.
Let Israel's foes suspect his heavenly call,
And rashly judge his writ apocryphal;
Our laws for such affronts have forfeits made;
He takes his life, who takes away his trade.
Were I myself in witness Corah's place,
The wretch who did me such a dire disgrace,
Should whet my memory, though once forgot,   670
To make him an appendix of my plot.
His zeal to heaven made him his prince despise,
And load his person with indignities;
But zeal peculiar privilege affords,
Indulging latitude to deeds and words;
And Corah might for Agag's murder call,°
In terms as coarse as Samuel used to Saul.
What others in his evidence did join
(The best that could be had for love or coin),
In Corah's own predicament will fall;   680
For witness is a common name to all.

   Surrounded thus with friends of every sort,
Deluded Absalom forsakes the court;
Impatient of high hopes, urged with renown,
And fired with near possession of a crown.
The admiring crowd are dazzled with surprise,
And on his goodly person feed their eyes.
His joy concealed, he sets himself to show,
On each side bowing popularly low;
His looks, his gestures, and his words he frames,   690
And with familiar ease repeats their names.
Thus formed by nature, furnished out with arts,
He glides unfelt into their secret hearts.
Then, with a kind compassionating look,
And sighs, bespeaking pity ere he spoke,
Few words he said; but easy those and fit,
More slow than Hybla-drops, and far more sweet.°

'I mourn, my countrymen, your lost estate;
Though far unable to prevent your fate:
Behold a banished man, for your dear cause°   700
Exposed a prey to arbitrary laws!
Yet O! that I alone could be undone,
Cut off from empire, and no more a son!
Now all your liberties a spoil are made;
Egypt and Tyrus intercept your trade,°
And Jebusites your sacred rites invade.
My father, whom with reverence yet I name,
Charmed into ease, is careless of his fame;
And, bribed with petty sums of foreign gold,
Is grown in Bathsheba's embraces old;°   710
Exalts his enemies, his friends destroys;
And all his power against himself employs.
He gives, and let him give, my right away;
But why should he his own and yours betray?
He, only he, can make the nation bleed,
And he alone from my revenge is freed.
Take then my tears' (with that he wiped his eyes),
''Tis all the aid my present power supplies:
No court-informer can these arms accuse;
These arms may sons against their fathers use,   720
And 'tis my wish, the next successor's reign
May make no other Israelite complain.'
    Youth, beauty, graceful action seldom fail;
But common interest always will prevail;
And pity never ceases to be shown
To him who makes the people's wrongs his own.
The crowd, that still believe their kings oppress,
With lifted hands their young messiah bless:
Who now begins his progress to ordain°
With chariots, horsemen, and a numerous train;   730
From east to west his glories he displays,
And, like the sun, the promised land surveys.
Fame runs before him as the morning star,
And shouts of joy salute him from afar:
Each house receives him as a guardian god,
And consecrates the place of his abode.
But hospitable treats did most commend
Wise Issachar, his wealthy western friend.°

This moving court, that caught the people's eyes,
And seemed but pomp, did other ends disguise: 740
Achitophel had formed it, with intent
To sound the depths, and fathom, where it went,
The people's hearts, distinguish friends from foes,
And try their strength, before they came to blows.
Yet all was coloured with a smooth pretence
Of specious love, and duty to their prince.
Religion, and redress of grievances,
Two names that always cheat and always please,
Are often urged; and good King David's life
Endangered by a brother and a wife.   750
Thus in a pageant show a plot is made,
And peace itself is war in masquerade.
O foolish Israel! never warned by ill!
Still the same bait, and circumvented still!
Did ever men forsake their present ease,
In midst of health imagine a disease;
Take pains contingent mischiefs to foresee,
Make heirs for monarchs, and for God decree?
What shall we think! Can people give away,
Both for themselves and sons, their native sway?   760
Then they are left defenceless to the sword
Of each unbounded, arbitrary lord:
And laws are vain, by which we right enjoy,
If kings unquestioned can those laws destroy.
Yet if the crowd be judge of fit and just,
And kings are only officers in trust,
Then this resuming covenant was declared
When kings were made, or is for ever barred.
If those who gave the sceptre could not tie
By their own deed their own posterity,   770
How then could Adam bind his future race?
How could his forfeit on mankind take place?
Or how could heavenly justice damn us all,
Who ne'er consented to our father's fall?
Then kings are slaves to those whom they command,
And tenants to their people's pleasure stand.
Add, that the power for property allowed
Is mischievously seated in the crowd;
For who can be secure of private right,

If sovereign sway may be dissolved by might?   780
Nor is the people's judgment always true:
The most may err as grossly as the few;
And faultless kings run down, by common cry,
For vice, oppression, and for tyranny.
What standard is there in a fickle rout,
Which, flowing to the mark, runs faster out?°
Nor only crowds but Sanhedrins may be
Infected with this public lunacy,
And share the madness of rebellious times,
To murder monarchs for imagined crimes.   790
If they may give and take whene'er they please,
Not kings alone (the Godhead's images)
But government itself at length must fall
To nature's state, where all have right to all,
Yet, grant our lords the people kings can make,
What prudent men a settled throne would shake?
For whatsoe'er their sufferings were before,
That change they covet makes them suffer more.
All other errors but disturb a state,
But innovation is the blow of fate.°   800
If ancient fabrics nod, and threat to fall,
To patch the flaws, and buttress up the wall,
Thus far 'tis duty. But here fix the mark
For all beyond it is to touch our ark.°
To change foundations, cast the frame anew,
Is work for rebels, who base ends pursue,
At once divine and human laws control,°
And mend the parts by ruin of the whole.
The tampering world is subject to this curse,
To physic their disease into a worse.   810
     Now what relief can righteous David bring?
How fatal 'tis to be too good a king!
Friends he has few, so high the madness grows:
Who dare be such, must be the people's foes.
Yet some there were, e'en in the worst of days;
Some let me name, and naming is to praise.
     In this short file Barzillai first appears;°
Barzillai, crowned with honour and with years:
Long since, the rising rebels he withstood
In regions waste, beyond the Jordan's flood;   820

Unfortunately brave to buoy the state,
But sinking underneath his master's fate.
In exile with his godlike prince he mourned,
For him he suffered, and with him returned.
The court he practised, not the courtier's art;
Large was his wealth, but larger was his heart;
Which well the noblest objects knew to choose,
The fighting warrior, and recording muse.
His bed could once a fruitful issue boast;
Now more than half a father's name is lost.°   830
His eldest hope, with every grace adorned,°
By me (so heaven will have it) always mourned,
And always honoured, snatched in manhood's prime
By unequal fates, and providence's crime;
Yet not before the goal of honour won,
All parts fulfilled of subject and of son:
Swift was the race, but short the time to run.
O narrow circle, but of power divine,
Scanted in space, but perfect in thy line!
By sea, by land, thy matchless worth was known,   840
Arms thy delight, and war was all thy own:
Thy force, infused, the fainting Tyrians propped;°
And haughty Pharaoh found his fortune stopped.
O ancient honour! O unconquered hand,
Whom foes unpunished never could withstand!
But Israel was unworthy of thy name;
Short is the date of all immoderate fame.
It looks as heaven our ruin had designed,
And durst not trust thy fortune and thy mind.
Now, free from earth, thy disencumbered soul   850
Mounts up, and leaves behind the clouds and starry pole:°
From thence thy kindred legions mayst thou bring,
To aid the guardian angel of thy king.
Here stop, my muse, here cease thy painful flight;
No pinions can pursue immortal height:
Tell good Barzillai thou canst sing no more,
And tell thy soul she should have fled before.
Or fled she with his life, and left this verse
To hang on her departed patron's hearse?
Now take thy steepy flight from heaven and see   860
If thou canst find on earth another he;

Another he would be too hard to find;
See then whom thou canst see not far behind.
Zadoc the priest, whom, shunning power and place,°
His lowly mind advanced to David's grace.
With him the Sagan of Jerusalem,°
Of hospitable soul, and noble stem;
Him of the western dome, whose weighty sense°
Flows in fit words and heavenly eloquence.
The prophets' sons, by such example led,°   870
To learning and to loyalty were bred:
For colleges on bounteous kings depend,
And never rebel was to arts a friend.
To these succeed the pillars of the laws;
Who best could plead, and best can judge a cause.
Next them a train of loyal peers ascend;
Sharp-judging Adriel, the muses' friend;°
Himself a muse—in Sanhedrin's debate
True to his prince, but not a slave of state:
Whom David's love with honours did adorn,   880
That from his disobedient son were torn.
Jotham of piercing wit, and pregnant thought;°
Endued by nature, and by learning taught
To move assemblies, who but only tried
The worse a while, then chose the better side:
Nor chose alone, but turned the balance too;
So much the weight of one brave man can do.
Hushai, the friend of David in distress;°
In public storms of manly steadfastness:
By foreign treaties he informed his youth,   890
And joined experience to his native truth.
His frugal care supplied the wanting throne;
Frugal for that, but bounteous of his own:
'Tis easy conduct when exchequers flow,
But hard the task to manage well the low;
For sovereign power is too depressed or high,
When kings are forced to sell, or crowds to buy.
Indulge one labour more, my weary muse,
For Amiel: who can Amiel's praise refuse?°
Of ancient race by birth, but nobler yet   900
In his own worth, and without title great:
The Sanhedrin long time as chief he ruled,

Their reason guided, and their passion cooled:
So dexterous was he in the crown's defence,
So formed to speak a loyal nation's sense,
That, as their band was Israel's tribes in small,
So fit was he to represent them all.
Now rasher charioteers the seat ascend,
Whose loose careers his steady skill commend:
They, like the unequal ruler of the day,°   910
Misguide the seasons, and mistake the way;
While he withdrawn at their mad labour smiles,
And safe enjoys the sabbath of his toils.

   These were the chief, a small but faithful band
Of worthies, in the breach who dared to stand,
And tempt the united fury of the land.
With grief they viewed such powerful engines bent
To batter down the lawful government:
A numerous faction, with pretended frights,
In Sanhedrins to plume the regal rights;   920
The true successor from the court removed;
The Plot, by hireling witnesses, improved.
These ills they saw, and, as their duty bound,
They showed the king the danger of the wound;
That no concessions from the throne would please,
But lenitives fomented the disease;
That Absalom, ambitious of the crown,
Was made the lure to draw the people down;
That false Achitophel's pernicious hate
Had turned the Plot to ruin Church and State;   930
The council violent, the rabble worse;
That Shimei taught Jerusalem to curse.

   With all these loads of injuries oppressed,
And long revolving in his careful breast
The event of things, at last, his patience tired,
Thus from his royal throne, by heaven inspired,
The godlike David spoke: with awful fear
His train their maker in their master hear.

   'Thus long have I, by native mercy swayed,
My wrongs dissembled, my revenge delayed: 940
So willing to forgive the offending age;
So much the father did the king assuage.
But now so far my clemency they slight,

The offenders question my forgiving right.°
That one was made for many, they contend;
But 'tis to rule; for that's a monarch's end.
They call my tenderness of blood, my fear;
Though manly tempers can the longest bear.
Yet, since they will divert my native course,
'Tis time to show I am not good by force.   950
Those heaped affronts that haughty subjects bring,
Are burdens for a camel, not a king,
Kings are the public pillars of the state,
Born to sustain and prop the nation's weight;
If my young Samson will pretend a call°
To shake the column, let him share the fall:
But O that yet he would repent and live!
How easy 'tis for parents to forgive!
With how few tears a pardon might be won
From nature, pleading for a darling son!   960
Poor pitied youth, by my paternal care
Raised up to all the height his frame could bear!
Had God ordained his fate for empire born,
He would have given his soul another turn:
Gulled with a patriot's name, whose modern sense
Is one that would by law supplant his prince;
The people's brave, the politician's tool;
Never was patriot yet, but was a fool.
Whence comes it that religion and the laws
Should more be Absalom's than David's cause?   970
His old instructor, ere he lost his place,
Was never thought indued with so much grace.
Good heavens, how faction can a patriot paint!
My rebel ever proves my people's saint.
Would *they* impose an heir upon the throne?
Let Sanhedrins be taught to give their own.
A king's at least a part of government,
And mine as requisite as their consent;
Without my leave a future king to choose,
Infers a right the present to depose.   980
True, they petititon me to approve their choice;
But Esau's hands suit ill with Jacob's voice.°
My pious subjects for my safety pray;
Which to secure, they take my power away.

From plots and treasons heaven preserve my years,
But save me most from my petitioners!
Unsatiate as the barren womb or grave;
God cannot grant so much as they can crave.
What then is left, but with a jealous eye
To guard the small remains of royalty?   990
The law shall still direct my peaceful sway,
And the same law teach rebels to obey:
Votes shall no more established power control—
Such votes as make a part exceed the whole:
No groundless clamours shall my friends remove,
Nor crowds have power to punish ere they prove;
For gods and godlike kings their care express,
Still to defend their servants in distress.
O that my power to saving were confined!
Why am I forced, like heaven, against my mind,   1000
To make examples of another kind?
Must I at length the sword of justice draw?
O cursed effects of necessary law!
How ill my fear they by my mercy scan!
Beware the fury of a patient man.
Law they require, let Law then show her face;
They could not be content to look on grace,
Her hinder parts, but with a daring eye°
To tempt the terror of her front and die.
By their own arts, 'tis righteously decreed,   1010
Those dire artificers of death shall bleed.
Against themselves their witnesses will swear,
Till viper-like their mother Plot they tear;
And suck for nutriment that bloody gore,
Which was their principle of life before.
Their Belial with their Belzebub will fight;
Thus on my foes, my foes shall do me right.
Nor doubt the event; for factious crowds engage,
In their first onset, all their brutal rage.
Then let 'em take an unresisted course;   1020
Retire, and traverse, and delude their force;
But, when they stand all breathless, urge the fight,
And rise upon 'em with redoubled might;
For lawful power is still superior found;
When long driven back, at length it stands the ground.'

He said. The almighty, nodding, gave consent;
And peals of thunder shook the firmament.
Henceforth a series of new time began,
The mighty years in long procession ran:
Once more the godlike David was restored,   1030
And willing nations knew their lawful lord.

From the Second Part of
Absalom and Achitophel

Now stop your noses, readers, all and some,
For here's a tun of midnight work to come,
Og, from a treason-tavern rolling home°
Round as a globe, and liquored every chink,
Goodly and great he sails behind his link.°
With all this bulk there's nothing lost in Og,
For every inch that is not fool is rogue:
A monstrous mass of foul corrupted matter,
As all the devils had spewed to make the batter.
When wine has given him courage to blaspheme,   10
He curses God, but God before cursed him;
And if man could have reason, none has more,
That made his paunch so rich, and him so poor.
With wealth he was not trusted, for heaven knew
What 'twas of old to pamper up a Jew;
To what would he on quail and pheasant swell,
That e'en on tripe and carrion could rebel?
But though heaven made him poor (with reverence
     speaking),
He never was a poet of God's making.
The midwife laid her hand on his thick skull,   20
With this prophetic blessing: *be thou dull*;
Drink, swear, and roar, forbear no lewd delight
Fit for thy bulk, do anything but write:
Thou art of lasting make, like thoughtless men,
A strong nativity—but for the pen;
Eat opium, mingle arsenic in thy drink,
Still thou mayst live, avoiding pen and ink.
I see, I see, 'tis counsel given in vain,
For treason botched in rhyme will be thy bane;

Rhyme is the rock on which thou art to wreck,   30
'Tis fatal to thy fame and to thy neck:
Why should thy metre good King David blast?
A psalm of his will surely be thy last.
Darest thou presume in verse to meet thy foes,
Thou whom the penny pamphlet foiled in prose?
Doeg, whom God for mankind's mirth has made,°
O'ertops thy talent in thy very trade;
Doeg to thee, thy paintings are so coarse,
A poet is, though he's the poets' horse.
A double noose thou on thy neck dost pull,   40
For writing treason, and for writing dull;
To die for faction is a common evil,
But to be hanged for nonsense is the devil:
Hadst thou the glories of thy king expressed,
Thy praises had been satire at the best;
But thou in clumsy verse, unlicked, unpointed,
Hast shamefully defied the Lord's anointed:
I will not rake the dunghill of thy crimes,
For who would read thy life that reads thy rhymes?
But of King David's foes, be this the doom,°   50
May all be like the young man Absalom;
And for my foes may this their blessing be,
To talk like Doeg, and to write like thee.

The Medal

A SATIRE AGAINST SEDITION

per Graium populos, mediaeque per Elidis urbem
ibat ovans, divumque sibi poscebat honorem.°

[he rode in triumph through the centre of the city of Elis, with
Greek nations about him, and claimed as his own the honours which
only gods may claim]

EPISTLE TO THE WHIGS

For to whom can I dedicate this poem, with so much justice as to you?
It is the representation of your own hero: it is the picture drawn at
length, which you admire and prize so much in little. None of your
ornaments are wanting; neither the landscape of the Tower, nor the
rising sun; nor the Anno Domini of your new sovereign's coronation.

This must needs be a grateful undertaking to your whole party; especially to those who have not been so happy as to purchase the original. I hear the graver has made a good market of it: all his kings are bought up already; or the value of the remainder so enhanced, that many a poor Polander° who would be glad to worship the image, is not able to go to the cost of him, but must be content to see him here. I must confess I am no great artist; but signpost painting will serve the turn to remember a friend by, especially when better is not to be had. Yet for your comfort the lineaments are true; and though he sat not five times to me, as he did to Bower,° yet I have consulted history, as the Italian painters do, when they would draw a Nero or a Caligula; though they have not seen the man, they can help their imagination by a statue of him, and find out the colouring from Suetonius and Tacitus. Truth is, you might have spared one side of your medal: the head would be seen to more advantage if it were placed on a spike of the Tower, a little nearer to the sun, which would then break out to better purpose.

You tell us in your preface to the *No-Protestant Plot*,° that you shall be forced hereafter to leave off your modesty: I suppose you mean that little which is left you; for it was worn to rags when you put out this medal. Never was there practised such a piece of notorious impudence in the face of an established government. I believe when he is dead you will wear him in thumb-rings, as the Turks did Scanderbeg,° as if there were virtue in his bones to preserve you against monarchy. Yet all this while you pretend not only zeal for the public good, but a due veneration for the person of the king. But all men who can see an inch before them may easily detect those gross fallacies. That it is necessary for men in your circumstances to pretend both, is granted you; for without them there could be no ground to raise a faction. But I would ask you one civil question. What right has any man among you, or any association of men (to come nearer to you), who, out of Parliament, cannot be considered in a public capacity, to meet as you daily do in factious clubs, to vilify the government in your discourses, and to libel it in all your writings? Who made you judges in Israel? Or how is it consistent with your zeal of the public welfare to promote sedition? Does your definition of loyal, which is to serve the king according to the laws, allow you the licence of traducing the executive power with which you own he is invested? You complain that his Majesty has lost the love and confidence of his people; and by your very urging it you endeavour, what in you lies, to make him lose them. All good subjects abhor the thought of arbitrary power, whether it be in one or many: if

you were the patriots you would seem, you would not at this rate
incense the multitude to assume it; for no sober man can fear it, either
from the king's disposition, or his practice, or even, where you would
odiously lay it, from his ministers. Give us leave to enjoy the
government and the benefit of laws under which we were born, and
which we desire to transmit to our posterity. You are not the trustees
of the public liberty; and if you have not right to petition in a crowd,°
much less have you to intermeddle in the management of affairs, or to
arraign what you do not like, which in effect is everything that is done
by the king and council. Can you imagine that any reasonable man will
believe you respect the person of his majesty, when it is apparent that
your seditious pamphlets are stuffed with particular reflections on
him? If you have the confidence to deny this, it is easy to be evinced
from a thousand passages, which I only forbear to quote, because I
desire they should die, and be forgotten. I have perused many of your
papers, and to show you that I have, the third part of your *No-
Protestant Plot* is much of it stolen from your dead author's°
pamphlet, called *The Growth of Popery*;° as manifestly as Milton's
Defence of the English People is from Buchanan,° *De Jure Regni apud
Scotos*; or your first Covenant° and new Association from the Holy
League° of the French Guisards. Anyone who reads Davila° may trace
your practices all along. There were the same pretences for reformation and loyalty, the same aspersions of the king, and the same
grounds of a rebellion. I know not whether you will take the
historian's word, who says it was reported that Poltrot, a Huguenot,
murdered Francis, Duke of Guise, by the instigations of Theodore
Beza,° or that it was a Huguenot minister, otherwise called a
Presbyterian (for our church abhors so devilish a tenet), who first
wrote a treatise of the lawfulness of deposing and murdering kings of
a different persuasion in religion; but I am able to prove, from the
doctrine of Calvin, and principles of Buchanan, that they set the
people above the magistrate; which, if I mistake not, is your own
fundamental, and which carries your loyalty no further than your
liking. When a vote of the House of Commons goes on your side, you
are as ready to observe it as if it were passed into a law; but when you
are pinched with any former, and yet unrepealed Act of Parliament,
you declare that in some cases you will not be obliged by it. The
passage is in the same third part of the *No-Protestant Plot*, and is too
plain to be denied. The late copy of your intended association, you
neither wholly justify nor condemn; but as the Papists, when they are
unopposed, fly out into all the pageantries of worship; but in times of

war, when they are hard pressed by arguments, lie close intrenched behind the Council of Trent: so now, when your affairs are in a low condition, you dare not pretend that to be a legal combination, but whensoever you are afloat, I doubt not but it will be maintained and justified to purpose. For indeed there is nothing to defend it but the sword: it is the proper time to say anything when men have all things in their power.

In the meantime, you would fain be nibbling at a parallel betwixt this association and that in the time of Queen Elizabeth. But there is this small difference betwixt them, that the ends of the one are directly opposite to the other: one with the queen's approbation and conjunction, as head of it; the other without either the consent or knowledge of the king, against whose authority it is manifestly designed. Therefore you do well to have recourse to your last evasion, that it was contrived by your enemies, and shuffled into the papers that were seized; which yet you see the nation is not so easy to believe as your own jury; but the matter is not difficult, to find twelve men in Newgate who would acquit a malefactor.

I have only one favour to desire of you at parting, that when you think of answering this poem, you would employ the same pens against it, who have combated with so much success against *Absalom and Achitophel*; for then you may assure yourselves of a clear victory, without the least reply. Rail at me abundantly; and, not to break a custom, do it without wit: by this method you will gain a considerable point, which is, wholly to waive the answer of my arguments. Never own the bottom of your principles, for fear they should be treason. Fall severely on the miscarriages of government; for if scandal be not allowed, you are no freeborn subjects. If God has not blessed you with the talent of rhyming, make use of my poor stock and welcome: let your verses run upon my feet; and, for the utmost refuge of notorious blockheads, reduced to the last extremity of sense, turn my own lines upon me, and, in utter despair of your own satire, make me satirise myself. Some of you have been driven to this bay already; but, above all the rest, commend me to the nonconformist parson, who wrote the *Whip and Key*. I am afraid it is not read so much as the piece deserves, because the bookseller is every week crying help at the end of his gazette, to get it off. You see I am charitable enough to do him a kindness, that it may be published as well as printed; and that so much skill in Hebrew derivations may not lie for waste paper in the shop. Yet I half suspect he went no further for his learning, than the index of Hebrew names and etymologies, which is printed at the end of some

English Bibles. If 'Achitophel' signify the brother of a fool, the author of that poem will pass with his readers for the next of kin. And perhaps it is the relation that makes the kindness. Whatever the verses are, buy 'em up, I beseech you, out of pity; for I hear the conventicle is shut up, and the brother of Achitophel out of service.

Now footmen, you know, have the generosity to make a purse for a member of their society, who has had his livery pulled over his ears; and even protestant socks are bought up among you, out of veneration to the name. A dissenter in poetry from sense and English will make as good a protestant rhymer, as a dissenter from the church of England a protestant parson. Besides, if you encourage a young beginner, who knows but he may elevate his style a little above the vulgar epithets of profane, and saucy Jack, and atheistic scribbler, with which he treats me, when the fit of enthusiasm is strong upon him; by which well-mannered and charitable expressions I was certain of his sect before I knew his name. What would you have more of a man? He has damned me in your cause from Genesis to the Revelations; and has half the texts of both the Testaments against me, if you will be so civil to yourselves as to take him for your interpreter, and not to take them for Irish witnesses. After all, perhaps you will tell me that you retained him only for the opening of your cause, and that your main lawyer is yet behind. Now if it so happen he meet with no more reply than his predecessors, you may either conclude that I trust to the goodness of my cause, or fear my adversary, or disdain him, or what you please, for the short on it is, it is indifferent to your humble servant, whatever your party says or thinks of him.

THE MEDAL

Of all our antic sights and pageantry,
Which English idiots run in crowds to see,
The Polish Medal bears the prize alone:°
A monster, more the favourite of the town
Than either fairs or theatres have shown.
Never did art so well with nature strive,
Nor ever idol seemed so much alive:
So like the man; so golden to the sight,
So base within, so counterfeit and light.
One side is filled with title and with face;   10
And, lest the king should want a regal place,
On the reverse, a tower the town surveys;

O'er which our mounting sun his beams displays.
The word, pronounced aloud by shrieval voice,°
Laetamur, which, in Polish, is *rejoice*.°
The day, month, year, to the great act are joined;
And a new canting holiday designed.
Five days he sat for every cast and look;
Four more than God to finish Adam took.
But who can tell what essence angels are,   20
Or how long heaven was making Lucifer?
O could the style that copied every grace,°
And plowed such furrows for a eunuch face,
Could it have formed his ever-changing will,
The various piece had tired the graver's skill!
A martial hero first, with early care,
Blown, like a pigmy by the winds, to war.
A beardless chief, a rebel, ere a man:
(So young his hatred to his prince began.)
Next this (how wildly will ambition steer!)   30
A vermin wriggling in the usurper's ear.°
Bartering his venal wit for sums of gold,
He cast himself into the saintlike mould;
Groaned, sighed, and prayed, while godliness was gain,
The loudest bagpipe of the squeaking train.
But, as 'tis hard to cheat a juggler's eyes,
His open lewdness he could ne'er disguise.
There split the saint; for hypocritic zeal
Allows no sins but those it can conceal.
Whoring to scandal gives too large a scope;   40
Saints must not trade, but they may interlope.°
The ungodly principle was all the same;
But a gross cheat betrays his partner's game.
Besides, their pace was formal, grave, and slack;
His nimble wit outran the heavy pack.
Yet still he found his fortune at a stay;
Whole droves of blockheads choking up his way:
They took, but not rewarded, his advice;
Villain and wit exact a double price.
Power was his aim; but, thrown from that pretence,   50
The wretch turned loyal in his own defence,
And malice reconciled him to his prince.
Him in the anguish of his soul he served,

Rewarded faster still than he deserved.
Behold him now exalted into trust;
His counsel's oft convenient, seldom just.
E'en in the most sincere advice he gave,
He had a grudging still to be a knave.
The frauds he learnt in his fanatic years
Made him uneasy in his lawful gears:°   60
At best as little honest as he could,
And, like white witches, mischievously good;
To his first bias longingly he leans,
And rather would be great by wicked means.
Thus, framed for ill, he loosed our triple hold:°
Advice unsafe, precipitous, and bold.
From hence those tears! that Ilium of our woe!°
Who helps a powerful friend, forearms a foe.
What wonder if the waves prevail so far,
When he cut down the banks that made the bar?   70
Seas follow but their nature to invade,
But he by art our native strength betrayed.
So Samson to his foe his force confessed;°
And, to be shorn, lay slumbering on her breast.
But when this fatal counsel, found too late,
Exposed its author to the public hate;
When his just sovereign by no impious way
Could be seduced to arbitrary sway;
Forsaken of that hope, he shifts the sail,
Drives down the current with a popular gale;   80
And shows the fiend confessed without a veil.
He preaches to the crowd that power is lent,
But not conveyed to kingly government;
That claims successive bear no binding force,
That coronation oaths are things of course;
Maintains the multitude can never err,
And sets the people in the papal chair.
The reason's obvious: *interest never lies*;
The most have still their interest in their eyes;
The power is always theirs, and power is ever wise.   90
Almighty crowd, thou shortenest all dispute;
Power is thy essence, wit thy attribute!
Nor faith nor reason make thee at a stay,
Thou leapest o'er all eternal truths in thy Pindaric way!°

Athens no doubt did righteously decide,
When Phocion and when Socrates were tried;°
As righteously they did those dooms repent;
Still they were wise, whatever way they went.
Crowds err not, though to both extremes they run;
To kill the father and recall the son.°   100
Some think the fools were most, as times went then;
But now the world's o'erstocked with prudent men.
The common cry is e'en religion's test:
The Turk's is at Constantinople best;
Idols in India; popery at Rome;
And our own worship only true at home.
And true, but for the time; 'tis hard to know
How long we please it shall continue so.
This side today, and that tomorrow burns;
So all are God almighties in their turns.   110
A tempting doctrine, plausible and new:
What fools our fathers were, if this be true!
Who, to destroy the seeds of civil war,
Inherent right in monarchs did declare;
And, that a lawful power might never cease,
Secured succession, to secure our peace.
Thus property and sovereign sway, at last,
In equal balances were justly cast:
But this new Jehu spurs the hot-mouthed horse;°
Instructs the beast to know his native force,   120
To take the bit between his teeth, and fly
To the next headlong steep of anarchy.
Too happy England, if our good we knew,
Would we possess the freedom we pursue!
The lavish government can give no more;
Yet we repine, and plenty makes us poor.
God tried us once: our rebel-fathers fought;
He glutted 'em with all the power they sought:
Till, mastered by their own usurping brave,°
The freeborn subject sunk into a slave.   130
We loathe our manna, and we long for quails;°
Ah, what is man, when his own wish prevails!
How rash, how swift to plunge himself in ill;
Proud of his power, and boundless in his will!
That kings can do no wrong we must believe;

None can they do, and must they all receive?
Help, heaven! or sadly we shall see an hour,
When neither wrong nor right are in their power!
Already they have lost their best defence,
The benefit of laws which they dispense: 140
No justice to their righteous cause allowed;
But baffled by an arbitrary crowd;
And medals graved, their conquest to record,
The stamp and coin of their adopted lord.

The man who laughed but once, to see an ass°
Mumbling to make the crossgrained thistles pass,
Might laugh again, to see a jury chaw
The prickles of unpalatable law.
The witnesses that, leech-like, lived on blood,
Sucking for them were medicinally good;   150
But when they fastened on their festered sore,
Then justice and religion they forswore;
Their maiden oaths debauched into a whore.
Thus men are raised by factions, and decried;
And rogue and saint distinguished by their side.
They rack e'en scripture to confess their cause,
And plead a call to preach in spite of laws.
But that's no news to the poor injured page:
It has been used as ill in every age:
And is constrained, with patience, all to take;   160
For what defence can Greek and Hebrew make?
Happy who can this talking trumpet seize;
They make it speak whatever sense they please!
'Twas framed at first our oracle to enquire;
But since our sects in prophecy grow higher,
The text inspires not them, but they the text inspire.

London, thou great emporium of our isle,°
O thou too bounteous, thou too fruitful Nile!
How shall I praise or curse to thy desert;
Or separate thy sound from thy corrupted part!   170
I called thee Nile; the parallel will stand:
Thy tides of wealth o'erflow the fattened land;
Yet monsters from thy large increase we find,
Engendered on the slime thou leav'st behind.
Sedition has not wholly seized on thee,
Thy nobler parts are from infection free.

Of Israel's tribes thou has a numerous band,
But still the Canaanite is in the land.°
Thy military chiefs are brave and true,
Nor are thy disenchanted burghers few.   180
The head is loyal which thy heart commands,°
But what's a head with two such gouty hands?
The wise and wealthy love the surest way,
And are content to thrive and to obey.
But wisdom is to sloth too great a slave;
None are so busy as the fool and knave.
Those let me curse; what vengeance will they urge,
Whose ordures neither plague nor fire can purge;
Nor sharp experience can to duty bring,
Nor angry heaven, nor a forgiving king!   190
In Gospel-phrase their chapmen they betray;
Their shops are dens, the buyer is their prey.
The knack of trades is living on the spoil;
They boast, e'en when each other they beguile.
Customs to steal is such a trivial thing,
That 'tis their charter to defraud their king.
All hands unite of every jarring sect;
They cheat the country first, and then infect.
They for God's cause their monarchs dare dethrone,
And they'll be sure to make his cause their own.   200
Whether the plotting Jesuit laid the plan
Of murdering kings, or the French puritan,°
Our sacrilegious sects their guides outgo,
And kings and kingly power would murder too.

   What means their traitorous combination less,
Too plain to evade, too shameful to confess!
But treason is not owned when 'tis descried:
Successful crimes alone are justified.
The men who no conspiracy would find,
Who doubts but, had it taken, they had joined—   210
Joined in a mutual covenant of defence,
At first without, at last against their prince.
If sovereign right by sovereign power they scan,
The same bold maxim holds in God and man:
God were not safe, his thunder could they shun,
He should be forced to crown another son.
Thus, when the heir was from the vineyard thrown,°

The rich possession was the murderers' own.
In vain to sophistry they have recourse:
By proving theirs no plot, they prove 'tis worse;   220
Unmasked rebellion, and audacious force;
Which though not actual, yet all eyes may see
'Tis working in the immediate power to be;
For from pretended grievances they rise,
First to dislike, and after to despise;
Then, cyclop-like, in human flesh to deal,
Chop up a minister at every meal;
Perhaps not wholly to melt down the king,
But clip his regal rights within the ring;°
From thence t'assume the power of peace and war;   230
And ease him by degrees of public care.
Yet, to consult his dignity and fame,
He should have leave to exercise the name,
And hold the cards, while commons played the game.
For what can power give more than food and drink,
To live at ease, and not be bound to think?
These are the cooler methods of their crime,
But their hot zealots think 'tis loss of time;
On utmost bounds of loyalty they stand,
And grin and whet like a Croatian band,   240
That waits impatient for the last command.
Thus outlaws open villainy maintain,
They steal not, but in squadrons scour the plain;
And, if their power the passengers subdue,
The most have right, the wrong is in the few.
Such impious axioms foolishly they show,
For in some soils republics will not grow:
Our temperate isle will no extremes sustain
Of popular sway or arbitrary reign,
But slides between them both into the best,   250
Secure in freedom, in a monarch blessed;
And though the climate, vexed with various winds,
Works through our yielding bodies on our minds,
The wholesome tempest purges what it breeds,
To recommend the calmness that succeeds.
  But thou, the pander of the people's hearts,
(O crooked soul, and serpentine in arts!)
Whose blandishments a loyal land have whored,°

And broke the bonds she plighted to her lord;
What curses on thy blasted name will fall!   260
Which age to age their legacy shall call;
For all must curse the woes that must descend on all.
Religion thou hast none; thy mercury°
Has passed through every sect, or theirs through thee.
But what thou givest, that venom still remains;
And the poxed nation feels thee in their brains.
What else inspires the tongues and swells the breasts
Of all thy bellowing renegado priests,°
That preach up thee for God, dispense thy laws,
And with thy stum ferment their fainting cause,   270
Fresh fumes of madness raise, and toil and sweat
To make the formidable cripple great?
Yet should thy crimes succeed, should lawless power
Compass those ends thy greedy hopes devour,
Thy canting friends thy mortal foes would be,
Thy God and theirs will never long agree;
For thine (if thou hast any) must be one
That lets the world and humankind alone;
A jolly God, that passes hours too well
To promise heaven, or threaten us with hell;   280
That unconcerned can at rebellion sit,
And wink at crimes he did himself commit.
A tyrant theirs; the heaven their priesthood paints
A conventicle of gloomy sullen saints;
A heaven like Bedlam, slovenly and sad,°
Foredoomed for souls with false religion mad.
    Without a vision poets can foreshow
What all but fools by common sense may know;
If true succession from our isle should fail,
And crowds profane with impious arms prevail,   290
Not thou, nor those thy factious arts engage,
Shall reap that harvest of rebellious rage,
With which thou flatterest thy decrepit age.
The swelling poison of the several sects,
Which, wanting vent, the nation's health infects,
Shall burst its bag; and, fighting out their way,
The various venoms on each other prey.
The presbyter, puffed up with spiritual pride,
Shall on the necks of the lewd nobles ride,

His brethren damn, the civil power defy,   300
And parcel out republic prelacy.
But short shall be his reign: his rigid yoke
And tyrant power will puny sects provoke;
And frogs and toads, and all the tadpole train,°
Will croak to heaven for help from this devouring crane.
The cut-throat sword and clamorous gown shall jar,
In sharing their ill-gotten spoils of war;
Chiefs shall be grudged the part which they pretend;
Lords envy lords, and friends with every friend
About their impious merit shall contend.   310
The surly Commons shall respect deny,°
And jostle peerage out with property.
Their general either shall his trust betray,
And force the crowd to arbitrary sway;
Or they, suspecting his ambitious aim,
In hate of kings shall cast anew the frame;
And thrust out Collatine that bore their name.°
    Thus inborn broils the factions would engage,
Or wars of exiled heirs, or foreign rage,
Till halting vengeance overtook our age;   320
And our wild labours wearied into rest,
Reclined us on a rightful monarch's breast.

        —*pudet haec opprobria, vobis*°
        et dici potuisse, et non potuisse refelli.

Prologue to the Duchess
on her Return from Scotland

When factious rage to cruel exile drove
The queen of beauty, and the court of love,
The muses drooped, with their forsaken arts,
And the sad Cupids broke their useless darts.
Our fruitful plains to wilds and deserts turned,
Like Eden's face, when banished man it mourned:
Love was no more, when loyalty was gone,
The great supporter of his awful throne.
Love could no longer after beauty stay,
But wandered northward to the verge of day,   10

As if the sun and he had lost their way.
But now the illustrious nymph, returned again,
Brings every grace triumphant in her train.
The wondering nereids, though they raised no storm,
Foreslowed her passage, to behold her form:
Some cried, 'a Venus'; some, 'a Thetis passed';°
But this was not so fair, nor that so chaste.
Far from her sight flew faction, strife, and pride;
And envy did but look on her, and died.
Whate'er we suffered from our sullen fate,   20
Her sight is purchased at an easy rate.
Three gloomy years against this day were set;
But this one mighty sum has cleared the debt:
Like Joseph's dream, but with a better doom,°
The famine past, the plenty still to come.
For her, the weeping heavens become serene;
For her, the ground is clad in cheerful green;
For her, the nightingales are taught to sing,
And nature has for her delayed the spring.
The muse resumes her long-forgotten lays,   30
And love restored his ancient realm surveys,
Recalls our beauties, and revives our plays,
His waste dominions peoples once again,
And from her presence dates his second reign.
But awful charms on her fair forehead sit,
Dispensing what she never will admit;
Pleasing, yet cold, like Cynthia's silver beam,
The people's wonder, and the poet's theme.
Distempered zeal, sedition, cankered hate,
No more shall vex the church, and tear the state;   40
No more shall faction civil discords move,
Or only discords of too tender love:
Discord, like that of music's various parts;
Discord, that makes the harmony of hearts;
Discord, that only this dispute shall bring,
Who best shall love the duke, and serve the king.

RELIGIO LAICI

OR, A LAYMAN'S FAITH

A POEM

Ornari res ipsa negat, contenta doceri.
[my subject is content to be taught, and scorns adornment]°

THE PREFACE

A poem with so bold a title,° and a name prefixed from which the handling of so serious a subject would not be expected, may reasonably oblige the author to say somewhat in defence both of himself and of his undertaking. In the first place, if it be objected to me that, being a layman, I ought not to have concerned myself with speculations which belong to the profession of divinity, I could answer that perhaps laymen with equal advantages of parts and knowledge, are not the most incompetent judges of sacred things; but, in the due sense of my own weakness and want of learning, I plead not this; I pretend not to make myself a judge of faith in others, but only to make a confession of my own; I lay no unhallowed hand upon the ark,° but wait on it, with the reverence that becomes me, at a distance. In the next place I will ingenuously confess that the helps I have used in this small treatise were many of them taken from the works of our own reverend divines of the Church of England; so that the weapons with which I combat irreligion are already consecrated; though I suppose they may be taken down as lawfully as the sword of Goliah° was by David, when they are to be employed for the common cause, against the enemies of piety. I intend not by this to entitle them to° any of my errors, which yet, I hope, are only those of charity to mankind; and such as my own charity has caused me to commit, that of others may more easily excuse. Being naturally inclined to scepticism in philosophy, I have no reason to impose my opinions in a subject which is above it; but whatever they are, I submit them with all reverence to my mother church, accounting them no further mine, than as they are authorised, or at least uncondemned by her. And indeed, to secure myself on this side, I have used the necessary precaution of showing this paper before it was published to a judicious and learned friend, a man indefatigably zealous in the service of the church and state; and whose writings have highly deserved of both. He was pleased to approve the body of the discourse, and I hope he is more my friend than to do it out of complaisance. It is true, he had too good a taste to

like it all; and amongst some other faults recommended to my second
view what I have written, perhaps too boldly, on St. Athanasius,°
which he advised me wholly to omit. I am sensible enough that I had
done more prudently to have followed his opinion; but then I could
not have satisfied myself that I had done honestly not to have written
what was my own. It has always been my thought that heathens who
never did, nor without miracle could, hear of the name of Christ, were
yet in a possibility of salvation. Neither will it enter easily into my
belief that, before the coming of our Saviour, the whole world,
excepting only the Jewish nation, should lie under the inevitable
necessity of everlasting punishment, for want of that revelation which
was confined to so small a spot of ground as that of Palestine. Among
the sons of Noah we read of one only who was accursed;° and if a
blessing in the ripeness of time was reserved for Japhet (of whose
progeny we are), it seems unaccountable to me why so many
generations of the same offspring, as preceded our Saviour in the flesh,
should be all involved in one common condemnation, and yet that
their posterity should be entitled to the hopes of salvation: as if a bill of
exclusion had passed only on the fathers, which debarred not the sons
from their succession. Or that so many ages had been delivered over to
hell, and so many reserved for heaven, and that the devil had the first
choice, and God the next. Truly, I am apt to think that the revealed
religion which was taught by Noah to all his sons might continue for
some ages in the whole posterity. That afterwards it was included
wholly in the family of Sem is manifest; but when the progenies of
Cham and Japhet swarmed into colonies, and those colonies were
subdivided into many others, in process of time their descendants lost
by little and little the primitive and purer rites of divine worship,
retaining only the notion of one deity; to which succeeding generations
added others; for men took their degrees in those ages from
conquerors to gods. Revelation being thus eclipsed to almost all
mankind, the light of nature, as the next in dignity, was substituted;
and that is it which St Paul concludes to be the rule of the heathens,°
and by which they are hereafter to be judged. If my supposition be
true, then the consequence which I have assumed in my poem may be
also true; namely, that Deism, or the principles of natural worship, are
only the faint remnants of dying flames of revealed religion in the
posterity of Noah: and that our modern philosophers, nay, and some
of our philosophising divines, have too much exalted the faculties of
our souls, when they have maintained that by their force mankind has
been able to find out that there is one supreme agent or intellectual

being which we call God; that praise and prayer are his due worship; and the rest of those deducements, which I am confident are the remote effects of revelation, and unattainable by our discourse; I mean as simply considered, and without the benefit of divine illumination. So that we have not lifted up ourselves to God by the weak pinions of our reason, but he has been pleased to descend to us; and what Socrates said of him, what Plato writ, and the rest of the heathen philosophers of several nations, is all no more than the twilight of revelation, after the sun of it was set in the race of Noah. That there is something above us, some principle of motion, our reason can apprehend, though it cannot discover what it is, by its own virtue. And indeed it is very improbable that we, who by the strength of our faculties cannot enter into the knowledge of any being, not so much as of our own, should be able to find out by them that supreme nature, which we cannot otherwise define than by saying it is infinite; as if infinite were definable, or infinity a subject for our narrow understanding. They who would prove religion by reason do but weaken the cause which they endeavour to support: it is to take away the pillars from our faith, and to prop it only with a twig; it is to design a tower like that of Babel, which, if it were possible (as it is not) to reach heaven, would come to nothing by the confusion of the workmen. For every man is building a several way, impotently conceited of his own model and his own materials: reason is always striving, and always at a loss; and of necessity it must so come to pass, while it is exercised about that which is not its proper object. Let us be content at last to know God by his own methods; at least, so much of him as he is pleased to reveal to us in the sacred scriptures; to apprehend them to be the word of God is all our reason has to do, for all beyond it is the work of faith, which is the seal of heaven impressed upon our human understanding.

And now for what concerns the holy bishop Athanasius, the preface of whose creed seems inconsistent with my opinion, which is, that heathens may possibly be saved. In the first place I desire it may be considered that it is the preface only, not the creed itself, which (till I am better informed) is of too hard a digestion for my charity. It is not that I am ignorant how many several texts of scripture seemingly support that cause; but neither am I ignorant how all those texts may receive a kinder and more mollified interpretation. Every man who is read in church history knows that belief was drawn up after a long contestation with Arius° concerning the divinity of our blessed saviour, and his being one substance with the father; and that, thus compiled, it was sent abroad among the Christian churches, as a kind of test, which whosoever took was looked on as an orthodox believer. It is manifest from hence that the heathen part of the empire was not concerned in it; for its business was not to distinguish betwixt pagans and Christians, but betwixt heretics and true believers. This, well considered, takes off the heavy weight of censure, which I would willingly avoid from so venerable a man; for if this proportion, 'whosoever will be saved', be restrained only to those to whom it was intended, and for whom it was composed, I mean the Christians; then the anathema reaches not the heathens, who had never heard of Christ, and were nothing interested in that dispute. After all, I am far from blaming even that prefatory addition to the creed, and as far from cavilling at the continuation of it in the liturgy of the church, where, on the days appointed, it is publicly read: for I suppose there is the same reason for it now, in opposition to the Socinians,° as there was then against the Arians; the one being a heresy which seems to have been refined out of the other; and with how much more plausibility of reason it combats our religion, with so much more caution to be avoided; and therefore the prudence of our church is to be commended, which has interposed her authority for the recommendation of this creed. Yet to such as are grounded in the true belief, those explanatory creeds, the Nicene° and this of Athanasius, might perhaps be spared; for what is supernatural will always be a mystery in spite of exposition, and, for my own part, the plain Apostles' Creed is most suitable to my weak understanding, as the simplest diet is the most easy of digestion.

I have dwelt longer on this subject than I intended, and longer than, perhaps, I ought; for having laid down as my foundation that the scripture is a rule, that in all things needful to salvation it is clear, sufficient, and ordained by God Almighty for that purpose, I have left myself no right to interpret obscure places, such as concern the possibility of eternal happiness to heathens; because whatsoever is obscure is concluded not necessary to be known.

But, by asserting the scripture to be the canon of our faith, I have unavoidably created to myself two sorts of enemies: the papists indeed, more directly, because they have kept the scripture from us, what they could; and have reserved to themselves a right of interpreting what they have delivered under the pretence of infalli-bility: and the fanatics more collaterally, because they have assumed what amounts to an infallibility in the private spirit; and have detorted those texts of scripture which are not necessary to salvation, to the damnable uses of sedition, disturbance, and destruction of the civil government. To begin with the papists, and to speak freely, I think them the less dangerous, at least in appearance, to our present state; for not only the penal laws are in force against them, and their number is contemptible, but also their peerage and commons are excluded from parliaments, and consequently those laws in no probability of being repealed. A general and uninterrupted plot of their clergy, ever since the Reformation, I suppose all protestants believe; for it is not reasonable to think but that so many of their orders as were outed from their fat possessions would endeavour a re-entrance against those whom they account heretics. As for the late design, Mr Coleman's letters,° for aught I know, are the best evidence; and what they discover, without wire-drawing their sense, or malicious glosses, all men of reason conclude credible. If there be anything more than this required of me, I must believe it as well as I am able, in spite of the witness, and out of a decent conformity to the votes of parliament; for I suppose the fanatics will not allow the private spirit in this case. Here the infallibility is at least in one part of the government; and our understandings as well as our wills are represented. But to return to the Roman Catholics, how can we be secure from the practice of jesuited papists in that religion? For not two or three of that order, as some of them would impose upon us, but almost the whole body of them, are of opinion that their infallible master has a right over kings, not only in spirituals but temporals. Not to name Mariana, Bellarmine, Emanuel Sa, Molina, Santarel, Simancha,° and at least twenty others of foreign countries; we can produce, of our own nation, Campion, and Doleman or Parsons,° besides many are named whom I have not read, who all of them attest this doctrine, that the Pope can depose and give away the right of any sovereign prince, *si vel paulum deflexerit*, 'if he shall never so little warp'; but if he once comes to be excommunicated, then the bond of obedience is taken off from subjects; and they may and ought to drive him, like another Nebuchadnezzar,° *ex hominum Christianorum dominatu*, 'from exercising dominion over Christians'; and to this they are bound by virtue of divine precept, and by all the ties of conscience under no less penalty than damnation. If they answer me (as a learned priest has lately written) that this doctrine of the Jesuits is not *de fide* [an article of faith]; and that consequently they are not obliged by it, they must pardon me if I think they have said nothing to the purpose; for it is a maxim in their Church, where points of faith are not decided, and that doctors are of contrary opinions, they may follow which part they please; but more safely the most received and most authorised. And their champion Bellarmine° has told the world, in his *Apology*, that the king of England is a vassal to the Pope, *ratione directi dominii* [after the manner of feudal tenure], and that he holds in villeinage° of his Roman landlord. Which is no new claim put in for England. Our chronicles are his authentic witnesses that King John was deposed by the same plea, and Philip Augustus admitted tenant. And which makes the more for Bellarmine, the French king was again ejected when our king submitted to the church, and the crown received under the sordid condition of a vassalage.

It is not sufficient for the more moderate and well-meaning papists (of which I doubt not there are many) to produce the evidences of their loyalty to the late king, and to declare their innocency in this Plot.° I will grant their behaviour in the first to have been as loyal and as brave as they desire; and will be willing to hold them excused as to the second (I mean when it comes to my turn, and after my betters; for it is a madness to be sober alone, while the nation continues drunk); but that saying of their Father Cres° is still running in my head, that they may be dispensed with in their obedience to a heretic prince, while the necessity of the times shall oblige them to it: for that (as another of them tells us) is only the effect of Christian prudence; but when once they shall get power to shake him off, a heretic is no lawful king, and consequently to rise against him is no rebellion. I should be glad, therefore, that they would follow the advice which was charitably given them by a reverend prelate of our church; namely, that they would join in a public act of disowning and detesting those jesuitic principles; and subscribe to all doctrines which deny the Pope's authority of deposing kings, and releasing subjects from their oath of allegiance; to which I should think they might easily be induced, if it be true that this present Pope° has condemned the doctrine of king-killing (a thesis of the Jesuits) amongst others, *ex cathedra* (as they call it) or in open consistory.

Leaving them, therefore, in so fair a way (if they please themselves) of satisfying all reasonable men of their sincerity and good meaning to the government, I shall make bold to consider that other extreme of our religion, I mean the fanatics, or schismatics, of the English church. Since the Bible has been translated into our tongue, they have used it so, as if their business was not to be saved but to be damned by its contents. If we consider only them, better had it been for the English nation that it had still remained in the original Greek and Hebrew, or at least in the honest Latin of St Jerome, than that several texts in it should have been prevaricated to the destruction of that government which put it into so ungrateful hands.°

How many heresies the first translation of Tyndal° produced in few years, let my Lord Herbert's *Henry the Eighth*° inform you; insomuch that for the gross errors in it and the great mischiefs it occasioned, a sentence passed on the first edition of the Bible, too shameful almost to be repeated. After the short reign of Edward the Sixth (who had continued to carry on the Reformation on other principles than it was begun) everyone knows that not only the chief promoters of that work, but many others whose consciences would not dispense with popery, were forced, for fear of persecution, to change climates: from whence returning at the beginning of Queen Elizabeth's reign, many of them who had been in France, and at Geneva, brought back the rigid opinions and imperious discipline of Calvin, to graft upon our Reformation. Which, though they cunningly concealed at first (as well knowing how nauseously that drug would go down in a lawful monarchy, which was prescribed for a rebellious commonwealth), yet they always kept it in reserve; and were never wanting to themselves either in court or parliament, when either they had any prospect of a numerous party of fanatic members in the one, or the encouragement of any favourite in the other, whose covetousness was gaping at the patrimony of the church. They who will consult the works of our venerable Hooker,° or the account of his life, or more particularly the letter written to him on this subject by George Cranmer,° may see by what gradations they proceeded: from the dislike of cap and surplice, the very next step was admonitions to the parliament against the whole government ecclesiastical; then came out volumes in English and Latin in defence of their tenets; and immediately practices were set on foot to erect their discipline without authority. Those not succeeding, satire and railing was the next; and Martin Mar-prelate° (the Marvell° of those times) was the first presbyterian scribbler who sanctified libels and scurrility to the use of the Good Old Cause. Which was done (says my author) upon this account: that (their serious treatises having been fully answered and refuted) they might compass by railing what they had lost by reasoning; and when their cause was sunk in court and parliament, they might at least hedge in a stake amongst the rabble: for to their ignorance all things are wit which are abusive. But if church and state were made the theme, then the doctoral degree of wit was to be taken at Billingsgate:° even the most saintlike of the party, though they durst not excuse this contempt and vilifying of the government, yet were pleased, and grinned at it with a pious smile, and called it a judgment of God against the hierarchy. Thus sectaries, we may see, were born with teeth, foulmouthed and scurrilous from their infancy; and if spiritual pride, venom, violence, contempt of superiors, and slander, had been the marks of orthodox belief, the presbytery and the rest of our schismatics, which are their spawn, were always the most visible church in the Christian world.

It is true, the government was too strong at that time for a rebellion; but to show what proficiency they had made in Calvin's school, even *then* their mouths watered at it; for two of their gifted brotherhood, Hacket° and Coppinger,° as the story tells us, got up into a pease-cart and harangued the people, to dispose them to an insurrection, and to establish their discipline by force: so that, however it comes about that now they celebrate Queen Elizabeth's birth-night as that of their saint and patroness, yet then they were for doing the work of the Lord by arms against her; and, in all probability, they wanted but a fanatic lord mayor and two sheriffs of their party, to have compassed it.

Our venerable Hooker, after many admonitions which he had given them, toward the end of his preface breaks out into this prophetic speech: 'There is in every one of these considerations most just cause to fear, lest our hastiness to embrace a thing of so perilous consequence' (meaning the presbyterian discipline) 'should cause posterity to feel those evils, which as yet are more easy for us to prevent, than they would be for them to remedy.'

How fatally this Cassandra has foretold, we know too well by sad experience: the seeds were sown in the time of Queen Elizabeth, the bloody harvest ripened in the reign of King Charles the Martyr; and, because all the sheaves could not be carried off without shedding some of the loose grains, another crop is too like to follow; nay, I fear it is unavoidable if the conventiclers° be permitted still to scatter.

A man may be suffered to quote an adversary to our religion, when he speaks truth; and it is the observation of Maimbourg, in his *History of Calvinism*, that wherever that discipline was planted and embraced, rebellion, civil war, and misery attended it. And how indeed should it happen otherwise? Reformation of church and state has always been the ground of our divisions in England. While we were papists, our Holy Father rid us, by pretending authority out of the scriptures to depose princes (a doctrine which, though some papists may reject, no Pope has hitherto denied, nor ever will); when we shook off his authority, the sectaries furnished themselves with the same weapons; and out of the same magazine, the Bible: so that the scriptures, which are in themselves the greatest security of governors, as commanding express obedience to them, are now turned to their destruction; and never since the Reformation has there wanted a text of their interpreting to authorise a rebel. And it is to be noted by the way that the doctrines of king-killing and deposing, which have been taken up only by the worst party of the papists, the most frontless flatterers of the Pope's authority, have been espoused, defended, and are still maintained by the whole body of non-conformists and republicans. It is but dubbing themselves the people of God, which it is the interest of their preachers to tell them they are, and their own interest to believe; and after that, they cannot dip into the Bible, but one text or another will turn up for their purpose; if they are under persecution, as they call it, then that is a mark of their election; if they flourish, then God works miracles for their deliverance, and the saints are to possess the earth.

They may think themselves to be too roughly handled in this paper; but I, who know best how far I could have gone on this subject, must be bold to tell them they are spared: though at the same time I am not ignorant that they interpret the mildness of a writer to them, as they do the mercy of the government; in the one they think it fear, and conclude it weakness in the other. The best way for them to confute me is, as I before advised the Papists, to disclaim their principles and renounce their practices. We shall all be glad to think them true Englishmen when they obey the king, and true protestants when they conform to the church discipline.

It remains that I acquaint the reader that the verses were written for an ingenious young gentleman, my friend, upon his translation of the *Critical History of the Old Testament*, composed by the learned Father Simon:° the verses therefore are addressed to the translator of that work, and the style of them is, what it ought to be, epistolary.

If anyone be so lamentable a critic as to require the smoothness, the numbers, and the turn of heroic poetry in this poem, I must tell him that if he has not read Horace, I have studied him, and hope the style of his *Epistles* is not ill imitated here. The expressions of a poem designed purely for instruction ought to be plain and natural, and yet majestic; for here the poet is presumed to be a kind of lawgiver, and those three qualities which I have named are proper to the legislative style. The florid, elevated, and figurative way is for the passions; for love and hatred, fear and anger, are begotten in the soul by showing their objects out of their true proportion, either greater than the life, or less; but instruction is to be given by showing them what they naturally are. A man is to be cheated into passion, but to be reasoned into truth.

RELIGIO LAICI

Dim as the borrowed beams of moon and stars
To lonely, weary, wandering travellers,
Is reason to the soul; and, as on high
Those rolling fires discover but the sky,
Not light us here, so reason's glimmering ray
Was lent, not to assure our doubtful way,
But guide us upward to a better day.
And as those nightly tapers disappear,
When day's bright lord ascends our hemisphere;
So pale grows reason at religion's sight;   10
So dies, and so dissolves in supernatural light.
Some few, whose lamp shone brighter, have been led
From cause to cause, to nature's secret head;
And found that one first principle must be:
But what, or who, that UNIVERSAL HE;
Whether some soul encompassing this ball,
Unmade, unmoved, yet making, moving all;
Or various atoms' interfering dance°
Leapt into form, the noble work of chance;
Or this great all was from eternity;   20
Not e'en the Stagirite himself could see,°
And Epicurus guessed as well as he.°
As blindly groped they for a future state;
As rashly judged of providence and fate:
But least of all could their endeavours find
What most concerned the good of humankind;
For happiness was never to be found,
But vanished from 'em like enchanted ground.
One thought content the good to be enjoyed;
This every little accident destroyed;   30
The wiser madmen did for virtue toil,°
A thorny, or at best a barren soil;
In pleasure some their glutton souls would steep,
But found their line too short, the well too deep,
And leaky vessels which no bliss could keep.
Thus anxious thoughts in endless circles roll,
Without a centre where to fix the soul;
In this wild maze their vain endeavours end:

Opinions of the
several sects of
philosophers
concerning the
summum bonum.

How can the less the greater comprehend?
Or finite reason reach infinity?   40
For what could fathom God were more than he.

System of Deism.   The Deist thinks he stands on firmer ground;
Cries: 'εὕρηκα, the mighty secret's found:°
God is that spring of good; supreme and best;
We, made to serve, and in that service blessed.'
If so, some rules of worship must be given,
Distributed alike to all by heaven:
Else God were partial, and to some denied
The means his justice should for all provide.
This general worship is to praise and pray,   50
One part to borrow blessings, one to pay;
And when frail nature slides into offence,
The sacrifice for crimes is penitence.
Yet, since the effects of providence, we find,
Are variously dispensed to humankind;
That vice triumphs, and virtue suffers here
(A brand that sovereign justice cannot bear);
Our reason prompts us to a future state,
The last appeal from fortune and from fate:
Where God's all-righteous ways will be declared,   60
The bad meet punishment, the good reward.

Of revealed   Thus man by his own strength to heaven would soar,
religion.   And would not be obliged to God for more.
Vain, wretched creature, how art thou misled
To think thy wit these godlike notions bred!
These truths are not the product of thy mind,
But dropped from heaven, and of a nobler kind.
Revealed religion first informed thy sight,
And reason saw not, till faith sprung the light.
Hence all thy natural worship takes the source: 70
'Tis revelation what thou think'st discourse.
Else, how com'st thou to see these truths so clear,
Which so obscure to heathens did appear?
Not Plato these, nor Aristotle found;
Socrates.   Nor he whose wisdom oracles renowned.
Hast thou a wit so deep, or so sublime,
Or canst thou lower dive, or higher climb?
Canst thou, by reason, more of Godhead know
Than Plutarch, Seneca, or Cicero?

Those giant wits, in happier ages born,   80
When arms and arts did Greece and Rome adorn,
Knew no such system; no such piles could raise
Of natural worship, built on prayer and praise,
To one sole GOD:
Nor did remorse to expiate sin prescribe,
But slew their fellow creatures for a bribe:
The guiltless victim groaned for their offence,
And cruelty and blood was penitence.
If sheep and oxen could atone for men,
Ah! at how cheap a rate the rich might sin!   90
And great oppressors might heaven's wrath beguile,
By offering his own creatures for a spoil!
   Dar'st thou, poor worm, offend infinity?
And must the terms of peace be given by thee?
Then thou art justice in the last appeal:
Thy easy God instructs thee to rebel;
And, like a king remote, and weak, must take
What satisfaction thou art pleased to make.
   But if there be a power too just and strong
To wink at crimes, and bear unpunished wrong;   100
Look humbly upward, see his will disclose
The forfeit first, and then the fine impose:°
A mulct thy poverty could never pay,
Had not eternal wisdom found the way,
And with celestial wealth supplied thy store:
His justice makes the fine, his mercy quits the score.
See God descending in thy human frame;
The offended suffering in the offender's name;
All thy misdeeds to him imputed see,
And all his righteousness devolved on thee.   110
   For granting we have sinned and that the offence
Of man is made against omnipotence,
Some price that bears proportion must be paid,
And infinite with infinite be weighed.
See then the Deist lost: remorse for vice,
Not paid; or paid, inadequate in price:
What farther means can reason now direct,
Or what relief from human wit expect?
That shows us sick; and sadly are we sure
Still to be sick, till heaven reveal the cure: 120

If then heaven's will must needs be understood
(Which must, if we want cure, and heaven be good),
Let all records of will revealed be shown;
With scripture all in equal balance thrown,
And our one sacred book will be that one.
   Proof needs not here, for whether we compare
That impious, idle superstitious ware
Of rites, lustrations, offerings, which before,°
In various ages, various countries bore,
With Christian faith and virtues, we shall find   130
None answering the great ends of humankind,
But this one rule of life, that shows us best
How God may be appeased, and mortals blessed.
Whether from length of time its worth we draw,
The world is scarce more ancient than the law:
Heaven's early care prescribed for every age;
First, in the soul, and after, in the page.
Or, whether more abstractedly we look,
Or on the writers, or the written book,
Whence, but from heaven, could men unskilled in arts,   140
In several ages born, in several parts,
Weave such agreeing truths? or how, or why,
Should all conspire to cheat us with a lie?
Unasked their pains, ungrateful their advice,
Starving their gain, and martyrdom their price.
   If on the book itself we cast our view,
Concurrent heathens prove the story true,
The doctrine, miracles; which must convince,
For heaven in them appeals to human sense:
And though they prove not, they confirm the cause,   150
When what is taught agrees with nature's laws.
   Then for the style; majestic and divine,
It speaks no less than God in every line:
Commanding words; whose force is still the same
As the first *Fiat* that produced our frame.
All faiths beside or did by arms ascend,
Or sense indulged has made mankind their friend:
This only doctrine does our lusts oppose,
Unfed by nature's soil, in which it grows;
Cross to our interests, curbing sense and sin;   160
Oppressed without, and undermined within,

It thrives through pain, its own tormentors tires;
And with a stubborn patience still aspires.
To what can reason such effects assign,
Transcending nature, but to laws divine?
Which in that sacred volume are contained;
Sufficient, clear, and for that use ordained.

Objection of the
Deist.

But stay: the Deist here will urge anew,
No supernatural worship can be true;
Because a general law is that alone   170
Which must to all, and everywhere, be known:
A style so large as not this book can claim,
Nor aught that bears revealed religion's name.
'Tis said the sound of a Messiah's birth
Is gone through all the habitable earth;
But still that text must be confined alone
To what was then inhabited, and known:
And what provision could from thence accrue
To Indian souls, and worlds discovered new?
In other parts it helps, that, ages past,   180
The scriptures there were known, and were embraced,
Till sin spread once again the shades of night:
What's that to these who never saw the light?

The objection
answered.

Of all objections this indeed is chief
To startle reason, stagger frail belief:
We grant, 'tis true, that heaven from human sense
Has hid the secret paths of providence;
But boundless wisdom, boundless mercy, may
Find e'en for those bewildered souls a way:
If from his nature foes may pity claim,   190
Much more may strangers who ne'er heard his name.
And though no name be for salvation known,
But that of his eternal son's alone;
Who knows how far transcending goodness can
Extend the merits of that son to man?
Who knows what reasons may his mercy lead,
Or ignorance invincible may plead?
Not only charity bids hope the best,
But more the great apostle has expressed:°
That if the Gentiles (whom no law inspired)   200
By nature did what was by law required;
They, who the written rule had never known,

Were to themselves both rule and law alone:
To nature's plain indictment they shall plead,
And by their conscience be condemned or freed.
Most righteous doom! because a rule revealed
Is none to those from whom it was concealed.
Then those who followed reason's dictates right,
Lived up, and lifted high their natural light;
With Socrates may see their maker's face,   210
While thousand rubric-martyrs want a place.°

    Nor does it balk my charity, to find
The Egyptian bishop of another mind:°
For though his creed eternal truth contains,
'Tis hard for man to doom to endless pains
All who believed not all his zeal required,
Unless he first could prove he was inspired.
Then let us either think he meant to say
This faith, where published, was the only way;
Or else conclude that, Arius to confute,°   220
The good old man, too eager in dispute,
Flew high; and, as his Christian fury rose,
Damned all for heretics who durst oppose.

Digression to the translator of Father Simon's Critical History of the Old Testament.

    Thus far my charity this path has tried;
(A much unskilful, but well-meaning guide:)
Yet what they are, e'en these crude thoughts were bred
By reading that which better thou hast read:
Thy matchless author's work; which thou, my friend,
By well translating better dost commend:
Those youthful hours which of thy equals most   230
In toys have squandered, or in vice have lost,
Those hours hast thou to nobler use employed;
And the severe delights of truth enjoyed.
Witness this weighty book, in which appears
The crabbèd toil of many thoughtful years,
Spent by thy author in the sifting care
Of rabbins' old sophisticated ware
From gold divine; which he who well can sort
May afterwards make algebra a sport:
A treasure, which if country curates buy,   240
They Junius and Tremellius may defy;°
Save pains in various readings and translations,
And without Hebrew make most learned quotations:

A work so full with various learning fraught,
So nicely pondered, yet so strongly wrought,
As nature's height and art's last hand required;
As much as man could compass, uninspired.
Where we may see what errors have been made
Both in the copier's and translator's trade;
How Jewish, popish interests have prevailed,   250
And where infallibility has failed.

    For some, who have his secret meaning guessed,
Have found our author not too much a priest:
For fashion sake he seems to have recourse
To Pope, and Councils, and tradition's force;
But he that old traditions could subdue,
Could not but find the weakness of the new:
If Scripture, though derived from heavenly birth,
Has been but carelessly preserved on earth;
If God's own people, who of God before   260
Knew what we know, and had been promised more,
In fuller terms, of heaven's assisting care,
And who did neither time nor study spare
To keep this book untainted, unperplexed,
Let in gross errors to corrupt the text,
Omitted paragraphs, embroiled the sense,
With vain traditions stopped the gaping fence,
Which every common hand pulled up with ease;
What safety from such brushwood-helps as these?°
If written words from time are not secured,   270
How can we think have oral sounds endured?
Which thus transmitted, if one mouth has failed,
Immortal lies on ages are entailed;
And that some such have been, is proved too plain;
If we consider interest, church, and gain.

Of the infallibility
of tradition in
general.
    'O, but,' says one, 'tradition set aside,
Where can we hope for an unerring guide?
For since the original scripture has been lost,
All copies disagreeing, maimed the most,
Or Christian faith can have no certain ground,   280
Or truth in church tradition must be found.'

    Such an omniscient church we wish indeed;
'Twere worth both Testaments, and cast in the Creed:°
But if this mother be a guide so sure,

As can all doubts resolve, all truth secure,
Then her infallibility as well
Where copies are corrupt or lame can tell;
Restore lost canon with as little pains,
As truly explicate what still remains;
Which yet no Council dare pretend to do,   290
Unless like Esdras they could write it new:°
Strange confidence, still to interpret true,
Yet not be sure that all they have explained,
Is in the blessed original contained.
More safe, and much more modest 'tis, to say
God would not leave mankind without a way;
And that the scriptures, though not everywhere
Free from corruption, or entire, or clear,
Are uncorrupt, sufficient, clear, entire,
In all things which our needful faith require.   300
If others in the same glass better see,
'Tis for themselves they look, but not for me:
For *my* salvation must its doom receive,
Not from what *others* but what *I* believe.

Objection in behalf of tradition, urged by Father Simon.

Must all tradition then be set aside?
This to affirm were ignorance or pride.
Are there not many points, some needful sure
To saving faith, that scripture leaves obscure?
Which every sect will wrest a several way
(For what one sect interprets, all sects may).   310
We hold, and say we prove from scripture plain,
That Christ is God; the bold Socinian
From the same scripture urges he's but man.
Now what appeal can end the important suit?
Both parts talk loudly, but the rule is mute.
Shall I speak plain, and in a nation free
Assume an honest layman's liberty?
I think (according to my little skill,
To my own mother church submitting still)
That many have been saved, and many may,   320
Who never heard this question brought in play.
The unlettered Christian, who believes in gross,
Plods on to heaven, and ne'er is at a loss;
For the strait gate would be made straiter yet,°
Were none admitted there but men of wit.

The few by nature formed, with learning fraught,
Born to instruct, as others to be taught,
Must study well the sacred page, and see
Which doctrine, this, or that, does best agree
With the whole tenor of the work divine,   330
And plainliest points to heaven's revealed design;
Which exposition flows from genuine sense,
And which is forced by wit and eloquence.
Not that tradition's parts are useless here,
When general, old, disinterested and clear:
That ancient Fathers thus expound the page
Gives truth the reverend majesty of age;
Confirms its force, by biding every test;
For best authority's next rules are best.°
And still the nearer to the spring we go,   340
More limpid, more unsoiled the waters flow.
Thus, first traditions were a proof alone,
Could we be certain such they were, so known;
But since some flaws in long descent may be,
They make not truth, but probability.
E'en Arius and Pelagius durst provoke°
To what the centuries preceeding spoke.
Such difference is there in an oft-told tale;
But truth by its own sinews will prevail.
Tradition written therefore more commends   350
Authority, than what from voice descends;
And this, as perfect as its kind can be,
Rolls down to us the sacred history,
Which, from the universal church received,
Is tried, and after for itself believed.

The second
objection.
     The partial papists would infer from hence
Their church, in last resort, should judge the sense;
Answer to the
objection.
But first they would assume, with wondrous art,
Themselves to be the whole, who are but part
Of that vast frame, the church; yet grant they were   360
The handers down, can they from thence infer
A right to interpret? or would they alone
Who brought the present, claim it for their own?
The book's a common largesse to mankind,
Not more for them than every man designed;
The welcome news is in the letter found;

The carrier's not commissioned to expound.
It speaks itself, and what it does contain,
In all things needful to be known, is plain.
　　In times o'ergrown with rust and ignorance,   370
A gainful trade their clergy did advance;
When want of learning kept the laymen low,
And none but priests were authorized to know;
When what small knowledge was, in them did dwell,
And he a god who could but read or spell:
Then mother church did mightily prevail;
She parcelled out the Bible by retail;
But still expounded what she sold or gave,
To keep it in her power to damn and save:
Scripture was scarce, and, as the market went,   380
Poor laymen took salvation on content;
As needy men take money, good or bad:
God's word they had not, but the priest's they had.
Yet, whate'er false conveyances they made,
The lawyer still was certain to be paid.
In those dark times they learned their knack so well,
That by long use they grew infallible:
At last, a knowing age began to enquire
If they the book, or that did them inspire;
And, making narrower search, they found, though late,   390
That what they thought the priest's was their estate,
Taught by the will produced (the written word)
How long they had been cheated on record.
Then every man who saw the title fair
Claimed a child's part, and put in for a share;
Consulted soberly his private good,
And saved himself as cheap as e'er he could.
　　'Tis true, my friend (and far be flattery hence),
This good had full as bad a consequence:
The book thus put in every vulgar hand,   400
Which each presumed he best could understand,
The common rule was made the common prey,
And at the mercy of the rabble lay.
The tender page with horny fists was galled,
And he was gifted most that loudest bawled:
The Spirit gave the doctoral degree;
And every member of a Company

Was of his trade and of the Bible free.
Plain truths enough for needful use they found,
But men would still be itching to expound: 410
Each was ambitious of the obscurest place,
No measure ta'en from knowledge, all from Grace.
Study and pains were now no more their care;
Texts were explained by fasting and by prayer:
This was the fruit the private spirit brought,
Occasioned by great zeal and little thought.
While crowds unlearned, with rude devotion warm,
About the sacred viands buzz and swarm,
The fly-blown text creates a crawling brood,
And turns to maggots what was meant for food.   420
A thousand daily sects rise up and die;
A thousand more the perished race supply:
So all we make of heaven's discovered will
Is, not to have it, or to use it ill.
The danger's much the same; on several shelves
If others wreck us, or we wreck ourselves.
    What then remains, but, waiving each extreme,
The tides of ignorance and pride to stem?
Neither so rich a treasure to forgo,
Nor proudly seek beyond our power to know.   430
Faith is not built on disquisitions vain;
The things we must believe are few and plain:
But since men will believe more than they need,
And every man will make himself a creed,
In doubtful questions 'tis the safest way
To learn what unsuspected ancients say;
For 'tis not likely we should higher soar
In search of heaven, than all the church before;
Nor can we be deceived, unless we see
The scripture and the Fathers disagree.   440
If, after all, they stand suspected still
(For no man's faith depends upon his will),
'Tis some relief that points not clearly known
Without much hazard may be let alone:
And after hearing what our church can say,
If still our reason runs another way,
That private reason 'tis more just to curb,
Than by disputes the public peace disturb.

For points obscure are of small use to learn;
But common quiet is mankind's concern.   450
    Thus have I made my own opinions clear;
Yet neither praise expect, nor censure fear:
And this unpolished, rugged verse, I chose,
As fittest for discourse, and nearest prose;
For while from sacred truth I do not swerve,
Tom Sternhold's, or Tom Shadwell's rhymes will serve.°

Ovid's Elegies. *Book II. The Nineteenth Elegy*

If for thyself thou wilt not watch thy whore,
Watch her for me that I may love her more;
What comes with ease we nauseously receive,
Who but a sot would scorn to love with leave?
With hopes and fears my flames are blown up higher,
Make me despair and then I can desire.
Give me a jilt to tease my jealous mind,
Deceits are virtues in the female kind.
Corinna my fantastic humour knew,
Played trick for trick and kept herself still new;   10
She, that next night I might the sharper come,
Fell out with me, and sent me fasting home;
Or some pretence to lie alone would take,
Whene'er she pleased her head and teeth would ache,
Till having won me to the highest strain,
She took occasion to be sweet again.
With what a gust, ye gods, we then embraced!
How every kiss was dearer than the last!
    Thou whom I now adore, be edified,
Take care that I may often be denied.   20
Forget the promised hour, or feign some fright,
Make me lie rough on bulks each other night.°
These are the arts that best secure thy reign,
And this the food that must my fires maintain.
Gross easy love does like gross diet pall,
In queasy stomachs honey turns to gall.
Had Danae not been kept in brazen towers,°
Jove had not thought her worth his golden showers.

When Juno to a cow turned Io's shape,
The watchman helped her to a second leap.   30
Let him who loves an easy Whetstone whore°
Pluck leaves from trees, and drink the common shore.
The jilting harlot strikes the surest blow,
A truth which I by sad experience know.
The kind poor constant creature we despise,
Man but pursues the quarry while it flies.
　　But thou dull husband of a wife too fair,
Stand on thy guard, and watch the precious ware;
If creaking doors or barking dogs thou hear,
Or windows scratched, suspect a rival there;   40
An orange-wench would tempt thy wife abroad,
Kick her, for she's a letter-bearing bawd;
In short be jealous as the devil in Hell,
And set my wit on work to cheat thee well.
The sneaking city cuckold is my foe,
I scorn to strike but when he wards the blow.
Look to thy hits, and leave off thy conniving,
I'll be no drudge to any wittol living;°
I have been patient and foreborn thee long,
In hope thou wouldst not pocket up thy wrong;°   50
If no affront can rouse thee, understand
I'll take no more indulgence at thy hand.
What, ne'er to be forbid thy house and wife!
Damn him who loves to lead so dull a life.
Now I can neither sigh, nor whine, nor pray,
All those occasions thou hast ta'en away.
Why art thou so incorrigibly civil?
Do somewhat I may wish thee at the devil.
For shame be no accomplice in my treason,
A pimping husband is too much in reason.   60
　　Once more wear horns before I quite forsake her,
In hopes whereof I rest thy cuckold-maker.

Prologue to The University of Oxford

SPOKEN BY MR HART
AT THE ACTING OF *THE SILENT WOMAN*

What Greece, when learning flourished, only knew,
Athenian judges, you this day renew.
Here, too, are annual rites to Pallas done,
And here poetic prizes lost or won.
Methinks I see you, crowned with olives, sit,
And strike a sacred horror from the pit.
A day of doom is this of your decree,
Where e'en the best are but by mercy free;
A day, which none but Jonson durst have wished to see.
Here they, who long have known the useful stage,   10
Come to be taught themselves to teach the age.
As your commissioners our poets go,
To cultivate the virtue which you sow;
In your Lyceum first themselves refined,°
And delegated thence to humankind.
But as ambassadors, when long from home,
For new instructions to their princes come,
So poets, who your precepts have forgot,
Return, and beg they may be better taught:
Follies and faults elsewhere by them are shown,   20
But by your manners they correct their own.
The illiterate writer, emp'ric-like, applies
To minds diseased, unsafe chance remedies:
The learned in schools, where knowledge first began.
Studies with care the anatomy of man;
Sees virtue, vice, and passions in their cause,
And fame from science, not from fortune, draws;
So poetry, which is in Oxford made
An art, in London only is a trade.
There haughty dunces, whose unlearned pen   30
Could ne'er spell grammar, would be reading men.
Such build their poems the Lucretian way;°
So many huddled atoms make a play;
And if they hit in order by some chance,
They call that nature, which is ignorance.

To such a fame let mere town-wits aspire,
And their gay nonsense their own cits admire.
Our poet, could he find forgiveness here,
Would wish it rather than a plaudit there.
He owns no crown from those Praetorian bands,°   40
But knows *that* right is in this senate's hands.
Not impudent enough to hope your praise,
Low at the muses' feet his wreath he lays,
And, where he took it up, resigns his bays.
Kings make their poets whom themselves think fit,
But 'tis your suffrage makes authentic wit.

Epilogue to Oxford

Oft has our poet wished this happy seat
Might prove his fading muse's last retreat:
I wondered at his wish, but now I find
He sought for quiet, and content of mind;
Which noiseful towns, and courts, can never know,
And only in the shades like laurels grow.
Youth, ere it sees the world, here studies rest,
And age returning thence concludes it best.
What wonder if we court that happiness
Yearly to share, which hourly you possess,   10
Teaching e'en you, while the vexed world we show,
Your peace to value more, and better know?
'Tis all we can return for favours past,°
Whose holy memory shall ever last,
For patronage from him whose care presides
O'er every noble art, and every science guides:
Bathurst, a name the learned with reverence know,°
And scarcely more to his own Virgil owe;
Whose age enjoys but what his youth deserved,
To rule those muses whom before he served.   20
His learning, and untainted manners too,
We find, Athenians, are derived to you:
Such ancient hospitality there rests
In yours, as dwelt in the first Grecian breasts,
Whose kindness was religion to their guests.

Such modesty did to our sex appear,
As, had there been no laws, we need not fear,
Since each of you was our protector here.
Converse so chaste, and so strict virtue shown,
As might Apollo with the muses own.   30
Till our return, we must despair to find
Judges so just, so knowing, and so kind.

Prologue to The University of Oxford

Though actors cannot much of learning boast,
Of all who want it, we admire it most.
We love the praises of a learned pit,
As we remotely are allied to wit.
We speak our poet's wit, and trade in ore,
Like those who touch upon the golden shore;
Betwixt our judges can distinction make,
Discern how much, and why our poems take;
Mark if the fools, or men of sense, rejoice;
Whether the applause be only sound or voice.   10
When our fop gallants, or our city folly,
Clap over-loud, it makes us melancholy:
We doubt that scene which does their wonder raise,
And, for their ignorance, contemn their praise.
Judge, then, if we who act, and they who write,
Should not be proud of giving you delight.
London likes grossly; but this nicer pit°
Examines, fathoms, all the depths of wit;
The ready finger lays on every blot;
Knows what should justly please, and what should not.   20
Nature herself lies open to your view;
You judge by her what draft of her is true,
Where outlines false, and colours seem too faint,
Where bunglers daub, and where true poets paint.
But by the sacred genius of this place,
By every muse, by each domestic grace,
Be kind to wit, which but endeavours well,
And, where you judge, presumes not to excel.
Our poets hither for adoption come,
As nations sued to be made free of Rome: 30

Not in the suffragating tribes to stand,
But in your utmost, last, provincial band.
If his ambition may those hopes pursue,
Who with religion loves your arts and you,
Oxford to him a dearer name shall be
Than his own mother-university.°
Thebes did his green, unknowing youth engage;
He chooses Athens in his riper age.

Prologue Spoken at the Opening
of the New House

26 March, 1674

A plain-built house, after so long a stay,
Will send you half unsatisfied away;
When, fallen from your expected pomp, you find
A bare convenience only is designed.
You who each day can theatres behold,
Like Nero's palace, shining all with gold,°
Our mean ungilded stage will scorn, we fear,
And, for the homely room, disdain the cheer.
Yet now cheap druggets to a mode are grown,°
And a plain suit, since we can make but one,   10
Is better than to be by tarnished gaudery known.
They who are by your favours wealthy made,
With mighty sums may carry on the trade;
We broken bankers half destroyed by fire,
With our small stock to humble roofs retire;
Pity our loss, while you their pomp admire.
For fame and honour we no longer strive;
We yield in both, and only beg to live;
Unable to support their vast expense,
Who build and treat with such magnificence,   20
That like the ambitious monarchs of the age,
They give the law to our provincial stage.
Great neighbours enviously promote excess,
While they impose their splendour on the less;
But only fools, and they of vast estate,
The extremity of modes will imitate,

The dangling knee-fringe, and the bib-cravat.
Yet if some pride with want may be allowed,
We in our plainness may be justly proud;
Our royal master willed it should be so;   30
Whate'er he's pleased to own, can need no show:
That sacred name gives ornament and grace,
And, like his stamp, makes basest metal pass.
'Twere folly now a stately pile to raise,
To build a playhouse while you throw down plays;
Whilst scenes, machines, and empty operas reign,
And for the pencil you the pen disdain;
While troupes of famished Frenchmen hither drive,
And laugh at those upon whose alms they live:
Old English authors vanish, and give place   40
To these new conquerors of the Norman race.
More tamely than your fathers you submit;
You're now grown vassals to them in your wit.
Mark, when they play, how our fine fops advance
The mighty merits of these men of France,
Keep time, cry *ben!* and humour the *cadence*.°
Well, please yourselves; but sure 'tis understood
That French machines have ne'er done England good.°
I would not prophesy our house's fate:
But while vain shows and scenes you overrate,   50
'Tis to be feared—
That, as a fire the former house o'erthrew,
Machines and tempests will destroy the new.°

To the Memory of Mr Oldham

Farewell, too little and too lately known,
Whom I began to think and call my own;
For sure our souls were near allied; and thine
Cast in the same poetic mould with mine.
One common note on either lyre did strike,
And knaves and fools we both abhorred alike:
To the same goal did both our studies drive,
The last set out the soonest did arrive.
Thus Nisus fell upon the slippery place,°
While his young friend performed and won the race.   10

O early ripe—to thy abundant store
What could advancing age have added more?
It might (what nature never gives the young)
Have taught the numbers of thy native tongue.°
But satire needs not those, and wit will shine
Through the harsh cadence of a rugged line:
A noble error, and but seldom made,
When poets are by too much force betrayed.
Thy generous fruits, though gathered ere their prime
Still showed a quickness; and maturing time   20
But mellows what we write to the dull sweets of rhyme.
Once more, hail and farewell; farewell thou young
But ah too short, Marcellus of our tongue;°
Thy brows with ivy, and with laurels bound;
But fate and gloomy night encompass thee around.

Preface to Sylvae, or The Second Part of Poetical Miscellanies

For this last half year I have been troubled with the disease (as I may call it) of translation; the cold prose fits of it (which are always the most tedious with me) were spent in the *History of the League;*° the hot (which succeeded them) in this volume of Verse Miscellanies. The truth is, I fancied to myself a kind of ease in the change of the paroxysm; never suspecting but that the humour would have wasted itself in two or three Pastorals of Theocritus, and as many Odes of Horace. But finding, or at least thinking I found, something that was more pleasing in them, than my ordinary productions, I encouraged myself to renew my old acquaintance with Lucretius and Virgil; and immediately fixed upon some parts of them which had most affected me in the reading. These were my natural impulses for the undertaking. But there was an accidental motive, which was full as forcible, and God forgive him who was the occasion of it. It was my Lord Roscommon's° *Essay on translated Verse,* which made me uneasy till I tried whether or no I was capable of following his rules, and of reducing the speculation into practice. For many a fair precept in poetry, is like a seeming demonstration in the mathematics; very specious° in the diagram, but failing in the mechanic operation. I think I have generally observed his instructions; I am sure my reason is sufficiently convinced both of their truth and usefulness; which, in other words, is to confess no less a vanity than to pretend that I have at least in some places made examples to his rules. Yet withal, I must acknowledge, that I have many times exceeded my commission; for I have both added and omitted, and even sometimes very boldly made such expositions of my authors, as no Dutch commentator will forgive me. Perhaps, in such particular passages, I have thought that I discovered some beauty yet undiscovered by those pedants, which none but a poet could have found. Where I have taken away some of their expressions, and cut them shorter, it may possibly be on this consideration, that what was beautiful in the Greek or Latin, would not appear so shining in the English. And where I have enlarged them, I desire the false critics would not always think that those thoughts are wholly mine, but that either they are secretly in the poet, or may be fairly deduced from him: or at least, if both those considerations should fail, that my own is of a piece with his, and that if he were living, and an Englishman, they are such, as he would probably have written.

For, after all, a translator is to make his author appear as charming as possibly he can, provided he maintains his character, and makes him not unlike himself. Translation is a kind of drawing after the life; where everyone will acknowledge there is a double sort of likeness, a good one and a bad. 'Tis one thing to draw the outlines true, the features like, the proportions exact, the colouring itself perhaps tolerable, and another thing to make all these graceful, by the posture, the shadowings, and chiefly by the spirit which animates the whole. I cannot without some indignation, look on an ill copy of an excellent original. Much less can I behold with patience Virgil, Homer, and some others, whose beauties I have been endeavouring all my life to imitate, so abused, as I may say to their faces by a botching interpreter. What English readers unacquainted with Greek or Latin will believe me or any other man, when we commend those authors, and confess we derive all that is pardonable in us from their fountains, if they take those to be the same poets, whom our Ogilbies° have translated? But I dare assure them, that a good poet is no more like himself in a dull translation, than his carcass would be to his living body. There are many who understand Greek and Latin, and yet are ignorant of their mother tongue. The proprieties and delicacies of the English are known to few; 'tis impossible even for a good wit to understand and practice them without the help of a liberal education, long reading, and digesting of those few good authors we have amongst us, the knowledge of men and manners, the freedom of habitudes and conversation with the best company of both sexes; and in short, without wearing off the rust which he contracted, while he was laying in a stock of learning. Thus difficult it is to understand the purity of English, and critically to discern not only good writers from bad, and a proper style from a corrupt, but also to distinguish that which is pure in a good author, from that which is vicious and corrupt in him. And for want of all these requisites, or the greatest part of them, most of our ingenious young men take up some cried up English poet for their model, adore him, and imitate him as they think, without knowing wherein he is defective, where he is boyish and trifling, wherein either his thoughts are improper to his subject, or his expressions unworthy of his thoughts, or the turn of both is unharmonious. Thus it appears necessary that a man should be a nice critic in his mother tongue, before he attempts to translate a foreign language. Neither is it sufficient that he be able to judge of words and style; but he must be a master of them too. He must perfectly understand his author's tongue, and absolutely command his own, so that to be a thorough translator, he must be a thorough poet. Neither is it enough to give his author's sense, in good English, in poetical expressions, and in musical numbers. For, though all these are exceeding difficult to perform, there yet remains a harder task; and 'tis a secret of which few translators have sufficiently thought. I have already hinted a word or two concerning it; that is, the maintaining the character of an author, which distinguishes him from all others, and makes him appear that individual poet whom you would interpret. For example, not only the thoughts, but the style and versification of Virgil and Ovid, are very different. Yet I see, even in our best poets, who have translated some parts of them, that they have confounded their several talents; and by endeavouring only at the sweetness and harmony of numbers, have made them both so much alike, that if I did not know the originals, I should never be able to judge by the copies, which was Virgil, and which was Ovid. It was objected against a late noble painter,° that he drew many graceful pictures, but few of them were like. And this happened to him, because he always studied himself more than those who sat to him. In such translators I can easily distinguish the hand which performed the work, but I cannot distinguish their poet from another. Suppose two authors are equally sweet, yet there is a great distinction to be made in sweetness, as in that of sugar, and that of honey. I can make the difference more plain, by giving you (if it be worth knowing) my own method of proceeding

in my translations out of four several poets in this volume; Virgil,
Theocritus, Lucretius and Horace. In each of these, before I
undertook them, I considered the genius and distinguishing character
of my author. I looked on Virgil, as a succinct and grave majestic
writer; one who weighed not only every thought, but every word and
syllable; who was still aiming to crowd his sense into as narrow a
compass as possibly he could, for which reason he is so very figurative,
that he requires, (I may almost say) a grammar apart to construe him.
His verse is everywhere sounding the very thing in your ears, whose
sense it bears, yet the numbers are perpetually varied, to increase the
delight of the reader; so that the same sounds are never repeated twice
together. On the contrary, Ovid and Claudian, though they write in
styles differing from each other, yet have each of them but one sort of
music in their verses. All the versification, and little variety of
Claudian, is included within the compass of four or five lines, and then
he begins again in the same tenor; perpetually closing his sense at the
end of a verse, and that verse commonly which they call golden, or two
substantives and two adjectives with a verb betwixt them to keep the
peace. Ovid with all his sweetness, has as little variety of numbers and
sound as he: he is always as it were upon the hand-gallop,° and his
verse runs upon carpet ground.° He avoids like the other all
synaloephas,° or cutting off one vowel when it comes before another in
the following word, so that minding only smoothness, he wants both
variety and majesty. But to return to Virgil, though he is smooth
where smoothness is required, yet he is so far from affecting it, that he
seems rather to disdain it; frequently makes use of synaloephas, and
concludes his sense in the middle of his verse. He is everywhere above
conceits of epigrammatic wit, and gross hyperboles. He maintains
majesty in the midst of plainness; he shines, but glares not; and is
stately without ambition, which is the vice of Lucan. I drew my
definition of poetical wit from my particular consideration of him. For
propriety of thoughts and words are only to be found in him; and
where they are proper, they will be delightful. Pleasure follows of
necessity, as the effect does the cause; and therefore is not to be put
into the definition. This exact propriety of Virgil I particularly
regarded as a great part of his character; but must confess to my
shame, that I have not been able to translate any part of him so well as
to make him appear wholly like himself. For where the original is
close, no version can reach it in the same compass. Hannibal Caro's in
the Italian,° is the nearest, the most poetical, and the most sonorous of
any translation of the *Aeneid*; yet, though he takes the advantage of

blank verse, he commonly allows two lines for one of Virgil, and does not always hit his sense. Tasso tells us in his letters,° that Sperone Speroni, a great Italian wit, who was his contemporary, observed of Virgil and Tully that the Latin orator endeavoured to imitate the copiousness of Homer the Greek poet; and that the Latin poet, made it his business to reach the conciseness of Demosthenes the Greek orator. Virgil therefore being so very sparing of his words, and leaving so much to be imagined by the reader, can never be translated as he ought, in any modern tongue. To make him copious is to alter his character; and to translate him line for line is impossible; because the Latin is naturally a more succinct language, than either the Italian, Spanish, French, or even than the English, (which by reason of its monosyllables is far the most compendious of them). Virgil is much the closest of any Roman poet, and the Latin hexameter, has more feet than the English heroic.

Besides all this, an author has the choice of his own thoughts and words, which a translator has not; he is confined by the sense of the inventor to those expressions which are the nearest to it, so that Virgil studying brevity, and having the command of his own language, could bring those words into a narrow compass, which a translator cannot render without circumlocutions. In short they who have called him the torture of grammarians, might also have called him the plague of translators; for he seems to have studied not to be translated. I own that endeavouring to turn his Nisus and Euryalus as close as I was able, I have performed that episode too literally; that giving more scope to Mezentius and Lausus,° that version which has more of the majesty of Virgil, has less of his conciseness; and all that I can promise for myself, is only that I have done both, better than Ogilby, and perhaps as well as Caro. So that methinks I come like a malefactor, to make a speech upon the gallows, and to warn all other poets, by my sad example, from the sacrilege of translating Virgil. Yet, by considering him so carefully as I did before my attempt, I have made some faint resemblance of him; and had I taken more time, might possibly have succeeded better; but never so well, as to have satisfied myself.

He who excels all other poets in his own language, were it possible to do him right, must appear above them in our tongue, which, as my Lord Roscommon justly observes,° approaches nearest to the Roman in its majesty: nearest indeed, but with a vast interval betwixt them. There is an inimitable grace in Virgil's words, and in them principally consists that beauty, which gives so unexpressible a pleasure to him who best understands their force; this diction of his, I must once again say, is never to be copied, and since it cannot, he will appear but lame in the best translation. The turns° of his verse, his breakings,° his propriety, his numbers, and his gravity, I have as far imitated, as the poverty of our language, and the hastiness of my performance would allow. I may seem sometimes to have varied from his sense; but I think the greatest variations may be fairly deduced from him; and where I leave his commentators, it may be I understand him better. At least I writ without consulting them in many places. But two particular lines in Mezentius and Lausus, I cannot so easily excuse; they are indeed remotely allied to Virgil's sense; but they are too like the trifling tenderness of Ovid; and were printed before I had considered them enough to alter them. The first of them I have forgotten, and cannot easily retrieve, because the copy is at the press. The second is this;

> When Lausus died, I was already slain.°

This appears pretty enough at first sight, but I am convinced for many reasons, that the expression is too bold, that Virgil would not have said it, though Ovid would. The reader may pardon it, if he please, for the freeness of the confession; and instead of that, and the former, admit these two lines which are more according to the author,

> Nor ask I life, nor fought with that design;
> As I had used my fortune, use thou thine.

Having with much ado got clear of Virgil, I have in the next place to consider the genius of Lucretius, whom I have translated more happily in those parts of him which I undertook. If he was not of the best age of Roman poetry, he was at least of that which preceded it; and he himself refined it to that degree of perfection, both in the language and the thoughts, that he left an easy task to Virgil, who as he succeeded him in time, so he copied his excellencies: for the method of the *Georgics* is plainly derived from him. Lucretius had chosen a subject naturally crabbed; he therefore adorned it with poetical descriptions, and precepts of morality, in the beginning and ending of his books. Which you see Virgil has imitated with great success, in those four books,° which in my opinion are more perfect in their kind, than even his divine *Aeneid*. The turn of his verse he has likewise followed, in those places which Lucretius has most laboured, and some of his very lines he has transplanted into his own works, without much variation. If I am not mistaken, the distinguishing character of Lucretius (I mean of his soul and genius) is a certain kind of noble

pride, and positive assertion of his opinions. He is everywhere confident of his own reason, and assuming an absolute command not only over his vulgar reader, but even his patron Memmius. For he is always bidding him attend, as if he had the rod over him; and using a magisterial authority, while he instructs him. From his time to ours, I know none so like him, as our poet and philosopher of Malmesbury.°
This is that perpetual dictatorship, which is exercised by Lucretius; who though often in the wrong, yet seems to deal *bona fide* [in good faith] with his reader, and tells him nothing but what he thinks; in which plain sincerity, I believe he differs from our Hobbes, who could not but be convinced, or at least doubt of some eternal truths which he has opposed. But for Lucretius, he seems to disdain all manner of replies, and is so confident of his cause, that he is beforehand with his antagonists; urging for them whatever he imagined they could say, and leaving them as he supposes, without an objection for the future. All this too, with so much scorn and indignation, as if he were assured of the triumph before he entered into the lists. From this sublime and daring genius of his, it must of necessity come to pass that his thoughts must be masculine, full of argumentation, and that sufficiently warm. From the same fiery temper proceeds the loftiness of his expressions, and the perpetual torrent of his verse, where the barrenness of his subject does not too much constrain the quickness of his fancy. For there is no doubt to be made, but that he could have been everywhere as poetical, as he is in his descriptions, and in the moral part of his philosophy, if he had not aimed more to instruct in his system of nature, than to delight. But he was bent upon making Memmius a materialist, and teaching him to defy an invisible power. In short, he was so much an atheist, that he forgot sometimes to be a poet. These are the considerations which I had of that author, before I attempted to translate some parts of him. And accordingly I laid by my natural diffidence and scepticism for a while, to take up that dogmatical way of his, which as I said, is so much his character, as to make him that individual poet. As for his opinions concerning the mortality of the soul, they are so absurd, that I cannot if I would believe them. I think a future state demonstrable even by natural arguments; at least to take away rewards and punishments is only a pleasing prospect to a man who resolves beforehand not to live morally. But on the other side, the thought of being nothing after death is a burden insupportable to a virtuous man, even though a heathen. We naturally aim at happiness, and cannot bear to have it confined to the shortness of our present being, especially when we

consider that virtue is generally unhappy in this world, and vice fortunate. So that 'tis hope of futurity alone, that makes this life tolerable, in expectation of a better. Who would not commit all the excesses to which he is prompted by his natural inclinations, if he may do them with security while he is alive, and be incapable of punishment after he is dead! If he be cunning and secret enough to avoid the laws, there is no band of morality to restrain him, for fame and reputation are weak ties; many men have not the least sense of them: powerful men are only awed by them, as they conduce to their interest, and that not always when a passion is predominant; and no man will be contained within the bounds of duty when he may safely transgress them. These are my thoughts abstractedly, and without entering into the notions of our Christian faith, which is the proper business of divines.

But there are other arguments in this poem (which I have turned into English), not belonging to the mortality of the soul, which are strong enough to a reasonable man, to make him less in love with life, and consequently in less apprehension of death. Such as are the natural satiety, proceeding from a perpetual enjoyment of the same things; the inconveniencies of old age, which make him incapable of corporeal pleasures; the decay of understanding and memory, which render him contemptible and useless to others; these and many other reasons so pathetically urged, so beautifully expressed, so adorned with examples, and so admirably raised by the prosopopeia° of nature, who is brought in speaking to her children, with so much authority and vigour, deserve the pains I have taken with them, which I hope have not been unsuccessful, or unworthy of my author. At least I must take the liberty to own, that I was pleased with my own endeavours, which but rarely happens to me, and that I am not dissatisfied upon the review, of anything I have done in this author.

'Tis true, there is something, and that of some moment, to be objected against my Englishing the nature of love, from the fourth book of Lucretius. And I can less easily answer why I translated it, than why I thus translated it. The objection arises from the obscenity of the subject; which is aggravated by the too lively and alluring delicacy of the verses. In the first place, without the least formality of an excuse, I own it pleased me: and let my enemies make the worst they can of this confession; I am not yet so secure from that passion, but that I want my author's antidotes against it. He has given the truest and most philosophical account both of the disease and remedy, which I ever found in any author: for which reasons I translated him.

But it will be asked why I turned him into this luscious English (for I will not give it a worse word). Instead of an answer, I would ask again of my supercilious adversaries, whether I am not bound when I translate an author, to do him all the right I can, and to translate him to the best advantage? If to mince his meaning, which I am satisfied was honest and instructive, I had either omitted some part of what he said, or taken from the strength of his expression, I certainly had wronged him; and that freeness of thought and words being thus cashiered,° in my hands he had no longer been Lucretius. If nothing of this kind be to be read, physicians must not study nature, anatomies must not be seen, and somewhat I could say of particular passages in books, which to avoid profaneness I do not name. But the intention qualifies the act; and both mine and my author's were to instruct as well as please. 'Tis most certain that bare-faced bawdry is the poorest pretence to wit imaginable: if I should say otherwise, I should have two great authorities against me. The one is the *Essay on Poetry*,° which I publicly valued before I knew the author of it, and with the commendation of which, my Lord Roscommon so happily begins° his *Essay on Translated Verse*. The other is no less than our admired Cowley; who says the same thing in other words: for in his Ode concerning Wit, he writes thus of it;

> Much less can that have any place
> At which a virgin hides her face:
> Such dross the fire must purge away; 'tis just
> The author blush, there where the reader must.°

Here indeed Mr Cowley goes further than the *Essay*; for he asserts plainly that obscenity has no place in wit; the other only says, 'tis a poor pretence to it, or an ill sort of wit, which has nothing more to support it than bare-faced ribaldry, which is both unmannerly in itself, and fulsome° to the reader. But neither of these will reach my case, for in the first place, I am only the translator, not the inventor; so that the heaviest part of the censure falls upon Lucretius, before it reaches me: in the next place, neither he nor I have used the grossest words; but the cleanliest metaphors we could find, to palliate the broadness of the meaning; and to conclude, have carried the poetical part no further, than the philosophical exacted.

There is one mistake of mine which I will not lay to the printer's charge, who has enough to answer for in false pointings: 'tis in the word 'viper': I would have the verse run thus,

> The scorpion, Love, must on the wound be bruised.°

There are a sort of blundering half-witted people, who make a great deal of noise about a verbal slip; though Horace would instruct them better in true criticism. *Non ego paucis, offendar maculis quas aut incuria fudit, aut humana parum cavit natura* [I shall not quarrel with those faults which were caused either by carelessness or human weakness].° True judgement in poetry, like that in painting, takes a view of the whole together, whether it be good or not; and where the beauties are more than the faults, concludes for the poet against the little judge; 'tis a sign that malice is hard driven, when 'tis forced to lay hold on a word or syllable; to arraign a man is one thing, and to cavil at him is another. In the midst of an ill-natured generation of scribblers, there is always justice enough left in mankind to protect good writers. And they too are obliged, both by humanity and interest, to espouse each other's cause against false critics, who are the common enemies. This last consideration puts me in mind of what I owe to the ingenious and learned translator of Lucretius;° I have not here designed to rob him of any part of that commendation, which he has so justly acquired by the whole author, whose fragments only fall to my portion. What I have now performed, is no more than I intended above twenty years ago. The ways of our translation are very different; he follows him more closely than I have done, which became an interpreter of the whole poem. I take more liberty, because it best suited with my design, which was to make him as pleasing as I could. He had been too voluminous had he used my method in so long a work, and I had certainly taken his, had I made it my business to translate the whole. The preference then is justly his, and I join with Mr Evelyn° in the confession of it, with this additional advantage to him; that his reputation is already established in this poet, mine is to make its fortune in the world. If I have been anywhere obscure in following our common author or if Lucretius himself is to be condemned, I refer myself to his excellent annotations, which I have often read, and always with some new pleasure.

My preface begins already to swell upon me, and looks as if I were afraid of my reader, by so tedious a bespeaking of him; and yet I have Horace and Theocritus upon my hands; but the Greek gentleman shall quickly be dispatched, because I have more business with the Roman.

That which distinguishes Theocritus from all other poets, both Greek and Latin, and which raises him even above Virgil in his *Eclogues*, is the inimitable tenderness of his passions; and the natural expression of them in words so becoming of a pastoral. A simplicity shines through all he writes: he shows his art and learning by disguising both. His shepherds never rise above their country education in their complaints of love. There is the same difference betwixt him and Virgil, as there is betwixt Tasso's *Aminta*, and the *Pastor Fido* of Guarini. Virgil's shepherds are too well read in the philosophy of Epicurus and of Plato; and Guarini's seem to have been bred in courts. But Theocritus and Tasso, have taken theirs from cottages and plains. It was said of Tasso, in relation to his similitudes, *mai esce dal bosco*:° that 'he never departed from the woods', that is, all his comparisons were taken from the country. The same may be said of our Theocritus; he is softer than Ovid, he touches the passions more delicately; and performs all this out of his own fund, without diving into the arts and sciences for a supply. Even his Doric dialect has an incomparable sweetness in its clownishness, like a fair shepherdess in her country russet,° talking in a Yorkshire tone. This was impossible for Virgil to imitate; because the severity of the Roman language denied him that advantage. Spenser has endeavoured it in his *Shepherd's Calendar*; but neither will it succeed in English, for which reason I forbore to attempt it. For Theocritus wrote to Sicilians, who spoke that dialect; and I direct this part of my translations to our ladies, who neither understand, nor will take pleasure in such homely expressions. I proceed to Horace.

Take him in parts, and he is chiefly to be considered in his three different talents, as he was a critic, a satirist, and a writer of odes. His morals are uniform, and run through all of them; for let his Dutch commentators say what they will, his philosophy was epicurean; and he made use of gods and providence only to serve a turn in poetry. But since neither his criticisms (which are the most instructive of any that are written in this art) nor his satires (which are incomparably beyond Juvenal's, if to laugh and rally is to be preferred to railing and declaiming,) are any part of my present undertaking, I confine myself wholly to his odes: these are also of several sorts; some of them are panegyrical, others moral, the rest jovial, or (if I may so call them) Bacchanalian. As difficult as he makes it, and as indeed it is, to imitate Pindar, yet in his most elevated flights, and in the sudden changes of his subject with almost imperceptible connexions, that Theban poet is his master. But Horace is of the more bounded fancy, and confines himself strictly to one sort of verse or stanza in every ode. That which will distinguish his style from all other poets, is the elegance of his words, and the numerousness° of his verse; there is nothing so delicately turned in all the Roman language. There appears in every part of his diction, or, (to speak English) in all his expressions, a kind of noble and bold purity. His words are chosen with as much exactness as Virgil's; but there seems to be a greater spirit in them. There is a secret happiness attends his choice, which in Petronius is called *curiosa felicitas*,° and which I suppose he had from the *feliciter audere*° of Horace himself. But the most distinguishing part of all his character, seems to me, to be his briskness, his jollity, and his good humour. And those I have chiefly endeavoured to copy; his other excellencies, I confess are above my imitation. One ode, which infinitely pleased me in the reading, I have attempted to translate in Pindaric verse: 'tis that which is inscribed to the present Earl of Rochester,° to whom I have particular obligations, which this small testimony of my gratitude can never pay. 'Tis his darling in the Latin, and I have taken some pains to make it my masterpiece in English, for which reason I took this kind of verse, which allows more latitude than any other. Everyone knows it was introduced into our language, in this age, by the happy genius of Mr Cowley.° The seeming easiness of it has made it spread; but it has not been considered enough, to be so well cultivated. It languishes in almost every hand but his, and some very few, whom (to keep the rest in countenance) I do not name. He, indeed, has brought it as near perfection as was possible in so short a time. But if I may be allowed to speak my mind modestly, and without injury to his sacred ashes, somewhat of the purity of English, somewhat of more equal of thoughts, somewhat of sweetness in the numbers, in one word, somewhat of a finer turn and more lyrical verse is yet wanting. As for the soul of it, which consists in the warmth and vigour of fancy, the masterly figures, and the copiousness of imagination, he has excelled all others in this kind. Yet, if the kind itself be capable of more perfection, though rather in the ornamental parts of it, than the essential, what rules of morality or respect have I broken, in naming the defects, that they may hereafter be amended? Imitation is a nice point, and there are few poets who deserve to be models in all they write. Milton's *Paradise Lost* is admirable; but am I therefore bound to maintain, that there are no flats amongst his elevations, when 'tis evident he creeps along sometimes, for above a hundred lines together? Cannot I admire the height of his invention, and the strength of his expression, without defending his antiquated words, and the perpetual harshness of their sound? 'Tis as much commendation as a man can bear, to own him excellent; all beyond it is idolatry. Since Pindar was the prince of lyric poets, let me have leave to say, that in imitating him, our numbers should for the most part be lyrical:

for variety, or rather where the majesty of the thought requires it, they may be stretched to the English heroic of five feet, and to the French Alexandrine of six. But the ear must preside, and direct the judgement to the choice of numbers. Without the nicety of this, the harmony of Pindaric verse can never be complete; the cadency of one line must be a rule to that of the next; and the sound of the former must slide gently into that which follows; without leaping from one extreme into another. It must be done like the shadowings of a picture, which fall by degrees into a darker colour. I shall be glad if I have so explained myself as to be understood, but if I have not, *quod nequeo dicere et sentio tantum* [what I cannot say and only feel],° must be my excuse.

There remains much more to be said on this subject, but to avoid envy, I will be silent. What I have said is the general opinion of the best judges, and in a manner has been forced from me, by seeing a noble sort of poetry so happily restored by one man, and so grossly copied by almost all the rest: a musical ear, and a great genius, if another Mr Cowley could arise, in another age may bring it to perfection. In the meantime,

> fungar vice cotis acutum
> reddere quae ferrum valet, expers ipsa secandi.°

[I shall act the whetstone, that sharpens steel, though it cannot itself cut]

I hope it will not be expected from me, that I should say anything of my fellow undertakers in this miscellany. Some of them are too nearly related° to me, to be commended without suspicion of partiality. Others I am sure need it not; and the rest I have not perused. To conclude, I am sensible that I have written this too hastily and too loosely; I fear I have been tedious, and which is worse, it comes out from the first draft, and uncorrected. This I grant is no excuse, for it may be reasonably urged, why did he not write with more leisure, or, if he had it not (which was certainly my case) why did he attempt to write on so nice a subject? The objection is unanswerable, but in part of recompense, let me assure the reader, that in hasty productions, he is sure to meet with an author's present sense, which cooler thoughts would possibly have disguised. There is undoubtedly more of spirit, though not of judgement in these incorrect essays, and consequently though my hazard be the greater, yet the reader's pleasure is not the less.

JOHN DRYDEN

The entire Episode of Nisus and Euryalus, translated from the Fifth and Ninth Books of Virgil's Aeneid

Connection of the first part of the episode in the fifth book, with the rest of the foregoing poem.

Aeneas, having buried his father Anchises in Sicily, and setting sail from thence in search of Italy, is driven by a storm on the same coasts from whence he departed. After a year's wandering, he is hospitably received by his friend Acestes, king of that part of the island, who was born of Trojan parentage. He applies himself to celebrate the memory of his father with divine honours, and accordingly institutes funeral games and appoints prizes for those who should conquer in them. One of these games was a foot race, in which Nisus and Euryalus were engaged amongst other Trojans and Sicilians.

From thence his way the Trojan hero bent,°
Into a grassy plain with mountains pent,
Whose brows were shaded with surrounding wood.
Full in the midst of this fair valley stood
A native theatre, which rising slow,
By just degrees, o'erlooked the ground below:
A numerous train attend in solemn state:
High on the new-raised turf their leader sate.
Here those who in the rapid race delight,
Desire of honour and the prize invite: 10
The Trojans and Sicilians mingled stand,
With Nisus and Euryalus the foremost of the band;
Euryalus with youth and beauty crowned,
Nisus for friendship to the boy renowned.
Diores next, of Priam's regal race,
Then Salius, joined with Patron, took his place:
But from Epirus one derived his birth,
The other owed it to Arcadian earth.
Then two Sicilian youths; the name of this
Was Helymus, of that was Panopes: 20
Two jolly huntsmen in the forest bred,
And owning old Acestes for their head.
With many others of obscurer name,
Whom time has not delivered o'er to fame.

To these Aeneas in the midst arose,
And pleasingly did thus his mind expose.
'Not one of you shall unrewarded go;
On each I will two Cretan spears bestow,
Pointed with polished steel; a battle-axe too,
With silver studded; these in common share,   30
The foremost three shall olive garlands wear;
The victor, who shall first the race obtain,
Shall for his prize a well-breathed courser gain,°
Adorned with trappings; to the next in fame,
The quiver of an Amazonian dame
With feathered Thracian arrows well supplied
Hung on a golden belt, and with a jewel tied;
The third this Grecian helmet must content.'
He said; to their appointed base they went:
With beating hearts the expected sign receive,   40
And starting all at once, the station leave.
Spread out, as on the wings of winds they flew
And seized the distant goal with eager view;
Shot from the crowd, swift Nisus all o'erpassed,
Not storms, nor thunder equal half his haste:
The next, but though the next, yet far disjoined,
Came Salius; then, a distant space behind,
Euryalus the third;
Next Helymus, whom young Diores plied,
Step after step, and almost side by side,   50
His shoulders pressing, and in longer space,
Had won, or left at least a doubtful race.
   Now spent, the goal they almost reach at last,
When eager Nisus, hapless in his haste,
Slipped first, and slipping, fell upon the plain,
Moist with the blood of oxen lately slain;
The careless victor had not marked his way
But treading where the treacherous puddle lay,
His heels flew up, and on the grassy floor,
He fell besmeared with filth and holy gore.   60
Nor mindless then Euryalus of thee,
Nor of the sacred bonds of amity,
He strove the immediate rival to oppose
And caught the foot of Salius as he rose;
So Salius lay extended on the plain:
Euryalus springs out the prize to gain,

And cuts the crowd; applauding peals attend°
The conqueror to the goal, who conquered through his friend.
Next Helymus and then Diores came,
By two misfortunes, now the third in fame.   70
    But Salius enters, and exclaiming loud
For justice, deafens and disturbs the crowd:
Urges his cause may in the court be heard,
And pleads the prize is wrongfully conferred.
But favour for Euryalus appears,
His blooming beauty and his graceful tears
Had bribed the judges to protect his claim:
Besides Diores does as loud exclaim,
Who vainly reaches at the last reward,
If the first palm on Salius be conferred.   80
Then thus the prince: 'Let not disputes arise;°
Where fortune placed it, I award the prize.
But give me leave her errors to amend,
At least to pity a deserving friend.'
Thus having said,
A lion's hide, amazing to behold,
Ponderous with bristles, and with paws of gold,
He gave the youth; which Nisus grieved to view:
'If such rewards to vanquished men are due,'
Said he, 'and falling is to rise by you,   90
What prize may Nisus from your bounty claim,
Who merited the first rewards and fame?
In falling did both equal fortune try,
Would fortune make me fall as happily.'
With this he pointed to his face, and showed
His hands and body all besmeared with blood;
The indulgent father of the people smiled,
And caused to be produced a massy shield
Of wonderous art by Didymaon wrought,
Long since from Neptune's bars in triumph brought;°   100
With this, the graceful youth he gratified,
Then the remaining presents did divide.

*Connection of the remaining part of the episode, translated out of the ninth
book of Virgil's Aeneid, with the foregoing part of the story.*

    *The war being now broken out betwixt the Trojans and Latins and
Aeneas being overmatched in numbers by his enemies who were aided by*

King Turnus, he fortifies his camp and leaves in it his young son Ascanius,
under the direction of his chief counsellors and captains while he goes in
person to beg succours from king Evander and the Tuscans. Turnus takes
advantage of his absence and assaults his camp. The Trojans in it are
reduced to great extremities, which gives the poet the occasion of
continuing this admirable episode, wherein he described the friendship, the
generosity, the adventures, and the death of Nisus and Euryalus.

The Trojan camp the common danger shared;
By turns they watched the walls, and kept the nightly guard.
To warlike Nisus fell the gate by lot,
(Whom Hyrtacus on huntress Ida got,
And sent to sea Aeneas to attend,)
Well could he dart the spear and shafts unerring send.
Beside him stood Euryalus, his ever-faithful friend.
No youth in all the Trojan host was seen
More beautiful in arms, or of a nobler mien;
Scarce was the down upon his chin begun;   10
One was their friendship, their desire was one;
With minds united in the field they warred
And now were both by choice upon the guard.
     Then Nisus thus:
'Or do the gods this warlike warmth inspire,
Or makes each man a god of his desire?
A noble ardour boils within my breast,
Eager of action, enemy of rest,
That urges me to fight, or undertake
Some deed that may my fame immortal make.   20
Thou seest the foe secure: how faintly shine
Their scattered fires; the most, in sleep supine,
Dissolved in ease, and drunk with victory;
The few awake the fuming flaggon ply,
All hushed around; now hear what I revolve
Within my mind, and what my labouring thoughts resolve.
Our absent lord both camp and council mourn;
By message both would hasten his return;
The gifts proposed if they confer on thee,
(For fame is recompense enough to me)
Methinks beneath yon hill, I have espied
A way that safely will my passage guide.'
Euryalus stood listening while he spoke,

With love of praise and noble envy struck;
Then to his ardent friend exposed his mind:
'All this alone, and leaving me behind!
Am I unworthy, Nisus, to be joined,
Thinkst thou my share of honour I will yield,
Or send thee unassisted to the field?
Not so my father taught my childhood arms,   40
Born in a siege, and bred amongst alarms;
Nor is my youth unworthy of my friend,
Or of the heaven-born hero I attend.
The thing called life with ease I can disclaim
And think it oversold to purchase fame.'
    To whom his friend:
'I could not think, alas, thy tender years
Would minister new matter to my fears;
Nor is it just thou shouldst thy wish obtain,
So Jove in triumph bring me back again   50
To those dear eyes, or if a god there be
To pious friends propitious more than he.
But if some one, as many sure there are,
Of adverse accidents in doubtful war,
If one should reach my head there let it fall,
And spare thy life, I would not perish all:
Thy youth is worthy of a longer date;
Do thou remain to mourn thy lover's fate,
To bear my mangled body from the foe,
Or buy it back, and funeral rites bestow,   60
Or if hard fortune shall my corpse deny
Those dues, with empty marble to supply.
O let not me the widow's tears renew,
Let not a mother's curse my name pursue;
Thy pious mother, who in love to thee
Left the fair coast of fruitful Sicily,
Her age committing to the seas and wind,
When every weary matron stayed behind.'
To this, Euryalus: 'Thou pleadst in vain,
And but delayst the cause thou canst not gain: 70
No more, 'tis loss of time.' With that he wakes
The nodding watch; each to his office takes.
    The guard relieved, in company they went
To find the council at the royal tent.

Now every living thing lay void of care,
And sleep, the common gift of nature, share;
Meantime the Trojan peers in council sate
And called their chief commanders to debate
The weighty business of the endangered state,
What next was to be done, who to be sent   80
To inform Aeneas of the foe's intent.
In midst of all the quiet camp they held
Nocturnal council; each sustains a shield
Which his o'erlaboured arm can hardly rear;
And leans upon a long projected spear.
　　Now Nisus and his friend approach the guard
And beg admittance, eager to be heard,
The affair important, not to be deferred.
Ascanius bids them be conducted in;
Then thus, commanded, Nisus does begin.   90
'Ye Trojan fathers lend attentive ears,
Nor judge our undertaking by our years.
The foes securely drenched in sleep and wine
Their watch neglect; their fires but thinly shine.
And where the smoke in thickening vapours flies
Covering the plain, and clouding all the skies,
Betwixt the spaces we have marked a way,
Close by the gate and coasting by the sea.
This passage undisturbed, and unespied
Our steps will safely to Aeneas guide,   100
Expect each hour to see him back again
Loaded with spoils of foes in battle slain.
Snatch we the lucky minute while we may,
Nor can we be mistaken in the way:
For, hunting in the vale, we oft have seen
The rising turrets with the stream between,
And know its winding course, with every ford.'
He paused, and old Alethes took the word.°
'Our country gods, in whom our trust we place,
Will yet from ruin save the Trojan race;   110
While we behold such springing worth appear,
In youth so brave, and breasts so void of fear.'
(With this he took the hand of either boy,
Embraced them closely both, and wept for joy:)
'Ye brave young men, what equal gifts can we,

What recompense for such desert, decree?
The greatest sure and best you can receive,
The gods your virtue and your fame will give:
The rest, our grateful general will bestow;
And young Ascanius, till his manhood, owe.'   120
'And I whose welfare in my father lies,'
(Ascanius adds) 'by all the deities,
By our great country, and our household gods,
By hoary Vesta's rites and dark abodes,
Adjure you both, on you my fortune stands,
That and my faith I plight into your hands,
Make me but happy in his safe return,
(For I no other loss but only his can mourn,)
Nisus your gift shall two large goblets be,
Of silver, wrought with curious imagery,   130
And high embossed, which, when old Priam reigned,
My conquering sire, at sacked Arisba gained.
And, more, two tripods cast in antique mould,
With two great talents of the finest gold:
Besides a bowl which Tyrian art did grave,
The present that Sidonian Dido gave.°
But if in conquered Italy we reign,
When spoils by lot the victors shall obtain—
Thou sawest the courser by proud Turnus pressed,
That, and his golden arms, and sanguine crest,   140
And shield, from lot exempted, thou shalt share;
With these, twelve captive damsels young and fair:
Male slaves as many; well appointed all
With vests and arms, shall to thy portion fall:
And, last, a fruitful field to thee shall rest,
The large demesnes the Latian king possessed.
But thou, whose years are more to mine allied,
No fate my vowed affection shall divide
From thee, O wondrous youth! be ever mine,
Take full possession; all my soul is thine: 150
My life's companion, and my bosom friend—
One faith, one fame, one fate shall both attend.
My peace shall be committed to thy care,
And to thy conduct my concerns in war.'
    Then thus the bold Euryalus replied:
'Whatever fortune, good or bad, betide,

The same shall be my age, as now my youth;
No time shall find me wanting to my truth.
This only from your bounty let me gain
(And this not granted, all rewards are vain)   160
Of Priam's royal race my mother came—
And sure the best that ever bore the name—
Whom neither Troy, nor Sicily could hold
From me departing; but o'erspent and old,
My fate she followed. Ignorant of this
Whatever danger, neither parting kiss,
Nor pious blessing taken, her I leave,
And in this only act of all my life deceive.
By this your hand and conscious night, I swear,
My youth so sad a farewell could not bear.   170
Be you her patron, fill my vacant place,
(Permit me to presume so great a grace)
Support her age forsaken and distressed,
That hope alone will fortify my breast,
Against the worst of fortunes and of fears.'
He said; the assistants shed presaging tears,
But above all, Ascanius moved to see
That image of paternal piety.
Then thus replied:
'So great beginnings in so green an age   180
Exact that faith which firmly I engage.
Thy mother all the privilege shall claim
Creusa had, and only want the name.
Whate'er event thy enterprise shall have,
'Tis merit to have borne a son so brave.
By this my head a sacred oath I swear,
(My father used it) what returning here
Crowned with success, I for thyself prepare,
Thy parent and thy family shall share.'
He said: and weeping while he spoke the word,   190
From his broad belt he drew a shining sword,
Magnificent with gold. Lycaon made,
And in an ivory scabbard sheathed the blade.
This was his gift. While Mnestheus did provide
For Nisus' arms a grizly lion's hide,
And true Alethes changed with him his helm of temper tried.
    Thus armed they went; the noble Trojans wait

Their going forth, and follow to the gate
With prayers and vows. Above the rest appears
Ascanius, manly far above his years,   200
And messages committed to their care;
Which all in winds were lost, and empty air.
The trenches first they passed, then took their way
Where their proud foes in pitched pavilions lay:
To many fatal, ere themselves were slain.
The careless host dispersed upon the plain
They found, who drunk with wine supinely snore.
Unharnessed chariots stand upon the shore
Midst wheels, and reins, and arms, the goblet by,
A medley of debauch and war, they lie.   210
Observing Nisus showed his friend the sight
Then thus: 'Behold a conquest without fight.
Occasion calls the sword to be prepared;
The way lies there; stand thou upon the guard,
And look behind, while I securely go
To cut an ample passage through the foe.'
Softly he spoke; then stalking took his way
With his drawn sword, where haughty Rhamnes lay,
His head raised high, on tapestry beneath,
And heaving from his breast, he puffed his breath——   220
A king and prophet by king Turnus loved;
But fate by prescience cannot be removed.
Three sleeping slaves he soon subdues, then spies
Where Remus with his proud retinue lies.
His armour-bearer first, and next he kills
His charioteer entrenched betwixt the wheels
And his loved horses; last invades their lord,
Full on his neck he aims the fatal sword;
The gasping head flies off. A purple flood
Flows from the trunk, that wallows in the blood   230
Which, by the spurning heels dispersed around
The bed besprinkles, and bedews the ground.
Then Lamyrus with Lamus and the young
Serranus, who with gaming did prolong
The night; oppressed with wine and slumber lay
The beauteous youth, and dreamt of lucky play—
More lucky, had it been protracted till the day.
    The famished lion thus with hunger bold,

O'erleaps the fences of the nightly fold,
The peaceful flock devours, and tears, and draws;   240
Wrapped up in silent fear, they lie and pant beneath his paws.
Nor with less rage Euryalus employs
The vengeful sword, nor fewer foes destroys;
But on the ignoble crowd his fury flew,
Which Fadus, Herbesus, and Rhoetus slew,
With Abaris; in sleep the rest did fall;
But Rhoetus waking, and observing all,
Behind a mighty jar he slunk for fear;
The sharp edged iron found and reached him there,
Full as he rose, he plunged it in his side—   250
The cruel sword returned in crimson dyed.
The wound a blended stream of wine and blood
Pours out; the purple soul comes floating in the flood.
   Now where Messapus quartered they arrive;
The fires were fainting there, and just alive;
The warlike horses, tied in order, fed;
Nisus the discipline observed, and said:
'Our eagerness of blood may both betray:
Behold the doubtful glimmering of the day,
Foe to these nightly thefts. No more, my friend,   260
Here let our glutted execution end;
A lane through slaughtered bodies we have made.'
The bold Euryalus, though loath, obeyed.
Rich arms and arras which they scattered find,
And plate, a precious load they leave behind.
Yet fond of gaudy spoils, the boy would stay
To make the proud caparisons his prey,
Which decked a neighbouring steed—
Nor did his eyes less longingly behold
The girdle studded o'er with nails of gold   270
Which Rhamnes wore. (This present long ago
On Remulus did Caedicus bestow,
And absent joined in hospitable ties;
He dying to this heir bequeathed the prize,
Till by the conquering Rutuli oppressed
He fell, and they the glorious gift possessed.)
These gaudy spoils Euryalus now bears,
And vainly on his brawny shoulders wears:
Messapus' helm, he found amongst the dead

Garnished with plumes, and fitted to his head.   280
They leave the camp and take the safest road.
    Meantime a squadron of their foes abroad,
Three hundred horse with bucklers armed, they spied,
Whom Volcens by the king's command did guide;
To Turnus these were from the city sent,
And to perform their message sought his tent.
Approaching near their utmost lines they draw;
When bending towards the left, their captain saw
The faithful pair (for through the doubtful shade
His glittering helm Euryalus betrayed,   290
On which the moon with full reflection played).
''Tis not for naught' cried Volcens from the crowd,
'These men go there,' then raised his voice aloud:
'Stand, stand! why thus in arms? and whither bent?
From whence, to whom, and on what errand sent?'
Silent they make away, and haste their flight
To neighbouring woods, and trust themselves to night.
The speedy horsemen spur their steeds to get
'Twixt them and home; and every path beset,
And all the windings of the well known wood;   300
(Black was the brake, and thick with oak it stood,
With fern all horrid, and perplexing thorn,°
Where tracks of bears had scarce a passage worn.)
The darkness of the shades, his heavy prey,
And fear, misled the younger from his way.
But Nisus hit the turns with happier haste,
Who now, unknowing, had the danger passed,
And Alban lakes, (from Alba's name so called)
Where king Latinus then his oxen stalled,
Till turning at the length, he stood his ground,   310
And vainly cast his longing eyes around
For his lost friend.
'Ah! wretch,' he cried, 'where have I left behind,
Where shall I hope, the unhappy youth to find?
Or what way take?' Again he ventures back
And treads the mazes of his former track
Through the wild wood; at last he hears the noise
Of trampling horses, and the riders' voice.
The sound approached; and suddenly he viewed
His foes enclosing, and his friend pursued,   320

Forelaid, and taken, while he strove in vain
The covert of the neighbouring wood to gain.
What should he next attempt? what arms employ
With fruitless force to free the captive boy?
Or tempt unequal numbers with the sword,
And die by him whom living he adored?
Resolved on death, his dreadful spear he shook,
And casting to the moon a mournful look,
'Fair queen,' said he, 'who dost in woods delight,
Grace of the stars, and goddess of the night,   330
Be present, and direct my dart aright.
If e'er my pious father for my sake
Did on the altars grateful offerings make,
Or I increased them with successful toils
And hung thy sacred roof with savage spoils—
Through the brown shadows guide my flying spear°
To reach this troop.' Then, poising from his ear°
The quivering weapon, with full force he threw;
Through the divided shades the deadly javelin flew;
On Sulmo's back it splits; the double dart   340
Drove deeper onward, and transfixed his heart.
He staggers round, his eyeballs roll in death,
And with short sobs, he gasps away his breath.
All stand amazed; a second javelin flies
From his stretched arm, and hisses through the skies.
The lance through Tagus' temples forced its way
And in his brain-pan warmly buried lay.
Fierce Volcens foams with rage; and gazing round
Descried no author of the fatal wound,
Nor where to fix revenge. 'But thou,' he cries   350
'Shalt pay for both,' and at the prisoner flies
With his drawn sword. Then struck with deep despair
That fatal sight the lover could not bear
But from the covert rushed in open view
And sent his voice before him as he flew,
'Me, me, employ your sword on me alone;
The crime confessed; the fact was all my own.
He neither could nor durst, the guiltless youth,
Ye moon and stars bear witness to the truth,
His only fault, if that be to offend,   360
Was too much loving his unhappy friend.'

Too late alas he speaks;
The sword which unrelenting fury guides
Driven with full force had pierced his tender sides;
Down fell the beauteous youth; the gaping wound
Gushed out a crimson stream and stained the ground;
His nodding neck reclines on his white breast
Like a fair flower, in furrowed fields oppressed
By the keen share, or poppy on the plain
Whose heavy head is overcharged with rain.   370

Disdain, despair, and deadly vengeance vowed,
Drove Nisus headlong on the hostile crowd.
Volcens he seeks; at him alone he bends;
Borne back, and pushed by his surrounding friends
He still pressed on, and kept him still in sight,
Then whirled aloft his sword with all his might.
The unerring weapon flew, and winged with death
Entered his gaping mouth, and stopped his breath.
Dying he slew, and staggering on the plain
Sought for the body of his lover slain.   380
Then quietly on his dear breast he fell,
Content in death to be revenged so well.

O happy pair! for if my verse can give
Eternity your fame shall ever live,
Fixed as the Capitol's foundation lies,
And spread where'er the Roman eagle flies.

TRANSLATIONS FROM LUCRETIUS

The Beginning of the FIRST BOOK

Delight of humankind, and gods above,
Parent of Rome, propitious queen of Love!
Whose vital power, air, earth, and sea supplies,
And breeds whate'er is born beneath the rolling skies;
For every kind, by thy prolific might,
Springs, and beholds the regions of the light.
Thee, goddess, thee the clouds and tempests fear,
And at thy pleasing presence disappear;
For thee the land in fragrant flowers is dressed;
For thee the ocean smiles, and smoothes her wavy breast,   10

And heaven itself with more serene and purer light is blest.
For when the rising spring adorns the mead,
And a new scene of nature stands displayed,
When teeming buds, and cheerful greens appear,
And western gales unlock the lazy year;
The joyous birds thy welcome first express,
Whose native songs thy genial fire confess;
Then savage beasts bound o'er their slighted food,
Struck with thy darts, and tempt the raging flood.
All nature is thy gift; earth, air, and sea;   20
Of all that breathes, the various progeny,
Stung with delight, is goaded on by thee.
O'er barren mountains, o'er the flowery plain,
The leafy forest, and the liquid main,
Extends thy uncontrolled and boundless reign;
Through all the living regions dost thou move,
And scatterest, where thou goest, the kindly seeds of love.
Since then the race of every living thing
Obeys thy power; since nothing new can spring
Without thy warmth, without thy influence bear,   30
Or beautiful, or lovesome can appear;
Be thou my aid, my tuneful song inspire,
And kindle with thy own productive fire;
While all thy province, nature, I survey,
And sing to Memmius an immortal lay°
Of heaven and earth, and everywhere thy wondrous power
    display;
To Memmius, under thy sweet influence born,
Whom thou with all thy gifts and graces dost adorn.
The rather then assist my muse and me,
Infusing verses worthy him and thee.   40
Meantime on land and sea let barbarous discord cease,
And lull the listening world in universal peace.
To thee mankind their soft repose must owe,
For thou alone that blessing canst bestow;
Because the brutal business of the war
Is managed by thy dreadful servant's care;°
Who oft retires from fighting fields, to prove
The pleasing pains of thy eternal love;
And panting on thy breast, supinely lies,
While with thy heavenly form he feeds his famished eyes;   50

Sucks in with open lips thy balmy breath,
By turns restored to life, and plunged in pleasing death.
There while thy curling limbs about him move,
Involved and fettered in the links of love,
When wishing all, he nothing can deny,
Thy charms in that auspicious moment try;
With winning eloquence our peace implore,
And quiet to the weary world restore.

The Beginning of the SECOND BOOK

'Tis pleasant, safely to behold from shore
The rolling ship, and hear the tempest roar;
Not that another's pain is our delight,
But pains unfelt produce the pleasing sight.
'Tis pleasant also to behold from far
The moving legions mingled in the war;
But much more sweet thy labouring steps to guide
To virtue's heights, with wisdom well supplied,
And all the magazines of learning fortified;
From thence to look below on humankind,   10
Bewildered in the maze of life, and blind;
To see vain fools ambitiously contend
For wit and power; their lost endeavours bend
To outshine each other, waste their time and health
In search of honour, and pursuit of wealth.
O wretched man! in what a mist of life,
Enclosed with dangers and with noisy strife,
He spends his little span; and overfeeds
His crammed desires, with more than nature needs!
For nature wisely stints our appetitite,°   20
And craves no more than undisturbed delight;
Which minds unmixed with cares and fears obtain;
A soul serene, a body void of pain.
So little this corporeal frame requires,
So bounded are our natural desires,
That wanting all, and setting pain aside,
With bare privation sense is satisfied.
If golden sconces hang not on the walls,°
To light the costly suppers and the balls;
If the proud palace shines not with the state   30
Of burnished bowls and of reflected plate;

If well-tuned harps, nor the more pleasing sound
Of voices, from the vaulted roofs rebound;
Yet on the grass beneath a poplar shade,
By the cool stream, our careless limbs are laid;
With cheaper pleasures innocently blest,
When the warm spring with gaudy flowers is dressed.
Nor will the raging fever's fire abate,
With golden canopies, and beds of state;
But the poor patient will as soon be sound   40
On the hard mattress, or the mother ground.
Then since our bodies are not eased the more
By birth, or power, or fortune's wealthy store,
'Tis plain, these useless toys of every kind
As little can relieve the labouring mind;
Unless we could suppose the dreadful sight
Of marshalled legions moving to the fight,
Could with their sound and terrible array
Expel our fears, and drive the thoughts of death away.
But, since the supposition vain appears,   50
Since clinging cares, and trains of inbred fears,
Are not with sounds to be affrighted thence,
But in the midst of pomp pursue the prince,
Not awed by arms, but in the presence bold,
Without respect to purple, or to gold;
Why should not we these pageantries despise,
Whose worth but in our want of reason lies?
For life is all in wandering errors led;°
And just as children are surprised with dread,
And tremble in the dark, so riper years,   60
E'en in broad daylight, are possessed with fears,
And shake at shadows fanciful and vain,
As those which in the breasts of children reign.
These bugbears of the mind, this inward hell,
No rays of outward sunshine can dispel;
But nature and right reason must display
Their beams abroad, and bring the darksome soul to day.

The Latter Part of the
THIRD BOOK

AGAINST THE FEAR OF DEATH

What has this bugbear, death, to frighten man,
If souls can die, as well as bodies can?
For, as before our birth we felt no pain,
When Punic arms infested land and main,°
When heaven and earth were in confusion hurled,
For the debated empire of the world,
Which awed with dreadful expectation lay,
Sure to be slaves, uncertain who should sway:
So, when our mortal frame shall be disjoined,
The lifeless lump uncoupled from the mind,   10
From sense of grief and pain we shall be free;
We shall not feel, because we shall not be.
Though earth in seas, and seas in heaven were lost,
We should not move, we only should be tossed.
Nay, e'en suppose when we have suffered fate,
The soul could feel in her divided state,
What's that to us? for we are only we,
While souls and bodies in one frame agree.
Nay, though our atoms should revolve by chance,
And matter leap into the former dance;   20
Though time our life and motion could restore,
And make our bodies what they were before;
What gain to us would all this bustle bring?
The new-made man would be another thing.
When once an interrupting pause is made,
That individual being is decayed.
We, who are dead and gone, shall bear no part
In all the pleasures, nor shall feel the smart,
Which to that other mortal shall accrue,
Whom of our matter time shall mould anew.   30
For backward if you look on that long space
Of ages past, and view the changing face
Of matter, tossed, and variously combined
In sundry shapes, 'tis easy for the mind
From thence to infer, that seeds of things have been
In the same order as they now are seen;

Which yet our dark remembrance cannot trace,
Because a pause of life, a gaping space,
Has come betwixt, where memory lies dead,
And all the wandering motions from the sense are fled.   40
For, whosoe'er shall in misfortunes live,
Must *be*, when those misfortunes shall arrive;
And since the man who *is* not, feels not woe,
(For death exempts him, and wards off the blow,
Which we, the living, only feel and bear,)
What is there left for us in death to fear?
When once that pause of life has come between,
'Tis just the same as we had never been.

   And therefore if a man bemoan his lot,
That after death his mouldering limbs shall rot,   50
Or flames, or jaws of beasts devour his mass,
Know, he's an unsincere, unthinking ass.
A secret sting remains within his mind;
The fool is to his own cast offals kind.°
He boasts no sense can after death remain,
Yet makes himself a part of life again,
As if some other he could feel the pain.
If, while he live, this thought molest his head,
What wolf or vulture shall devour me dead?
He wastes his days in idle grief, nor can   60
Distinguish 'twixt the body and the man;
But thinks himself can still himself survive,
And what when dead he feels not, feels alive.
Then he repines that he was born to die,
Nor knows in death there is no other he,
No living he remains his grief to vent,
And o'er his senseless carcase to lament.
If, after death, 'tis painful to be torn
By birds and beasts, then why not so to burn,
Or drenched in floods of honey to be soaked,   70
Embalmed to be at once preserved and choked;
Or on an airy mountain's top to lie,
Exposed to cold and heaven's inclemency;
Or crowded in a tomb, to be oppressed
With monumental marble on thy breast?
But to be snatched from all thy household joys,
From thy chaste wife, and thy dear prattling boys,

Whose little arms about thy legs are cast,
And climbing for a kiss prevent their mother's haste,
Inspiring secret pleasure through thy breast;   80
All these shall be no more; thy friends oppressed,
Thy care and courage now no more shall free;
'Ah! wretch,' thou criest, 'ah! miserable me!
One woeful day sweeps children, friends, and wife,
And all the brittle blessings of my life!'
Add one thing more, and all thou say'st is true;
Thy want and wish of them is vanished too;
Which well considered were a quick relief
To all thy vain imaginary grief.
For thou shalt sleep, and never wake again,   90
And, quitting life, shalt quit thy living pain.
But we, thy friends, shall all those sorrows find,
Which in forgetful death thou leav'st behind;
No time shall dry our tears, nor drive thee from our mind.
The worst that can befall thee, measured right,
Is a sound slumber, and a long goodnight.
Yet thus the fools, that would be thought the wits,
Disturb their mirth with melancholy fits;
When healths go round, and kindly brimmers flow,°
Till the fresh garlands on their foreheads glow,   100
They whine, and cry: 'Let us make haste to live,
Short are the joys that human life can give.'
Eternal preachers, that corrupt the draught,
And pall the god that never thinks with thought;°
Idiots with all that thought, to whom the worst
Of death, is want of drink, and endless thirst,
Or any fond desire as vain as these.
For, e'en in sleep, the body, wrapped in ease,
Supinely lies, as in the peaceful grave;
And wanting nothing, nothing can it crave.   110
Were that sound sleep eternal, it were death;
Yet the first atoms then, the seeds of breath,
Are moving near to sense; we do but shake
And rouse that sense, and straight we are awake.
Then death to us, and death's anxiety,
Is less than nothing, if a less could be;
For then our atoms, which in order lay,
Are scattered from their heap, and puffed away,

And never can return into their place,
When once the pause of life has left an empty space.   120
    And, last, suppose great nature's voice should call
To thee, or me, or any of us all,
'What dost thou mean, ungrateful wretch, thou vain,
Thou mortal thing, thus idly to complain,
And sigh and sob, that thou shalt be no more?
For if thy life were pleasant heretofore,
If all the bounteous blessings I could give
Thou hast enjoyed, if thou hast known to live,
And pleasure not leaked through thee like a sieve,
Why dost thou not give thanks as at a plenteous feast,   130
Crammed to the throat with life, and rise and take thy rest?
But if my blessings thou hast thrown away,
If undigested joys passed through, and would not stay,
Why dost thou wish for more to squander still?
If life be grown a load, a real ill,
And I would all thy cares and labours end,
Lay down thy burden, fool, and know thy friend.
To please thee, I have emptied all my store;
I can invent, and can supply no more,
But run the round again, the round I ran before.   140
Suppose thou art not broken yet with years,
Yet still the selfsame scene of things appears,
And would be ever, couldst thou ever live;
For life is still but life, there's nothing new to give.'
What can we plead against so just a bill?
We stand convicted, and our cause goes ill.
But if a wretch, a man oppressed by fate,
Should beg of nature to prolong his date,
She speaks aloud to him with more disdain,
'Be still, thou martyr fool, thou covetous of pain.'   150
But if an old decrepit sot lament,
'What, thou,' she cries, 'who hast outlived content!
Dost thou complain, who hast enjoyed my store?
But this is still the effect of wishing more.
Unsatisfied with all that nature brings;
Loathing the present, liking absent things;
From hence it comes, thy vain desires, at strife
Within themselves, have tantalised thy life,
And ghastly death appeared before thy sight,

Ere thou hadst gorged thy soul and senses with delight.   160
Now leave those joys unsuiting to thy age
To a fresh comer, and resign the stage.'
   Is nature to be blamed if thus she chide?
No, sure; for 'tis her business to provide
Against this ever-changing frame's decay,
New things to come, and old to pass away.
One being, worn, another being makes;
Changed, but not lost; for nature gives and takes:
New matter must be found for things to come,
And these must waste like those, and follow nature's doom.   170
All things, like thee, have time to rise and rot,
And from each other's ruin are begot:
For life is not confined to him or thee;
'Tis given to all for use, to none for property.
Consider former ages past and gone,
Whose circles ended long ere thine begun,
Then tell me, fool, what part in them thou hast?
Thus mayst thou judge the future by the past.
What horror seest thou in that quiet state,
What bugbear dreams to fright thee after fate?   180
No ghost, no goblins, that still passage keep;
But all is there serene, in that eternal sleep.
For all the dismal tales that poets tell
Are verified on earth, and not in hell.
No Tantalus looks up with fearful eye,°
Or dreads the impending rock to crush him from on high;
But fear of chance on earth disturbs our easy hours,
Or vain imagined wrath of vain imagined powers.
No Tityus torn by vultures lies in hell;°
Nor could the lobes of his rank liver swell   190
To that prodigious mass for their eternal meal;
Not though his monstrous bulk had covered o'er
Nine spreading acres, or nine thousand more;
Not though the globe of earth had been the giant's floor;
Nor in eternal torments could he lie,
Nor could his corpse sufficient food supply.
But he's the Tityus, who, by love oppressed,
Or tyrant passion preying on his breast,
And ever anxious thoughts, is robbed of rest.
The Sisyphus is he, whom noise and strife°   200

Seduce from all the soft retreats of life,
To vex the government, disturb the laws;
Drunk with the fumes of popular applause,
He courts the giddy crowd to make him great,
And sweats and toils in vain, to mount the sovereign seat.
For still to aim at power, and still to fail,
Ever to strive, and never to prevail,
What is it but in reason's true account,
To heave the stone against the rising mount?
Which urged, and laboured, and forced up with pain,   210
Recoils, and rolls impetuous down, and smokes along the plain.°
Then still to treat thy ever-craving mind
With every blessing, and of every kind,
Yet never fill thy ravening appetite,
Though years and seasons vary thy delight,
Yet nothing to be seen of all the store,
But still the wolf within thee barks for more;
This is the fable's moral, which they tell
Of fifty foolish virgins damned in hell°
To leaky vessels, which the liquor spill;   220
To vessels of their sex, which none could ever fill.
As for the dog, the furies, and their snakes,°
The gloomy caverns, and the burning lakes,
And all the vain infernal trumpery,
They neither are, nor were, nor e'er can be.
But here on earth the guilty have in view
The mighty pains to mighty mischiefs due:
Racks, prisons, poisons, the Tarpeian rock,°
Stripes, hangmen, pitch, and suffocating smoke;
And last, and most, if these were cast behind,   230
The avenging horror of a conscious mind,
Whose deadly fear anticipates the blow,
And sees no end of punishment and woe,
But looks for more, at the last gasp of breath:
This makes a hell on earth, and life a death.
    Meantime, when thoughts of death disturb thy head,
Consider, Ancus, great and good, is dead;°
Ancus, thy better far, was born to die,
And thou, dost thou bewail mortality?
So many monarchs with their mighty state,   240
Who ruled the world, were overruled by fate.

That haughty king, who lorded o'er the main,°
And whose stupendous bridge did the wild waves restrain,
(In vain they foamed, in vain they threatened wreck,
While his proud legions marched upon their back):
Him death, a greater monarch, overcame;
Nor spared his guards the more, for their immortal name.
The Roman chief, the Carthaginian dread,
Scipio, the thunderbolt of war, is dead,
And like a common slave, by fate in triumph led.   250
The founders of invented arts are lost,
And wits who made eternity their boast.
Where now is Homer, who possessed the throne?
The immortal work remains, the mortal author's gone.
Democritus, perceiving age invade,°
His body weakened, and his mind decayed,
Obeyed the summons with a cheerful face;
Made haste to welcome death, and met him half the race.
That stroke e'en Epicurus could not bar,
Though he in wit surpassed mankind, as far   260
As does the midday sun the midnight star.
And thou, dost thou disdain to yield thy breath,
Whose very life is little more than death?
More than one half by lazy sleep possessed;
And when awake, thy soul but nods at best,
Day-dreams and sickly thoughts revolving in thy breast,
Eternal troubles haunt thy anxious mind,
Whose cause and cure thou never hop'st to find;
But still uncertain, with thyself at strife,
Thou wanderest in the labyrinth of life.   270
      Oh, if the foolish race of man, who find
A weight of cares still pressing on their mind,
Could find as well the cause of this unrest,
And all this burden lodged within the breast;
Sure they would change their course, nor live as now,
Uncertain what to wish or what to vow.
Uneasy both in country and in town,
They search a place to lay their burden down.
One, restless in his palace, walks abroad,
And vainly thinks to leave behind the load,   280
But straight returns; for he's as restless there,
And finds there's no relief in open air.

Another to his villa would retire,
And spurs as hard as if it were on fire;
No sooner entered at his country door,
But he begins to stretch, and yawn, and snore,
Or seeks the city, which he left before.
Thus every man o'erworks his weary will,
To shun himself, and to shake off his ill;
The shaking fit returns, and hangs upon him still.   290
No prospect of repose, nor hope of ease,
The wretch is ignorant of his disease;
Which known, would all his fruitless trouble spare,
For he would know the world not worth his care:
Then would he search more deeply for the cause,
And study nature well, and nature's laws:
For in this moment lies not the debate,
But on our future, fixed, eternal state;
That never-changing state, which all must keep,
Whom death has doomed to everlasting sleep.   300
    Why are we then so fond of mortal life,
Beset with dangers, and maintained with strife?
A life, which all our care can never save;
One fate attends us, and one common grave.
Besides we tread but a perpetual round;
We ne'er strike out, but beat the former ground,
And the same mawkish joys in the same track are found.
For still we think an absent blessing best,
Which cloys, and is no blessing when possessed;
A new arising wish expels it from the breast.   310
The feverish thirst of life increases still;
We call for more and more, and never have our fill;
Yet know not what tomorrow we shall try,
What dregs of life in the last draught may lie.
Nor, by the longest life we can attain,
One moment from the length of death we gain;
For all behind belongs to his eternal reign.
When once the fates have cut the mortal thread,
The man as much to all intents is dead,
Who dies today, and will as long be so,   320
As he who died a thousand years ago.

THE FOURTH BOOK

CONCERNING THE NATURE OF LOVE

Thus therefore, he who feels the fiery dart
Of strong desire transfix his amorous heart,
Whether some beauteous boy's alluring face,
Or lovelier maid, with unresisting grace,
From her each part the winged arrow sends,
From whence he first was struck he thither tends;
Restless he roams, impatient to be freed,
And eager to inject the sprightly seed;
For fierce desire does all his mind employ,
And ardent love assures approaching joy.   10
Such is the nature of that pleasing smart,
Whose burning drops distil upon the heart,
The fever of the soul shot from the fair,
And the cold ague of succeeding care.
If absent, her idea still appears,
And her sweet name is chiming in your ears.
But strive those pleasing phantoms to remove,
And shun the aerial images of love
That feed the flame: when one molests thy mind,
Discharge thy loins on all the leaky kind;°   20
For that's a wiser way, than to restrain
Within thy swelling nerves that hoard of pain.
For every hour some deadlier symptom shows,
And by delay the gathering venom grows,
When kindly applications are not used;
The scorpion, love, must on the wound be bruised.
On that one object 'tis not safe to stay,
But force the tide of thought some other way;
The squandered spirits prodigally throw,
And in the common glebe of nature sow.   30
Nor wants he all the bliss that lovers feign,
Who takes the pleasure, and avoids the pain;
For purer joys in purer health abound,
And less affect the sickly than the sound.
    When love its utmost vigour does employ,
E'en then 'tis but a restless wandering joy;
Nor knows the lover in that wild excess,

With hands or eyes, what first he would possess:
But strains at all, and fastening where he strains,
Too closely presses with his frantic pains;   40
With biting kisses hurts the twining fair,
Which shows his joys imperfect, insincere:
For, stung with inward rage, he flings around,
And strives to avenge the smart on that which gave the wound.
But love those eager bitings does restrain,
And mingling pleasure mollifies the pain.
For ardent hope still flatters anxious grief,
And sends him to his foe to seek relief:
Which yet the nature of the thing denies;
For love, and love alone of all our joys,   50
By full possession does but fan the fire;
The more we still enjoy, the more we still desire.
Nature for meat and drink provides a space,
And when received, they fill their certain place;
Hence thirst and hunger may be satisfied,
But this repletion is to love denied:
Form, feature, colour, whatsoe'er delight
Provokes the lover's endless appetite,
These fill no space, nor can we thence remove
With lips, or hands, or all our instruments of love: 60
In our deluded grasp we nothing find,
But thin aerial shapes, that fleet before the mind.
As he, who in a dream with drought is cursed,
And finds no real drink to quench his thirst,
Runs to imagined lakes his heat to steep,
And vainly swills and labours in his sleep;
So love with phantoms cheats our longing eyes,
Which hourly seeing never satisfies:
Our hands pull nothing from the parts they strain,
But wander o'er the lovely limbs in vain.   70
Nor when the youthful pair more closely join,
When hands in hands they lock, and thighs in thighs
          they twine,
Just in the raging foam of full desire,
When both press on, both murmur, both expire,
They grip, they squeeze, their humid tongues they dart,
As each would force their way to t'other's heart:
In vain; they only cruise about the coast;

For bodies cannot pierce, nor be in bodies lost,
As sure they strive to be, when both engage
In that tumultuous momentary rage;   80
So tangled in the nets of love they lie,
Till man dissolves in that excess of joy.
Then, when the gathered bag has burst its way,
And ebbing tides the slackened nerves betray,
A pause ensues; and nature nods awhile,
Till with recruited rage new spirits boil;
And then the same vain violence returns,
With flames renewed the erected furnace burns;
Again they in each other would be lost,
But still by adamantine bars are crossed.   90
All ways they try, successless all they prove,
To cure the secret sore of lingering love.
  Besides—
They waste their strength in the venereal strife,
And to a woman's will enslave their life;
The estate runs out, and mortgages are made,
All offices of friendship are decayed,
Their fortune ruined, and their fame betrayed.
Assyrian ointment from their temples flows,
And diamond buckles sparkle at their shoes;   100
The cheerful emerald twinkles on their hands,
With all the luxury of foreign lands;
And the blue coat, that with embroidery shines,
Is drunk with sweat of their o'er-laboured loins.
Their frugal father's gains they misemploy,
And turn to point, and pearl, and every female toy.°
French fashions, costly treats are their delight;
The park by day, and plays and balls by night.
In vain;—
For in the fountain, where their sweets are sought,   110
Some bitter bubbles up, and poisons all the draught.
First, guilty conscience does the mirror bring,
Then sharp remorse shoots out her angry sting;
And anxious thoughts, within themselves at strife,
Upbraid the long misspent, luxurious life.
Perhaps the fickle fair one proves unkind,
Or drops a doubtful word, that pains his mind,
And leaves a rankling jealousy behind.

Perhaps he watches close her amorous eyes,
And in the act of ogling does surprise,   120
And thinks he sees upon her cheeks the while
The dimpled tracks of some foregoing smile;
His raging pulse beats thick, and his pent spirits boil.
This is the product e'en of prosperous love;
Think then what pangs disastrous passions prove;
Innumerable ills; disdain, despair,
With all the meagre family of care.
    Thus, as I said, 'tis better to prevent,
Than flatter the disease, and late repent;
Because to shun the allurement is not hard   130
To minds resolved, forewarned, and well prepared;
But wondrous difficult, when once beset,
To struggle through the straits, and break the involving net.
    Yet thus ensnared thy freedom thou may'st gain,
If, like a fool, thou dost not hug thy chain;
If not to ruin obstinately blind,
And wilfully endeavouring not to find
Her plain defects of body and of mind.
For thus the bedlam train of lovers use°
To enhance the value, and the faults excuse;   140
And therefore 'tis no wonder if we see
They dote on dowdies and deformity.
E'en what they cannot praise, they will not blame,
But veil with some extenuating name.
The sallow skin is for the swarthy put,
And love can make a slattern of a slut;
If cat-eyed, then a Pallas is their love;
If freckled, she's a particoloured dove;
If little, then she's life and soul all o'er;
An Amazon, the large two-handed whore.°   150
She stammers; oh, what grace in lisping lies!
If she says nothing, to be sure she's wise.
If shrill, and with a voice to drown a choir,
Sharp-witted she must be, and full of fire;
The lean, consumptive wench, with coughs decayed,
Is called a pretty, tight, and slender maid;
The o'ergrown, a goodly Ceres is expressed,
A bed-fellow for Bacchus at the least;
Flat-nose the name of Satyr never misses,

And hanging blubber lips but pout for kisses.   160
The task were endless all the rest to trace;
Yet grant she were a Venus for her face
And shape, yet others equal beauty share,
And time was you could live without the fair;
She does no more, in that for which you woo,
Than homelier women full as well can do.
Besides, she daubs, and stinks so much of paint,
Her own attendants cannot bear the scent,
But laugh behind, and bite their lips to hold.
Meantime excluded, and exposed to cold,   170
The whining lover stands before the gates,
And there with humble adoration waits;
Crowning with flowers the threshold and the floor,
And printing kisses on the obdurate door;
Who, if admitted in that nick of time,
If some unsavoury whiff betray the crime,
Invents a quarrel straight, if there be none,
Or makes some faint excuses to be gone;
And calls himself a doting fool to serve,
Ascribing more than woman can deserve;   180
Which well they understand, like cunning queans,°
And hide their nastiness behind the scenes,
From him they have allured, and would retain;
But to a piercing eye 'tis all in vain:
For common sense brings all their cheats to view,
And the false light discovers by the true;
Which a wise harlot owns, and hopes to find
A pardon for defects, that run through all the kind.
   Nor always do they feign the sweets of love,
When round the panting youth their pliant limbs they move,   190
And cling, and heave, and moisten every kiss;
They often share, and more than share the bliss:
From every part, e'en to their inmost soul,
They feel the trickling joys, and run with vigour to the goal.
Stirred with the same impetuous desire,
Birds, beasts, and herds, and mares, their males require;
Because the throbbing nature in their veins
Provokes them to assuage their kindly pains.
The lusty leap the expecting female stands,
By mutual heat compelled to mutual bands.   200

Thus dogs with lolling tongues by love are tied,
Nor shouting boys nor blows their union can divide;
At either end they strive the link to loose,
In vain, for stronger Venus holds the noose;
Which never would those wretched lovers do,
But that the common heats of love they know;
The pleasure therefore must be shared in common too:
And when the woman's more prevailing juice
Sucks in the man's, the mixture will produce,
The mother's likeness; when the man prevails,   210
His own resemblance in the seed he seals.
But when we see the new-begotten race
Reflect the features of each parent's face,
Then of the father's and the mother's blood
The justly tempered seed is understood;
When both conspire, with equal ardour bent,
From every limb the due proportion sent,
When neither party foils, when neither foiled,
This gives the blended features of the child.
Sometimes the boy the grandsire's image bears;   220
Sometimes the more remote progenitor he shares;
Because the genial atoms of the seed
Lie long concealed ere they exert the breed;
And after sundry ages past, produce
The tardy likeness of the latent juice.
Hence families such different figures take,
And represent their ancestors in face, and hair, and make;
Because of the same seed, the voice, and hair,
And shape, and face, and other members are,
And the same antique mould the likeness does prepare.   230
Thus oft the father's likeness does prevail
In females, and the mother's in the male;
For since the seed is of a double kind,
From that where we the most resemblance find,
We may conclude the strongest tincture sent,
And that was in conception prevalent.
    Nor can the vain decrees of powers above
Deny production to the act of love,
Or hinder fathers of that happy name,
Or with a barren womb the matron shame;   240
As many think, who stain with victims' blood

The mournful altars, and with incense load,
To bless the showery seed with future life,
And to impregnate the well-laboured wife.
In vain they weary heaven with prayer, or fly
To oracles, or magic numbers try;
For barrenness of sexes will proceed
Either from too condensed, or watery, seed:
The watery juice too soon dissolves away,
And in the parts projected will not stay;   250
The too condensed, unsouled, unwieldy mass,°
Drops short, nor carries to the destined place;
Nor pierces to the parts, nor, though injected home,
Will mingle with the kindly moisture of the womb.
For nuptials are unlike in their success;
Some men with fruitful seed some women bless,
And from some men some women fruitful are,
Just as their constitutions join or jar:
And many seeming barren wives have been,
Who after matched with more prolific men,   260
Have filled a family with prattling boys;
And many, not supplied at home with joys,
Have found a friend abroad to ease their smart,
And to perform the sapless husband's part.
So much it does import that seed with seed
Should of the kindly mixture make the breed;
And thick with thin, and thin with thick should join,
So to produce and propagate the line.
Of such concernment too is drink and food,
To incrassate, or attenuate the blood.°   270
    Of like importance is the posture too,
In which the genial feat of love we do;
For as the females of the four-foot kind
Receive the leapings of their males behind,
So the good wives, with loins uplifted high,
And leaning on their hands, the fruitful stroke may try:
For in that posture will they best conceive;
Not when, supinely laid, they frisk and heave;
For action motions only break the blow,
And more of strumpets than of wives they show,   280
When answering stroke with stroke, the mingled liquors flow.
Endearments eager, and too brisk a bound,

Throws off the plough-share from the furrowed ground;
But common harlots in conjunction heave,
Because 'tis less their business to conceive
Than to delight, and to provoke the deed;
A trick which honest wives but little need.
Nor is it from the gods, or Cupid's dart,
That many a homely woman takes the heart,
But wives well-humoured, dutiful, and chaste,   290
And clean, will hold their wandering husbands fast;
Such are the links of love, and such a love will last.
For what remains, long habitude, and use,
Will kindness in domestic bands produce;
For custom will a strong impression leave.
Hard bodies, which the lightest stroke receive,
In length of time will moulder and decay,
And stones with drops of rain are washed away.

From the Fifth Book

Thus, like a sailor by the tempest hurled
Ashore, the babe is shipwrecked on the world.
Naked he lies, and ready to expire,
Helpless of all that human wants require;
Exposed upon unhospitable earth,
From the first moment of his hapless birth.
Straight with foreboding cries he fills the room,
(Too true presages of his future doom).
But flocks and herds, and every savage beast,
By more indulgent nature are increased:
They want no rattles for their froward mood,   10
Nor nurse to reconcile them to their food,
With broken words; nor winter blasts they fear,
Nor change their habits with the changing year;
Nor, for their safety, citadels prepare,
Nor forge the wicked instruments of war;
Unlaboured earth her bounteous treasure grants,
And nature's lavish hands supply their common wants.

Daphnis

From The
Twenty-seventh Idyll of Theocritus

Daphnis

The shepherd Paris bore the Spartan bride
By force away, and then by force enjoyed;
But I by free consent can boast a bliss,
A fairer Helen, and a sweeter kiss.

Chloris

Kisses are empty joys, and soon are o'er.

Daphnis

A kiss betwixt the lips is something more.

Chloris

I wipe my mouth, and where's your kissing then?

Daphnis

I swear you wipe it to be kissed again.

Chloris

Go tend your herd, and kiss your cows at home;
I am a maid, and in my beauty's bloom.   10

Daphnis

'Tis well remembered; do not waste your time,
But wisely use it ere you pass your prime.

Chloris

Blown roses hold their sweetness to the last,
And raisins keep their luscious native taste.

Daphnis

The sun's too hot; those olive shades are near;
I fain would whisper something in your ear.

Chloris

'Tis honest talking where we may be seen;
God knows what secret mischief you may mean;
I doubt you'll play the wag, and kiss again.

Daphnis

At least beneath yon elm you need not fear;   20
My pipe's in tune, if you're disposed to hear.

Chloris

Play by yourself, I dare not venture thither;
You, and your naughty pipe, go hang together.

Daphnis

Coy nymph, beware, lest Venus you offend.

Chloris

I shall have chaste Diana still to friend.

Daphnis

You have a soul, and Cupid has a dart.

Chloris

Diana will defend, or heal my heart.
Nay, fie, what mean you in this open place?
Unhand me, or I swear I'll scratch your face.
Let go for shame; you make me mad for spite;   30
My mouth's my own; and if you kiss, I'll bite.

Daphnis

Away with your dissembling female tricks;
What, would you scape the fate of all your sex?

Chloris

I swear, I'll keep my maidenhead till death.
And die as pure as Queen Elizabeth.

Daphnis

Nay, mum for that, but let me lay thee down;
Better with me, than with some nauseous clown.

Chloris
I'd have you know, if I were so inclined,
I have been woo'd by many a wealthy hind;
But never found a husband to my mind.   40

Daphnis
But they are absent all, and I am here.

Chloris
The matrimonial yoke is hard to bear,
And marriage is a woeful word to hear.

Daphnis
A scarecrow, set to frighten fools away;
Marriage has joys, and you shall have a say.

Chloris
Sour sauce is often mixed with our delight;
You kick by day more than you kiss by night.

Daphnis
Sham stories all; but say the worst you can,
A very wife fears neither God nor man.

Chloris
But childbirth is, they say, a deadly pain;   50
It costs at least a month to knit again.

Daphnis
Diana cures the wounds Lucina made;
Your goddess is a midwife by her trade.

Chloris
But I shall spoil my beauty, if I bear.

Daphnis
But Mam and Dad are pretty names to hear.

Chloris
But there's a civil question used of late;
Where lies my jointure, where your own estate?

Daphnis

My flocks, my fields, my wood, my pastures take,
With settlement as good as law can make.

Chloris

Swear then you will not leave me on the common,   60
But marry me, and make an honest woman.

Daphnis

I swear by Pan, though he wears horns you'll say,
Cudgelled and kicked, I'll not be forced away.

Chloris

I bargain for a wedding-bed at least,
A house, and handsome lodging for a guest.

Daphnis

A house well furnished shall be thine to keep;
And, for a flock-bed, I can shear my sheep.

Chloris

What tale shall I to my old father tell?

Daphnis

'Twill make him chuckle thou art bestowed so well.

Chloris

But, after all, in truth I am to blame   70
To be so loving, ere I know your name;
A pleasant sounding name's a pretty thing.

Daphnis

Faith, mine's a very pretty name to sing.
They call me Daphnis; Lycidas my sire;
Both sound as well as woman can desire.
Nomaea bore me; farmers in degree;
He a good husband, a good housewife she.

Chloris

Your kindred is not much amiss, 'tis true;
Yet I am somewhat better born than you.

Daphnis

I know your father, and his family;   80
And, without boasting, am as good as he,
Menalcas; and no master goes before.

Chloris

Hang both our pedigrees! not one word more;
But if you love me, let me see your living,
Your house, and home; for seeing is believing.

Daphnis

See first yon cypress grove, a shade from noon.

Chloris

Browse on, my goats; for I'll be with you soon.

Daphnis

Feed well, my bulls, to whet your appetite,
That each may take a lusty leap at night.

Chloris

What do you mean, uncivil as you are,   90
To touch my breasts, and leave my bosom bare?

Daphnis

These pretty bubbies first I make my own.

Chloris

Pull out your hand, I swear, or I shall swoon.

Daphnis

Why does thy ebbing blood forsake thy face?

Chloris

Throw me at least upon a cleaner place;
My linen ruffled, and my waistcoat soiling—
What, do you think new clothes were made for spoiling?

Daphnis

I'll lay my lambskins underneath thy back.

Chloris

My head-gear's off; what filthy work you make!

Daphnis

To Venus, first, I lay these offerings by.   100

Chloris

Nay, first look round, that nobody be nigh:
Methinks I hear a whispering in the grove.

Daphnis

The cypress trees are telling tales of love.

Chloris

You tear off all behind me, and before me;
And I'm as naked as my mother bore me.

Daphnis

I'll buy thee better clothes than these I tear,
And lie so close I'll cover thee from air.

Chloris

You're liberal now; but when your turn is sped,
You'll wish me choked with every crust of bread.

Daphnis

I'll give thee more, much more than I have told;   110
Would I could coin my very heart to gold!

Chloris

Forgive thy handmaid, huntress of the wood!
I see there's no resisting flesh and blood!

Daphnis

The noble deed is done!—my herds I'll cull;
Cupid, be thine a calf; and Venus, thine a bull.

Chloris

A maid I came in an unlucky hour,
But hence return without my virgin flower.

Daphnis

A maid is but a barren name at best;
If thou canst hold, I bid for twins at least.

    Thus did this happy pair their love dispense   120
With mutual joys, and gratified their sense;
The god of love was there, a bidden guest,
And present at his own mysterious feast.
His azure mantle underneath he spread,
And scattered roses on the nuptial bed;
While folded in each other's arms they lay,
He blew the flames, and furnished out the play,
And from their foreheads wiped the balmy sweat away.
First rose the maid, and with a glowing face,
Her downcast eyes beheld her print upon the grass;   130
Thence to her herd she sped herself in haste:
The bridegroom started from his trance at last,
And piping homeward jocundly he passed.

A New Song

I

Sylvia the fair, in the bloom of fifteen,
Felt an innocent warmth as she lay on the green;
She had heard of a pleasure, and something she guessed
By the tousing, and tumbling, and touching her breast.°
She saw the men eager, but was at a loss,
What they meant by their sighing, and kissing so close;
      By their praying and whining,
      And clasping and twining,
      And panting and wishing,
      And sighing and kissing,   10
      And sighing and kissing so close.

II

Ah! she cried, ah for a languishing maid,
In a country of Christians, to die without aid!
Not a Whig, or a Tory, or Trimmer at least,°
Or a Protestant parson, or Catholic priest,

To instruct a young virgin, that is at a loss,
What they meant by their sighing, and kissing so close!
    By their praying and whining,
    And clasping and twining,
    And panting and wishing.   20
    And sighing and kissing,
    And sighing and kissing so close.

III

Cupid, in shape of a swain, did appear,
He saw the sad wound, and in pity drew near;
Then showed her his arrow, and bid her not fear,
For the pain was no more than a maiden may bear.
When the balm was infused, she was not at a loss,
What they meant by their sighing, and kissing so close;
    By their praying and whining,
    And clasping and twining,   30
    And panting and wishing,
    And sighing and kissing,
    And sighing and kissing so close.

Song

I

Go tell Amynta, gentle swain,
I would not die, nor dare complain:
Thy tuneful voice with numbers join,
Thy words will more prevail than mine.
To souls oppressed, and dumb with grief,
The gods ordain this kind relief,
That music should in sounds convey,
What dying lovers dare not say.

II

A sigh or tear perhaps she'll give,
But love on pity cannot live.   10
Tell her that hearts for hearts were made,
And love with love is only paid.

Tell her my pains so fast increase,
That soon they will be past redress;
But ah! the wretch that speechless lies,
Attends but death to close his eyes.

The Third Ode of the First Book of Horace

Inscribed to the Earl of Roscommon on
his Intended Voyage to Ireland°

So may the auspicious queen of Love
And the twin stars (the seed of Jove),°
And he, who rules the raging wind°
To thee, O sacred ship, be kind,
And gentle breezes fill thy sails,
Supplying soft Etesian gales,°
As thou to whom the muse commends,
The best of poets and of friends,
Dost thy committed pledge restore,
And land him safely on the shore;   10
And save the better part of me,
From perishing with him at sea.
Sure he, who first the passage tried,
In hardened oak his heart did hide,
And ribs of iron armed his side!
Or his at least, in hollow wood,
Who tempted first the briny flood,
Nor feared the winds' condending roar,
Nor billows beating on the shore;
Nor Hyades portending rain,°   20
Nor all the tyrants of the main.
What form of death could him affright,
Who unconcerned with steadfast sight
Could view the surges mounting steep,
And monsters rolling in the deep?
Could through the ranks of ruin go,
With storms above, and rocks below!
In vain did nature's wise command,
Divide the waters from the land,
If daring ships, and men profane,   30
Invade the inviolable main:

The eternal fences over-leap
And pass at will the boundless deep.
No toil, no hardship can restrain
Ambitious man inured to pain:
The more confined, the more he tries,
And at forbidden quarry flies.
Thus bold Prometheus did aspire,
And stole from heaven the seed of fire:
A train of ills, a ghastly crew,   40
The robber's blazing track pursue:
Fierce famine, with her meagre face,
And fevers of the fiery race,
In swarms the offending wretch surround,
All brooding on the blasted ground:
And limping death lashed on by fate
Comes up to shorten half our date.
This made not Daedalus beware,
With borrowed wings to sail in air;
To hell Alcides forced his way,   50
Plunged through the lake, and snatched the prey.
Nay scarce the gods, or heavenly climes
Are safe from our audacious crimes:
We reach at Jove's imperial crown,
And pull the unwilling thunder down.°

The Ninth Ode of the First Book of Horace

I

Behold yon mountain's hoary height,
    Made higher with new mounts of snow;
Again behold the winter's weight
    Oppress the labouring woods below;
And streams, with icy fetters bound,
Benumbed and cramped to solid ground.

II

With well-heaped logs dissolve the cold,
    And feed the genial hearth with fires;
Produce the wine, that makes us bold,
    And sprightly wit and love inspires: 10

For what hereafter shall betide,
God, if 'tis worth his care, provide.

III

Let him alone, with what he made,
   To toss and turn the world below;
At his command the storms invade;
   The winds by his commission blow;
Till with a nod he bids them cease,
And then the calm returns, and all is peace.

IV

Tomorrow and her works defy,
   Lay hold upon the present hour,   20
And snatch the pleasures passing by,
   To put them out of fortune's power:
Nor love, nor love's delights disdain;
Whate'er thou gettest today is gain.

V

Secure those golden early joys
   That youth unsoured with sorrow bears,
Ere withering time the taste destroys,
   With sickness and unwieldy years.
For active sports, for pleasing rest,
This is the time to be possessed;   30
The best is but in season best.

VI

The pointed hour of promised bliss,°
   The pleasing whisper in the dark,
The half-unwilling willing kiss,
   The laugh that guides thee to the mark,
When the kind nymph would coyness feign,
And hides but to be found again;
These, these are joys the gods for youth ordain.

The Twenty-ninth Ode of the
Third Book of Horace

Paraphrased in Pindaric Verse, and Inscribed to
the Right Honourable Laurence Earl of Rochester

I

Descended of an ancient line,
   That long the Tuscan sceptre swayed,°
Make haste to meet the generous wine,
   Whose piercing is for thee delayed:
The rosy wreath is ready made,
   And artful hands prepare
The fragrant Syrian oil that shall perfume thy hair.

II

When the wine sparkles from afar,
   And the well-natured friend cries 'Come away,'
Make haste and leave thy business and thy care,   10
   No mortal interest can be worth thy stay.

III

Leave for a while thy costly country seat;
   And, to be great indeed, forget
The nauseous pleasures of the great:
   Make haste and come:
Come, and forsake thy cloying store;
   Thy turret that surveys from high
The smoke and wealth and noise of Rome,
   And all the busy pageantry
That wise men scorn and fools adore: 20
Come, give thy soul a loose, and taste the pleasures of
   the poor.°

Sometimes 'tis grateful to the rich, to try
A short vicissitude, and fit of poverty:°
   A savoury dish, a homely treat,
   Where all is plain, where all is neat,

Without the stately spacious room,
The Persian carpet, or the Tyrian loom,°
Clear up the cloudy foreheads of the great.

V

    The sun is in the Lion mounted high;
        The Syrian star°   30
        Barks from afar,
    And with his sultry breath infects the sky;
The ground below is parched, the heavens above us fry;
    The shepherd drives his fainting flock,
    Beneath the covert of a rock
    And seeks refreshing rivulets nigh.
    The sylvans to their shades retire,°
Those very shades and streams new shades and streams require,
And want a cooling breeze of wind to fan the raging fire.

VI

    Thou, what befits the new Lord Mayor,°   40
    And what the city faction dare,
    And what the Gallic arms will do,
    And what the quiver-bearing foe,
    Art anxiously inquisitive to know:
But God has wisely hid from human sight
    The dark decrees of future fate,
    And sown their seeds in depth of night:
He laughs at all the giddy turns of state,
When mortals search too soon and fear too late.

VII

    Enjoy the present smiling hour,   50
    And put it out of fortune's power;
The tide of business, like a running stream,
    Is sometimes high and sometimes low,
A quiet ebb or a tempestuous flow,
    And always in extreme.
    Now with a noiseless gentle course
    It keeps within the middle bed;
    Anon it lifts aloft the head,

And bears down all before it, with impetuous force;
  And trunks of trees come rolling down,
  Sheep and their folds together drown,
 Both house and homestead into seas are borne,
 And rocks are from their old foundations torn,
And woods made thin with winds, their scattered honours
 mourn.

VIII

 Happy the man, and happy he alone,
 He who can call today his own;
 He who, secure within, can say
'Tomorrow do thy worst, for I have lived today:
 Be fair or foul or rain or shine,
The joys I have possessed in spite of fate are mine.   70
 Not heaven itself upon the past has power;
But what has been has been, and I have had my hour.'

IX

 Fortune, that with malicious joy
  Does man her slave oppress,
 Proud of her office to destroy,
  Is seldom pleased to bless;
 Still various and unconstant still,
But with an inclination to be ill,
 Promotes, degrades, delights in strife,
 And makes a lottery of life.   80
 I can enjoy her while she's kind;
 But when she dances in the wind,
 And shakes her wings, and will not stay,
 I puff the prostitute away.
The little or the much she gave is quietly resigned;
 Content with poverty my soul I arm,
 And virtue, though in rags, will keep me warm.

X

    What is't to me,
Who never sail in her unfaithful sea,
 If storms arise and clouds grow black;   90
 If the mast split and threaten wreck?

Then let the greedy merchant fear
    For his ill-gotten gain,
And pray to gods that will not hear,
While the debating winds and billows bear
    His wealth into the main.
For me secure from fortune's blows,
(Secure of what I cannot lose,)
In my small pinnace I can sail,
    Contemning all the blustering roar,   100
    And running with a merry gale,
With friendly stars my safety seek,
Within some little winding creek;
    And see the storm ashore.

The Second Epode of Horace

'How happy in his low degree,
How rich in humble poverty, is he,
Who leads a quiet country life,
Discharged of business, void of strife,
And from the griping scrivener free.
(Thus, ere the seeds of vice were sown,
Lived men in better ages born
Who ploughed with oxen of their own
Their small paternal field of corn.)
Nor trumpets summon him to war,   10
    Nor drums disturb his morning sleep,
Nor knows he merchants' gainful care,
    Nor fears the dangers of the deep.
The clamours of contentious law,
    And court and state, he wisely shuns,
Nor bribed with hopes nor dared with awe
    To servile salutations runs;
But either to the clasping vine
    Does the supporting poplar wed,
Or with his pruning-hook disjoin   20
    Unbearing branches from their head,
    And grafts more happy in their stead;
Or climbing to a hilly steep
    He views his herds in vales afar,
Or shears his overburdened sheep,

Or mead for cooling drink prepares,
  Of virgin honey in the jars.
Or in the now declining year,
  When bounteous Autumn rears his head,
He joys to pull the ripened pear,   30
  And clustering grapes with purple spread.
The fairest of his fruit he serves,
  Priapus, thy rewards:°
Sylvanus too his part deserves,°
  Whose care the fences guards.
Sometimes beneath an ancient oak
  Or on the matted grass he lies:
No god of sleep he need invoke;
  The stream that o'er the pebbles flies,
  With gentle slumber crowns his eyes.   40
The wind that whistles through the sprays
  Maintains the consort of the song;
And hidden birds, with native lays
  The golden sleep prolong.
But when the blast of winter blows,
  And hoary frost inverts the year,
Into the naked woods he goes
  And seeks the tusky boar to rear,°
  With well-mouthed hounds and pointed spear;
Or spreads his subtle nets from sight   50
  With twinkling glasses to betray
The larks that in the meshes light,
  Or makes the fearful hare his prey.
Amidst his harmless easy joys
  No anxious care invades his health,
Nor love his peace of mind destroys,
  Nor wicked avarice of wealth.
But if a chaste and pleasing wife,
To ease the business of his life,
Divides with him his household care,   60
Such as the Sabine matrons were,°
Such as the swift Apulian's bride,°
  Sunburnt and swarthy though she be,
Will fire for winter nights provide,
  And without noise will oversee
  His children and his family,

And order all things till he come
Sweaty and overlaboured home;
If she in pens his flocks will fold,
   And then produce her dairy store,   70
With wine to drive away the cold
   And unbought dainties of the poor;
Not oysters of the Lucrine lake
   My sober appetite would wish,
   Nor turbot, or the foreign fish
That rolling tempests overtake,
   And hither waft the costly dish.
Not heath-poult, or the rarer bird°
   Which Phasis or Ionia yields,
More pleasing morsels would afford   80
     Than the fat olives of my fields;
Than chards or mallows for the pot,°
   That keep the loosened body sound,
Or than the lamb, that falls by lot
   To the just guardian of my ground.
Amidst these feasts of happy swains,
   The jolly shepherd smiles to see
His flock returning from the plains;
   The farmer is as pleased as he
To view his oxen, sweating smoke,   90
Bear on their necks the loosened yoke:
To look upon his menial crew
   That sit around his cheerful hearth,
And bodies spent in toil renew
   With wholesome food and country mirth.'
This Morecraft said within himself,°
   Resolved to leave the wicked town,
   And live retired upon his own.
He called his money in;
   But the prevailing love of pelf   100
   Soon split him on the former shelf,
And put it out again.

To Sir George Etherege

To you who live in chill degree,°
(As map informs of fifty-three,)
And do not much for cold atone,
By bringing thither fifty-one,
Methinks all climes should be alike,
From tropic e'en to pole arctic;
Since you have such a constitution
As nowhere suffers diminution.
You can be old in grave debate,
And young in love's affairs of state;　　　　　　　　10
And both to wives and husbands show
The vigour of a plenipo.
Like mighty missioner you come
Ad partes infidelium—°
A work of wondrous merit sure,
So far to go, so much endure;
And all to preach to German dame,
Where sound of Cupid never came.
Less had you done, had you been sent
As far as Drake or Pinto went,°　　　　　　　　20
For cloves or nutmegs to the line-a,
Or even for oranges to China.
That had indeed been charity,
Where lovesick ladies helpless lie,
Chapped, and for want of liquor dry.
But you have made your zeal appear
Within the circle of the Bear.°
What region of the earth so dull,
That is not of your labours full?
Triptolemus (so sing the nine)°　　　　　　　　30
Strewed plenty from his cart divine;
But spite of all those fable-makers,
He never sowed on Almain acres.
No, that was left by fate's decree
To be performed and sung by thee.
Thou breakst through forms with as much ease
As the French king through articles.°
In grand affairs thy days are spent,

Of waging weighty compliment,
With such as monarchs represent.   40
They who such vast fatigues attend,
Want some soft minutes to unbend,
To show the world that now and then
Great ministers are mortal men.
Then Rhenish rummers walk the round;°
In bumpers every king is crowned;
Besides three holy mitred hectors,°
And the whole college of Electors.
No health of potentate is sunk,
That pays to make his envoy drunk.   50
These Dutch delights, I mentioned last,
Suit not, I know, your English taste:
For wine to leave a whore or play,
Was ne'er your Excellency's way.
Nor need the title give offence,
For here you were his Excellence;
For gaming, writing, speaking, keeping,
His Excellence for all, but sleeping.
Now if you tope in form, and treat,
'Tis the sour sauce to the sweet meat,   60
The fine you pay for being great.
Nay, there's a harder imposition,
Which is indeed the court petition,
That, setting worldly pomp aside,
(Which poet has at font defied,)°
You would be pleased in humble way
To write a trifle called a play.
This truly is a degradation,
But would oblige the crown and nation
Next to your wise negotiation.   70
If you pretend (as well you may)
Your high degree, your friends will say,
That Duke St. Aignan made a play.°
If Gallic peer convince you scarce,
His Grace of Bucks has made a farce,°
And you, whose comic wit is terse all,
Can hardly fall below Rehearsal.
Then finish what you here began,
But scribble faster if you can;

For yet no George, to our discerning,   80
Has writ without a ten years' warning.

To the Pious Memory
of the Accomplished Young Lady
Mrs Anne Killigrew

Excellent in the Two Sister-arts of
Poesy and Painting

AN ODE

I

Thou youngest virgin-daughter of the skies,
Made in the last promotion of the blest;
Whose palms, new plucked from paradise,
In spreading branches more sublimely rise,
Rich with immortal green above the rest:
Whether, adopted to some neighbouring star,°
Thou roll'st above us, in thy wandering race,
    Or, in procession fixed and regular,°
    Moved with the heaven's majestic pace;
    Or, called to more superior bliss,   10
Thou tread'st with seraphims the vast abyss:
Whatever happy region is thy place,
Cease thy celestial song a little space;
Thou wilt have time enough for hymns divine,
    Since heaven's eternal year is thine.
Hear, then, a mortal muse thy praise rehearse,°
        In no ignoble verse;
But such as thy own voice did practise here,
When thy first fruits of poesy were given,
To make thyself a welcome inmate there;   20
        While yet a young probationer,
        And candidate of heaven.

II

    If by traduction came thy mind,°
    Our wonder is the less to find
A soul so charming from a stock so good;
Thy father was transfused into thy blood:

So wert thou born into a tuneful strain,
An early, rich, and inexhausted vein.
   But if thy pre-existing soul
   Was formed, at first, with myriads more,   30
It did through all the mighty poets roll,
   Who Greek or Latin laurels wore,
And was that Sappho last, which once it was before.°
   If so, then cease thy flight, O heaven-born mind!
   Thou hast no dross to purge from thy rich ore:
   Nor can thy soul a fairer mansion find,
   Than was the beauteous frame she left behind:
Return to fill or mend the choir of thy celestial kind.

III
   May we presume to say that at thy birth,
New joy was sprung in heaven as well as here on earth.   40
   For sure the milder planets did combine
   On thy auspicious horoscope to shine,
   And e'en the most malicious were in trine.°
      Thy brother-angels at thy birth
      Strung each his lyre, and tuned it high,
      That all the people of the sky
   Might know a poetess was born on earth
      And then, if ever, mortal ears
      Had heard the music of the spheres.
      And if no clustering swarm of bees°   50
   On thy sweet mouth distilled their golden dew,
      'Twas that such vulgar miracles
      Heaven had not leisure to renew:
   For all the blest fraternity of love
Solemnised there thy birth, and kept thy holiday above.

IV
   O gracious God! how far have we
Profaned thy heavenly gift of poesy?
Made prostitute and profligate the muse,
Debased to each obscene and impious use,
Whose harmony was first ordained above   60
For tongues of angels, and for hymns of love?

O wretched we! why were we hurried down
   This lubric and adulterate age,
   (Nay, added fat pollutions of our own)
   To increase the steaming ordures of the stage?
   What can we say to excuse our second fall?
   Let this thy vestal, heaven, atone for all:
   Her Arethusian stream remains unsoiled,°
   Unmixed with foreign filth, and undefiled;
Her wit was more than man, her innocence a child.   70

V

   Art she had none, yet wanted none;
   For nature did that want supply:
   So rich in treasures of her own,
   She might our boasted stores defy:
   Such noble vigour did her verse adorn,
   That it seemed borrowed where 'twas only born.
   Her morals, too, were in her bosom bred,
      By great examples daily fed,
What in the best of books, her father's life, she read:
   And to be read herself she need not fear;   80
   Each test, and every light, her muse will bear,
   Though Epictetus with his lamp were there.°
   E'en love (for love sometimes her muse expressed)
Was but a lambent flame which played about her breast,
   Light as the vapours of a morning dream,
   So cold herself, whilst she such warmth expressed,
   'Twas Cupid bathing in Diana's stream.

VI

   Born to the spacious empire of the Nine,°
   One would have thought she should have been content
   To manage well that mighty government;   90
   But what can young ambitious souls confine?
      To the next realm she stretched her sway,
      For painture near adjoinging lay,°
   A plenteous province, and alluring prey.
   A chamber of dependencies was framed,
   (As conquerors will never want pretence,
      When armed, to justify the offence,)
And the whole fief, in right of poetry, she claimed.

The country open lay without defence;
For poets frequent inroads there had made,   100
    And perfectly could represent
The shape, the face, with every lineament,
And all the large domains which the dumb sister swayed;
    All bowed beneath her government,
    Received in triumph wheresoe'er she went.
Her pencil drew whate'er her soul designed,
And oft the happy draft surpassed the image in her mind.
    The sylvan scenes of herds and flocks,
    And fruitful plains and barren rocks,
    Of shallow brooks that flowed so clear,   110
    The bottom did the top appear;
    Of deeper too and ampler floods,
    Which, as in mirrors, showed the woods;
    Of lofty trees, with sacred shades,
    And perspectives of pleasant glades,
    Where nymphs of brightest form appear,
    And shaggy satyrs standing near,
    Which them at once admire and fear.
    The ruins, too, of some majestic piece,
    Boasting the power of ancient Rome or Greece,   120
    Whose statues, friezes, columns, broken lie,
    And, though defaced, the wonder of the eye;
    What nature, art, bold fiction, e'er durst frame,
    Her forming hand gave feature to the name.
    So strange a concourse ne'er was seen before,
But when the peopled ark the whole creation bore.

VII

The scene then changed; with bold erected look
Our martial king the sight with reverence strook:°
For not content to express his outward part,
Her hand called out the image of his heart: 130
His warlike mind, his soul devoid of fear,
His high-designing thoughts were figured there,
As when, by magic, ghosts are made appear.
    Our phoenix-queen was portrayed too so bright,°
Beauty alone could beauty take so right:
Her dress, her shape, her matchless grace,
Were all observed, as well as heavenly face.

With such a peerless majesty she stands,
As in that day she took the crown from sacred hands:
Before a train of heroines was seen,   140
In beauty foremost, as in rank, the queen.
   Thus nothing to her genius was denied,
But like a ball of fire the further thrown,
   Still with a greater blaze she shone,
And her bright soul broke out on every side.
What next she had designed, heaven only knows:
To such immoderate growth her conquest rose,
That fate alone its progress could oppose.

VIII

   Now all those charms, that blooming grace,
The well-proportioned shape, and beauteous face,   150
Shall never more be seen by mortal eyes;
In earth the much-lamented virgin lies.
   Not wit, nor piety, could fate prevent;
   Nor was the cruel destiny content°
To finish all the murder at a blow,
To sweep at once her life and beauty too;
But, like a hardened felon, took a pride
   To work more mischievously slow,
   And plundered first, and then destroyed.
O double sacrilege on things divine,   160
To rob the relic, and deface the shrine!
   But thus Orinda died;°
Heaven, by the same disease, did both translate,
As equal were their souls, so equal were their fate.

IX

Meantime her warlike brother on the seas°
His waving streamers to the winds displays,
And vows for his return, with vain devotion, pays.
   Ah, generous youth! that wish forbear,
   The winds too soon will waft thee here:
Slack all thy sails, and fear to come;   170
Alas, thou know'st not, thou art wrecked at home!
No more shalt thou behold thy sister's face,
Thou hast already had her last embrace.

But look aloft, and if thou ken'st from far
Among the Pleiads a new-kindled star,°
If any sparkles than the rest more bright,
'Tis she that shines in that propitious light.

X

When in mid-air the golden trump shall sound,
   To raise the nations under ground;
     When in the valley of Jehosaphat,°   180
The judging God shall close the book of fate,
   And there the last assizes keep,
     For those who wake, and those who sleep;
       When rattling bones together fly,
From the four corners of the sky;
When sinews o'er the skeletons are spread,
Those clothed with flesh, and life inspires the dead;
The sacred poets first shall hear the sound,
   And foremost from the tomb shall bound,
For they are covered with the lightest ground;   190
And straight, with inborn vigour, on the wing,
Like mounting larks, to the new morning sing.
There thou, sweet saint, before the choir shalt go,
As harbinger of heaven, the way to show,
The way which thou so well hast learned below.

To my Ingenious Friend, Mr Henry Higden, Esq. on his Translation of the Tenth Satire of Juvenal

The Grecian wits, who satire first began,
Were pleasant pasquins on the life of man;°
At mighty villains, who the state oppressed,
They durst not rail: perhaps they laughed at least
And turned them out of office with a jest.
No fool could peep abroad, but ready stand
The drolls to clap a bauble in his hand.°
Wise legislators never yet could draw
A fop within the reach of common law;
For posture, dress, grimace, and affectation,   10
Though foes to sense, are harmless to the nation.

Our last redress is dint of verse to try,
And satire is our Court of Chancery.
This way took Horace to reform an age,
Not bad enough to need an author's rage:
But yours, who lived in more degenerate times,
Was forced to fasten deep, and worry crimes.
Yet you, my friend, have tempered him so well,
You make him smile in spite of all his zeal;
An art peculiar to yourself alone,   20
To join the virtues of two styles in one.

  Oh! were your author's principle received,
Half of the labouring world would be relieved;
For not to wish is not to be deceived.
Revenge would into charity be changed,
Because it costs too dear to be revenged;
It costs our quiet and content of mind,
And when 'tis compassed leaves a sting behind.
Suppose I had the better end o' the staff,
Why should I help the ill-natured world to laugh?   30
'Tis all alike to them, who get the day;
They love the spite and mischief of the fray.
No; I have cured myself of that disease;
Nor will I be provoked, but when I please:
But let me half that cure to you restore;°
You give the salve, I laid it to the sore.

  Our kind relief against a rainy day,
Beyond a tavern, or a tedious play,
We take your book, and laugh our spleen away.
If all your tribe, too studious of debate,   40
Would cease false hopes and titles to create,
Led by the rare example you begun,
Clients would fail, and lawyers be undone.

A Song for St Cecilia's Day, 1687

I

From harmony, from heavenly harmony,
  This universal frame began:°
    When nature underneath a heap
      Of jarring atoms lay,

And could not heave her head,
The tuneful voice was heard from high,
    'Arise, ye more than dead.'
Then cold, and hot, and moist, and dry,
In order to their stations leap,
    And Music's power obey.   10
From harmony, from heavenly harmony,
    This universal frame began;
    From harmony to harmony
Through all the compass of the notes it ran,°
The diapason closing full in man.°

II

What passion cannot music raise and quell?
    When Jubal struck the corded shell,°
    His listening brethren stood around,
    And, wondering, on their faces fell
    To worship that celestial sound: 20
Less than a God they thought there could not dwell
    Within the hollow of that shell
    That spoke so sweetly and so well.
What passion cannot music raise the quell?

III

    The trumpet's loud clangour
      Excites us to arms,
    With shrill notes of anger
      And mortal alarms.
The double double double beat
    Of the thundering drum,   30
Cries 'hark! the foes come:
Charge, charge! 'tis too late to retreat.'

IV

The soft complaining flute,
In dying notes discovers
The woes of hopeless lovers;
Whose dirge is whispered by the warbling lute.

V

Sharp violins proclaim
Their jealous pangs, and desperation,
Fury, frantic indignation,
Depth of pains, and height of passion,   40
    For the fair, disdainful dame.

VI

But, oh! what art can teach,
    What human voice can reach,
The sacred organ's praise?
Notes inspiring holy love,
Notes that wing their heavenly ways
        To mend the choirs above.

VII

Orpheus could lead the savage race;
And trees unrooted left their place,
    Sequacious of the lyre:°   50
But bright Cecilia raised the wonder higher;
When to her organ vocal breath was given,
An angel heard, and straight appeared,
    Mistaking earth for heaven.

GRAND CHORUS

As from the power of sacred lays
    The spheres began to move,
And sung the great creator's praise
    To all the blessed above;
So when the last and dreadful hour
This crumbling pageant shall devour,   60
The trumpet shall be heard on high,
The dead shall live, the living die,
And Music shall untune the sky.°

Prologue to Don Sebastian

Spoken by a woman

The judge removed, though he's no more My Lord,
May plead at bar or at the Council-Board:
So may cast poets write; there's no pretension°
To argue loss of wit from loss of pension.
Your looks are cheerful; and in all this place
I see not one that wears a damning face.
The British nation is too brave to show
Ignoble vengeance on a vanquished foe;
At least be civil to the wretch imploring,
And lay your paws upon him without roaring.   10
Suppose our poet was your foe before;
Yet now the business of the field is o'er;
'Tis time to let your civil wars alone
When troops are into winter-quarters gone.
Jove was alike to Latian and to Phrygian;°
And you well know a play's of no religion.
Take good advice and please yourselves this day;
No matter from what hands you have the play.
Among good fellows every health will pass
That serves to carry round another glass: 20
When with full bowls of burgundy you dine,
Though at the mighty monarch you repine,
You grant him still most christian in his wine.
   Thus far the poet, but his brains grow addle;
And all the rest is purely from this noddle.
You've seen young ladies at the senate door
Prefer petitions and your grace implore;
However grave the legislators were,
Their cause went ne'er the worse for being fair.
Reasons as weak as theirs perhaps I bring,   30
But I could bribe you with as good a thing.
I heard him make advances of good nature,
That he for once would sheath his cutting satire;
Sign but his peace, he vows he'll ne'er again
The sacred names of fops and beaus profane.

Strike up the bargain quickly; for I swear,
As times go now, he offers very fair.
Be not too hard on him with statutes neither;
Be kind, and do not set your teeth together
To stretch the laws, as cobblers do their leather.   40
Horses by papists are not to be ridden;
But sure the muses' horse was ne'er forbidden.
For in no rate-book it was ever found
That Pegasus was valued at five pound.°
Fine him to daily drudging and inditing;
And let him pay his taxes out, in writing.

Prologue to The Mistakes

Save ye, sirs, save ye! I am in a hopeful way.
I should speak something, in rhyme, now, for the play;
But the deuce take me, if I know what to say.
I'll stick to my friend the author, that I can tell ye,
To the last drop of claret in my belly.
So far I'm sure 'tis rhyme—that needs no granting;
And, if my verses' feet stumble—you see my own are wanting.
Our young poet has brought a piece of work,
In which though much of art there does not lurk,
It may hold out three days—and that's as long as Cork.°   10
But, for this play—(which till I have done, we show not)
What may be its fortune—by the Lord—I know not.
This I dare swear, no malice here is writ;
'Tis innocent of all things—e'en of wit.
He's no high-flyer—he makes no sky-rockets,
His squibs are only levelled at your pockets;
And if his crackers light among your pelf,
You are blown up; if not, then he's blown up himself.
By this time, I'm something recovered of my flustered
      madness;
And now, a word or two in sober sadness.   20
Ours is a common play; and you pay down
A common harlot's price—just half a crown.
You'll say, I play the pimp, on my friend's score;
But, since 'tis for a friend, your gibes give o'er,
For many a mother has done that before.

How's this? you cry: an actor write?—we know it;
But Shakespeare was an actor, and a poet.
Has not great Jonson's learning often failed?
But Shakespeare's greater genius still prevailed.
Have not some writing actors, in this age,⁣   30
Deserved and found success upon the stage?
To tell the truth, when our old wits are tired,
Not one of us but means to be inspired.
Let your kind presence grace our homely cheer;
Peace and the butt is all our business here;°
So much for that—and the devil take small beer.

The Lady's Song

I

A choir of bright beauties in spring did appear,
To choose a May-lady to govern the year:°
All the nymphs were in white, and the shepherds in green,
The garland was given, and Phyllis was queen;
But Phyllis refused it, and sighing did say,
I'll not wear a garland while Pan is away.

II

While Pan and fair Syrinx are fled from our shore,°
The Graces are banished, and Love is no more;
The soft god of pleasure, that warmed our desires,
Has broken his bow, and extinguished his fires,⁣   10
And vows that himself and his mother will mourn,
Till Pan and fair Syrinx in triumph return.

III

Forbear your addresses, and court us no more,
For we will perform what the deity swore:
But, if you dare think of deserving our charms,
Away with your sheep-hooks, and take to your arms;
Then laurels and myrtles your brows shall adorn,
When Pan, and his son, and far Syrinx, return.

To Mr Southerne on his Comedy called
The Wives' Excuse

Sure there's a fate in plays, and 'tis in vain
To write, while these malignant planets reign.
Some very foolish influence rules the pit,
Not always kind to sense, or just to wit;
And whilst it lasts, let buffoonery succeed,
To make us laugh, for never was more need.
Farce, in itself, is of a nasty scent;
But the gain smells not of the excrement.
The Spanish nymph, a wit and beauty too,°
With all her charms, bore but a single show;   10
But let a monster Muscovite appear,
He draws a crowded audience round the year.
Maybe thou hast not pleased the box and pit;
Yet those who blame thy tale commend thy wit:
So Terence plotted, but so Terence writ.°
Like his, thy thoughts are true, thy language clean;
E'en lewdness is made moral in thy scene.°
The hearers may for want of Nokes repine;°
But rest secure, the readers will be thine.
Nor was thy laboured drama damned or hissed,   20
But with a kind civility dismissed;
With such good manners as the wife did use,°
Who, not accepting, did but just refuse.
There was a glance at parting; such a look,
As bids thee not give o'er for one rebuke.
But if thou wouldst be seen, as well as read,
Copy one living author, and one dead.
The standard of thy style let Etherege be;°
For wit, the immortal spring of Wycherley.°
Learn, after both, to draw some just design,   30
And the next age will learn to copy thine.

Eleonora

A PANEGYRIC POEM DEDICATED TO THE MEMORY OF THE LATE COUNTESS OF ABINGDON

> superas evadere ad auras,°
> hoc opus, hic labor est. pauci, quos aequus amavit
> Iuppiter, aut ardens evexit ad aethera virtus,
> dis geniti potuere

[to pass out to the upper air, this is the task, this the toil. Some few, whom kindly Jupiter has loved, or shining worth uplifted to heaven, sons of the gods, have availed]

To
the Right Honourable
the Earl of Abingdon°

MY LORD,

The commands, with which you honoured me some months ago, are now performed: they had been sooner, but betwixt ill health, some business, and many troubles, I was forced to defer them till this time. Ovid, going to his banishment, and writing from on shipboard to his friends, excused the faults of his poetry by his misfortunes;° and told them, that good verses never flow, but from a serene and composed spirit. Wit, which is a kind of Mercury, with wings fastened to his head and heels, can fly but slowly in a damp air. I therefore chose rather to obey you late than ill; if at least I am capable of writing anything, at any time, which is worthy your perusal and your patronage. I cannot say that I have escaped from a shipwreck; but have only gained a rock by hard swimming, where I may pant awhile and gather breath; for the doctors give me a sad assurance, that my disease° never took its leave of any man but with a purpose to return. However, my lord, I have laid hold on the interval, and managed the small stock, which age has left me, to the best advantage, in performing this inconsiderable service to my lady's memory. We, who are priests of Apollo, have not the inspiration when we please; but must wait till the god comes rushing on us, and invades us with a fury which we are not able to resist; which gives us double strength while the fit continues, and leaves us languishing and spent, at its departure. Let me not seem to boast, my lord, for I have really felt it on this occasion, and prophesied beyond my natural power. Let me add, and hope to be believed, that the excellency of the subject contributed much to the happiness of the execution; and that the weight of thirty years was taken off me while I was writing. I swam with the tide, and the water under me was buoyant. The reader will easily observe that I was transported by the multitude and variety of my similitudes;° which are generally the product of a luxuriant fancy, and the wantonness of wit. Had I called in my judgment to my assistance, I had certainly retrenched many of them. But I defend them not; let them pass for beautiful faults amongst the better sort of critics; for the whole poem, though written in that which they call heroic verse, is of the pindaric nature, as well in the thought as the expression; and, as such, requires the same grains of allowance for it. It was intended, as your lordship sees in the title, not for an elegy, but a panegyric: a kind of apotheosis, indeed, if a heathen word may be applied to a Christian use. And on all occasions of praise, if we take the ancients for our patterns, we are bound by prescription to employ the magnificence of words, and the force of figures,° to adorn the sublimity of thoughts. Isocrates amongst the Grecian orators, and Cicero, and the younger Pliny, amongst the Romans, have left us their precedents for our security; for I think I need not mention the inimitable Pindar, who stretches on these pinions out of sight, and is carried upward, as it were, into another world.

This, at least, my lord, I may justly plead, that, if I have not performed so well as I think I have, yet I have used my best endeavours to excel myself. One disadvantage I have had, which is never to have known or seen my lady; and to draw the lineaments of her mind from the description which I have received from others, is for a painter to set himself at work without the living original before him; which, the more beautiful it is, will be so much the more difficult for him to conceive, when he has only a relation given him of such and such features by an acquaintance or a friend, without the nice touches, which give the best resemblance, and make the graces of the picture. Every artist is apt enough to flatter himself, and I amongst the rest, that their own ocular observations would have discovered more perfections, at least others, than have been delivered to them; though I have received mine from the best hands, that is, from persons who neither want a just understanding of my lady's worth, nor a due veneration for her memory.

Doctor Donne, the greatest wit, though not the best poet of our nation, acknowledges that he had never seen Mrs Drury,° whom he has made immortal in his admirable 'Anniversaries'. I have had the same fortune, though I have not succeeded to the same genius. However, I have followed his footsteps in the design of his panegyric; which was to raise an emulation in the living, to copy out the example of the dead. And therefore it was, that I once intended to have called this poem 'The Pattern'; and though, on a second consideration, I changed the title into the name of that illustrious person, yet the design continues, and Eleonora is still the pattern of charity, devotion, and humility; of the best wife, the best mother, and the best of friends.

And now, my lord, though I have endeavoured to answer your commands, yet I could not answer it to the world, nor to my conscience, if I gave not your lordship my testimony of being the best husband now living: I say my testimony only; for the praise of it is given you by yourself. They, who despise the rules of virtue both in their practice and their morals, will think this a very trivial commendation. But I think it the peculiar happiness of the Countess of Abingdon, to have been so truly loved by you, while she was living, and so gratefully honoured, after she was dead. Few there are who have either had, or could have, such a loss; and yet fewer, who carried their love and constancy beyond the grave. The exteriors of mourning, a decent funeral, and black habits, are the usual stints of common husbands; and perhaps their wives deserve no better than to be mourned with hypocrisy, and forgot with ease. But you have distinguished yourself from ordinary lovers, by a real and lasting grief for the deceased; and by endeavouring to raise for her the most durable monument, which is that of verse. And so it would have proved, if the workman had been equal to the work, and your choice of the artificer as happy as your design. Yet, as Phidias,° when he had made the statue of Minerva, could not forbear to engrave his own name, as author of the piece; so give me leave to hope, that by subscribing mine to this poem, I may live by the goddess, and transmit my name to posterity by the memory of hers. It is no flattery to assure your lordship, that she is remembered in the present age by all who have had the honour of her conversation and acquaintance; and that I have never been in any company since the news of her death was first brought me where they have not extolled her virtues, and even spoken the same things of her in prose, which I have done in verse.

I therefore think myself obliged to thank your lordship for the commission which you have given me: how I have acquitted myself of it, must be left to the opinion of the world, in spite of any protestation which I can enter against the present age, as incompetent or corrupt judges. For my comfort they are but Englishmen; and, as such, if they think ill of me today, they are inconstant enough to think well of me tomorrow. And after all, I have not much to thank my fortune that I was born amongst them. The good of both sexes are so few, in England, that they stand like exceptions against general rules; and though one of them has deserved a greater commendation than I could give her, they have taken care that I should not tire my pen with frequent exercise on the like subjects; that praises, like taxes, should be appropriated, and left almost as individual as the person. They say my talent is satire; if it be so, it is a fruitful age, and there is an extraordinary crop to gather. But a single hand is insufficient for such a harvest: they° have sown the dragon's teeth° themselves, and it is but just they should reap each other in lampoons. You, my lord, who have the character of honour, though it is not my happiness to know you, may stand aside, with the small remainders of the English nobility, truly such, and, unhurt yourselves, behold the mad combat. If I have pleased you, and some few others, I have obtained my end. You see I have disabled myself, like an elected Speaker of the House;° yet, like him, I have undertaken the charge, and find the burden sufficiently recompensed by the honour. Be pleased to accept of these my unworthy labours, this paper monument; and let her pious memory, which I am sure is sacred to you, not only plead the pardon of my many faults, but gain me your protection, which is ambitiously sought by,

> MY LORD,
> Your Lordship's
> Most obedient Servant,
> JOHN DRYDEN

The introduction As when some great and gracious monarch dies,
Soft whispers first and mournful murmurs rise
Among the sad attendants; then the sound
Soon gathers voice and spreads the news around,
Through town and country till the dreadful blast
Is blown to distant colonies at last,
Who then, perhaps, were offering vows in vain,
For his long life and for his happy reign:
So slowly, by degrees, unwilling fame
Did matchless Eleonora's fate proclaim,   10
Till public as the loss the news became.
Of her charity   The nation felt it in the extremest parts,
With eyes o'erflowing, and with bleeding hearts;

But most the poor, whom daily she supplied,
Beginning to be such, but when she died.
For while she lived they slept in peace by night,
Secure of bread, as of returning light,
And with such firm dependence on the day,
That need grew pampered, and forgot to pray;
So sure the dole, so ready at their call,   20
They stood prepared to see the manna fall.
   Such multitudes she fed, she clothed, she nursed,
That she herself might fear her wanting first.
Of her five talents, other five she made;°
Heaven, that had largely given, was largely paid;
And in few lives, in wondrous few, we find
A fortune better fitted to the mind.
Nor did her alms from ostentation fall,
Or proud desire of praise—the soul gave all:
Unbribed it gave; or, if a bribe appear,   30
No less than heaven, to heap huge treasures there.
   Want passed for merit at her open door:
Heaven saw, he safely might increase his poor,
And trust their sustenance with her so well,
As not to be at charge of miracle.
None could be needy, whom she saw or knew;
All in the compass of her sphere she drew:
He, who could touch her garment, was as sure,°
As the first Christians of the apostles' cure.
The distant heard, by fame, her pious deeds,   40
And laid her up for their extremest needs;
A future cordial for a fainting mind;
For what was ne'er refused, all hoped to find,
Each in his turn: the rich might freely come,
As to a friend; but to the poor, 'twas home.
As to some holy house the afflicted came,
The hunger-starved, the naked, and the lame,
Want and diseases fled before her name.
For zeal like hers her servants were too slow;
She was the first, where need required, to go;   50
Herself the foundress and attendant too.
   Sure she had guests sometimes to entertain,
Guests in disguise, of her great master's train:

Her lord himself might come, for aught we know,
Since in a servant's form he lived below:
Beneath her roof he might be pleased to stay;
Or some benighted angel, in his way,
Might ease his wings, and seeing heaven appear
In its best work of mercy, think it there;
Where all the deeds of charity and love   60
Were in as constant method, as above,
All carried on; all of a piece with theirs;
As free her alms, as diligent her cares;
As loud her praises, and as warm her prayers.

Of her prudent   Yet was she not profuse; but feared to waste,
management   And wisely managed, that the stock might last;
That all might be supplied, and she not grieve,
When crowds appeared, she had not to relieve:
Which to prevent, she still increased her store;
Laid up, and spared, that she might give the more.   70
So Pharaoh, or some greater king than he,°
Provided for the seventh necessity:
Taught from above his magazines to frame,
That famine was prevented ere it came.
Thus heaven, though all-sufficient, shows a thrift°
In his economy, and bounds his gift;
Creating for our day one single light,
And his reflection too supplies the night.
Perhaps a thousand other worlds, that lie
Remote from us, and latent in the sky,°   80
Are lightened by his beams, and kindly nursed,
Of which our earthly dunghill is the worst.

    Now, as all virtues keep the middle line,
Yet somewhat more to one extreme incline,
Such was her soul; abhorring avarice,
Bounteous, but almost bounteous to a vice:
Had she given more, it had profusion been,
And turned the excess of goodness into sin.

Of her humility   These virtues raised her fabric to the sky;
For that which is next heaven is charity.   90
But as high turrets for their airy steep°
Require foundations in proportion deep,
And lofty cedars as far upward shoot
As to the nether heavens they drive the root;

So low did her secure foundation lie,
She was not humble, but humility.
Scarcely she knew that she was great, or fair,
Or wise, beyond what other women are,
Or, which is better, knew, but never durst compare.°
For to be conscious of what all admire,   100
And not be vain, advances virtue higher.
But still she found, or rather thought she found,
Her own worth wanting, others' to abound;
Ascribed above their due to everyone,
Unjust and scanty to herself alone.

Of her piety   Such her devotion was, as might give rules
Of speculation to disputing schools,
And teach us equally the scales to hold
Betwixt the two extremes of hot and cold;
That pious heat may moderately prevail,   110
And we be warmed, but not be scorched with zeal.
Business might shorten, not disturb, her prayer;
Heaven had the best, if not the greater share.
An active life long orisons forbids;°
Yet still she prayed, for still she prayed by deeds.

Her every day was Sabbath; only free
From hours of prayer, for hours of charity.
Such as the Jews from servile toil released,
Where works of mercy were a part of rest;
Such as blest angels exercise above,   120
Varied with sacred hymns and acts of love;
Such Sabbaths as that one she now enjoys,
E'en that perpetual one, which she employs
(For such vicissitudes in heaven there are)°
In praise alternate, and alternate prayer.
All this she practised here, that when she sprung
Amidst the choirs, at the first sight she sung;
Sung, and was sung herself in angels' lays;
For, praising her, they did her maker praise.
All offices of heaven so well she knew,°   130
Before she came, that nothing there was new;
And she was so familiarly received,
As one returning, not as one arrived.

Of her various   Muse, down again precipitate thy flight;
virtues   For how can mortal eyes sustain immortal light?°

But as the sun in water we can bear,
Yet not the sun, but his reflection there,
So let us view her here in what she was,
And take her image in this watery glass:°
Yet look not every lineament to see;   140
Some will be cast in shades, and some will be
So lamely drawn, you scarcely know 'tis she.
For where such various virtues we recite,
'Tis like the milky way, all over bright,
But sown so thick with stars, 'tis undistinguished light.

Her virtue, not her virtues, let us call;
For one heroic comprehends them all:°
One, as a constellation is but one,
Though 'tis a train of stars, that, rolling on,
Rise in their turn, and in the zodiac run,   150
Ever in motion; now 'tis faith ascends,
Now hope, now charity, that upward tends,
And downwards with diffusive good descends.

As in perfumes composed with art and cost,
'Tis hard to say what scent is uppermost;
Nor this part musk or civet can we call,
Or amber, but a rich result of all;
So she was all a sweet, whose every part,
In due proportion mixed, proclaimed the maker's
        art.

No single virtue we could most commend,   160
Whether the wife, the mother, or the friend;
For she was all, in that supreme degree,
That as no one prevailed, so all was she.
The several parts lay hidden in the piece;
The occasion but exerted that, or this.

Of her conjugal   A wife as tender, and as true withal,
virtues   As the first woman was before her fall:
Made for the man of whom she was a part;
Made to attract his eyes, and keep his heart.
A second Eve, but by no crime accursed;   170
As beauteous, not as brittle as the first.
Had she been first, still paradise had been,
And death had found no entrance by her sin.
So she not only had preserved from ill
Her sex and ours, but lived their pattern still.

Love and obedience to her lord she bore;
She much obeyed him, but she loved him more:
Not awed to duty by superior sway,
But taught by his indulgence to obey.
Thus we love God, as author of our good;   180
So subjects love just kings, or so they should.
Nor was it with ingratitude returned;
In equal fires the blissful couple burned;
One joy possessed them both, and in one grief they
    mourned.
His passion still improved; he loved so fast,
As if he feared each day would be her last.
Too true a prophet to foresee the fate
That should so soon divide their happy state;
When he to heaven entirely must restore
That love, that heart, where he went halves before.   190
Yet as the soul is all in every part,
So God and he might each have all her heart.

Of her love to   So had her children too; for charity
her children Was not more fruitful, or more kind, than she:
Each under other by degrees they grew;
A goodly perspective of distant view.
Anchises looked not with so pleased a face,°
In numbering o'er his future Roman race,
And marshalling the heroes of his name,
As, in their order, next to light they came;   200
Nor Cybele, with half so kind an eye,°
Surveyed her sons and daughters of the sky;
Proud, shall I say, of her immortal fruit,
As far as pride with heavenly minds may suit.

Her care of   Her pious love excelled to all she bore;
their education New objects only multiplied it more.
And as the chosen found the pearly grain°
As much as every vessel could contain;
As in the blissful vision each shall share
As much of glory as his soul can bear;   210
So did she love, and so dispense her care.
Her eldest thus, by consequence, was best,
As longer cultivated than the rest.
The babe had all that infant care beguiles,
And early knew his mother in her smiles:

But when dilated organs let in day
To the young soul, and gave it room to play,
At his first aptness, the maternal love
Those rudiments of reason did improve:
The tender age was pliant to command;   220
Like wax it yielded to the forming hand:
True to the artificer, the laboured mind
With ease was pious, generous, just, and kind;
Soft for impression, from the first prepared,
Till virtue with long exercise grew hard:
With every act confirmed, and made at last
So durable as not to be effaced,
It turned to habit; and, from vices free,
Goodness resolved into necessity.

Thus fixed she virtue's image, (that's her own,)   230
Till the whole mother in the children shone;
For that was their perfection: she was such,
They never could express her mind too much.
So unexhausted her perfections were,
That, for more children, she had more to spare;
For souls unborn, whom her untimely death
Deprived of bodies, and of mortal breath;
And could they take the impressions of her mind,
Enough still left to sanctify her kind.

Of her friendship   Then wonder not to see this soul extend   240
The bounds, and seek some other self, a friend:
As swelling seas to gentle rivers glide,
To seek repose, and empty out the tide;
So this full soul, in narrow limits pent,
Unable to contain her, sought a vent
To issue out, and in some friendly breast
Discharge her treasures, and securely rest;
To unbosom all the secrets of her heart,
Take good advice, but better to impart.
For 'tis the bliss of friendship's holy state,   250
To mix their minds, and to communicate;
Though bodies cannot, souls can penetrate:
Fixed to her choice, inviolably true,
And wisely choosing, for she chose but few.
Some she must have; but in no one could find
A tally fitted for so large a mind.

The souls of friends like kings in progress are,
Still in their own, though from the palace far:
Thus her friend's heart her country dwelling was,
A sweet retirement to a coarser place;   260
Where pomp and ceremonies entered not,
Where greatness was shut out, and business well
    forgot.

This is the imperfect draft; but short as far
As the true height and bigness of a star
Exceeds the measures of the astronomer.
She shines above, we know; but in what place,
How near the throne, and heaven's imperial face,
By our weak optics is but vainly guessed;
Distance and altitude conceal the rest.

Reflections on the   Though all these rare endowments of the mind   270
shortness of her   Were in a narrow space of life confined,
life   The figure was with full perfection crowned;
Though not so large an orb, as truly round.

As when in glory, through the public place,
The spoils of conquered nations were to pass,
And but one day for triumph was allowed,
The consul was constrained his pomp to crowd;°
And so the swift procession hurried on,
That all, though not distinctly, might be shown:
So in the straitened bounds of life confined,   280
She gave but glimpses of her glorious mind;
And multitudes of virtues passed along,
Each pressing foremost in the mighty throng,
Ambitious to be seen, and then make room
For greater multitudes that were to come.

Yet unemployed no minute slipped away;
Moments were precious in so short a stay.
The haste of heaven to have her was so great,
That some were single acts, though each complete;
But every act stood ready to repeat.   290

Her fellow-saints with busy care will look
For her blessed name in fate's eternal book;
And, pleased to be outdone, with joy will see
Numberless virtues, endless charity:
But more will wonder at so short an age,
To find a blank beyond the thirtieth page;

And with a pious fear begin to doubt
The piece imperfect, and the rest torn out.

*She died in
her thirty-third
year*
But 'twas her saviour's time; and, could there be
A copy near the original, 'twas she.   300
    As precious gums are not for lasting fire,
They but perfume the temple, and expire;
So was she soon exhaled, and vanished hence;
A short sweet odour of a vast expense.
She vanished, we can scarcely say she died;
For but a now did heaven and earth divide:
She passed serenely with a single breath;
This moment perfect health, the next was death:

*The manner of
her death*
One sigh did her eternal bliss assure;
So little penance needs, when souls are almost   310
      pure.
As gentle dreams our waking thoughts pursue,
Or one dream passed we slide into a new;
So close they follow, such wild order keep,
We think ourselves awake, and are asleep;
So softly death succeeded life in her,
She did but dream of heaven, and she was there.
    No pains she suffered, nor expired with noise;
Her soul was whispered out with God's still voice;
As an old friend is beckoned to a feast,
And treated like a long-familiar guest.   320

*Her preparedness
to die*
He took her as he found, but found her so,
As one in hourly readiness to go;
E'en on that day, in all her trim prepared,
As early notice she from heaven had heard,
And some descending courtier from above
Had given her timely warning to remove;
Or counselled her to dress the nuptial room,

*She died on
Whitsunday
night*
For on that night the bridegroom was to come.
He kept his hour, and found her where she lay,
Clothed all in white, the livery of the day.°   330
Scarce had she sinned in thought, or word, or act,
Unless omissions were to pass for fact;
That hardly death a consequence could draw,
To make her liable to nature's law.
And that she died, we only have to show
The mortal part of her she left below;

The rest, so smooth, so suddenly she went,
Looked like translation through the firmament,
Or like the fiery car on the third errand sent.

Apostrophe to
her soul

    O happy soul! if thou canst view from high,   340
Where thou art all intelligence, all eye,
If looking up to God, or down to us,
Thou findst, that any way be pervious,°
Survey the ruins of thy house, and see
Thy widowed and thy orphan family;
Look on thy tender pledges left behind;
And if thou canst a vacant minute find
From heavenly joys, that interval afford
To thy sad children, and thy mourning lord.
See how they grieve, mistaken in their love,   350
And shed a beam of comfort from above;
Give them, as much as mortal eyes can bear,
A transient view of thy full glories there;
That they with moderate sorrow may sustain,
And mollify their losses in thy gain.
Or else divide the grief; for such thou wert,
That should not all relations bear a part,
It were enough to break a single heart.

Epiphonema, or
close of the poem

    Let this suffice: nor thou, great saint, refuse
This humble tribute of no vulgar muse;   360
Who, not by cares, or wants, or age depressed,
Stems a wild deluge with a dauntless breast;
And dares to sing thy praises in a clime
Where vice triumphs, and virtue is a crime;
Where e'en to draw the picture of thy mind,
Is satire on the most of humankind:
Take it, while yet 'tis praise; before my rage,
Unsafely just, break loose on this bad age;
So bad, that thou thyself hadst no defence
From vice, but barely by departing hence.   370
    Be what, and where thou art; to wish thy place
Were, in the best, presumption more than grace.
Thy relics (such thy works of mercy are)
Have, in this poem, been my holy care.
As earth thy body keeps, thy soul the sky,
So shall this verse preserve thy memory;
For thou shalt make it live, because it sings of thee.

The Sixth Satire of Juvenal

ARGUMENT

This Satire, of almost double length to any of the rest, is a bitter invective against the fair sex. It is indeed a commonplace, from whence all the moderns have notoriously stolen their sharpest railleries. In his other satires, the poet has only glanced on some particular women, and generally scourged the men; but this he reserved wholly for the ladies. How they had offended him, I know not; but, upon the whole matter, he is not to be excused for imputing to all the vices of some few amongst them. Neither was it generously done of him to attack the weakest, as well as the fairest, part of the creation; neither do I know what moral he could reasonably draw from it. It could not be to avoid the whole sex, if all had been true which he alleges against them; for that had been to put an end to humankind. And to bid us beware of their artifices, is a kind of silent acknowledgment that they have more wit than men; which turns the satire upon us, and particularly upon the poet, who thereby makes a compliment where he meant a libel. If he intended only to exercise his wit, he has forfeited his judgment, by making the one half of his readers his mortal enemies; and amongst the men, all the happy lovers, by their own experience, will disprove his accusations. The whole world must allow this to be the wittiest of his satires; and truly he had need of all his parts to maintain with so much violence so unjust a charge. I am satisfied he will bring but few over to his opinion; and on that consideration chiefly I ventured to translate him. Though there wanted not another reason, which was, that no one else would undertake it; at least Sir Charles Sedley, who could have done more right to the author, after a long delay, at length absolutely refused so ungrateful an employment; and everyone will grant, that the work must have been imperfect and lame, if it had appeared without one of the principal members belonging to it. Let the poet, therefore, bear the blame of his own invention; and let me satisfy the world that I am not of his opinion. Whatever his Roman ladies were, the English are free from all his imputations. They will read with wonder and abhorrence the vices of an age which was the most infamous of any on record. They will bless themselves when they behold those examples related of Domitian's time;° they will give back to antiquity those monsters it produced, and believe, with reason, that the species of those women is extinguished, or, at least, that they were never here propagated. I may safely, therefore, proceed to the argument of a satire which is no way relating to them; and first observe, that my author makes their lust the most heroic of their vices; the rest are in a manner but digression. He skims them over, but he dwells on this; when he seems to have taken his last leave of it, on the sudden he returns to it: it is one branch of it in Hippia, another in Messalina, but lust is the main body of the tree. He begins with this text in the first line, and takes it up, with intermissions, to the end of the chapter. Every vice is a loader,° but that is a ten. The fillers, or intermediate parts, are their revenge; their contrivances of secret crimes; their arts to hide them; their wit to excuse them; and their impudence to own them, when they can no

*longer be kept secret. Then the persons to whom they are most addicted, and on
whom they commonly bestow the last favours, as stage-players, fiddlers, singing-
boys, and fencers. Those who pass for chaste amongst them are not really so; but
only, for their vast doweries, are rather suffered, than loved, by their own husbands. That they are imperious, domineering, scolding wives; set up for learning,
and criticism in poetry; but are false judges: love to speak Greek (which was then
the fashionable tongue, as French is now with us). That they plead causes at the
bar, and play prizes° at the beargarden: that they are gossips and newsmongers;
wrangle with their neighbours abroad, and beat their servants at home. That they
lie-in for new faces once a month; are sluttish with their husbands in private, and
paint and dress in public for their lovers: that they deal with Jews, diviners, and
fortune-tellers; learn the arts of miscarrying and barrenness; buy children, and
produce them for their own; murder their husband's sons, if they stand in their way
to his estate, and make their adulterers his heirs. From hence the poet proceeds to
show the occasions of all these vices, their original, and how they were introduced
in Rome by peace, wealth, and luxury. In conclusion, if we will take the word of
our malicious author, bad women are the general standing rule; and the good, but
some few exceptions to it.*

In Saturn's reign, at nature's early birth,°
There was that thing called chastity on earth;
When in a narrow cave, their common shade,
The sheep, the shepherds, and their gods were laid;
When reeds, and leaves, and hides of beasts, were spread,
By mountain-housewives, for their homely bed,
And mossy pillows raised, for the rude husband's head.
Unlike the niceness of our modern dames,
(Affected nymphs, with new-affected names,)
The Cynthias, and the Lesbias of our years,   10
Who for a sparrow's death dissolve in tears.
Those first unpolished matrons, big and bold,
Gave suck to infants of gigantic mould;
Rough as their savage lords, who ranged the wood,
And, fat with acorns, belched their windy food.°
For when the world was buxom, fresh, and young,
Her sons were undebauched, and therefore strong;
And whether born in kindly beds of earth,
Or struggling from the teeming oaks to birth,
Or from what other atoms they begun,   20
No sires they had, or, if a sire, the sun.
Some thin remains of chastity appeared
E'en under Jove, but Jove without a beard;°

Before the servile Greeks had learned to swear
By heads of kings; while yet the bounteous year
Her common fruits in open plains exposed;
Ere thieves were feared, or gardens were enclosed.
At length uneasy justice upwards flew,°
And both the sisters to the stars withdrew;
From that old era whoring did begin,   30
So venerably ancient is the sin.
Adulterers next invade the nuptial state,
And marriage-beds creaked with a foreign weight;
All other ills did iron times adorn,
But whores and silver in one age were born.
    Yet thou, they say, for marriage dost provide;
Is this an age to buckle with a bride?
They say thy hair the curling art is taught,
The wedding-ring perhaps already bought;
A sober man like thee to change his life!   40
What fury would possess thee with a wife?
Art thou of every other death bereft,
No knife, no ratsbane, no kind halter left?
(For every noose compared to hers is cheap.)
Is there no city bridge from whence to leap?
Wouldst thou become her drudge, who dost enjoy
A better sort of bedfellow, thy boy?
He keeps thee not awake with nightly brawls,
Nor, with a begged reward, thy pleasure palls;
Nor, with insatiate heavings, calls for more,   50
When all thy spirits were drained out before.
But still Ursidius courts the marriage-bait,
Longs for a son to settle his estate,
And takes no gifts, though every gaping heir
Would gladly grease the rich old bachelor.
What revolution can appear so strange,
As such a lecher such a life to change?
A rank, notorious whoremaster, to choose
To thrust his neck into the marriage-noose?
He who so often in a dreadful fright   60
Had in a coffer scaped the jealous cuckold's sight,
That he to wedlock dotingly betrayed,
Should hope, in this lewd town, to find a maid!

The man's grown mad! to ease his frantic pain,
Run for the surgeon, breathe the middle vein;
But let a heifer with gilt horns be led
To Juno, regent of the marriage-bed;
And let him every deity adore,
If his new bride prove not an arrant whore,
In head and tail and every other pore.   70
On Ceres' feast, restrained from their delight,°
Few matrons there, but curse the tedious night;
Few whom their fathers dare salute, such lust
Their kisses have, and come with such a gust.
With ivy now adorn thy doors, and wed;
Such is thy bride, and such thy genial bed.
Thinkst thou one man is for one woman meant?
She sooner with one eye would be content.

    And yet, 'tis noised, a maid did once appear
In some small village, though fame says not where.   80
'Tis possible; but sure no man she found;
'Twas desert all about her father's ground.
And yet some lustful god might there make bold;
Are Jove and Mars grown impotent and old?°
Many a fair nymph has in a cave been spread,
And much good love without a feather-bed.
Whither wouldst thou to choose a wife resort,
The park, the mall, the playhouse, or the court?
Which way soever thy adventures fall,
Secure alike of chastity in all.°   90
    One sees a dancing-master capering high,
And raves, and pisses, with pure ecstasy;
Another does with all his motions move,
And gapes, and grins, as in the feat of love;
A third is charmed with the new opera notes,
Admires the song, but on the singer dotes.
The country lady in the box appears,
Softly she warbles over all she hears,
And sucks in passion both at eyes and ears.
    The rest (when now the long vacation's come,   100
The noisy hall and theatres grown dumb)
Their memories to refresh, and cheer their hearts,
In borrowed breeches act the players' parts.

The poor, that scarce have wherewithal to eat,
Will pinch, to make the singing-boy a treat;
The rich, to buy him, will refuse no price,
And stretch his quail-pipe, till they crack his voice.°
Tragedians, acting love, for lust are sought,
(Though but the parrots of a poet's thought.)
The pleading lawyer, though for counsel used,   110
In chamber-practice often is refused.
Still thou wilt have a wife, and father heirs,
(The product of concurring theatres.)
Perhaps a fencer did thy brows adorn,
And a young swordsman to thy lands is born.
   Thus Hippia loathed her old patrician lord,
And left him for a brother of the sword.
To wondering Pharos with her love she fled,°
To show one monster more than Afric bred;
Forgetting house and husband left behind,   120
E'en children too, she sails before the wind;
False to them all, but constant to her kind.
But, stranger yet, and harder to conceive,
She could the playhouse and the players leave.
Born of rich parentage, and nicely bred,
She lodged on down, and in a damask bed;
Yet daring now the dangers of the deep,
On a hard mattress is content to sleep.
Ere this, 'tis true, she did her fame expose;
But that great ladies with great ease can lose.   130
The tender nymph could the rude ocean bear,
So much her lust was stronger than her fear.
But had some honest cause her passage pressed,
The smallest hardship had disturbed her breast.
Each inconvenience makes their virtue cold;
But womankind in ills is ever bold.
Were she to follow her own lord to sea,
What doubts and scruples would she raise to stay?
Her stomach sick, and her head giddy grows,
The tar and pitch are nauseous to her nose;   140
But in love's voyage nothing can offend,
Women are never sea-sick with a friend.
Amidst the crew she walks upon the board,
She eats, she drinks, she handles every cord;

And if she spew, 'tis thinking of her lord.
Now ask, for whom her friends and fame she lost?
What youth, what beauty, could the adulterer boast?
What was the face, for which she could sustain
To be called mistress to so base a man?
The gallant of his days had known the best;   150
Deep scars were seen indented on his breast,
And all his battered limbs required their needful rest;
A promontory wen, with grisly grace,
Stood high upon the handle of his face:
His blear eyes ran in gutters to his chin;
His beard was stubble, and his cheeks were thin.
But 'twas his fencing did her fancy move;
'Tis arms and blood and cruelty they love.
But should he quit his trade, and sheathe his sword,
Her lover would begin to be her lord.   160
    This was a private crime; but you shall hear
What fruits the sacred brows of monarchs bear:
The good old sluggard but began to snore,°
When, from his side, up rose the imperial whore;
She, who preferred the pleasures of the night
To pomps, that are but impotent delight,
Strode from the palace, with an eager pace,
To cope with a more masculine embrace.
Muffled she marched, like Juno in a cloud,
Of all her train but one poor wench allowed;   170
One whom in secret service she could trust,
The rival and companion of her lust.
To the known brothel-house she takes her way,
And for a nasty room gives double pay;
That room in which the rankest harlot lay.
Prepared for fight, expectingly she lies,
With heaving breasts, and with desiring eyes.
The fair unbroken belly lay displayed
Where once the brave Britannicus was laid.°
Bare was her bosom, bare the field of lust,   180
Eager to swallow every sturdy thrust.
Still as one drops, another takes his place,
And baffled, still succeeds to like disgrace.
At length, when friendly darkness is expired,
And every strumpet from her cell retired,

She lags behind and lingering at the gate,
With a repining sigh submits to fate;
All filth without, and all afire within,
Tired with the toil, unsated with the sin.
Old Caesar's bed the modest matron seeks,   190
The steam of lamps still hanging on her cheeks
In ropy smut; thus foul, and thus bedight,
She brings him back the product of the night.
   Now should I sing what poisons they provide,
With all their trumpery of charms beside,
And all their arts of death,—it would be known,
Lust is the smallest sin the sex can own.
Caesinia still, they say, is guiltless found
Of every vice, by her own lord renowned;
And well she may, she brought ten thousand pound.   200
She brought him wherewithal to be called chaste;
His tongue is tied in golden fetters fast:
He sighs, adores, and courts her every hour;
Who would not do as much for such a dower?
She writes love-letters to the youth in grace,
Nay, tips the wink before the cuckold's face;
And might do more, her portion makes it good;
Wealth has the privilege of widowhood.°
   These truths with his example you disprove,
Who with his wife is monstrously in love: 210
But know him better; for I heard him swear,
'Tis not that she's his wife, but that she's fair.
Let her but have three wrinkles in her face,
Let her eyes lessen, and her skin unbrace,
Soon you will hear the saucy steward say
'Pack up with all your trinkets, and away;
You grow offensive both at bed and board;
Your betters must be had to please my lord.'
   Meantime she's absolute upon the throne,
And knowing time is precious, loses none.   220
She must have flocks of sheep, with wool more fine
Than silk, and vineyards of the noblest wine;
Whole droves of pages for her train she craves,
And sweeps the prisons for attending slaves.
In short, whatever in her eyes can come,
Or others have abroad, she wants at home.

When winter shuts the seas, and fleecy snows
Make houses white, she to the merchant goes;
Rich crystals of the rock she takes up there,
Huge agate vases, and old china ware;   230
Then Berenice's ring her finger proves,°
More precious made by her incestuous loves,
And infamously dear; a brother's bribe,
E'en God's anointed, and of Judah's tribe;
Where barefoot they approach the sacred shrine,
And think it only sin to feed on swine.
　　But is none worthy to be made a wife
In all this town? Suppose her free from strife,
Rich, fair, and fruitful, of unblemished life;
Chaste as the Sabines, whose prevailing charms,   240
Dismissed their husbands' and their brothers' arms;
Grant her, besides, of noble blood, that ran
In ancient veins, ere heraldry began;
Suppose all these, and take a poet's word,
A black swan is not half so rare a bird.
A wife, so hung with virtues, such a freight,
What mortal shoulders could support the weight!
Some country girl, scarce to a curtsey bred,
Would I much rather than Cornelia wed;°
If supercilious, haughty, proud, and vain,   250
She brought her father's triumphs in her train.
Away with all your Carthaginian state;
Let vanquished Hannibal without doors wait,
Too burly, and too big, to pass my narrow gate.
　　'O Paean!' cries Amphion, 'bend thy bow°
Against my wife, and let my children go!'
But sullen Paean shoots at sons and mothers too.
His Niobe and all his boys he lost;
E'en her, who did her numerous offspring boast,
As fair and fruitful as the sow that carried   260
The thirty pigs, at one large litter farrowed.°
　　What beauty or what chastity, can bear
So great a price, if stately and severe,
She still insults, and you must still adore?
Grant that the honey's much, the gall is more.
Upbraided with the virtues she displays,
Seven hours in twelve you loathe the wife you praise.

Some faults, though small, intolerable grow;
For what so nauseous and affected too,
As those that think they due perfection want,   270
Who have not learnt to lisp the Grecian cant?°
In Greece, their whole accomplishments they seek:
Their fashion, breeding, language, must be Greek;
But, raw in all that does to Rome belong,
They scorn to cultivate their mother tongue.
In Greek they flatter, all their fears they speak;
Tell all their secrets, nay, they scold in Greek:
E'en in the feat of love, they use that tongue.
Such affectations may become the young;
But thou, old hag, of threescore years and three,   280
Is showing of thy parts in Greek for thee?
Ζωὴ καὶ ψυχή! All those tender words°
The momentary trembling bliss affords;
The kind soft murmurs of the private sheets
Are bawdy, while thou speakst in public streets.
Those words have fingers; and their force is such,
They raise the dead, and mount him with a touch.
But all provocatives from thee are vain;
No blandishment the slackened nerve can strain:
It looks but on thy face and falls again.   290
   If then thy lawful spouse thou canst not love,
What reason should thy mind to marriage move?
Why all the charges of the nuptial feast,
Wine and desserts, and sweetmeats to digest?
The endowing gold that buys the dear delight,
Given for thy first and only happy night?
If thou art thus uxoriously inclined,
To bear thy bondage with a willing mind,
Prepare thy neck, and put it in the yoke;
But for no mercy from thy woman look.   300
For though, perhaps, she loves with equal fires,
To absolute dominion she aspires,
Joys in the spoils, and triumphs o'er thy purse;
The better husband makes the wife the worse.
Nothing is thine to give, or sell, or buy,
All offices of ancient friendship die,
Nor hast thou leave to make a legacy.°

By thy imperious wife thou art bereft
A privilege, to pimps and panders left;
Thy testament's her will; where she prefers   310
Her ruffians, drudges, and adulterers,
Adopting all thy rivals for thy heirs.
   'Go drag that slave to death!'—'Your reason? why
Should the poor innocent be doomed to die?
What proofs? For, when man's life is in debate,
The judge can ne'er too long deliberate.'
'Call'st thou that slave a man?' the wife replies;
'Proved, or unproved the crime, the villain dies.
I have the sovereign power to save or kill
And give no other reason but my will.'   320
   Thus the she-tyrant reigns, till pleased with change,
Her wild affections to new empires range;
Another subject-husband she desires;
Divorced from him, she to the first retires,
While the last wedding-feast is scarcely o'er,
And garlands hang yet green upon the door.
So still the reckoning rises; and appears
In total sum, eight husbands in five years.
The title for a tombstone might be fit,
But that it would too commonly be writ.   330
   Her mother living, hope no quiet day;
She sharpens her, instructs her how to flay
Her husband bare, and then divides the prey.
She takes love-letters, with a crafty smile,
And, in her daughter's answer, mends the style.
In vain the husband sets his watchful spies;
She cheats their cunning, or she bribes their eyes.
The doctor's called; the daughter, taught the trick,
Pretends to faint, and in full health is sick.
The panting stallion in the closet stands,   340
Hears all, and thinks, and loves, and helps it with his hands.
Canst thou, in reason, hope, a bawd so known,
Should teach her other manners than her own?
Her interest is in all the advice she gives
'Tis on the daughter's rents the mother lives.
   No cause is tried at the litigious bar
But women plaintiffs or defendants are;

They form the process, all the briefs they write,
The topics furnish, and the pleas indict,
And teach the toothless lawyer how to bite.   350
 They turn viragoes too; the wrestler's toil
They try, and smear the naked limbs with oil;
Against the post their wicker shields they crush,
Flourish the sword, and at the plastron push.°
Of every exercise the mannish crew
Fulfils the parts, and oft excels us too;
Prepared not only in feigned fights to engage,
But rout the gladiators on the stage.
What sense of shame in such a breast can lie,
Inured to arms, and her own sex to fly?   360
Yet to be wholly man she would disclaim;
To quit her tenfold pleasure at the game,
For frothy praises and an empty name.
Oh what a decent sight 'tis to behold
All thy wife's magazine by auction sold!°
The belt, the crested plume, the several suits
Of armour, and the Spanish leather boots!

Yet these are they, that cannot bear the heat
Of figured silks, and under sarcenet sweat.°
Behold the strutting Amazonian whore,   370
She stands in guard with her right foot before;
Her coats tucked up, and all her motions just,
She stamps, and then cries, 'Hah!' at every thrust;
But laugh to see her, tired with many a bout,
Call for the pot, and like a man piss out.
The ghosts of ancient Romans, should they rise,
Would grin to see their daughters play a prize.°
 Besides, what endless brawls by wives are bred!
The curtain-lecture makes a mournful bed.°
Then, when she has thee sure within the sheets,   380
Her cry begins, and the whole day repeats.
Conscious of crimes herself, she teases first;
Thy servants are accused; thy whore is cursed;
She acts the jealous, and at will she cries;
For women's tears are but the sweat of eyes.
Poor cuckold fool! thou think'st that love sincere,
And suckst between her lips the falling tear;
But search her cabinet, and thou shalt find

Each tiller there with love-epistles lined.°
Suppose her taken in a close embrace,   390
This you would think so manifest a case,
No rhetoric could defend, no impudence outface;
And yet e'en then she cries, 'The marriage-vow
A mental reservation must allow;
And there's a silent bargain still implied,
The parties should be pleased on either side,
And both may for their private needs provide.
Though men yourselves, and women us you call,
Yet *homo* is a common name for all.'
There's nothing bolder than a woman caught;   400
Guilt gives them courage to maintain their fault.
   You ask, from whence proceed these monstrous crimes?
Once poor, and therefore chaste, in former times
Our matrons were; no luxury found room,
In low-roofed houses, and bare walls of loam;
Their hands with labour hardened while 'twas light,
And frugal sleep supplied the quiet night;
While pinched with want, their hunger held them straight,
When Hannibal was hovering at the gate:
But wanton now, and lolling at our ease,   410
We suffer all the inveterate ills of peace,
And wasteful riot; whose destructive charms,
Revenge the vanquished world of our victorious arms.
No crime, no lustful postures are unknown,
Since poverty, our guardian god, is gone;
Pride, laziness, and all luxurious arts,
Pour like a deluge in from foreign parts:
Since gold obscene, and silver found the way,
Strange fashions, with strange bullion, to convey,
And our plain simple manners to betray.   420
   What care our drunken dames to whom they spread?
Wine no distinction makes of tail or head,
Who lewdly dancing at a midnight ball,
For hot eringoes and fat oysters call:°
Full brimmers to their fuddled noses thrust,
Brimmers, the last provocatives of lust;
When vapours to their swimming brains advance,
And double tapers on the table dance.
   Now think what bawdy dialogues they have,

What Tullia talks to her confiding slave,   430
At Modesty's old statue; when by night
They make a stand, and from their litters light;
They straighten with their hands the nameless place
And spouting thence bepiss her venerable face;
Before the conscious moon they get astride,
By turns are ridden, and by turns they ride.
The good man early to the levee goes,
And treads the nasty puddle of his spouse.
    The secrets of the goddess named the good°
Are e'en by boys and barbers understood: 440
Where the rank matrons, dancing to the pipe,
Jig with their bums, and are for action ripe;
With music raised, they spread abroad their hair
And toss their heads like an enamoured mare,
Confess their itching in their ardent eyes
While tears of wine run trickling down their thighs.
Laufella lays her garland by, and proves
The mimic lechery of manly loves,
Provokes to flats some battered household whore,°
And heaving up the rubster does adore,°   450
Ranked with the lady the cheap sinner lies;
For here not blood, but virtue, gives the prize.
Nothing is feigned in this venereal strife,
'Tis downright lust, and acted to the life.
So full, so fierce, so vigorous, and so strong,
That looking on would make old Nestor young.°
Impatient of delay, a general sound,
A universal groan of lust goes round;
For then, and only then, the sex sincere is found.
'Now is the time of action; now begin,'   460
They cry, 'and let the lusty lovers in.
The whoresons are asleep; then bring the slaves
And watermen, a race of strong-backed knaves;
Bring anything that's man: if none be nigh,
Asses have better parts their places to supply.'
    I wish, at least, our sacred rites were free
From those pollutions of obscenity:
But 'tis well known what singer, how disguised,°
A lewd audacious action enterprised;
Into the fair, with women mixed, he went,   470

Armed with a huge two-handed instrument;
A grateful present to those holy choirs,
Where the mouse, guilty of his sex, retires,
And e'en male pictures modestly are veiled:
Yet no profaneness on that age prevailed;
No scoffers at religious rites were found,
Though now at every altar they abound.

    I hear your cautious council; you would say,
'Keep close your women under lock and key:'
But, who shall keep those keepers? Women, nursed   480
In craft; begin with those, and bribe them first.
The sex is turned all whore; they love the game,
And mistresses and maids are both the same.

    The poor Ogulnia, on the poet's day,
Will borrow clothes and chair to see the play;
She, who before had mortgaged her estate,
And pawned the last remaining piece of plate.
Some are reduced their utmost shifts to try;
But women have no shame of poverty.
They live beyond their stint, as if their store,   490
The more exhausted, would increase the more:
Some men, instructed by the labouring ant,
Provide against the extremities of want;
But womankind, that never knows a mean,
Down to the dregs their sinking fortune drain:
Hourly they give, and spend, and waste, and wear,
And think no pleasure can be bought too dear.

    There are, who in soft eunuchs place their bliss,
To shun the scrubbing of a bearded kiss,
And scape abortion; but their solid joy   500
Is when the page, already past a boy,
Is caponed late, and to the gelder shown,
With his two-pounders to perfection grown;
When all the navel-string could give, appears;°
All but the beard, and that's the barber's loss, not theirs.
Seen from afar, and famous for his ware,
He struts into the bath among the fair;
The admiring crew to their devotions fall,
And, kneeling, on their new Priapus call.
Carved for my lady's use, with her he lies;°   510
And let him drudge for her, if thou art wise,

Rather than trust him with thy favourite boy;
He proffers death, in proffering to enjoy.
    If songs they love, the singer's voice they force
Beyond his compass, till his quail-pipe's hoarse.
His lute and lyre with their embrace is worn;
With knots they trim it, and with gems adorn;
Run over all the strings, and kiss the case,
And make love to it in the master's place.
    A certain lady once, of high degree,   520
To Janus vowed, and Vesta's deity,
That Pollio might in singing win the prize;°
Pollio, the dear, the darling of her eyes:
She prayed, and bribed; what could she more have done
For a sick husband, or an only son?
With her face veiled, and heaving up her hands,
The shameless suppliant at the altar stands;
The forms of prayer she solemnly pursues,
And, pale with fear, the offered entrails views.
Answer, ye powers; for if you heard her vow,   530
Your godships sure had little else to do.
    This is not all; for actors they implore;°
An impudence unknown to heaven before.
The haruspex, tired with this religious rout,°
Is forced to stand so long, he gets the gout.
But suffer not thy wife abroad to roam:
If she love singing, let her sing at home;
Not strut in streets with Amazonian pace,
For that's to cuckold thee before thy face.
    Their endless itch of news comes next in play;   540
They vent their own, and hear what others say;
Know what in Thrace, or what in France is done;
The intrigues betwixt the stepdame and the son;
Tell who loves who, what favours some partake,
And who is jilted for another's sake;
What pregnant widow in what month was made;
How oft she did, and, doing, what she said.
    She first beholds the raging comet rise,
Knows whom it threatens, and what lands destroys.
Still for the newest news she lies in wait,   550
And takes reports just entering at the gate.
Wrecks, floods, and fires, whatever she can meet,

She spreads, and is the fame of every street.
    This is a grievance; but the next is worse;
A very judgment, and her neighbours' curse;
For if their barking dog disturb her ease,
No prayer can bend her, no excuse appease.
The unmannered malefactor is arraigned;
But first the master, who the cur maintained,
Must feel the scourge. By night she leaves her bed,   560
By night her bathing equipage is led,
That marching armies a less noise create;
She moves in tumult, and she sweats in state.
Meanwhile, her guests their appetites must keep;
Some gape for hunger, and some gasp for sleep.
At length she comes, all flushed; but ere she sup,
Swallows a swingeing preparation-cup,
And then, to clear her stomach, spews it up.
The deluge-vomit all the floor o'erflows,
And the sour savour nauseates every nose.   570
She drinks again, again she spews a lake;
Her wretched husband sees, and dares not speak;
But mutters many a curse against his wife,
And damns himself for choosing such a life.
    But of all plagues, the greatest is untold;
The book-learned wife, in Greek and Latin bold;
The critic-dame, who at her table sits,
Homer and Virgil quotes, and weighs their wits,
And pities Dido's agonising fits.
She has so far the ascendant of the board,   580
The prating pedant puts not in one word;
The man of law is nonplussed in his suit,
Nay, every other female tongue is mute.
Hammers and beating anvils you would swear,
And Vulcan, with his whole militia, there.
Tabors and trumpets, cease; for she alone°
Is able to redeem the labouring moon.
E'en wit's a burden, when it talks too long;
But she, who has no continence of tongue,
Should walk in breeches, and should wear a beard,   590
And mix among the philosophic herd.
O what a midnight curse has he, whose side
Is pestered with a mood and figure bride!°

Let mine, ye Gods! (if such must be my fate,)
No logic learn, nor history translate,
But rather be a quiet, humble fool;
I hate a wife to whom I go to school,
Who climbs the grammar-tree, distinctly knows
Where noun, and verb, and participle grows;
Corrects her country-neighbour; and, abed,   600
For breaking Priscian's breaks her husband's head.°

    The gaudy gossip, when she's set agog,
In jewels dressed, and at each ear a bob,
Goes flaunting out, and in her trim of pride,
Thinks all she says or does is justified.
When poor, she's scarce a tolerable evil;
But rich, and fine, a wife's a very devil.

    She duly, once a month, renews her face;
Meantime it lies in daub and hid in grease.
Those are the husband's nights; she craves her due,   610
He takes fat kisses, and is stuck in glue.
But to the loved adulterer when she steers,
Fresh from the bath, in brightness she appears:
For him the rich Arabia sweats her gum,
And precious oils from distant Indies come,
How haggardly soe'er she looks at home.
The eclipse then vanishes, and all her face
Is opened, and restored to every grace;
The crust removed, her cheeks as smooth as silk
Are polished with a wash of asses' milk;   620
And should she to the furthest north be sent,
A train of these attend her banishment.°
But hadst thou seen her plastered up before,
'Twas so unlike a face, it seemed a sore.

    'Tis worth our while, to know what all the day
They do, and how they pass their time away;
For, if o'ernight the husband has been slack,
Or counterfeited sleep, and turned his back,
Next day, be sure, the servants go to wrack.
The chambermaid and dresser are called whores,   630
The page is stripped, and beaten out of doors;
The whole house suffers for the master's crime,
And he himself is warned to wake another time.

    She hires tormentors by the year; she treats

Her visitors, and talks, but still she beats.
Beats while she paints her face, surveys her gown,
Casts up the day's account, and still beats on:
Tired out, at length, with an outrageous tone,
She bids them in the devil's name be gone.
Compared with such a proud, insulting dame,   640
Sicilian tyrants may renounce their name.°

   For if she hastes abroad to take the air,
Or goes to Isis' church, (the bawdy house of prayer,)
She hurries all her handmaids to the task;
Her head, alone, will twenty dressers ask.
Psecas, the chief, with breast and shoulders bare,
Trembling, considers every sacred hair;
If any straggler from his rank be found,
A pinch must for the mortal sin compound.
Psecas is not in fault; but in the glass,   650
The dame's offended at her own ill face.
That maid is banished; and another girl
More dexterous manages the comb and curl.
The rest are summoned on a point so nice,
And first the grave old woman gives advice;
The next is called, and so the turn goes round,
As each for age, or wisdom, is renowned:
Such counsel, such deliberate care they take,
As if her life and honour lay at stake:
With curls on curls, they build her head before,   660
And mount it with a formidable tower.
A giantess she seems; but look behind,
And then she dwindles to the pigmy kind.
Duck-legged, short-waisted, such a dwarf she is,
That she must rise on tiptoes for a kiss.
Meanwhile her husband's whole estate is spent;
He may go bare, while she receives his rent.
She minds him not; she lives not as a wife,
But like a bawling neighbour, full of strife:
Near him in this alone, that she extends   670
Her hate to all his servants and his friends.

   Bellona's priests, a eunuch at their head,°
About the streets a mad procession lead:
The venerable gelding, large, and high,
O'erlooks the herd of his inferior fry.

His awkward clergymen about him prance,
And beat the timbrels to their mystic dance;
Guiltless of testicles, they tear their throats,
And squeak in treble their unmanly notes.
Meanwhile his cheeks the mitred prophet swells,   680
And dire presages of the year foretells;
Unless with eggs (his priestly hire) they haste
To expiate, and avert the autumnal blast;
And add beside a murrey-coloured vest,°
Which, in their places, may receive the pest,
And thrown into the flood, their crimes may bear,
To purge the unlucky omens of the year.
The astonished matrons pay, before the rest;
That sex is still obnoxious to the priest.

    Through ice they beat, and plunge into the stream,   690
If so the god has warned them in a dream.
Weak in their limbs, but in devotion strong,
On their bare hands and feet they crawl along
A whole field's length, the laughter of the throng.
Should Io (Io's priest, I mean) command
A pilgrimage to Meroe's burning sand,°
Through deserts they would seek the secret spring,
And holy water for lustration bring.
How can they pay their priests too much respect,
Who trade with heaven, and earthly gains neglect?   700
With him domestic gods discourse by night;
By day, attended by his choir in white,
The bald-pate tribe runs madding through the street,
And smile to see with how much ease they cheat.
The ghostly sire forgives the wife's delights,
Who sins, through frailty, on forbidden nights,
And tempts her husband in the holy time,
When carnal pleasure is a mortal crime.
The sweating image shakes its head, but he,
With mumbled prayers, atones the deity.   710
The pious priesthood the fat goose receive,
And they once bribed, the godhead must forgive.

    No sooner these remove, but full of fear,
A gipsy Jewess whispers in your ear,
And begs an alms; a high-priest's daughter she,
Versed in their Talmud, and divinity,

And prophesies beneath a shady tree.
Her goods a basket, and old hay her bed,
She strolls, and telling fortunes, gains her bread:
Farthings and some small moneys are her fees;   720
Yet she interprets all your dreams for these.
Foretells the estate when the rich uncle dies
And sees a sweetheart in the sacrifice.
Such toys, a pigeon's entrails can disclose,
Which yet the Armenian augur far outgoes;
In dogs, a victim more obscene, he rakes;
And murdered infants for inspection takes:
For gain his impious practice he pursues,
For gain will his accomplices accuse.

    More credit yet is to Chaldeans given;°   730
What they foretell is deemed the voice of heaven.
Their answers, as from Hammon's altar, come;
Since now the Delphian oracles are dumb,
And mankind, ignorant of future fate,
Believes what fond astrologers relate.

    Of these the most in vogue is he, who, sent
Beyond seas, is returned from banishment;
His art who to aspiring Otho sold,°
And sure succession to the crown foretold;
For his esteem is in his exile placed;   740
The more believed, the more he was disgraced.
No astrologic wizard honour gains,
Who has not oft been banished, or in chains.
He gets renown, who, to the halter near,
But narrowly escapes, and buys it dear.

    From him your wife inquires the planets' will,
When the black jaundice shall her mother kill;
Her sister's and her uncle's end would know,
But first consults his art, when you shall go;
And what's the greatest gift that heaven can give,   750
If after her the adulterer shall live.
She neither knows, nor cares to know, the rest,
If Mars and Saturn shall the world infest;°
Or Jove and Venus, with their friendly rays,
Will interpose, and bring us better days.

    Beware the woman too, and shun her sight,
Who in these studies does herself delight,

By whom a greasy almanac is borne,
With often handling, like chafed amber worn:
Not now consulting, but consulted, she   760
Of the twelve houses, and their lords, is free.°
She, if the scheme a fatal journey show,
Stays safe at home, but lets her husband go.
If but a mile she travel out of town,
The planetary hour must first be known,
And lucky moment; if her eye but aches,
Or itches, its decumbiture she takes.°
No nourishment receives in her disease,
But what the stars and Ptolemy shall please.°

    The middle sort, who have not much to spare,   770
To chiromancers' cheaper art repair,
Who clap the pretty palm, to make the lines more fair.
But the rich matron who has more to give,
Her answers from the Brahmin will receive;
Skilled in the globe and sphere, he gravely stands,
And with his compass measures seas and lands.

    The poorest of the sex have still an itch
To know their fortunes, equal to the rich.
The dairymaid inquires, if she shall take
The trusty tailor, and the cook forsake,   780

    Yet these, though poor, the pain of childbed bear,
And without nurses their own infants rear:
You seldom hear of the rich mantle spread
For the babe, born in the great lady's bed:
Such is the power of herbs, such arts they use
To make them barren, or their fruit to lose.
But thou, whatever slops she will have bought,
Be thankful and supply the deadly draught;
Help her to make manslaughter; let her bleed,
And never want for savin at her need.°   790
For if she holds till her nine months be run,
Thou may'st be father to an Ethiop's son;°
A boy who, ready gotten to thy hands,
By law is to inherit all thy lands;
One of that hue, that should he cross the way,
His omen would discolour all the day.°

    I pass the foundling by, a race unknown,
At doors exposed, whom matrons make their own;

And into noble families advance
A nameless issue, the blind work of chance.   800
Indulgent fortune does her care employ,
And smiling, broods upon the naked boy:
Her garment spreads, and laps him in the fold,
And covers with her wings from nightly cold:
Gives him her blessing, puts him in a way,
Sets up the farce, and laughs at her own play.
Him she promotes; she favours him alone,
And makes provision for him as her own.
    The craving wife the force of magic tries,
And philtres for the unable husband buys;   810
The potion works not on the part designed,
But turns his brain, and stupefies his mind.
The sotted moon-calf gapes, and staring on,
Sees his own business by another done:
A long oblivion, a benumbing frost,
Constrains his head, and yesterday is lost.
Some nimbler juice would make him foam and rave,
Like that Caesinia to her Caius gave,°
Who plucking from the forehead of the foal
His mother's love, infused it in the bowl;°   820
The boiling blood ran hissing in his veins,
Till the mad vapour mounted to his brains.
The thunderer was not half so much on fire,
When Juno's girdle kindled his desire.
What woman will not use the poisoning trade,
When Caesar's wife the precedent has made?
Let Agrippina's mushroom be forgot,°
Given to a slavering, old, unuseful sot;
That only closed the drivelling dotard's eyes,
And sent his godhead downward to the skies;   830
But this fierce potion calls for fire and sword,
Nor spares the commons, when it strikes the lord.
So many mischiefs were in one combined;
So much one single poisoner cost mankind.
    If step-dames seek their sons-in-law to kill,
'Tis venial trespass—let them have their will;
But let the child, entrusted to the care
Of his own mother, of her bread beware;
Beware the food she reaches with her hand;

The morsel is intended for thy land.   840
Thy tutor be thy taster, ere thou eat;
There's poison in thy drink and in thy meat.
   You think this feigned; the satire in a rage
Struts in the buskins of the tragic stage;
Forgets his business is to laugh and bite,
And will of deaths and dire revenges write.
Would it were all a fable that you read!
But Drymon's wife pleads guilty to the deed.°
'I', she confesses, 'in the fact was caught,
Two sons dispatching at one deadly draught.'   850
'What, two! two sons, thou viper, in one day!'
'Yes, seven,' she cries, 'if seven were in my way.'
Medea's legend is no more a lie,°
Our age adds credit to antiquity.
Great ills, we grant, in former times did reign,
And murders then were done, but not for gain.
Less admiration to great crimes is due,
Which they through wrath, or through revenge pursue;
For, weak of reason, impotent of will,
The sex is hurried headlong into ill;   860
And like a cliff, from its foundations torn
By raging earthquakes, into seas is borne.
But those are fiends, who crimes from thought begin,
And cool in mischief meditate the sin.
They read the example of a pious wife,
Redeeming, with her own, her husband's life;
Yet if the laws did that exchange afford,
Would save their lap-dog sooner than their lord.
   Where'er you walk the Belides you meet,°
And Clytemnestras grow in every street;   870
But here's the difference,—Agamemnon's wife
Was a gross butcher with a bloody knife;
But murder now is to perfection grown,
And subtle poisons are employed alone;
Unless some antidote prevents their arts,
And lines with balsam all the noble parts.
In such a case, reserved for such a need,
Rather than fail, the dagger does the deed.

The Tenth Satire of Juvenal

ARGUMENT

The poet's design, in this divine satire, is to represent the various wishes and desires of mankind, and to set out the folly of them. He runs through all the several heads, of riches, honours, eloquence, fame for martial achievements, long life, and beauty; and gives instances in each, how frequently they have proved the ruin of those that owned them. He concludes, therefore, that, since we generally choose so ill for ourselves, we should do better to leave it to the gods to make the choice for us. All we can safely ask of heaven lies within a very small compass—it is but 'health of body and mind'; and if we have these, it is not much matter what we want besides; for we have already enough to make us happy.

Look round the habitable world, how few
Know their own good, or, knowing it, pursue.
How void of reason are our hopes and fears!
What in the conduct of our life appears
So well designed, so luckily begun,
But when we have our wish, we wish undone?
    Whole houses, of their whole desires possessed,
Are often ruined at their own request.
In wars and peace things hurtful we require,
When made obnoxious to our own desire.   10
    With laurels some have fatally been crowned;
Some, who the depths of eloquence have found,
In that unnavigable stream were drowned.
    The brawny fool, who did his vigour boast,°
In that presuming confidence was lost;
But more have been by avarice oppressed,
And heaps of money crowded in the chest:
Unwieldy sums of wealth, which higher mount
Than files of marshalled figures can account;
To which the stores of Croesus in the scale   20
Would look like little dolphins, when they sail
In the vast shadow of the British whale.
    For this, in Nero's arbitrary time,
When virtue was a guilt, and wealth a crime,
A troop of cut-throat guards were sent to seize
The rich men's goods, and gut their palaces:
The mob, commissioned by the government,
Are seldom to an empty garret sent.

The fearful passenger who travels late,
Charged with the carriage of a paltry plate,   30
Shakes at the moonshine shadow of a rush,
And sees a red-coat rise from every bush:
The beggar sings, e'en when he sees the place
Beset with thieves, and never mends his pace.
    Of all the vows, the first and chief request
Of each, is to be richer than the rest:
And yet no doubts the poor man's draught control,
He dreads no poison in his homely bowl;
Then fear the deadly drug, when gems divine
Enchase the cup, and sparkle in the wine.   40
    Will you not now the pair of sages praise,
Who the same end pursued by several ways?
One pitied, one contemned, the woeful times;
One laughed at follies, one lamented crimes.
Laughter is easy; but the wonder lies,
What stores of brine supplied the weeper's eyes.
Democritus could feed his spleen, and shake
His sides and shoulders, till he felt them ache;
Though in his country town no lictors were,°
Nor rods, nor axe, nor tribune, did appear;   50
Nor all the foppish gravity of show,
Which cunning magistrates on crowds bestow.
    What had he done, had he beheld on high
Our praetor seated in mock majesty;
His chariot rolling o'er the dusty place,
While, with dumb pride, and a set formal face,
He moves, in the dull ceremonial track,
With Jove's embroidered coat upon his back!
A suit of hangings had not more oppressed
His shoulders, than that long laborious vest;   60
A heavy gewgaw, (called a crown,) that spread°
About his temples, drowned his narrow head,
And would have crushed it with the massy freight,
But that a sweating slave sustained the weight;
A slave in the same chariot seen to ride
To mortify the mighty madman's pride.
Add now the imperial eagle, raised on high,
With golden beak, (the mark of majesty)
Trumpets before, and on the left and right

A cavalcade of nobles, all in white;   70
In their own natures false; and flattering tribes
But made his friends by places and by bribes.°

In his own age, Democritus could find
Sufficient cause to laugh at humankind:
Learn from so great a wit; a land of bogs
With ditches fenced, a heaven fat with fogs,
May form a spirit fit to sway the state,
And make the neighbouring monarchs fear their fate.

He laughs at all the vulgar cares and fears;
At their vain triumphs, and their vainer tears: 80
An equal temper in his mind he found,
When fortune flattered him and when she frowned.
'Tis plain from hence that what our vows request
Are hurtful things, or useless at the best.

Some ask for envied power; which public hate
Pursues, and hurries headlong to their fate:
Down go the titles; and the statue crowned,
Is by base hands in the next river drowned.
The guiltless horses, and the chariot wheel,
The same effects of vulgar fury feel: 90
The smith prepares his hammer for the stroke,
While the lunged bellows hissing fire provoke.
Sejanus, almost first of Roman names,°
The great Sejanus crackles in the flames:
Formed in the forge, the pliant brass is laid
On anvils; and of head and limbs are made,
Pans, cans, and piss-pots, a whole kitchen trade.

Adorn your doors with laurels; and a bull,
Milk white, and large, lead to the Capitol;
Sejanus with a rope is dragged along,   100
The sport and laughter of the giddy throng;
'Good lord!' they cry, 'what Ethiop lips he has;
How foul a snout, and what a hanging face!°
By heaven, I never could endure his sight!
But say, how came his monstrous crimes to light?
What is the charge, and who the evidence,
(The saviour of the nation and the prince?)'
'Nothing of this; but our old Caesar sent
A noisy letter to his parliament.'
'Nay, sirs, if Caesar writ, I ask no more;   110

He's guilty, and the question's out of door.'
How goes the mob? (for that's a mighty thing,)
When the king's trump, the mob are for the king;
They follow fortune, and the common cry
Is still against the rogue condemned to die.

    But the same very mob, that rascal crowd,
Had cried 'Sejanus,' with a shout as loud,
Had his designs (by fortune's favour blest)
Succeeded, and the prince's age oppressed.
But long, long since, the times have changed their face,   120
The people grown degenerate and base;
Not suffered now the freedom of their choice
To make their magistrates, and sell their voice.

    Our wise forefathers, great by sea and land,
Had once the power and absolute command;
All offices of trust themselves disposed;
Raised whom they pleased, and whom they pleased deposed:
But we who give our native rights away
And our enslaved posterity betray
Are now reduced to beg an alms, and go   130
On holidays to see a puppet-show.

    'There was a damned design,' cries one, 'no doubt,
For warrants are already issued out:
I met Brutidius in a mortal fright,
He's dipped for certain, and plays least in sight;°
I fear the rage of our offended prince,
Who thinks the senate slack in his defence.
Come, let us haste, our loyal zeal to show,
And spurn the wretched corpse of Caesar's foe:
But let our slaves be present there; lest they   140
Accuse their masters, and for gain betray.'
Such were the whispers of those jealous times,
About Sejanus' punishment and crimes.

    Now tell me truly, wouldst thou change thy fate,
To be, like him, first minister of state?
To have thy levees crowded with resort,
Of a depending, gaping, servile court;
Dispose all honours of the sword and gown,
Grace with a nod, and ruin with a frown;
To hold thy prince in pupilage, and sway   150
That monarch, whom the mastered world obey?

While he, intent on secret lusts alone,
Lives to himself, abandoning the throne;
Cooped in a narrow isle, observing dreams°
With flattering wizards, and erecting schemes!
   I well believe thou wouldst be great as he,
For every man's a fool to that degree:
All wish the dire prerogative to kill;
E'en they would have the power, who want the will:
But wouldst thou have thy wishes understood,   160
To take the bad together with the good?
Wouldst thou not rather choose a small renown,
To be the mayor of some poor paltry town;
Bigly to look, and barbarously to speak;
To pound false weights, and scanty measures break?
Then, grant we that Sejanus went astray
In every wish, and knew not how to pray;
For he who grasped the world's exhausted store
Yet never had enough, but wished for more,
Raised a top-heavy tower, of monstrous height,   170
Which, mouldering, crushed him underneath the weight.
   What did the mighty Pompey's fall beget,
And ruined him, who, greater than the great,°
The stubborn pride of Roman nobles broke,
And bent their haughty necks beneath his yoke?
What else but his immoderate lust of power,
Prayers made and granted in a luckless hour,
For few usurpers to the shades descend
By a dry death, or with a quiet end.
   The boy who scarce has paid his entrance down   180
To his proud pedant, or declined a noun,
(So small an elf, that, when the days are foul,
He and his satchel must be borne to school,)
Yet prays, and hopes, and aims at nothing less,
To prove a Tully, or Demosthenes:°
But both those orators, so much renowned,
In their own depths of eloquence were drowned:
The hand and head were never lost of those
Who dealt in doggrel, or who punned in prose.
'Fortune foretuned the dying notes of Rome,°   190
Till I, thy consul sole, consoled thy doom.'
His fate had crept below the lifted swords,

Had all his malice been to murder words.
I rather would be Maevius, thrash for rhymes
Like his, the scorn and scandal of the times,
Than that Philippic, fatally divine,°
Which is inscribed the second, should be mine.

Nor he, the wonder of the Grecian throng,
Who drove them with the torrent of his tongue,
Who shook the theatres, and swayed the state   200
Of Athens, found a more propitious fate.
Whom, born beneath a boding horoscope,
His sire, the blear-eyed Vulcan of a shop,
From Mars's forge, sent to Minerva's schools,
To learn the unlucky art of wheedling fools.

With itch of honour, and opinion vain,
All things beyond their native worth we strain;
The spoils of war, brought to Feretrian Jove,°
An empty coat of armour hung above
The conqueror's chariot, and in triumph borne,   210
A streamer from a boarded galley torn,
A chapfallen beaver loosely hanging by
The cloven helm, an arch of victory;
On whose high convex sits a captive foe,
And, sighing, casts a mournful look below;
Of every nation each illustrious name,
Such toys as these have cheated into fame;
Exchanging solid quiet, to obtain
The windy satisfaction of the brain.

So much the thirst of honour fires the blood;   220
So many would be great, so few be good:
For who would Virtue for herself regard,
Or wed, without the portion of reward?
Yet this mad chase of fame, by few pursued,
Has drawn destruction on the multitude;
This avarice of praise in times to come,
Those long inscriptions crowded on the tomb;
Should some wild fig-tree take her native bent,
And heave below the gaudy monument,
Would crack the marble titles, and disperse   230
The characters of all the lying verse.
For sepulchres themselves must crumbling fall
In time's abyss, the common grave of all.

Great Hannibal within the balance lay,
And tell how many pounds his ashes weigh;
Whom Afric was not able to contain,
Whose length runs level with the Atlantic main,
And wearies fruitful Nilus, to convey
His sun-beat waters by so long a way;
Which Ethiopia's double clime divides,   240
And elephants in other mountains hides.
Spain first he won, the Pyreneans past,
And steepy Alps, the mounds that nature cast;
And with corroding juices, as he went,
A passage through the living rocks he rent:
Then, like a torrent rolling from on high,
He pours his headlong rage on Italy,
In three victorious battles overrun;
Yet still uneasy, cries, 'There's nothing done,
Till level with the ground their gates are laid,   250
And Punic flags on Roman towers displayed.'
   Ask what a face belonged to this high fame,
His picture scarcely would deserve a frame:
A sign-post dauber would disdain to paint
The one-eyed hero on his elephant.
Now, what's his end, O charming glory say,
What rare fifth act to crown this huffing play?°
In one deciding battle overcome,
He flies, is banished from his native home;
Begs refuge in a foreign court, and there   260
Attends, his mean petition to prefer;
Repulsed by surly grooms, who wait before
The sleeping tyrant's interdicted door.
   What wondrous sort of death has heaven designed,
Distinguished from the herd of humankind,
For so untamed, so turbulent a mind?
Nor swords at hand, nor hissing darts afar,
Are doomed to avenge the tedious bloody war;
But poison, drawn through a ring's hollow plate,
Must finish him—a sucking infant's fate.   270
Go, climb the rugged Alps, ambitious fool,
To please the boys, and be a theme at school.
   One world sufficed not Alexander's mind;
Cooped up, he seemed in earth and seas confined,

And, struggling, stretched his restless limbs about
The narrow globe, to find a passage out:
Yet entered in the brick-built town, he tried°
The tomb, and found the strait dimensions wide.
Death only this mysterious truth unfolds,
The mighty soul how small a body holds   280
   Old Greece a tale of Athos would make out,°
Cut from the continent, and sailed about;
Seas hid with navies, chariots passing o'er
The channel, on a bridge from shore to shore:
Rivers, whose depth no sharp beholder sees,
Drunk at an army's dinner to the lees;
With a long legend of romantic things,
Which in his cups the boozy poet sings.
But how did he return, this haughty brave,
Who whipped the winds, and made the sea his slave?   290
(Though Neptune took unkindly to be bound,
And Eurus never such hard usage found
In his Aeolian prisons under ground;)
What god so mean, e'en he who points the way,°
So merciless a tyrant to obey?
But how returned he, let us ask again?
In a poor skiff he passed the bloody main,
Choked with the slaughtered bodies of his train.
For fame he prayed, but let the event declare
He had no mighty penn'worth of his prayer: 300
   'Jove, grant me length of life, and years' good store
Heap on my bending back! I ask no more.'
Both sick and healthful, old and young, conspire
In this one silly mischievous desire.
Mistaken blessing, which old age they call;
'Tis a long, nasty, darksome hospital:
A ropy chain of rheums; a visage rough,
Deformed, unfeatured, and a skin of buff;
A stitch-fallen cheek, that hangs below the jaw;
Such wrinkles as a skilful hand would draw   310
For an old grandame ape, when, with a grace,
She sits at squat, and scrubs her leathern face.
   In youth, distinctions infinite abound;
No shape, or feature, just alike are found;
The fair, the black, the feeble, and the strong:

But the same foulness does to age belong,
The self-same palsy, both in limbs and tongue;
The skull and forehead one bald barren plain,
And gums unarmed to mumble meat in vain;°
Besides the eternal drivel that supplies   320
The dropping beard, from nostrils, mouth, and eyes.
His wife and children loathe him, and, what's worse,
Himself does his offensive carrion curse!
Flatterers forsake him too; for who would kill
Himself, to be remembered in a will?
His taste not only palled to wine and meat,
But to the relish of a nobler treat.
The limber nerve, in vain provoked to rise,
Inglorious from the field of battle flies;
Poor feeble dotard, how could he advance   330
With his blue head-piece, and his broken lance?
Add that endeavouring still without effect,
A lust more sordid justly we suspect.
    Those senses lost, behold a new defeat,
The soul dislodging from another seat.
What music, or enchanting voice, can cheer
A stupid, old, impenetrable ear?
No matter in what place, or what degree
Of the full theatre he sits to see;
Cornets and trumpets cannot reach his ear;   340
Under an actor's nose he's never near.
    His boy must bawl, to make him understand
The hour o' the day, or such a lord's at hand;
The little blood that creeps within his veins,
Is but just warmed in a hot fever's pains.
In fine, he wears no limb about him sound,
With sores and sicknesses beleaguered round;
Ask me their names, I sooner could relate
How many drudges on salt Hippia wait;°
What crowds of patients the town doctor kills,   350
Or how, last fall, he raised the weekly bills;
What provinces by Basilus were spoiled;
What herds of heirs by guardians are beguiled;
How many bouts a day that bitch has tried;
How many boys that pedagogue can ride;
What lands and lordships for their owner know

My quondam barber, but his worship now.°
   This dotard of his broken back complains;
One his legs fail, and one his shoulder pains:
Another is of both his eyes bereft,   360
And envies who has one for aiming left,
A fifth, with trembling lips expecting stands
As in his childhood, crammed by others' hands;
One, who at sight of supper opened wide
His jaws before, and whetted grinders tried,
Now only yawns, and waits to be supplied;
Like a young swallow, when, with weary wings,
Expected food her fasting mother brings.
   His loss of members is a heavy curse,
But all his faculties decayed, a worse.   370
His servants' names he has forgotten quite;
Knows not his friend who supped with him last night:
Not e'en the children he begot and bred;
Or his will knows them not; for, in their stead,
In form of law, a common hackney jade,°
Sole heir, for secret services, is made:
So lewd, and such a battered brothel whore,
That she defies all comers at her door.
Well, yet suppose his senses are his own,
He lives to be chief mourner for his son: 380
Before his face his wife and brother burns;
He numbers all his kindred in their urns.
These are the fines he pays for living long,
And dragging tedious age in his own wrong;
Grief always green, a household still in tears,
Sad pomps, a threshold thronged with daily biers,
And liveries of black for length of years.
   Next to the raven's age, the Pylian king°
Was longest lived of any two-legged thing.
Blest, to defraud the grave so long, to mount   390
His numbered years, and on his right hand count.°
Three hundred seasons, guzzling must of wine.
But hold a while, and hear himself repine
At fate's unequal laws, and at the clew
Which, merciless in length, the midmost sister drew.°
When his brave son upon the funeral pyre
He saw extended, and his beard on fire,

He turned, and weeping, asked his friends, what crime
Had cursed his age to this unhappy time?
Thus mourned old Peleus for Achilles slain,   400
And thus Ulysses' father did complain.
   How fortunate an end had Priam made,
Among his ancestors a mighty shade,
While Troy yet stood; when Hector, with the race
Of royal bastards, might his funeral grace;
Amidst the tears of Trojan dames inurned,
And by his loyal daughters truly mourned.
Had heaven so blest him, he had died before
The fatal fleet to Sparta Paris bore:
But mark what age produced, he lived to see   410
His town in flames, his falling monarchy.
In fine, the feeble sire, reduced by fate
To change his sceptre for a sword too late,
His last effort before Jove's altar tries,°
A soldier half, and half a sacrifice:
Falls like an ox that waits the coming blow,
Old and unprofitable to the plough.
   At least he died a man: his queen survived,
To howl, and in a barking body lived.
   I hasten to our own; nor will relate   420
Great Mithridates, and rich Croesus' fate;°
Whom Solon wisely counselled to attend
The name of happy, till he knew his end.
   That Marius was an exile, that he fled,°
Was ta'en, in ruined Carthage begged his bread;
All these were owing to a life too long:
For whom had Rome beheld so happy, young?
High in his chariot, and with laurel crowned,
When he had led the Cimbrian captives round°
The Roman streets, descending from his state,   430
In that blest hour he should have begged his fate;
Then, then, he might have died of all admired,
And his triumphant soul with shouts expired.
   Campania, Fortune's malice to prevent,
To Pompey an indulgent fever sent;°
But public prayers imposed on heaven to give
Their much loved leader an unkind reprieve.
The city's fate and his conspired to save

The head reserved for an Egyptian slave.

Cethegus, though a traitor to the State,°   440
And tortured, scaped this ignominious fate;
And Sergius, who a bad cause bravely tried,°
All of a piece, and undiminished, died.

To Venus, the fond mother makes a prayer,
That all her sons and daughters may be fair:
True, for the boys a mumbling vow she sends,
But for the girls the vaulted temple rends:
They must be finished pieces; 'tis allowed
Diana's beauty made Latona proud,
And pleased to see the wondering people pray   450
To the new-rising sister of the day.

And yet Lucretia's fate would bar that vow;
And fair Virginia would her fate bestow°
On Rutila, and change her faultless make°
For the foul rumple of her camel back.

But, for his mother's boy, the beau, what frights
His parents have by day, what anxious nights!
Form joined with virtue is a sight too rare;
Chaste is no epithet to suit with fair.
Suppose the same traditionary strain   460
Of rigid manners in the house remain;
Inveterate truth, an old plain Sabine's heart;
Suppose that nature too has done her part,
Infused into his soul a sober grace,
And blushed a modest blood into his face,
(For nature is a better guardian far
Than saucy pedants, or dull tutors are;)
Yet still the youth must ne'er arrive at man,
(So much almighty bribes and presents can;)
E'en with a parent, where persuasions fail,   470
Money is impudent, and will prevail.

We never read of such a tyrant king,
Who gelt a boy deformed, to hear him sing;
Nor Nero, in his more luxurious rage,
E'er made a mistress of an ugly page:
Sporus, his spouse, nor crooked was, nor lame,
With mountain back, and belly, from the game
Cross-barred; but both his sexes well became.
Go, boast your springald, by his beauty cursed°

To ills, nor think I have declared the worst;   480
His form procures him journey-work; a strife
Betwixt town-madams, and the merchant's wife:
Guess, when he undertakes this public war,
What furious beasts offended cuckolds are.

    Adulterers are with dangers round beset;
Born under Mars, they cannot scape the net;
And from revengeful husbands oft have tried
Worse handling than severest laws provide:
One stabs, one slashes, one, with cruel art,
Makes colon suffer for the peccant part.   490

    But your Endymion, your smooth smock-faced boy,
Unrivalled, shall a beauteous dame enjoy.
Not so: one more salacious, rich, and old,
Outbids, and buys her pleasure for her gold:
Now, he must moil, and drudge, for one he loathes;°
She keeps him high in equipage and clothes;
She pawns her jewels, and her rich attire,
And thinks the workman worthy of his hire.
In all things else immoral, stingy, mean,
But, in her lusts, a conscionable quean.   500

    'She may be handsome, yet be chaste,' you say,
Good observator, not so fast away;
Did it not cost the modest youth his life,°
Who shunned the embraces of his father's wife?
And was not t' other stripling forced to fly,°
Who coldly did his patron's queen deny,
And pleaded laws of hospitality?
The ladies charged them home, and turned the tale;
With shame they reddened, and with spite grew pale.
'Tis dangerous to deny the longing dame;   510
She loses pity, who has lost her shame.

    Now Silius wants thy counsel, give advice;
Wed Caesar's wife, or die—the choice is nice.°
Her comet-eyes she darts on every grace,
And takes a fatal liking to his face.
Adorned with bridal pomp, she sits in state;
The public notaries and haruspex wait;
The genial bed is in the garden dressed,
The portion paid, and every rite expressed,
Which in a Roman marriage is professed.   520

'Tis no stolen wedding this; rejecting awe,
She scorns to marry, but in form of law:
In this moot case, your judgment to refuse
Is present death, besides the night you lose.
If you consent, 'tis hardly worth your pain,
A day or two of anxious life you gain;
Till loud reports through all the town have past,
And reach the prince—for cuckolds hear the last.
Indulge thy pleasure, youth, and take thy swing,
For not to take is but the self-same thing;   530
Inevitable death before thee lies,
But looks more kindly through a lady's eyes.

    What then remains? are we deprived of will?
Must we not wish, for fear of wishing ill?
Receive my counsel, and securely move—
Intrust thy fortune to the powers above;
Leave them to manage for thee, and to grant
What their unerring wisdom sees thee want:
In goodness, as in greatness, they excel;
Ah that we loved ourselves but half so well!   540
We, blindly by our headstrong passions led
Are hot for action, and desire to wed;
Then wish for heirs; but to the gods alone
Our future offspring, and our wives, are known;
The audacious strumpet, and ungracious son.

    Yet not to rob the priests of pious gain,
That altars be not wholly built in vain,
Forgive the gods the rest, and stand confined
To health of body and content of mind;
A soul that can securely death defy   550
And count it nature's privilege to die;
Serene and manly, hardened to sustain
The load of life, and exercised in pain;
Guiltless of hate and proof against desire,
That all things weighs and nothing can admire;
That dares prefer the toils of Hercules,
To dalliance, banquets, and ignoble ease.

    The path to peace is virtue: what I show,
Thyself may freely on thyself bestow;
Fortune was never worshipped by the wise,   560
But set aloft by fools usurps the skies.°

The First Satire of Persius

ARGUMENT OF THE PROLOGUE TO THE FIRST SATIRE

The design of the author was to conceal his name and quality. He lived in the dangerous times of the tyrant Nero, and aims particularly at him in most of his satires. For which reason, though he was a Roman knight, and of a plentiful fortune, he would appear in this prologue but a beggarly poet, who writes for bread. After this, he breaks into the business of the First Satire; which is chiefly to decry the poetry then in fashion, and the impudence of those who were endeavouring to pass their stuff upon the world.

PROLOGUE

I never did on cleft Parnassus dream°
Nor taste the sacred Heliconian stream;
Nor can remember when my brain, inspired,
Was by the muses into madness fired.
My share in pale Pyrene I resign,°
And claim no part in all the mighty nine.
Statues with winding ivy crowned belong
To nobler poets, for a nobler song;
Heedless of verse, and hopeless of the crown,
Scarce half a wit, and more than half a clown,   10
Before the shrine I lay my rugged numbers down.°
Who taught the parrot human notes to try,
Or with a voice endued the chattering pie?°
'Twas witty want, fierce hunger to appease;
Want taught their masters, and their masters these.
Let gain, that gilded bait, be hung on high,
The hungry witlings have it in their eye;
Pies, crows, and daws, poetic presents bring;
You say they squeak, but they will swear they sing.

ARGUMENT OF THE FIRST SATIRE

I need not repeat that the chief aim of the author is against bad poets in this Satire. But I must add, that he includes also bad orators who began at that time (as Petronius in the beginning of his book tells us) to enervate manly eloquence by tropes and figures, ill placed, and worse applied. Amongst the poets, Persius covertly strikes at Nero, some of whose verses he recites with scorn and indignation. He also takes notice of the noblemen,

*and their abominable poetry, who, in the luxury of their fortune, set up
for wits and judges. The Satire is in dialogue betwixt the author and his
friend, or monitor; who dissuades him from this dangerous attempt of
exposing great men. But Persius, who is of a free spirit, and has not forgotten that Rome was once a commonwealth, breaks through all those difficulties, and boldly arraigns the false judgment of the age in which he
lives. The reader may observe, that our poet was a Stoic philosopher; and
that all his moral sentences, both here and in all the rest of his satires, are
drawn from the dogmas of that sect.*

Persius

How anxious are our cares, and yet how vain
The bent of our desires!

Friend

               Thy spleen contain;
For none will read thy satires

Persius

This to me?

Friend

None, or what's next to none, but two or three.
'Tis hard, I grant.

Persius

              'Tis nothing; I can bear,
That paltry scribblers have the public ear;
That this vast universal fool, the town,
Should cry up Labeo's stuff, and cry me down.°
They damn themselves; nor will my muse descend
To clap with such, who fools and knaves commend: 10
Their smiles and censures are to me the same;
I care not what they praise, or what they blame.
In full assemblies let the crowd prevail;
I weigh no merit by the common scale.
The conscience is the test of every mind;
'Seek not thyself, without thyself, to find.'
But where's that Roman?—Somewhat I would say,
But fear—let fear, for once, to truth give way.
Truth lends the Stoic courage: when I look

On human acts, and read in nature's book,   20
From the first pastimes of our infant age,
To elder cares, and man's severer page;
When stern as tutors, and as uncles hard,
We lash the pupil, and defraud the ward,
Then, then I say—or would say, if I durst—
But, thus provoked, I must speak out, or burst.

Friend
Once more forbear.

Persius
           I cannot rule my spleen;
My scorn rebels, and tickles me within.
  First, to begin at home: our authors write
In lonely rooms, secured from public sight;   30
Whether in prose, or verse, 'tis all the same,
The prose is fustian, and the numbers lame:
All noise, and empty pomp, a storm of words,
Labouring with sound, that little sense affords.
They comb, and then they order every hair;°
A gown, or white, or scoured to whiteness, wear,
A birthday jewel bobbing at their ear;
Next, gargle well their throats; and thus prepared,
They mount, a-God's name, to be seen and heard
From their high scaffold; with a trumpet cheek,   40
And ogling all their audience ere they speak.
The nauseous nobles, e'en the chief of Rome,
With gaping mouths to these rehearsals come,
And pant with pleasure, when some lusty line
The marrow pierces, and invades the chine;°
At open fulsome bawdry they rejoice,
And slimy jests applaud with broken voice.
Base prostitute! thus dost thou gain thy bread?
Thus dost thou feed their ears, and thus art fed?
At his own filthy stuff he grins and brays,   50
And gives the sign where he expects their praise.
  Why have I learned, sayst thou, if thus confined,
I choke the noble vigour of my mind?
Know, my wild fig-tree, which in rocks is bred,
Will split the quarry, and shoot out the head.

Fine fruits of learning! old ambitious fool,
Darest thou apply that adage of the school,
As if 'tis nothing worth that lies concealed,
And 'science is not science till revealed.'
Oh, but 'tis brave to be admired, to see   60
The crowd, with pointing fingers, cry, 'That's he;
That's he, whose wondrous poem is become
A lecture for the noble youth of Rome!
Who, by their fathers, is at feasts renowned,
And often quoted when the bowls go round.'
Full gorged and flushed, they wantonly rehearse.
And add to wine the luxury of verse.
One, clad in purple, not to lose his time,
Eats and recites some lamentable rhyme,
Some senseless Phyllis, in a broken note,   70
Snuffling at nose, or croaking in his throat.
Then graciously the mellow audience nod;
Is not the immortal author made a god?
Are not his manes blest, such praise to have?
Lies not the turf more lightly on his grave?
And roses (while his loud applause they sing)
Stand ready from his sepulchre to spring?
    All these, you cry, but light objections are,
Mere malice, and you drive the jest too far:
For does there breathe a man, who can reject   80
A general fame, and his own lines neglect?
In cedar tablets worthy to appear,°
That need not fish, or frankincense, to fear?
    Thou, whom I make the adverse part to bear,
Be answered thus: If I by chance succeed
In what I write (and that's a chance indeed),
Know, I am not so stupid, or so hard,
Not to feel praise, or fame's deserved reward;
But this I cannot grant, that thy applause
Is my work's ultimate, or only cause.   90
Prudence can ne'er propose so mean a prize;
For mark what vanity within it lies.
Like Labeo's Iliads, in whose verse is found
Nothing but trifling care, and empty sound;
Such little elegies as nobles write,
Who would be poets, in Apollo's spite.
Them and their woeful works the Muse defies;

Products of citron beds, and golden canopies.°
To give thee all thy due, thou hast the heart
To make a supper, with a fine dessert,   100
And to thy thread-bare friend a cast old suit impart.

   Thus bribed, thou thus bespeak'st him, 'Tell me, friend,
(For I love truth, nor can plain speech offend,)
What says the world of me and of my muse?'

   The poor dare nothing tell but flattering news;
But shall I speak? Thy verse is wretched rhyme,
And all thy labours are but loss of time.
Thy strutting belly swells, thy paunch is high;
Thou writ'st not, but thou pissest poetry.

   All authors to their own defects are blind;   110
Hadst thou but, Janus-like, a face behind,°
To see the people, what splay-mouths they make;
To mark their fingers, pointed at thy back,
Their tongues lolled out, a foot beyond the pitch,
When most athirst, of an Apulian bitch:°
But noble scribblers are with flattery fed,
For none dare find their faults, who eat their bread.
To pass the poets of patrician blood,
What is 't the common reader takes for good?
The verse in fashion is, when numbers flow,   120
Soft without sense, and without spirit slow;
So smooth and equal, that no sight can find
The rivet, where the polished piece was joined;
So even all, with such a steady view,
As if he shut one eye to level true.
Whether the vulgar vice his satire stings,
The people's riots, or the rage of kings,°
The gentle poet is alike in all;
His reader hopes no rise, and fears no fall.

Friend

Hourly we see some raw pin-feathered thing°   130
Attempt to mount, and fights and heroes sing;
Who for false quantities was whipped at school
But t' other day, and breaking grammar-rule;
Whose trivial art was never tried above°
The bare description of a native grove;
Who knows not how to praise the country store,

The feasts, the baskets, nor the fatted boar,
Nor paint the flowery fields that paint themselves before;
Where Romulus was bred, and Quintius born,°
Whose shining plough-share was in furrows worn,   140
Met by his trembling wife returning home,
And rustically joyed, as chief of Rome:
She wiped the sweat from the dictator's brow,
And o'er his back his robe did rudely throw;
The lictors bore in state their lord's triumphant plough.

   Some love to hear the fustian poet roar,
And some on antiquated authors pore;
Rummage for sense, and think those only good
Who labour most, and least are understood.
When thou shalt see the blear-eyed fathers teach   150
Their sons this harsh and mouldy sort of speech,
Or others new affected ways to try,
Of wanton smoothness, female poetry;
One would inquire from whence this motley style
Did first our Roman purity defile.
For our old dotards cannot keep their seat,
But leap and catch at all that's obsolete.

   Others, by foolish ostentation led,
When called before the bar, to save their head,
Bring trifling tropes, instead of solid sense,   160
And mind their figures more than their defence;
Are pleased to hear their thick-skulled judges cry,
'Well moved, oh finely said, and decently!'
'Theft' (says the accuser) 'to thy charge I lay,
O Pedius'. What does gentle Pedius say?
Studious to please the genius of the times,
With periods, points, and tropes, he slurs his crimes:°
'He robbed not, but he borrowed from the poor,
And took but with intention to restore.'
He lards with flourishes his long harangue;   170
'Tis fine, say'st thou; what, to be praised, and hang?
Effeminate Roman, shall such stuff prevail
To tickle thee, and make thee wag thy tail?
Say, should a shipwrecked sailor sing his woe,
Wouldst thou be moved to pity, or bestow
An alms? What's more preposterous than to see
A merry beggar, mirth in misery?

Persius

He seems a trap for charity to lay,
And cons, by night, his lesson for the day.

Friend

But to raw numbers and unfinished verse,   180
Sweet sound is added now, to make it terse:
"Tis tagged with rhyme, like Berecyntian Attis,°
The mid-part chimes with art, which never flat is.
The dolphin brave, that cut the liquid wave,
Or he who in his line can chine the long-ribbed Apennine.'°

Persius

All this is doggrel stuff.

Friend
                    What if I bring
A nobler verse? 'Arms and the man I sing.'°

Persius

Why name you Virgil with such fops as these?
He's truly great, and must for ever please:
Not fierce, but awful, is his manly page;   190
Bold is his strength, but sober is his rage.

Friend

What poems think you soft, and to be read
With languishing regards, and bending head?

Persius

'Their crooked horns the Mimallonian crew°
With blasts inspired; and Bassaris, who slew
The scornful calf, with sword advanced on high,
Made from his neck his haughty head to fly:
And Maenas, when with ivy bridles bound,
She led the spotted lynx, then Evion rung around;
Evion from woods and floods repairing echoes sound.'   200
Could such rude lines a Roman mouth become,
Were any manly greatness left in Rome?
Maenas and Attis in the mouth were bred,°
And never hatched within the labouring head;

No blood from bitten nails those poems drew,
But churned, like spittle, from the lips they flew.

Friend

'Tis fustian all; 'tis execrably bad;
But if they will be fools, must you be mad?
Your satires, let me tell you, are too fierce;
The great will never bear so blunt a verse.   210
Their doors are barred against a bitter flout;°
Snarl, if you please, but you shall snarl without.
Expect such pay as railing rhymes deserve;
You're in a very hopeful way to starve.

Persius

Rather than so, uncensured let them be;
All, all is admirably well, for me.
My harmless rhyme shall scape the dire disgrace
Of common-shores, and every pissing-place.°
Two painted serpents shall on high appear;
'Tis holy ground; you must not urine here.   220
This shall be writ, to fright the fry away,
Who draw their little baubles when they play.

   Yet old Lucilius never feared the times,°
But lashed the city, and dissected crimes.
Mutius and Lupus both by name he brought;
He mouthed them, and betwixt his grinders caught.
Unlike in method, with concealed design,
Did crafty Horace his low numbers join;
And, with a sly insinuating grace,
Laughed at his friend, and looked him in the face;   230
Would raise a blush where secret vice he found,
And tickle while he gently probed the wound,
With seeming innocence the crowd beguiled,
But made the desperate passes when he smiled.

   Could he do this, and is my muse controlled
By servile awe? Born free, and not be bold?
At least, I'll dig a hole within the ground,
And to the trusty earth commit the sound;
The reeds shall tell you what the poet fears,
'King Midas has a snout, and ass's ears.'°   240

This mean conceit, this darling mystery,
Which thou think'st nothing, friend, thou shalt not buy;
Nor will I change for all the flashy wit,
That flattering Labeo in his Iliads writ.
   Thou, if there be a thou in this base town,
Who dares, with angry Eupolis, to frown;°
He who, with bold Cratinus, is inspired
With zeal, and equal indignation fired;
Who at enormous villainy turns pale,
And steers against it with a full-blown sail,   250
Like Aristophanes, let him but smile
On this my honest work, though writ in homely style;
And if two lines or three in all the vein
Appear less drossy, read those lines again.
May they perform their author's just intent,
Glow in thy ears, and in thy breast ferment.
But from the reading of my book and me,
Be far, ye foes of virtuous poverty;
Who fortune's fault upon the poor can throw,
Point at the tattered coat, and ragged shoe;   260
Lay nature's failings to their charge, and jeer
The dim weak eyesight, when the mind is clear;
When thou thyself, thus insolent in state,
Art but, perhaps, some country magistrate,
Whose power extends no further than to speak
Big on the bench, and scanty weights to break.
   Him also for my censor I disdain,
Who thinks all science, as all virtue, vain;
Who counts geometry, and numbers, toys,
And with his foot the sacred dust destroys;°   270
Whose pleasure is to see a strumpet tear
A cynic's beard, and lug him by the hair.
Such all the morning to the pleadings run;
But when the business of the day is done,
On dice, and drink, and drabs, they spend their afternoon.°

The First Book of
Ovid's Metamorphoses

Of bodies changed to various forms I sing:
Ye gods, from whom these miracles did spring,
Inspire my numbers with celestial heat,
Till I my long laborious work complete;
And add perpetual tenor to my rhymes,°
Deduced from nature's birth to Caesar's times.°

  Before the seas, and this terrestrial ball,
And heaven's high canopy, that covers all,
One was the face of nature, if a face;
Rather a rude and indigested mass;°   10
A lifeless lump, unfashioned, and unframed,
Of jarring seeds, and justly chaos named.
No sun was lighted up the world to view;
No moon did yet her blunted horns renew;
Nor yet was earth suspended in the sky,
Nor, poised, did on her own foundations lie;
Nor seas about the shores their arms had thrown;
But earth, and air, and water, were in one.
Thus air was void of light, and earth unstable,
And water's dark abyss unnavigable.   20
No certain form on any was imprest;
All were confused, and each disturbed the rest:
For hot and cold were in one body fixed;
And soft with hard, and light with heavy, mixed.

  But god, or nature, while they thus contend,
To these intestine discords put an end.°
Then earth from air, and seas from earth, were driven,
And grosser air sunk from ethereal heaven.
Thus disembroiled, they take their proper place;
The next of kin contiguously embrace;   30
And foes are sundered by a larger space.
The force of fire ascended first on high,
And took its dwelling in the vaulted sky.
Then air succeeds, in lightness next to fire,
Whose atoms from unactive earth retire.
Earth sinks beneath, and draws a numerous throng,
Of ponderous, thick, unwieldy seeds along.

About her coasts unruly waters roar,
And, rising on a ridge, insult the shore.
Thus when the god, whatever god was he,   40
Had formed the whole, and made the parts agree,
That no unequal portions might be found,
He moulded earth into a spacious round;
Then, with a breath, he gave the winds to blow,
And bade the congregated waters flow:
He adds the running springs, and standing lakes,
And bounding banks for winding rivers makes.°
Some part in earth are swallowed up, the most
In ample oceans, disembogued, are lost:
He shades the woods, the valleys he restrains   50
With rocky mountains, and extends the plains.
    And as five zones the ethereal regions bind,
Five, correspondent, are to earth assigned;
The sun, with rays directly darting down,
Fires all beneath, and fries the middle zone:
The two beneath the distant poles complain
Of endless winter, and perpetual rain.
Betwixt the extremes, two happier climates hold
The temper that partakes of hot and cold.
The fields of liquid air, enclosing all,   60
Surround the compass of this earthly ball:
The lighter parts lie next the fires above;
The grosser near the watery surface move:
Thick clouds are spread, and storms engender there,
And thunder's voice, which wretched mortals fear,
And winds that on their wings cold winter bear.
Nor were those blustering brethren left at large,
On seas and shores their fury to discharge:
Bound as they are, and circumscribed in place,
They rend the world, resistless, where they pass,   70
And mighty marks of mischief leave behind;
Such is the rage of their tempestuous kind.
First, Eurus to the rising morn is sent,
(The regions of the balmy continent,)
And eastern realms, where early Persians run,
To greet the blest appearance of the sun.
Westward the wanton Zephyr wings his flight,
Pleased with the remnants of departing light;

Fierce Boreas with his offspring issues forth,
To invade the frozen waggon of the north;°   80
While frowning Auster seeks the southern sphere,
And rots, with endless rain, the unwholesome year.°

   High o'er the clouds, and empty realms of wind,
The god a clearer space for heaven designed;
Where fields of light and liquid ether flow,
Purged from the ponderous dregs of earth below.

   Scarce had the power distinguished these, when straight
The stars, no longer overlaid with weight,
Exert their heads from underneath the mass,
And upward shoot, and kindle as they pass,   90
And with diffusive light adorn their heavenly place.
Then, every void of nature to supply,
With forms of gods he fills the vacant sky:
New herds of beasts he sends, the plains to share;
New colonies of birds, to people air;
And to their oozy beds the finny fish repair.
A creature of a more exalted kind
Was wanting yet, and then was man designed;
Conscious of thought, of more capacious breast,
For empire formed, and fit to rule the rest: 100
Whether with particles of heavenly fire
The god of nature did his soul inspire;
Or earth, but new divided from the sky,
And pliant still, retained the ethereal energy;
Which wise Prometheus tempered into paste,
And, mixed with living streams, the god-like image cast.
Thus, while the mute creation downward bend
Their sight, and to their earthly mother tend,
Man looks aloft, and, with erected eyes,
Beholds his own hereditary skies.   110
From such rude principles our form began,
And earth was metamorphosed into man.

THE GOLDEN AGE

The Golden Age was first; when man yet new
No rule but uncorrupted reason knew;
And, with a native bent, did good pursue.
Unforced by punishment, unawed by fear,
His words were simple, and his soul sincere.

Needless was written law, where none oppressed;
The law of man was written in his breast.
No suppliant crowds before the judge appeared;   120
No court erected yet, nor cause was heard;
But all was safe, for conscience was their guard.
The mountain trees in distant prospect please,
Ere yet the pine descended to the seas;
Ere sails were spread, new oceans to explore;
And happy mortals, unconcerned for more,
Confined their wishes to their native shore.
No walls were yet, nor fence, nor moat, nor mound;
Nor drum was heard, nor trumpet's angry sound;
Nor swords were forged; but, void of care and crime,   130
The soft creation slept away their time.
The teeming earth, yet guiltless of the plough,°
And unprovoked, did fruitful stores allow:
Content with food, which nature freely bred,
On wildings and on strawberries they fed;
Cornels and bramble-berries gave the rest,
And falling acorns furnished out a feast.
The flowers, unsown, in fields and meadows reigned;
And western winds immortal spring maintained.
In following years the bearded corn ensued   140
From earth unasked, nor was that earth renewed.
From veins of valleys milk and nectar broke,
And honey sweating through the pores of oak.

THE SILVER AGE

But when good Saturn, banished from above,
Was driven to hell, the world was under Jove.
Succeeding times a Silver Age behold,
Excelling brass, but more excelled by gold.
Then Summer, Autumn, Winter did appear,
And Spring was but a season of the year.
The sun his annual course obliquely made,   150
Good days contracted, and enlarged the bad.
Then air with sultry heats began to glow,
The winds of winds were clogged with ice and snow;
And shivering mortals, into houses driven,
Sought shelter from the inclemency of heaven.

Those houses, then, were caves, or homely sheds,
With twining osiers fenced, and moss their beds.
Then ploughs for seed the fruitful furrows broke,
And oxen laboured first beneath the yoke.

THE BRAZEN AGE

To this came next, in course the Brazen Age: 160
A warlike offspring prompt to bloody rage,
Not impious yet—

THE IRON AGE

                 Hard steel succeeded then;
And stubborn as the metal were the men.
Truth, modesty, and shame, the world forsook;
Fraud, avarice, and force, their places took.
Then sails were spread to every wind that blew;
Raw were the sailors, and the depths were new:
Trees, rudely hollowed, did the waves sustain,
Ere ships in triumph ploughed the watery plain.
  Then landmarks limited to each his right;   170
For all before was common as the light.
Nor was the ground alone required to bear
Her annual income to the crooked share;
But greedy mortals, rummaging her store,°
Dug from her entrails the precious ore;
Which next to hell the prudent gods had laid,
And that alluring ill to sight displayed.
Thus cursed steel, and more accursed gold,
Gave mischief birth, and made that mischief bold;
And double death did wretched man invade,   180
By steel assaulted, and by gold betrayed.
Now (brandished weapons glittering in their hands)
Mankind is broken loose from moral bands:
No rights of hospitality remain,
The guest, by him who harboured him, is slain;
The son-in-law pursues the father's life;
The wife her husband murders, he the wife;
The step-dame poison for the son prepares;
The son inquires into his father's years.
Faith flies, and piety in exile mourns: 190
And justice, here oppressed, to heaven returns.

THE GIANTS' WAR

Nor were the Gods themselves more safe above;
Against beleaguered heaven the Giants move.
Hills piled on hills, on mountains mountains lie,
To make their mad approaches to the sky:
Till Jove, no longer patient, took his time
To avenge with thunder their audacious crime;
Red lightning played along the firmament,
And their demolished works to pieces rent.
Singed with the flames, and with the bolts transfixed,   200
With native earth their blood the monsters mixed;
The blood, endued with animating heat,
Did in the impregnant earth new sons beget;
They, like the seed from which they sprung, accursed,
Against the gods immortal hatred nursed;
An impious, arrogant, and cruel brood,
Expressing their original from blood;
Which when the king of gods beheld from high,
(Withal revolving in his memory
What he himself had found on earth of late,   210
Lycaon's guilt, and his inhuman treat,)°
He sighed, nor longer with his pity strove,
But kindled to a wrath becoming Jove.
Then called a general council of the gods;
Who, summoned, issue from their blest abodes,
And fill the assembly with a shining train.
A way there is in heaven's expanded plain,
Which, when the skies are clear, is seen below,
And mortals by the name of Milky know.
The ground-work is of stars; through which the road   220
Lies open to the thunderer's abode.
The gods of greater nations dwell around,
And on the right and left the palace bound;
The commons where they can; the nobler sort,
With winding doors wide open, front the court.
This place, as far as earth with heaven may vie,
I dare to call the Louvre of the sky.
When all were placed, in seats distinctly known,
And he, their father, had assumed the throne,
Upon his ivory sceptre first he leant,   230
Then shook his head, that shook the firmament;

Air, earth, and seas obeyed the almighty nod,
And with a general fear confessed the god.
At length, with indignation, thus he broke
His awful silence, and the powers bespoke.
  'I was not more concerned in that debate
Of empire, when our universal state
Was put to hazard, and the giant race
Our captive skies were ready to embrace:
For though the foe was fierce, the seeds of all   240
Rebellion sprung from one original;
Now wheresoever ambient waters glide,
All are corrupt, and all must be destroyed.
Let me this holy protestation make,
By hell, and hell's inviolable lake,
I tried whatever in the godhead lay;
But gangrened members must be lopped away,
Before the nobler parts are tainted to decay.°
There dwells below a race of demi-gods,
Of nymphs in waters, and of fauns in woods;   250
Who, though not worthy yet in heaven to live,
Let them at least enjoy that earth we give.
Can these be thought securely lodged below,
When I myself, who no superior know,
I, who have heaven and earth at my command,
Have been attempted by Lycaon's hand?'°
  At this a murmur through the synod went,
And with one voice they vote his punishment.
Thus when conspiring traitors dared to doom
The fall of Caesar, and in him of Rome,   260
The nations trembled with a pious fear,
All anxious for their earthly thunderer:
Nor was their care, O Caesar, less esteemed
By thee, than that of heaven for Jove was deemed;
Who with his hand and voice did first restrain
Their murmurs, then resumed his speech again.
The gods to silence were composed, and sate
With reverence due to his superior state.
  'Cancel your pious cares; already he
Has paid his debt to justice, and to me.   270
Yet what his crimes, and what my judgments were,
Remains for me thus briefly to declare.

The clamours of this vile degenerate age,
The cries of orphans, and the oppressor's rage,
Had reached the stars; "I will descend," said I,
"In hope to prove this loud complaint a lie."
Disguised in human shape, I travelled round
The world, and more than what I heard, I found.
O'er Maenalus I took my steepy way,
By caverns infamous for beasts of prey;   280
Then crossed Cyllene, and the piny shade,
More infamous by cursed Lycaon made.
Dark night had covered heaven and earth, before
I entered his unhospitable door.
Just at my entrance, I displayed the sign
That somewhat was approaching of divine.
The prostrate people pray; the tyrant grins;
And, adding profanation to his sins,
"I'll try," said he, "and if a God appear,
To prove his deity shall cost him dear."   290
'Twas late; the graceless wretch my death prepares,
When I should soundly sleep, oppressed with cares:
This dire experiment he chose, to prove
If I were mortal, or undoubted Jove.
But first he had resolved to taste my power:
Not long before, but in a luckless hour,
Some legates, sent from the Molossian state,°
Were on a peaceful errand come to treat;
Of these he murders one, he boils the flesh,
And lays the mangled morsels in a dish;   300
Some part he roasts; then serves it up so dressed,
And bids me welcome to this human feast.
Moved with disdain, the table I o'erturned,
And with avenging flames the palace burned.
The tyrant, in a fright, for shelter gains
The neighbouring fields, and scours along the plains.
Howling he fled, and fain he would have spoke,
But human voice his brutal tongue forsook.
About his lips the gathered foam he churns,
And, breathing slaughter, still with rage he burns,   310
But on the bleating flock his fury turns.
His mantle, now his hide, with rugged hairs
Cleaves to his back; a famished face he bears;°

His arms descend, his shoulders sink away,
To multiply his legs for chase of prey.
He grows a wolf, his hoariness remains,
And the same rage in other members reigns.
His eyes still sparkle in a narrower space,
His jaws retain the grin, and violence of his face.
   'This was a single ruin, but not one   320
Deserves so just a punishment alone.
Mankind's a monster, and the ungodly times,
Confederate into guilt, are sworn to crimes.
All are alike involved in ill, and all
Must by the same relentless fury fall.'
   Thus ended he; the greater gods assent,
By clamours urging his severe intent;
The less fill up the cry for punishment.
Yet still with pity they remember man,
And mourn as much as heavenly spirits can.   330
They ask, when those were lost of human birth,
What he would do with all this waste of earth.
If his dispeopled world he would resign
To beasts, a mute, and more ignoble line.
Neglected altars must no longer smoke,
If none were left to worship and invoke.
To whom the father of the gods replied:
'Lay that unnecessary fear aside;
Mine be the care new people to provide.
I will from wondrous principles ordain   340
A race unlike the first, and try my skill again.'
   Already had he tossed the flaming brand,
And rolled the thunder in his spacious hand,
Preparing to discharge on seas and land;
But stopped, for fear, thus violently driven,
The sparks should catch his axle-tree of heaven;°
Remembering, in the fates, a time, when fire
Should to the battlements of heaven aspire,
And all his blazing worlds above should burn,
And all the inferior globe to cinders turn.   350
His dire artillery thus dismissed, he bent
His thoughts to some securer punishment;
Concludes to pour a watery deluge down,
And what he durst not burn, resolves to drown.

The northern breath, that freezes floods, he binds,
With all the race of cloud-dispelling winds;
The south he loosed, who night and horror brings,
And fogs are shaken from his flaggy wings.°
From his divided beard two streams he pours;
His head and rheumy eyes distil in showers;   360
With rain his robe and heavy mantle flow,
And lazy mists are louring on his brow.
Still as he swept along, with his clenched fist,
He squeezed the clouds; the imprisoned clouds resist;
The skies, from pole to pole, with peals resound,
And showers enlarged come pouring on the ground.
Then clad in colours of a various dye,
Junonian Iris breeds a new supply°
To feed the clouds: impetuous rain descends;
The bearded corn beneath the burden bends;   370
Defrauded clowns deplore their perished grain,
And the long labours of the year are vain.

    Nor from his patrimonial heaven alone
Is Jove content to pour his vengeance down;
Aid from his brother of the seas he craves,
To help him with auxiliary waves.
The watery tyrant calls his brooks and floods,°
Who roll from mossy caves, their moist abodes;
And with perpetual urns his palace fill:
To whom, in brief, he thus imparts his will.   380
    'Small exhortation needs; your powers employ,
And this bad world (so Jove requires) destroy.
Let loose the reins to all your watery store;
Bear down the dams, and open every door.'
    The floods, by nature enemies to land,
And proudly swelling with their new command,
Remove the living stones that stopped their way,
And gushing from their source, augment the sea.
Then with his mace their monarch struck the ground;
With inward trembling earth received the wound,   390
And rising streams a ready passage found.
The expanded waters gather on the plain,
They float the fields, and overtop the grain;
Then rushing onwards, with a sweepy sway,
Bear flocks and folds and labouring hinds away.

Nor safe their dwellings were; for sapped by floods,
Their houses fell upon their household gods.
The solid piles, too strongly built to fall,
High o'er their heads behold a watery wall.
Now seas and earth were in confusion lost;   400
A world of waters, and without a coast.
　　One climbs a cliff; one in his boat is borne,
And ploughs above, where late he sowed his corn.
Others o'er chimney-tops and turrets row,
And drop their anchors on the meads below;
Or, downward driven, they bruise the tender vine,
Or, tossed aloft, are knocked against a pine;
And where of late the kids had cropped the grass,
The monsters of the deep now take their place.
Insulting nereids on the cities ride,   410
And wondering dolphins o'er the palace glide;
On leaves and masts of mighty oaks they browse;
And their broad fins entangle in the boughs.
The frighted wolf now swims amongst the sheep;
The yellow lion wanders in the deep;
His rapid force no longer helps the boar;
The stag swims faster than he ran before.
The fowls, long beating on their wings in vain,
Despair of land and drop into the main.
Now hills and vales no more distinction know,   420
And levelled nature lies oppressed below.
The most of mortals perish in the flood,
The small remainder dies for want of food.
　　A mountain of stupendous height there stands
Betwixt the Athenian and Boeotian lands,
The bound of fruitful fields, while fields they were,°
But then a field of waters did appear:
Parnassus is its name, whose forky rise
Mounts through the clouds and mates the lofty skies.
High on the summit of this dubious cliff,   430
Deucalion wafting moored his little skiff.
He with his wife were only left behind
Of perished man; they two were humankind.
The mountain-nymphs and Themis they adore,
And from her oracles relief implore.
The most upright of mortal men was he;

The most sincere and holy woman, she.
    When Jupiter, surveying earth from high,
Beheld it in a lake of water lie,
That where so many millions lately lived,   440
But two, the best of either sex, survived,
He loosed the northern wind; fierce Boreas flies
To puff away the clouds, and purge the skies;
Serenely, while he blows, the vapours driven
Discover heaven to earth, and earth to heaven.
The billows fall, while Neptune lays his mace
On the rough sea, and smooths its furrowed face.
Already Triton at his call appears
Above the waves; a Tyrian robe he wears;°
And in his hand a crooked trumpet bears.   450
The sovereign bids him peaceful sounds inspire,
And give the waves the signal to retire.
His writhen shell he takes, whose narrow vent
Grows by degrees into a large extent,
Then gives it breath; the blast, with doubling sound,
Runs the wide circuit of the world around.
The sun first heard it in his early east
And met the rattling echoes in the west.
The waters, listening to the trumpet's roar,
Obey the summons, and forsake the shore.   460
    A thin circumference of land appears;
And earth, but not at once, her visage rears,
And peeps upon the seas from upper grounds:
The streams, but just contained within their bounds,
By slow degrees into their channels crawl,
And earth increases as the waters fall.
In longer time the tops of trees appear,
Which mud on their dishonoured branches bear.
    At length the world was all restored to view,
But desolate, and of a sickly hue: 470
Nature beheld herself, and stood aghast,
A dismal desert, and a silent waste.
    Which when Deucalion, with a piteous look,
Beheld, he wept, and thus to Pyrrha spoke:
'O wife, O sister, oh, of all thy kind,
The best and only creature left behind,
By kindred, love, and now by dangers joined;

Of multitudes, who breathed the common air,
We two remain, a species in a pair:
The rest the seas have swallowed; nor have we   480
E'en of this wretched life a certainty.
The clouds are still above; and while I speak,
A second deluge o'er our heads may break.
Should I be snatched from hence, and thou remain,
Without relief, or partner of thy pain,
How couldst thou such a wretched life sustain?
Should I be left, and thou be lost, the sea,
That buried her I loved, should bury me.
Oh could our father his old arts inspire,°
And make me heir of his informing fire,   490
That so I might abolished man retrieve,
And perished people in new souls might live!
But heaven is pleased, nor ought we to complain,
That we, the examples of mankind, remain.'
   He said; the careful couple join their tears,
And then invoke the gods with pious prayers.
Thus in devotion having eased their grief,
From sacred oracles they seek relief,
And to Cephisus' brook their way pursue;
The stream was troubled, but the ford they knew.   500
With living waters in the fountain bred,
They sprinkle first their garments, and their head,
Then took the way which to the temple led.
The roofs were all defiled with moss and mire,
The desert altars void of solemn fire.
Before the gradual prostrate they adored,°
The pavement kissed, and thus the saint implored.
   'O righteous Themis, if the powers above
By prayers are bent to pity and to love;
If human miseries can move their mind;   510
If yet they can forgive, and yet be kind;
Tell how we may restore, by second birth,
Mankind, and people desolated earth.'
Then thus the gracious goddess, nodding, said:
'Depart, and with your vestments veil your head:
And stooping lowly down, with loosened zones,
Throw each behind your backs your mighty mother's bones.'
Amazed the pair, and mute with wonder, stand,

Till Pyrrha first refused the dire command.
'Forbid it heaven,' said she, 'that I should tear   520
Those holy relics from the sepulchre.'
They pondered the mysterious words again,
For some new sense; and long they sought in vain.
At length Deucalion cleared his cloudy brow,
And said: 'The dark enigma will allow
A meaning which, if well I understand,
From sacrilege will free the god's command:
This earth our mighty mother is, the stones
In her capacious body are her bones.
These we must cast behind.' With hope, and fear,   530
The woman did the new solution hear:
The man diffides in his own augury,°
And doubts the gods; yet both resolve to try.
Descending from the mount, they first unbind
Their vests, and, veiled, they cast the stones behind:
The stone (a miracle to mortal view,
But long tradition makes it pass for true,)
Did first the rigour of their kind expel,
And suppled into softness as they fell;
Then swelled, and, swelling, by degrees grew warm,   540
And took the rudiments of human form;
Imperfect shapes, in marble such are seen,
When the rude chisel does the man begin,
While yet the roughness of the stone remains,
Without the rising muscles, and the veins.
The sappy parts, and next resembling juice,
Were turned to moisture, for the body's use;
Supplying humours, blood, and nourishment:
The rest, too solid to receive a bent,
Converts to bones; and what was once a vein,   550
Its former name and nature did retain.
By help of power divine, in little space,
What the man threw, assumed a manly face;
And what the wife, renewed the female race.
Hence we derive our nature, born to bear
Laborious life, and hardened into care.

The rest of animals, from teeming earth
Produced, in various forms received their birth.
The native moisture, in its close retreat,

Digested by the sun's ethereal heat,   560
As in a kindly womb, began to breed;
Then swelled, and quickened by the vital seed:
And some in less, and some in longer space,
Were ripened into form, and took a several face.
Thus when the Nile from Pharian fields is fled,°
And seeks with ebbing tides his ancient bed,
The fat manure with heavenly fire is warmed,
And crusted creatures, as in wombs, are formed:°
These, when they turn the glebe, the peasants find:
Some rude, and yet unfinished in their kind;   570
Short of their limbs, a lame imperfect birth;
One half alive, and one of lifeless earth.

For heat and moisture, when in bodies joined,
The temper that results from either kind,°
Conception makes; and fighting till they mix,
Their mingled atoms in each other fix.
Thus nature's hand the genial bed prepares,
With friendly discord, and with fruitful wars.

From hence the surface of the ground, with mud
And slime besmeared, (the faeces of the flood,)   580
Received the rays of heaven; and sucking in
The seeds of heat, new creatures did begin.
Some were of several sorts produced before;
But of new monsters earth created more.

Unwillingly, but yet she brought to light
Thee, Python, too, the wondering world to fright,°
And the new nations with so dire a sight;
So monstrous was his bulk, so large a space
Did his vast body and long train embrace:
Whom Phoebus basking on a bank espied.   590
Ere now the god his arrows had not tried
But on the trembling deer, or mountain-goat;
At this new quarry he prepares to shoot.
Though every shaft took place, he spent the store
Of his full quiver; and 'twas long before
The expiring serpent wallowed in his gore.
Then to preserve the fame of such a deed,
For Python slain, he Pythian games decreed,
Where noble youths for mastership should strive,
To quoit, to run, and steeds and chariots drive.   600

The prize was fame. In witness of renown,
An oaken garland did the victor crown.
The laurel was not yet for triumphs borne;
But every green alike, by Phoebus worn,
Did, with promiscuous grace, his flowing locks adorn.°

THE TRANSFORMATION OF DAPHNE INTO A LAUREL

The first and fairest of his loves was she,
Whom not blind fortune, but the dire decree
Of angry Cupid, forced him to desire;
Daphne her name, and Peneus was her sire.
Swelled with the pride that new success attends,   610
He sees the stripling, while his bow he bends,
And thus insults him: 'Thou lascivious boy,
Are arms like these for children to employ?
Know, such achievements are my proper claim,
Due to my vigour and unerring aim:
Resistless are my shafts, and Python late,
In such a feathered death, has found his fate.
Take up thy torch, and lay my weapons by;
With that the feeble souls of lovers fry.'
To whom the son of Venus thus replied: 620
'Phoebus, thy shafts are sure on all beside;
But mine on Phoebus; mine the fame shall be
Of all thy conquests, when I conquer thee.'

He said, and soaring swiftly winged his flight;
Nor stopped but on Parnassus' airy height.
Two different shafts he from his quiver draws;
One to repel desire, and one to cause.
One shaft is pointed with refulgent gold,
To bribe the love, and make the lover bold;
One blunt, and tipped with lead, whose base allay   630
Provokes disdain, and drives desire away.
The blunted bolt against the nymph he dressed;
But with the sharp transfixed Apollo's breast.

The enamoured deity pursues the chase;
The scornful damsel shuns his loathed embrace:
In hunting beasts of prey her youth employs,
And Phoebe rivals in her rural joys.
With naked neck she goes, and shoulders bare,
And with a fillet binds her flowing hair.

By many suitors sought, she mocks their pains,   640
And still her vowed virginity maintains.
Impatient of a yoke, the name of bride
She shuns, and hates the joys she never tried.
On wilds and woods she fixes her desire;
Nor knows what youth and kindly love inspire.
Her father chides her oft: 'Thou ow'st,' says he,°
'A husband to thyself, a son to me.'°
She, like a crime, abhors the nuptial bed;
She glows with blushes, and she hangs her head.
Then casting round his neck her tender arms,   650
Soothes him with blandishments, and filial charms:
'Give me, my lord,' she said, 'to live and die
A spotless maid, without the marriage-tie.
'Tis but a small request; I beg no more
Than what Diana's father gave before.'
The good old sire was softened to consent;
But said her wish would prove her punishment;
For so much youth, and so much beauty joined,
Opposed the state which her desires designed.
    The god of light, aspiring to her bed,   660
Hopes what he seeks, with flattering fancies fed,
And is by his own oracles misled.
And as in empty fields the stubble burns,
Or nightly travellers, when day returns,
Their useless torches on dry hedges throw,
That catch the flames, and kindle all the row;
So burns the god, consuming in desire,
And feeding in his breast a fruitless fire:
Her well-turned neck he viewed, (her neck was bare,)
And on her shoulders her dishevelled hair: 670
'Oh, were it combed,' said he, 'with what a grace
Would every waving curl become her face!'
He viewed her eyes, like heavenly lamps that shone;
He viewed her lips, too sweet to view alone;
Her taper fingers, and her panting breast:
He praises all he sees; and for the rest,
Believes the beauties yet unseen are best.
Swift as the wind, the damsel fled away,
Nor did for these alluring speeches stay.
'Stay, nymph,' he cried; 'I follow, not a foe:°   680

Thus from the lion trips the trembling doe;
Thus from the wolf the frightened lamb removes,
And from pursuing falcons fearful doves;
Thou shunn'st a god, and shunn'st a god that loves.
Ah! lest some thorn should pierce thy tender foot,
Or thou shouldst fall in flying my pursuit,
To sharp uneven ways thy steps decline,
Abate thy speed, and I will bate of mine.
Yet think from whom thou dost so rashly fly;
Nor basely born, nor shepherd's swain am I.   690
Perhaps thou know'st not my superior state,
And from that ignorance proceeds thy hate.
Me Claros, Delphos, Tenedos, obey;°
These hands the Patareian sceptre sway.°
The king of gods begot me: what shall be,
Or is, or ever was, in fate, I see.
Mine is the invention of the charming lyre;
Sweet notes, and heavenly numbers, I inspire.
Sure is my bow, unerring is my dart;
But ah! more deadly his, who pierced my heart.   700
Medicine is mine; what herbs and simples grow
In fields and forests, all their powers I know,
And am the great physician called below.
Alas, that fields and forests can afford
No remedies to heal their love-sick lord!
To cure the pains of love, no plant avails,
And his own physic the physician fails.'
     She heard not half, so furiously she flies,
And on her ear the imperfect accent dies.
Fear gave her wings; and as she fled, the wind   710
Increasing spread her flowing hair behind;
And left her legs and thighs exposed to view,
Which made the god more eager to pursue.
The god was young, and was too hotly bent
To lose his time in empty compliment;
But led by love, and fired with such a sight,
Impetuously pursued his near delight.
     As when the impatient greyhound, slipped from far,°
Bounds o'er the glebe, to course the fearful hare,
She in her speed does all her safety lay,   720
And he with double speed pursues the prey;

O'erruns her at the sitting turn, and licks°
His chaps in vain, and blows upon the flix;
She scapes, and for the neighbouring covert strives,
And gaining shelter doubts if yet she lives.
If little things with great we may compare,
Such was the god, and such the flying fair:
She, urged by fear, her feet did swiftly move,
But he more swiftly, who was urged by love.
He gathers ground upon her in the chase;   730
Now breathes upon her hair, with nearer pace,
And just is fastening on the wished embrace.
The nymph grew pale, and in a mortal fright,
Spent with the labour of so long a flight,
And now despairing, cast a mournful look
Upon the streams of her paternal brook:°
'Oh, help,' she cried, 'in this extremest need,
If water-gods are deities indeed!
Gape, earth, and this unhappy wretch entomb,
Or change my form, whence all my sorrows come.'   740
Scarce had she finished, when her feet she found
Benumbed with cold, and fastened to the ground;
A filmy rind about her body grows,
Her hair to leaves, her arms extend to boughs;
The nymph is all into a laurel gone,
The smoothness of her skin remains alone.

    Yet Phoebus loves her still, and, casting round
Her bole his arms, some little warmth he found.
The tree still panted in the unfinished part,
Not wholly vegetive, and heaved her heart.   750
He fixed his lips upon the trembling rind;
It swerved aside, and his embrace declined.
To whom the god: 'Because thou canst not be
My mistress, I espouse thee for my tree:
Be thou the prize of honour and renown;
The deathless poet, and the poem, crown.
Thou shalt the Roman festivals adorn,
And, after poets, be by victors worn;
Thou shalt returning Caesar's triumph grace,
When pomps shall in a long procession pass;   760
Wreathed on the posts before his palace wait,
And be the sacred guardian of the gate:

Secure from thunder, and unharmed by Jove,
Unfading as the immortal powers above;
And as the locks of Phoebus are unshorn,
So shall perpetual green thy boughs adorn.'
The grateful tree was pleased with what he said,
And shook the shady honours of her head.

THE TRANSFORMATION OF IO INTO A HEIFER

An ancient forest in Thessalia grows,
Which Tempe's pleasing valley does enclose;   770
Through this the rapid Peneus takes his course,
From Pindus rolling with impetuous force;
Mists from the river's mighty fall arise,
And deadly damps enclose the cloudy skies;
Perpetual fogs are hanging o'er the wood,
And sounds of waters deaf the neighbourhood.
Deep in a rocky cave he makes abode;
A mansion proper for a mourning god.
Here he gives audience; issuing out decrees
To rivers, his dependent deities.   780
On this occasion hither they resort,
To pay their homage, and to make their court;
All doubtful, whether to congratulate
His daughter's honour, or lament her fate.
Spercheus, crowned with poplar, first appears;
Then old Apidanus came, crowned with years;
Enipeus turbulent, Amphrysos tame,
And Aeas, last, with lagging waters came.
Then of his kindred brooks a numerous throng
Condole his loss, and bring their urns along: 790
Not one was wanting of the watery train,
That filled his flood, or mingled with the main,
But Inachus, who, in his cave alone,
Wept not another's losses, but his own.
For his dear Io, whether strayed, or dead,
To him uncertain, doubtful tears he shed.
He sought her through the world, but sought in vain;
And nowhere finding, rather feared her slain.
  Her, just returning from her father's brook,
Jove had beheld with a desiring look;   800

And, 'Oh, fair daughter of the flood,' he said,
'Worthy alone of Jove's imperial bed,
Happy whoever shall those charms possess!
The king of gods (nor is thy lover less,)
Invites thee to yon cooler shades, to shun
The scorching rays of the meridian sun.
Nor shalt thou tempt the dangers of the grove
Alone without a guide; thy guide is Jove.
No puny power, but he, whose high command
Is unconfined, who rules the seas and land,   810
And tempers thunder in his awful hand.
Oh, fly not!'—for she fled from his embrace.
O'er Lerna's pastures he pursued the chase,
Along the shades of the Lyrcaean plain.
At length the god, who never asks in vain,
Involved with vapours, imitating night,
Both air and earth; and then suppressed her flight,
And, mingling force with love, enjoyed the full delight.
    Meantime the jealous Juno, from on high,
Surveyed the fruitful fields of Arcady;   820
And wondered that the mist should overrun
The face of daylight and obscure the sun.
No natural cause she found, from brooks or bogs,
Or marshy lowlands, to produce the fogs:
Then round the skies she sought for Jupiter,
Her faithless husband; but no Jove was there.
Suspecting now the worst, 'Or I,' she said,
'Am much mistaken, or am much betrayed.'
With fury she precipitates her flight,
Dispels the shadows of dissembled night,   830
And to the day restores his native light.
The almighty lecher, careful to prevent
The consequence, foreseeing her descent,
Transforms his mistress in a trice; and now,
In Io's place, appears a lovely cow.
So sleek her skin, so faultless was her make,
E'en Juno did unwilling pleasure take
To see so fair a rival of her love;
And what she was, and whence, inquired of Jove,
Of what fair herd, and from what pedigree?   840
The god, half-caught, was forced upon a lie,

And said she sprang from earth. She took the word,
And begged the beauteous heifer of her lord.
What should he do? 'twas equal shame to Jove,
Or to relinquish, or betray his love;
Yet to refuse so slight a gift, would be
But more to increase his consort's jealousy.
Thus fear, and love, by turns his heart assailed;
And stronger love had sure at length prevailed,
But some faint hope remained, his jealous queen   850
Had not the mistress through the heifer seen.
The cautious goddess, of her gift possessed,
Yet harboured anxious thoughts within her breast;
As she who knew the falsehood of her Jove,
And justly feared some new relapse of love;
Which to prevent, and to secure her care,
To trusty Argus she commits the fair.

The head of Argus (as with stars the skies,)
Was compassed round, and wore a hundred eyes.
But two by turns their lids in slumber steep;   860
The rest on duty still their station keep;
Nor could the total constellation sleep.
Thus, ever present to his eyes and mind,
His charge was still before him, though behind.
In fields he suffered her to feed by day;
But when the setting sun to night gave way,
The captive cow he summoned with a call,
And drove her back, and tied her to the stall.
On leaves of trees and bitter herbs she fed,
Heaven was her canopy, bare earth her bed,   870
So hardly lodged; and, to digest her food,°
She drank from troubled streams, defiled with mud.
Her woeful story fain she would have told,
With hands upheld, but had no hands to hold.
Her head to her ungentle keeper bowed,
She strove to speak; she spoke not, but she lowed;
Affrighted with the noise, she looked around,
And seemed to inquire the author of the sound.

Once on the banks where often she had played,
(Her father's banks,) she came, and there surveyed   880
Her altered visage, and her branching head;
And starting, from herself she would have fled.

Her fellow-nymphs, familiar to her eyes,
Beheld, but knew her not in this disguise.
E'en Inachus himself was ignorant;
And in his daughter did his daughter want.
She followed where her fellows went, as she
Were still a partner of the company:
They stroke her neck; the gentle heifer stands,
And her neck offers to their stroking hands.   890
Her father gave her grass; the grass she took,
And licked his palms, and cast a piteous look,
And in the language of her eyes she spoke.
She would have told her name, and asked relief,
But wanting words, in tears she tells her grief;
Which with her foot she makes him understand,
And prints the name of Io in the sand.
    'Ah, wretched me!' her mournful father cried;
She, with a sigh, to 'wretched me!' replied.
About her milk-white neck his arms he threw,   900
And wept, and then these tender words ensue.
'And art thou she, whom I have sought around
The world, and have at length so sadly found?
So found, is worse than lost: with mutual words
Thou answerest not, no voice thy tongue affords;
But sighs are deeply drawn from out thy breast,
And speech denied, by lowing is expressed.
Unknowing, I prepared thy bridal bed;
With empty hopes of happy issue fed.
But now the husband of a herd must be   910
Thy mate, and bellowing sons thy progeny.
Oh, were I mortal, death might bring relief!
But now my godhead but extends my grief;
Prolongs my woes, of which no end I see,
And makes me curse my immortality.'
More had he said, but fearful of her stay,
The starry guardian drove his charge away
To some fresh pasture; on a hilly height
He sat himself, and kept her still in sight.

THE EYES OF ARGUS TRANSFORMED INTO A
PEACOCK'S TRAIN

Now Jove no longer could her sufferings bear;   920
But called in haste his airy messenger,°
The son of Maia, with severe decree
To kill the keeper, and to set her free.
With all his harness soon the god was sped;
His flying hat was fastened on his head;
Wings on his heels were hung, and in his hand
He holds the virtue of the snaky wand.
The liquid air his moving pinions wound,
And, in the moment, shoot him on the ground.
Before he came in sight, the crafty god   930
His wings dismissed, but still retained his rod:
That sleep-procuring wand wise Hermes took,
But made it seem to sight a shepherd's hook.
With this he did a herd of goats control;
Which by the way he met, and slyly stole.
Clad like a country swain, he piped and sung;
And, playing, drove his jolly troop along.

   With pleasure Argus the musician heeds;
But wonders much at those new vocal reeds.
And, 'Whosoe'er thou art, my friend,' said he,   940
'Up hither drive thy goats, and play by me;
This hill has browse for them, and shade for thee.'
The god, who was with ease induced to climb,
Began discourse to pass away the time;
And still, betwixt, his tuneful pipe he plies,
And watched his hour, to close the keeper's eyes.
With much ado, he partly kept awake;
Not suffering all his eyes repose to take;
And asked the stranger, who did reeds invent,
And whence began so rare an instrument.   950

THE TRANSFORMATION OF SYRINX INTO REEDS

Then Hermes thus; 'A nymph of late there was,
Whose heavenly form her fellows did surpass;
The pride and joy of fair Arcadia's plains,
Beloved by deities, adored by swains;
Syrinx her name, by sylvans oft pursued,
As oft she did the lustful gods delude:
The rural and the woodland powers disdained;
With Cynthia hunted, and her rites maintained;
Like Phoebe clad, e'en Phoebe's self she seems,
So tall, so straight, such well-proportioned limbs: 960
The nicest eye did no distinction know,
But that the goddess bore a golden bow;
Distinguished thus, the sight she cheated too.
Descending from Lycaeus, Pan admires
The matchless nymph, and burns with new desires.
A crown of pine upon his head he wore;
And thus began her pity to implore.
But ere he thus began, she took her flight
So swift, she was already out of sight;
Nor stayed to hear the courtship of the god,   970
But bent her course to Ladon's gentle flood;
There by the river stopped, and, tired before,
Relief from water-nymphs her prayers implore.
   'Now while the lustful god, with speedy pace,
Just thought to strain her in a strict embrace,
He filled his arms with reeds, new rising on the place.
And while he sighs his ill success to find,
The tender canes were shaken by the wind;
And breathed a mournful air, unheard before,
That, much surprising Pan, yet pleased him more.   980
Admiring this new music, "Thou," he said,
"Who canst not be the partner of my bed,
At least shalt be the consort of my mind,
And often, often, to my lips be joined."
He formed the reeds, proportioned as they are;
Unequal in their length, and waxed with care,
They still retain the name of his ungrateful fair.'
   While Hermes piped, and sung, and told his tale,
The keeper's winking eyes began to fail,

And drowsy slumber on the lids to creep,   990
Till all the watchman was at length asleep.
Then soon the god his voice and song suppressed,
And with his powerful rod confirmed his rest;
Without delay his crooked falchion drew,
And at one fatal stroke the keeper slew.
Down from the rock fell the dissevered head,
Opening its eyes in death, and falling bled;
And marked the passage with a crimson trail:
Thus Argus lies in pieces, cold and pale;
And all his hundred eyes, with all their light,   1000
Are closed at once, in one perpetual night.
These Juno takes, that they no more may fail,
And spreads them in her peacock's gaudy tail.
   Impatient to revenge her injured bed,
She wreaks her anger on her rival's head;
With furies frights her from her native home,
And drives her gadding round the world to roam:
Nor ceased her madness and her flight, before
She touched the limits of the Pharian shore.
At length, arriving on the banks of Nile,   1010
Wearied with length of ways, and worn with toil,
She laid her down; and leaning on her knees,
Invoked the cause of all her miseries;
And cast her languishing regards above,
For help from heaven, and her ungrateful Jove.
She sighed, she wept, she lowed; 'twas all she could;
And with unkindness seemed to tax the god.
Last, with a humble prayer, she begged repose,
Or death at least to finish all her woes.
Jove heard her vows, and with a flattering look,   1020
In her behalf to jealous Juno spoke.
He cast his arms about her neck, and said,
'Dame, rest secure; no more thy nuptial bed
This nymph shall violate; by Styx I swear,
And every oath that binds the thunderer.'
The goddess was appeased; and at the word
Was Io to her former shape restored.
The rugged hair began to fall away;
The sweetness of her eyes did only stay,

Though not so large; her crooked horns decrease;   1030
The wideness of her jaws and nostrils cease;
Her hoofs to hands return, in little space:
The five long taper fingers take their place;
And nothing of the heifer now is seen,
Beside the native whiteness of the skin.
Erected on her feet, she walks again,
And two the duty of the four sustain.
She tries her tongue, her silence softly breaks,
And fears her former lowings when she speaks:
A goddess now through all the Egyptian state,   1040
And served by priests, who in white linen wait.
    Her son was Epaphus, at length believed
The son of Jove, and as a god received.
With sacrifice adored, and public prayers,
He common temples with his mother shares.
Equal in years, and rival in renown
With Epaphus, the youthful Phaeton
Like honour claims, and boasts his sire the sun.
His haughty looks, and his assuming air,
The son of Isis could no longer bear;   1050
'Thou tak'st thy mother's word too far,' said he,
'And hast usurped thy boasted pedigree.
Go, base pretender to a borrowed name!'
Thus taxed, he blushed with anger, and with shame;
But shame repressed his rage: the daunted youth°
Soon seeks his mother, and inquires the truth.
'Mother,' said he, 'this infamy was thrown
By Epaphus on you, and me your son.
He spoke in public, told it to my face,
Nor durst I vindicate the dire disgrace: 1060
E'en I, the bold, the sensible of wrong,
Restrained by shame, was forced to hold my tongue.
To hear an open slander, is a curse;
But not to find an answer, is a worse.
If I am heaven-begot, assert your son
By some sure sign, and make my father known,
To right my honour, and redeem your own.'
He said, and, saying, cast his arms about
Her neck, and begged her to resolve the doubt.
    'Tis hard to judge if Clymene were moved   1070

More by his prayer, whom she so dearly loved,
Or more with fury fired, to find her name
Traduced, and made the sport of common fame.
She stretched her arms to heaven, and fixed her eyes
On that fair planet that adorns the skies;
'Now by those beams,' said she, 'whose holy fires
Consume my breast, and kindle my desires;
By him who sees us both, and cheers our sight,
By him, the public minister of light,
I swear that sun begot thee; if I lie   1080
Let him his cheerful influence deny;
Let him no more this perjured creature see,
And shine on all the world but only me.
If still you doubt your mother's innocence,
His eastern mansion is not far from hence;
With little pains you to his levée go,°
And from himself your parentage may know.'
With joy the ambitious youth his mother heard.
And, eager for the journey, soon prepared.
He longs the world beneath him to survey,   1090
To guide the chariot, and to give the day.
From Meroe's burning sands he bends his course,
Nor less in India feels his father's force;
His travel urging, till he came in sight,
And saw the palace by the purple light.

Ovid's Amours

BOOK I. ELEGY 1

For mighty wars I thought to tune my lute,
And make my measures to my subject suit.
Six feet for every verse the muse designed;°
But Cupid, laughing, when he saw my mind,
From every second verse a foot purloined.°
Who gave thee, boy, this arbitrary sway,
On subjects, not thy own, commands to lay,
Who Phoebus only and his laws obey?
'Tis more absurd than if the queen of love
Should in Minerva's arms to battle move;   10

Or manly Pallas from that queen should take
Her torch, and o'er the dying lover shake:
In fields as well may Cynthia sow the corn,
Or Ceres wind in woods the bugle-horn:
As well may Phoebus quit the trembling string
For sword and shield; and Mars may learn to sing.
Already thy dominions are too large;
Be not ambitious of a foreign charge.
If thou wilt reign o'er all, and everywhere,
The god of music for his harp may fear,   20
Thus, when with soaring wings I seek renown,
Thou pluck'st my pinions, and I flutter down.
Could I on such mean thoughts my muse employ,
I want a mistress, or a blooming boy.
  Thus I complained; his bow the stripling bent,
And chose an arrow fit for his intent.
The shaft his purpose fatally pursues:
'Now, poet, there's a subject for thy muse!'
He said. Too well, alas! he knows his trade;
For in my breast a mortal wound he made.   30
Far hence, ye proud hexameters, remove,
My verse is paced and trammelled into love.
With myrtle wreaths my thoughtful brows enclose,
While in unequal verse I sing my woes.°

Ovid's Amours

BOOK I. ELEGY 4

*To his Mistress, whose husband is invited to a feast with them. The Poet instructs
her how to behave herself in his company.*

Your husband will be with us at the treat;
May that be the last supper he shall eat!
And am poor I, a guest invited there,
Only to see, while he may touch the fair?
To see you kiss and hug your nauseous lord,
While his lewd hand descends below the board?
Now wonder not that Hippodamia's charms,°
At such a sight, the centaurs urged to arms;
That in a rage they threw their cups aside,
Assailed the bridegroom, and would force the bride.   10

I am not half a horse, (I would I were!)
Yet hardly can from you my hands forbear.
Take then my counsel; which, observed, may be
Of some importance both to you and me.
Be sure to come before your man be there;
There's nothing can be done; but come, howe'er.
Sit next him, (that belongs to decency,)
But tread upon my foot in passing by;
Read in my looks what silently they speak,
And slily, with your eyes, your answer make.   20
My lifted eyebrow shall declare my pain;
My right hand to his fellow shall complain,
And on the back a letter shall design,
Besides a note that shall be writ in wine.
Whene'er you think upon our last embrace,
With your forefinger gently touch your face;
If any word of mine offend my dear,
Pull with your hand the velvet of your ear;
If you are pleased with what I do or say,
Handle your rings, or with your fingers play.   30
As suppliants use at altars, hold the board,
Whene'er you wish the devil may take your lord.
When he fills for you, never touch the cup,
But bid the officious cuckold drink it up.
The waiter on those services employ;
Drink you, and I will snatch it from the boy,
Watching the part where your sweet mouth hath been,
And thence with eager lips will suck it in.
If he, with clownish manners, thinks it fit
To taste, and offer you the nasty bit,   40
Reject his greasy kindness, and restore
The unsavoury morsel he had chewed before.
Nor let his arms embrace your neck, nor rest
Your tender cheek upon his hairy breast;
Let not his hand within your bosom stray,
And rudely with your pretty bubbies play;
But above all let him no kiss receive!
That's an offence I never can forgive.
Do not, O do not that sweet mouth resign,
Lest I rise up in arms and cry, ''Tis mine.'   50
I shall thrust in betwixt, and, void of fear,

The manifest adulterer will appear.
These things are plain to sight; but more I doubt
What you conceal beneath your petticoat.
Take not his leg between your tender thighs,
Nor with your hand provoke my foe to rise.
How many love inventions I deplore,
Which I myself have practised all before!
How oft have I been forced the robe to lift
In company; to make a homely shift   60
For a bare bout, ill huddled o'er in haste,
While o'er my side the fair her mantle cast!
You to your husband shall not be so kind;
But, lest you should, your mantle leave behind.
Encourage him to tope; but kiss him not,
Nor mix one drop of water in his pot.
If he be fuddled well, and snores apace,
Then we may take advice from time and place.
When all depart, while compliments are loud,
Be sure to mix among the thickest crowd;   70
There I will be, and there we cannot miss,
Perhaps to grubble, or at least to kiss.°
Alas! what length of labour I employ,
Just to secure a short and transient joy!
For night must part us; and when night is come,
Tucked underneath his arms he leads you home.
He locks you in; I follow to the door,
His fortune envy, and my own deplore.
He kisses you, he more than kisses too;
The outrageous cuckold thinks it all his due.   80
But add not to his joy by your consent,
And let it not be given, but only lent.
Return no kiss, nor move in any sort;
Make it a dull and a malignant sport.
Had I my wish, he should no pleasure take,
But slubber o'er your business for my sake;
And whate'er fortune shall this night befall,
Coax me tomorrow, by forswearing all.

OVID'S *ART OF LOVE*
BOOK I

In Cupid's school whoe'er would take degree,
Must learn his rudiments, by reading me.
Seamen with sailing arts their vessels move,
Art guides the chariot, art instructs to love.
Of ships and chariots others know the rule,
But I am master in love's mighty school.
Cupid indeed is obstinate and wild,
A stubborn god, but yet the god's a child:
Easy to govern in his tender age,
Like fierce Achilles in his pupillage: 10
That hero, born for conquest, trembling stood
Before the centaur, and received the rod.
As Chiron mollified his cruel mind°
With art, and taught his warlike hands to wind
The silver strings of his melodious lyre,
So love's fair goddess does my soul inspire,
To teach her softer arts, to soothe the mind,
And smooth the rugged breasts of humankind.

    Yet Cupid and Achilles, each with scorn
And rage were filled, and both were goddess-born.   20
The bull, reclaimed and yoked, the burden draws,
The horse receives the bit within his jaws;
And stubborn love shall bend beneath my sway,
Though struggling oft he strives to disobey.
He shakes his torch, he wounds me with his darts,
But vain his force, and vainer are his arts.
The more he burns my soul, or wounds my sight,
The more he teaches to revenge the spite.

    I boast no aid the Delphian god affords,°
Nor auspice from the flight of chattering birds;   30
Nor Clio, nor her sisters, have I seen,
As Hesiod saw them on the shady green:°
Experience makes my work a truth so tried,
You may believe, and Venus be my guide.

    Far hence you vestals be, who bind your hair;
And wives, who gowns below your ankles wear.

I sing the brothels loose and unconfined,
The unpunishable pleasures of the kind;
Which all alike, for love, or money, find.

You who in Cupid's rolls inscribe your name,   40
First seek an object worthy of your flame;
Then strive with art your lady's mind to gain;
And last provide your love may long remain.
On these three precepts all my work shall move;
These are the rules and principles of love.

Before your youth with marriage is oppressed,
Make choice of one who suits your humour best;
And such a damsel drops not from the sky,
She must be sought for with a curious eye.

The wary angler, in the winding brook,   50
Knows what the fish, and where to bait his hook.
The fowler and the huntsman know by name
The certain haunts and harbour of their game.
So must the lover beat the likeliest grounds;
The assemblies where his quarry most abounds.
Nor shall my novice wander far astray;
These rules shall put him in the ready way.
Thou shalt not sail around the continent,
As far as Perseus, or as Paris went,
For Rome alone affords thee such a store,   60
As all the world can hardly show thee more:
The face of heaven with fewer stars is crowned,
Than beauties in the Roman sphere are found.

Whether thy love is bent on blooming youth,
On dawning sweetness in unartful truth,
Or courts the juicy joys of riper growth;
Here may'st thou find thy full desires in both.
Or if autumnal beauties please thy sight,
(An age that knows to give, and take delight,)
Millions of matrons of the graver sort,   70
In common prudence, will not balk the sport.

In summer heats thou need'st but only go
To Pompey's cool and shady portico;°
Or Concord's fane; or that proud edifice,
Whose turrets near the bawdy suburb rise;
Or to that other portico, where stands°
The cruel father urging his commands,

And fifty daughters wait the time of rest,
To plunge their poniards in the bridegroom's breast;
Or Venus' temple, where, on annual nights,   80
They mourn Adonis with Assyrian rites.
Nor shun the Jewish walk, where the foul drove,
On Sabbaths, rest from everything but love;
Nor Isis' temple; for that sacred whore
Makes others what to Jove she was before.
And if the hall itself be not belied,°
E'en there the cause of love is often tried;
Near it at least, or in the palace-yard,
From whence the noisy combatants are heard,
The crafty counsellors, in formal gown,   90
There gain another's cause, but lose their own.
There eloquence is nonplussed in the suit,
And lawyers, who had words at will, are mute.
Venus, from her adjoining temple, smiles
To see them caught in their litigious wiles.
Grave senators lead home the youthful dame,
Returning clients, when they patrons came.
But, above all, the playhouse is the place,
There's choice of quarry in that narrow chase.°
There take thy stand and sharply looking out,   100
Soon may'st thou find a mistress in the rout,
For length of time, or for a single bout.
The theatres are burrows for the fair,
Like ants on mole-hills thither they repair;
Like bees to hives, so numerously they throng,
It may be said, they to that place belong.
Thither they swarm, who have the public voice;
There choose, if plenty not distracts thy choice.
To see, and to be seen, in heaps they run;
Some to undo, and some to be undone.   110

   From Romulus the rise of plays began,
To his new subjects a commodious man;
Who his unmarried soldiers to supply
Took care the commonwealth should multiply.
Providing Sabine women for his braves
Like a true king to get a race of slaves.
His playhouse not of Parian marble made,°
Nor was it spread with purple sails for shade;

The stage with rushes, or with leaves they strewed,
No scenes in prospect, no machining god.   120
On rows of homely turf they sat to see,
Crowned with the wreaths of every common tree.
There, while they sat in rustic majesty,
Each lover had his mistress in his eye;
And whom he saw most suiting to his mind,
For joys of matrimonial rape designed.
Scarce could they wait the plaudit in their haste;
But ere the dances and the song were past,
The monarch gave the signal from his throne,
And rising, bade his merry men fall on.   130
The martial crew, like soldiers ready pressed,
Just at the word, (the word too was, 'The best,')°
With joyful cries each other animate;
Some choose, and some at hazard seize their mate.
As doves from eagles, or from wolves the lambs,
So from their lawless lovers fly the dames.
Their fear was one, but not one face of fear;
Some rend the lovely tresses of their hair,
Some shriek, and some are struck with dumb despair.
Her absent mother one invokes in vain,   140
One stands amazed not daring to complain,
The nimbler trust their feet, the slow remain.
But nought availing, all are captives led,
Trembling and blushing, to the genial bed.
She who too long resisted, or denied,
The lusty lover made by force a bride,
And with superior strength, compelled her to his side.
Then soothed her thus: 'My soul's far better part,
Cease weeping, nor afflict thy tender heart;
For what thy father to thy mother was,   150
That faith to thee, that solemn vow, I pass.'
   Thus Romulus became so popular;
This was the way to thrive in peace and war.
To pay his army, and fresh whores to bring:
Who would not fight for such a gracious king?
   Thus love in theatres did first improve,
And theatres are still the scene of love.
Nor shun the chariot's and the courser's race,
The circus is no inconvenient place.

No need is there of talking on the hand, 160
Nor nods, nor signs, which lovers understand:
But boldly next the fair your seat provide,
Close as you can to hers, and side by side.
Pleased or unpleased, no matter, crowding sit,
For so the laws of public shows permit.
Then find occasion to begin discourse,
Inquire, whose chariot this, and whose that horse?
To whatsoever side she is inclined
Suit all your inclinations to her mind;
Like what she likes; from thence your court begin, 170
And whom she favours, wish that he may win.
But when the statues of the deities,
In chariots rolled, appear before the prize,
When Venus comes, with deep devotion rise.
If dust be on her lap, or grains of sand,
Brush both away with your officious hand;
If none be there, yet brush that nothing thence,
And still to touch her lap make some pretence.
Touch anything of hers; and if her train
Sweep on the ground, let it not sweep in vain, 180
But gently take it up, and wipe it clean,
And while you wipe it, with observing eyes,
Who knows but you may see her naked thighs!
Observe who sits behind her; and beware
Lest his encroaching knee should press the fair.
Light service takes light minds; for some can tell°
Of favours won, by laying cushions well:
By fanning faces, some their fortune meet,
And some by laying footstools for their feet.
These overtures of love the circus gives, 190
Nor at the sword-play less the lover thrives;
For there the son of Venus fights his prize,
And deepest wounds are oft received from eyes.
One, while the crowd their acclamations make,
Or while he bets, and puts his ring to stake,
Is struck from far, and feels the flying dart,
And of the spectacle is made a part.
　　Caesar would represent a naval fight,
For his own honour, and for Rome's delight;
From either sea the youths and maidens come, 200

And all the world was then contained in Rome.
In this vast concourse, in this choice of game,
What Roman heart but felt a foreign flame?
Once more our prince prepares to make us glad,
And the remaining East to Rome will add.
Rejoice, you Roman soldiers, in your urn,
Your ensigns from the Parthians shall return,
And the slain Crassi shall no longer mourn.°
A youth is sent those trophies to demand,°
And bears his father's thunder in his hand;   210
Doubt not the imperial boy in wars unseen,
In childhood all of Caesar's race are men;
Celestial seeds shoot out before their day,
Prevent their years, and brook no dull delay:
Thus infant Hercules the snakes did press,
And in his cradle did his sire confess;°
Bacchus, a boy, yet like a hero fought,
And early spoils from conquered India brought.
Thus you your father's troops shall lead to fight,
And thus shall vanquish in your father's right.
These rudiments you to your lineage owe,
Born to increase your titles, as you grow.
Brethren you had, revenge your brethren slain;
You have a father, and his rights maintain,°
Armed by your country's parent, and your own,
Redeem your country, and restore his throne.
Your enemies assert an impious cause,
You fight both for divine and human laws.
Already in their cause they are o'ercome,
Subject them too, by force of arms, to Rome.   230
Great father Mars with greater Caesar join,
To give a prosperous omen to your line;
One of you is, and one shall be divine.
I prophesy you shall, you shall o'ercome,
My verse shall bring you back in triumph home.
Speak in my verse, exhort to loud alarms;
Oh were my numbers equal to your arms!
Then will I sing the Parthians' overthrow,
Their shot averse sent from a flying bow:°
The Parthians, who already flying fight,   240
Already give an omen of their flight.

Oh when will come the day, by heaven designed,
When thou, the best and fairest of mankind,°
Drawn by white horses shalt in triumph ride,
With conquered slaves attending on thy side;
Slaves, that no longer can be safe in flight;
O glorious object, O surprising sight,
O day of public joy, too good to end in night!
On such a day, if thou, and next to thee,°
Some beauty sits, the spectacle to see;   250
If she inquire the names of conquered kings,
Of mountains, rivers, and their hidden springs,
Answer to all thou knowest; and if need be,
Of things unknown seem to speak knowingly.
This is Euphrates crowned with reeds, and there
Flows the swift Tigris with his sea-green hair.
Invent new names of things unknown before,
Call this Armenia, that the Caspian shore;
Call this a Mede, and that a Parthian youth;
Talk probably, no matter for the truth.   260
   In feasts as at our shows new means abound,
More pleasure there than that of wine is found.
The Paphian goddess there her ambush lays;°
And love betwixt the horns of Bacchus plays;
Desires increase at every swilling draught;
Brisk vapours add new vigour to the thought.
There Cupid's purple wings no flight afford,
But wet with wine, he flutters on the board;
He shakes his pinions, but he cannot move,
Fixed he remains, and turns a maudlin love.   270
Wine warms the blood, and makes the spirits flow;
Care flies and wrinkles from the forehead go;
Exalts the poor, invigorates the weak,
Gives mirth and laughter, and a rosy cheek.
Bold truths it speaks, and, spoken, dares maintain,
And brings our old simplicity again.
Love sparkles in the cup and fills it higher,
Wine feeds the flames, and fuel adds to fire.
But choose no mistress in thy drunken fit,
Wine gilds too much their beauties and their wit.   280
Nor trust thy judgment when the tapers dance,
But sober, and by day, thy suit advance.

By daylight Paris judged the beauteous three,
And for the fairest did the prize decree.
Night is a cheat and all deformities
Are hid or lessened in her dark disguise.
The sun's fair light each error will confess,
In face, in shape, in jewels, and in dress.

Why name I every place where youths abound?
'Tis loss of time, and a too fruitful ground.   290
The Baian baths, where ships at anchor ride,°
And wholesome streams from sulphur fountains glide,
Where wounded youths are by experience taught,
The waters are less healthful than they thought;
Or Dian's fane, which near the suburb lies,
Where priests, for their promotion, fight a prize.°
That maiden goddess is love's mortal foe,
And much from her his subjects undergo.

Thus far the sportful muse with myrtle bound°
Has sung where lovely lasses may be found.   300
Now let me sing how she who wounds your mind
With art may be to cure your wounds inclined.
Young nobles, to my laws attention lend;
And all you vulgar of my school, attend.

First then believe, all women may be won,
Attempt with confidence, the work is done.
The grasshopper shall first forbear to sing
In summer season or the birds in spring,
Than women can resist your flattering skill,
E'en she will yield who swears she never will.   310
To secret pleasure both the sexes move,
But women most, who most dissemble love.
'Twere best for us, if they would first declare,
Avow their passion, and submit to prayer.
The cow by lowing tells the bull her flame,
The neighing mare invites her stallion to the game.
Man is more temperate in his lust than they,
And more than women can his passion sway.
Byblis we know did first her love declare,°
And had recourse to death in her despair.   320
Her brother she, her father Myrrha sought,°
And loved, but loved not as a daughter ought.
Now from a tree she stills her odorous tears

Which yet the name of her who shed them bears.
  In Ida's shady vale a bull appeared°
White as the snow, the fairest of the herd;
A beauty-spot of black there only rose,
Betwixt his equal horns and ample brows,
The love and wish of all the Cretan cows.
The queen beheld him as his head he reared,°   330
And envied every leap he gave the herd;
A secret fire she nourished in her breast,
And hated every heifer he caressed.
A story known, and known for true, I tell,
Nor Crete, though lying, can the truth conceal.
She cut him grass; (so much can love command,)
She stroked, she fed him with her royal hand,
Was pleased in pastures with the herd to roam;
And Minos by the bull was overcome.°
  Cease, queen, with gems to adorn thy beauteous brows;   340
The monarch of thy heart no jewel knows.
Nor in thy glass compose thy looks and eyes;
Secure from all thy charms thy lover lies;
Yet trust thy mirror, when it tells thee true,
Thou art no heifer to allure his view.
Soon wouldst thou quit thy royal diadem
To thy fair rivals, to be horned like them.
If Minos please, no lover seek to find;
If not, at least seek one of human kind.
  The wretched queen the Cretan court forsakes,   350
In woods and wilds her habitation makes:
She curses every beauteous cow she sees,
'Ah, why dost thou my lord and master please?
And think'st, ungrateful creature as thou art,
With frisking awkwardly, to gain his heart!'
She said, and straight commands, with frowning look
To put her, undeserving, to the yoke,
Or feigns some holy rites of sacrifice,
And sees her rival's death with joyful eyes:
Then, when the bloody priest has done his part,   360
Pleased, in her hand she holds the beating heart,
Nor from a scornful taunt can scarce refrain,
'Go, fool, and strive to please my love again.'
  Now she would be Europa, Io now;

(One bore a bull, and one was made a cow.)
Yet she at last her brutal bliss obtained,
And in a wooden cow the bull sustained;
Filled with his seed, accomplished her desire,
Till by his form the son betrayed the sire.°

    If Atreus' wife to incest had not run,°   370
(But, ah, how hard it is to love but one!)
His coursers Phoebus had not driven away,
To shun that sight, and interrupt the day.

    Thy daughter, Nisus, pulled thy purple hair,°
And barking sea-dogs yet her bowels tear.
At sea and land Atrides saved his life,
Yet fell a prey to his adulterous wife.
Who knows not what revenge Medea sought,
When the slain offspring bore the father's fault?
Thus Phoenix did a woman's love bewail,°   380
And thus Hippolytus by Phaedra fell.
These crimes revengeful matrons did commit,
Hotter their lust, and sharper is their wit.
Doubt not from them an easy victory,
Scarce of a thousand dames will one deny.
All women are content that men should woo,
She who complains, and she who will not do.
Rest then secure, whate'er thy luck may prove,
Not to be hated for declaring love.
And yet how canst thou miss, since womankind   390
Is frail and vain and still to change inclined?
Old husbands and stale gallants they despise;
And more another's, than their own, they prize.
A larger crop adorns our neighbour's field,
More milk his kine from swelling udders yield.

    First gain the maid; by her thou shalt be sure
A free access and easy to procure:
Who knows what to her office does belong,
Is in the secret, and can hold her tongue,
Bribe her with gifts, with promises, and prayers;   400
For her good word goes far in love-affairs.
The time and fit occasion leave to her,
When she most aptly can thy suit prefer.
The time for maids to fire their lady's blood
Is when they find her in a merry mood.

When all things at her wish and pleasure move,
Her heart is open then, and free to love;
Then mirth and wantonness to lust betray,
And smooth the passage to the lover's way.
Troy stood the siege, when filled with anxious care;   410
One merry fit concluded all the war.
    If some fair rival vex her jealous mind,
Offer thy service to revenge in kind.
Instruct the damsel, while she combs her hair,
To raise the choler of that injured fair;
And sighing, make her mistress understand,
She has the means of vengeance in her hand:
Then, naming thee, thy humble suit prefer,
And swear thou languishest and diest for her.
Then let her lose no time, but push at all;   420
For women soon are raised, and soon they fall.
Give their first fury leisure to relent,
They melt like ice, and suddenly repent.
    To enjoy the maid, will that thy suit advance?
'Tis a hard question, and a doubtful chance.
One maid, corrupted, bawds the better for 't;
Another for herself would keep the sport.
Thy business may be furthered or delayed;
But, by my counsel, let alone the maid;
E'en though she should consent to do the feat,   430
The profit's little, and the danger great.
I will not lead thee through a rugged road,
But, where the way lies open, safe, and broad.
Yet if thou find'st her very much thy friend,
And her good face her diligence commend,
Let the fair mistress have thy first embrace,
And let the maid come after in her place.
    But this I will advise, and mark my words,
For 'tis the best advice my skill affords:
If needs thou with the damsel wilt begin,   440
Before the attempt is made, make sure to win;
For then the secret better will be kept,
And she can tell no tales when once she's dipped.°
'Tis for the fowler's interest to beware,
The bird entangled should not scape the snare.
The fish, once pricked, avoids the bearded hook,

And spoils the sport of all the neighbouring brook.
But if the wench be thine, she makes thy way,
And for thy sake, her mistress will betray,
Tell all she knows, and all she hears her say.   450
Keep well the counsel of thy faithful spy,
So shalt thou learn whene'er she treads awry.

   All things the stations of their seasons keep,
And certain times there are to sow and reap.
Ploughmen and sailors for the season stay,
One to plough land, and one to plough the sea;
So should the lover wait the lucky day.
Then stop thy suit, it hurts not thy design;
But think, another hour she may be thine.
And when she celebrates her birth at home,   460
Or when she views the public shows of Rome,
Know, all thy visits then are troublesome.
Defer thy work, and put not then to sea,
For that's a boding and a stormy day.
Else take thy time, and when thou canst begin,
To break a Jewish Sabbath, think no sin:
Nor e'en on superstitious days abstain;
Not when the Romans were at Allia slain.°
Ill omens in her frowns are understood,
When she's in humour, every day is good.   470
But than her birthday seldom comes a worse,
When bribes and presents must be sent of course,
And that's a bloody day, that costs thy purse.
Be staunch, yet parsimony will be vain,
The craving sex will still the lover drain.
No skill can shift them off, nor art remove,
They will be begging, when they know we love.
The merchant comes upon the appointed day,
Who shall before thy face his wares display;
To choose for her she craves thy kind advice,   480
Then begs again, to bargain for the price:
But when she has her purchase in her eye,
She hugs thee close, and kisses thee to buy:
''Tis what I want, and 'tis a penn'orth too;
In many years I will not trouble you.'
If you complain you have no ready coin;
No matter, 'tis but writing of a line,

A little bill, not to be paid at sight;
Now curse the time when thou wert taught to write!
She keeps her birthday; you must send the cheer,   490
And she'll be born a hundred times a year.
With daily lies she dribs thee into cost;°
That earring dropped a stone, that ring is lost.
They often borrow what they never pay,
Whate'er you lend her, think it thrown away.
Had I ten mouths and tongues to tell each art,
All would be wearied ere I told a part.

   By letters, not by words, thy love begin,
And ford the dangerous passage with thy pen.
If to her heart thou aim'st to find the way,   500
Extremely flatter, and extremely pray.
Priam by prayers did Hector's body gain,
Nor is an angry god invoked in vain.
With promised gifts her easy mind bewitch,
For e'en the poor in promise may be rich.
Vain hopes awhile her appetite will stay,
'Tis a deceitful, but commodious way.
Who gives is mad; but make her still believe
'Twill come, and that's the cheapest way to give.
E'en barren lands fair promises afford,   510
But the lean harvest cheats the starving lord.
Buy not thy first enjoyment, lest it prove
Of bad example to thy future love:
But get it gratis, and she'll give thee more,
For fear of losing what she gave before.
The losing gamester shakes the box in vain,
And bleeds, and loses on, in hopes to gain.

   Write then, and in thy letter, as I said,
Let her with mighty promises be fed.
Cydippe by a letter was betrayed,°   520
Writ on an apple to the unwary maid.
She read herself into a marriage-vow,
(And every cheat in love the gods allow.)
Learn eloquence, ye noble youth of Rome;
It will not only at the bar o'ercome:
Sweet words the people and the senate move,
But the chief end of eloquence is love.
But in thy letter hide thy moving arts,

Affect not to be thought a man of parts.
None but vain fools to simple women preach,   530
A learned letter oft has made a breach.
In a familiar style your thoughts convey,
And write such things as present you would say;
Such words as from the heart may seem to move;
'Tis wit enough, to make her think you love.
If sealed she sends it back, and will not read,
Yet hope in time the business may succeed.
In time the steer will to the yoke submit,
In time the restive horse will bear the bit;
E'en the hard ploughshare use will wear away,   540
And stubborn steel in length of time decay.
Water is soft, and marble hard; and yet
We see soft water through hard marble eat.
Though late, yet Troy at length in flames expired,
And ten years more Penelope had tired.
Perhaps thy lines unanswered she retained;
No matter, there's a point already gained,
For she who reads in time will answer too:
Things must be left by just degrees to grow.
Perhaps she writes, but answers with disdain,   550
And sharply bids you not to write again:
What she requires, she fears you should accord;
The jilt would not be taken at her word.
   Meantime if she be carried in her chair,
Approach, but do not seem to know she's there.
Speak softly, to delude the standers-by;
Or if aloud, then speak ambiguously.
If sauntering in the portico she walk,
Move slowly too, for that's a time for talk;
And sometimes follow, sometimes be her guide,   560
But when the crowd permits, go side by side.
Nor in the playhouse let her sit alone;
For she's the playhouse, and the play, in one.
There thou may'st ogle, or by signs advance
Thy suit, and seem to touch her hand by chance.
Admire the dancer who her liking gains,
And pity in the play the lover's pains:
For her sweet sake the loss of time despise;
Sit while she sits, and when she rises, rise.

But dress not like a fop, nor curl your hair,   570
Nor with a pumice make your body bare.
Leave those effeminate and useless toys
To eunuchs, who can give no solid joys.
Neglect becomes a man; this Theseus found,
Uncurled, uncombed, the nymph his wishes crowned.
The rough Hippolytus was Phaedra's care,
And Venus thought the rude Adonis fair.
Be not too finical; but yet be clean,
And wear well-fashioned clothes, like other men.
Let not your teeth be yellow, or be foul,   580
Nor in wide shoes your feet too loosely roll;
Of a black muzzle and long beard beware,
And let a skilful barber cut your hair;
Your nails be picked from filth, and even pared,
Nor let your nasty nostrils bud with beard;
Cure your unsavoury breath, gargle your throat,
And free your armpits from the ram and goat:
Dress not, in short, too little or too much;
And be not wholly French, nor wholly Dutch.
Now Bacchus calls me to his jolly rites;   590
Who would not follow, when a god invites?
He helps the poet, and his pen inspires,
Kind and indulgent to his former fires.

   Fair Ariadne wandered on the shore,
Forsaken now, and Theseus loves no more:
Loose was her gown, dishevelled was her hair,
Her bosom naked, and her feet were bare;
Exclaiming, in the water's brink she stood;
Her briny tears augment the briny flood.
She shrieked, and wept, and both became her face;   600
No posture could that heavenly form disgrace.
She beat her breast: 'The traitor's gone,' said she;
'What shall become of poor forsaken me?
What shall become'—she had not time for more,
The sounding cymbals rattled on the shore.
She swoons for fear, she falls upon the ground;
No vital heat was in her body found.
The Mimallonian dames about her stood,°
And scudding satyrs ran before their god.
Silenus on his ass did next appear,   610

And held upon the mane (the god was clear:)°
The drunken sire pursues, the dames retire,
Sometimes the drunken dames pursue the drunken sire.
At last he topples over on the plain;
The satyrs laugh, and bid him rise again.
And now the god of wine came driving on,
High on his chariot by swift tigers drawn.
Her colour, voice, and sense forsook the fair,
Thrice did her trembling feet for flight prepare,
And thrice, affrighted, did her flight forbear,   620
She shook, like leaves of corn when tempests blow,
Or slender reeds that in the marshes grow.
To whom the god: 'Compose thy fearful mind;
In me a truer husband thou shalt find.
With heaven I will endow thee, and thy star
Shall with propitious light be seen afar,
And guide on seas the doubtful mariner.'
He said, and from his chariot leaping light,
Lest the grim tigers should the nymph affright,
His brawny arms around her waist he threw;   630
(For Gods, whate'er they will, with ease can do)
And swiftly bore her thence: the attending throng
Shout at the sight, and sing the nuptial song.
Now in full bowls her sorrow she may steep,
The bridegroom's liquor lays the bride asleep.

    But thou, when flowing cups in triumph ride,
And the loved nymph is seated by thy side,
Invoke the god, and all the mighty powers,
That wine may not defraud thy genial hours.
Then in ambiguous words thy suit prefer,   640
Which she may know were all addressed to her.
In liquid purple letters write her name,
Which she may read, and, reading, find the flame.
Then may your eyes confess your mutual fires
(For eyes have tongues, and glances tell desires.)
Whene'er she drinks, be first to take the cup,
And, where she laid her lips, the blessing sup.
When she to carving does her hand advance,
Put out thy own, and touch it as by chance.
Thy service e'en her husband must attend: 650
(A husband is a most convenient friend.)

Seat the fool cuckold in the highest place,
And with thy garland his dull temples grace.
Whether below or equal in degree,
Let him be lord of all the company,
And what he says, be seconded by thee.
'Tis common to deceive through friendship's name;
But common though it be 'tis still to blame.
Thus factors frequently their trust betray,
And to themselves their masters' gains convey.   660
Drink to a certain pitch, and then give o'er;
Thy tongue and feet may stumble, drinking more.
Of drunken quarrels in her sight beware;
Pot-valour only serves to fright the fair.
Eurytion justly fell, by wine oppressed,°
For his rude riot at a wedding-feast.
Sing, if you have a voice; and show your parts
In dancing, if endued with dancing arts.
Do anything within your power to please,
Nay, e'en affect a seeming drunkenness: 670
Clip every word, and if by chance you speak
Too home, or if too broad a jest you break,
In your excuse the company will join,
And lay the fault upon the force of wine.
True drunkenness is subject to offend,
But when 'tis feigned, 'tis oft a lover's friend.
Then safely you may praise her beauteous face,
And call him happy, who is in her grace.
Her husband thinks himself the man designed,
But curse the cuckold in your secret mind.   680
When all are risen, and prepare to go,
Mix with the crowd, and tread upon her toe.
This is the proper time to make thy court;
For now she's in the vein, and fit for sport.
Lay bashfulness, that rustic virtue, by,
To manly confidence thy thoughts apply.
On fortune's foretop timely fix thy hold,
Now speak and speed, for Venus loves the bold.
No rules of rhetoric here I need afford;
Only begin, and trust the following word—   690
It will be witty of its own accord.

    Act well the lover; let thy speech abound

In dying words, that represent thy wound;
Distrust not her belief, she will be moved,
All women think they merit to be loved.

   Sometimes a man begins to love in jest,
And after feels the torment he professed.
For your own sakes be pitiful, ye fair,
For a feigned passion may a true prepare.
By flatteries we prevail on womankind,   700
As hollow banks by streams are undermined.
Tell her her face is fair, her eyes are sweet,
Her taper fingers praise, and little feet.
Such praises e'en the chaste are pleased to hear,
Both maids and matrons hold their beauty dear.
   Once naked Pallas with Jove's queen appeared,°
And still they grieve that Venus was preferred.
Praise the proud peacock, and he spreads his train;
Be silent, and he pulls it in again.
Pleased is the courser in his rapid race;   710
Applaud his running, and he mends his pace.
But largely promise, and devoutly swear,
And if need be call every god to hear.
Jove sits above, forgiving with a smile
The perjuries that easy maids beguile.
He swore to Juno by the Stygian lake;
Forsworn, he dares not an example make,
Or punish falsehood, for his own dear sake.
'Tis for our interest that the gods should be;
Let us believe them; I believe, they see,   720
And both reward, and punish equally.
Not that they live above like lazy drones,
Or kings below, supine upon their thrones.
Lead then your lives as present in their sight;
Be just in dealings, and defend the right,
By fraud betray not, nor oppress by might.
But 'tis a venial sin to cheat the fair,
All men have liberty of conscience there.
On cheating nymphs a cheat is well designed,
'Tis a profane and a deceitful kind.   730
   'Tis said, that Egypt for nine years was dry,
Nor Nile did floods, nor heaven did rain supply.
A foreigner at length informed the king,

That slaughtered guests would kindly moisture bring.
The king replied: 'On thee the lot shall fall;
Be thou, my guest, the sacrifice for all.'
Thus Phalaris Perillus taught to low,°
And made him season first the brazen cow.°
A rightful doom, the laws of nature cry,
'Tis, the artificers of death should die: 740
Thus justly women suffer by deceit,
Their practice authorises us to cheat.
Beg her with tears thy warm desires to grant,
For tears will pierce a heart of adamant.
If tears will not be squeezed, then rub your eye,
Or 'noint the lids and seem at least to cry.
Kiss, if you can; resistance if she make,
And will not give you kisses, let her take.
'Fie, fie, you naughty man,' are words of course,
She struggles but to be subdued by force.   750
Kiss only soft, I charge you, and beware,
With your hard bristles not to brush the fair.
He who has gained a kiss, and gains no more,
Deserves to lose the bliss he got before.
If once she kiss, her meaning is expressed,
There wants but little pushing for the rest,
Which if thou dost not gain, by strength or art,
The name of clown then suits with thy desert;
'Tis downright dulness, and a shameful part.
Perhaps she calls it force; but if she scape,   760
She will not thank you for the omitted rape.
The sex is cunning to conceal their fires,
They would be forced e'en to their own desires.
They seem to accuse you, with a downcast sight,
But in their souls confess you did them right.
Who might be forced, and yet untouched depart,
Thank with their tongues, but curse you with their heart.
Fair Phoebe and her sister did prefer°
To their dull mates the noble ravisher.
   What Deidamia did, in days of yore,°   770
The tale is old, but worth the telling o'er.
When Venus had the golden apple gained,
And the just judge fair Helen had obtained;
When she with triumph was at Troy received,

The Trojans joyful, while the Grecians grieved,
They vowed revenge of violated laws,
And Greece was arming in the cuckold's cause:
Achilles, by his mother warned from war,
Disguised his sex, and lurked among the fair.
What means Aeacides to spin and sow?°   780
With spear and sword in field thy valour show,
And leaving this the nobler Pallas know.
Why dost thou in that hand the distaff wield,
Which is more worthy to sustain the shield?
Or with that other draw the woolly twine,
The same the fates for Hector's thread assign?
Brandish thy falchion in thy powerful hand,
Which can alone the ponderous lance command.
In the same room by chance the royal maid
Was lodged, and, by his seeming sex betrayed,   790
Close to her side the youthful hero laid.
I know not how his courtship he began;
But to her cost she found it was a man.
'Tis thought she struggled; but withal 'tis thought,
Her wish was to be conquered when she fought.
For when disclosed, and hastening to the field,
He laid his distaff down, and took the shield;
With tears her humble suit she did prefer,
And thought to stay the grateful ravisher.
She sighs, she sobs, she begs him not to part;   800
And now 'tis nature, what before was art.
She strives by force her lover to detain,
And wishes to be ravished once again.
This is the sex; they will not first begin,
But when compelled are pleased to suffer sin.
Is there, who thinks that women first should woo?
Lay by thy self-conceit, thou foolish beau!
Begin, and save their modesty the shame,
'Tis well for thee if they receive thy flame.
'Tis decent for a man to speak his mind;   810
They but expect the occasion to be kind.
Ask that thou may'st enjoy; she waits for this,
And on thy first advance depends thy bliss,
E'en Jove himself was forced to sue for love,
None of the nymphs did first solicit Jove.

But if you find your prayers increase her pride,
Strike sail awhile and wait another tide.
They fly when we pursue; but make delay,
And when they see you slacken they will stay.
Sometimes it profits to conceal your end;   820
Name not yourself her lover, but her friend.
How many skittish girls have thus been caught!
He proved a lover, who a friend was thought.
Sailors by sun and wind are swarthly made;
A tanned complexion best becomes their trade:
'Tis a disgrace to ploughmen to be fair;
Bluff cheeks they have, and weather-beaten hair.
The ambitious youth, who seeks an olive crown.
Is sunburnt with his daily toil, and brown,
But if the lover hopes to be in grace,   830
Wan be his looks, and meagre be his face.
That colour from the fair compassion draws;
She thinks you sick, and thinks herself the cause.
Orion wandered in the woods for love,°
His paleness did the nymphs to pity move,
His ghastly visage argued hidden love.
Nor fail a nightcap, in full health, to wear;
Neglect thy dress, and discompose thy hair.
All things are decent that in love avail.
Read long by night, and study to be pale,   840
Forsake your food, refuse your needful rest,
Be miserable, that you may be blest.

　　Shall I complain, or shall I warn you most?
Faith, truth, and friendship in the world are lost;
A little and an empty name they boast.
Trust not thy friend, much less thy mistress praise;
If he believe, thou may'st a rival raise.
'Tis true, Patroclus by no lust misled,
Sought not to stain his dear companion's bed,
Nor Pylades Hermione embraced,   850
E'en Phaedra to Pirithous still was chaste.
But hope not thou in this vile age to find
Those rare examples of a faithful mind;
The sea shall sooner with sweet honey flow,
Or from the furzes pears and apples grow.
We sin with gust, we love by fraud to gain,

And find a pleasure in our fellow's pain.
From rival foes you may the fair defend;
But would you ward the blow, beware your friend:
Beware your brother, and your next of kin,   860
But from your bosom friend your care begin.

   Here I had ended, but experience finds
That sundry women are of sundry minds,
With various crotchets filled, and hard to please;
They therefore must be caught by various ways.
All things are not produced in any soil,
This ground for wine is proper, that for oil.
So 'tis in men, but more in womankind,
Different in face, in manners, and in mind,
But wise men shift their sails with every wind.   870
As changeful Proteus varied oft his shape,
And did in sundry forms and figures scape,
A running stream, a standing tree became,
A roaring lion, or a bleating lamb.
Some fish with harpoons, some with darts are struck,
Some drawn with nets, some hang upon the hook;
So turn thyself; and imitating them,
Try several tricks, and change thy stratagem.
One rule will not for different ages hold,
The jades grow cunning, as they grow more old.   880
Then talk not bawdy to the bashful maid,°
Bug-words will make her innocence afraid:°
Nor to an ignorant girl of learning speak,
She thinks you conjure, when you talk in Greek.
And hence 'tis often seen, the simple shun
The learned, and into vile embraces run.

   Part of my task is done, and part to do,
But here 'tis time to rest myself and you.

The Fable of Iphis and Ianthe°

From the Ninth Book of
Ovid's *Metamorphoses*

The fame of this, perhaps, through Crete had flown;°
But Crete had newer wonders of her own,
In Iphis changed; for near the Gnossian bounds,°

As loud report the miracle resounds,
At Phaestus dwelt a man of honest blood,
But meanly born, and not so rich as good,
Esteemed and loved by all the neighbourhood;
Who to his wife, before the time assigned
For childbirth came, thus bluntly spoke his mind:
'If heaven', said Ligdus, 'will vouchsafe to hear,   10
I have but two petitions to prefer;
Short pains for thee, for me a son and heir.
Girls cost as many throes in bringing forth;
Beside, when born, the tits are little worth;°
Weak puling things, unable to sustain
Their share of labour, and their bread to gain.
If therefore thou a creature shalt produce,
Of so great charges, and so little use,
Bear witness, heaven, with what reluctancy,
Her hapless innocence I doom to die.'   20
He said, and tears the common grief display,
Of him who bade, and her who must obey.
    Yet Telethusa still persists to find
Fit arguments to move a father's mind;
To extend his wishes to a larger scope,
And in one vessel not confine his hope.
Ligdus continues hard; her time drew near,
And she her heavy load could scarcely bear;
When slumbering in the latter shades of night
Before the approaches of returning light,   30
She saw, or thought she saw, before her bed,
A glorious train, and Isis at their head;
Her moony horns were on her forehead placed,
And yellow sheaves her shining temples graced;
A mitre, for a crown, she wore on high;
The dog and dappled bull were waiting by;
Osiris, sought along the banks of Nile;°
The silent god; the sacred crocodile;°
And last a long procession moving on,
With timbrels, that assist the labouring moon.   40
Her slumbers seemed dispelled, and, broad awake,
She heard a voice, that thus distinctly spake:
'My votary, thy babe from death defend,
Nor fear to save whate'er the gods will send;

Delude with art thy husband's dire decree;
When danger calls, repose thy trust on me;
And know thou hast not served a thankless deity.'
This promise made, with night the goddess fled;
With joy the woman wakes, and leaves her bed;
Devoutly lifts her spotless hands on high,   50
And prays the powers their gift to ratify.

   Now grinding pains proceed to bearing throes,
Till its own weight the burden did disclose.
'Twas of the beauteous kind, and brought to light
With secrecy, to shun the father's sight.
The indulgent mother did her care employ,
And passed it on her husband for a boy.
The nurse was conscious of the fact alone;
The father paid his vows as for a son;
And called him Iphis, by a common name,   60
Which either sex with equal right may claim.
Iphis his grandsire was; the wife was pleased,
Of half the fraud by fortune's favour eased;
The doubtful name was used without deceit,
And truth was covered with a pious cheat.
The habit showed a boy, the beauteous face
With manly fierceness mingled female grace.

   Now thirteen years of age were swiftly run,
When the fond father thought the time drew on
Of settling in the world his only son.   70
Ianthe was his choice; so wondrous fair,
Her form alone with Iphis could compare;
A neighbour's daughter of his own degree,
And not more blessed with fortune's goods than he.

   They soon espoused; for they with ease were joined,
Who were before contracted in the mind.
Their age the same, their inclinations too,
And bred together in one school, they grew.
Thus, fatally disposed to mutual fires,
They felt, before they knew, the same desires.   80
Equal their flame, unequal was their care;
One loved with hope, one languished in despair.
The maid accused the lingering days alone;
For whom she thought a man, she thought her own.
But Iphis bends beneath a greater grief;

As fiercely burns, but hopes for no relief.
E'en her despair adds fuel to her fire;
A maid with madness does a maid desire.
  And, scarce refraining tears, 'Alas,' said she,
'What issue of my love remains for me!   90
How wild a passion works within my breast!
With what prodigious flames am I possessed!
Could I the care of providence deserve,
Heaven must destroy me, if it would preserve.
And that's my fate, or sure it would have sent
Some usual evil for my punishment;
Not this unkindly curse; to rage and burn,
Where nature shows no prospect of return.
Nor cows for cows consume with fruitless fire;
Nor mares, when hot, their fellow-mares desire;   100
The father of the fold supplies his ewes;
The stag through secret woods his hind pursues;
And birds for mates the males of their own species choose.
Her females nature guards from female flame,
And joins two sexes to preserve the game;
Would I were nothing, or not what I am!
Crete, famed for monsters, wanted of her store,
Till my new love produced one monster more.
The daughter of the sun a bull desired;°
And yet e'en then a male a female fired: 110
Her passion was extravagantly new;
But mine is much the madder of the two.
To things impossible she was not bent,
But found the means to compass her intent.
To cheat his eyes she took a different shape;
Yet still she gained a lover, and a leap.
Should all the wit of all the world conspire,
Should Daedalus assist my wild desire,
What art can make me able to enjoy,
Or what can change Ianthe to a boy?   120
Extinguish then thy passion, hopeless maid,
And recollect thy reason for thy aid.
Know what thou art, and love as maidens ought,
And drive these golden wishes from thy thought.
Thou canst not hope thy fond desires to gain;
Where hope is wanting, wishes are in vain.

And yet no guards against our joys conspire;
No jealous husband hinders our desire;
My parents are propitious to my wish,
And she herself consenting to the bliss.   130
All things concur to prosper our design;
All things to prosper any love but mine.
And yet I never can enjoy the fair;
'Tis past the power of heaven to grant my prayer.
Heaven has been kind, as far as heaven can be;
Our parents with our own desires agree;
But nature, stronger than the gods above,
Refuses her assistance to my love:
She sets the bar that causes all my pain;
One gift refused makes all their bounty vain.   140
And now the happy day is just at hand,
To bind our hearts in Hymen's holy band;
Our hearts, but not our bodies; thus accursed,
In midst of water I complain of thirst.
Why comest thou, Juno, to these barren rites,
To bless a bed defrauded of delights?
And why should Hymen lift his torch on high,
To see two brides in cold embraces lie?'

    Thus love-sick Iphis her vain passion mourns;
With equal ardour fair Ianthe burns;   150
Invoking Hymen's name, and Juno's power,
To speed the work, and haste the happy hour.

    She hopes, while Telethusa fears the day,
And strives to interpose some new delay;
Now feigns a sickness, now is in a fright
For this bad omen, or that boding sight.
But having done whate'er she could devise,
And emptied all her magazine of lies,
The time approached; the next ensuing day
The fatal secret must to light betray.   160
Then Telethusa had recourse to prayer,
She and her daughter with dishevelled hair;
Trembling with fear, great Isis they adored,
Embraced her altar, and her aid implored.

    'Fair queen, who dost on fruitful Egypt smile,
Who sway'st the sceptre of the Pharian isle,
And seven-fold falls of disemboguing Nile;

Relieve, in this our last distress,' she said,
'A suppliant mother, and a mournful maid.
Thou, goddess, thou wert present to my sight;   170
Revealed I saw thee by thy own fair light;
I saw thee in my dream, as now I see,
With all thy marks of awful majesty;
The glorious train that compassed thee around;
And heard the hollow timbrel's holy sound.
Thy words I noted, which I still retain;
Let not thy sacred oracles be vain.
That Iphis lives, that I myself am free
From shame and punishment, I owe to thee.
On thy protection all our hopes depend;   180
Thy counsel saved us, let thy power defend.'
   Her tears pursued her words, and while she spoke,
The goddess nodded, and her altar shook;
The temple doors, as with a blast of wind,
Were heard to clap; the lunar horns, that bind
The brows of Isis, cast a blaze around;
The trembling timbrel made a murmuring sound.
   Some hopes these happy omens did impart;
Forth went the mother with a beating heart,
Not much in fear, nor fully satisfied;   190
But Iphis followed with a larger stride:
The whiteness of her skin forsook her face;
Her looks emboldened with an awful grace;
Her features and her strength together grew,
And her long hair to curling locks withdrew.
Her sparkling eyes with manly vigour shone;
Big was her voice, audacious was her tone.
The latent parts, at length revealed, began
To shoot, and spread, and burnish into man.°
The maid becomes a youth; no more delay   200
Your vows, but look, and confidently pay.
Their gifts the parents to the temple bear;
The votive tables this inscription wear;—
'Iphis, the man, has to the goddess paid
The vows, that Iphis offered when a maid.'
   Now when the star of day had shown his face,
Venus and Juno with their presence grace
The nuptial rites, and Hymen from above

Descending to complete their happy love;
The gods of marriage lend their mutual aid,   210
And the warm youth enjoys the lovely maid.

The Fable of Acis, Polyphemus, and Galatea

From the Thirteenth Book of
Ovid's *Metamorphoses*

Galatea relates the story

Acis, the lovely youth, whose loss I mourn,
From Faunus and the nymph Symaethis born,
Was both his parents' pleasure; but to me
Was all that love could make a lover be.
The gods our minds in mutual bonds did join;
I was his only joy, and he was mine.
Now sixteen summers the sweet youth had seen,
And doubtful down began to shade his chin;
When Polyphemus first disturbed our joy,
And loved me fiercely, as I loved the boy.   10
Ask not which passion in my soul was higher,
My last aversion, or my first desire;
Nor this the greater was, nor that the less,
Both were alike, for both were in excess.
Thee, Venus, thee both heaven and earth obey;
Immense thy power, and boundless is thy sway.
The cyclops, who defied the ethereal throne,
And thought no thunder louder than his own,
The terror of the woods, and wilder far
Than wolves in plains, or bears in forests are;   20
The inhuman host, who made his bloody feasts
On mangled members of his butchered guests,
Yet felt the force of love, and fierce desire,
And burnt for me with unrelenting fire;
Forgot his caverns, and his woolly care,
Assumed the softness of a lover's air,
And combed with teeth of rakes his rugged hair.
Now with a crooked scythe his beard he sleeks,
And mows the stubborn stubble of his cheeks;
Now in the crystal stream he looks, to try   30
His simagres, and rolls his glaring eye.°

His cruelty and thirst of blood are lost;
And ships securely sail along the coast.
   The prophet Telemus (arrived by chance
Where Etna's summits to the seas advance,
Who marked the tracks of every bird that flew.
And sure presages from their flying drew,)
Foretold the cyclops, that Ulysses' hand
In his broad eye should thrust a flaming brand.
The giant, with a scornful grin, replied.   40
'Vain augur, thou hast falsely prophesied:
Already love his flaming brand has tossed;
Looking on two fair eyes, my sight I lost.'
Thus warned in vain, with stalking pace he strode,
And stamped the margin of the briny flood
With heavy steps, and weary sought again
The cool retirement of his gloomy den.
   A promontory sharpening by degrees
Ends in a wedge, and overlooks the seas;
On either side below the water flows: 50
This airy walk the giant-lover chose;
Here on the midst he sat; his flocks, unled,
Their shepherd followed, and securely fed.
A pine so burly and of length so vast
That sailing ships required it for a mast,
He wielded for a staff, his steps to guide;
But laid it by, his whistle while he tried.
A hundred reeds, of a prodigious growth,
Scarce made a pipe proportioned to his mouth;
Which when he gave it wind, the rocks around,   60
And watery plains, the dreadful hiss resound.
I heard the ruffian shepherd rudely blow,
Where, in a hollow cave, I sat below.
On Acis' bosom I my head reclined;
And still preserve the poem in my mind.
   'O lovely Galatea, whiter far
Than falling snows, and rising lilies are;
More flowery than the meads, as crystal bright,
Erect as alders, and of equal height;
More wanton than a kid; more sleek thy skin,   70
Than orient shells, that on the shores are seen;
Than apples fairer, when the boughs they lade;

Pleasing, as winter suns, or summer shade;
More grateful to the sight than goodly planes,
And softer to the touch than down of swans,
Or curds new turned; and sweeter to the taste.
Than swelling grapes, that to the vintage haste;
More clear than ice, or running streams, that stray
Through garden plots, but ah! more swift than they.

   'Yet, Galatea, harder to be broke   80
Than bullocks, unreclaimed to bear the yoke,
And far more stubborn than the knotted oak;
Like sliding streams, impossible to hold,
Like them fallacious, like their fountains cold;
More warping than the willow, to decline
My warm embrace; more brittle than the vine;
Immovable, and fixed in thy disdain;
Rough as these rocks, and of a harder grain:
More violent than is the rising flood;
And the praised peacock is not half so proud;   90
Fierce as the fire, and sharp as thistles are,
And more outrageous than a mother bear;
Deaf as the billows to the vows I make,
And more revengeful than a trodden snake;
In swiftness fleeter than the flying hind,
Or driven tempests, or the driving wind.
All other faults with patience I can bear;
But swiftness is the vice I only fear.

   'Yet, if you knew me well, you would not shun
My love, but to my wished embraces run,   100
Would languish in your turn and court my stay,
And much repent of your unwise delay.

   'My palace in the living rock is made
By nature's hand; a spacious pleasing shade,
Which neither heat can pierce, nor cold invade.
My garden filled with fruits you may behold,
And grapes in clusters, imitating gold;
Some blushing bunches of a purple hue;
And these, and those, are all reserved for you.
Red strawberries in shades expecting stand,   110
Proud to be gathered by so white a hand.
Autumnal cornels later fruit provide,
And plums, to tempt you, turn their glossy side;

Not those of common kinds, but such alone,
As in Phaeacian orchards might have grown.°
Nor chestnuts shall be wanting to your food,
Nor garden-fruits, nor wildings of the wood.
The laden boughs for you alone shall bear,
And yours shall be the product of the year.

   'The flocks you see are all my own, beside   120
The rest that woods and winding valleys hide,
And those that folded in the caves abide.
Ask not the numbers of my growing store;
Who knows how many, knows he has no more.
Nor will I praise my cattle; trust not me,
But judge yourself, and pass your own decree.
Behold their swelling dugs; the sweepy weight°
Of ewes that sink beneath the milky freight;
In the warm folds their tender lambkins lie;
Apart from kids that call with human cry.   130
New milk in nut-brown bowls is duly served
For daily drink, the rest for cheese reserved.
Nor are these household dainties all my store;
The fields and forests will afford us more;
The deer, the hare, the goat, the savage boar,
All sorts of venison, and of birds the best;
A pair of turtles taken from the nest.°
I walked the mountains, and two cubs I found,
Whose dam had left them on the naked ground;
So like, that no distinction could be seen;   140
So pretty, they were presents for a queen:
And so they shall; I took them both away,
And keep, to be companions of your play.

   'Oh raise, fair nymph, your beauteous face above
The waves; nor scorn my presents, and my love.
Come, Galatea, come, and view my face;
I late beheld it in the watery glass,
And found it lovelier than I feared it was.
Survey my towering stature, and my size:
Not Jove, the Jove you dream, that rules the skies,   150
Bears such a bulk, or is so largely spread.
My locks (the plenteous harvest of my head,)
Hang o'er my manly face, and dangling down,
As with a shady grove, my shoulders crown.

Nor think, because my limbs and body bear
A thick-set underwood of bristling hair,
My shape deformed; what fouler sight can be,
Than the bald branches of a leafless tree?
Foul is the steed without a flowing main;
And birds, without their feathers, and their train: 160
Wool decks the sheep; and man receives a grace
From bushy limbs, and from a bearded face.
My forehead with a single eye is filled,
Round as a ball, and ample as a shield.
The glorious lamp of heaven, the radiant sun,
Is nature's eye; and she's content with one.
Add, that my father sways your seas, and I,
Like you, am of the watery family.
I make you his, in making you my own;
You I adore, and kneel to you alone;   170
Jove, with his fabled thunder, I despise,
And only fear the lightning of your eyes.
Frown not, fair nymph! yet I could bear to be
Disdained, if others were disdained with me.
But to repulse the cyclops, and prefer
The love of Acis,—heavens! I cannot bear.
But let the stripling please himself; nay more,
Please you, though that's the thing I most abhor;
The boy shall find, if e'er we cope in fight,
These giant limbs endued with giant might.   180
His living bowels from his belly torn,
And scattered limbs, shall on the flood be borne,
Thy flood, ungrateful nymph; and fate shall find
That way for thee and Acis to be joined.
For oh! I burn with love, and thy disdain
Augments at once my passion, and my pain.
Translated Etna flames within my heart,
And thou, inhuman, wilt not ease my smart.'
    Lamenting thus in vain, he rose, and strode
With furious paces to the neighbouring wood;   190
Restless his feet, distracted was his walk,
Mad were his motions, and confused his talk;
Mad as the vanquished bull, when forced to yield
His lovely mistress, and forsake the field.
    Thus far unseen I saw; when, fatal chance

His looks directing, with a sudden glance,
Acis and I were to his sight betrayed;
Where, naught suspecting, we securely played.
From his wide mouth a bellowing cry he cast,
'I see, I see, but this shall be your last.'   200
A roar so loud made Etna to rebound,
And all the cyclops laboured in the sound.
Affrighted with his monstrous voice, I fled,
And in the neighbouring ocean plunged my head.
Poor Acis turned his back, and, 'Help,' he cried,
'Help, Galatea! help, my parent gods,
And take me, dying, to your deep abodes!'
The cyclops followed; but he sent before
A rib, which from the living rock he tore;
Though but an angle reached him of the stone,   210
The mighty fragment was enough alone,
To crush all Acis; 'twas too late to save,
But what the fates allowed to give, I gave;
That Acis to his lineage should return,
And roll among the river gods his urn.
Straight issued from the stone a stream of blood,
Which lost the purple mingling with the flood;
Then like a troubled torrent it appeared;
The torrent too in little space was cleared;
The stone was cleft, and through the yawning chink   220
New reeds arose, on the new river's brink.
The rock, from out its hollow womb, disclosed
A sound like water in its course opposed:
When (wondrous to behold!) full in the flood,
Up starts a youth, and navel-high he stood.
Horns from his temples rise; and either horn
Thick wreaths of reeds (his native growth) adorn.
Were not his stature taller than before,
His bulk augmented, and his beauty more,
His colour blue, for Acis he might pass;   230
And Acis, changed into a stream, he was.
But mine no more, he rolls along the plains
With rapid motion, and his name retains.

A Song to a Fair Young Lady Going Out of Town in the Spring

1

Ask not the cause why sullen spring
    So long delays her flowers to bear;
Why warbling birds forget to sing,
    And winter storms invert the year.
Chloris is gone, and fate provides
To make it spring where she resides.

2

Chloris is gone, the cruel fair;
    She cast not back a pitying eye;
But left her lover in despair,
    To sigh, to languish, and to die.
Ah, how can those fair eyes endure,
To give the wounds they will not cure!

3

Great god of love, why hast thou made
    A face that can all hearts command,
That all religions can invade,
    And change the laws of every land?
Where thou hadst placed such power before,
Thou shouldst have made her mercy more.

4

When Chloris to the temple comes,
    Adoring crowds before her fall;
She can restore the dead from tombs,
    And every life but mine recall.
I only am, by love, designed
To be the victim for mankind.

Prologue

Gallants, a bashful poet bids me say,
He's come to lose his maidenhead to-day.
Be not too fierce; for he's but green of age,
And ne'er, till now, debauched upon the stage.
He wants the suffering part of resolution,
And comes with blushes to his execution.
Ere you deflower his muse, he hopes the pit
Will make some settlement upon his wit.
Promise him well, before the play begin;
For he would fain be cozened into sin.   10
'Tis not but that he knows you mean to fail;
But if you leave him after being frail,
He'll have at least a fair pretence to rail;
To call you base, and swear you used him ill,
And put you in the new Deserters' Bill.°
Lord, what a troop of perjured men we see;
Enough to fill another Mercury!°
But this the ladies may with patience brook;
Theirs are not the first colours you forsook.
He would be loath the beauties to offend;   20
But if he should, he's not too old to mend.
He's a young plant, in his first year of bearing;
But his friend swears he will be worth the rearing.
His gloss is still upon him; though 'tis true
He's yet unripe, yet take him for the blue.°
You think an apricot half green is best;
There's sweet and sour, and one side good at least.
Mangos and limes, whose nourishment is little,
Though not for food, are yet preserved for pickle,
So this green writer may pretend, at least,   30
To whet your stomachs for a better feast.
He makes this difference in the sexes too;
He sells to men, he gives himself to you.
To both he would contribute some delight;
A mere poetical hermaphrodite.
Thus he's equipped, both to be wooed, and woo;
With arms offensive, and defensive too;
'Tis hard, he thinks, if neither part will do.

Veni, Creator Spiritus

Translated in Paraphrase

Creator Spirit, by whose aid
The world's foundations first were laid,
Come visit every pious mind;
Come pour thy joys on humankind;
From sin and sorrow set us free,
And make thy temples worthy thee.

O source of uncreated light,
The Father's promised Paraclete!°
Thrice holy fount, thrice holy fire,
Our hearts with heavenly love inspire;   10
Come, and thy sacred unction bring
To sanctify us, while we sing.

Plenteous of grace, descend from high,
Rich in thy sevenfold energy!°
Thou strength of his almighty hand,
Whose power does heaven and earth command.
Proceeding Spirit, our defence,
Who dost the gifts of tongues dispense,
And crown'st thy gift with eloquence.

Refine and purge our earthly parts;   20
But, oh, inflame and fire our hearts!
Our frailties help, our vice control,
Submit the senses to the soul;
And when rebellious they are grown,
Then lay thy hand, and hold them down.

Chase from our minds the infernal foe;
And peace, the fruit of love, bestow;
And, lest our feet should step astray,
Protect and guide us in the way.

Make us eternal truths receive,   30
And practise all that we believe;
Give us thyself, that we may see
The Father, and the Son, by thee.

Immortal honour, endless fame,
Attend the almighty Father's name;
The saviour Son be glorified,
Who for lost man's redemption died;
And equal adoration be,
Eternal Paraclete, to thee.

Rondelay

1

Chloe found Amyntas lying
    All in tears, upon the plain,
Sighing to himself, and crying,
    'Wretched I, to love in vain!
Kiss me, dear, before my dying;
    Kiss me once, and ease my pain.'

2

Sighing to himself, and crying,
    'Wretched I, to love in vain!
Ever scorning, and denying
    To reward your faithful swain.
Kiss me, dear, before my dying;
    Kiss me once, and ease my pain.'

3

'Ever scorning, and denying
    To reward your faithful swain,'
Chloe, laughing at his crying,
    Told him, that he loved in vain.
'Kiss me, dear, before my dying;
    Kiss me once, and ease my pain.'

4

Chloe, laughing at his crying,
    Told him, that he loved in vain;
But, repenting, and complying,
    When he kissed, she kissed again:
Kissed him up, before his dying;
    Kissed him up, and eased his pain.

The Last Parting of Hector
and Andromache

From the Sixth Book of Homer's *Iliad*

ARGUMENT

Hector returning from the field of battle to visit Helen, his sister-in-law, and his brother Paris, who had fought unsuccessfully, hand to hand with Menelaus, from thence goes to his own palace to see his wife Andromache, and his infant son Astyanax. The description of that interview is the subject of this translation.

Thus having said, brave Hector went to see
His virtuous wife, the fair Andromache.
He found her not at home; for she was gone,
Attended by her maid and infant son,
To climb the steepy tower of Ilion;
From whence, with heavy heart, she might survey
The bloody business of the dreadful day.
Her mournful eyes she cast around the plain,
And sought the lord of her desires in vain.

But he, who thought his peopled palace bare,   10
When she, his only comfort, was not there,
Stood in the gate, and asked of every one,
Which way she took, and whither she was gone;
If to the court, or with his mother's train,
In long procession to Minerva's fane?
The servants answered, neither to the court,
Where Priam's sons and daughters did resort;
Nor to the temple was she gone, to move
With prayers the blue-eyed progeny of Jove;°
But more solicitous for him alone,   20
Than all their safety, to the tower was gone,
There to survey the labours of the field,
Where the Greeks conquer, and the Trojans yield;
Swiftly she passed, with fear and fury wild;
The nurse went lagging after with the child.

This heard, the noble Hector made no stay,
The admiring throng divide to give him way;
He passed through every street, by which he came,
And at the gate he met the mournful dame.

His wife beheld him; and with eager pace,   30

Flew to his arms, to meet a dear embrace.
His wife, who brought in dower Cilicia's crown.
And in herself a greater dower alone;
Eetion's heir, who, on the woody plain°
Of Hippoplacus, did in Thebe reign.°
Breathless she flew, with joy and passion wild;
The nurse came lagging after with her child.

   The royal babe upon her breast was laid,
Who, like the morning star, his beams displayed.
Scamandrius was his name, which Hector gave,   40
From that fair flood which Ilion's wall did lave;
But him Astyanax the Trojans call,
From his great father who defends the wall.

   Hector beheld him with a silent smile,
His tender wife stood weeping by the while;
Pressed in her own, his warlike hand she took,
Then sighed, and thus prophetically spoke:
   'Thy dauntless heart, which I foresee too late,
Too daring man, will urge thee to thy fate.
Nor dost thou pity, with a parent's mind,   50
This helpless orphan, whom thou leav'st behind;
Nor me, the unhappy partner of thy bed,
Who must in triumph by the Greeks be led.
They seek thy life; and in unequal fight
With many, will oppress thy single might.
Better it were for miserable me
To die before the fate which I foresee;
For, ah! what comfort can the world bequeath
To Hector's widow, after Hector's death!

   'Eternal sorrow and perpetual tears   60
Began my youth, and will conclude my years;
I have no parents, friends, nor brothers left,
By stern Achilles all of life bereft.
Then, when the walls of Thebes he o'erthrew,
His fatal hand my royal father slew;
He slew Eetion, but despoiled him not,
Nor in his hate the funeral rites forgot;
Armed as he was he sent him whole below,
And reverenced thus the manes of his foe.
A tomb he raised; the mountain-nymphs around   70
Enclosed, with planted elms, the holy ground.

'My seven brave brothers in one fatal day
To death's dark mansions took the mournful way;
Slain by the same Achilles, while they keep
The bellowing oxen, and the bleating sheep.
My mother, who the royal sceptre swayed,
Was captive to the cruel victor made,
And hither led; but, hence redeemed with gold,
Her native country did again behold,
And but beheld; for soon Diana's dart   80
In an unhappy chase transfixed her heart.
   'But thou, my Hector, art thyself alone
My parents, brothers, and my lord, in one.
O kill not all my kindred o'er again,
Nor tempt the dangers of the dusty plain,
But in this tower, for our defence, remain.
Thy wife and son are in thy ruin lost;
This is a husband's and a father's post.
The Scaean gate commands the plains below;
Here marshal all thy soldiers as they go;   90
And hence, with other hands, repel the foe.
By yon wild fig-tree lies their chief ascent,
And thither all their powers are daily bent.
The two Ajaxes have I often seen,
And the wronged husband of the Spartan queen:
With him his greater brother; and, with these,
Fierce Diomede, and bold Meriones;
Uncertain if by augury, or chance,
But by this easy rise they all advance;
Guard well that pass, secure of all beside.'   100
To whom the noble Hector thus replied:
   'That and the rest are in my daily care;
But should I shun the dangers of the war,
With scorn the Trojans would reward my pains.
And their proud ladies, with their sweeping trains;
The Grecian swords and lances I can bear,
But loss of honour is my only fear.
Shall Hector, born to war, his birthright yield,
Belie his courage, and forsake the field?
Early in rugged arms I took delight,   110
And still have been the foremost in the fight;
With dangers dearly have I bought renown,

And am the champion of my father's crown.
    And yet my mind forebodes, with sure presage,
That Troy shall perish by the Grecian rage:
The fatal day draws on, when I must fall,
And universal ruin cover all.
Not Troy itself, though built by hands divine,
Nor Priam, nor his people, nor his line,
My mother, nor my brothers of renown,   120
Whose valour yet defends the unhappy town,
Not these, nor all their fates which I foresee,
Are half of that concern I have for thee.
I see, I see thee in that fatal hour,
Subjected to the victor's cruel power;
Led hence a slave to some insulting sword,
Forlorn and trembling at a foreign lord;
A spectacle in Argos, at the loom,
Gracing with Trojan fights, a Grecian room,
Or from deep wells the living stream to take,   130
And on thy weary shoulders bring it back.
While, groaning under this laborious life,
They insolently call thee Hector's wife;
Upbraid thy bondage with thy husband's name,
And from my glory propagate thy shame.
This when they say, thy sorrows will increase
With anxious thoughts of former happiness;
That he is dead who could thy wrongs redress.
But I, oppressed with iron sleep before,
Shall hear thy unavailing cries no more.'   140
    He said;
Then holding forth his arms, he took his boy,
The pledge of love and other hope of Troy.
The fearful infant turned his head away,
And on his nurse's neck reclining lay,
His unknown father shunning with affright,
And looking back on so uncouth a sight;
Daunted to see a face with steel o'erspread,
And his high plume that nodded o'er his head.
His sire and mother smiled with silent joy,   150
And Hector hastened to relieve his boy;
Dismissed his burnished helm, that shone afar,
The pride of warriors, and the pomp of war;

The illustrious babe, thus reconciled, he took,
Hugged in his arms, and kissed, and thus he spoke:
   'Parent of gods and men, propitious Jove!
And you, bright synod of the powers above!
On this my son your gracious gifts bestow;
Grant him to live, and great in arms to grow,
To reign in Troy, to govern with renown,   160
To shield the people, and assert the crown;
That when hereafter he from war shall come,
And bring his Trojans peace and triumph home,
Some aged man, who lives this act to see,
And who in former times remembered me,
May say the son in fortitude and fame
Outgoes the mark, and drowns his father's name:
That, at these words, his mother may rejoice,
And add her suffrage to the public voice.'
   Thus having said;   170
He first, with suppliant hands, the Gods adored;
Then to the mother's arms the child restored.
With tears and smiles she took her son, and pressed
The illustrious infant to her fragrant breast.
He, wiping her fair eyes, indulged her grief,
And eased her sorrows with this last relief:
   'My wife and mistress, drive thy fears away,
Nor give so bad an omen to the day;
Think not it lies in any Grecian's power
To take my life, before the fatal hour.   180
When that arrives, nor good nor bad can fly
The irrevocable doom of destiny.
Return; and to divert thy thoughts at home
There task thy maids, and exercise the loom,
Employed in works that womankind become.
The toils of war and feats of chivalry
Belong to men; and, most of all, to me.'
At this, for new replies he did not stay,
But laced his crested helm, and strode away.
   His lovely consort to her house returned,   190
And, looking often back, in silence mourned.
Home when she came, her secret woe she vents,
And fills the palace with her loud laments;

Those loud laments her echoing maids restore,°
And Hector, yet alive, as dead deplore.

To my Dear Friend Mr Congreve on his Comedy called The Double Dealer

Well then, the promised hour is come at last,
The present age of wit obscures the past.
Strong were our sires, and as they fought they writ,
Conquering with force of arms and dint of wit:
Theirs was the giant race before the flood;°
And thus, when Charles returned, our empire stood.
Like Janus he the stubborn soil manured,°
With rules of husbandry the rankness cured;
Tamed us to manners when the stage was rude,
And boisterous English wit with art endued.   10
Our age was cultivated thus at length;
But what we gained in skill we lost in strength.
Our builders were with want of genius cursed;
The second temple was not like the first;°
Till you, the best Vitruvius, come at length,°
Our beauties equal, but excel our strength.
Firm doric pillars found your solid base;
The fair corinthian crowns the higher space:
Thus all below is strength, and all above is grace.
In easy dialogue is Fletcher's praise;   20
He moved the mind, but had not power to raise.
Great Jonson did by strength of judgment please;
Yet doubling Fletcher's force, he wants his ease.
In differing talents both adorned their age;
One for the study, t' other for the stage.
But both to Congreve justly shall submit,
One matched in judgment, both o'ermatched in wit.
In him all beauties of this age we see,
Etherege's courtship, Southerne's purity,°
The satire, wit, and strength, of manly Wycherley.°   30
All this in blooming youth you have achieved,
Nor are your foiled contemporaries grieved.°
So much the sweetness of your manners move,
We cannot envy you, because we love.

Fabius might joy in Scipio, when he saw°
A beardless consul made against the law,
And join his suffrage to the votes of Rome,
Though he with Hannibal was overcome.
Thus old Romano bowed to Raphael's fame,°
And scholar to the youth he taught became.   40

   O that your brows my laurel had sustained,
Well had I been deposed, if you had reigned!
The father had descended for the son;
For only you are lineal to the throne.
Thus when the state one Edward did depose,°
A greater Edward in his room arose:
But now not I, but poetry, is cursed;
For Tom the second reigns like Tom the first.°
But let them not mistake my patron's part,°
Nor call his charity their own desert   50
Yet this I prophesy; thou shalt be seen,
(Though with some short parenthesis between,)
High on the throne of wit, and, seated there,°
Not mine (that's little) but thy laurel wear.
Thy first attempt an early promise made;°
That early promise this has more than paid.
So bold, yet so judiciously you dare
That your least praise is to be regular.
Time, place, and action may with pains be wrought,
But genius must be born and never can be taught.   60
This is your portion, this your native store.
Heaven, that but once was prodigal before,
To Shakespeare gave as much, he could not give him more.

   Maintain your post; that's all the fame you need;
For 'tis impossible you should proceed.
Already I am worn with cares and age,
And just abandoning the ungrateful stage;
Unprofitably kept at heaven's expense,
I live a rent-charge on his providence:
But you, whom every muse and grace adorn,   70
Whom I foresee to better fortune born,
Be kind to my remains; and oh, defend,°
Against your judgment, your departed friend!
Let not the insulting foe my fame pursue,
But shade those laurels which descend to you;

And take for tribute what these lines express:
You merit more, nor could my love do less.

To Sir Godfrey Kneller

Once I beheld the fairest of her kind,
And still the sweet idea charms my mind:
True, she was dumb; for nature gazed so long,
Pleased with her work, that she forgot her tongue;
But, smiling, said—She still shall gain the prize;
I only have transferred it to her eyes.
Such are thy pictures, Kneller, such thy skill,
That nature seems obedient to thy will;
Comes out, and meets thy pencil in the draft,°
Lives there, and wants but words to speak her thought. 10
At least thy pictures look a voice; and we
Imagine sounds, deceived to that degree,
We think 'tis somewhat more than just to see.

 Shadows are but privations of the light;
Yet when we walk, they shoot before the sight;
With us approach, retire, arise, and fall;
Nothing themselves, and yet expressing all.
Such are thy pieces, imitating life
So near, they almost conquered in the strife;
And from their animated canvas came,   20
Demanding souls, and loosened from the frame.

 Prometheus, were he here, would cast away°
His Adam, and refuse a soul to clay;
And either would thy noble work inspire,
Or think it warm enough, without his fire.

 But vulgar hands may vulgar likeness raise;
This is the least attendant on thy praise:
From hence the rudiments of art began;
A coal, or chalk, first imitated man:
Perhaps the shadow, taken on a wall,   30
Gave outlines to the rude original; —
Ere canvas yet was strained, before the grace
Of blended colours found their use and place,
Or cypress tablets first received a face.

 By slow degrees the godlike art advanced;
As man grew polished, picture was enhanced:°

Greece added posture, shade, and perspective,
And then the mimic piece began to live.
Yet perspective was lame, no distance true,
But all came forward in one common view: 40
No point of light was known, no bounds of art;
When light was there, it knew not to depart,
But glaring on remoter objects played;
Not languished and insensibly decayed.

  Rome raised not art, but barely kept alive,
And with old Greece unequally did strive;
Till Goths and Vandals, a rude northern race,
Did all the matchless monuments deface.
Then all the muses in one ruin lie,
And rhyme began to enervate poetry.   50
Thus in a stupid military state,
The pen and pencil find an equal fate.
Flat faces, such as would disgrace a screen,
Such as in Bantam's embassy were seen,°
Unraised, unrounded, were the rude delight
Of brutal nations, only born to fight.

  Long time the sister arts, in iron sleep,°
A heavy sabbath did supinely keep;
At length, in Raphael's age, at once they rise,
Stretch all their limbs, and open all their eyes.   60

  Thence rose the Roman, and the Lombard line;
One coloured best, and one did best design.
Raphael's, like Homer's, was the nobler part,
But Titian's painting looked like Virgil's art.

  Thy genius gives thee both; where true design,
Postures unforced, and lively colours join,
Likeness is ever there; but still the best,
(Like proper thoughts in lofty language dressed,)
Where light, to shades descending, plays, not strives,
Dies by degrees and by degrees revives.   70
Of various parts a perfect whole is wrought;
Thy pictures think, and we divine their thought.

  Shakespeare, thy gift, I place before my sight;°
With awe I ask his blessing ere I write;
With reverence look on his majestic face;
Proud to be less, but of his godlike race.

His soul inspires me, while thy praise I write,
And I, like Teucer, under Ajax fight;°
Bids thee, through me, be bold; with dauntless breast
Contemn the bad, and emulate the best.   80
Like his, thy critics in the attempt are lost;°
When most they rail, know then, they envy most.
In vain they snarl aloof; a noisy crowd,
Like women's anger, impotent and loud.
While they their barren industry deplore,
Pass on secure, and mind the goal before,
Old as she is, my muse shall march behind,
Bear off the blast, and intercept the wind.
Our arts are sisters, though not twins in birth,
For hymns were sung in Eden's happy earth: 90
By the first pair, while Eve was yet a saint,
Before she fell with pride and learned to paint.
Forgive the allusion: 'twas not meant to bite,
But satire will have room where'er I write.
For oh, the painter muse, though last in place,
Has seized the blessing first, like Jacob's race.°
Apelles' art an Alexander found,°
And Raphael did with Leo's gold abound;°
But Homer was with barren laurel crowned.
Thou hadst thy Charles a while, and so had I;   100
But pass we that unpleasing image by.
Rich in thyself, and of thyself divine,
All pilgrims come and offer at the shrine.
A graceful truth thy pencil can command;
The fair themselves go mended from thy hand.
Likeness appears in every lineament,
But likeness in thy work is eloquent.
Though nature there her true resemblance bears,
A nobler beauty in thy piece appears.
So warm thy work, so glows the generous frame,   110
Flesh looks less living in the lovely dame.
     Thou paint'st as we describe, improving still,
When on wild nature we engraft our skill,
But not creating beauties at our will.
     Some other hand perhaps may reach a face,
But none like thee a finished figure place:

None of this age, for that's enough for thee,
The first of these inferior times to be,
Not to contend with heroes' memory.°

Due honours to those mighty names we grant,   120
But shrubs may live beneath the lofty plant:
Sons may succeed their greater parents gone;
Such is thy lot, and such I wish my own.

But poets are confined in narrower space,
To speak the language of their native place;
The painter widely stretches his command,
Thy pencil speaks the tongue of every land.
From hence, my friend, all climates are your own,
Nor can you forfeit, for you hold of none.
All nations all immunities will give   130
To make you theirs, where'er you please to live;
And not seven cities, but the world, would strive.°

Sure some propitious planet then did smile,
When first you were conducted to this isle;
Our genius brought you here, to enlarge our fame,
For your good stars are everywhere the same.
Thy matchless hand, of every region free,
Adopts our climate, not our climate thee.

Great Rome and Venice early did impart°
To thee the examples of their wondrous art.   140
Those masters, then but seen, not understood,
With generous emulation fired thy blood;
For what in nature's dawn the child admired,
The youth endeavoured, and the man acquired.

That yet thou hast not reached their high degree,
Seems only wanting to this age, not thee.
Thy genius, bounded by the times, like mine,
Drudges on petty drafts, nor dare design
A more exalted work, and more divine.
For what a song, or senseless opera,   150
Is to the living labour of a play;
Or what a play to Virgil's work would be,
Such is a single piece to history.

But we, who life bestow, ourselves must live;
Kings cannot reign, unless their subjects give;
And they who pay the taxes bear the rule:
Thus thou sometimes art forced to draw a fool;

But so his follies in thy posture sink,°
The senseless idiot seems at last to think.
Good heaven! that sots and knaves should be so vain   160
To wish their vile resemblance may remain
And stand recorded, at their own request,
To future days, a libel or a jest!
     Meantime, whilst just encouragement you want,
You only paint to live, not live to paint.
     Else should we see your noble pencil trace
Our unities of action, time, and place;
A whole composed of parts, and those the best,
With every various character expressed;
Heroes at large, and at a nearer view;   170
Less, and at distance, an ignobler crew;
While all the figures in one action join,
As tending to complete the main design.
     More cannot be by mortal art expressed,
But venerable age shall add the rest:
For time shall with his ready pencil stand,
Retouch your figures with his ripening hand,
Mellow your colours, and embrown the taint,°
Add every grace, which time alone can grant;
To future ages shall your fame convey,   180
And give more beauties than he takes away.

An Ode on the Death of Mr Henry Purcell

Late Servant of his Majesty and Organist of the Chapel Royal and of
St Peter's Westminster

I

Mark how the lark and linnet sing;
          With rival notes
     They strain their warbling throats,
          To welcome in the spring.
          But in the close of night,°
     When Philomel begins her heavenly lay,°
          They cease their mutual spite,
     Drink in her music with delight,
And, listening and silent and silent and listening,
          and listening and silent obey.

II

So ceased the rival crew when Purcell came;   10
They sang no more, or only sang his fame.
Struck dumb, they all admired the godlike man:
   The godlike man,
    Alas! too soon retired,
    As he too late began.
We beg not hell our Orpheus to restore;
  Had he been there,
  Their sovereigns' fear°
  Had sent him back before.
The power of harmony too well they know: 20
He long ere this had tuned their jarring sphere,
   And left no hell below.

III

The heavenly choir, who heard his notes from high,
Let down the scale of music from the sky;
   They handed him along,°
And all the way he taught, and all the way they sung.
  Ye brethren of the lyre, and tuneful voice,
  Lament his lot, but at your own rejoice:
  Now live secure, and linger out your days;
  The gods are pleased alone with Purcell's lays,   30
   Nor know to mend their choice.

Virgil's *Georgics*

BOOK I

THE ARGUMENT

The poet, in the beginning of this book, propounds the general design of each Georgic: and, after a solemn invocation of all the gods who are any way related to his subject, he addresses himself, in particular, to Augustus, whom he compliments with divinity; and after strikes into his business. He shows the different kinds of tillage proper to different soils; traces out the original of agriculture; gives a catalogue of the husbandman's tools; specifies the employments peculiar to each season; describes the changes of the weather, with the signs in heaven and earth that forebode them; instances many of the prodigies that happened near the time of Julius Caesar's death; and shuts up all with a supplication to the gods for the safety of Augustus, and the preservation of Rome.

What makes a plenteous harvest, when to turn
The fruitful soil, and when to sow the corn;
The care of sheep, of oxen, and of kine,
And how to raise on elms the teeming vine;
The birth and genius of the frugal bee,
I sing, Maecenas, and I sing to thee.°
   Ye deities! who fields and plains protect,
Who rule the seasons, and the year direct,
Bacchus and fostering Ceres, powers divine,
Who gave us corn for mast, for water, wine,   10
Ye fauns, propitious to the rural swains,
Ye nymphs, that haunt the mountains and the plains,
Join in my work, and to my numbers bring
Your needful succour; for your gifts I sing.
And thou, whose trident struck the teeming earth,°
And made a passage for the courser's birth;
And thou, for whom the Caean shore sustains°
Thy milky herds, that graze thy flowery plains;
And thou, the shepherds' tutelary god,
Leave, for a while, O Pan! thy loved abode;   20
And if Arcadian fleeces be thy care,
From fields and mountains to my song repair.
Inventor, Pallas, of the fattening oil,°
Thou founder of the plough, and ploughman's toil;°
And thou, whose hands the shroud-like cypress rear,

Come, all ye gods and goddesses, that wear
The rural honours, and increase the year,
You, who supply the ground with seeds of grain;
And you, who swell those seeds with kindly rain;
And chiefly thou, whose undetermined state°　　　　　　30
Is yet the business of the gods' debate,
Whether in after times to be declared
The patron of the world, and Rome's peculiar guard,
Or o'er the fruits and seasons to preside,
And the round circuit of the year to guide,
Powerful of blessings, which thou strew'st around,
And with thy goddess-mother's myrtle crowned.
Or wilt thou, Caesar, choose the watery reign,°
To smooth the surges, and correct the main?
Then mariners in storms to thee shall pray;　　　　　　40
Even utmost Thule shall thy power obey;
And Neptune shall resign the fasces of the sea.
The watery virgins for thy bed shall strive,
And Tethys all her waves in dowry give.
Or wilt thou bless our summers with thy rays,
And, seated near the Balance, poise the days°
Where in the void of heaven, a space is free,
Betwixt the Scorpion and the Maid, for thee?
The Scorpion, ready to receive thy laws,
Yields half his region, and contracts his claws.　　　　　　50
Whatever part of heaven thou shalt obtain
(For let not hell presume of such a reign;
Nor let so dire a thirst of empire move
Thy mind, to leave thy kindred gods above;)
Though Greece admires Elysium's blest retreat,
Though Proserpine affects her silent seat,
And, importuned by Ceres to remove,
Prefers the fields below to those above:
But thou propitious Caesar! guide my course,
And to my bold endeavours add thy force:　　　　　　60
Pity the poet's and the ploughman's cares;
Interest thy greatness in our mean affairs,
And use thyself betimes to hear and grant our prayers.

　　While yet the spring is young, while earth unbinds
Her frozen bosom to the western winds;
While mountain snows dissolve against the sun,

And streams, yet new, from precipices run;
E'en in this early dawning of the year,
Produce the plough, and yoke the sturdy steer,
And goad him till he groans beneath his toil,   70
Till the bright share is buried in the soil.
That crop rewards the greedy peasant's pains,
Which twice the sun, and twice the cold sustains,
And bursts the crowded barns with more than promised gains.
But ere we stir the yet unbroken ground,
The various course of seasons must be found;
The weather, and the setting of the winds,
The culture suiting to the several kinds
Of seeds and plants, and what will thrive and rise,
And what the genius of the soil denies.   80
This ground with Bacchus, that with Ceres, suits:
That other loads the trees with happy fruits:
A fourth, with grass unbidden, decks the ground.
Thus Tmolus is with yellow saffron crowned:
India black ebon and white ivory bears;
And soft Idume weeps her odorous tears.°
Thus Pontus sends her beaver-stones from far;°
And naked Spaniards temper steel for war:
Epirus, for the Elean chariot, breeds°
(In hopes of palms) a race of running steeds.   90
This is the original contract; these the laws
Imposed by nature, and by nature's cause,
On sundry places, when Deucalion hurled°
His mother's entrails on the desert world;
Whence men, a hard laborious kind, were born.
Then borrow part of winter for thy corn;
And early with thy team the glebe in furrows turn;
That while the turf lies open and unbound,
Succeeding suns may bake the mellow ground.
But if the soil be barren, only scar   100
The surface, and but lightly print the share,
When cold Arcturus rises with the sun;°
Lest wicked weeds the corn should overrun
In watery soils; or lest the barren sand
Should suck the moisture from the thirsty land.
Both these unhappy soils the swain forbears,
And keeps a sabbath of alternate years,

That the spent earth may gather heart again,
And, bettered by cessation, bear the grain.
At least where vetches, pulse and tares, have stood,   110
And stalks of lupins grew (a stubborn wood),
The ensuing season in return may bear
The bearded product of the golden year.
For flax and oats will burn the tender field,
And sleepy poppies harmful harvests yield.
But sweet vicissitudes of rest and toil
Make easy labour, and renew the soil.
Yet sprinkle sordid ashes all around,
And load with fattening dung thy fallow ground.
Thus change of seeds for meagre soils is best;   120
And earth manured, not idle, though at rest.

    Long practice has a sure improvement found,
With kindled fires to burn the barren ground,
When the light stubble, to the flames resigned,
Is driven along, and crackles in the wind.
Whether from hence the hollow womb of earth
Is warmed with secret strength for better birth;
Or when the latent vice is cured by fire,
Redundant humours through the pores expire;
Or that the warmth distends the chinks, and makes   130
New breathings, whence new nourishment she takes;
Or that the heat the gaping ground constrains,
New knits the surface, and new strings the veins;
Lest soaking showers should pierce her secret seat,
Or freezing Boreas chill her genial heat,
Or scorching suns too violently beat.

    Nor is the profit small the peasant makes,
Who smooths with harrows, or who pounds with rakes,
The crumbling clods: nor Ceres from on high
Regards his labours with a grudging eye: 140
Nor his, who ploughs across the furrowed grounds,
And on the back of earth inflicts new wounds;
For he, with frequent exercise, commands
The unwilling soil, and tames the stubborn lands.

    Ye swains, invoke the powers who rule the sky,
For a moist summer, and a winter dry;
For winter drought rewards the peasant's pain,
And broods indulgent on the buried grain.

Hence Mysia boasts her harvests, and the tops°
Of Gargarus admire their happy crops.   150
When first the soil receives the fruitful seed,
Make no delay, but cover it with speed:
So fenced from cold, the pliant furrows break,
Before the surly clod resists the rake;
And call the floods from high, to rush amain
With pregnant streams, to swell the teeming grain.
Then when the fiery suns too fiercely play
And shrivelled herbs on withering stems decay,
The wary ploughman on the mountain's brow,
Undams his watery stores, huge torrents flow,   160
And rattling down the rocks, large moisture yield,
Tempering the thirsty fever of the field;
And, lest the stem, too feeble for the freight,
Should scarce sustain the head's unwieldy weight,
Sends in his feeding flocks betimes, to invade
The rising bulk of the luxuriant blade,
Ere yet the aspiring offspring of the grain
O'ertops the ridges of the furrowed plain;
And drains the standing waters, when they yield
Too large a beverage to the drunken field.   170
But most in autumn, and the showery spring,
When dubious months uncertain weather bring;
When fountains open, when impetuous rain
Swells hasty brooks, and pours upon the plain;
When earth with slime and mud is covered o'er,
Or hollow places spew their watery store.
Nor yet the ploughman, nor the labouring steer,
Sustain alone the hazards of the year:
But glutton geese, and the Strymonian crane,°
With foreign troops invade the tender grain;   180
And towering weeds malignant shadows yield;
And spreading chicory chokes the rising field.
The sire of gods and men, with hard decrees,°
Forbids our plenty to be bought with ease,
And wills that mortal men, inured to toil,
Should exercise, with pains, the grudging soil;
Himself invented first the shining share,
And whetted human industry by care;
Himself did handicrafts and arts ordain,

Nor suffered sloth to rust his active reign.   190
Ere this, no peasant vexed the peaceful ground,
Which only turfs and greens for altars found:
No fences parted fields, nor marks nor bounds
Distinguished acres of litigious grounds;
But all was common, and the fruitful earth
Was free to give her unexacted birth.
Jove added venom to the viper's brood,
And swelled with raging storms the peaceful flood;
Commissioned hungry wolves to infest the fold,
And shook from oaken leaves the liquid gold;°   200
Removed from human reach the cheerful fire,
And from the rivers bade the wine retire;
That studious need might useful arts explore;
From furrowed fields to reap the foodful store,
And force the veins of clashing flints to expire
The lurking seeds of their celestial fire.
Then first on seas the hollowed alder swam;
Then sailors quartered heaven, and found a name
For every fixed and every wandering star:
The Pleiads, Hyads, and the Northern Car.°   210
Then toils for beasts, and lime for birds, were found.
And deep-mouthed dogs did forest-walks surround;
And casting-nets were spread in shallow brooks,
Drags in the deep, and baits were hung on hooks.°
Then saws were toothed, and sounding axes made
(For wedges first did yielding wood invade);
And various arts in order did succeed,
(What cannot endless labour, urged by need?)
    First Ceres taught, the ground with grain to sow,
And armed with iron shares the crooked plough;   220
When now Dodonian oaks no more supplied°
Their mast, and trees their forest-fruit denied.
Soon was his labour doubled to the swain,
And blasting mildews blackened all his grain:
Tough thistles choked the fields, and killed the corn,
And an unthrifty crop of weeds was born:
Then burs and brambles, an unbidden crew
Of graceless guests, the unhappy field subdue;
And oats unblest, and darnel domineers,
And shoots its head above the shining ears;   230

So that unless the land with daily care
Is exercised, and, with an iron war
Of rakes and harrows, the proud foes expelled,
And birds with clamours frighted from the field,
Unless the boughs are lopped that shade the plain,
And heaven invoked with vows for fruitful rain,
On other crops you may with envy look,
And shake for food the long-abandoned oak.
Nor must we pass untold what arms they wield,
Who labour tillage and the furrowed field;   240
Without whose aid the ground her corn denies,
And nothing can be sown, and nothing rise.
The crooked plough, the share, the towering height
Of waggons, and the cart's unwieldy weight,
The sled, the tumbril, hurdles, and the flail,
The fan of Bacchus, with the flying sail:
These all must be prepared, if ploughmen hope
The promised blessing of a bounteous crop.
Young elms, with early force, in copses bow,
Fit for the figure of the crooked plough.   250
Of eight foot long a fastened beam prepare:
On either side the head, produce an ear;
And sink a socket for the shining share.
Of beech the plough-tail, and the bending yoke,
Or softer linden hardened in the smoke.
I could be long in precepts; but I fear
So mean a subject might offend your ear.
Delve of convenient depth your threshing-floor:
With tempered clay then fill and face it o'er;
And let the weighty roller run the round,   260
To smooth the surface of the unequal ground;
Lest cracked with summer heats the flooring flies,
Or sinks, and through the crannies weeds arise.
For sundry foes the rural realm surround:
The field-mouse builds her garner under ground;
For gathered grain the blind laborious mole
In winding mazes works her hidden hole.
In hollow caverns vermin make abode,
The hissing serpent, and the swelling toad:
The corn-devouring weevil here abides,   270
And the wise ant her wintry store provides.

Mark well the flowering almonds in the wood:
If odorous blooms the bearing branches load,
The glebe will answer to the sylvan reign;
Great heats will follow, and large crops of grain.
But if a wood of leaves o'ershade the tree,
Such and so barren will thy harvest be:
In vain the hind shall vex the threshing-floor;
For empty chaff and straw will be thy store.
Some steep their seed, and some in cauldrons boil,   280
With vigorous nitre and with lees of oil,
O'er gentle fires, the exuberant juice to drain,
And swell the flattering husks with fruitful grain.
Yet is not the success for years assured,
Though chosen is the seed, and fully cured,
Unless the peasant, with his annual pain,
Renews his choice, and culls the largest grain.
Thus all below, whether by nature's curse,
Or fate's decree, degenerate still to worse.
So the boat's brawny crew the current stem,   290
And slow advancing struggle with the stream:
But, if they slack their hands, or cease to strive,
Then down the flood with headlong haste they drive.
    Nor must the ploughman less observe the skies,
When the Kids, Dragon, and Arcturus rise,°
Than sailors homeward bent, who cut their way
Through Helle's stormy straits, and oyster-breeding sea.°
But, when Astraea's balance, hung on high,°
Betwixt the nights and days divides the sky,
Then yoke your oxen, sow your winter grain,   300
Till cold December comes with driving rain.
Linseed and fruitful poppy bury warm,
In a dry season, and prevent the storm.
Sow beans and clover in a rotten soil,
And millet rising from your annual toil,
When with his golden horns, in full career,
The Bull beats down the barriers of the year,°
And Argo and the Dog forsake the northern sphere.°
    But, if your care to wheat alone extend,
Let Maia with her sisters first descend,°   310
And the bright Gnossian diadem downward bend,°
Before you trust in earth your future hope:

Or else expect a listless lazy crop.
Some swains have sown before; but most have found
A husky harvest from the grudging ground.
Vile vetches would you sow, or lentils lean,
The growth of Egypt, or the kidney-bean?
Begin when the slow waggoner descends;°
Nor cease your sowing till mid-winter ends.
For this, through twelve bright signs Apollo guides°   320
The year, and earth in several climes divides.
Five girdles bind the skies: the torrid zone
Glows with the passing and repassing sun:
Far on the right and left, the extremes of heaven
To frosts and snows and bitter blasts are given:
Betwixt the midst and these, the gods assigned
Two habitable seats for human kind,
And 'cross their limits, cut a sloping way,
Which the twelve signs in beauteous order sway.
Two poles turn round the globe; one seen to rise   330
O'er Scythian hills, and one in Libyan skies;
The first sublime in heaven, the last is whirled
Below the regions of the nether world.
Around our pole the spiry Dragon glides,
And, like a winding stream, the Bears divides—
The less and greater, who, by fate's decree,
Abhor to dive beneath the southern sea.
There, as they say, perpetual night is found
In silence brooding on the unhappy ground:
Or, when Aurora leaves our northern sphere,   340
She lights the downward heaven, and rises there;
And, when on us she breathes the living light,
Red Vesper kindles there the tapers of the night.
From hence uncertain seasons we may know,
And when to reap the grain, and when to sow;
Or when to fell the furzes; when 'tis meet°
To spread the flying canvas for the fleet.
Observe what stars arise, or disappear;
And the four quarters of the rolling year.
But, when cold weather and continued rain   350
The labouring husband in his house restrain,
Let him forecast his work with timely care,
Which else is huddled, when the skies are fair:

Then let him mark the sheep, or whet the shining share.
Or hollow trees for boats, or number o'er
His sacks, or measure his increasing store,
Or sharpen stakes, or head the forks, or twine
The sallow twigs to tie the straggling vine;
Or wicker baskets weave, or air the corn,
Or grinded grain betwixt two marbles turn.   360
No laws, divine or human, can restrain
From necessary works the labouring swain.
E'en holidays and feasts permission yield
To float the meadows, or to fence the field,
To fire the brambles, snare the birds, and steep
In wholesome waterfalls the woolly sheep.
And oft the drudging ass is driven, with toil,
To neighbouring towns with apples and with oil;
Returning, late and loaden, home with gain
Of bartered pitch, and hand mills for the grain.   370
    The lucky days, in each revolving moon,
For labour choose: the fifth be sure to shun;
That gave the furies and pale Pluto birth,
And armed against the skies the sons of earth.°
With mountains piled on mountains, thrice they strove
To scale the steepy battlements of Jove;
And thrice his lightening and red thunder played,
And their demolished works in ruin laid.
The seventh is, next the tenth, the best to join
Young oxen to the yoke, and plant the vine.   380
Then weavers stretch your stays upon the weft:
The ninth is good for travel, bad for theft.
Some works in dead of night are better done,
Or when the morning dew prevents the sun.
Parched meads and stubble mow by Phoebe's light,°
Which both require the coolness of the night;
For moisture then abounds, and pearly rains
Descend in silence to refresh the plains.
The wife and husband equally conspire
To work by night, and rake the winter fire: 390
He sharpens torches in the glimmering room;
She shoots the flying shuttle through the loom,
Or boils in kettles must of wine, and skims
With leaves the dregs that overflow the brims.

And till the watchful cock awakes the day,
She sings to drive the tedious hours away.
But in warm weather, when the skies are clear,
By daylight reap the product of the year;
And in the sun your golden grain display,
And thrash it out, and winnow it by day.   400
Plough naked, swain, and naked sow the land;
For lazy winter numbs the labouring hand.
In genial winter, swains enjoy their store,
Forget their hardships, and recruit for more.
The farmer to full bowls invites his friends,
And what he got with pains with pleasure spends.
So sailors, when escaped from stormy seas,
First crown their vessels, then indulge their ease.
Yet that's the proper time to thresh the wood
For mast of oak, your fathers' homely food;   410
To gather laurel-berries, and the spoil
Of bloody-myrtles, and to press your oil;
For stalking cranes to set the guileful snare;
To enclose the stags in toils, and hunt the hare.
With Balearic slings, or Gnossian bow,
To persecute from far the flying doe,
Then, when the fleecy skies new clothe the wood,
And cakes of rustling ice come rolling down the flood.
   Now sing we stormy stars, when autumn weighs
The year, and adds to nights, and shortens days,   420
And suns declining shine with feeble rays:
What cares must then attend the toiling swain;
Or when the louring spring, with lavish rain,
Beats down the slender stem and bearded grain,
While yet the head is green, or, lightly swelled
With milky moisture, overlooks the field.
E'en when the farmer, now secure of fear,
Sends in the swains to spoil the finished year,
E'en while the reaper fills his greedy hands,
And binds the golden sheaves in brittle bands,   430
Oft have I seen a sudden storm arise,
From all the warring winds that sweep the skies:
The heavy harvest from the root is torn,
And whirled aloft the lighter stubble borne:
With such a force the flying rack is driven,

And such a winter wears the face of heaven:
And oft whole sheets descend of sluicy rain,
Sucked by the spongy clouds from off the main:
The lofty skies at once come pouring down,
The promised crop and golden labours drown.   440
The dykes are filled; and with a roaring sound
The rising rivers float the nether ground,
And rocks the bellowing voice of boiling seas rebound.
The father of the gods his glory shrouds,
Involved in tempests, and a night of clouds;
And from the middle darkness flashing out,
By fits he deals his fiery bolts about.
Earth feels the motions of her angry god;
Her entrails tremble, and her mountains nod,
And flying beasts in forests seek abode: 450
Deep horror seizes every human breast;
Their pride is humbled and their fear confessed,
While he from high his rolling thunder throws,
And fires the mountains with repeated blows:
The rocks are from their old foundations rent;
The winds redouble, and the rains augment:
The waves on heaps are dashed against the shore;
And now the woods, and now the billows, roar.
   In fear of this, observe the starry signs,
Where Saturn houses, and where Hermes joins.°   460
But first to heaven thy due devotions pay,
And annual gifts on Ceres' altars lay.
When winter's rage abates, when cheerful hours
Awake the spring, and spring awakes the flowers,
On the green turf thy careless limbs display,
And celebrate the mighty Mother's day.°
For then the hills with pleasing shades are crowned,
And sleeps are sweeter on the silken ground:
With milder beams the sun securely shines;
Fat are the lambs, and luscious are the wines.   470
Let every swain adore her power divine,
And milk and honey mix with sparkling wine:
Let all the choir of clowns attend the show,
In long procession, shouting as they go;
Invoking her to bless their yearly stores,
Inviting plenty to their crowded floors.

Thus in the spring, and thus in summer's heat,
Before the sickles touch the ripening wheat,
On Ceres call; and let the labouring hind
With oaken wreaths his hollow temples bind: 480
On Ceres let him call, and Ceres praise,
With uncouth dances, and with country lays.
    And that by certain signs we may presage
Of heats and rains, and wind's impetuous rage,
The sovereign of the heavens has set on high
The moon, to mark the changes of the sky;
When southern blasts should cease, and when the swain
Should near their folds his feeding flocks restrain.
For ere the rising winds begin to roar,
The working seas advance to wash the shore;   490
Soft whispers run along the leafy woods,
And mountains whistle to the murmuring floods.
E'en then the doubtful billows scarce abstain
From the tossed vessel on the troubled main;
When crying cormorants forsake the sea,
And stretching to the covert, wing their way;
When sportful coots run skimming o'er the strand;
When watchful herons leave their watery stand,
And mounting upward with erected flight,
Gain on the skies, and soar above the sight.   500
And oft, before tempestuous winds arise,
The seeming stars fall headlong from the skies,
And shooting through the darkness, gild the night
With sweeping glories, and long trails of light;
And chaff with eddy-winds is whirled around,
And dancing leaves are lifted from the ground;
And floating feathers on the waters play.
But when the winged thunder takes his way
From the cold north, and east and west engage
And at their frontiers meet with equal rage,   510
The clouds are crushed; a glut of gathered rain
The hollow ditches fills, and floats the plain;
And sailors furl their dropping sheets amain.
Wet weather seldom hurts the most unwise;
So plain the signs, such prophets are the skies.
The wary crane foresees it first, and sails
Above the storm, and leaves the lowly vales;

The cow looks up, and from afar can find
The change of heaven, and snuffs it in the wind.
The swallow skims the river's watery face: 520
The frogs renew the croaks of their loquacious race.
The careful ant her secret cell forsakes,
And drags her eggs along the narrow tracks.
At either horn the rainbow drinks the flood;
Huge flocks of rising rooks forsake their food,
And crying seek the shelter of the wood.
Besides, the several sorts of watery fowls,
That swim the seas, or haunt the standing pools,
The swans that sail along the silver flood,
And dive with stretching necks to search their food,   530
Then lave their backs with sprinkling dews in vain,
And stem the stream to meet the promised rain.
The crow with clamorous cries the shower demands,
And single stalks along the desert sands.
The nightly virgin, while her wheel she plies,
Foresees the storm impending in the skies,
When sparkling lamps their sputtering light advance,
And in the sockets oily bubbles dance.
     Then, after showers, 'tis easy to descry
Returning suns, and a serener sky: 540
The stars shine smarter; and the moon adorns,
As with unborrowed beams, her sharpened horns.
The filmy gossamer now flits no more,
Nor halcyons bask on the short sunny shore:
Their litter is not tossed by sows unclean;
But a blue droughty mist descends upon the plain;
And owls, that mark the setting sun, declare
A starlight evening, and a morning fair.
Towering aloft, avenging Nisus flies,°
While, dared, below the guilty Scylla lies.°   550
Wherever frighted Scylla flies away,
Swift Nisus follows, and pursues his prey:
Where injured Nisus takes his airy course,
Thence trembling Scylla flies, and shuns his force.
This punishment pursues the unhappy maid,
And thus the purple hair is dearly paid:°
Then thrice the ravens rend the liquid air,
And croaking notes proclaim the settled fair.

Then round their airy palaces they fly,
To greet the sun; and, seized with secret joy,   560
When storms are over-blown, with food repair
To their forsaken nests, and callow care.
Not that I think their breasts with heavenly souls
Inspired, as man, who destiny controls;
But, with the changeful temper of the skies,
As rains condense, and sunshine rarefies,
So turn the species in their altered minds,
Composed by calms, and discomposed by winds.
From hence proceeds the birds' harmonious voice;
From hence the crows exult, and frisking lambs rejoice.   570
Observe the daily circle of the sun,
And the short year of each revolving moon:
By them thou shalt foresee the following day,
Nor shall a starry night thy hopes betray.
When first the moon appears, if then she shrouds
Her silver crescent tipped with sable clouds,
Conclude she bodes a tempest on the main,
And brews for fields impetuous floods of rain.
Or if her face with fiery flushing glow,
Expect the rattling winds aloft to blow.   580
But four nights old (for that's the surest sign),
With sharpened horns if glorious then she shine,
Next day, nor only that, but all the moon,
Till her revolving race be wholly run,
Are void of tempests, both by land and sea,
And sailors in the port their promised vow shall pay.
Above the rest, the sun, who never lies,
Foretells the change of weather in the skies:
For if he rise unwilling to his race,
Clouds on his brow, and spots upon his face,   590
Or if through mists he shoots his sullen beams,
Frugal of light, in loose and straggling streams;
Suspect a drizzling day, with southern rain,
Fatal to fruits, and flocks, and promised grain.
Or if Aurora, with half-opened eyes,
And a pale sickly cheek, salute the skies;
How shall the vine, with tender leaves, defend
Her teeming clusters, when the storms descend,
When ridgy roofs and tiles can scarce avail

To bar the ruin of the rattling hail?   600
But more than all, the setting sun survey,
When down the steep of heaven he drives the day.
For oft we find him finishing his race,
With various colours erring on his face.°
If fiery red his glowing globe descends,
High winds and furious tempests he portends:
But if his cheeks are swollen with livid blue,
He bodes wet weather by his watery hue.
If dusky spots are varied on his brow,
And, streaked with red, a troubled colour show;   610
That sullen mixture shall at once declare
Winds, rain, and storms, and elemental war.
What desperate madman then would venture o'er
The firth, or haul his cables from the shore?°
But if with purple rays he brings the light,
And a pure heaven resigns to quiet night,
No rising winds or falling storms are nigh;
But northern breezes through the forest fly,
And drive the rack, and purge the ruffled sky.
The unerring sun by certain signs declares   620
What the late even or early morn prepares,
And when the south projects a stormy day,
And when the clearing north will puff the clouds away.

   The sun reveals the secrets of the sky;
And who dares give the source of light the lie?
The change of empires often he declares,
Fierce tumults, hidden treasons, open wars.
He first the fate of Caesar did foretell,
And pitied Rome when Rome in Caesar fell.
In iron clouds concealed the public light,   630
And impious mortals feared eternal night.

   Nor was the fact foretold by him alone:
Nature herself stood forth, and seconded the sun.
Earth, air, and seas, with prodigies were signed;
And birds obscene, and howling dogs, divined.
What rocks did Etna's bellowing mouth expire
From her torn entrails! and what floods of fire!
What clanks were heard, in German skies afar,
Of arms, and armies rushing to the war!
Dire earthquakes rent the solid Alps below,   640

And from their summits shook the eternal snow.
Pale spectres in the close of night were seen,
And voices heard, of more than mortal men,
In silent groves: dumb sheep and oxen spoke;
And streams ran backward, and their beds forsook:
The yawning earth disclosed the abyss of hell,
The weeping statues did the wars foretell,
And holy sweat from brazen idols fell.
Then, rising in his might, the king of floods
Rushed through the forests, tore the lofty woods,   650
And rolling onward with a sweepy sway,
Bore houses, herds, and labouring hinds away.
Blood sprang from wells, wolves howled in towns by night,
And boding victims did the priests affright.
Such peals of thunder never poured from high,
Nor lightning flashed from so serene a sky.
Red meteors ran across the ethereal space;
Stars disappeared and comets took their place.
For this, the Emathian plains once more were strewed°
With Roman bodies, and just heaven thought good   660
To fatten twice those fields with Roman blood.°
Then, after length of time, the labouring swains,
Who turn the turfs of those unhappy plains,
Shall rusty piles from the ploughed furrows take,
And over empty helmets pass the rake.
Amazed at antique titles on the stones,
And mighty relics of gigantic bones.
  Ye home-born deities, of mortal birth!
Thou father Romulus, and mother earth,
Goddess unmoved! whose guardian arms extend   670
O'er Tuscan Tiber's course, and Roman towers defend;
With youthful Caesar your joint powers engage,°
Nor hinder him to save the sinking age.
O! let the blood, already spilt, atone
For the past crimes of cursed Laomedon!°
Heaven wants thee there; and long the gods, we know,
Have grudged thee, Caesar, to the world below,
Where fraud and rapine right and wrong confound,
Where impious arms from every part resound,
And monstrous crimes in every shape are crowned.   680
The peaceful peasant to the wars is pressed;

The fields lie fallow in inglorious rest;
The plain no pasture to the flock affords;
The crooked scythes are straightened into swords:
And there Euphrates her soft offspring arms,
And here the Rhine rebellows with alarms;
The neighbouring cities range on several sides,
Perfidious Mars long-plighted leagues divides,
And o'er the wasted world in triumph rides.
So four fierce coursers, starting to the race,   690
Scour through the plain, and lengthen every pace;
Nor reins, nor curbs, nor threatening cries, they fear,
But force along the trembling charioteer.

BOOK II

THE ARGUMENT

The subject of the following book is planting: in handling of which argument the poet shows all the different methods of raising trees, describes their variety, and gives rules for the management of each in particular. He then points out the soils in which the several plants thrive best, and thence takes occasion to run out into the praises of Italy: after which he gives some directions for discovering the nature of every soil, prescribes rules for the dressing of vines, olives, etc., and concludes the Georgic with a panegyric on a country life.

Thus far of tillage, and of heavenly signs:
Now sing, my muse, the growth of generous vines,
The shady groves, the woodland progeny,
And the slow product of Minerva's tree.°
  Great father Bacchus! to my song repair;
For clustering grapes are thy peculiar care:
For thee, large bunches load the bending vine,
And the last blessings of the year are thine.
To thee his joys the jolly autumn owes,
When the fermenting juice the vat o'erflows.   10
Come, strip with me, my god! come drench all o'er
Thy limbs in must of wine, and drink at every pore.
  Some trees their birth to bounteous Nature owe;
For some without the pains of planting grow.
With osiers thus the banks of brooks abound,
Sprung from the watery genius of the ground.
From the same principles grey willows come,

Herculean poplar, and the tender broom.
But some from seeds enclosed in earth arise;
For thus the mastful chestnut mates the skies.°   20
Hence rise the branching beech and vocal oak,
Where Jove of old oraculously spoke.
Some from the root a rising wood disclose:
Thus elms, and thus the savage cherry grows:
Thus the green bays, that binds the poet's brows,
Shoots, and is sheltered by the mother's boughs.

   These ways of planting nature did ordain,
For trees and shrubs, and all the sylvan reign.
Others there are, by late experience found:
Some cut the shoots, and plant in furrowed ground;   30
Some cover rooted stalks in deeper mould;
Some, cloven stakes; and (wondrous to behold)
Their sharpened ends in earth their footing place;
And the dry poles produce a living race.
Some bow their vines, which buried in the plain,
Their tops in distant arches rise again.
Others no root require; the labourer cuts
Young slips, and in the soil securely puts.
E'en stumps of olives, bared of leaves, and dead,
Revive, and oft redeem their withered head.   40
'Tis usual now an inmate graft to see
With insolence invade a foreign tree:
Thus pears and quinces from the crabtree come,
And thus the ruddy cornel bears the plum.

   Then let the learned gardener mark with care
The kinds of stocks, and what those kinds will bear;
Explore the nature of each several tree,
And known, improve with artful industry:
And let no spot of idle earth be found,
But cultivate the genius of the ground: 50
For open Ismarus will Bacchus please;°
Taburnus loves the shade of olive trees.°

   The virtues of the several soils I sing.
Maecenas, now thy needful succour bring!
O thou! the better part of my renown,
Inspire thy poet, and thy poem crown:
Embark with me, while I new tracts explore,
With flying sails and breezes from the shore:

Not that my song, in such a scanty space,
So large a subject fully can embrace: 60
Not though I were supplied with iron lungs,
A hundred mouths, filled with as many tongues:
But steer my vessel with a steady hand,
And coast along the shore in sight of land.
Nor will I tire thy patience with a train
Of preface, or what ancient poets feign.
The trees, which of themselves advance in air,
Are barren kinds, but strongly built and fair,
Because the vigour of the native earth
Maintains the plant, and makes a manly birth.   70
Yet these, receiving grafts of other kind,
Or thence transplanted, change their savage mind,
Their wildness lose, and quitting nature's part
Obey the rules and discipline of art.
The same do trees, that, sprung from barren roots,
In open fields transplanted bear their fruits.
For where they grow the native energy
Turns all into the substance of the tree,
Starves and destroys the fruit, is only made
For brawny bulk, and for a barren shade.   80
The plant that shoots from seed, a sullen tree,
At leisure grows for late posterity;
The generous flavour lost, the fruits decay,
And savage grapes are made the birds' ignoble prey.
Much labour is required in trees to tame
Their wild disorder, and in ranks reclaim.
Well must the ground be dug, and better dressed,
New soil to make, and meliorate the rest.
Old stakes of olive trees in plants revive;
By the same method Paphian myrtles live;°   90
But nobler vines by propagation thrive.
From roots hard hazels, and from scions rise
Tall ash, and taller oak that mates the skies;
Palm, poplar, fir, descending from the steep
Of hills, to try the dangers of the deep.
The thin-leaved arbute hazel-grafts receives;°
And planes huge apples bear, that bore but leaves.
Thus mastful beech the bristly chestnut bears,
And the wild ash is white with blooming pears,

And greedy swine from grafted elms are fed   100
With falling acorns, that on oaks are bred.
   But various are the ways to change the state
Of plants, to bud, to graft, to inoculate.
For, where the tender rinds of trees disclose
Their shooting gems, a swelling knot there grows;
Just in that space a narrow slit we make,
Then other buds from bearing trees we take;
Inserted thus, the wounded rind we close,
In whose moist womb the admitted infant grows.
But, when the smoother bole from knots is free,   110
We make a deep incision in the tree,
And in the solid wood the slip enclose.
The battening bastard shoots again and grows;
And in short space the laden boughs arise,
With happy fruit advancing to the skies.
The mother plant admires the leaves unknown
Of alien trees, and apples not her own.
   Of vegetable woods are various kinds,
And the same species are of several minds.
Lotes, willows, elms, have different forms allowed;°   120
So funeral cypress, rising like a shroud.
Fat olive trees of sundry sorts appear,
Of sundry shapes their unctuous berries bear.
Radii long olives, orchites round produce,°
And bitter pausia, pounded for the juice.°
Alcinous' orchard various apples bears:°
Unlike are bergamots and pounder pears.
Nor our Italian vines produce the shape,
Or taste, or flavour, of the Lesbian grape.
The Thasian vines in richer soils abound;   130
The Mareotic grow in barren ground.
The psythian grape we dry: Lagean juice
Will stammering tongues and staggering feet produce.
Rathe ripe are some, and some of later kind,
Of golden some, and some of purple rind.
How shall I praise the Rhaetian grape divine,
Which yet contends not with Falernian wine?
The Aminean many a consulship survives,
And longer than the Lydian vintage lives,
Or high Phanaeus, king of Chian growth: 140

But, for large quantities and lasting, both,
The less Argitis bears the prize away.
The Rhodian, sacred to the solemn day,
In second services is poured to Jove,
And best accepted by the gods above.
Nor must Bumastus his old honours lose,°
In length and largeness like the dugs of cows.
I pass the rest, whose every race, and name,
And kinds, are less material to my theme;
Which who would learn, as soon may tell the sands,   150
Driven by the western wind on Libyan lands,
Or number, when the blustering Eurus roars,°
The billows beating on Ionian shores.

   Nor every plant on every soil will grow:
The sallow loves the watery ground, and low;
The marshes, alders: nature seems to ordain
The rocky cliff for the wild ash's reign;
The baleful yew to northern blasts assigns,
To shores the myrtles, and to mounts the vines.

   Regard the extremest cultivated coast,   160
From hot Arabia to the Scythian frost:
All sorts of trees their several countries know;
Black ebon only will in India grow,
And odorous frankincense on the Sabaean bough.
Balm slowly trickles through the bleeding veins
Of happy shrubs in Idumaean plains.
The green Egyptian thorn, for medicine good,
With Ethiops' hoary trees and woolly wood,
Let others tell; and how the Seres spin°
Their fleecy forests in a slender twine;°   170
With mighty trunks of trees on Indian shores,
Whose height above the feathered arrow soars,
Shot from the toughest bow, and, by the brawn
Of expert archers, with vast vigour drawn.
Sharp-tasted citrons Median climes produce,
(Bitter the rind, but generous is the juice,)
A cordial fruit, a present antidote
Against the direful stepdame's deadly draught,°
Who, mixing wicked weeds with words impure,
The fate of envied orphans would procure.   180
Large is the plant, and like a laurel grows,

And did it not a different scent disclose,
A laurel were: the fragrant flowers contemn
The stormy winds, tenacious of their stem.
With this, the Medes to labouring age bequeath
New lungs, and cure the sourness of the breath,
    But neither Median woods (a plenteous land),
Fair Ganges, Hermus' rolling golden sand,
Nor Bactria, nor the richer Indian fields,
Nor all the gummy stores Arabia yields,   190
Nor any foreign earth of greater name,
Can with sweet Italy contend in fame.
No bulls, whose nostrils breathe a living flame,
Have turned our turf; no teeth of serpents here
Were sown, an armed host and iron crop to bear.
But fruitful vines, and the fat olive's freight,
And harvests heavy with their fruitful weight,
Adorn our fields; and on the cheerful green
The grazing flocks and lowing herds are seen.
The warrior horse, here bred, is taught to train: 200
There flows Clitumnus through the flowery plain,
Whose waves, for triumphs after prosperous war,
The victim ox and snowy sheep prepare.
Perpetual spring our happy climate sees:
Twice breed the cattle, and twice bear the trees;
And summer suns recede by slow degrees.
    Our land is from the rage of tigers freed,
Nor nourishes the lion's angry seed;
Nor poisonous aconite is here produced,°
Or grows unknown, or is, when known, refused;   210
Nor in so vast a length our serpents glide,
Or raised on such a spiry volume ride.
    Next add our cities of illustrious name,
Their costly labour, and stupendous frame;
Our forts on steepy hills, that far below
See wanton streams in winding valleys flow;
Our twofold seas, that, washing either side,
A rich recruit of foreign stores provide;
Our spacious lakes; thee, Larius, first; and next°
Benacus, with tempestuous billows vexed.°   220
Or shall I praise thy ports, or mention make
Of the vast mound that binds the Lucrine lake?

Or the disdainful sea, that shut from thence
Roars round the structure, and invades the fence;
There, where secure the Julian waters glide,°
Or where Avernus' jaws admit the Tyrrhene tide?
Our quarries, deep in earth, were famed of old
For veins of silver, and for ore of gold.
The inhabitants themselves their country grace:
Hence rose the Marsian and Sabellian race,   230
Strong-limbed and stout, and to the wars inclined,
And hard Ligurians, a laborious kind,
And Volscians armed with iron-headed darts.
Besides—an offspring of undaunted hearts—
The Decii, Marii, great Camillus, came
From hence, and greater Scipio's double name,°
And mighty Caesar, whose victorious arms°
To furthest Asia carry fierce alarms,
Avert unwarlike Indians from his Rome,
Triumph abroad, secure our peace at home.   240
    Hail, sweet Saturnian soil! of fruitful grain°
Great parent, greater of illustrious men!
For thee my tuneful accents will I raise,
And treat of arts disclosed in ancient days,
Once more unlock for thee the sacred spring,
And old Ascraean verse in Roman cities sing.°
    The nature of the several soils now see,
Their strength their colour, their fertility:
And first for heath, and barren hilly ground,
Where meagre clay and flinty stones abound,   250
Where the poor soil all succour seems to want—
Yet this suffices the Palladian plant.°
Undoubted signs of such a soil are found;
For here wild olive-shoots o'erspread the ground,
And heaps of berries strew the fields around.
But where the soil with fattening moisture filled
Is clothed with grass, and fruitful to be tilled,
Such as in cheerful vales we view from high,
Which dripping rocks with rolling streams supply,
And feed with ooze; where rising hillocks run   260
In length, and open to the southern sun;
Where fern succeeds, ungrateful to the plough:
That gentle ground to generous grapes allow.

Strong stocks of vines it will in time produce,
And overflow the vats with friendly juice,
Such as our priests in golden goblets pour
To gods, the givers of the cheerful hour,
Then when the bloated Tuscan blows his horn,
And reeking entrails are in chargers borne.
    If herds or fleecy flocks be more thy care,   270
Or goats that graze the field, and burn it bare,
Then seek Tarentum's lawns, and furthest coast,
Or such a field as hapless Mantua lost,°
Where silver swans sail down the watery road,
And graze the floating herbage of the flood.
There crystal streams perpetual tenor keep,
Nor food nor springs are wanting to thy sheep;
For what the day devours the nightly dew
Shall to the morn in pearly drops renew.
Fat crumbling earth is fitter for the plough,   280
Putrid and loose above, and black below;
For ploughing is an imitative toil,
Resembling nature in an easy soil.
No land for seed like this; no fields afford
So large an income to the village lord:
No toiling teams from harvest-labour come
So late at night, so heavy-laden home.
The like of forest land is understood,
From whence the surly ploughman grubs the wood,
Which had for length of ages idle stood.   290
Then birds forsake the ruins of their seat,
And, flying from their nests, their callow young forget.
The coarse lean gravel, on the mountain sides,
Scarce dewy beverage for the bees provides;
Nor chalk nor crumbling stones, the food of snakes,
That work in hollow earth their winding tracks.
The soil exhaling clouds of subtle dews,
Imbibing moisture which with ease she spews,
Which rusts not iron, and whose mould is clean,
Well clothed with cheerful grass, and ever green,   300
Is good for olives, and aspiring vines,
Embracing husband elms in amorous twines;
Is fit for feeding cattle, fit to sow,
And equal to the pasture and the plough.

Such is the soil of fat Campanian fields;
Such large increase the land that joins Vesuvius yields;
And such a country could Acerra boast,
Till Clanius overflowed the unhappy coast.
  I teach thee next the differing soils to know,
The light for vines, the heavier for the plough.   310
Choose first a place for such a purpose fit:
There dig the solid earth, and sink a pit;
Next fill the hole with its own earth again,
And trample with thy feet, and tread it in:
Then, if it rise not to the former height°
Of superfice, conclude that soil is light,
A proper ground for pasturage and vines.
But, if the sullen earth, so pressed, repines
Within its native mansion to retire,
And stays without, a heap of heavy mire,   320
'Tis good for arable, a glebe that asks
Tough teams of oxen, and laborious tasks.
  Salt earth and bitter are not fit to sow,
Nor will be tamed or mended with the plough.
Sweet grapes degenerate there; and fruits, declined
From their first flavorous taste, renounce their kind.
This truth by sure experiment is tried;
For first an osier colander provide
Of twigs thick wrought (such toiling peasants twine,
When through strait passages they strain their wine):   330
In this close vessel place that earth accursed,
But filled brimful with wholesome water first;
Then run it through; the drops will rope around,°
And, by the bitter taste, disclose the ground.
The fatter earth by handling we may find,
With ease distinguished from the meagre kind:
Poor soil will crumble into dust; the rich
Will to the fingers cleave like clammy pitch:
Moist earth produces corn and grass, but both
Too rank and too luxuriant in their growth.   340
Let not my land so large a promise boast,
Lest the lank ears in length of stem be lost.
The heavier earth is by her weight betrayed;
The lighter in the poising hand is weighed.
'Tis easy to distinguish by the sight,

The colour of the soil, and black from white.
But the cold ground is difficult to know;
Yet this the plants, that prosper there, will show;
Black ivy, pitch-trees, and the baleful yew.
These rules considered well, with early care   350
The vineyard destined for thy vines prepare:
But, long before the planting, dig the ground,
With furrows deep that cast a rising mound.
The clods, exposed to winter winds, will bake;
For putrid earth will best in vineyards take;
And hoary frosts, after the painful toil
Of delving hinds, will rot the mellow soil.

　Some peasants, not to omit the nicest care,
Of the same soil their nursery prepare,
With that of their plantation; lest the tree,   360
Translated, should not with the soil agree.
Beside, to plant it as it was, they mark
The heaven's four quarters on the tender bark,
And to the north or south restore the side,
Which at their birth did heat or cold abide:
So strong is custom; such effects can use
In tender souls of pliant plants produce.

　Choose next a province for thy vineyard's reign,
On hills above, or in the lowly plain.
If fertile fields or valleys be thy choice,   370
Plant thick; for bounteous Bacchus will rejoice
In close plantations there; but, if the vine
On rising ground be placed, or hills supine,
Extend thy loose battalions largely wide,
Opening thy ranks and files on either side,
But marshalled all in order as they stand;
And let no soldier straggle from his band.
As legions in the field their front display,
To try the fortune of some doubtful day,
And move to meet their foes with sober pace,   380
Strict to their figure, though in wider space,
Before the battle joins, while from afar
The field yet glitters with the pomp of war,
And equal Mars, like an impartial lord,
Leaves all to fortune, and the dint of sword:
So let thy vines in intervals be set,

But not their rural discipline forget;
Indulge their width, and add a roomy space,
That their extremest lines may scarce embrace.
Nor this alone to indulge a vain delight,   390
And make a pleasing prospect for the sight;
But, for the ground itself, this only way
Can equal vigour to the plants convey,
Which crowded, want the room their branches to display.
    How deep they must be planted, wouldst thou know?
In shallow furrows vines securely grow.
Not so the rest of plants; for Jove's own tree,°
That holds the woods in awful sovereignty,
Requires a depth of lodging in the ground,
And, next the lower skies, a bed profound: 400
High as his topmost boughs to heaven ascend,
So low his roots to hell's dominion tend.
Therefore, nor winds, nor winter's rage o'erthrows
His bulky body, but unmoved he grows;
For length of ages lasts his happy reign,
And lives of mortal man contend in vain.
Full in the midst of his own strength he stands,
Stretching his brawny arms, and leafy hands;
His shade protects the plains, his head the hills commands.
    The hurtful hazel in thy vineyard shun;   410
Nor plant it to receive the setting sun;
Nor break the topmost branches from the tree;
Nor prune, with blunted knife, the progeny.
Root up wild olives from thy laboured lands;
For sparkling fire, from hinds' unwary hands,
Is often scattered o'er their unctuous rinds,
And after spread abroad by raging winds:
For first the smouldering flame the trunk receives;
Ascending thence, it crackles in the leaves;
At length victorious to the top aspires,   420
Involving all the wood in smoky fires;
But most, when driven by winds, the flaming storm
Of the long files destroys the beauteous form.
In ashes then the unhappy vineyard lies;
Nor will the blasted plants from ruin rise,
Nor will the withered stock be green again;
But the wild olive shoots, and shades the ungrateful plain.

Be not seduced with wisdom's empty shows,
To stir the peaceful ground when Boreas blows.
When winter frosts constrain the field with cold,   430
The fainty root can take no steady hold.
But when the golden spring reveals the year,
And the white bird returns, whom serpents fear,°
That season deem the best to plant thy vines:
Next that, is when autumnal warmth declines,
Ere heat is quite decayed, or cold begun,
Or Capricorn admits the winter sun.
    The spring adorns the woods, renews the leaves;
The womb of earth the genial seed receives:
For then almighty Jove descends, and pours   440
Into his buxom bride his fruitful showers;
And mixing his large limbs with hers, he feeds
Her births with kindly juice, and fosters teeming seeds.
Then joyous birds frequent the lonely grove,
And beasts, by nature stung, renew their love.
Then fields the blades of buried corn disclose;
And while the balmy western spirit blows,
Earth to the breath her bosom dares expose.
With kindly moisture then the plants abound;
The grass securely springs above the ground;   450
The tender twig shoots upward to the skies,
And on the faith of the new sun relies.
The swerving vines on the tall elms prevail;
Unhurt by southern showers or northern hail
They spread their gems the genial warmth to share,
And boldly trust their buds in open air.
In this soft season (let me dare to sing),
The world was hatched by heaven's imperial king
In prime of all the year, and holidays of spring.
Then did the new creation first appear;   460
Nor other was the tenor of the year,
When laughing heaven did the great birth attend;
And eastern winds their wintry breath suspend:
Then sheep first saw the sun in open fields,
And savage beasts were sent to stock the wilds;
And golden stars flew up to light the skies,
And man's relentless race from stony quarries rise.
Nor could the tender new creation bear

The excessive heats or coldness of the year,
But, chilled by winter, or by summer fired,   470
The middle temper of the spring required,
When warmth and moisture did at once abound,
And heaven's indulgence brooded on the ground.

For what remains, in depth of earth secure
Thy covered plants, and dung with hot manure;
And shells and gravel in the ground enclose;
For through their hollow chinks the water flows,
Which, thus imbibed, returns in misty dews,
And steaming up, the rising plant renews.
Some husbandmen of late, have found the way,   480
A hilly heap of stones above to lay,
And press the plants with shards of potters' clay.
This fence against immoderate rain they found,
Or when the Dog-star cleaves the thirsty ground.°

Be mindful, when thou hast entombed the shoot,
With store of earth around to feed the root;
With iron teeth of rakes and prongs, to move
The crusted earth, and loosen it above.
Then exercise thy sturdy steers to plough
Betwixt thy vines, and teach thy feeble row   490
To mount on reeds, and wands, and, upward led,
On ashen poles to raise their forky head.
On these new crutches let them learn to walk,
Till swerving upwards with a stronger stalk,
They brave the winds, and, clinging to their guide,
On tops of elms at length triumphant ride.
But in their tender nonage, while they spread
Their springing leaves, and lift their infant head,
And upward while they shoot in open air,
Indulge their childhood, and the nursling spare,   500
Nor exercise thy rage on new-born life,
But let thy hand supply the pruning-knife,
And crop luxuriant stragglers, nor be loath
To strip the branches of their leafy growth.
But when the rooted vines, with steady hold,
Can clasp their elms, then, husbandman, be bold
To lop the disobedient boughs that strayed
Beyond their ranks; let crooked steel invade
The lawless troops, which discipline disclaim,

And their superfluous growth with rigour tame.   510
Next, fenced with hedges and deep ditches round,
Exclude the encroaching cattle from thy ground,
While yet the tender gems but just appear,
Unable to sustain the uncertain year;
Whose leaves are not alone foul winter's prey,
But oft by summer suns are scorched away,
And worse than both, become the unworthy browse
Of buffaloes, salt goats, and hungry cows.
For not December's frost that burns the boughs,
Nor dog-days' parching heat that splits the rocks,   520
Are half so harmful as the greedy flocks,
Their venomed bite, and scars indented on the stocks.
For this the malefactor goat was laid
On Bacchus' altar, and his forfeit paid.
At Athens thus old comedy began,
When round the streets the reeling actors ran,
In country villages, and crossing ways,
Contending for the prizes of their plays;
And glad with Bacchus, on the grassy soil,°
Leaped o'er the skins of goats besmeared with oil.   530
Thus Roman youth, derived from ruined Troy,
In rude Saturnian rhymes express their joy;°
With taunts, and laughter loud, their audience please,
Deformed with visors cut from barks of trees:
In jolly hymns they praise the god of wine,
Whose earthen images adorn the pine,
And there are hung on high, in honour of the vine:
A madness so devout the vineyard fills.
In hollow valleys and on rising hills,
On whate'er side he turns his honest face,°   540
And dances in the wind, those fields are in his grace.
To Bacchus therefore let us tune our lays,
And in our mother tongue resound his praise.
Thin cakes in chargers, and a guilty goat,
Dragged by the horns, be to his altars brought;
Whose offered entrails shall his crime reproach,
And drip their fatness from the hazel broach.°
To dress thy vines new labour is required;
Nor must the painful husbandman be tired:
For thrice, at least, in compass of the year,   550

Thy vineyard must employ the sturdy steer
To turn the glebe, besides thy daily pain
To break the clods, and make the surface plain,
To unload the branches, or the leaves to thin,
That suck the vital moisture of the vine.
Thus in a circle runs the peasant's pain,
And the year rolls within itself again.
E'en in the lowest months, when storms have shed
From vines the hairy honours of their head,
Not then the drudging hind his labour ends,   560
But to the coming year his care extends.
E'en then the naked vine he persecutes;
His pruning knife at once reforms and cuts.
Be first to dig the ground; be first to burn
The branches lopped; and first the props return
Into thy house, that bore the burdened vines;
But last to reap the vintage of thy wines.
Twice in the year luxuriant leaves o'ershade
The encumbered vine; rough brambles twice invade:
Hard labour both! Commend the large excess   570
Of spacious vineyards; cultivate the less.
Besides, in woods the shrubs of prickly thorn,
Sallows and reeds on banks of rivers born,
Remain to cut; for vineyards useful found,
To stay thy vines, and fence thy fruitful ground.
Nor when thy tender trees at length are bound,°
When peaceful vines from pruning hooks are free,
When husbands have surveyed the last degree,
And utmost files of plants, and ordered every tree;
E'en when they sing at ease in full content,   580
Insulting o'er the toils they underwent,
Yet still they find a future task remain,
To turn the soil, and break the clods again;
And after all, their joys are unsincere,
While falling rains on ripening grapes they fear.
Quite opposite to these are olives found:
No dressing they require, and dread no wound,
Nor rakes nor harrows need; but fixed below,
Rejoice in open air, and unconcernedly grow.
The soil itself due nourishment supplies: 590
Plough but the furrows, and the fruits arise,

Content with small endeavours till they spring.
Soft peace they figure, and sweet plenty bring,
Then olives plant, and hymns to Pallas sing.
   Thus apple trees, whose trunks are strong to bear
Their spreading boughs, exert themselves in air,
Want no supply, but stand secure alone,
Not trusting foreign forces, but their own,
Till with the ruddy freight the bending branches groan.
   Thus trees of nature, and each common bush,   600
Uncultivated thrive, and with red berries blush.
Vile shrubs are shorn for browse; the towering height
Of unctuous trees are torches for the night.
And shall we doubt (indulging easy sloth),
To sow, to set, and to reform their growth?
To leave the lofty plants—the lowly kind
Are for the shepherd or the sheep designed.
E'en humble broom and osiers have their use,
And shade for sleep, and food for flocks, produce;
Hedges for corn, and honey for the bees,   610
Besides the pleasing prospect of the trees.
How goodly looks Cytorus, ever green°
With boxen groves, with what delight are seen
Narycian woods of pitch, whose gloomy shade°
Seems for retreat of heavenly muses made!
But much more pleasing are those fields to see,
That need not ploughs, nor human industry.
E'en cold Caucasian rocks with trees are spread,
And wear green forests on their hilly head.
Though bending from the blast of eastern storms,   620
Though shent their leaves, and shattered are their arms,
Yet heaven their various plants for use designs,
For houses, cedars, and for shipping, pines;
Cypress provides for spokes and wheels of wains,
And all for keels of ships that scour the watery plains.
Willows in twigs are fruitful, elms in leaves;
The war, from stubborn myrtle, shafts receives,
From cornels, javelins; and the tougher yew
Receives the bending figure of a bow.
Nor box, nor limes, without their use are made,   630
Smooth-grained, and proper for the turner's trade;
Which curious hands may carve, and steel with ease invade.

Light alder stems the Po's impetuous tide,
And bees in hollow oaks their honey hide.
Now balance with these gifts the fumy joys
Of wine, attended with eternal noise.
Wine urged to lawless lust the centaurs' train;
Through wine they quarrelled, and through wine were slain.
　O happy, if he knew his happy state,
The swain, who, free from business and debate,　　　　640
Receives his easy food from nature's hand,
And just returns of cultivated land!
No palace, with a lofty gate, he wants,
To admit the tide of early visitants,
With eager eyes devouring, as they pass,
The breathing figures of Corinthian brass.
No statues threaten from high pedestals;
No Persian arras hides his homely walls,
With antic vests which through their shady fold
Betray the streaks of ill-dissembled gold:　　　　650
He boasts no wool whose native white is dyed
With purple poison of Assyrian pride;
No costly drugs of Araby defile
With foreign scents the sweetness of his oil:
But easy quiet, a secure retreat,
A harmless life that knows not how to cheat,
With home-bred plenty, the rich owner bless,
And rural pleasures crown his happiness.
Unvexed with quarrels, undisturbed with noise,
The country king his peaceful realm enjoys,　　　　660
Cool grots, and living lakes, the flowery pride
Of meads, and streams that through the valley glide,
And shady groves that easy sleep invite,
And after toilsome days, a soft repose at night.
Wild beasts of nature in his woods abound;
And youth, of labour patient, plough the ground,
Inured to hardship, and to homely fare.
Nor venerable age is wanting there,
In great examples to the youthful train;
Nor are the gods adored with rites profane.　　　　670
From hence Astraea took her flight; and here
The prints of her departing steps appear.
　Ye sacred muses! with whose beauty fired,

My soul is ravished, and my brain inspired
(Whose priest I am, whose holy fillets wear)
Would you your poet's first petition hear:
Give me the ways of wandering stars to know,
The depths of heaven above, and earth below.
Teach me the various labours of the moon,
And whence proceed the eclipses of the sun.   680
Why flowing tides prevail upon the main,
And in what dark recess they shrink again;
What shakes the solid earth; what cause delays
The summer nights, and shortens winter days.
But if my heavy blood restrain the flight
Of my free soul, aspiring to the height
Of nature, and unclouded fields of light;
My next desire is, void of care and strife,
To lead a soft, secure, inglorious life.
A country cottage near a crystal flood,   690
A winding valley, and a lofty wood.
Some god conduct me to the sacred shades,
Where Bacchanals are sung by Spartan maids,
Or lift me high to Haemus' hilly crown,
Or in the plains of Tempe lay me down,
Or lead me to some solitary place,
And cover my retreat from human race.

  Happy the man who studying nature's laws,
Through known effects can trace the secret cause.
His mind possessing in a quiet state,   700
Fearless of fortune, and resigned to fate!
And happy too is he, who decks the bowers
Of sylvans, and adores the rural powers,
Whose mind, unmoved, the bribes of courts can see,
Their glittering baits, and purple slavery,
Nor hopes the people's praise, nor fears their frown,
Nor, when contending kindred tear the crown,
Will set up one, or pull another down.

  Without concern he hears, but hears from far,
Of tumults, and descents, and distant war;   710
Nor with a superstitious fear is awed,
For what befalls at home, or what abroad.
Nor envies he the rich their heapy store,
Nor his own peace disturbs with pity for the poor.

He feeds on fruits which of their own accord
The willing ground and laden trees afford.
From his loved home no lucre him can draw;
The senate's mad decrees he never saw;
Nor heard, at bawling bars, corrupted law.
Some to the seas and some to camps resort   720
And some with impudence invade the court:
In foreign countries others seek renown;
With wars and taxes, others waste their own.
And houses burn, and household gods deface,
To drink in bowls which glittering gems enchase;
To loll on couches, rich with citron steads,
And lay their guilty limbs in Tyrian beds.
This wretch in earth entombs his golden ore,
Hovering and brooding on his buried store.
Some patriot fools to popular praise aspire   730
Or public speeches, which worse fools admire.
While from both benches, with redoubled sounds,
The applause of lords and commoners abounds.
Some through ambition, or through thirst of gold,
Have slain their brothers, or their country sold,
And leaving their sweet homes, in exile run
To lands that lie beneath another sun.

    The peasant, innocent of all these ills,
With crooked ploughs the fertile fallows tills,
And the round year with daily labour fills.   740
From hence the country markets are supplied:
Enough remains for household charge beside,
His wife and tender children to sustain,
And gratefully to feed his dumb deserving train.
Nor cease his labours, till the yellow field
A full return of bearded harvest yield,
A crop so plenteous, as the land to load,
O'ercome the crowded barns, and lodge on ricks abroad.
Thus every several season is employed,
Some spent in toil, and some in ease enjoyed.   750
The yeaning ewes prevent the springing year:
The laded boughs their fruits in autumn bear:
'Tis then the vine her liquid harvest yields,
Baked in the sunshine of ascending fields.
The winter comes; and then the falling mast

For greedy swine provides a full repast:
Then olives, ground in mills, their fatness boast,
And winter fruits are mellowed by the frost.
His cares are eased with intervals of bliss;
His little children, climbing for a kiss,   760
Welcome their father's late return at night;
His faithful bed is crowned with chaste delight.
His kine with swelling udders ready stand,
And lowing for the pail, invite the milker's hand.
His wanton kids, with budding horns prepared,
Fight harmless battles in his homely yard:
Himself, in rustic pomp, on holidays,
To rural powers a just oblation pays,
And on the green his careless limbs displays.
The hearth is in the midst; the herdsmen round   770
The cheerful fire, provoke his health in goblets crowned.
He calls on Bacchus, and propounds the prize;
The groom his fellow-groom at butts defies,
And bends his bow, and levels with his eyes.
Or stripped for wrestling, smears his limbs with oil,
And watches, with a trip his foe to foil.
Such was the life the frugal Sabines led;
So Remus and his brother god were bred,
From whom the austere Etrurian virtue rose;
And this rude life our homely fathers chose.   780
Old Rome from such a race derived her birth
(The seat of empire, and the conquered earth),
Which now on seven high hills triumphant reigns,
And in that compass all the world contains.
Ere Saturn's rebel son usurped the skies,
When beasts were only slain for sacrifice,
While peaceful Crete enjoyed her ancient lord,
Ere sounding hammers forged the inhuman sword,
Ere hollow drums were beat, before the breath
Of brazen trumpets rung the peals of death,   790
The good old god his hunger did assuage
With roots and herbs, and gave the golden age.
But, over-laboured with so long a course,
'Tis time to set at ease the smoking horse.

BOOK III

THE ARGUMENT

This book begins with the invocation of some rural deities, and a compliment to Augustus; after which Virgil directs himself to Maecenas, and enters on his subject. He lays down rules for the breeding and management of horses, oxen, sheep, goats, and dogs; and interweaves several pleasant descriptions of a chariot-race, of the battle of the bulls, of the force of love, and of the Scythian winter. In the latter part of the book, he relates the diseases incident to cattle; and ends with the description of a fatal murrain that formerly raged among the Alps.

Thy fields, propitious Pales, I rehearse,°
And sing thy pastures in no vulgar verse,
Amphrysian shepherd! the Lycaean woods,°
Arcadia's flowery plains, and pleasing floods.
   All other themes that careless minds invite
Are worn with use, unworthy me to write.
Busiris' altars, and the dire decrees°
Of hard Eurystheus, every reader sees:°
Hylas the boy, Latona's erring isle,°
And Pelops' ivory shoulder, and his toil°   10
For fair Hippodame, with all the rest
Of Grecian tales, by poets are expressed.
New ways I must attempt, my grovelling name
To raise aloft, and wing my flight to fame.
   I, first of Romans, shall in triumph come
From conquered Greece and bring her trophies home,
With foreign spoils adorn my native place,
And with Idume's palms my Mantua grace.
Of Parian stone a temple will I raise,°
Where the slow Mincius through the valley strays,°   20
Where cooling streams invite the flocks to drink,
And reeds defend the winding water's brink.
Full in the midst shall mighty Caesar stand,°
Hold the chief honours, and the dome command.
Then I, conspicuous in my Tyrian gown°
(Submitting to his godhead my renown),
A hundred coursers from the goal will drive:
The rival chariots in the race shall strive.
All Greece shall flock from far, my games to see;

The whirlbat, and the rapid race, shall be°   30
Reserved for Caesar, and ordained by me.
Myself, with olive crowned, the gifts will bear.
E'en now methinks the public shouts I hear;
The passing pageants, and the pomps appear.
I to the temple will conduct the crew,
The sacrifice and sacrificers view,
From thence return, attended with my train,
Where the proud theatres disclose the scene,
Which interwoven Britons seem to raise,°
And show the triumph which their shame displays.   40
High o'er the gate, in elephant and gold,°
The crowd shall Caesar's Indian war behold:°
The Nile shall flow beneath; and, on the side,
His shattered ships on brazen pillars ride.°
Next him Niphates, with inverted urn,°
And dropping sedge, shall his Armenia mourn;
And Asian cities in our triumph borne.
With backward bows the Parthian shall be there,°
And spurring from the fight, confess their fear.
A double wreath shall crown our Caesar's brows: 50
Two differing trophies, from two different foes.
Europe with Afric in his fame shall join;
But neither shore his conquest shall confine.
The Parian marble there shall seem to move
In breathing statues, not unworthy Jove,
Resembling heroes, whose ethereal root
Is Jove himself, and Caesar is the fruit.
Tros and his race the sculptor shall employ;°
And he—the god who built the walls of Troy.°
Envy herself at last, grown pale and dumb   60
(By Caesar combated and overcome),
Shall give her hands, and fear the curling snakes
Of lashing furies, and the burning lakes;
The pains of famished Tantalus shall feel,
And Sisyphus, that labours up the hill
The rolling rock in vain; and cursed Ixion's wheel.
Meantime we must pursue the sylvan lands
(The abode of nymphs), untouched by former hands:
For such, Maecenas, are thy hard commands.
Without thee, nothing lofty can I sing.   70

Come then, and with thyself, thy genius bring,
With which inspired, I brook no dull delay:
Cithaeron loudly calls me to my way;°
Thy hounds, Taygetus, open and pursue their prey.°
High Epidaurus urges on my speed,°
Famed for his hills, and for his horses' breed:
From hills and dales the cheerful cries rebound;
For echo hunts along, and propagates the sound.

    A time will come, when my maturer muse,
In Caesar's wars, a nobler theme shall choose,   80
And through more ages bear my sovereign's praise,
Than have from Tithon passed to Caesar's days.°

    The generous youth, who, studious of the prize,
The race of running coursers multiplies,
Or to the plough the sturdy bullocks breeds,
May know that from the dam the worth of each proceeds.
The mother cow must wear a louring look,
Sour-headed, strongly necked, to bear the yoke.
Her double dew-lap from her chin descends,
And at her thighs the ponderous burden ends.   90
Long are her sides and large; her limbs are great;
Rough are her ears, and broad her horny feet.
Her colour shining black, but flecked with white;
She tosses from the yoke; provokes the fight:
She rises in her gait, is free from fears,
And in her face a bull's resemblance bears:
Her ample forehead with a star is crowned,
And with her length of tail she sweeps the ground.
The bull's insult at four she may sustain;°
But after ten from nuptial rites refrain.   100
Six seasons use; but then release the cow,
Unfit for love, and for the labouring plough.

    Now while their youth is filled with kindly fire,
Submit thy females to the lusty sire:
Watch the quick motions of the frisking tail;
Then serve their fury with the rushing male,
Indulging pleasure, lest the breed should fail.

    In youth alone unhappy mortals live;
But ah! the mighty bliss is fugitive:
Discoloured sickness, anxious labours, come,   110
And age, and death's inexorable doom.

Yearly thy herds in vigour will impair.
Recruit and mend them with thy yearly care:
Still propagate; for still they fall away:
'Tis prudence to prevent the entire decay.
   Like diligence requires the courser's race,
In early choice, and for a longer space.
The colt, that for a stallion is designed,
By sure presages shows his generous kind,
Of able body, sound of limb and wind.   120
Upright he walks, on pasterns firm and straight;
His motions easy; prancing in his gait;
The first to lead the way, to tempt the flood,
To pass the bridge unknown, nor fear the trembling wood;
Dauntless at empty noises; lofty necked;
Sharp-headed, barrel-bellied, broadly backed;
Brawny his chest, and deep; his colour gray;
For beauty, dappled, or the brightest bay:
Faint white and dun will scarce the rearing pay.
   The fiery courser, when he hears from far   130
The sprightly trumpets, and the shouts of war,
Pricks up his ears; and trembling with delight,
Shifts place, and paws, and hopes the promised fight.
On his right shoulder his thick mane reclined,
Ruffles at speed, and dances in the wind.
His horny hoofs are jetty black and round;
His chine is double; starting with a bound
He turns the turf, and shakes the solid ground.
Fire from his eyes, clouds from his nostrils flow:
He bears his rider headlong on the foe.   140
Such was the steed in Grecian poets famed,
Proud Cyllarus, by Spartan Pollux tamed:
Such coursers bore to fight the god of Thrace;°
And such, Achilles, was thy warlike race.
In such a shape, grim Saturn did restrain°
His heavenly limbs, and flowed with such a mane,
When, half-surprised, and fearing to be seen,
The lecher galloped from his jealous queen,
Ran up the ridges of the rocks amain,
And with shrill neighings filled the neighbouring plain. 150
   But worn with years, when dire diseases come,
Then hide his not ignoble age at home,

In peace to enjoy his former palms and pains;
And gratefully be kind to his remains.
For when his blood no youthful spirits move,
He languishes and labours in his love;
And when the sprightly seed should swiftly come,
Dribbling he drudges, and defrauds the womb.
In vain he burns, like hasty stubble-fires,
And in himself his former self requires.   160

   His age and courage weigh; nor those alone,
But note his father's virtues and his own:
Observe if he disdains to yield the prize,
Of loss impatient, proud of victories.

   Hast thou beheld, when from the goal they start,
The youthful charioteers with heaving heart
Rush to the race; and panting, scarcely bear
The extremes of feverish hope and chilling fear;
Stoop to the reins and lash with all their force;
The flying chariot kindles in the course: 170
And now alow, and now aloft, they fly,
As borne through air, and seem to touch the sky.
No stop, no stay: but clouds of sand arise,
Spurned, and cast backward on the followers' eyes.
The hindmost blows the foam upon the first:
Such is the love of praise, an honourable thirst.

   Bold Erichthonius was the first who joined°
Four horses for the rapid race designed,
And o'er the dusty wheels presiding sate:
The Lapithae, to chariots, add the state°   180
Of bits and bridles; taught the steed to bound,
To run the ring, and trace the mazy round;
To stop, to fly, the rules of war to know;
To obey the rider, and to dare the foe.

   To choose a youthful steed with courage fired,
To breed him, break him, back him, are required
Experienced masters; and in sundry ways,
Their labours equal, and alike their praise.
But once again the battered horse beware:
The weak old stallion will deceive thy care,   190
Though famous in his youth for force and speed,
Or was of Argos or Epirian breed,
Or did from Neptune's race, or from himself, proceed.

These things premised, when now the nuptial time
Approaches for the stately steed to climb,
With food enable him to make his court;
Distend his chine, and pamper him for sport:
Feed him with herbs, whatever thou canst find,
Of generous warmth, and of salacious kind:°
Then water him, and (drinking what he can)   200
Encourage him to thirst again, with bran.
Instructed thus, produce him to the fair,
And join in wedlock to the longing mare.
For if the sire be faint, or out of case,°
He will be copied in his famished race,
And sink beneath the pleasing task assigned
(For all's too little for the craving kind).°
      As for the females, with industrious care
Take down their mettle; keep them lean and bare:
When conscious of their past delight, and keen   210
To take the leap, and prove the sport again,
With scanty measure then supply their food;
And when athirst restrain them from the flood;
Their bodies harass; sink them when they run;
And fry their melting marrow in the sun.
Starve them, when barns beneath their burden groan,
And winnowed chaff by western winds is blown;
For fear the rankness of the swelling womb
Should scant the passage, and confine the room;
Lest the fat furrows should the sense destroy   220
Of genial lust, and dull the seat of joy.
But let them suck the seed with greedy force,
And close involve the vigour of the horse.
      The male has done: thy care must now proceed°
To teeming females, and the promised breed.
First let them run at large, and never know
The taming yoke, or draw the crooked plough.
Let them not leap the ditch, or swim the flood,
Or lumber o'er the meads, or cross the wood;
But range the forest, by the silver side   230
Of some cool stream, where nature shall provide
Green grass and fattening clover for their fare,
And mossy caverns for their noontide lair,
With rocks above, to shield the sharp nocturnal air.

About the Alburnian groves, with holly green,°
Of winged insects mighty swarms are seen:
This flying plague (to mark its quality)
Oestros the Grecians call: Asylus, we;°
A fierce loud-buzzing breeze. Their stings draw blood,
And drive the cattle gadding through the wood.   240
Seized with unusual pains, they loudly cry:
Tanagrus hastens thence, and leaves his channel dry.°
This curse the jealous Juno did invent,
And first employed for Io's punishment.°
To shun this ill, the cunning leech ordains,
In summer's sultry heats (for then it reigns)
To feed the females ere the sun arise,
Or late at night, when stars adorn the skies.

When she has calved, then set the dam aside,
And for the tender progeny provide.   250
Distinguish all betimes with branding fire,
To note the tribe, the lineage, and the sire.
Whom to reserve for husband of the herd,
Or who shall be to sacrifice preferred;
Or whom thou shalt to turn thy glebe allow,
To smooth the furrows, and sustain the plough:
The rest, for whom no lot is yet decreed,
May run in pastures, and at pleasure feed.
The calf, by nature and by genius made
To turn the glebe, breed to the rural trade.   260
Set him betimes to school; and let him be
Instructed there in rules of husbandry,
While yet his youth is flexible and green,
Nor bad examples of the world has seen.
Early begin the stubborn child to break;
For his soft neck, a supple collar make
Of bending osiers; and (with time and care
Inured that easy servitude to bear)
Thy flattering method on the youth pursue:
Joined with his school-fellows by two and two,   270
Persuade them first to lead an empty wheel,
That scarce the dust can raise, or they can feel:
In length of time produce the labouring yoke
And shining shares, that make the furrow smoke.
Ere the licentious youth be thus restrained,

Or moral precepts on their minds have gained,
Their wanton appetites not only feed
With delicates of leaves, and marshy weed,
But with thy sickle reap the rankest land,
And minister the blade with bounteous hand: 280
Nor be with harmful parsimony won
To follow what our homely sires have done,
Who filled the pail with beestings of the cow,°
But all her udder to the calf allow.

   If to the warlike steed thy studies bend,
Or for the prize in chariots to contend,
Near Pisa's flood the rapid wheels to guide.°
Or in Olympian groves aloft to ride,
The generous labours of the courser, first,
Must be with sight of arms and sounds of trumpets nursed; 290
Inured the groaning axle-tree to bear,
And let him clashing whips in stables hear.
Soothe him with praise, and make him understand
The loud applauses of his master's hand:
This, from his weaning, let him well be taught;
And then betimes in a soft snaffle wrought,
Before his tender joints with nerves are knit,
Untried in arms, and trembling at the bit.
But when to four full springs his years advance,
Teach him to run the round, with pride to prance,   300
And (rightly managed) equal time to beat,
To turn, to bound in measure, and curvet.
Let him to this with easy pains be brought,
And seem to labour, when he labours not.
Thus formed for speed, he challenges the wind,
And leaves the Scythian arrow far behind:
He scours along the field, with loosened reins,
And treads so light, he scarcely prints the plains;
Like Boreas in his race, when, rushing forth,
He sweeps the skies, and clears the cloudy north: 310
The waving harvest bends beneath his blast,
The forest shakes, the groves their honours cast;
He flies aloft, and with impetuous roar
Pursues the foaming surges to the shore.
Thus, o'er the Elean plains, thy well-breathed horse
Impels the flying car and wins the course,

Or, bred to Belgian waggons, leads the way,
Untired at night, and cheerful all the day.
    When once he's broken, feed him full and high;
Indulge his growth, and his gaunt sides supply.   320
Before his training keep him poor and low;
For his stout stomach with his food will grow:
The pampered colt will discipline disdain,
Impatient of the lash, and restive to the rein.
    Wouldst thou their courage and their strength improve?
Too soon they must not feel the stings of love.
Whether the bull or courser be thy care,
Let him not leap the cow, nor mount the mare.
The youthful bull must wander in the wood
Behind the mountain, or beyond the flood,   330
Or in the stall at home his fodder find,
Far from the charms of that alluring kind.
With two fair eyes his mistress burns his breast:
He looks, and languishes, and leaves his rest,
Forsakes his food, and, pining for the lass,
Is joyless of the grove, and spurns the growing grass.
The soft seducer, with enticing looks,
The bellowing rivals to the fight provokes.
    A beauteous heifer in the wood is bred:
The stooping warriors, aiming head to head,   340
Engage their clashing horns: with dreadful sound
The forest rattles, and the rocks rebound.
They fence, they push, and, pushing, loudly roar:
Their dewlaps and their sides are bathed in gore.
Nor when the war is over, is it peace;
Nor will the vanquished bull his claim release;
But feeding in his breast his ancient fires,
And cursing fate, from his proud foe retires.
Driven from his native land to foreign grounds,
He with a generous rage resents his wounds,   350
His ignominious flight, the victor's boast,
And more than both, the loves which unrevenged he lost.
Often he turns his eyes, and, with a groan,
Surveys the pleasing kingdoms, once his own:
And therefore to repair his strength he tries,
Hardening his limbs with painful exercise,
And rough upon the flinty rock he lies.

On prickly leaves and on sharp herbs he feeds,
Then to the prelude of a war proceeds.
His horns, yet sore, he tries against a tree,   360
And meditates his absent enemy.
He snuffs the wind; his heels the sand excite;
But when he stands collected in his might,
He roars, and promises a more successful fight.
Then, to redeem his honour at a blow,
He moves his camp, to meet his careless foe.
Not with more madness, rolling from afar,
The spumy waves proclaim the watery war,
And mounting upwards, with a mighty roar,
March onwards, and insult the rocky shore.   370
They mate the middle region with their height,
And fall no less than with a mountain's weight;
The waters boil, and belching from below
Black sands, as from a forceful engine, throw.

   Thus every creature, and of every kind,
The secret joys of sweet coition find.
Not only man's imperial race, but they
That wing the liquid air, or swim the sea,
Or haunt the desert, rush into the flame:
For love is lord of all, and is in all the same.   380
   'Tis with this rage, the mother lion stung,
Scours o'er the plain, regardless of her young:
Demanding rites of love, she sternly stalks,
And hunts her lover in his lonely walks.
'Tis then the shapeless bear his den forsakes;
In woods and fields a wild destruction makes:
Boars whet their tusks; to battle tigers move,
Enraged with hunger, more enraged with love.
Then woe to him, that in the desert land
Of Libya, travels o'er the burning sand.   390
The stallion snuffs the well-known scent afar,
And snorts and trembles for the distant mare;
Nor bits nor bridles can his rage restrain,
And rugged rocks are interposed in vain:
He makes his way o'er mountains, and contemns
Unruly torrents, and unforded streams.
The bristled boar, who feels the pleasing wound,
New grinds his arming tusks, and digs the ground.

The sleepy lecher shuts his little eyes;
About his churning chaps the frothy bubbles rise: 400
He rubs his sides against a tree; prepares
And hardens both his shoulders for the wars.

   What did the youth, when Love's unerring dart°
Transfixed his liver, and inflamed his heart?
Alone, by night, his watery way he took;
About him, and above, the billows broke;
The sluices of the sky were open spread,
And rolling thunder rattled o'er his head;
The raging tempest called him back in vain
And every boding omen of the main: 410
Nor could his kindred, nor the kindly force
Of weeping parents, change his fatal course;
No, not the dying maid, who must deplore
His floating carcase on the Sestian shore.°

   I pass the wars that spotted lynxes make
With their fierce rivals for the female's sake,
The howling wolves', the mastiffs' amorous rage;
When e'en the fearful stag dares for his hind engage.
But, far above the rest, the furious mare,
Barred from the male, is frantic with despair: 420
For, when her pouting vent declares her pain,
She tears the harness, and she rends the rein.
For this (when Venus gave them rage and power)
Their master's mangled members they devour,
Of love defrauded in their longing hour.
For love they force through thickets of the wood,
They climb the steepy hills, and stem the flood.

   When at the spring's approach, their marrow burns
(For with the spring their genial warmth returns),
The mares to cliffs of rugged rocks repair,   430
And with wide nostrils snuff the western air:
When (wondrous to relate!) the parent wind,
Without the stallion, propagates the kind.
Then fired with amorous rage, they take their flight
Through plains, and mount the hills' unequal height;
Nor to the north, nor to the rising sun,
Nor southward to the rainy regions run,
But boring to the west, and hovering there,°
With gaping mouths, they draw prolific air;

With which impregnate, from their groins they shed   440
A slimy juice, by false conception bred.
The shepherd knows it well, and calls by name
Hippomanes, to note the mother's flame.°
This, gathered in the planetary hour,
With noxious weeds, and spelled with words of power,
Dire stepdames in the magic bowl infuse,
And mix, for deadly draughts, the poisonous juice.
But time is lost, which never will renew,
While we too far the pleasing path pursue,
Surveying nature with too nice a view.   450
    Let this suffice for herds: our following care
Shall woolly flocks and shaggy goats declare.
Nor can I doubt what oil I must bestow,°
To raise my subject from a ground so low;
And the mean matter which my theme affords,
To embellish with magnificence of words.
But the commanding muse my chariot guides,
Which o'er the dubious cliff securely rides;
And pleased I am, no beaten road to take,
But first the way to new discoveries make.   460
    Now, sacred Pales, in a lofty strain
I sing the rural honours of thy reign.
First, with assiduous care from winter keep
Well foddered in the stalls thy tender sheep:
Then spread with straw the bedding of thy fold,
With fern beneath, to fend the bitter cold;
That free from gouts thou mayst preserve thy care,
And clear from scabs, produced by freezing air.
Next let thy goats officiously be nursed,
And led to living streams, to quench their thirst.   470
Feed them with winter-browse; and, for their lair,
A cote that opens to the south prepare;
Where basking in the sunshine they may lie,
And the short remnants of his heat enjoy.
This during winter's drizzly reign be done,
Till the new Ram receives the exalted sun.°
For hairy goats of equal profit are
With woolly sheep, and ask an equal care.
'Tis true, the fleece, when drunk with Tyrian juice,°
Is dearly sold; but not for needful use: 480

For the salacious goat increases more,
And twice as largely yields her milky store.
The still distended udders never fail,
But when they seem exhausted, swell the pail.
Meantime the pastor shears their hoary beards,
And eases of their hair the loaden herds.
Their camlets warm in tents, the soldier hold,°
And shield the shivering mariner from cold.
  On shrubs they browse, and on the bleaky top
Of rugged hills, the thorny bramble crop.   490
Attended with their bleating kids they come
At night unasked and mindful of their home;
And scarce their swelling bags the threshold overcome.
So much the more thy diligence bestow
In depth of winter, to defend the snow,°
By how much less the tender helpless kind,
For their own ills, can fit provision find.
Then minister the browse with bounteous hand,
And open let thy stacks all winter stand.
But when the western winds with vital power   500
Call forth the tender grass and budding flower,
Then, at the last, produce in open air
Both flocks; and send them to their summer fare.
Before the sun while Hesperus appears,°
First let them sip from herbs the pearly tears
Of morning dews, and after break their fast
On greensward ground—a cool and grateful taste.
But when the day's fourth hour has drawn the dews,
And the sun's sultry heat their thirst renews;
When creaking grasshoppers on shrubs complain,   510
Then lead them to their watering-troughs again.
In summer's heat, some bending valley find,
Closed from the sun, but open to the wind;
Or seek some ancient oak, whose arms extend
In ample breadth, thy cattle to defend,
Or solitary grove, or gloomy glade,
To shield them with its venerable shade.
Once more to watering lead; and feed again
When the low sun is sinking to the main.
When rising Cynthia sheds her silver dews,   520
And the cool evening-breeze the meads renews,

When linnets fill the woods with tuneful sound,
And hollow shores the halcyon's voice rebound.

Why should my muse enlarge on Libyan swains,
Their scattered cottages, and ample plains,
Where oft the flocks without a leader stray,
Or through continued deserts take their way,
And feeding, add the length of night to day?
Whole months they wander, grazing as they go;
Nor folds nor hospitable harbour know.   530
Such an extent of plains, so vast a space
Of wilds unknown, and of untasted grass,
Allures their eyes: the shepherd last appears,
And with him all his patrimony bears,
His house and household gods, his trade of war,
His bow and quiver, and his trusty cur.
Thus, under heavy arms, the youth of Rome
Their long laborious marches overcome,
Cheerly their tedious travels undergo,
And pitch their sudden camp before the foe.   540

Not so the Scythian shepherd tends his fold,°
Nor he who bears in Thrace the bitter cold,
Nor he who treads the bleak Maeotian strand,°
Or where proud Hister rolls his yellow sand.°
Early they stall their flocks and herds; for there
No grass the fields, no leaves the forests, wear:
The frozen earth lies buried there below
A hilly heap, seven cubits deep in snow;
And all the west allies of stormy Boreas blow.

The sun from far peeps with a sickly face,   550
Too weak the clouds and mighty fogs to chase,
When up the skies he shoots his rosy head,
Or in the ruddy ocean seeks his bed.
Swift rivers are with sudden ice constrained;
And studded wheels are on its back sustained,
A hostry now for waggons, which before°
Tall ships of burden on its bosom bore.
The brazen cauldrons with the frost are flawed;
The garment, stiff with ice, at hearths is thawed;
With axes first they cleave the wine; and thence,   560
By weight, the solid portions they dispense.
From locks uncombed and from the frozen beard,

Long icicles depend and crackling sounds are heard.
Meantime perpetual sleet, and driving snow,
Obscure the skies, and hang on herds below.
The starving cattle perish in their stalls;
Huge oxen stand enclosed in wintry walls
Of snow congealed; whole herds are buried there
Of mighty stags, and scarce their horns appear.
The dexterous huntsman wounds not these afar   570
With shafts or darts, or makes a distant war
With dogs, or pitches toils to stop their flight,
But close engages in unequal fight;
And while they strive in vain to make their way
Through hills of snow, and pitifully bray,
Assaults with dint of sword, or pointed spears,
And homeward, on his back, the joyful burden bears.
The men to subterranean caves retire,
Secure from cold and crowd the cheerful fire:
With trunks of elm and oaks the hearth they load,   580
Nor tempt the inclemency of heaven abroad.
Their jovial nights in frolics and in play
They pass, to drive the tedious hours away,
And their cold stomachs with crowned goblets cheer
Of windy cider, and of barmy beer.°
Such are the cold Rhipaean race, and such°
The savage Scythian, and unwarlike Dutch.
Where skins of beasts the rude barbarians wear,
The spoils of foxes, and the furry bear.

   Is wool thy care? Let not thy cattle go   590
Where bushes are, where burs and thistles grow;
Nor in too rank a pasture let them feed;
Then of the purest white select thy breed.
E'en though a snowy ram thou shalt behold,
Prefer him not in haste for husband to thy fold:
But search his mouth; and, if a swarthy tongue
Is underneath his humid palate hung,
Reject him, lest he darken all the flock,
And substitute another from thy stock.
'Twas thus, with fleeces milky white (if we   600
May trust report), Pan, god of Arcady,
Did bribe thee, Cynthia; nor didst thou disdain,°
When called in woody shades, to cure a lover's pain.

If milk be thy design, with plenteous hand
Bring clover-grass; and from the marshy land
Salt herbage for the foddering rack provide,
To fill their bags, and swell the milky tide.
These raise their thirst, and to the taste restore
The savour of the salt, on which they fed before.

Some, when the kids their dams too deeply drain,   610
With gags and muzzles their soft mouths restrain.
Their morning milk the peasants press at night;
Their evening meal, before the rising light,
To market bear; or sparingly they steep
With seasoning salt, and stored for winter keep.

Nor last, forget thy faithful dogs; but feed
With fattening whey the mastiffs' generous breed,
And Spartan race, who, for the fold's relief,
Will prosecute with cries the nightly thief,
Repulse the prowling wolf, and hold at bay   620
The mountain robbers rushing to the prey.
With cries of hounds, thou mayst pursue the fear
Of flying hares, and chase the fallow deer,
Rouse from their desert dens the bristled rage
Of boars, and beamy stags in toils engage.°

With smoke of burning cedars scent thy walls,
And fume with stinking galbanum thy stalls,°
With that rank odour from thy dwelling-place
To drive the viper's brood, and all the venomed race.
For often under stalls unmoved they lie   630
Obscure in shades, and shunning heaven's broad eye:
And snakes, familiar, to the hearth succeed,
Disclose their eggs, and near the chimney breed.
Whether to roofy houses they repair,
Or sun themselves abroad in open air,
In all abodes of pestilential kind
To sheep and oxen, and the painful hind.
Take, shepherd, take a plant of stubborn oak,
And labour him with many a sturdy stroke,
Or with hard stones demolish from afar   640
His haughty crest, the seat of all the war.
Invade his hissing throat, and winding spires;
Till stretched in length, the unfolded foe retires.
He drags his tail, and for his head provides,

And in some secret cranny slowly glides;
But leaves exposed to blows his back and battered sides.
   In fair Calabria's woods a snake is bred
With curling crest, and with advancing head:
Waving he rolls, and makes a winding track;
His belly spotted, burnished is his back.   650
While springs are broken, while the southern air
And dropping heavens the moistened earth repair,
He lives on standing lakes and trembling bogs,
And fills his maw with fish, or with loquacious frogs:
But when, in muddy pools, the water sinks,
And the chapped earth is furrowed o'er with chinks,
He leaves the fens, and leaps upon the ground,
And, hissing, rolls his glaring eyes around.
With thirst inflamed impatient of the heats,
He rages in the fields, and wide destruction threats.   660
Oh! let not sleep my closing eyes invade
In open plains, or in the secret shade,
When he, renewed in all the speckled pride
Of pompous youth, has cast his slough aside,
And in his summer livery rolls along,
Erect, and brandishing his forky tongue,
Leaving his nest, and his imperfect young,
And thoughtless of his eggs, forgets to rear
The hopes of poison for the following year.
   The causes and the signs shall next be told,   670
Of every sickness that infects the fold.
A scabby tetter on their pelts will stick,°
When the raw rain has pierced them to the quick,
Or searching frosts have eaten through the skin,
Or burning icicles are lodged within;
Or when the fleece is shorn, if sweat remains
Unwashed, and soaks into their empty veins;
When their defenceless limbs the brambles tear,
Short of their wool, and naked from the shear.
Good shepherds, after shearing, drench their sheep: 680
And their flock's father (forced from high to leap)
Swims down the stream, and plunges in the deep.
They oint their naked limbs with mothered oil;°
Or from the founts where living sulphurs boil,
They mix a medicine to foment their limbs,

With scum that on the molten silver swims.
Fat pitch, and black bitumen, add to these,
Besides the waxen labour of the bees,
And hellebore, and squills deep-rooted in the seas.°
Receipts abound; but searching all thy store,   690
The best is still at hand to lance the sore,
And cut the head; for till the core be found,
The secret vice is fed, and gathers ground.
While making fruitless moan the shepherd stands,
And, when the lancing-knife requires his hands,
Vain help, with idle prayers, from heaven demands.
Deep in their bones when fevers fix their seat,
And rack their limbs, and lick the vital heat,
The ready cure to cool the raging pain
Is underneath the foot to breathe a vein.   700
This remedy the Scythian shepherds found:
The inhabitants of Thracia's hilly ground,
And Gelons use it, when for drink and food°
They mix their curdled milk with horses' blood.
    But where thou seest a single sheep remain
In shades aloof, or couched upon the plain,
Or listlessly to crop the tender grass,
Or late to lag behind with truant pace;
Revenge the crime, and take the traitor's head,
Ere in the faultless flock the dire contagion spread.   710
    On winter seas we fewer storms behold,
Than foul diseases that infect the fold.
Nor do those ills on single bodies prey,
But oftener bring the nation to decay,
And sweep the present stock and future hope away.
    A dire example of this truth appears,
When after such a length of rolling years,
We see the naked Alps, and thin remains
Of scattered cots, and yet unpeopled plains,
Once filled with grazing flocks, the shepherds' happy reigns.   720
    Here from the vicious air and sickly skies,
A plague did on the dumb creation rise:
During the autumnal heats the infection grew,
Tame cattle and the beasts of nature slew,
Poisoning the standing lakes, and pools impure;
Nor was the foodful grass in fields secure.

Strange death! for when the thirsty fire had drunk
Their vital blood, and the dry nerves were shrunk,
When the contracted limbs were cramped, e'en then
A waterish humour swelled and oozed again,   730
Converting into bane the kindly juice,
Ordained by nature for a better use.
The victim ox, that was for altars pressed,
Trimmed with white ribbons, and with garlands drest,
Sunk of himself, without the god's command,
Preventing the slow sacrificer's hand.
Or by the holy butcher if he fell,
The inspected entrails could no fates foretell;
Nor, laid on altars, did pure flames arise;
But clouds of smouldering smoke forbade the sacrifice.   740
Scarcely the knife was reddened with his gore,
Or the black poison stained the sandy floor.°
The thriven calves in meads their food forsake,°
And render their sweet souls before the plenteous rack.
The fawning dog runs mad; the wheezing swine
With coughs is choked, and labours from the chine:°
The victor horse, forgetful of his food,
The palm renounces, and abhors the flood,
He paws the ground; and on his hanging ears
A doubtful sweat in clammy drops appears: 750
Parched is his hide, and rugged are his hairs.
Such are the symptoms of the young disease;
But in time's process, when his pains increase,
He rolls his mournful eyes; he deeply groans
With patient sobbing, and with manly moans.
He heaves for breath; which from his lungs supplied,
And fetched from far, distends his labouring side.
To his rough palate his dry tongue succeeds;
And ropy gore he from his nostrils bleeds.°
A drench of wine has with success been used,   760
And through a horn the generous juice infused,
Which timely taken oped his closing jaws,
But, if too late, the patient's death did cause:
For the too vigorous dose too fiercely wrought,
And added fury to the strength it brought.
Recruited into rage, he grinds his teeth
In his own flesh, and feeds approaching death.

Ye gods, to better fate good men dispose,
And turn that impious error on our foes!
   The steer, who to the yoke was bred to bow   770
(Studious of tillage, and the crooked plough),
Falls down and dies; and dying, spews a flood
Of foamy madness, mixed with clotted blood.
The clown, who cursing providence, repines,
His mournful fellow from the team disjoins;
With many a groan forsakes his fruitless care,
And in the unfinished furrow leaves the share.
The pining steer, no shades of lofty woods,
Nor flowery meads can ease, nor crystal floods
Rolled from the rock: his flabby flanks decrease;   780
His eyes are settled in a stupid peace;
His bulk too weighty for his thighs is grown,
And his unwieldy neck hangs drooping down.
Now what avails his well-deserving toil
To turn the glebe, or smooth the rugged soil?
And yet he never supped in solemn state
(Nor undigested feasts did urge his fate),
Nor day to night luxuriously did join,
Nor surfeited on rich Campanian wine.
Simple his beverage, homely was his food,   790
The wholesome herbage, and the running flood:
No dreadful dreams awaked him with affright;
His pains by day secured his rest by night.
   'Twas then that buffaloes, ill paired, were seen
To draw the car of Jove's imperial queen,
For want of oxen; and the labouring swain
Scratched with a rake, a furrow for his grain,
And covered with his hand the shallow seed again.
He yokes himself, and up the hilly height,
With his own shoulders, draws the waggon's weight.   800
   The nightly wolf, that round the enclosure prowled
To leap the fence, now plots not on the fold,
Tamed with a sharper pain. The fearful doe,
And flying stag, amidst the greyhounds go,
And round the dwellings roam of man, their fiercer foe.
The scaly nations of the sea profound,
Like shipwrecked carcases, are driven aground,
And mightly phocae, never seen before°

In shallow streams, are stranded on the shore.
The viper dead within her hole is found: 810
Defenceless was the shelter of the ground.
The water-snake, whom fish and paddocks fed,°
With staring scales lies poisoned in his bed:
To birds their native heavens contagious prove;
From clouds they fall, and leave their souls above.

    Besides, to change their pasture 'tis in vain,
Or trust to physic; physic is their bane.
The learned leeches in despair depart,
And shake their heads, desponding of their art.

    Tisiphone, let loose from under ground,°   820
Majestically pale, now treads the round,
Before her drives diseases and affright,
And every moment rises to the sight,
Aspiring to the skies, encroaching on the light.
The rivers, and their banks, and hills around,
With lowings and with dying bleats resound.
At length, she strikes a universal blow;
To death at once whole herds of cattle go;
Sheep, oxen, horses, fall; and heaped on high,
The differing species in confusion lie,   830
Till warned by frequent ills, the way they found
To lodge their loathsome carrion under ground:
For useless to the currier were their hides;
Nor could their tainted flesh with ocean tides
Be freed from filth; nor could Vulcanian flame
The stench abolish, or the savour tame.
Nor safely could they shear their fleecy store
(Made drunk with poisonous juice, and stiff with gore).
Or touch the web: but, if the vest they wear,
Red blisters rising on their paps appear,   840
And flaming carbuncles, and noisome sweat,
And clammy dews, that loathsome lice beget;
Till the slow-creeping evil eats his way,
Consumes the parching limbs, and makes the life his prey.

BOOK IV

THE ARGUMENT

Virgil has taken care to raise the subject of each Georgic. In the first, he has only dead matter on which to work. In the second, he just steps on the world of life, and describes that degree of it which is to be found in vegetables. In the third, he advances to animals: and, in the last, singles out the Bee, which may be reckoned the most sagacious of them, for his subject.

  In this Georgic he shows us what station is most proper for the bees, and when they begin to gather honey; how to call them home when they swarm; and how to part them when they are engaged in battle. From hence he takes occasion to discover their different kinds; and, after an excursion, relates their prudent and politic administration of affairs, and the several diseases that often rage in their hives, with the proper symptoms and remedies of each disease. In the last place, he lays down a method of repairing their kind, supposing their whole breed lost; and gives at large the history of its invention.

> The gifts of heaven my following song pursues,
> Aerial honey, and ambrosial dews.°
> Maecenas, read this other part, that sings
> Embattled squadrons, and adventurous kings:
> A mighty pomp, though made of little things.
> Their arms, their arts, their manners, I disclose,
> And how they war, and whence the people rose.
> Slight is the subject, but the praise not small,
> If heaven assist, and Phoebus hear my call.
>
>   First, for thy bees a quiet station find,   10
> And lodge them under covert of the wind
> (For winds, when homeward they return, will drive
> The loaded carriers from their evening hive),
> Far from the cows' and goats' insulting crew,
> That trample down the flowers, and brush the dew.
> The painted lizard, and the birds of prey,
> Foes of the frugal kind, be far away:
> The titmouse, and the pecker's hungry brood,
> And Procne, with her bosom stained in blood:°
> These rob the trading citizens, and bear   20
> The trembling captives through the liquid air,
> And for their callow young a cruel feast prepare.
> But near a living stream their mansion place,

Edged round with moss and tufts of matted grass:
And plant (the winds' impetuous rage to stop)
Wild olive-trees, or palms, before the busy shop;
That when the youthful prince, with proud alarm,
Calls out the venturous colony to swarm;
When first their way through yielding air they wing,
New to the pleasures of their native spring;   30
The banks of brooks may make a cool retreat
For the raw soldiers from the scalding heat,
And neighbouring trees with friendly shade invite
The troops, unused to long laborious flight.
Then o'er the running stream, or standing lake,
A passage for thy weary people make;
With osier floats the standing water strow;
Of massy stones make bridges, if it flow;
That basking in the sun thy bees may lie,
And resting there, their flaggy pinions dry,   40
When, late returning home, the laden host
By raging winds is wrecked upon the coast.
Wild thyme and savory set around their cell,
Sweet to the taste, and fragrant to the smell:
Set rows of rosemary with flowering stem,
And let the purple violets drink the stream.

   Whether thou build the palace of thy bees
With twisted osiers, or with barks of trees,
Make but a narrow mouth: for as the cold
Congeals into a lump the liquid gold,   50
So 'tis again dissolved by summer's heat,
And the sweet labours both extremes defeat.
And therefore not in vain the industrious kind
With dauby wax and flowers the chinks have lined,
And with their stores of gathered glue, contrive
To stop the vents and crannies of their hive.
Not birdlime, or Idaean pitch, produce°
A more tenacious mass of clammy juice.

   Nor bees are lodged in hives alone, but found
In chambers of their own beneath the ground;   60
Their vaulted roofs are hung in pumices,
And in the rotten trunks of hollow trees.
   But plaster thou the chinky hives with clay,
And leafy branches o'er their lodgings lay:

Nor place them where too deep a water flows,
Or where the yew, their poisonous neighbour, grows;
Nor roast red crabs, to offend the niceness of their nose;°
Nor near the steaming stench of muddy ground;
Nor hollow rocks that render back the sound,
And doubled images of voice rebound.   70

   For what remains, when golden suns appear,
And under earth have driven the winter year,
The winged nation wanders through the skies,
And o'er the plains and shady forest flies;
Then stooping on the meads and leafy bowers,
They skim the floods, and sip the purple flowers.
Exalted hence, and drunk with secret joy,
Their young succession all their cares employ:
They breed, they brood, instruct and educate,
And make provision for the future state;   80
They work their waxen lodgings in their hives,
And labour honey to sustain their lives.
But when thou seest a swarming cloud arise,
That sweeps aloft, and darkens all the skies,
The motions of their hasty flight attend;
And know, to floods or woods, their airy march they bend.
Then milfoil beat, and honeysuckles pound;°
With these alluring savours strew the ground;
And mix with tinkling brass the cymbal's droning sound.
Straight to their ancient cells, recalled from air,   90
The reconciled deserters will repair.
But if intestine broils alarm the hive
(For two pretenders oft for empire strive),
The vulgar in divided factions jar;
And murmuring sounds proclaim the civil war.
Inflamed with ire, and trembling with disdain,
Scarce can their limbs their mighty souls contain.
With shouts, the coward's courage they excite,
And martial clangours call them out to fight;
With hoarse alarms the hollow camp rebounds,   100
That imitates the trumpet's angry sounds;
Then to their common standard they repair;
The nimble horsemen scour the fields of air.
In form of battle drawn, they issue forth,
And every knight is proud to prove his worth.

Pressed for their country's honour, and their king's,°
On their sharp beaks they whet their pointed stings,
And exercise their arms, and tremble with their wings.
Full in the midst the haughty monarchs ride;
The trusty guards come up, and close the side;   110
With shouts the daring foe to battle is defied.
Thus in the season of unclouded spring,
To war they follow their undaunted king,
Crowd through their gates, and, in the fields of light,
The shocking squadrons meet in mortal fight.°
Headlong they fall from high, and wounded, wound,
And heaps of slaughtered soldiers bite the ground.
Hard hailstones lie not thicker on the plain,
Nor shaken oaks such showers of acorns rain.
With gorgeous wings, the marks of sovereign sway,   120
The two contending princes make their way;
Intrepid through the midst of danger go,
Their friends encourage and amaze the foe.
With mighty souls in narrow bodies pressed,
They challenge, and encounter breast to breast;
So fixed on fame, unknowing how to fly,
And obstinately bent to win or die,
That long the doubtful combat they maintain,
Till one prevails—for one can only reign.
Yet all these dreadful deeds, this deadly fray,   130
A cast of scattered dust will soon allay,
And undecided leave the fortune of the day.
When both the chiefs are sundered from the fight,
Then to the lawful king restore his right;
And let the wasteful prodigal be slain,
That he, who best deserves, alone may reign.
With ease distinguished is the regal race:
One monarch wears an honest open face;
Shaped to his size and godlike to behold,
His royal body shines with specks of gold,   140
And ruddy scales; for empire he designed,
Is better born, and of a nobler kind.
That other looks like nature in disgrace:
Gaunt are his sides, and sullen is his face;
And like their grisly prince appear his gloomy race,
Grim, ghastly, rugged, like a thirsty train

That long have travelled through a desert plain,
And spit from their dry chaps the gathered dust again.
The better brood, unlike the bastard crew,
Are marked with royal streaks of shining hue;   150
Glittering and ardent, though in body less:
From these at pointed seasons hope to press°
Huge heavy honeycombs of golden juice,
Not only sweet, but pure, and fit for use,
To allay the strength and hardness of the wine,
And with old Bacchus new metheglin join.°
    But when their swarms are eager of their play,
And loathe their empty hives, and idly stray,
Restrain the wanton fugitives, and take
A timely care to bring the truants back.   160
The task is easy: but to clip the wings
Of their high-flying arbitrary kings.
At their command, the people swarm away;
Confine the tyrant, and the slave will stay.
    Sweet gardens, full of saffron flowers, invite
The wandering gluttons, and retard their flight.
Besides the god obscene, who frights away,°
With his lath sword, the thieves and birds of prey.°
With his own hand, the guardian of the bees,
For slips of pines may search the mountain trees,   170
And with wild thyme and savory plant the plain,
Till his hard horny fingers ache with pain;
And deck with fruitful trees the fields around,
And with refreshing waters drench the ground.
    Now, did I not so near my labours end,
Strike sail, and hastening to the harbour tend,
My song to flowery gardens might extend.
To teach the vegetable arts, to sing
The Paestan roses, and their double spring;°
How chicory drinks the running streams, and how   180
Green beds of parsley near the river grow;
How cucumbers along the surface creep
With crooked bodies, and with bellies deep:
The late narcissus, and the winding trail
Of bear's-foot, myrtles green, and ivy pale:°
For where with stately towers Tarentum stands,
And deep Galaesus soaks the yellow sands,

I chanced an old Corycian swain to know,°
Lord of few acres, and those barren too,
Unfit for sheep or vines, and more unfit to sow;   190
Yet labouring well his little spot of ground,
Some scattering pot-herbs here and there he found,
Which cultivated with his daily care,
And bruised with vervain, were his frugal fare.°
Sometimes white lilies did their leaves afford,
With wholesome poppy-flowers, to mend his homely board;
For late returning home, he supped at ease,
And wisely deemed the wealth of monarchs less;
The little of his own, because his own, did please.
To quit his care, he gathered first of all   200
In spring the roses, apples in the fall;
And when cold winter split the rocks in twain,
And ice the running rivers did restrain,
He stripped the bear's-foot of its leafy growth,
And, calling western winds, accused the Spring of sloth.
He therefore first among the swains was found
To reap the product of his laboured ground,
And squeeze the combs with golden liquor crowned.
His limes were first in flowers; his lofty pines,
With friendly shade, secured his tender vines.   210
For every bloom his trees in spring afford,
An autumn apple was by tale restored.°
He knew to rank his elms in even rows,
For fruit the grafted pear-tree to dispose,
And tame to plums the sourness of the sloes.
With spreading planes he made a cool retreat,
To shade good fellows from the summer's heat.°
But straitened in my space, I must forsake
This task, for others afterwards to take.

   Describe we next the nature of the bees,   220
Bestowed by Jove for secret services,
When by the tinkling sound of timbrels led,
The king of heaven in Cretan caves they fed.
Of all the race of animals, alone
The bees have common cities of their own,
And common sons; beneath one law they live,
And with one common stock their traffic drive.°
Each has a certain home, a several stall;

All is the state's, the state provides for all.
Mindful of coming cold, they share the pain,   230
And hoard, for winter's use, the summer's gain.
Some o'er the public magazines preside,
And some are sent new forage to provide;
These drudge in fields abroad, and those at home
Lay deep foundations for the laboured comb,
With dew, narcissus-leaves, and clammy gum.
To pitch the waxen flooring some contrive;
Some nurse the future nation of the hive;
Sweet honey some condense; some purge the grout;°
The rest in cells apart the liquid nectar shut: 240
All, with united force, combine to drive
The lazy drones from the laborious hive:
With envy stung, they view each other's deeds;
With diligence the fragrant work proceeds.
As when the cyclops, at the almighty nod,
New thunder hasten for their angry god,
Subdued in fire the stubborn metal lies;
One brawny smith the puffing bellows plies,
And draws and blows reciprocating air:
Others to quench the hissing mass prepare;   250
With lifted arms they order every blow,
And chime their sounding hammers in a row;
With laboured anvils Etna groans below.
Strongly they strike; huge flakes of flames expire;
With tongs they turn the steel, and vex it in the fire.
If little things with great we may compare,
Such are the bees, and such their busy care;
Studious of honey, each in his degree,
The youthful swain, the grave experienced bee—
That in the field; this, in affairs of state   260
Employed at home, abides within the gate,
To fortify the combs, to build the wall,
To prop the ruins, lest the fabric fall:
But late at night, with weary pinions come
The labouring youth, and heavy laden, home.
Plains, meads, and orchards, all the day he plies,
The gleans of yellow thyme distend his thighs:
He spoils the saffron flowers; he sips the blues
Of violets, wilding blooms, and willow dews.

Their toil is common, common is their sleep;   270
They shake their wings when morn begins to peep,
Rush through the city gates without delay,
Nor ends their work, but with declining day.
Then having spent the last remains of light,
They give their bodies due repose at night,
When hollow murmurs of their evening bells
Dismiss the sleepy swains, and toll them to their cells.
When once in beds their weary limbs they steep,
No buzzing sounds disturb their golden sleep:
'Tis sacred silence all. Nor dare they stray,   280
When rain is promised, or a stormy day;
But near the city walls their watering take,
Nor forage far, but short excursions make.

    And as when empty barks on billows float,
With sandy ballast sailors trim the boat;
So bees bear gravel-stones, whose poising weight
Steers through the whistling winds their steady flight.

    But (what's more strange) their modest appetites,
Averse from Venus, fly the nuptial rites.
No lust enervates their heroic mind,   290
Nor wastes their strength on wanton womankind.
But in their mouths reside their genial powers:
They gather children from the leaves and flowers.
Thus make they kings to fill the regal seat,
And thus their little citizens create,
And waxen cities build, the palaces of state.
And oft on rocks their tender wings they tear,
And sink beneath the burdens which they bear:
Such rage of honey in their bosom beats,
And such a zeal they have for flowery sweets.   300

    Thus though the race of life they quickly run,
Which in the space of seven short years is done,
The immortal line in sure succession reigns
The fortune of the family remains,
And grandsires, grandsons, the long list contains.
    Besides, not Egypt, India, Media, more,°
With servile awe, their idol king adore:
While he survives, in concord and content
The commons live, by no divisions rent;
But the great monarch's death dissolves the government.   310

All goes to ruin; they themselves contrive
To rob the honey, and subvert the hive.
The king presides, his subjects' toil surveys,
The servile rout their careful Caesar praise:
Him they extol; they worship him alone;
They crowd his levees, and support his throne:
They raise him on their shoulders with a shout;
And when their sovereign's quarrel calls them out,
His foes to mortal combat they defy,
And think it honour at his feet to die.   320
 Induced by such examples, some have taught
That bees have portions of ethereal thought,
Endued with particles of heavenly fires;
For God the whole created mass inspires.
Through heaven, and earth, and ocean's depth, he throws
His influence round, and kindles as he goes.
Hence flocks, and herds, and men, and beasts, and fowls,
With breath are quickened, and attract their souls;
Hence take the forms his prescience did ordain,
And into him at length resolve again.   330
No room is left for death: they mount the sky,
And to their own congenial planets fly.°
 Now, when thou hast decreed to seize their stores,
And by prerogative to break their doors,
With sprinkled water first the city choke,
And then pursue the citizens with smoke.
Two honey-harvests fall in every year:
First, when the pleasing Pleiades appear,°
And, springing upward, spurn the briny seas;
Again, when their affrighted choir surveys   340
The watery Scorpion mend his pace behind,°
With a black train of storms, and winter wind,
They plunge into the deep, and safe protection find.
Prone to revenge, the bees, a wrathful race,
When once provoked, assault the aggressor's face,
And through the purple veins a passage find;
There fix their stings, and leave their souls behind.
 But, if a pinching winter thou foresee,
And wouldst preserve thy famished family;
With fragrant thyme the city fumigate,   350
And break the waxen walls to save the state.

For lurking lizards often lodge, by stealth
Within the suburbs, and purloin their wealth;
And beetles shunning light, a dark retreat
Have found in combs, and undermined the seat.
Or lazy drones, without their share of pain,
In winter-quarters free, devour the gain;
Or wasps infest the camp with loud alarms,
And mix in battle with unequal arms;
Or secret moths are there in silence fed;   360
Or spiders in the vault their snary webs have spread.

    The more oppressed by foes, or famine-pined,
The more increase thy care to save the sinking kind:
With greens and flowers recruit their empty hives,
And seek fresh forage to sustain their lives.

    But since they share with man one common fate
In health and sickness, and in turns of state,
Observe the symptoms when they fall away
And languish with insensible decay.
They change their hue; with haggard eyes they stare;   370
Lean are their looks, and shagged is their hair:°
And crowds of dead, that never must return
To their loved hives, in decent pomp are borne:
Their friends attend the hearse; the next relations mourn.
The sick, for air, before the portal gasp,
Their feeble legs within each other clasp,
Or idle in their empty hives remain,
Benumbed with cold, and listless of their gain.
Soft whispers then, and broken sounds, are heard,
As when the woods by gentle winds are stirred;   380
Such stifled noise as the close furnace hides,
Or dying murmurs of departing tides.
This when thou seest, galbanean odours use,°
And honey in the sickly hive infuse.
Through reeden pipes convey the golden flood,
To invite the people to their wonted food.
Mix it with thickened juice of sodden wines,°
And raisins from the grapes of psithian vines:
To these add pounded galls, and roses dry,°
And, with Cecropian thyme, strong-scented centaury.°   390

    A flower there is, that grows in meadow-ground,
Amellus called, and easy to be found;°

For from one root the rising stem bestows
A wood of leaves, and violet-purple boughs:
The flower itself is glorious to behold,
And shines on altars like refulgent gold,
Sharp to the taste; by shepherds near the stream
Of Mella found; and thence they gave the name.
Boil this restoring root in generous wine,
And set beside the door, the sickly stock to dine.   400
But if the labouring kind be wholly lost,
And not to be retrieved with care or cost;
'Tis time to touch the precepts of an art,
The Arcadian master did of old impart;°
And how he stocked his empty hives again,
Renewed with putrid gore of oxen slain.
An ancient legend I prepare to sing,
And upward follow fame's immortal spring.
   For where with seven-fold horns mysterious Nile
Surrounds the skirts of Egypt's fruitful isle,   410
And where in pomp the sun-burnt people ride,
On painted barges, o'er the teeming tide,
Which, pouring down from Ethiopian lands,
Makes green the soil with slime, and black prolific sands,
That length of region, and large tract of ground,
In this one art a sure relief have found.
First, in a place by nature close, they build
A narrow flooring, guttered, walled, and tiled.
In this, four windows are contrived, that strike
To the four winds opposed, their beams oblique.   420
A steer of two years old they take, whose head
Now first with burnished horns begins to spread:°
They stop his nostrils, while he strives in vain
To breathe free air, and struggles with his pain.
Knocked down, he dies: his bowels, bruised within,
Betray no wound on his unbroken skin.
Extended thus, in this obscene abode
They leave the beast; but first sweet flowers are strowed
Beneath his body, broken boughs and thyme,
And pleasing cassia just renewed in prime.   430
This must be done ere spring makes equal day,
When western winds on curling waters play;
Ere painted meads produce their flowery crops,

Or swallows twitter on the chimney-tops.
The tainted blood, in this close prison pent,
Begins to boil and through the bones ferment.
Then (wondrous to behold) new creatures rise,
A moving mass at first, and short of thighs;
'Till, shooting out with legs, and imped with wings,
The grubs proceed to bees with pointed stings;   440
And, more and more affecting air, they try
Their tender pinions, and begin to fly:
At length, like summer storms from spreading clouds,
That burst at once and pour impetuous floods,
Or flights of arrows from the Parthian bows,
When from afar they gall embattled foes,°
With such a tempest through the skies they steer,
And such a form the winged squadrons bear.
  What god, O muse! this useful science taught?
Or by what man's experience was it brought?   450
  Sad Aristaeus from fair Tempe fled,°
His bees with famine or diseases dead:
On Peneus' banks he stood, and near his holy head;
And, while his falling tears the streams supplied,
Thus, mourning, to his mother goddess cried:
'Mother Cyrene! mother, whose abode°
Is in the depth of this immortal flood!
What boots it, that from Phoebus' loins I spring,
The third, by him and thee, from heaven's high king?
O! where is all thy boasted pity gone,   460
And promise of the skies to thy deluded son?
Why didst thou me, unhappy me, create,
Odious to gods, and born to bitter fate?
Whom scarce my sheep, and scarce my painful plough,
The needful aids of human life allow:
So wretched is thy son, so hard a mother thou!
Proceed, inhuman parent, in thy scorn;
Root up my trees; with blights destroy my corn;
My vineyards ruin, and my sheepfolds burn.
Let loose thy rage; let all thy spite be shown,   470
Since thus thy hate pursues the praises of thy son.'
But from her mossy bower below the ground,
His careful mother heard the plaintive sound,
Encompassed with her sea-green sisters round.

One common work they plied; their distaffs full°
With carded locks of blue Milesian wool.°
Spio, with Drymo brown, and Xantho fair,
And sweet Phyllodoce with long dishevelled hair;
Cydippe with Lycorias, one a maid,
And one that once had called Lucina's aid;°   480
Clio and Beroe, from one father both;
Both girt with gold, and clad in particoloured cloth.
Opis the meek, and Deiopeia proud;
Nisaea softly, with Ligea loud;
Thalia joyous, Ephyre the sad,
And Arethusa, once Diana's maid,
But now (her quiver left) to love betrayed.
To these Clymene the sweet theft declares
Of Mars, and Vulcan's unavailing cares,°
And all the rapes of gods, and every love,   490
From ancient Chaos down to youthful Jove.
     Thus while she sings, the sisters turn the wheel,
Empty the woolly rock, and fill the reel.
A mournful sound again the mother hears;
Again the mournful sound invades the sisters' ears.
Starting at once from their green seats, they rise
Fear in their heart, amazement in their eyes.
But Arethusa, leaping from her bed,
First lifts above the waves her beauteous head,
And crying from afar thus to Cyrene said: 500
'O sister, not with causeless fear possessed!
No stranger voice disturbs thy tender breast.
'Tis Aristaeus, 'tis thy darling son,
Who to his careless mother makes his moan.
Near his paternal stream he sadly stands,°
With downcast eyes, wet cheeks, and folded hands,
Upbraiding heaven from whence his lineage came,
And cruel calls the gods, and cruel thee, by name.'
     Cyrene, moved with love, and seized with fear,
Cries out, 'Conduct my son, conduct him here: 510
'Tis lawful for the youth, derived from gods,
To view the secrets of our deep abodes.'
At once she waved her hand on either side,
At once the ranks of swelling streams divide.
Two rising heaps of liquid crystal stand,

And leave a space betwixt of empty sand.
Thus safe received, the downward track he treads,
Which to his mother's watery palace leads.
With wondering eyes he views the secret store
Of lakes, that, pent in hollow caverns, roar;   520
He hears the crackling sounds of coral woods,
And sees the secret source of subterranean floods.
And where, distinguished in their several cells,
The fount of Phasis and of Lycus dwells;
Where swift Enipeus in his bed appears,
And Tiber his majestic forehead rears;
Whence Anio flows, and Hypanis profound
Breaks through the opposing rocks with raging sound.
Where Po first issues from his dark abodes,
And awful in his cradle, rules the floods: 530
Two golden horns on his large front he wears,
And his grim face a bull's resemblance bears;
With rapid course he seeks the sacred main,
And fattens, as he runs, the fruitful plain.°

    Now to the court arrived, the admiring son
Beholds the vaulted roofs of pory stone,°
Now to his mother goddess tells his grief,
Which she with pity hears, and promises relief.
The officious nymphs, attending in a ring,
With waters drawn from their perpetual spring,   540
From earthly dregs his body purify,
And rub his temples, with fine towels, dry;
Then load the tables with a liberal feast,
And honour with full bowls their friendly guest.
The sacred altars are involved in smoke,
And the bright choir their kindred gods invoke.
Two bowls the mother fills with Lydian wine;
Then thus: 'Let these be poured, with rites divine,
To the great authors of our watery line.
To father Ocean, this; and this,' she said,   550
'Be to the nymphs his sacred sisters paid,
Who rule the watery plains, and hold the woodland shade.'
She sprinkled thrice, with wine, the Vestal fire;°
Thrice to the vaulted roof the flames aspire.
Raised with so blest an omen, she begun,

With words, like these, to cheer her drooping son:
'In the Carpathian bottom, makes abode°
The shepherd of the seas, a prophet and a god.
High o'er the main in watery pomp he rides,
His azure car and finny coursers guides;   560
Proteus his name. To his Pallenian port°
I see from far the weary god resort.
Him not alone we river gods adore,
But aged Nereus hearkens to his lore.°
With sure foresight, and with unerring doom,
He sees what is, and was, and is to come.
This Neptune gave him, when he gave to keep
His scaly flocks, that graze the watery deep.
Implore his aid; for Proteus only knows
The secret cause, and cure, of all thy woes.   570
But first the wily wizard must be caught;
For unconstrained, he nothing tells for nought;
Nor is with prayers, or bribes, or flattery bought.
Surprise him first, and with hard fetters bind;
Then all his frauds will vanish into wind.
I will myself conduct thee on thy way:
When next the southing sun inflames the day,°
When the dry herbage thirsts for dews in vain,
And sheep, in shades, avoid the parching plain;
Then will I lead thee to his secret seat,   580
When weary with his toil, and scorched with heat,
The wayward sire frequents his cool retreat.
His eyes with heavy slumber overcast;
With force invade his limbs, and bind him fast.
Thus surely bound, yet be not over bold:
The slippery god will try to loose his hold,
And various forms assume, to cheat thy sight,
And with vain images of beasts affright;
With foamy tusks will seem a bristly boar,
Or imitate the lion's angry roar;   590
Break out in crackling flames to shun thy snare,
Or hiss a dragon, or a tiger stare;
Or with a wile thy caution to betray,
In fleeting streams attempt to slide away.
But thou, the more he varies forms, beware

To strain his fetters with a stricter care.
Till, tiring all his arts, he turns again
To his true shape, in which he first was seen.'
    This said, with nectar she her son anoints,
Infusing vigour through his mortal joints: 600
Down from his head the liquid odours ran;
He breathed of heaven, and looked above a man.
    Within a mountain's hollow womb, there lies
A large recess, concealed from human eyes,
Where heaps of billows, driven by wind and tide,
In form of war, their watery ranks divide,
And there, like sentries set, without the mouth abide:
A station safe for ships, when tempests roar,
A silent harbour, and a covered shore.
Secure within resides the various god, 610
And draws a rock upon his dark abode.
Hither with silent steps, secure from sight,
The goddess guides her son, and turns him from the light:
Herself, involved in clouds, precipitates her flight.
    'Twas noon; the sultry Dog-star from the sky
Scorched Indian swains; the rivelled grass was dry;
The sun with flaming arrows pierced the flood,
And darting to the bottom, baked the mud;
When weary Proteus, from the briny waves,
Retired for shelter to his wonted caves. 620
His finny flocks about their shepherd play,
And rolling round him, spurt the bitter sea.
Unwieldily they wallow first in ooze,
Then in the shady covert seek repose.
Himself, their herdsman, on the middle mount,
Takes of his mustered flocks a just account.
So, seated on a rock, a shepherd's groom
Surveys his evening flocks returning home,
When lowing calves and bleating lambs, from far,
Provoke the prowling wolf to nightly war. 630
The occasion offers, and the youth complies:
For scarce the weary god had closed his eyes,
When rushing on with shouts he binds in chains
The drowsy prophet, and his limbs constrains.
He, not unmindful of his usual art,
First in dissembled fire attempts to part:°

Then roaring beasts and running streams he tries,
And wearies all his miracles of lies;
But, having shifted every form to scape,
Convinced of conquest, he resumed his shape,   640
And thus at length in human accent spoke:
'Audacious youth! what madness could provoke
A mortal man to invade a sleeping god?
What business brought thee to my dark abode?'
   To this the audacious youth: 'Thou know'st full well
My name and business, god, nor need I tell.
No man can Proteus cheat; but, Proteus, leave
Thy fraudful arts, and do not thou deceive.
Following the gods' command, I come to implore
Thy help, my perished people to restore.'   650
   The seer, who could not yet his wrath assuage,
Rolled his green eyes, that sparkled with his rage
And gnashed his teeth, and cried, 'No vulgar god
Pursues thy crimes, nor with a common rod.
Thy great misdeeds have met a due reward;
And Orpheus' dying prayers at length are heard.
For crimes not his, the lover lost his life,°
And at thy hands requires his murdered wife:
Nor (if the fates assist not) canst thou scape
The just revenge of that intended rape.   660
To shun thy lawless lust, the dying bride,
Unwary, took along the river's side,
Nor at her heels perceived the deadly snake,
That kept the bank, in covert of the brake.
But all her fellow-nymphs and mountains tear
With loud laments, and break the yielding air:
The realms of Mars remurmured all around,
And echoes to the Athenian shores rebound.
The unhappy husband, husband now no more,
Did on his tuneful harp his loss deplore,   670
And sought his mournful mind with music to restore.
On thee, dear wife, in deserts all alone
He called, sighed, sung: his griefs with day begun,
Nor were they finished with the setting sun.
E'en to the dark dominions of the night
He took his way, through forests void of light,
And dared amidst the trembling ghosts to sing,

And stood before the inexorable king.°
The infernal troops like passing shadows glide,
And listening, crowd the sweet musician's side.   680
Not flocks of birds, when driven by storms or night,
Stretch to the forest with so thick a flight.
Men, matrons, children, and the unmarried maid,
The mighty hero's more majestic shade,
And youths, on funeral piles before their parents laid.
All these Cocytus bounds with squalid reeds,°
With muddy ditches, and with deadly weeds;
And baleful Styx encompasses around,°
With nine slow circling streams, the unhappy ground.
E'en from the depths of hell the damned advance;   690
The infernal mansions, nodding, seem to dance;
The gaping three-mouthed dog forgets to snarl;
The Furies hearken, and their snakes uncurl;
Ixion seems no more his pains to feel,
But leans attentive on his standing wheel.
    All dangers past, at length the lovely bride
In safety goes, with her melodious guide,
Longing the common light again to share,
And draw the vital breath of upper air—
He first; and close behind him followed she;   700
For such was Proserpine's severe decree.
When strong desires the impatient youth invade,
By little caution and much love betrayed:
A fault which easy pardon might receive,
Were lovers judges, or could hell forgive:
For near the confines of ethereal light,
And longing for the glimmering of a sight,
The unwary lover cast his eyes behind,
Forgetful of the law, nor master of his mind.
Straight all his hopes exhaled in empty smoke,   710
And his long toils were forfeit for a look.
Three flashes of blue lightning gave the sign
Of covenants broke; three peals of thunder join.
Then thus the bride: "What fury seized on thee,
Unhappy man! to lose thyself and me?
Dragged back again by cruel destinies,
An iron slumber shuts my swimming eyes.
And now, farewell! Involved in shades of night,

For ever I am ravished from thy sight.
In vain I reach my feeble hands, to join   720
In sweet embraces—ah! no longer thine!"
She said; and from his eyes the fleeting fair
Retired like subtle smoke dissolved in air,
And left her hopeless lover in despair.
In vain, with folding arms, the youth essayed
To stop her flight, and strain the flying shade:
He prays, he raves, all means in vain he tries,
With rage inflamed, astonished with surprise;
But she returned no more, to bless his longing eyes.
Nor would the infernal ferryman once more   730
Be bribed to waft him to the further shore.
What should he do, who twice had lost his love?
What notes invent? what new petitions move?
Her soul already was consigned to fate,
And shivering in the leaky sculler sate.
For seven continued months, if fame say true,
The wretched swain his sorrows did renew:
By Strymon's freezing streams he sat alone:°
The rocks were moved to pity with his moan;
Trees bent their heads to hear him sing his wrongs: 740
Fierce tigers couched around, and lolled their fawning tongues.

So, close in poplar shades, her children gone,
The mother nightingale laments alone,
Whose nest some prying churl had found, and thence,
By stealth, conveyed the unfeathered innocence.
But she supplies the night with mournful strains;
And melancholy music fills the plains.

Sad Orpheus thus his tedious hours employs,
Averse from Venus, and from nuptial joys.
Alone he tempts the frozen floods, alone   750
The unhappy climes, where spring was never known:
He mourned his wretched wife, in vain restored,
And Pluto's unavailing boon deplored.

The Thracian matrons—who the youth accused
Of love disdained, and marriage rites refused—
With furies and nocturnal orgies fired,
At length against his sacred life conspired.
Whom e'en the savage beasts had spared, they killed,
And strewed his mangled limbs about the field.

Then, when his head, from his fair shoulders torn,   760
Washed by the waters, was on Hebrus borne,°
E'en then his trembling tongue invoked his bride;
With his last voice, "Eurydice," he cried.
"Eurydice," the rocks and river-banks replied.'
This answer Proteus gave; nor more he said,
But in the billows plunged his hoary head;
And where he leaped the waves in circles widely spread.

   The nymph returned her drooping son to cheer,
And bade him banish his superfluous fear:
'For now,' said she, 'the cause is known, from whence   770
Thy woe succeeded, and for what offence.
The nymphs, companions of the unhappy maid,
This punishment upon thy crimes have laid;
And sent a plague among thy thriving bees.—
With vows and suppliant prayers their powers appease:
The soft Napaean race will soon repent°
Their anger, and remit the punishment.
The secret in an easy method lies;
Select four brawny bulls for sacrifice,
Which on Lycaeus graze without a guide;   780
Add four fair heifers yet in yoke untried.
For these, four altars in their temple rear,
And then adore the woodland powers with prayer.
From the slain victims pour the streaming blood,
And leave their bodies in the shady wood:
Nine mornings thence, Lethaean poppy bring,
To appease the manes of the poets' king:°
And to propitiate his offended bride,
A fatted calf and a black ewe provide:
This finished, to the former woods repair.'   790
His mother's precepts he performs with care;
The temple visits, and adores with prayer;
Four altars raises; from his herd he culls
For slaughter four the fairest of his bulls:
Four heifers from his female store he took,
All fair, and all unknowing of the yoke.
Nine mornings thence, with sacrifice and prayers,
The powers atoned, he to the grove repairs.
Behold a prodigy! for from within
The broken bowels, and the bloated skin,   800

A buzzing noise of bees his ears alarms:
Straight issue through the sides assembling swarms:
Dark as a cloud they make a wheeling flight,
Then on a neighbouring tree, descending, light:
Like a large cluster of black grapes they show,
And make a large dependence from the bough.
    Thus have I sung of fields, and flocks, and trees,
And of the waxen work of labouring bees;
While mighty Caesar, thundering from afar,°
Seeks on Euphrates' banks the spoils of war;   810
With conquering arms asserts his country's cause,
With arts of peace the willing people draws;
On the glad earth the golden age renews,
And his great father's path to heaven pursues;
While I at Naples pass my peaceful days,
Affecting studies of less noisy praise;
And, bold through youth, beneath the beechen shade,
The lays of shepherds and their loves, have played.

Postscript to the Reader appended to the Aeneid

What Virgil wrote in the vigour of his age, in plenty and at ease, I have undertaken to translate in my declining years; struggling with wants, oppressed with sickness, curbed in my genius, liable to be misconstrued in all I write; and my judges, if they are not very equitable, already prejudiced against me by the lying character which has been given them of my morals. Yet steady to my principles, and not dispirited with my afflictions, I have, by the blessing of God on my endeavours, overcome all difficulties and, in some measure, acquitted myself of the debt which I owed the public when I undertook this work. In the first place, therefore, I thankfully acknowledge to the almighty power the assistance he had given me in the beginning, the prosecution, and conclusion of my present studies, which are more happily performed than I could have promised to myself, when I laboured under such discouragements. For, what I have done, imperfect as it is for want of health and leisure to correct it, will be judged in after-ages, and possibly in the present, to be no dishonour to my native country, whose language and poetry would be more esteemed abroad, if they were better understood. Somewhat (give me leave to say) I have added to both of them in the choice of words, and

harmony of numbers, which were wanting, especially the last, in all
our poets, even in those who, being endued with genius, yet have not
cultivated their mother-tongue with sufficient care; or, relying on the
beauty of their thoughts, have judged the ornament of words, and
sweetness of sound, unnecessary. One is for raking in Chaucer (our
English Ennius°) for antiquated words, which are never to be revived,
but when sound or significance is wanting in the present language. But
many of his deserve not this redemption, any more than the crowds of
men who daily die, or are slain for sixpence in a battle, merit to be
restored to life, if a wish could revive them. Others have no ear for
verse, nor choice of words, nor distinction of thoughts; but mingle
farthings° with their gold to make up the sum. Here is a field of satire
opened to me: but since the Revolution,° I have wholly renounced that
talent. For who would give physic to the great, when he is uncalled?
To do his patient no good, and endanger himself for his prescription?
Neither am I ignorant but I may justly be condemned for many of
those faults of which I have too liberally arraigned others.

Cynthius aurem

vellit et admonuit.°

[the Cynthian plucked my ear and warned me]

'Tis enough for me, if the government will let me pass unquestioned. In the meantime, I am obliged, in gratitude, to return my
thanks to many of them, who have not only distinguished me from
others of the same party, by a particular exception of grace, but,
without considering the man, have been bountiful to the poet: have
encouraged Virgil to speak such English as I could teach him, and
rewarded his interpreter for the pains he has taken in bringing him
over into Britain, by defraying the charges of his voyage. Even
Cerberus, when he had received the sop, permitted Aeneas to pass
freely to Elysium.° Had it been offered me, and I had refused it, yet
still some gratitude is due to such who were willing to oblige me; but
how much more to those from whom I have received the favours
which they have offered to one of a different persuasion! Amongst
whom I cannot omit naming the Earls of Derby and of Peterborough.°
To the first of these I have not the honour to be known; and therefore
his liberality was as much unexpected as it was undeserved. The
present Earl of Peterborough has been pleased long since to accept the
tenders of my service: his favours are so frequent to me that I receive
them almost by prescription. No difference of interests° or opinion
have been able to withdraw his protection from me; and I might justly

be condemned for the most unthankful of mankind, if I did not always preserve for him a most profound respect and inviolable gratitude. I must also add that if the last *Aeneid*° shine amongst its fellows, 'tis owing to the commands of Sir William Trumball,° one of the principal secretaries of state, who recommended it, as his favourite, to my care; and for his sake particularly I have made it mine. For who would confess weariness when he enjoined a fresh labour? I could not but invoke the assistance of a muse, for this last office:

> Extremum hunc, Arethusa . . .
> neget quis carmina Gallo?°

[my last task, Arethusa . . . who would refuse Gallus songs?]

Neither am I to forget the noble present which was made me by Gilbert Dolben,° esq., the worthy son of the late Archbishop of York, who, when I began this work, enriched me with all the several editions of Virgil, and all the commentaries of those editions in Latin; amongst which, I could not but prefer the Delphin,° as the last, the shortest, and the most judicious. Fabrini° I had also sent me from Italy; but either he understands Virgil very imperfectly, or I have no knowledge of my author.

Being invited by that worthy gentleman, Sir William Bowyer,° to Denham Court, I translated the first *Georgic* at his house, and the greatest part of the last *Aeneid*. A more friendly entertainment no man ever found. No wonder, therefore, if both those versions surpass the rest, and own the satisfaction I received in his converse, with whom I had the honour to be bred in Cambridge, and in the same college. The seventh *Aeneid* was made English at Burleigh, the magnificent abode of the Earl of Exeter.° In a village belonging to his family I was born; and under his roof I endeavoured to make that *Aeneid* appear in English with as much lustre as I could; though my author has not given the finishing strokes either to it, or to the eleventh, as I perhaps could prove in both, if I durst presume to criticise my master.

By a letter from William Walsh° of Abberley, esq. (who has so long honoured me with his friendship and who, without flattery, is the best critic of our nation), I have been informed that his grace the Duke of Shrewsbury° has procured a printed copy of the *Pastorals*, *Georgics*, and six first *Aeneid*s, from my bookseller, and has read them in the country, together with my friend. This noble person having been pleased to give them a commendation, which I presume not to insert, has made me vain enough to boast of so great a favour, and to think I have succeeded beyond my hopes; the character of his excellent judgment, the acuteness of his wit, and his general knowledge of good letters,° being known as well to all the world as the sweetness of his disposition, his humanity, his easiness of access, and desire of obliging those who stand in need of his protection, are known to all who have approached him; and to me in particular, who have formerly had the honour of his conversation. Whoever has given the world the translation of part of the third *Georgic*, which he calls 'The Power of Love',° has put me to sufficient pains to make my own not inferior to his; as my Lord Roscommon's 'Silenus'° had formerly given me the same trouble. The most ingenious Mr Addison° of Oxford has also been as troublesome to me as the other two, and on the same account. After his 'Bees', my latter swarm is scarcely worth the hiving. Mr Cowley's praise of a country life° is excellent, but it is rather an imitation of Virgil than a version. That I have recovered, in some measure, the health which I had lost by too much application to this work, is owing, next to God's mercy, to the skill and care of Dr Gibbons and Dr Hobbs, the two ornaments of their profession, whom I can only pay by this acknowledgment. The whole faculty has always been ready to oblige me; and the only one of them° who endeavoured to defame me had it not in his power. I desire pardon from my readers for saying so much in relation to myself, which concerns not them; and with my acknowledgments to all my subscribers, have only to add that the few notes which follow are *par manière d'acquit* [for form's sake], because I had obliged myself by articles to do somewhat of that kind. These scattering observations are rather guesses at my author's meaning in some passages than proofs that so he meant. The unlearned may have recourse to any poetical dictionary in English for the names of persons, places, or fables, which the learned need not: but that little which I say is either new or necessary. And the first of these qualifications never fails to invite a reader, if not to please him.

Alexander's Feast
or The Power of Music:
an Ode in honour of St Cecilia's Day

I

'Twas at the royal feast, for Persia won
          By Philip's warlike son:
          Aloft, in awful state,
          The godlike hero sate
          On his imperial throne.
    His valiant peers were placed around;
Their brows with roses and with myrtles bound:
      (So should desert in arms be crowned.)
The lovely Thais, by his side,°
Sat like a blooming eastern bride,   10
In flower of youth and beauty's pride.
          Happy, happy, happy pair!
          None but the brave,
          None but the brave,
          None but the brave deserves the fair.

CHORUS

Happy, happy, happy pair!
None but the brave,
None but the brave,
None but the brave deserves the fair.

II

    Timotheus, placed on high°   20
          Amid the tuneful choir,
          With flying fingers touched the lyre:
    The trembling notes ascend the sky,
              And heavenly joys inspire.
The song began from Jove,
Who left his blissful seats above,
(Such is the power of mighty love.)
A dragon's fiery form belied the god;°
Sublime on radiant spires he rode,°
      When he to fair Olympia pressed,°   30
      And while he sought her snowy breast;

Then, round her slender waist he curled,
And stamped an image of himself, a sovereign of the world.
The listening crowd admire the lofty sound,
'A present deity' they shout around;
'A present deity' the vaulted roofs rebound.
   With ravished ears,
   The monarch hears;
   Assumes the god,
   Affects to nod,   40
  And seems to shake the spheres.

CHORUS

With ravished ears,
The monarch hears;
Assumes the god,
Affects to nod,
And seems to shake the spheres.

III

The praise of Bacchus, then, the sweet musician sung;
 Of Bacchus ever fair, and ever young.
  The jolly god in triumph comes;
  Sound the trumpets, beat the drums;   50
   Flushed with a purple grace°
   He shows his honest face:°
Now give the hautboys breath; he comes, he comes.°
  Bacchus, ever fair and young,
   Drinking joys did first ordain;
  Bacchus' blessings are a treasure,
  Drinking is the soldier's pleasure;
   Rich the treasure,
   Sweet the pleasure,
  Sweet is pleasure after pain.   60

CHORUS

Bacchus' blessings are a treasure,
Drinking is the soldier's pleasure;
* Rich the treasure,*
* Sweet the pleasure,*
Sweet is pleasure after pain.

IV

    Soothed with the sound, the king grew vain:
        Fought all his battles o'er again;
And thrice he routed all his foes, and thrice he slew the slain.
    The master saw the madness rise,°
    His glowing cheeks, his ardent eyes;°   70
    And while he heaven and earth defied,
    Changed his hand, and checked his pride.
        He chose a mournful muse,
        Soft pity to infuse,
    He sung Darius great and good,
        By too severe a fate,
    Fallen, fallen, fallen, fallen,
        Fallen from his high estate,
            And weltering in his blood:
    Deserted at his utmost need,°   80
    By those his former bounty fed;
    On the bare earth exposed he lies,
    With not a friend to close his eyes.

With downcast looks the joyless victor sate,
    Revolving in his altered soul
        The various turns of chance below;
    And now and then, a sigh he stole,
        And tears began to flow.

CHORUS

Revolving in his altered soul
    The various turns of chance below;   90
And now and then, a sigh he stole,
    And tears began to flow.

V

    The mighty master smiled, to see
    That love was in the next degree;
    'Twas but a kindred-sound to move,
    For pity melts the mind to love.
        Softly sweet, in Lydian measures,°
        Soon he soothed his soul to pleasures:
        War, he sung, is toil and trouble;
        Honour, but an empty bubble;   100

Never ending, still beginning,
Fighting still, and still destroying:
  If the world be worth thy winning,
Think, O think it worth enjoying;
  Lovely Thais sits beside thee,
  Take the good the gods provide thee.

The many rend the skies with loud applause;
So love was crowned, but music won the cause.
  The prince, unable to conceal his pain,
      Gazed on the fair,   110
      Who caused his care,
    And sighed and looked, sighed and looked,
  Sighed and looked, and sighed again;
At length, with love and wine at once oppressed,
The vanquished victor sunk upon her breast.

CHORUS

  The prince, unable to conceal his pain,
      Gazed on the fair,
      Who caused his care,
    And sighed and looked, sighed and looked,
  Sighed and looked, and sighed again;   120
At length, with love and wine at once oppressed,
The vanquished victor sunk upon her breast.

VI

Now strike the golden lyre again;
A louder yet, and yet a louder strain.
Break his bonds of sleep asunder,
And rouse him, like a rattling peal of thunder.
      Hark, hark! the horrid sound
      Has raised up his head;
      As awaked from the dead,
      And amazed, he stares around.   130
Revenge, revenge! Timotheus cries,
  See the furies arise;
    See the snakes, that they rear,
    How they hiss in their hair,
  And the sparkles that flash from their eyes!

      Behold a ghastly band,
      Each a torch in his hand!
Those are Grecian ghosts, that in battle were slain,
        And, unburied, remain
        Inglorious on the plain: 140
      Give the vengeance due
      To the valiant crew.
Behold how they toss their torches on high,
      How they point to the Persian abodes,
And glittering temples of their hostile gods.
The princes applaud, with a furious joy,
And the king seized a flambeau with zeal to destroy;°
      Thais led the way,
      To light him to his prey,
And, like another Helen, fired another Troy.   150

CHORUS

And the king seized a flambeau with zeal to destroy;
      Thais led the way,
      To light him to his prey,
And, like another Helen, fired another Troy.

VII

      Thus, long ago,
    Ere heaving bellows learned to blow,
      While organs yet were mute,
    Timotheus, to his breathing flute,
      And sounding lyre,
Could swell the soul to rage, or kindle soft desire.   160
      At last divine Cecilia came,
      Inventress of the vocal frame;°
The sweet enthusiast, from her sacred store,
      Enlarged the former narrow bounds,
      And added length to solemn sounds,
With nature's mother-wit, and arts unknown before.
      Let old Timotheus yield the prize,
        Or both divide the crown;
      He raised a mortal to the skies,
        She drew an angel down.   170

GRAND CHORUS

At last divine Cecilia came,
Inventress of the vocal frame:
The sweet enthusiast, from her sacred store,
   Enlarged the former narrow bounds,
   And added length to solemn sounds,
With nature's mother-wit, and arts unknown before.
   Let old Timotheus yield the prize,
   Or both divide the crown;
   He raised a mortal to the skies,
   She drew an angel down.   180

To Mr Motteux on his Tragedy called Beauty in Distress

'Tis hard, my friend, to write in such an age,
As damns not only poets, but the stage.°
That sacred art, by heaven itself infused,
Which Moses, David, Solomon, have used,
Is now to be no more: the muses' foes
Would sink their maker's praises into prose.
Were they content to prune the lavish vine
Of straggling branches, and improve the wine,
Who but a madman, would his faults defend?
All would submit; for all but fools will mend.   10
But when to common sense they give the lie,
And turn distorted words to blasphemy,
They give the scandal; and the wise discern,
Their glosses teach an age too apt to learn.
What I have loosely, or profanely, writ,
Let them to fires (their due desert) commit:
Nor, when accused by me, let *them* complain;
Their faults, and not their function, I arraign.
Rebellion, worse than witchcraft, they pursued;°
The pulpit preached the crime, the people rued. 20
The stage was silenced; for the saints would see
In fields performed their plotted tragedy.
But let us first reform, and then so live,
That we may teach our teachers to forgive;
Our desk be placed below their lofty chairs,
Ours be the practice, as the precept theirs.

The moral part, at least we may divide,
Humility reward, and punish pride;
Ambition, interest, avarice, accuse;
These are the province of the tragic muse. 30
These hast thou chosen; and the public voice
Has equalled thy performance with thy choice.
Time, action, place, are so preserved by thee,
That e'en Corneille might with envy see°
The alliance of his tripled unity.°
Thy incidents, perhaps, too thick are sown,
But too much plenty is thy fault alone.
At least but two can that good crime commit,
Thou in design, and Wycherley in wit.°
Let thy own Gauls condemn thee, if they dare, 40
Contented to be thinly regular:
Born there, but not for them, our fruitful soil
With more increase rewards thy happy toil.
Their tongue, enfeebled, is refined so much,
That, like pure gold, it bends at every touch.
Our sturdy Teuton yet will art obey,
More fit for manly thought, and strengthened with allay.°
But whence art thou inspired, and thou alone,
To flourish in an idiom not thy own?
It moves our wonder, that a foreign guest 50
Should overmatch the most, and match the best.
In under-praising, thy deserts I wrong;
Here find the first deficience of our tongue:
Words, once my stock, are wanting, to commend
So great a poet, and so good a friend.

FABLES ANCIENT AND MODERN

PREFACE

'Tis with a poet as with a man who designs to build, and is very exact, as he supposes, in casting up the cost beforehand; but, generally speaking, he is mistaken in his account, and reckons short of the expense he first intended. He alters his mind as the work proceeds, and will have this or that convenience more, of which he had not thought when he began. So has it happened to me; I have built a house where I intended but a lodge. Yet with better success than a certain nobleman,° who beginning with a dog-kennel never lived to finish the palace he had contrived.

From translating the first of Homer's *Iliad* (which I intended as an essay to the whole work) I proceeded to the translation of the twelfth book of Ovid's *Metamorphoses*, because it contains, among other things, the causes, the beginning, and ending of the Trojan War. Here I ought in reason to have stopped, but the speeches of Ajax and Ulysses lying next in my way, I could not balk 'em. When I had compassed them, I was so taken with the former part of the fifteenth book (which is the masterpiece of the whole *Metamorphoses*) that I enjoined myself the pleasing task of rendering it into English. And now I found, by the number of my verses, that they began to swell into a little volume, which gave me an occasion of looking backward on some beauties of my author in his former books. There occurred to me the *Hunting of the Boar*, *Cinyras and Myrrha*, the good-natured story of *Baucis and Philemon*, with the rest, which I hope I have translated closely enough and given them the same turn of verse which they had in the original; and this, I may say without vanity, is not the talent of every poet. He who has arrived the nearest to it is the ingenious and learned Sandys,° the best versifier of the former age; if I may properly call it by that name, which was the former part of this concluding century. For Spenser and Fairfax both flourished in the reign of Queen Elizabeth: great masters in our language, and who saw much further into the beauties of our numbers than those who immediately followed them. Milton was the poetical son of Spenser, and Mr Waller° of Fairfax;° for we have our lineal descents and clans as well as other families. Spenser more than once insinuates° that the soul of Chaucer was transfused into his body, and that he was begotten by him two hundred years after his decease. Milton has acknowledged to me that Spenser was his original; and many besides myself have heard our famous Waller own that he derived the harmony of his numbers from the *Godfrey of Bulloigne* which was turned into English by Mr Fairfax. But to return: having done with Ovid for this time, it came into my mind that our old English poet Chaucer in many things resembled him, and that with no disadvantage on the side of the modern author, as I shall endeavour to prove when I compare them. And as I am, and always have been studious to promote the honour of my native country, so I soon resolved to put their merits to the trial by turning some of the *Canterbury Tales* into our language, as it is now refined; for by this means both the poets being set in the same light, and dressed in the same English habit, story to be compared with story, a certain judgement may be made betwixt them by the reader, without obtruding my opinion on him. Or if I seem partial to my countryman, and predecessor in the laurel, the friends of antiquity are not few; and besides many of the learned, Ovid has almost all the *beaux* and the whole fair sex his declared patrons. Perhaps I have assumed somewhat more to myself than they allow me, because I have adventured to sum up the evidence; but the readers are the jury, and their privilege remains entire to decide according to the merits of the cause; or, if they please, to bring it to another hearing before some other court. In the meantime, to follow the thread of my discourse (as thoughts, according to Mr Hobbes, have always some connexion°), so from Chaucer I was led to think on Boccace,° who was not only his contemporary but also pursued the same studies; wrote novels in prose, and many works in verse; particularly is said to have invented the octave rhyme,° or stanza of eight lines, which ever since has been maintained by the practice of all Italian writers, who are, or at least assume the title of, heroic poets. He and Chaucer, among other things, had this in common, that they refined their mother tongues; but with this difference, that Dante had begun to file their language, at least in verse, before the time of Boccace, who likewise received no little help from his master Petrarch. But the reformation of their prose was wholly owing to Boccace himself, who is yet the standard of purity in the Italian tongue, though many of his phrases are become obsolete, as in process of time it must needs happen. Chaucer (as you have formerly been told by our learned Mr Rymer°) first adorned and amplified our barren tongue from the Provençal, which was then the most polished of all the modern languages; but this subject has been copiously treated by that great critic, who deserves no little commendation from us his country-

men. For these reasons of time, and resemblance of genius in Chaucer
and Boccace, I resolved to join them in my present work; to which I
have added some original papers of my own, which whether they are
equal or inferior to my other poems, an author is the most improper
judge; and therefore I leave them wholly to the mercy of the reader. I
will hope the best, that they will not be condemned; but if they should,
I have the excuse of an old gentleman, who mounting on horseback
before some ladies, when I was present, got up somewhat heavily
but desired of the fair spectators that they would count forescore
and eight before they judged him. By the mercy of God, I am already
come within twenty years of his number, a cripple in my limbs, but
what decays are in my mind, the reader must determine. I think myself as vigorous as ever in the faculties of my soul, excepting only my
memory, which is not impaired to any great degree; and if I lose not
more of it, I have no great reason to complain. What judgement I had
increases rather than diminishes, and thoughts, such as they are, come
crowding in so fast upon me that my only difficulty is to choose or to
reject, to run them into verse or to give them the other harmony of
prose; I have so long studied and practised both that they are grown
into a habit and become familiar to me. In short, though I may lawfully plead some part of the old gentleman's excuse, yet I will reserve
it till I think I have greater need and ask no grains of allowance for the
faults of this my present work but those which are given of course to
human frailty. I will not trouble my reader with the shortness of time
in which I writ it, or the several intervals of sickness. They who think
too well of their own performances are apt to boast in their prefaces
how little time their works have cost them, and what other business of
more importance interfered, but the reader will be as apt to ask the
question, why they allowed not a longer time to make their works
more perfect and why they had so despicable an opinion of their
judges as to thrust their indigested stuff upon them, as if they
deserved no better.

With this account of my present undertaking, I conclude the first
part of this discourse. In the second part, as at a second sitting, though
I alter not the draft, I must touch the same features over again and
change the dead-colouring° of the whole. In general I will only say
that I have written nothing which savours of immorality or profaneness; at least, I am not conscious to myself of any such intention. If
there happen to be found an irreverent expression, or a thought too
wanton, they are crept into my verses through my inadvertency. If the
searchers find any in the cargo, let them be staved° or forfeited like

contraband goods; at least, let their authors be answerable for them, as being but imported merchandise and not of my own manufacture. On the other side, I have endeavoured to choose such fables, both ancient and modern, as contain in each of them some instructive moral, which I could prove by induction, but the way is tedious; and they leap foremost into sight, without the reader's trouble of looking after them. I wish I could affirm with a safe conscience that I had taken the same care in all my former writings; for it must be owned, that supposing verses are never so beautiful or pleasing, yet if they contain anything which shocks religion or good manners, they are at best what Horace says of good numbers without good sense, *versus inopes rerum, nugaeque canorae* [lines lacking substance, melodious trifles].° Thus far, I hope, I am right in court, without renouncing to my other right of self-defence where I have been wrongfully accused and my sense wire-drawn into blasphemy or bawdry, as it has often been by a religious lawyer in a late pleading against the stage, in which he mixes truth with falsehood, and has not forgotten the old rule of calumniating strongly that something may remain.°

I resume the thread of my discourse with the first of my translations, which was the first *Iliad* of Homer. If it shall please God to give me longer life and moderate health, my intentions are to translate the whole *Iliad*; provided still that I meet with those encouragements from the public which may enable me to proceed in my undertaking with some cheerfulness. And this I dare assure the world beforehand, that I have found by trial Homer a more pleasing task than Virgil (though I say not the translation will be less laborious); for the Grecian is more according to my genius than the Latin poet. In the works of the two authors we may read their manners and natural inclinations, which are wholly different. Virgil was of a quiet, sedate temper; Homer was violent, impetuous, and full of fire. The chief talent of Virgil was propriety of thoughts and ornament of words; Homer was rapid in his thoughts, and took all the liberties both of numbers and of expressions which his language and the age in which he lived allowed him. Homer's invention was more copious, Virgil's more confined; so that if Homer had not led the way, it was not in Virgil to have begun heroic poetry; for nothing can be more evident than that the Roman poem is but the second part of the *Iliad*, a continuation of the same story and the persons already formed. The manners of Aeneas are those of Hector superadded to those which Homer gave him. The adventures of Ulysses in the *Odyssey*, are imitated in the first six books of Virgil's *Aeneis*; and though the accidents are not the same (which would have argued him of a servile, copying, and total barrenness of invention), yet the seas were the same in which both the heroes wandered; and Dido° cannot be denied to be the poetical daughter of Calypso.° The six latter Books of Virgil's poem, are the four-and-twenty *Iliads* contracted; a quarrel occasioned by a lady, a single combat, battles fought, and a town besieged. I say not this in derogation to Virgil, neither do I contradict anything which I have formerly said in his just praise; for his episodes are almost wholly of his own invention, and the form which he has given to the telling makes the tale his own, even though the original story had been the same. But this proves, however, that Homer taught Virgil to design; and if invention be the first virtue of an epic poet, then the Latin poem can only be allowed the second place. Mr Hobbes,° in the preface to his own bald translation of the *Iliad* (studying poetry as he did mathematics, when it was too late), Mr Hobbes, I say, begins the praise of Homer where he should have ended it. He tells us that the first beauty of an epic poem consists in diction, that is, in the choice of words, and harmony of numbers. Now, the words are the colouring of the work, which in the order of nature is last to be considered. The design, the disposition, the manners, and the thoughts are all before it. Where any of those are wanting or imperfect, so much wants or is imperfect in the imitation of human life, which is in the very definition of a poem. Words indeed, like glaring colours, are the first beauties that arise and strike the sight; but if the draft be false or lame, the figures ill disposed, the manners obscure or inconsistent, or the thoughts unnatural, then the finest colours are but daubing, and the piece is a beautiful monster at the best. Neither Virgil nor Homer were deficient in any of the former beauties, but in this last, which is expression, the Roman poet is at least equal to the Grecian, as I have said elsewhere; supplying the poverty of his language by his musical ear and by his diligence. But to return: our two great poets, being so different in their tempers; one choleric and sanguine, the other phlegmatic and melancholic; that which makes them excel in their several ways is that each of them has followed his own natural inclination, as well in forming the design as in the execution of it. The very heroes show their authors. Achilles is hot, impatient, revengeful, *impiger, iracundus, inexorabilis, acer,* etc. [restless, hot-tempered, unyielding, and fierce].° Aeneas patient, considerate, careful of his people, and merciful to his enemies; ever submissive to the will of heaven, *quo fata trahunt retrahuntque, sequamur* [let us follow where the fates convey us, be it forth or back].° I could please myself with enlarging on this subject, but am forced to defer it to a fitter time. From all I have said, I will only draw this inference, that the action of Homer being more full of vigour than that of Virgil, according to the temper of the writer, is of consequence more pleasing to the reader. One warms you by degrees; the other sets you on fire all at once, and never intermits his heat. 'Tis the same difference which Longinus° makes betwixt the effects of eloquence in Demosthenes and Tully. One persuades; the other commands. You never cool while you read Homer, even not in the second book (a graceful flattery to his countrymen); but he hastens from the ships, and concludes not that book till he has made you an amends by the violent playing of a new machine.° From thence he hurries on his action with variety of events, and ends it in less compass than two months. This vehemence of his, I confess, is more suitable to my temper, and therefore I have translated his first book with greater pleasure than any part of Virgil. But it was not a pleasure without pains. The continual agitations of the spirits must needs be a weakening of any constitution, especially in age; and many pauses are required for refreshment betwixt the heats; the *Iliad* of itself being a third part longer than all Virgil's works together.

This is what I thought needful in this place to say of Homer. I proceed to Ovid and Chaucer; considering the former only in relation to the latter. With Ovid ended the golden age of the Roman tongue; from Chaucer the purity of the English tongue began. The manners of the poets were not unlike; both of them were well-bred, well-natured, amorous, and libertine, at least in their writings, it may be also in their lives. Their studies were the same, philosophy and philology.° Both of them were knowing in astronomy, of which Ovid's books of the Roman Feasts, and Chaucer's Treatise of the Astrolabe, are sufficient witnesses. But Chaucer was likewise an astrologer, as were Virgil, Horace, Persius, and Manilius. Both writ with wonderful facility and clearness; neither were great inventors: for Ovid only copied the Grecian fables, and most of Chaucer's stories were taken from his Italian contemporaries, or their predecessors. Boccace's *Decameron* was first published, and from thence our Englishman has borrowed many° of his *Canterbury Tales*; yet that of Palamon and Arcite° was written in all probability by some Italian wit in a former age, as I shall prove hereafter. The tale of Grizild was the invention of Petrarch; by him sent to Boccace, from whom it came to Chaucer.° *Troilus and Cressida* was also written by a Lombard author,° but much amplified by our English translator, as well as beautified; the genius of our countrymen in general being rather to improve an invention than to invent themselves, as is evident not only in our poetry but in many of our manufactures. I find I have anticipated already and taken up from Boccace before I come to him, but there is so much less behind, and I am of the temper of most kings, *who love to be in debt*, are all for present money no matter how they pay it afterwards. Besides, the nature of a preface is rambling; never wholly out of the way, nor in it. This I have learned from the practice of honest Montaigne, and return at my pleasure to Ovid and Chaucer, of whom I have little more to say. Both of them built on the inventions of other men; yet since Chaucer had something of his own, as *The Wife of Bath's Tale*, *The Cock and the Fox*,° which I have translated, and some others, I may justly give our countryman the precedence in that part, since I can remember nothing of Ovid which was wholly his. Both of them understood the manners; under which name I comprehend the passions and, in a larger sense, the descriptions of persons and their very habits. For an example, I see Baucis and Philemon as perfectly before me as if some ancient painter had drawn them; and all the pilgrims in the *Canterbury Tales*, their humours, their features, and the very dress, as distinctly as if I had supped with them at the Tabard in Southwark. Yet even there too the figures of Chaucer are much more lively and set in a better light; which though I have not time to prove, yet I appeal to the reader, and am sure he will clear me from partiality. The thoughts and words remain to be considered, in the comparison of the two poets, and I have saved myself one-half of that labour by owning that Ovid lived when the Roman tongue was in its meridian; Chaucer, in the dawning of our language. Therefore that part of the comparison stands not on an equal foot, any more than the diction of Ennius° and Ovid, or of Chaucer and our present English. The words are given up as a post not to be defended in our poet, because he wanted the modern art of fortifying. The thoughts remain to be considered: and they are to be measured only by their propriety; that is, as they flow more or less naturally from the persons described, on such and such occasions. The vulgar judges, which are nine parts in ten of all nations, who call conceits and jingles wit, who see Ovid full of them, and Chaucer altogether without them, will think me little less than mad for preferring the Englishman to the Roman. Yet, with their leave, I must presume to say that the things they admire are only glittering trifles, and so far from being witty that in a serious poem they are nauseous, because they are unnatural. Would any man who is ready to die for love, describe his passion like Narcissus? Would he think of *inopem me copia fecit* [plenty has made me poor],° and a dozen more of such expressions, poured on the neck of one another and signifying all the same thing? If this were wit, was this a time to be witty, when the poor wretch was in the agony of death? This is just John Littlewit in *Bartholomew Fair*, who had a conceit (as he tells you) left him in his misery; a miserable conceit.° On these occasions the poet should endeavour to raise pity; but instead of this, Ovid is tickling you to laugh. Virgil never made use of such machines when he was moving you to commiserate the death of Dido;° he would not destroy what he was building. Chaucer makes Arcite violent in his love, and unjust in the pursuit of it; yet when he came to die, he made him think more reasonably. He repents not of his love, for that had altered his character; but acknowledges the injustice of his proceedings, and resigns Emilia to Palamon. What would Ovid have done on this occasion? He would certainly have made Arcite witty on his deathbed. He had complained he was farther off from possession, by being so near, and a thousand such boyisms, which Chaucer rejected as below the dignity of the subject. They who think otherwise would by the same reason prefer Lucan and Ovid to Homer and Virgil, and Martial to all four of them. As for the turn of words, in which Ovid particularly excels all poets, they are sometimes a fault and sometimes a beauty, as they are used properly or improperly; but in strong passions always to be shunned, because passions are serious and will admit no playing. The French have a high value for them; and I confess they are often what they call delicate, when they are introduced with judgement; but Chaucer writ with more simplicity, and followed nature more closely, than to use them. I have thus far, to the best of my knowledge, been an upright judge betwixt the parties in competition, not meddling with the design nor the disposition of it; because the design was not their own, and in the disposing of it they were equal. It remains that I say somewhat of Chaucer in particular.

In the first place, as he is the father of English poetry, so I hold him in the same degree of veneration as the Grecians held Homer, or the Romans Virgil. He is a perpetual fountain of good sense; learned in all sciences; and therefore speaks properly on all subjects. As he knew what to say, so he knows also when to leave off; a continence which is practised by few writers, and scarcely by any of the ancients, excepting Virgil and Horace. One of our late great poets° is sunk in his reputation because he could never forgive° any conceit which came in his way, but swept like a drag-net, great and small. There was plenty enough, but the dishes were ill sorted; whole pyramids of sweetmeats, for boys and women, but little of solid meat, for men. All this proceeded not from any want of knowledge, but of judgement; neither did he want that in discerning the beauties and faults of other poets, but only indulged himself in the luxury of writing, and perhaps knew it was a fault but hoped the reader would not find it. For this reason, though he must always be thought a great poet, he is no longer esteemed a good writer. And for ten impressions which his works have had in so many successive years, yet at present a hundred books are scarcely purchased once a twelvemonth; for, as my last Lord Rochester said, though somewhat profanely, *Not being of God, he could not stand.*°

Chaucer followed nature everywhere; but was never so bold to go beyond her. And there is a great difference of being *Poeta* and *nimis Poeta* [a poet ... too much of a poet], if we may believe Catullus,° as much as betwixt a modest behaviour and affectation. The verse of Chaucer, I confess, is not harmonious to us; but 'tis like the eloquence of one whom Tacitus° commends, it was *auribus istius temporis accommodata* [suited to the ears of that time]. They who lived with him, and some time after him, thought it musical; and it continues so even in our judgement if compared with the numbers of Lydgate and Gower his contemporaries. There is the rude sweetness of a Scotch tune in it, which is natural and pleasing, though not perfect. 'Tis true, I cannot go so far as he who published the last edition of him;° for he would make us believe the fault is in our ears, and that there were really ten syllables in a verse where we find but nine. But this opinion is not worth confuting; 'tis so gross and obvious an error, that common sense (which is a rule in everything but matters of faith and revelation) must convince the reader that equality of numbers in every verse which we call heroic° was either not known, or not always practised in Chaucer's age. It were an easy matter to produce some thousands of his verses which are lame for want of half a foot, and sometimes a whole one, and which no pronunciation can make otherwise. We can only say that he lived in the infancy of our poetry, and that nothing is brought to perfection at the first. We must be children before we grow men. There was an Ennius, and in process of time a Lucilius, and a Lucretius, before Virgil and Horace; even after Chaucer there was a Spenser, a Harrington, a Fairfax, before Waller and Denham were in being; and our numbers were in their nonage till these last appeared. I need say little of his parentage, life, and fortunes; they are to be found at large in all the editions of his works. He was employed abroad, and favoured by Edward the Third, Richard the Second, and Henry the Fourth, and was poet, as I suppose, to all three of them. In Richard's time, I doubt, he was a little dipped° in the rebellion of the commons; and being brother-in-law to John of Gaunt, it was no wonder if he followed the fortunes of that family, and was well with Henry the Fourth when he had deposed his predecessor. Neither is it to be admired that Henry, who was a wise as well as a valiant prince, who claimed by succession, and was sensible that his title was not sound but was rightfully in Mortimer, who had married the heir of York; it was not to be admired, I say, if that great politician should be pleased to have the greatest wit of those times in his interests, and to be the trumpet of his praises. Augustus had given him the example, by the advice of Maecenas, who recommended Virgil and Horace to him; whose praises helped to make him popular while he was alive, and after his death have made him precious to posterity. As for the religion of our poet, he seems to have some little bias towards the opinions of Wyclif, after John of Gaunt his patron; somewhat of which appears in the tale of *Piers Plowman.*° Yet I cannot blame him for inveighing so sharply against the vices of the clergy in his age. Their pride, their ambition, their pomp, their avarice, their worldly interest, deserved the lashes which he gave them, both in that, and in most of his *Canterbury Tales.* Neither has his contemporary Boccace spared them. Yet both those poets lived in much esteem with good and holy men in orders; for the scandal which is given by particular priests, reflects not on the sacred function. Chaucer's Monk, his Canon, and his Friar, took° not from the character of his Good Parson. A satirical poet is the check of the laymen on bad priests. We are only to take care that we involve not the innocent with the guilty in the same condemnation. The good cannot be too much honoured, nor the bad too coarsely used; for the corruption of the best becomes the worst. When a clergyman is whipped, his gown is first taken off, by which the dignity of his order is secured. If he be wrongfully accused, he has his action of slander; and 'tis at the poet's peril if he transgress the law. But they will tell us that all kind of satire, though never so well deserved by particular priests, yet brings the whole order into contempt. Is then the peerage of England anything dishonoured when a peer suffers for his treason? If he be libelled, or any way defamed, he has his *scandalum magnatum*° to punish the offender. They who use this kind of argument, seem to be conscious to themselves of somewhat which has deserved the poet's lash, and are less concerned for their public capacity, than for their private; at least, there is pride at the bottom of their reasoning. If the faults of men in orders are only to be judged among themselves, they are all in some sort parties; for since they say the honour of their order is concerned in every member of it, how can we be sure that they will be impartial judges? How far I may be allowed to speak my opinion in this case, I know not; but I am sure a dispute of this nature caused mischief in abundance betwixt a king of England and an Archbishop of Canterbury; one standing up for the laws of his land, and the other for the honour (as he called it) of God's church; which ended in the murder of the prelate, and in the whipping of his majesty from post to pillar for his penance. The learned and ingenious Dr Drake° has saved me the labour of inquiring into the esteem and reverence which the priests have had of old; and I would rather extend than diminish any part of it. Yet I must needs say, that when a priest provokes me without any occasion given him, I have no reason, unless it be the charity of a Christian, to forgive him: *prior laesit* [he hit first]° is justification sufficient in the civil law. If I answer him in his own language, self-defence, I am sure, must be allowed me; and if I carry it further, even to a sharp recrimination, somewhat may be indulged to human frailty. Yet my resentment has not wrought so far but that I have followed Chaucer in his character of a holy man, and have enlarged on that subject with some pleasure, reserving to myself the right, if I shall think fit hereafter, to describe another sort of priests, such as are more easily to be found than the good Parson; such as have given the last blow to Christianity in this age by a practice so contrary to their doctrine. But this will keep cold till another time. In the meanwhile, I take up Chaucer where I left him. He must have been a man of a most wonderful comprehensive nature, because, as it has been truly observed of him, he has taken into the compass of his *Canterbury Tales* the various manners and humours (as we now call them) of the whole English nation in his age. Not a single character has escaped him. All his Pilgrims are severally distinguished from each other; and not only in their inclinations but in their very physiognomies and persons. Baptisa Porta° could not have described their natures better, than by the marks which the poet gives them. The matter and manner of their tales, and of their telling, are so suited to their different educations, humours, and callings, that each of them would be improper in any other mouth. Even the grave and serious characters are distinguished by their several sorts of gravity. Their discourses are such as belong to their age, their calling, and their breeding; such as are becoming of them, and of them only. Some of his persons are vicious, and some virtuous; some are unlearned, or (as Chaucer calls them) lewd, and some are learned. Even the ribaldry of the low characters is different. The Reeve, the Miller, and the Cook are several men, and distinguished from each other as much as the mincing Lady Prioress, and the broad-speaking gap-toothed Wife of Bath. But enough of this: there is such a variety of game springing up before me that I am distracted in my choice, and know not which to follow. 'Tis sufficient to say according to the proverb, that here is God's plenty. We have our forefathers and great grandames all before us as they were in Chaucer's days; their general characters are still remaining in mankind, and even in England, though they are called by other names than those of monks, and friars, and canons, and lady abbesses, and nuns; for mankind is ever the same, and nothing lost out of nature, though everything is altered. May I have leave to do myself the justice (since my enemies will do me none, and are so far from granting me to be a good poet that they will not allow me so much as to be a Christian, or a moral man), may I have leave, I say, to inform my reader that I have confined my choice to such tales of Chaucer as savour nothing of immodesty. If I had desired more to please than to instruct, the Reeve, the Miller, the Shipman, the Merchant, the Sumner, and, above all, the Wife of Bath, in the Prologue to her Tale, would have procured me as many friends and readers as there are beaux and ladies of pleasure in the town. But I will no more offend against good manners. I am sensible as I ought to be of the scandal I have given by my loose writings, and make what reparation I am able by this public acknowledgement. If anything of this nature, or of profaneness, be crept into these poems, I am so far from defending it, that I disown it. *Totum hoc indictum volo* [I would have all this unsaid]. Chaucer makes another manner of apology for his broad-speaking, and Boccace makes the like; but I will follow neither of them. Our countryman, in the end of his characters, before the *Canterbury Tales*, thus excuses the ribaldry, which is very gross, in many of his novels:

> But first, I pray you, of your courtesy,
> That ye ne arrete it not my villany,
> Though that I plainly speak in this mattere
> To tellen you her words, and eke her chere:
> Ne though I speak her words properly,
> For this ye knowen as well as I,
> Who shall tellen a tale after a man
> He mote rehearse as nye, as ever he can:
> Everich word of it been in his charge,
> *All speke, he, never so rudely, ne large.*
> Or else he mote tellen his tale untrue,
> Or feine things, or find words new:

He may not spare, altho he were his brother,
He mote as well say o word as another.
Christ spake himself full broad in holy Writ,
And well I wote no villany is it.
Eke *Plato* saith, who so can him rede,
The words mote been cousin to the dede.°

Yet if a man should have inquired of Boccace or of Chaucer what need they had of introducing such characters, where obscene words were proper in their mouths, but very undecent to be heard; I know not what answer they could have made. For that reason, such tales shall be left untold by me. You have here a specimen of Chaucer's language, which is so obsolete that his sense is scarce to be understood; and you have likewise more than one example of his unequal numbers, which were mentioned before. Yet many of his verses consist of ten syllables, and the words not much behind our present English; as for example, these two lines in the description of the Carpenter's young wife:

Wincing, she was as is a jolly Colt,
Long as a Mast, and upright as a Bolt.°

I have almost done with Chaucer, when I have answered some objections relating to my present work. I find some people are offended that I have turned these tales into modern English; because they think them unworthy of my pains, and look on Chaucer as a dry, old-fashioned wit, not worth receiving. I have often heard the late Earl of Leicester° say that Mr Cowley himself was of that opinion, who having read him over at my Lord's request, declared he had no taste of him. I dare not advance my opinion against the judgement of so great an author; but I think it fair, however, to leave the decision to the public. Mr Cowley was too modest to set up for a dictator; and being shocked perhaps with his old style, never examined into the depth of his good sense. Chaucer, I confess, is a rough diamond, and must first be polished ere he shines. I deny not likewise that, living in our early days of poetry, he writes not always of a piece, but sometimes mingles trivial things with those of greater moment. Sometimes also, though not often, he runs riot, like Ovid, and knows not when he has said enough. But there are more great wits beside Chaucer whose fault is their excess of conceits, and those ill sorted. An author is not to write all he can, but only all he ought. Having observed this redundancy in Chaucer (as it is an easy matter for a man of ordinary parts to find a fault in one of greater), I have not tied myself to a literal translation; but have often omitted what I judged unnecessary or not of dignity enough to appear in the company of better thoughts. I have presumed further in some places, and added somewhat of my own where I thought my author was deficient, and had not given his thoughts their true lustre for want of words in the beginning of our language. And to this I was the more emboldened, because (if I may be permitted to say it of myself) I found I had a soul congenial to his, and that I had been conversant in the same studies. Another poet in another age may take the same liberty with my writings; if at least they live long enough to deserve correction. It was also necessary sometimes to restore the sense of Chaucer, which was lost or mangled in the errors of the press. Let this example suffice at present; in the story of Palamon and Arcite, where the temple of Diana is described, you find these verses, in all the editions of our author:

> There saw I *Danè* turned unto a Tree,
> I mean not the Goddess *Diane*,
> But *Venus* Daughter, which that hight *Danè*.°

Which after a little consideration I knew was to be reformed into this sense, that Daphne the daughter of Peneus was turned into a tree. I durst not make thus bold with Ovid, lest some future Milbourne° should arise and say, I varied from my author because I understood him not.

But there are other judges who think I ought not to have translated Chaucer into English, out of a quite contrary notion. They suppose there is a certain veneration due to his old language; and that it is little less than profanation and sacrilege to alter it. They are further of opinion, that somewhat of his good sense will suffer in this transfusion, and much of the beauty of his thoughts will infallibly be lost, which appear with more grace in their old habit. Of this opinion was that excellent person, whom I mentioned, the late Earl of Leicester, who valued Chaucer as much as Mr Cowley despised him. My Lord dissuaded me from this attempt (for I was thinking of it some years before his death) and his authority prevailed so far with me as to defer my undertaking while he lived in deference to him; yet my reason was not convinced with what he urged against it. If the first end of a writer be to be understood, then as his language grows obsolete, his thoughts must grow obscure, *multa renascuntur, quae nunc cecidere; cadentque quae nunc sunt in honore vocabula, si volet usus, quem penes arbitrium est et jus et norma loquendi* [Many words which now have fallen out of use will be reborn, and words now in esteem will become obsolete, if usage, which owns the right to decide, by law and standard of speaking, so wills it].° When an ancient word for its sounds and significancy deserves to be revived, I have that reasonable veneration for antiquity to restore it. All beyond this is superstition. Words are not like landmarks, so sacred as never to be removed. Customs are changed, and even statutes are silently repealed, when the reason ceases for which they were enacted. As for the other part of the argument, that his thoughts will lose of their original beauty by the innovation of words; in the first place, not only their beauty but their being is lost where they are no longer understood, which is the present case. I grant that something must be lost in all transfusion, that is, in all translations; but the sense will remain, which would otherwise be lost, or at least be maimed, when it is scarce intelligible, and that but to a few. How few are there who can read Chaucer so as to understand him perfectly! And if imperfectly, then with less profit and no pleasure. 'Tis not for the use of some old Saxon friends that I have taken these pains with him. Let them neglect my version, because they have no need of it. I made it for their sakes who understand sense and poetry as well as they; when that poetry and sense is put into words which they understand. I will go further, and dare to add, that what beauties I lose in some places, I give to others which had them not originally. But in this I may be partial to myself; let the reader judge, and I submit to his decision. Yet I think I have just occasion to complain of them, who because they understand Chaucer, would deprive the greater part of their countrymen of the same advantage, and hoard him up, as misers do their grandam gold,° only to look on it themselves and hinder others from making use of it. In sum, I seriously protest that no man ever had, or can have, a greater veneration for Chaucer than myself. I have translated some parts of his works only that I might perpetuate his memory, or at least refresh it, amongst my countrymen. If I have altered him anywhere for the better, I must at the same time acknowledge that I could have done nothing without him: *facile est inventis addere* [it is easy to add to what has already been discovered] is no great commendation; and I am not so vain to think I have deserved a greater. I will conclude what I have to say of him singly with this one remark: a lady of my acquaintance, who keeps a kind of correspondence with some authors of the fair sex in France, has been informed by them that Mademoiselle de Scudéry,° who is as old as Sibyl and inspired like her by the same god of poetry, is at this time translating Chaucer into modern French. From which I gather that he has been formerly translated into the old Provençal° (for how she should come to understand old English I know not). But the matter of fact being true, it makes me think that there is something in it like fatality; that, after certain periods of time, the fame and memory of great wits should be renewed, as Chaucer is both in France and England. If this be wholly chance, 'tis extraordinary; and I dare not call it more for fear of being taxed with superstition.

Boccace comes last to be considered, who living in the same age with Chaucer, had the same genius and followed the same studies. Both writ novels, and each of them cultivated his mother tongue. But the greatest resemblance of our two modern authors being in their familiar style and pleasing way of relating comical adventures, I may pass it over because I have translated nothing from Boccace of that nature. In the serious part of poetry, the advantage is wholly on Chaucer's side, for though the Englishman has borrowed many tales from the Italian, yet it appears that those of Boccace were not generally of his own making but taken from authors of former ages and by him only modelled; so that what there was of invention in either of them may be judged equal. But Chaucer has refined on Boccace, and has mended the stories which he has borrowed, in his way of telling; though prose allows more liberty of thought, and the expression is more easy when unconfined by numbers. Our countryman carries weight, and yet wins the race at disadvantage. I desire not the reader should take my word; and therefore I will set two of their discourses on the same subject in the same light, for every man to judge betwixt them. I translated Chaucer first, and amongst the rest, pitched on the Wife of Bath's Tale; not daring, as I have said, to adventure on her Prologue, because 'tis too licentious. There Chaucer introduces an old woman of mean parentage, whom a youthful knight of noble blood was forced to marry, and consequently loathed her. The crone being in bed with him on the wedding night, and finding his aversion, endeavours to win his affection by reason, and speaks a good word for herself (as who could blame her?) in hope to mollify the sullen bridegroom. She takes her topics from the benefits of poverty, the advantages of old age and ugliness, the vanity of youth, and the silly pride of ancestry and titles without inherent virtue, which is the true nobility. When I had closed Chaucer, I returned to Ovid and translated some more of his fables; and by this time had so far forgotten the Wife of Bath's Tale, that when I took up Boccace, unawares I fell on the same argument of preferring virtue to nobility of blood and titles, in the story of Sigismonda, which I had certainly avoided for the resemblance of the two discourses if my memory had not failed me. Let the reader weigh them both; and if he thinks me partial to Chaucer, 'tis in him to right Boccace.

I prefer in our countryman, far above all his other stories, the noble poem of Palamon and Arcite, which is of the epic kind, and perhaps not much inferior to the *Iliad* or the *Aeneid*. The story is more pleasing than either of them, the manners as perfect, the diction as poetical, the learning as deep and various, and the disposition full as artful: only it includes a greater length of time, as taking up seven years at least; but Aristotle has left undecided the duration of the action; which yet is easily reduced into the compass of a year by a narration of what preceded the return of Palamon to Athens. I had thought for the honour of our nation, and more particularly for his whose laurel, though unworthy, I have worn after him, that this story was of English growth and Chaucer's own. But I was undeceived by Boccace; for casually looking on the end of his seventh *Giornata*, I found Dioneo (under which name he shadows himself) and Fiametta (who represents his mistress, the natural daughter of Robert, King of Naples) of whom these words are spoken: *Dioneo e Fiametta gran pezza cantarono insieme d'Arcita, e di Palemone* [Dioneo and Fiametta together told a long tale of Arcite and Palamon]. By which it appears that this story was written before the time of Boccace, but the name of its author being wholly lost, Chaucer is now become an original; and I question not but the poem has received many beauties by passing through his noble hands. Besides this tale, there is another of his own invention, after the manner of the Provençals, called *The Flower and the Leaf*,° with which I was so particularly pleased, both for the invention and the moral, that I cannot hinder myself from recommending it to the reader.

As a corollary to this preface, in which I have done justice to others, I owe somewhat to myself: not that I think it worth my time to enter the lists with one Milbourne, and one Blackmore,° but barely to take notice that such men there are who have written scurrilously against me without any provocation. Milbourne, who is in orders, pretends amongst the rest this quarrel to me, that I have fallen foul on priesthood. If I have, I am only to ask pardon of good priests, and am afraid his part of the reparation will come to little. Let him be satisfied that he shall not be able to force himself upon me for an adversary. I contemn him too much to enter into competition with him. His own translations of Virgil have answered his criticisms on mine. If (as they say he has declared in print) he prefers the version of Ogilby° to mine, the world has made him the same compliment; for 'tis agreed on all hands that he writes even below Ogilby. That, you will say, is not easily to be done; but what cannot Milbourne bring about? I am satisfied, however, that while he and I live together, I shall not be thought the worst poet of the age. It looks as if I had desired him underhand to write so ill against me; but upon my honest word I have not bribed him to do me this service and am wholly guiltless of his pamphlet. 'Tis true I should be glad if I could persuade him to continue his good offices and write such another critique on anything of mine; for I find by experience he has a great stroke with the reader when he condemns any of my poems to make the world have a better opinion of them. He has taken some pains with my poetry; but nobody will be persuaded to take the same with his. If I had taken to the church (as he affirms, but which was never in my thoughts) I should have had more sense, if not more grace, than to have turned myself out of my benefice by writing libels on my parishioners. But his account of my manners and my principles are of a piece with his cavils and his poetry: and so I have done with him for ever.

As for the City Bard, or Knight Physician, I hear his quarrel to me is that I was the author of *Absalom and Achitophel*, which he thinks is a little hard on his fanatic patrons in London.

But I will deal the more civilly with his two poems,° because nothing ill is to be spoken of the dead: and therefore peace be to the *manes* [shades] of his *Arthurs*. I will only say that it was not for this noble Knight that I drew the plan of an epic poem on King Arthur in my preface to the translation of Juvenal. The guardian angels of kingdoms were machines too ponderous for him to manage; and therefore he rejected them as Dares did the whirlbats of Eryx when they were thrown before him by Entellus.° Yet from that preface he plainly took his hint; for he began immediately upon the story, though he had the baseness not to acknowledge his benefactor, but instead of it to traduce me in a libel.

I shall say the less of Mr Collier, because in many things he has taxed me justly; and I have pleaded guilty to all thoughts and expressions of mine which can be truly argued of obscenity, profaneness, or immorality, and retract them. If he be my enemy, let him triumph; if he be my friend, as I have given him no personal occasion to be otherwise, he will be glad of my repentance. It becomes me not to draw my pen in the defence of a bad cause when I have so often drawn it for a good one. Yet it were not difficult to prove that in many places he has perverted my meaning by his glosses; and interpreted my words into blasphemy and bawdry of which they were not guilty. Besides that, he is too much given to horse-play in his raillery; and comes to battle like a dictator from the plough. I will not say, *the zeal of God's house has eaten him up;*° but I am sure it has devoured some part of his good manners and civility. It might also be doubted whether it were altogether zeal which prompted him to this rough manner of proceeding; perhaps it became not one of his function to rake into the rubbish of ancient and modern plays; a divine might have employed his pains to better purpose than in the nastiness of Plautus and Aristophanes, whose examples, as they excuse not me, so it might be possibly supposed that he read them not without some pleasure. They who have written commentaries on those poets, or on Horace, Juvenal, and Martial, have explained some vices, which without their interpretation had been unknown to modern times. Neither has he judged impartially betwixt the former age and us.

There is more bawdry in one play of Fletcher's, called *The Custom of the Country*, than in all ours together. Yet this has been often acted on the stage in my remembrance. Are the times so much more reformed now than they were five-and-twenty years ago? If they are, I congratulate the amendment of our morals. But I am not to prejudice the cause of my fellow-poets, though I abandon my own defence. They have some of them answered for themselves, and neither they nor I can think Mr Collier so formidable an enemy that we should shun him. He has lost ground at the latter end of the day by pursuing his point too far, like the Prince of Condé at the battle of Senneph° from immoral plays, to no plays, *ab abusu ad usum, non valet consequentia* [no valid inference can be made from the abuse of a thing to its use]. But being a party, I am not to erect myself into a judge. As for the rest of those who have written against me, they are such scoundrels that they deserve not the least notice to be taken of them. Blackmore and Milbourne are only distinguished from the crowd by being remembered to their infamy:

> Demetri, teque Tigelli
> Discipularum inter jubeo plorare cathedras.°

[As for you, Demetrius and Tigellius, I bid you go lament among your women pupils]

To Her Grace the Duchess of Ormonde

Madam,
The bard who first adorned our native tongue°
Tuned to his British lyre this ancient song;°
Which Homer might without a blush rehearse,
And leaves a doubtful palm in Virgil's verse:°
He matched their beauties, where they most excel;
Of love sung better, and of arms as well.

   Vouchsafe, illustrious Ormonde, to behold
What power the charms of beauty had of old;
Nor wonder if such deeds of arms were done,
Inspired by two fair eyes, that sparkled like your own.   10

   If Chaucer by the best idea wrought,
And poets can divine each other's thought,
The fairest nymph before his eyes he set;
And then the fairest was Plantagenet,°
Who three contending princes made her prize,
And ruled the rival nations with her eyes:
Who left immortal trophies of her fame,
And to the noblest order gave the name.

   Like her, of equal kindred to the throne,
You keep her conquests, and extend your own: 20
As when the stars, in their ethereal race,
At length have rolled around the liquid space,
At certain periods they resume their place,
From the same point of heaven their course advance,
And move in measures of their former dance;
Thus, after length of age, she returns,
Restored in you, and the same place adorns;
Or you perform her office in the sphere,
Born of her blood, and make a new Platonic year.°

   O true Plantagenet, O race divine,   30
(For beauty still is fatal to the line,)
Had Chaucer lived that angel-face to view,
Sure he had drawn his Emily from you:
Or had you lived, to judge the doubtful right,
Your noble Palamon had been the knight:
And conquering Theseus from his side had sent
Your generous lord, to guide the Theban government.

Time shall accomplish that; and I shall see
A Palamon in him, in you an Emily.
    Already have the fates your path prepared,   40
And sure presage your future sway declared:
When westward, like the sun, you took your way,°
And from benighted Britain bore the day,
Blue Triton gave the signal from the shore,
The ready nereids heard, and swam before
To smooth the seas; a soft etesian gale°
But just inspired, and gently swelled the sail;
Portunus took his turn, whose ample hand°
Heaved up the lightened keel, and sunk the sand,
And steered the sacred vessel safe to land.   50
The land, if not restrained, had met your way,
Projected out a neck, and jutted to the sea.
Hibernia, prostrate at your feet, adored°
In you, the pledge of her expected lord,°
Due to her isle; a venerable name;
His father and his grandsire known to fame:°
Awed by that house, accustomed to command,
The sturdy kerns in due subjection stand;
Nor hear the reins in any foreign hand.
    At your approach, they crowded to the port;   60
And scarcely landed, you create a court:
As Ormonde's harbinger, to you they run;
For Venus is the promise of the sun.
    The waste of civil wars, their towns destroyed,
Pales unhonoured, Ceres unemployed,
Were all forgot; and one triumphant day
Wiped all the tears of three campaigns away.
Blood, rapines, massacres, were cheaply bought,
So mighty recompense your beauty brought.
    As when the dove returning, bore the mark   70
Of earth restored to the long-labouring ark,°
The relics of mankind, secure of rest,
Oped every window to receive the guest,
And the fair bearer of the message blessed;
So, when you came, with loud repeated cries,
The nation took an omen from your eyes,
And God advanced his rainbow in the skies,

To sign inviolable peace restored;
The saints with solemn shouts proclaimed the new accord.
    When at your second coming you appear,   80
(For I foretell that millenary year)
The sharpened share shall vex the soil no more,
But earth unbidden shall produce her store:
The land shall laugh, the circling ocean smile,
And heaven's indulgence bless the holy isle.
    Heaven from all ages has reserved for you
That happy clime which venom never knew;
Or if it had been there, your eyes alone
Have power to chase all poison, but their own.
    Now in this interval, which fate has cast   90
Betwixt your future glories and your past,
This pause of power, 'tis Ireland's hour to mourn;
While England celebrates your safe return,
By which you seem the seasons to command,
And bring our summers back to their forsaken land.
    The vanquished isle our leisure must attend,
Till the fair blessing we vouchsafe to send;
Nor can we spare you long, though often we may lend.
The dove was twice employed abroad, before°
The world was dried; and she returned no more.   100
Nor dare we trust so soft a messenger,
New from her sickness, to that northern air;
Rest here awhile, your lustre to restore,
That they may see you as you shone before:
For yet, the eclipse not wholly past, you wade
Through some remains, and dimness of a shade.
    A subject in his prince may claim a right,
Nor suffer him with strength impaired to fight;
Till force returns, his ardour we restrain,
And curb his warlike wish to cross the main.   110
    Now past the danger, let the learned begin
The enquiry, where disease could enter in;
How those malignant atoms forced their way,
What in the faultless frame they found to make their prey,
Where every element was weighed so well,
That heaven alone, who mixed the mass, could tell
Which of the four ingredients could rebel;°

And where, imprisoned in so sweet a cage,
A soul might well be pleased to pass an age.
　　And yet the fine materials made it weak;　　　　120
Porcelain by being pure, is apt to break:
E'en to your breast the sickness durst aspire;
And forced from that fair temple to retire,
Profanely set the holy place on fire.
In vain your lord like young Vespasian mourned,°
When the fierce flames the sanctuary burned:
And I prepared to pay in verses rude
A most detested act of gratitude;
E'en this had been your elegy, which now
Is offered for your health, the table of my vow.°　　130
　　Your angel sure our Morley's mind inspired,°
To find the remedy your ill required;
As once the Macedon, by Jove's decree,°
Was taught to dream a herb for Ptolemy:
Or heaven, which had such over-cost bestowed,
As scarce it could afford to flesh and blood,
So liked the frame, he would not work anew,
To save the charges of another you.
Or by his middle science did he steer,
And saw some great contingent good appear,　　　140
Well worth a miracle to keep you here:
And for that end, preserved the precious mould,
Which all the future Ormondes was to hold;
And meditated in his better mind
An heir from you, who may redeem the failing kind.
　　Blessed be the power which has at once restored
The hopes of lost succession to your lord;
Joy to the first, and last of each degree,
Virtue to courts, and what I longed to see,
To you the graces, and the muse to me.　　　　　150
　　O daughter of the rose, whose cheeks unite
The differing titles of the red and white;
Who heaven's alternate beauty well display,
The blush of morning, and the milky way;
Whose face is paradise, but fenced from sin:
For God in either eye has placed a cherubin.
　　All is your lord's alone; e'en absent, he
Employs the care of chaste Penelope.°

For him you waste in tears your widowed hours,
For him your curious needle paints the flowers: 160
Such works of old imperial dames were taught;
Such for Ascanius, fair Elissa wrought.°

The soft recesses of your hours improve
The three fair pledges of your happy love:
All other parts of pious duty done,
You owe your Ormonde nothing but a son:
To fill in future times his father's place,
And wear the garter of his mother's race.

Palamon and Arcite:

OR THE KNIGHT'S TALE

In three books

BOOK I

In days of old there lived, of mighty fame,
A valiant prince, and Theseus was his name;
A chief, who more in feats of arms excelled,
The rising nor the setting sun beheld.
Of Athens he was lord; much land he won,
And added foreign countries to his crown.
In Scythia with the warrior queen he strove,°
Whom first by force he conquered, then by love;
He brought in triumph back the beauteous dame,
With whom her sister fair Emilia, came.   10
With honour to his home let Theseus ride,
With love to friend, and fortune for his guide,
And his victorious army at his side.
I pass their warlike pomp, their proud array,
Their shouts, their songs, their welcome on the way;
But, were it not too long, I would recite
The feats of Amazons, the fatal fight
Betwixt the hardy queen and hero knight;
The town besieged, and how much blood it cost
The female army, and the Athenian host;   20
The spousals of Hippolyta the queen;

What tilts and tourneys at the feast were seen;°
The storm at their return, the ladies' fear:
But these and other things I must forbear.
The field is spacious I design to sow
With oxen far unfit to draw the plough:
The remnant of my tale is of a length
To tire your patience, and to waste my strength;
And trivial accidents shall be forborn,
That others may have time to take their turn,   30
As was at first enjoined us by mine host,
That he, whose tale is best and pleases most,
Should win his supper at our common cost.

    And therefore where I left, I will pursue
This ancient story, whether false or true,
In hope it may be mended with a new.
The prince I mentioned, full of high renown,
In this array drew near the Athenian town;
When in his pomp and utmost of his pride
Marching, he chanced to cast his eye aside,   40
And saw a choir of mourning dames, who lay
By two and two across the common way:
At his approach they raised a rueful cry,
And beat their breasts, and held their hands on high,
Creeping and crying, till they seized at last
His courser's bridle and his feet embraced.
'Tell me,' said Theseus, 'what and whence you are,
And why this funeral pageant you prepare?
Is this the welcome of my worthy deeds,
To meet my triumph in ill-omened weeds?   50
Or envy you my praise, and would destroy
With grief my pleasures, and pollute my joy?
Or are you injured, and demand relief?
Name your request, and I will ease your grief.'

    The most in years of all the mourning train
Began; but sounded first away for pain;°
Then scarce recovered spoke: 'Nor envy we
Thy great renown, nor grudge thy victory;
'Tis thine, O king, the afflicted to redress,
And fame has filled the world with thy success: 60
We wretched women sue for that alone,
Which of thy goodness is refused to none;

Let fall some drops of pity on our grief,
If what we beg be just, and we deserve relief;
For none of us, who now thy grace implore,
But held the rank of sovereign queen before;
Till, thanks to giddy chance, which never bears°
That mortal bliss should last for length of years,
She cast us headlong from our high estate,
And here in hope of thy return we wait,   70
And long have waited in the temple nigh,
Built to the gracious goddess Clemency.
But reverence thou the power whose name it bears,
Relieve the oppressed, and wipe the widows' tears.
I, wretched I, have other fortune seen,
The wife of Capaneus, and once a queen;°
At Thebes he fell; cursed be the fatal day!
And all the rest thou seest in this array
To make their moan their lords in battle lost,
Before that town besieged by our confederate host.   80
'But Creon, old and impious, who commands°
The Theban city, and usurps the lands,
Denies the rites of funeral fires to those
Whose breathless bodies yet he calls his foes.
Unburned, unburied, on a heap they lie;
Such is their fate, and such his tyranny;
No friends has leave to bear away the dead,
But with their lifeless limbs his hounds are fed.'
At this she screaked aloud; the mournful train°
Echoed her grief, and grovelling on the plain,   90
With groans, and hands upheld, to move his mind,
Besought his pity to their helpless kind.

   The prince was touched, his tears began to flow,
And, as his tender heart would break in two,
He sighed; and could not but their fate deplore,
So wretched now, so fortunate before.
Then lightly from his lofty steed he flew,
And rising one by one the suppliant crew,
To comfort each, full solemnly he swore,
That by the faith which knights to knighthood bore,   100
And whate'er else to chivalry belongs,
He would not cease, till he revenged their wrongs;
That Greece should see performed what he declared,

And cruel Creon find his just reward.
He said no more, but shunning all delay
Rode on, nor entered Athens on his way;
But left his sister and his queen behind,°
And waved his royal banner in the wind,
Where in an argent field the god of war°
Was drawn triumphant on his iron car;   110
Red was his sword, and shield, and whole attire,
And all the godhead seemed to glow with fire;
E'en the ground glittered where the standard flew,
And the green grass was dyed to sanguine hue.°
High on his pointed lance his pennon bore
His Cretan fight, the conquered Minotaur:°
The soldiers shout around with generous rage,
And in that victory their own presage.
He praised their ardour, inly pleased to see
His host, the flower of Grecian chivalry.   120
All day he marched, and all the ensuing night,
And saw the city with returning light.°
The process of the war I need not tell,
How Theseus conquered, and how Creon fell;
Or after, how by storm the walls were won,
Or how the victor sacked and burned the town;
How to the ladies he restored again
The bodies of their lords in battle slain;
And with what ancient rites they were interred;
All these to fitter time shall be deferred;   130
I spare the widows' tears, their woeful cries,
And howling at their husbands' obsequies;
How Theseus at these funerals did assist,
And with what gifts the mourning dames dismissed.
  Thus when the victor chief had Creon slain,
And conquered Thebes, he pitched upon the plain
His mighty camp, and when the day returned,
The country wasted and the hamlets burned,
And left the pillagers, to rapine bred,
Without control to strip and spoil the dead.   140
  There, in a heap of slain, among the rest
Two youthful knights they found beneath a load oppressed
Of slaughtered foes, whom first to death they sent,
The trophies of their strength, a bloody monument.°

Both fair, and both of royal blood they seemed,
Whom kinsmen to the crown the heralds deemed;
That day in equal arms they fought for fame;
Their swords, their shields, their surcoats were the same:
Close by each other laid they pressed the ground,
Their manly bosoms pierced with many a grisly wound;   150
Nor well alive nor wholly dead they were,
But some faint signs of feeble life appear;
The wandering breath was on the wing to part,
Weak was the pulse, and hardly heaved the heart.
These two were sisters' sons; and Arcite one,
Much famed in fields, with valiant Palamon.
From these their costly arms the spoilers rent,
And softly both conveyed to Theseus' tent:
Whom, known of Creon's line and cured with care,
He to his city sent as prisoners of the war;   160
Hopeless of ransom, and condemned to lie
In durance, doomed a lingering death to die.
    This done, he marched away with warlike sound,
And to his Athens turned with laurels crowned,
Where happy long he lived, much loved, and more renowned.
But in a tower, and never to be loosed,
The woeful captive kinsmen are enclosed.
    Thus year by year they pass, and day by day,
Till once ('twas on the morn of cheerful May)
The young Emilia, fairer to be seen   170
Than the fair lily on the flowery green,
More fresh than May herself in blossoms new,
(For with the rosy colour strove her hue,)
Waked, as her custom was before the day,
To do the observance due to sprightly May;
For sprightly May commands our youth to keep
The vigils of her night, and breaks their sluggard sleep;
Each gentle breast with kindly warmth she moves;
Inspires new flames, revives extinguished loves.
In this remembrance Emily ere day   180
Arose, and dressed herself in rich array;
Fresh as the month, and as the morning fair,
Adown her shoulders fell her length of hair:
A ribbon did the braided tresses bind,
The rest was loose, and wantoned in the wind:

Aurora had but newly chased the night,
And purpled o'er the sky with blushing light,
When to the garden walk she took her way,
To sport and trip along in cool of day,
And offer maiden vows in honour of the May.   190
At every turn she made a little stand,
And thrust among the thorns her lily hand
To draw the rose; and every rose she drew,
She shook the stalk, and brushed away the dew;
Then particoloured flowers of white and red
She wove, to make a garland for her head.
This done, she sung and carolled out so clear,
That men and angels might rejoice to hear;
E'en wondering Philomel forgot to sing,
And learned from her to welcome in the spring.   200
The tower, of which before was mention made,
Within whose keep the captive knights were laid,
Built of a large extent, and strong withal,
Was one partition of the palace wall;°
The garden was enclosed within the square,
Where young Emilia took the morning air.
   It happened Palamon, the prisoner knight,
Restless for woe, arose before the light,
And with his gaoler's leave desired to breathe
An air more wholesome than the damps beneath.   210
This granted, to the tower he took his way,
Cheered with the promise of a glorious day;
Then cast a languishing regard around,
And saw with hateful eyes the temples crowned
With golden spires, and all the hostile ground.
He sighed, and turned his eyes, because he knew
'Twas but a larger gaol he had in view;
Then looked below, and from the castle's height
Beheld a nearer and more pleasing sight.
The garden, which before he had not seen,   220
In spring's new livery clad of white and green,
Fresh flowers in wide parterres, and shady walks between.°
This viewed, but not enjoyed, with arms across°
He stood, reflecting on his country's loss;
Himself an object of the public scorn,
And often wished he never had been born.

At last (for so his destiny required),
With walking giddy, and with thinking tired,
He through a little window cast his sight,
Though thick of bars, that gave a scanty light;   230
But e'en that glimmering served him to descry
The inevitable charms of Emily.

   Scarce had he seen, but, seized with sudden smart,
Stung to the quick, he felt it at his heart;
Struck blind with overpowering light he stood,
Then started back amazed, and cried aloud.

   Young Arcite heard; and up he ran with haste,
To help his friend, and in his arms embraced;
And asked him why he looked so deadly wan,
And whence, and how, his change of cheer began?   240
Or who had done the offence? 'But if', said he,
'Your grief alone is hard captivity,
For love of heaven with patience undergo
A cureless ill, since fate will have it so:
So stood our horoscope in chains to lie,
And Saturn in the dungeon of the sky,
Or other baleful aspect, ruled our birth,°
When all the friendly stars were under earth;
Whate'er betides, by destiny 'tis done;
And better bear like men than vainly seek to shun.'   250
'Nor of my bonds,' said Palamon again,
'Nor of unhappy planets I complain;
But when my mortal anguish caused my cry,
That moment I was hurt through either eye;°
Pierced with a random shaft, I faint away,
And perish with insensible decay;
A glance of some new goddess gave the wound,°
Whom, like Actaeon, unaware I found.°
Look how she walks along yon shady space;
Not Juno moves with more majestic grace,   260
And all the Cyprian queen is in her face.°
If thou art Venus (for thy charms confess
That face was formed in heaven), nor art thou less,
Disguised in habit, undisguised in shape,
O help us captives from our chains to scape!
But if our doom be passed in bonds to lie
For life, and in a loathsome dungeon die,

Then be thy wrath appeased with our disgrace,
And show compassion to the Theban race,
Oppressed by tyrant power!'——While yet he spoke,   270
Arcite on Emily had fixed his look;
The fatal dart a ready passage found
And deep within his heart infixed the wound:
So that if Palamon were wounded sore,
Arcite was hurt as much as he or more:
Then from his inmost soul he sighed, and said
'The beauty I behold has struck me dead:
Unknowingly she strikes, and kills by chance;
Poison is in her eyes, and death in every glance.
O, I must ask; nor ask alone, but move   280
Her mind to mercy, or must die for love.'

    Thus Arcite: and thus Palamon replies,
(Eager his tone, and ardent were his eyes):
'Speakst thou in earnest or in jesting vein?'
'Jesting', said Arcite, 'suits but ill with pain.'
'It suits far worse', said Palamon again,
And bent his brows, 'with men who honour weigh
Their faith to break, their friendship to betray.
But worst with thee, of noble lineage born,
My kinsman, and in arms my brother sworn.   290
Have we not plighted each our holy oath,
That one should be the common good of both?
One soul should both inspire, and neither prove
His fellow's hindrance in pursuit of love?
To this before the gods we gave our hands,
And nothing but our death can break the bands.
This binds thee, then, to further my design,
As I am bound by vow to further thine.
Nor canst, nor darst thou, traitor, on the plain
Appeach my honour, or thy own maintain,   300
Since thou art of my counsel, and the friend
Whose faith I trust, and on whose care depend.
And wouldst thou court my lady's love, which I
Much rather than release, would choose to die?
But thou, false Arcite, never shalt obtain
Thy bad pretence: I told thee first my pain;°
For first my love began ere thine was born.
Thou, as my counsel and my brother sworn,

Art bound to assist my eldership of right,
Or justly to be deemed a perjured knight.'   310
   Thus Palamon. But Arcite with disdain
In haughty language thus replied again:
'Forsworn thyself: the traitor's odious name
I first return, and then disprove thy claim.
If love be passion, and that passion nursed
With strong desires, I loved the lady first.
Canst thou pretend desire, whom zeal inflamed
To worship, and a power celestial named?
Thine was devotion to the blessed above;
I saw the woman, and desired her love: 320
First owned my passion, and to thee commend
The important secret, as my chosen friend.
Suppose (which yet I grant not) thy desire
A moment elder than my rival fire.
Can chance of seeing first thy title prove?
And knowst thou not, no law is made for love?
Law is to things which to free choice relate:
Love is not in our choice, but in our fate.
Laws are but positive. Love's power we see°
Is nature's sanction, and her first decree.   330
Each day we break the bond of human laws
For love, and vindicate the common cause.
Laws for defence of civil rights are placed,
Love throws the fences down, and makes a general waste.
Maids, widows, wives, without distinction fall:
The sweeping deluge, love, comes on, and covers all.
If then the laws of friendship I transgress,
I keep the greater, while I break the less:
And both are mad alike, since neither can possess.
Both hopeless to be ransomed; never more   340
To see the sun, but as he passes o'er.
Like Aesop's hounds contending for the bone,°
Each pleaded right, and would be lord alone.
The fruitless fight continued all the day;
A cur came by, and snatched the prize away.
As courtiers therefore jostle for a grant,
And when they break their friendship, plead their want,
So thou, if fortune will thy suit advance,
Love on; nor envy me my equal chance,

For I must love, and am resolved to try   350
My fate, or failing in the adventure die.'
    Great was their strife, which hourly was renewed,
Till each with mortal hate his rival viewed.
Now friends no more, nor walking hand in hand:
But when they met, they made a surly stand
And glared like angry lions as they passed,
And wished that every look might be their last.
    It chanced at length, Pirithous came, to attend
This worthy Theseus, his familiar friend.
Their love in early infancy began,   360
And rose as childhood ripened into man.
Companions of the war: and loved so well
That when one died, as ancient stories tell,
His fellow to redeem him went to hell.
    But to pursue my tale. To welcome home
His warlike brother is Pirithous come.
Arcite of Thebes was known in arms long since,
And honoured by this young Thessalian prince.
Theseus, to gratify his friend and guest,
Who made our Arcite's freedom his request,   370
Restored to liberty the captive knight,
But on these hard conditions I recite:
That if hereafter Arcite should be found
Within the compass of Athenian ground,
By day or night, or on whate'er pretence,
His head should pay the forfeit of the offence.
To this Pirithous for his friend agreed,
And on his promise was the prisoner freed.
    Unpleased and pensive hence he takes his way,
At his own peril; for his life must pay.   380
Who now but Arcite mourns his bitter fate,
Finds his dear purchase, and repents too late?
'What have I gained,' he said, 'in prison pent,
If I but change my bonds for banishment?
And banished from her sight, I suffer more
In freedom than I felt in bonds before;
Forced from her presence and condemned to live,
Unwelcome freedom and unthanked reprieve:
Heaven is not but where Emily abides,
And where she's absent, all is hell besides.   390

Next to my day of birth, was that accurst
Which bound my friendship to Pirithous first:
Had I not known that prince, I still had been
In bondage, and had still Emilia seen:
For though I never can her grace deserve,
'Tis recompense enough to see and serve.
O Palamon, my kinsman and my friend,
How much more happy fates thy love attend!
Thine is the adventure, thine the victory,
Well has thy fortune turned the dice for thee: 400
Thou on that angel's face mayst feed thy eyes,
In prison, no; but blissful paradise!
Thou daily seest that sun of beauty shine,
And lov'st at least in love's extremest line.°
I mourn in absence, love's eternal night;
And who can tell but since thou hast her sight,
And art a comely, young, and valiant knight,
Fortune (a various power) may cease to frown,
And by some ways unknown thy wishes crown?
But I, the most forlorn of human kind,   410
Nor help can hope nor remedy can find;
But doomed to drag my loathsome life in care,
For my reward, must end it in despair.
Fire, water, air, and earth, and force of Fates°
That governs all, and heaven that all creates,
Nor art, nor nature's hand can ease my grief;
Nothing but death, the wretch's last relief:
Then farewell youth, and all the joys that dwell
With youth and life, and life itself, farewell!

'But why, alas! do mortal men in vain   420
Of Fortune, Fate, or Providence complain?
God gives us what he knows our wants require,
And better things than those which we desire:
Some pray for riches; riches they obtain;
But, watched by robbers, for their wealth are slain;
Some pray from prison to be freed; and come,
When guilty of their vows, to fall at home;°
Murdered by those they trusted with their life,
A favoured servant or a bosom wife.
Such dear-bought blessings happen every day,   430
Because we know not for what things to pray.

Like drunken sots about the streets we roam:
Well knows the sot he has a certain home,
Yet know not how to find the uncertain place,
And blunders on, and staggers every pace.
Thus all seek happiness; but few can find,
For far the greater part of men are blind.
This is my case, who thought our utmost good
Was in one word of freedom understood:
The fatal blessing came: from prison free,
I starve abroad, and lose the sight of Emily.'   440

   Thus Arcite: but if Arcite thus deplore
His sufferings, Palamon yet suffers more.
For when he knew his rival freed and gone,
He swells with wrath; he makes outrageous moan;
He frets, he fumes, he stares, he stamps the ground;
The hollow tower with clamours rings around:
With briny tears he bathed his fettered feet,
And dropped all o'er with agony of sweat.
'Alas!' he cried, 'I, wretch, in prison pine,   450
Too happy rival, while the fruit is thine:
Thou livest at large, thou drawest thy native air,
Pleased with thy freedom, proud of my despair:
Thou mayest, since thou hast youth and courage joined,
A sweet behaviour and a solid mind,
Assemble ours, and all the Theban race,
To vindicate on Athens thy disgrace;
And after (by some treaty made) possess
Fair Emily, the pledge of lasting peace.
So thine shall be the beauteous prize, while I   460
Must languish in despair, in prison die.
Thus all the advantage of the strife is thine,
Thy portion double joys, and double sorrows mine.'

   The rage of jealousy then fired his soul,
And his face kindled like a burning coal:
Now cold despair, succeeding in her stead,
To livid paleness turns the glowing red.
His blood, scarce liquid, creeps within his veins,
Like water which the freezing wind constrains.
Then thus he said: 'Eternal deities,   470
Who rule the world with absolute decrees,
And write whatever time shall bring to pass

With pens of adamant on plates of brass;°
What is the race of human kind your care
Beyond what all his fellow-creatures are?
He with the rest is liable to pain,
And like the sheep, his brother-beast, is slain.
Cold, hunger, prisons, ills without a cure,
All these he must, and guiltless oft, endure;
Or does your justice, power, or prescience fail   480
When the good suffer and the bad prevail?
What worse to wretched virtue could befall,
If fate or giddy fortune governed all?
Nay, worse than other beasts' is our estate:
Them, to pursue their pleasures, you create;
We, bound by harder laws, must curb our will,
And your commands, not our desires, fulfil:
Then, when the creature is unjustly slain,
Yet, after death at least, he feels no pain;
But man in life surcharged with woe before,   490
Not freed when dead, is doomed to suffer more.
A serpent shoots his sting at unaware;
An ambushed thief forelays a traveller;
The man lies murdered, while the thief and snake,
One gains the thickets, and one threads the brake.°
This let divines decide; but well I know,
Just or unjust, I have my share of woe:
Through Saturn seated in a luckless place,°
And Juno's wrath that persecutes my race;
Or Mars and Venus in a quartile move°   500
My pangs of jealousy for Arcite's love.'
   Let Palamon oppressed in bondage mourn,
While to his exiled rival we return.
By this the sun, declining from his height,
The day had shortened to prolong the night:
The lengthened night gave length of misery,
Both to the captive lover and the free:
For Palamon in endless prison mourns,
And Arcite forfeits life if he returns;
The banished never hopes his love to see,   510
Nor hopes the captive lord his liberty.
'Tis hard to say who suffers greater pains;
One sees his love, but cannot break his chains;

One free, and all his motions uncontrolled,
Beholds whate'er he would but what he would behold.
Judge as you please, for I will haste to tell
What fortune to the banished knight befell.
When Arcite was to Thebes returned again,
The loss of her he loved renewed his pain;
What could be worse than never more to see   520
His life, his soul, his charming Emily?
He raved with all the madness of despair,
He roared, he beat his breast, he tore his hair.
Dry sorrow in his stupid eyes appears,
For wanting nourishment, he wanted tears;
His eyeballs in their hollow sockets sink,
Bereft of sleep; he loathes his meat and drink;
He withers at his heart, and looks as wan
As the pale spectre of a murdered man:
That pale turns yellow, and his face receives   530
The faded hue of sapless boxen leaves;
In solitary groves he makes his moan,
Walks early out and ever is alone;
Nor, mixed in mirth, in youthful pleasure shares,
But sighs when songs and instruments he hears.
His spirits are so low, his voice is drowned,
He hears as from afar, or in a swound,
Like the deaf murmurs of a distant sound:
Uncombed his locks, and squalid his attire,
Unlike the trim of love and gay desire;   540
But full of museful mopings, which presage
The loss of reason and conclude in rage.

    This when he had endured a year and more,
Now wholly changed from what he was before,
It happened once, that, slumbering as he lay,
He dreamt (his dream began at break of day)
That Hermes o'er his head in air appeared,
And with soft words his drooping spirits cheered;
His hat adorned with wings disclosed the god,
And in his hand he bore the sleep-compelling rod;   550
Such as he seemed, when, at his sire's command,°
On Argus' head he laid the snaky wand.
'Arise,' he said, 'to conquering Athens go;
There fate appoints an end of all thy woe.'

The fright awakened Arcite with a start,
Against his bosom bounced his heaving heart;
But soon he said, with scarce recovered breath,
'And thither will I go to meet my death,
Sure to be slain; but death is my desire,
Since in Emilia's sight I shall expire.'   560
By chance he spied a mirror while he spoke,
And gazing there beheld his altered look;
Wondering, he saw his features and his hue
So much were changed, that scarce himself he knew.
A sudden thought then starting in his mind,
'Since I in Arcite cannot Arcite find,
The world may search in vain with all their eyes,
But never penetrate through this disguise.
Thanks to the change which grief and sickness give,
In low estate I may securely live,   570
And see, unknown, my mistress day by day.'
He said, and clothed himself in coarse array,
A labouring hind in show; then forth he went,
And to the Athenian towers his journey bent:
One squire attended in the same disguise,
Made conscious of his master's enterprise.
Arrived at Athens, soon he came to court,
Unknown, unquestioned in that thick resort:°
Proffering for hire his service at the gate,
To drudge, draw water, and to run or wait.   580
   So fair befell him, that for little gain
He served at first Emilia's chamberlain;
And, watchful all advantages to spy,
Was still at hand, and in his master's eye;
And as his bones were big, and sinews strong,
Refused no toil that could to slaves belong;
But from deep wells with engines water drew,
And used his noble hands the wood to hew.
He passed a year at least attending thus
On Emily, and called Philostratus.°   590
But never was there man of his degree
So much esteemed, so well beloved as he.
So gentle of condition was he known,
That through the court his courtesy was blown:
All think him worthy of a greater place,

And recommend him to the royal grace;
That exercised within a higher sphere,
His virtues more conspicuous might appear.
Thus by the general voice was Arcite praised,
And by great Theseus to high favour raised;   600
Among his menial servants first enrolled,
And largely entertained with sums of gold:
Besides what secretly from Thebes was sent,
Of his own income and his annual rent.
This well employed, he purchased friends and fame,
But cautiously concealed from whence it came.
Thus for three years he lived with large increase
In arms of honour, and esteem in peace;
To Theseus' person he was ever near,
And Theseus for his virtues held him dear.   610

BOOK II

While Arcite lives in bliss, the story turns
Where hopeless Palamon in prison mourns.
For six long years immured, the captive knight
Had dragged his chains, and scarcely seen the light:
Lost liberty and love at once he bore;
His prison pained him much, his passion more:
Nor dares he hope his fetters to remove,
Nor ever wishes to be free from love.

  But when the sixth revolving year was run,
And May within the Twins received the sun,°   10
Were it by chance, or forceful destiny,
Which forms in causes first whate'er shall be,°
Assisted by a friend one moonless night,
This Palamon from prison took his flight:
A pleasant beverage he prepared before
Of wine and honey mixed, with added store
Of opium; to his keeper this he brought,
Who swallowed unaware the sleepy draught,
And snored secure till morn, his senses bound
In slumber, and in long oblivion drowned.   20
Short was the night, and careful Palamon
Sought the next covert ere the rising sun.
A thick-spread forest near the city lay,
To this with lengthened strides he took his way,

(For far he could not fly, and feared the day)
Safe from pursuit, he meant to shun the light,
Till the brown shadows of the friendly night
To Thebes might favour his intended flight.
When to his country come, his next design
Was all the Theban race in arms to join,   30
And war on Theseus, till he lost his life,
Or won the beauteous Emily to wife.
Thus while his thoughts the lingering day beguile,
To gentle Arcite let us turn our style;°
Who little dreamt how nigh he was to care,
Till treacherous fortune caught him in the snare.
The morning-lark, the messenger of day,
Saluted in her song the morning gray;
And soon the sun arose with beams so bright,
That all the horizon laughed to see the joyous sight;   40
He with his tepid rays the rose renews,
And licks the dropping leaves, and dries the dews;
When Arcite left his bed, resolved to pay
Observance to the month of merry May,
Forth on his fiery steed betimes he rode,
That scarcely prints the turf on which he trod:
At ease he seemed, and prancing o'er the plains,
Turned only to the grove his horse's reins,
The grove I named before, and, lighting there,
A woodbine garland sought to crown his hair;   50
Then turned his face against the rising day,°
And raised his voice to welcome in the May:
    'For thee, sweet month, the groves green liveries wear,
If not the first, the fairest of the year:
For thee the Graces lead the dancing hours,
And nature's ready pencil paints the flowers:
When thy short reign is past, the feverish sun
The sultry tropic fears, and moves more slowly on.
So may thy tender blossoms fear no blight,
Nor goats with venomed teeth thy tendrils bite,   60
As thou shalt guide my wandering feet to find
The fragrant greens I seek, my brows to bind.'
    His vows addressed, within the grove he strayed,
Till fate or fortune near the place conveyed
His steps where secret Palamon was laid.

Full little thought of him the gentle knight,
Who flying death had there concealed his flight,
In brakes and brambles hid, and shunning mortal sight;
And less he knew him for his hated foe,
But feared him as a man he did not know.   70
But as it has been said of ancient years,
That fields are full of eyes and woods have ears,
For this the wise are ever on their guard,
For unforeseen, they say, is unprepared.
Uncautious Arcite thought himself alone,
And less than all suspected Palamon,
Who, listening, heard him, while he searched the grove,
And loudly sung his roundelay of love:
But on the sudden stopped, and silent stood,
(As lovers often muse, and change their mood);   80
Now high as heaven, and then as low as hell,
Now up, now down, as buckets in a well:
For Venus, like her day, will change her cheer,°
And seldom shall we see a Friday clear.
Thus Arcite, having sung, with altered hue
Sunk on the ground, and from his bosom drew
A desperate sigh, accusing heaven and fate,
And angry Juno's unrelenting hate:
'Cursed be the day when first I did appear;
Let it be blotted from the calendar,   90
Lest it pollute the month, and poison all the year.
Still will the jealous queen pursue our race?
Cadmus is dead, the Theban city was:
Yet ceases not her hate; for all who come
From Cadmus are involved in Cadmus' doom.°
I suffer for my blood: unjust decree,
That punishes another's crime on me.
In mean estate I serve my mortal foe,
The man who caused my country's overthrow.
This is not all; for Juno, to my shame,   100
Has forced me to forsake my former name:
Arcite I was, Philostratus I am.
That side of heaven is all my enemy:
Mars ruined Thebes; his mother ruined me.
Of all the royal race remains but one
Beside myself, the unhappy Palamon,

Whom Theseus holds in bonds and will not free;
Without a crime, except his kin to me.
Yet these and all the rest I could endure;
But love's a malady without a cure: 110
Fierce love has pierced me with his fiery dart,
He fries within, and hisses at my heart.
Your eyes, fair Emily, my fate pursue;
I suffer for the rest, I die for you.
Of such a goddess no time leaves record,
Who burned the temple where she was adored:
And let it burn, I never will complain,
Pleased with my sufferings, if you knew my pain.'
    At this a sickly qualm his heart assailed,
His ears ring inward, and his senses failed.   120
No word missed Palamon of all he spoke;
But soon to deadly pale he changed his look:
He trembled every limb, and felt a smart,
As if cold steel had glided through his heart;
Nor longer stayed, but starting from his place,
Discovered stood, and showed his hostile face:
    'False traitor, Arcite, traitor to thy blood,
Bound by thy sacred oath to seek my good,
Now art thou found forsworn for Emily,
And darst attempt her love, for whom I die.   130
So hast thou cheated Theseus with a wile,
Against thy vow, returning to beguile
Under a borrowed name: as false to me,
So false thou art to him who set thee free.
But rest assured, that either thou shalt die,
Or else renounce thy claim in Emily;
For though unarmed I am, and, freed by chance,
Am here without my sword or pointed lance,
Hope not, base man, unquestioned hence to go,
For I am Palamon, thy mortal foe.'   140
    Arcite, who heard his tale and knew the man,
His sword unsheathed, and fiercely thus began:
'Now, by the gods who govern heaven above,
Wert thou not weak with hunger, mad with love,
That word had been thy last; or in this grove
This hand should force thee to renounce thy love;
The surety which I gave thee I defy:

Fool, not to know that love endures no tie,
And Jove but laughs at lovers' perjury.
Know, I will serve the fair in thy despite;   150
But since thou art my kinsman and a knight,
Here, have my faith, tomorrow in this grove
Our arms shall plead the titles of our love:
And heaven so help my right, as I alone
Will come, and keep the cause and quarrel both unknown,
With arms of proof both for myself and thee;°
Choose thou the best, and leave the worst to me.
And, that at better ease thou mayest abide,
Bedding and clothes I will this night provide,
And needful sustenance, that thou mayst be   160
A conquest better won, and worthy me.'
His promise Palamon accepts; but prayed,
To keep it better than the first he made.
Thus fair they parted till the morrow's dawn;
For each had laid his plighted faith to pawn.
Oh love! thou sternly dost thy power maintain,
And wilt not bear a rival in thy reign!
Tyrants and thou all fellowship disdain.
This was in Arcite proved and Palamon:
Both in despair, yet each would love alone.   170
Arcite returned, and, as in honour tied,
His foe with bedding and with food supplied;
Then, ere the day, two suits of armour sought,
Which borne before him on his steed he brought:
Both were of shining steel, and wrought so pure
As might the strokes of two such arms endure.
Now, at the time, and in the appointed place,
The challenger and challenged, face to face,
Approach; each other from afar they knew,
And from afar their hatred changed their hue.   180
So stands the Thracian herdsman with his spear,
Full in the gap, and hopes the hunted bear,
And hears him rustling in the wood, and sees
His course at distance by the bending trees:
And thinks, here comes my mortal enemy,
And either he must fall in fight, or I:
This while he thinks, he lifts aloft his dart;

A generous chillness seizes every part,
The veins pour back the blood, and fortify the heart.
  Thus pale they meet; their eyes with fury burn;   190
None greets, for none the greeting will return;
But in dumb surliness each armed with care
His foe professed, as brother of the war;
Then both, no moment lost, at once advance
Against each other, armed with sword and lance:
They lash, they foin, they pass, they strive to bore
Their corslets, and the thinnest parts explore.
Thus two long hours in equal arms they stood,
And wounded wound, till both were bathed in blood
And not a foot of ground had either got,   200
As if the world depended on the spot.
Fell Arcite like an angry tiger fared,
And like a lion Palamon appeared:
Or, as two boars whom love to battle draws,
With rising bristles and with frothy jaws,
Their adverse breasts with tusks oblique they wound
With grunts and groans the forest rings around.
So fought the knights, and fighting must abide,
Till fate an umpire sends their difference to decide.
The power that ministers to God's decrees,   210
And executes on earth what heaven foresees,
Called providence, or chance, or fatal sway,
Comes with resistless force, and finds or makes her way.
Nor kings, nor nations, nor united power
One moment can retard the appointed hour;
And some one day, some wondrous chance appears,
Which happened not in centuries of years:
For sure, whate'er we mortals hate or love
Or hope or fear depends on powers above:
They move our appetites to good or ill,   220
And by foresight necessitate the will.
In Theseus this appears, whose youthful joy
Was beasts of chase in forests to destroy;
This gentle knight, inspired by jolly May,
Forsook his easy couch at early day,
And to the wood and wilds pursued his way.
Beside him rode Hippolyta the queen,

And Emily attired in lively green,
With horns and hounds and all the tuneful cry,
To hunt a royal hart within the covert nigh: 230
And, as he followed Mars before, so now
He serves the goddess of the silver bow.°
The way that Theseus took was to the wood,
Where the two knights in cruel battle stood:
The laund on which they fought, the appointed place
In which the uncoupled hounds began the chase.
Thither forth-right he rode to rouse the prey,
That shaded by the fern in harbour lay;
And thence dislodged, was wont to leave the wood
For open fields, and cross the crystal flood.   240
Approached, and looking underneath the sun,
He saw proud Arcite and fierce Palamon,
In mortal battle doubling blow on blow;
Like lightning flamed their falchions to and fro,
And shot a dreadful gleam; so strong they strook,
There seemed less force required to fell an oak.
He gazed with wonder on their equal might,
Looked eager on, but knew not either knight.
Resolved to learn, he spurred his fiery steed
With goring rowels to provoke his speed.   250
The minute ended that began the race,
So soon he was betwixt them on the place;
And with his sword unsheathed, on pain of life
Commands both combatants to cease their strife;
Then with imperious tone pursues his threat:
'What are you? Why in arms together met?
How dares your pride presume against my laws,
As in a listed field to fight your cause,°
Unasked the royal grant; no marshal by,
As knightly rites require, nor judge to try?'   260
Then Palamon, with scarce recovered breath,
Thus hasty spoke: 'We both deserve the death,
And both would die; for look the world around,
A pair so wretched is not to be found.
Our life's a load; encumbered with the charge,
We long to set the imprisoned soul at large.
Now, as thou art a sovereign judge, decree
The rightful doom of death to him and me;

Let neither find thy grace, for grace is cruelty.
Me first, O kill me first, and cure my woe;   270
Then sheath the sword of justice on my foe;
Or kill him first, for when his name is heard,
He foremost will receive his due reward.
Arcite of Thebes is he, thy mortal foe,
On whom thy grace did liberty bestow;
But first contracted, that, if ever found
By day or night upon the Athenian ground,
His head should pay the forfeit; see returned
The perjured knight, his oath and honour scorned:
For this is he, who, with a borrowed name   280
And proffered service, to thy palace came,
Now called Philostratus; retained by thee,
A traitor trusted, and in high degree,
Aspiring to the bed of beauteous Emily.
My part remains, from Thebes my birth I own,
And call myself the unhappy Palamon.
Think me not like that man; since no disgrace
Can force me to renounce the honour of my race.
Know me for what I am: I broke thy chain,
Nor promised I thy prisoner to remain: 290
The love of liberty with life is given,
And life itself the inferior gift of heaven.
Thus without crime I fled; but further know,
I, with this Arcite, am thy mortal foe:
Then give me death, since I thy life pursue;
For safeguard of thyself, death is my due.
More wouldst thou know? I love bright Emily,
And for her sake and in her sight will die:
But kill my rival too, for he no less
Deserves; and I thy righteous doom will bless,   300
Assured that what I lose he never shall possess.'
To this replied the stern Athenian prince,
And sourly smiled: 'In owning your offence
You judge yourself, and I but keep record
In place of law, while you pronounce the word.
Take your desert, the death you have decreed;
I seal your doom, and ratify the deed:
By Mars, the patron of my arms, you die.'
     He said; dumb sorrow seized the standers-by.

The queen, above the rest, by nature good,   310
(The pattern formed of perfect womanhood)
For tender pity wept: when she began,
Through the bright choir the infectious virtue ran.
All dropt their tears, e'en the contended maid;
And thus among themselves they softly said:
'What eyes can suffer this unworthy sight!
Two youths of royal blood, renowned in fight,
The mastership of heaven in face and mind,
And lovers, far beyond their faithless kind:
See their wide streaming wounds; they neither came   320
From pride of empire nor desire of fame:
Kings fight for kingdoms, madmen for applause;
But love for love alone, that crowns the lover's cause.'
This thought, which ever bribes the beauteous kind,°
Such pity wrought in every lady's mind,
They left their steeds, and prostrate on the place,
From the fierce king implored the offenders' grace.
    He paused a while, stood silent in his mood;
(For yet his rage was boiling in his blood;)
But soon his tender mind the impression felt   330
(As softest metals are not slow to melt
And pity soonest runs in gentle minds:)
Then reasons with himself; and first he finds
His passion cast a mist before his sense,
And either made or magnified the offence.
Offence? Of what? To whom? Who judged the cause?
The prisoner freed himself by nature's laws;
Born free, he sought his right; the man he freed
Was perjured, but his love excused the deed:
Thus pondering, he looked under with his eyes,   340
And saw the women's tears, and heard their cries,
Which moved compassion more; he shook his head,
And softly sighing to himself he said:
    'Curse on the unpardoning prince, whom tears can draw
To no remorse, who rules by lion's law;
And deaf to prayers, by no submission bowed,
Rends all alike, the penitent and proud!'
At this with look serene he raised his head;
Reason resumed her place, and passion fled:
Then thus aloud he spoke:—'The power of love,   350

In earth, and seas, and air, and heaven above,
Rules, unresisted, with an awful nod,
By daily miracles declared a god;
He blinds the wise, gives eyesight to the blind;
And moulds and stamps anew the lover's mind.
Behold that Arcite, and this Palamon,
Freed from my fetters, and in safety gone,
What hindered either in their native soil
At ease to reap the harvest of their toil?
But love, their lord, did otherwise ordain,   360
And brought them, in their own despite again,
To suffer death deserved; for well they know
'Tis in my power, and I their deadly foe.
The proverb holds, that to be wise and love,°
Is hardly granted to the gods above.
See how the madmen bleed! behold the gains
With which their master, love, rewards their pains!
For seven long years, on duty every day,
Lo! their obedience, and their monarch's pay!
Yet, as in duty bound, they serve him on;   370
And ask the fools, they think it wisely done;
Nor ease nor wealth nor life itself regard,
For 'tis their maxim, love is love's reward.
This is not all; the fair, for whom they strove,
Nor knew before, nor could suspect their love,
Nor thought, when she beheld the fight from far,
Her beauty was the occasion of the war.
But sure a general doom on man is past,
And all are fools and lovers first or last:
This both by others and myself I know,   380
For I have served their sovereign long ago;
Oft have been caught within the winding train°
Of female snares, and felt the lover's pain,
And learned how far the god can human hearts constrain.
To this remembrance, and the prayers of those
Who for the offending warriors interpose,
I give their forfeit lives, on this accord,
To do me homage as their sovereign lord;
And as my vassals, to their utmost might,
Assist my person and assert my right.'   390
This freely sworn, the knights their grace obtained;

Then thus the king his secret thoughts explained:
'If wealth or honour or a royal race,
Or each or all, may win a lady's grace,
Then either of you knights may well deserve
A princess born; and such is she you serve:
For Emily is sister to the crown,
And but too well to both her beauty known:
But should you combat till you both were dead,
Two lovers cannot share a single bed.   400
As, therefore, both are equal in degree,
The lot of both be left to destiny.
Now hear the award, and happy may it prove
To her, and him who best deserves her love.
Depart from hence in peace, and free as air,
Search the wide world, and where you please repair;
But on the day when this returning sun
To the same point through every sign has run,
Then each of you his hundred knights shall bring
In royal lists, to fight before the king;   410
And then the knight, whom fate or happy chance
Shall with his friends to victory advance,
And grace his arms so far in equal fight,
From out the bars to force his opposite,°
Or kill, or make him recreant on the plain,
The prize of valour and of love shall gain;
The vanquished party shall their claim release,
And the long jars conclude in lasting peace.
The charge be mine to adorn the chosen ground,
The theatre of war, for champions so renowned: 420
And take the patron's place of either knight,
With eyes impartial to behold the fight;
And heaven of me so judge as I shall judge aright.
If both are satisfied with this accord,
Swear by the laws of knighthood on my sword.'
    Who now but Palamon exults with joy?
And ravished Arcite seems to touch the sky.
The whole assembled troop was pleased as well,
Extolled the award, and on their knees they fell
To bless the gracious king. The knights, with leave   430
Departing from the place, his last commands receive;
On Emily with equal ardour look,

And from her eyes their inspiration took:
From thence to Thebes' old walls pursue their way,
Each to provide his champions for the day.
   It might be deemed, on our historian's part,
Or too much negligence or want of art,
If he forgot the vast magnificence
Of royal Theseus, and his large expense.
He first enclosed for lists a level ground,   440
The whole circumference a mile around;
The form was circular; and all without
A trench was sunk, to moat the place about.
Within, an amphitheatre appeared,
Raised in degrees, to sixty paces reared:°
That when a man was placed in one degree,
Height was allowed for him above to see.
   Eastward was built a gate of marble white;
The like adorned the western opposite.
A nobler object than this fabric was   450
Rome never saw, nor of so vast a space:
For, rich with spoils of many a conquered land,
All arts and artists Theseus could command,
Who sold for hire, or wrought for better fame;
The master-painters and the carvers came.
So rose within the compass of the year
An age's work, a glorious theatre.
Then o'er its eastern gate was raised above
A temple, sacred to the Queen of Love;
An altar stood below; on either hand   460
A priest with roses crowned, who held a myrtle wand.°
   The dome of Mars was on the gate opposed,°
And on the north a turret was enclosed
Within the wall of alabaster white
And crimson coral, for the Queen of Night,
Who takes in sylvan sports her chaste delight.
   Within these oratories might you see
Rich carvings, portraitures, and imagery;
Where every figure to the life expressed
The godhead's power to whom it was addressed.   470
In Venus' temple on the sides were seen
The broken slumbers of enamoured men;
Prayers that e'en spoke, and pity seemed to call,

And issuing sighs that smoked along the wall;
Complaints and hot desires, the lover's hell,
And scalding tears that wore a channel where they fell;
And all around were nuptial bonds, the ties
Of love's assurance, and a train of lies,
That, made in lust, conclude in perjuries;
Beauty, and Youth, and Wealth and Luxury,   480
And sprightly Hope, and short-enduring Joy,
And sorceries, to raise the infernal powers,
And sigils framed in planetary hours;°
Expense, and After-thought, and idle Care,
And Doubts of motley hue, and dark Despair;
Suspicions and fantastical Surmise,
And Jealousy suffused, with jaundice in her eyes,
Discolouring all she viewed, in tawny dressed,
Down-looked, and with a cuckoo on her fist.°
Opposed to her, on t'other side advance   490
The costly feast, the carol, and the dance,°
Minstrels and music, poetry and play,
And balls by night, and tournaments by day.
All these were painted on the wall, and more;
With acts and monuments of times before
And others added by prophetic doom,
And lovers yet unborn, and loves to come:
For there the Idalian mount, and Citheron,°
The court of Venus, was in colours drawn;
Before the palace gate, in careless dress   500
And loose array, sat portress Idleness;
There by the fount Narcissus pined alone;°
There Samson was; with wiser Solomon,°
And all the mighty names by love undone.
Medea's charms were there; Circean feasts,°
With bowls that turned enamoured youth to beasts.
Here might be seen, that beauty, wealth, and wit,
And prowess to the power of love submit;
The spreading snare for all mankind is laid,
And lovers all betray, and are betrayed.   510
The goddess' self some noble hand had wrought;
Smiling she seemed, and full of pleasing thought;
From ocean as she first began to rise,°
And smoothed the ruffled seas, and cleared the skies,

She trod the brine, all bare below the breast,
And the green waves but ill concealed the rest:
A lute she held; and on her head was seen
A wreath of roses red and myrtles green;
Her turtles fanned the buxom air above;°
And by his mother stood an infant Love,°   520
With wings unfledged; his eyes were banded o'er,
His hands a bow, his back a quiver bore,
Supplied with arrows bright and keen, a deadly store.

    But in the dome of mighty Mars the red
With different figures all the sides were spread;
This temple, less in form, with equal grace,
Was imitative of the first in Thrace;
For that cold region was the loved abode
And sovereign mansion of the warrior god.
The landscape was a forest wide and bare,   530
Where neither beast nor human kind repair,
The fowl that scent afar the borders fly,
And shun the bitter blast, and wheel about the sky.
A cake of scurf lies baking on the ground,°
And prickly stubs, instead of trees, are found;
Or woods with knots and knares deformed and old,
Headless the most, the hideous to behold;
A rattling tempest through the branches went,
That stripped them bare, and one sole way they bent.
Heaven froze above severe, the clouds congeal,   540
And through the crystal vault appeared the standing hail.
Such was the face without: a mountain stood
Threatening from high, and overlooked the wood:
Beneath the louring brow, and on a bent,°
The temple stood of Mars armipotent;
The frame of burnished steel, that cast a glare
From far, and seemed to thaw the freezing air
A strait long entry to the temple led,
Blind with high walls, and horror over head;°
Thence issued such a blast, and hollow roar,   550
As threatened from the hinge to heave the door;
In through that door a northern light there shone;
'Twas all it had, for windows there were none.
The gate was adamant; eternal frame,
Which, hewed by Mars himself, from Indian quarries came,

The labour of a god; and all along
Tough iron plates were clenched to make it strong.
A tun about was every pillar there;°
A polished mirror shone not half so clear.
There saw I how the secret felon wrought,   560
And treason labouring in the traitor's thought,
And midwife Time the ripened plot to murder brought.
There the red Anger dared the pallid Fear;
Next stood Hypocrisy, with holy leer,
Soft smiling, and demurely looking down,
But hid the dagger underneath the gown;
The assassinating wife, the household fiend;
And far the blackest there, the traitor-friend.
On t'other side there stood Destruction bare,
Unpunished Rapine, and a waste of war;   570
Contest with sharpened knives in cloisters drawn,
And all with blood bespread the holy lawn.
Loud menaces were heard, and foul disgrace,
And bawling infamy, in language base;
Till sense was lost in sound, and silence fled the place.
The slayer of himself yet saw I there,
The gore congealed was clottered in his hair;
With eyes half closed and gaping mouth he lay,
And grim as when he breathed his sullen soul away.
In midst of all the dome, Misfortune sate,   580
And gloomy Discontent, and fell Debate,
And Madness laughing in his ireful mood;
And armed Complaint on theft; and cries of blood.
There was the murdered corpse in covert laid,
And violent death in thousand shapes displayed:
The city to the soldier's rage resigned;
Successless wars, and poverty behind:
Ships burnt in fight, or forced on rocky shores,
And the rash hunter strangled by the boars:°
The new-born babe by nurses overlaid;°   590
And the cook caught within the raging fire he made.
All ills of Mars's nature, flame and steel;
The gasping charioteer beneath the wheel
Of his own car; the ruined house that falls
And intercepts her lord betwixt the walls:°
The whole division that to Mars pertains,

All trades of death that deal in steel for gains
Were there: the butcher, armourer, and smith,
Who forges sharpened falchions, or the scythe.
The scarlet Conquest on a tower was placed,   600
With shouts and soldiers' acclamations graced:
A pointed sword hung threatening o'er his head,
Sustained but by a slender twine of thread.°
There saw I Mars's ides, the Capitol,°
The seer in vain foretelling Caesar's fall;
The last Triumvirs, and the wars they move,
And Antony, who lost the world for love.
These, and a thousand more, the fane adorn;
Their fates were painted ere the men were born,
All copied from the heavens, and ruling force   610
Of the red star, in his revolving course.
The form of Mars high on a chariot stood,
All sheathed in arms, and gruffly looked the god;
Two geomantic figures were displayed°
Above his head, a warrior and a maid,°
One when direct, and one when retrograde.
   Tired with deformities of death, I haste
To the third temple of Diana chaste.
A sylvan scene with various greens was drawn,
Shades on the sides, and on the midst a lawn;   620
The silver Cynthia, with her nymphs around,
Pursued the flying deer, the woods with horns resound:
Callisto there stood manifest of shame,°
And, turned a bear, the northern star became:
Her son was next, and, by peculiar grace,
In the cold circle held the second place;
The stag Actaeon in the stream had spied°
The naked huntress, and for seeing died;
His hounds, unknowing of his change, pursue
The chase, and their mistaken master slew.   630
Peneian Daphne too was there to see,°
Apollo's love before, and now his tree.
The adjoining fane the assembled Greeks expressed,
And hunting of the Calydonian beast:°
Oenides' valour, and his envied prize;°
The fatal power of Atalanta's eyes;°
Diana's vengeance on the victor shown,

The murderess mother, and consuming son;
The Volscian queen extended on the plain,°
The treason punished, and the traitor slain,   640
The rest were various huntings, well designed,
And savage beasts destroyed, of every kind.
The graceful goddess was arrayed in green;
About her feet were little beagles seen,
That watched with upward eyes the motions of their queen.
Her legs were buskined, and the left before,
In act to shoot; a silver bow she bore,
And at her back a painted quiver wore.
She trod a waxing moon, that soon would wane,
And, drinking borrowed light, be filled again;   650
With downcast eyes, as seeming to survey
The dark dominions, her alternate sway.
Before her stood a woman in her throes,
And called Lucina's aid, her burden to disclose.°
All these the painter drew with such command,
That nature snatched the pencil from his hand,
Ashamed and angry that his art could feign,
And mend the tortures of a mother's pain.
Theseus beheld the fanes of every god,
And thought his mighty cost was well bestowed.   660
So princes now their poets should regard;
But few can write, and fewer can reward.
    The theatre thus raised, the lists enclosed,
And all with vast magnificence disposed,
We leave the monarch pleased, and haste to bring
The knights to combat, and their arms to sing.

BOOK III

The day approached when fortune should decide
The important enterprise, and give the bride;
For now the rivals round the world had sought,
And each his number, well appointed, brought.
The nations far and near contend in choice,
And send the flower of war by public voice;°
That after or before were never known
Such chiefs, as each an army seemed alone:
Beside the champions, all of high degree,
Who knighthood loved, and deeds of chivalry,   10

Thronged to the lists, and envied to behold
The names of others, not their own, enrolled.
Nor seems it strange; for every noble knight
Who loves the fair, and is endued with might,
In such a quarrel would be proud to fight.
There breathes not scarce a man on British ground
(An isle for love and arms of old renowned)
But would have sold his life to purchase fame,
To Palamon or Arcite sent his name;
And had the land selected of the best,   20
Half had come hence, and let the world provide the rest.
A hundred knights with Palamon there came,
Approved in fight, and men of mighty name;°
Their arms were several, as their nations were,
But furnished all alike with sword and spear.
Some wore coat armour, imitating scale,
And next their skins were stubborn shirts of mail;
Some wore a breastplate and a light jupon,
Their horses clothed with rich caparison;
Some for defence would leathern bucklers use   30
Of folded hides, and others shields of Pruce.
One hung a pole-axe at his saddle-bow,
And one a heavy mace to stun the foe;
One for his legs and knees provided well,
With jambeaux armed, and double plates of steel;
This on his helmet wore a lady's glove,
And that a sleeve embroidered by his love.
    With Palamon above the rest in place,
Lycurgus came, the surly king of Thrace;
Black was his beard, and manly was his face,   40
The balls of his broad eyes rolled in his head,
And glared betwixt a yellow and a red;
He looked a lion with a gloomy stare,
And o'er his eyebrows hung his matted hair;
Big-boned and large of limbs, with sinews strong,
Broad-shouldered, and his arms were round and long.
Four milk-white bulls (the Thracian use of old)
Were yoked to draw his car of burnished gold.
Upright he stood, and bore aloft his shield,
Conspicuous from afar, and overlooked the field.   50
His surcoat was a bear-skin on his back;

His hair hung long behind, and glossy raven-black.
His ample forehead bore a coronet,
With sparkling diamonds and with rubies set.
Ten brace, and more, of greyhounds, snowy fair,
And tall as stags, ran loose, and coursed around his chair,
A match for pards in flight, in grappling, for the bear;
With golden muzzles all their mouths were bound,
And collars of the same their necks surround.
Thus through the fields Lycurgus took his way;   60
His hundred knights attend in pomp and proud array.

     To match this monarch, with strong Arcite came
Emetrius, king of Inde, a mighty name,
On a bay courser, goodly to behold,
The trappings of his horse embossed with barbarous gold.
Not Mars bestrode a steed with greater grace;
His surcoat o'er his arms was cloth of Thrace,
Adorned with pearls, all orient, round, and great;
His saddle was of gold, with emeralds set;
His shoulders large a mantle did attire,   70
With rubies thick, and sparkling as the fire;
His amber-coloured locks in ringlets run,
With graceful negligence, and shone against the sun.
His nose was aquiline, his eyes were blue,
Ruddy his lips, and fresh and fair his hue;
Some sprinkled freckles on his face were seen,
Whose dusk set off the whiteness of the skin.
His awful presence did the crowd surprise,
Nor durst the rash spectator meet his eyes;
Eyes that confessed him born for kingly sway,   80
So fierce, they flashed intolerable day.
His age in nature's youthful prime appeared,
And just began to bloom his yellow beard.
Whene'er he spoke, his voice was heard around,
Loud as a trumpet, with a silver sound;
A laurel wreathed his temples, fresh, and green,
And myrtle sprigs, the marks of love, were mixed between.
Upon his fist he bore, for his delight,
An eagle well reclaimed, and lily white.°

     His hundred knights attend him to the war,   90
All armed for battle; save their heads were bare.
Words and devices blazed on every shield,

And pleasing was the terror of the field.
For kings, and dukes, and barons you might see,
Like sparkling stars, though different in degree,
All for the increase of arms, and love of chivalry.°
Before the king tame leopards led the way,
And troops of lions innocently play.
So Bacchus through the conquered Indies rode,°
And beasts in gambols frisked before their honest god.   100
   In this array the war of either side
Through Athens passed with military pride.
At prime, they entered on the Sunday morn;
Rich tapestry spread the streets, and flowers the posts adorn.
The town was all a jubilee of feasts;
So Theseus willed in honour of his guests;
Himself with open arms the kings embraced,
Then all the rest in their degrees were graced.
No harbinger was needful for the night,°
For every house was proud to lodge a knight.   110
   I pass the royal treat, nor must relate
The gifts bestowed, nor how the champions sate;
Who first, who last, or how the knights addressed
Their vows, or who was fairest at the feast;
Whose voice, whose graceful dance did most surprise,
Soft amorous sighs, and silent love of eyes.
The rivals call my muse another way,
To sing their vigils for the ensuing day.
   'Twas ebbing darkness, past the noon of night:
And Phosphor, on the confines of the light,°   120
Promised the sun; ere day began to spring,
The tuneful lark already stretched her wing,
And flickering on her nest, made short essays to sing:
When wakeful Palamon, preventing day,°
Took to the royal lists his early way,
To Venus at her fane, in her own house, to pray.
There, falling on his knees before her shrine,
He thus implored with prayers her power divine:
'Creator Venus, genial power of love,°
The bliss of men below, and gods above!   130
Beneath the sliding sun thou runst thy race,
Dost fairest shine, and best become thy place.
For thee the winds their eastern blasts forbear,

Thy month reveals the spring, and opens all the year.°
Thee, goddess, thee the storms of winter fly;
Earth smiles with flowers renewing, laughs the sky,
And birds to lays of love their tuneful notes apply.
For thee the lion loathes the taste of blood,
And roaring hunts his female through the wood;
For thee the bulls rebellow through the groves,   140
And tempt the stream, and snuff their absent loves.
'Tis thine, whate'er is pleasant, good, or fair;
All nature is thy province, life thy care;
Thou mad'st the world, and dost the world repair.
Thou gladder of the mount of Cytheron,
Increase of Jove, companion of the sun,°
If e'er Adonis touched thy tender heart,°
Have pity, goddess, for thou knowst the smart!
Alas! I have not words to tell my grief;
To vent my sorrow would be some relief;   150
Light sufferings give us leisure to complain;
We groan, but cannot speak, in greater pain.
O goddess, tell thyself what I would say!
Thou knowst it, and I feel too much to pray.
So grant my suit, as I enforce my might,
In love to be thy champion and thy knight,
A servant to thy sex, a slave to thee,
A foe professed to barren chastity:
Nor ask I fame or honour of the field,
Nor choose I more to vanquish than to yield: 160
In my divine Emilia make me blest,
Let fate or partial chance dispose the rest:
Find thou the manner, and the means prepare;
Possession, more than conquest, is my care.
Mars is the warrior's god; in him it lies
On whom he favours to confer the prize;
With smiling aspect you serenely move
In your fifth orb, and rule the realm of love.°
The Fates but only spin the coarser clue,°
The finest of the wool is left for you: 170
Spare me but one small portion of the twine,
And let the Sisters cut below your line:°
The rest among the rubbish may they sweep,
Or add it to the yarn of some old miser's heap.

But if you this ambitious prayer deny,
(A wish, I grant, beyond mortality,)
Then let me sink beneath proud Arcite's arms,
And, I once dead, let him possess her charms.'
    Thus ended he; then, with observance due,
The sacred incense on her altar threw: 180
The curling smoke mounts heavy from the fires;
At length it catches flame, and in a blaze expires;
At once the gracious goddess gave the sign,
Her statue shook, and trembled all the shrine:
Pleased, Palamon the tardy omen took;
For since the flames pursued the trailing smoke,
He knew his boon was granted, but the day
To distance driven, and joy adjourned with long delay.
    Now morn with rosy light had streaked the sky,
Up rose the sun, and up rose Emily;   190
Addressed her early steps to Cynthia's fane,
In state attended by her maiden train,
Who bore the vests that holy rites require,°
Incense, and odorous gums, and covered fire.
The plenteous horns with pleasant mead they crown,
Nor wanted aught besides in honour of the moon.
Now, while the temple smoked with hallowed steam,
They wash the virgin in a living stream;
The secret ceremonies I conceal,
Uncouth, perhaps unlawful to reveal: 200
But such they were as pagan use required,
Performed by women when the men retired,
Whose eyes profane their chaste mysterious rites
Might turn to scandal or obscene delights.
Well-meaners think no harm; but for the rest,
Things sacred they pervert, and silence is the best.
Her shining hair, uncombed, was loosely spread,
A crown of mastless oak adorned her head:°
When to the shrine approached, the spotless maid
Had kindling fires on either altar laid;   210
(The rites were such as were observed of old,
By Statius in his Theban story told.)°
Then kneeling with her hands across her breast,
Thus lowly she preferred her chaste request.
    'O goddess, haunter of the woodland green,

To whom both heaven and earth and seas are seen;
Queen of the nether skies, where half the year°
Thy silver beams descend, and light the gloomy sphere;°
Goddess of maids, and conscious of our hearts,°
So keep me from the vengeance of thy darts,   220
(Which Niobe's devoted issue felt,°
When hissing through the skies the feathered deaths were dealt,)
As I desire to live a virgin life,
Nor know the name of mother or of wife.
Thy votress from my tender years I am,
And love, like thee, the woods and sylvan game.
Like death, thou knowest, I loathe the nuptial state,
And man, the tyrant of our sex, I hate,
A lowly servant, but a lofty mate;
Where love is duty on the female side,   230
On theirs mere sensual gust, and sought with surly pride.°
Now by thy triple shape, as thou art seen°
In heaven, earth, hell, and everywhere a queen,
Grant this my first desire; let discord cease,
And make betwixt the rivals lasting peace;
Quench their hot fire, or far from me remove
The flame, and turn it on some other love;
Or if my frowning stars have so decreed,
That one must be rejected, one succeed,
Make him my lord, within whose faithful breast   240
Is fixed my image, and who loves me best.
But oh! e'en that avert! I choose it not,
But take it as the least unhappy lot.
A maid I am, and of thy virgin train;
Oh, let me still that spotless name retain!
Frequent the forests, thy chaste will obey,
And only make the beasts of chase my prey!'
    The flames ascend on either altar clear,
While thus the blameless maid addressed her prayer.
When lo! the burning fire that shone so bright   250
Flew off, all sudden, with extinguished light,
And left one altar dark, a little space,
Which turned self-kindled, and renewed the blaze;
That other victor-flame a moment stood,
Then fell, and lifeless left the extinguished wood;
For ever lost, the irrevocable light

Forsook the blackening coals, and sunk to night:
At either end it whistled as it flew,
And as the brands were green, so dropped the dew,
Infected as it fell with sweat of sanguine hue.   260
　　The maid from that ill omen turned her eyes,
And with loud shrieks and clamours rent the skies;
Nor knew what signified the boding sign,
But found the powers displeased, and feared the wrath divine.
　　Then shook the sacred shrine, and sudden light
Sprung through the vaulted roof, and made the temple bright.
The power, behold! the power in glory shone,
By her bent bow and her keen arrows known;
The rest, a huntress issuing from the wood,
Reclining on her cornel spear she stood.°   270
Then gracious thus began: 'Dismiss thy fear,
And heaven's unchanged decrees attentive hear:
More powerful gods have torn thee from my side,
Unwilling to resign, and doomed a bride;
The two contending knights are weighed above;
One Mars protects, and one the queen of love:
But which the man is in the thunderer's breast;°
This he pronounced, "'Tis he who loves thee best."
The fire that, once extinct, revived again
Foreshows the love allotted to remain.   280
Farewell!' she said, and vanished from the place;
The sheaf of arrows shook, and rattled in the case.
Aghast at this, the royal virgin stood,
Disclaimed, and now no more a sister of the wood:°
But to the parting goddess thus she prayed:
'Propitious still, be present to my aid,
Nor quite abandon your once favoured maid.'
Then sighing she returned; but smiled betwixt,
With hopes, and fears, and joys with sorrows mixed.
　　The next returning planetary hour°   290
Of Mars, who shared the heptarchy of power,°
His steps bold Arcite to the temple bent,
To adore with pagan rites the power armipotent:
Then prostrate, low before his altar lay,
And raised his manly voice, and thus began to pray:
'Strong god of arms, whose iron sceptre sways
The freezing north, and hyperborean seas,°

And Scythian colds, and Thracia's wintry coast,
Where stand thy steeds, and thou art honoured most:
There most, but everywhere thy power is known,   300
The fortune of the fight is all thy own:
Terror is thine, and wild amazement, flung
From out thy chariot, withers e'en the strong;
And disarray and shameful rout ensue,
And force is added to the fainting crew.
Acknowledged as thou art, accept my prayer!
If aught I have achieved deserve thy care,
If to my utmost power with sword and shield
I dared the death, unknowing how to yield,
And falling in my rank, still kept the field;   310
Then let my arms prevail, by thee sustained,
That Emily by conquest may be gained.
Have pity on my pains; nor those unknown
To Mars, which, when a lover, were his own.
Venus, the public care of all above,
Thy stubborn heart has softened into love:
Now, by her blandishments and powerful charms,
When yielded she lay curling in thy arms,
Even by thy shame, if shame it may be called,
When Vulcan had thee in his net enthralled:°   320
O envied ignominy, sweet disgrace,
When every god that saw thee wished thy place!
By those dear pleasures, aid my arms in fight,
And make me conquer in my patron's right:
For I am young, a novice in the trade,
The fool of love, unpractised to persuade,
And want the soothing arts that catch the fair,
But, caught myself, lie struggling in the snare;
And she I love or laughs at all my pain
Or knows her worth too well, and pays me with disdain.   330
For sure I am, unless I win in arms,
To stand excluded from Emilia's charms:
Nor can my strength avail, unless by thee
Endued with force I gain the victory;
Then for the fire which warmed thy generous heart,
Pity thy subject's pains and equal smart.
So be the morrow's sweat and labour mine,
The palm and honour of the conquest thine:

Then shall the war, and stern debate, and strife
Immortal be the business of my life;   340
And in thy fane, the dusty spoils among,
High on the burnished roof, my banner shall be hung,
Ranked with my champion's bucklers; and below,
With arms reversed, the achievements of my foe;°
And while these limbs the vital spirit feeds;
While day to night and night to day succeeds,
Thy smoking altar shall be fat with food
Of incense and the grateful steam of blood;
Burnt-offerings morn and evening shall be thine,
And fires eternal in thy temple shine.   350
This bush of yellow beard, this length of hair,
Which from my birth inviolate I bear,
Guiltless of steel, and from the razor free,
Shall fall a plenteous crop, reserved for thee.
So may my arms with victory be blest,
I ask no more; let fate dispose the rest.'

    The champion ceased; there followed in the close°
A hollow groan; a murmuring wind arose;
The rings of iron, that on the doors were hung,
Sent out a jarring sound and harshly rung: 360
The bolted gates flew open at the blast,
The storm rushed in, and Arcite stood aghast:
The flames were blown aside, yet shone they bright,
Fanned by the wind, and gave a ruffled light.°

    Then from the ground a scent began to rise,
Sweet smelling as accepted sacrifice:
This omen pleased, and as the flames aspire,
With odorous incense Arcite heaps the fire:
Nor wanted hymns to Mars or heathen charms:°
At length the nodding statue clashed his arms,   370
And with a sullen sound and feeble cry,
Half sunk and half pronounced the word of victory.
For this, with soul devout, he thanked the god,
And, of success secure, returned to his abode.

    These vows thus granted raised a strife above
Betwixt the god of war and queen of love.
She, granting first, had right of time to plead;°
But he had granted too, nor would recede.
Jove was for Venus, but he feared his wife,°

And seemed unwilling to decide the strife;   380
Till Saturn from his leaden throne arose,
And found a way the difference to compose:
Though sparing of his grace, to mischief bent,
He seldom does a good with good intent.
Wayward, but wise; by long experience taught,
To please both parties, for ill ends, he sought:
For this advantage age from youth has won,
As not to be outridden, though outrun.°
By fortune he was now to Venus trined,°
And with stern Mars in Capricorn was joined:°   390
Of him disposing in his own abode,
He soothed the goddess, while he gulled the god:
'Cease, daughter, to complain, and stint the strife;
Thy Palamon shall have his promised wife:
And Mars, the lord of conquest, in the fight
With palm and laurel shall adorn his knight.
Wide is my course, not turn I to my place°
Till length of time, and move with tardy pace,
Man feels me, when I press the ethereal plains;
My hand is heavy, and the wound remains.   400
Mine is the shipwreck in a watery sign;°
And in an earthy the dark dungeon mine.°
Cold shivering agues, melancholy care,
And bitter blasting winds, and poisoned air,
Are mine, and wilful death, resulting from despair.°
The throttling quinsy 'tis my star appoints,
And rheumatisms I send to rack the joints:
When churls rebel against their native prince,
I arm their hands, and furnish the pretence;
And housing in the lion's hateful sign,°   410
Bought senates and deserting troops are mine.
Mine is the privy poisoning; I command
Unkindly seasons and ungrateful land.°
By me the king's palaces are pushed to ground,
And miners crushed beneath their mines are found.
'Twas I slew Samson, when the pillared hall
Fell down, and crushed the many with the fall.
My looking is the sire of pestilence,
That sweeps at once the people and the prince.
Now weep no more, but trust thy grandsire's art.°   420

Mars shall be pleased, and thou perform thy part.
'Tis ill, though different your complexions are,°
The family of heaven for men should war.'
The expedient pleased, where neither lost his right;
Mars had the day, and Venus had the night.
The management they left to Chronos' care.°
Now turn we to the effect, and sing the war.

   In Athens all was pleasure, mirth, and play,
All proper to the spring, and sprightly May:
Which every soul inspired with such delight,   430
'Twas jousting all the day, and love at night.
Heaven smiled, and gladded was the heart of man;
And Venus had the world as when it first began.
At length in sleep their bodies they compose,
And dreamt the future fight, and early rose.

   Now scarce the dawning day began to spring,
As at a signal given, the streets with clamours ring:
At once the crowd arose; confused and high,
E'en from the heaven was heard a shouting cry,
For Mars was early up, and roused the sky.   440
The gods came downward to behold the wars,
Sharpening their sights, and leaning from their stars.
The neighing of the generous horse was heard,
For battle by the busy groom prepared:
Rustling of harness, rattling of the shield,
Clattering of armour, furbished for the field.
Crowds to the castle mounted up the street;
Battering the pavement with their coursers' feet:
The greedy sight might there devour the gold
Of glittering arms, too dazzling to behold: 450
And polished steel that cast the view aside,
And crested morions, with their plumy pride.
Knights, with a long retinue of their squires,
In gaudy liveries march, and quaint attires.
One laced the helm, another held the lance;
A third the shining buckler did advance.
The courser pawed the ground with restless feet,
And snorting foamed, and champed the golden bit.
The smiths and armourers on palfreys ride,°
Files in their hands, and hammers at their side,   460
And nails for loosened spears, and thongs for shields provide.

The yeomen guard the streets, in seemly bands;
And clowns come crowding on, with cudgels in their hands.
   The trumpets, next the gate, in order placed,
Attend the sign to sound the martial blast:
The palace-yard is filled with floating tides,
And the last comers bear the former to the sides.
The throng is in the midst; the common crew
Shut out, the hall admits the better few.
In knots they stand, or in a rank they walk,   470
Serious in aspect, earnest in their talk:
Factious, and favouring this or t'other side,
As their strong fancies and weak reason guide.
Their wagers back their wishes; numbers hold
With the fair freckled king, and beard of gold:
So vigorous are his eyes, such rays they cast,
So prominent his eagle's beak is placed.
But most their looks on the black monarch bend,
His rising muscles, and his brawn commend;
His double-biting axe, and beamy spear,   480
Each asking a gigantic force to rear.
All spoke as partial favour moved the mind;
And safe themselves, at others' cost divined.
   Waked by the cries, the Athenian chief arose,
The knightly forms of combat to dispose;
And passing through the obsequious guards, he sate
Conspicuous on a throne, sublime in state;
There, for the two contending knights he sent;
Armed cap-a-pe, with reverence low they bent;°
He smiled on both, and with superior look   490
Alike their offered adoration took.
The people press on every side to see
Their awful prince, and hear his high decree.
Then signing to the heralds with his hand,
They gave his orders from their lofty stand.
Silence is thrice enjoined; then thus aloud
The king at arms bespeaks the knights and listening crowd:°
   'Our sovereign lord has pondered in his mind
The means to spare the blood of gentle kind;
And of his grace, and inborn clemency,   500
He modifies his first severe decree,
The keener edge of battle to rebate,

The troops for honour fighting, not for hate.
He wills, not death should terminate their strife,
And wounds, if wounds ensue, be short of life;
But issues ere the fight his dread command,
That slings afar, and poniards hand to hand,
Be banished from the field; that none shall dare
With shortened sword to stab in closer war;
But in fair combat fight with manly strength,   510
Nor push with biting point, but strike at length.
The tourney is allowed but one career,°
Of the tough ash, with the sharp-grinded spear;°
But knights unhorsed may rise from off the plain,
And fight on foot their honour to regain;
Nor, if at mischief taken, on the ground°
Be slain, but prisoners to the pillar bound,
At either barrier placed; nor (captives made)
Be freed, or armed anew the fight invade.
The chief of either side, bereft of life,   520
Or yielded to his foe, concludes the strife.
Thus dooms the lord; now, valiant knights and young,
Fight each his fill with swords and maces long.'
    The herald ends: the vaulted firmament
With loud acclaims and vast applause is rent:
Heaven guard a prince so gracious and so good,
So just, and yet so provident of blood!
This was the general cry. The trumpets sound,
And warlike symphony is heard around.
The marching troops through Athens take their way,   530
The great earl-marshal orders their array.°
The fair from high the passing pomp behold;
A rain of flowers is from the windows rolled.
The casements are with golden tissue spread,
And horses' hoofs, for earth, on silken tapestry tread.
The king goes midmost, and the rivals ride
In equal rank, and close his either side.
Next after these, there rode the royal wife,
With Emily the cause and the reward of strife.
The following cavalcade, by three and three,   540
Proceed by titles marshalled in degree.°
Thus through the southern gate they take their way,
And at the lists arrived ere prime of day.

There, parting from the king, the chiefs divide,
And wheeling east and west, before their many ride.
The Athenian monarch mounts his throne on high,
And after him the queen and Emily:
Next these, the kindred of the crown are graced
With nearer seats, and lords by ladies placed.
Scarce were they seated, when with clamours loud   550
In rushed at once a rude promiscuous crowd:°
The guards, and then each other overbear,
And in a moment throng the spacious theatre.
Now changed the jarring noise to whispers low,
As winds forsaking seas more softly blow;
When at the western gate, on which the car
Is placed aloft, that bears the god of war,
Proud Arcite, entering armed before his train,
Stops at the barrier, and divides the plain.
Red was his banner, and displayed abroad   560
The bloody colours of his patron god.
    At that self moment enters Palamon
The gate of Venus, and the rising sun;
Waved by the wanton winds, his banner flies,
All maiden white, and shares the people's eyes.
From east to west, look all the world around,
Two troops so matched were never to be found;
Such bodies built for strength, of equal age,
In stature sized; so proud an equipage:
The nicest eye could no distinction make,   570
Where lay the advantage, or what side to take.
    Thus ranged, the herald for the last proclaims
A silence, while they answered to their names:
For so the king decreed, to shun with care
The fraud of musters false, the common bane of war.
The tale was just, and then the gates were closed;
And chief to chief, and troop to troop opposed.
The heralds last retired, and loudly cried,
'The fortune of the field be fairly tried!'
    At this, the challenger, with fierce defy,   580
His trumpet sounds; the challenged makes reply.
With clangour rings the field, resounds the vaulted sky.
Their visors closed, their lances in the rest,
Or at the helmet pointed, or the crest,

They vanish from the barrier, speed the race,
And spurring see decrease the middle space.
A cloud of smoke envelops either host,
And all at once the combatants are lost:
Darkling they join adverse, and shock unseen,°
Coursers with coursers jostling, men with men:°   590
As labouring in eclipse, a while they stay,
Till the next blast of wind restores the day.
They look anew; the beauteous form of fight
Is changed, and war appears a grisly sight.
Two troops in fair array one moment showed,
The next, a field with fallen bodies strowed:
Not half the number in their seats are found;
But men and steeds lie grovelling on the ground.
The points of spears are stuck within the shield,
The steeds without their riders scour the field.   600
The knights, unhorsed, on foot renew the fight;
The glittering falchions cast a gleaming light;
Hauberks and helms are hewed with many a wound;°
Out spins the streaming blood, and dyes the ground.
The mighty maces with such haste descend,
They break the bones, and make the solid armour bend.
This thrusts amid the throng with furious force;
Down goes, at once, the horseman and the horse:
That courser stumbles on the fallen steed,
And floundering throws the rider o'er his head.   610
One rolls along, a football to his foes;
One with a broken truncheon deals his blows.°
This halting, this disabled with his wound,°
In triumph led, is to the pillar bound,
Where by the king's award he must abide;
There goes a captive led on t'other side.
By fits they cease; and leaning on the lance,°
Take breath awhile, and to new fight advance.
    Full oft the rivals met, and neither spared
His utmost force, and each forgot to ward.   620
The head of this was to the saddle bent,
That other backward to the crupper sent:°
Both were by turns unhorsed; the jealous blows
Fall thick and heavy, when on foot they close.
So deep their falchions bite, that every stroke

Pierced to the quick, and equal wounds they gave and took.
Borne far asunder by the tides of men,
Like adamant and steel they meet again.
    So when a tiger sucks the bullock's blood,
A famished lion issuing from the wood   630
Roars lordly fierce, and challenges the food.°
Each claims possession, neither will obey,
But both their paws are fastened on the prey;
They bite, they tear; and while in vain they strive,
The swains come armed between, and both to distance drive.
    At length, as fate foredoomed, and all things tend
By course of time to their appointed end;
So when the sun to west was far declined,
And both afresh in mortal battle joined,
The strong Emetrius came in Arcite's aid,   640
And Palamon with odds was overlaid:
For turning short, he struck with all his might°
Full on the helmet of the unwary knight.
Deep was the wound; he staggered with the blow,
And turned him to his unexpected foe;
Whom with such force he struck, he felled him down,
And cleft the circle of his golden crown.
But Arcite's men, who now prevailed in fight,
Twice ten at once surround the single knight:
O'erpowered, at length, they force him to the ground,   650
Unyielded as he was, and to the pillar bound;°
And king Lycurgus, while he fought in vain
His friend to free, was tumbled on the plain.
    Who now laments but Palamon, compelled
No more to try the fortune of the field!
And, worse than death, to view with hateful eyes
His rival's conquest, and renounce the prize!
    The royal judge on his tribunal placed,°
Who had beheld the fight from first to last,
Bade cease the war; pronouncing from on high   660
Arcite of Thebes had won the beauteous Emily.
The sound of trumpets to the voice replied,
And round the royal lists the heralds cried,
'Arcite of Thebes has won the beauteous bride!'
    The people rend the skies with vast applause;
All own the chief, when fortune owns the cause.

Arcite is owned e'en by the gods above,
And conquering Mars insults the Queen of Love.
So laughed he, when the rightful Titan failed,°
And Jove's usurping arms in heaven prevailed.   670
Laughed all the powers who favour tyranny,
And all the standing army of the sky.
But Venus with dejected eyes appears,
And, weeping, on the lists distilled her tears;
Her will refused, which grieves a woman most,
And, in her champion foiled, the cause of love is lost.
Till Saturn said: 'Fair daughter, now be still:
The blustering fool has satisfied his will;°
His boon is given; his knight has gained the day,
But lost the prize; the arrears are yet to pay.   680
Thy hour is come, and mine the care shall be
To please thy knight, and set thy promise free.'
    Now while the heralds run the lists around,
And 'Arcite, Arcite,' heaven and earth resound;
A miracle (nor less it could be called)
Their joy with unexpected sorrow palled.
The victor knight had laid his helm aside,
(Part for his ease, the greater part for pride,)
Bare-headed, popularly low he bowed,°
And paid the salutations of the crowd;   690
Then spurring, at full speed ran endlong on
Where Theseus sat on his imperial throne;
Furious he drove, and upward cast his eye,
Where next the queen was placed his Emily;
Then passing, to the saddle-bow he bent;
A sweet regard the gracious virgin lent;
(For women, to the brave an easy prey,
Still follow Fortune where she leads the way;)
Just then from earth sprung out a flashing fire,°
By Pluto sent, at Saturn's bad desire: 700
The startling steed was seized with sudden fright,
And, bounding, o'er the pommel cast the knight:
Forward he flew, and, pitching on his head,
He quivered with his feet, and lay for dead.
Black was his countenance in a little space,
For all the blood was gathered in his face.
Help was at hand: they reared him from the ground,

And from his cumbrous arms his limbs unbound;
Then lanced a vein, and watched returning breath;
It came, but clogged with symptoms of his death.   710
The saddle-bow the noble parts had pressed,°
All bruised and mortified his manly breast.°
Him still entranced, and in a litter laid,
They bore from field, and to his bed conveyed.
At length he waked, and with a feeble cry,
The word he first pronounced was Emily.

    Meantime the king, though inwardly he mourned,
In pomp triumphant to the town returned,
Attended by the chiefs, who fought the field,
(Now friendly mixed, and in one troop compelled,)°   720
Composed his looks to counterfeited cheer,
And bade them not for Arcite's life to fear,
But that which gladded all the warrior train,
Though most were sorely wounded, none were slain.
The surgeons soon despoiled them of their arms,°
And some with salves they cure, and some with charms;
Foment the bruises, and the pains assuage,°
And heal their inward hurts with sovereign draughts of sage.
The king, in person, visits all around,
Comforts the sick, congratulates the sound;   730
Honours the princely chiefs, rewards the rest,
And holds, for thrice three days, a royal feast.
None was disgraced, for falling is no shame,
And cowardice alone is loss of fame.
The venturous knight is from the saddle thrown;
But 'tis the fault of fortune, not his own:
If crowns and palms the conquering side adorn,
The victor under better stars was born:
The brave man seeks not popular applause,
Nor, overpowered with arms, deserts his cause;   740
Unshamed, though foiled, he does the best he can;
Force is of brutes, but honour is of man.

    Thus Theseus smiled on all with equal grace,
And each was set according to his place;
With ease were reconciled the differing parts,°
For envy never dwells in noble hearts.
At length they took their leave, the time expired,
Well pleased, and to their several homes retired.

  Meanwhile the health of Arcite still impairs;°
From bad proceeds to worse, and mocks the leeches' cares: 750
Swollen is his breast, his inward pains increase,
All means are used, and all without success.
The clotted blood lies heavy on his heart,
Corrupts, and there remains in spite of art;°
Nor breathing veins, nor cupping, will prevail;°
All outward remedies and inward fail:
The mould of nature's fabric is destroyed,
Her vessels discomposed, her virtue void:
The bellows of his lungs begins to swell;
All out of frame is every secret cell,   760
Nor can the good receive, nor bad expel.
Those breathing organs, thus within oppressed,
With venom soon distend the sinews of his breast.
Naught profits him to save abandoned life,
Nor vomits upward aid, nor downward laxative.
The midmost region battered and destroyed,
When nature cannot work, the effect of art is void;
For physic can but mend our crazy state,
Patch an old building, not a new create.
Arcite is doomed to die in all his pride,   770
Must leave his youth, and yield his beauteous bride,
Gained hardly, against right, and unenjoyed.°
When 'twas declared all hope of life was past,
Conscience (that of all physic works the last)
Caused him to send for Emily in haste.
With her, at his desire, came Palamon;
Then, on his pillow raised, he thus began:
'No language can express the smallest part
Of what I feel, and suffer in my heart,
For you whom best I love and value most: 780
But to your service I bequeath my ghost;
Which, from this mortal body when untied,
Unseen, unheard, shall hover at your side;
Nor fright you waking, nor your sleep offend,
But wait officious, and your steps attend.
How I have loved, excuse my faltering tongue,
My spirits feeble, and my pains are strong:
This I may say, I only grieve to die,
Because I lose my charming Emily.

To die, when heaven had put you in my power!   790
Fate could not choose a more malicious hour.
What greater curse could envious fortune give,
Than just to die, when I began to live!
Vain men! how vanishing a bliss we crave,
Now warm in love, now withering in the grave!
Never, O never more to see the sun!
Still dark, in a damp vault, and still alone!
This fate is common; but I lose my breath
Near bliss, and yet not blessed, before my death.   800
Farewell! but take me, dying, in your arms,
'Tis all I can enjoy of all your charms:
This hand I cannot but in death resign;
Ah, could I live! but while I live 'tis mine.
I feel my end approach, and, thus embraced,
Am pleased to die; but hear me speak my last:
Ah, my sweet foe! for you, and you alone,
I broke my faith with injured Palamon.
But love the sense of right and wrong confounds;
Strong love and proud ambition have no bounds.
And much I doubt, should heaven my life prolong,   810
I should return to justify my wrong;
For, while my former flames remain within,
Repentance is but want of power to sin.
With mortal hatred I pursued his life,
Nor he, nor you, were guilty of the strife;
Nor I, but as I loved; yet all combined,
Your beauty, and my impotence of mind;°
And his concurrent flame, that blew my fire;°
For still our kindred souls had one desire.
He had a moment's right, in point of time;   820
Had I seen first, then his had been the crime.
Fate made it mine, and justified his right;
Nor holds this earth a more deserving knight,
For virtue, valour, and for noble blood,
Truth, honour, all that is comprised in good;
So help me heaven, in all the world is none
So worthy to be loved as Palamon.
He loves you too, with such a holy fire,
As will not, cannot, but with life expire:
Our vowed affections both have often tried,   830

Nor any love but yours could ours divide.
Then, by my love's inviolable band,
By my long suffering, and my short command,
If e'er you plight your vows when I am gone,
Have pity on the faithful Palamon.'
   This was his last; for death came on amain,
And exercised below his iron reign;
Then upward to the seat of life he goes;°
Sense fled before him, what he touched he froze:
Yet could he not his closing eyes withdraw,°   840
Though less and less of Emily he saw;
So, speechless, for a little space he lay;
Then grasped the hand he held, and sighed his soul away.
   But whither went his soul, let such relate
Who search the secrets of the future state:
Divines can say but what themselves believe;
Strong proofs they have, but not demonstrative;°
For, were all plain, then all sides must agree,
And faith itself be lost in certainty.
To live uprightly, then, is sure the best;   850
To save ourselves, and not to damn the rest.
The soul of Arcite went where heathens go,
Who better live than we, though less they know.
   In Palamon a manly grief appears;
Silent he wept, ashamed to show his tears.
Emilia shrieked but once; and then, oppressed
With sorrow, sunk upon her lover's breast:
Till Theseus in his arms conveyed, with care,
Far from so sad a sight, the swooning fair.
'Twere loss of time her sorrow to relate;   860
Ill bears the sex a youthful lover's fate,°
When just approaching to the nuptial state:
But, like a low-hung cloud, it rains so fast,
That all at once it falls, and cannot last.
The face of things is changed, and Athens now,
That laughed so late, becomes the scene of woe:
Matrons and maids, both sexes, every state,
With tears lament the knight's untimely fate.
Not greater grief in falling Troy was seen
For Hector's death, but Hector was not then.°   870
Old men with dust deformed their hoary hair;

The women beat their breasts, their cheeks they tear.
'Why wouldst thou go,' (with one consent they cry,)
'When thou hadst gold enough, and Emily!'
   Theseus himself, who should have cheered the grief
Of others, wanted now the same relief:
Old Egeus only could revive his son,
Who various changes of the world had known,
And strange vicissitudes of human fate,
Still altering, never in a steady state: 880
Good after ill, and after pain, delight,
Alternate, like the scenes of day and night.
Since every man who lives, is born to die,
And none can boast sincere felicity,
With equal mind, what happens, let us bear,
Nor joy, nor grieve too much, for things beyond our care.
Like pilgrims, to the appointed place we tend;
The world's an inn, and death the journey's end.
E'en kings but play; and, when their part is done,
Some other, worse or better, mount the throne.   890
With words like these the crowd was satisfied,
And so they would have been had Theseus died.
   But he, their king, was labouring in his mind,
A fitting place for funeral pomps to find,
Which were in honour of the dead designed.
And, after long debate, at last he found
(As love itself had marked the spot of ground)
That grove, for ever green, that conscious laund,°
Where he with Palamon fought hand to hand;
That where he fed his amorous desires   900
With soft complaints, and felt his hottest fires,
There other flames might waste his earthly part,
And burn his limbs, where love had burned his heart.
   This once resolved, the peasants were enjoined,
Sere-wood, and firs, and doddered oaks to find.
With sounding axes to the grove they go,
Fell, split, and lay the fuel on a row;
Vulcanian food: a bier is next prepared,°
On which the lifeless body should be reared,
Covered with cloth of gold; on which was laid   910
The corpse of Arcite, in like robes arrayed.
White gloves were on his hands, and on his head

A wreath of laurel, mixed with myrtle, spread.°
A sword, keen-edged, within his right he held,
The warlike emblem of the conquered field.
Bare was his manly visage on the bier;
Menaced his countenance, e'en in death severe.
Then to the palace-hall they bore the knight,
To lie in solemn state, a public sight:
Groans, cries, and howlings, fill the crowded place,   920
And unaffected sorrow sat on every face.
Sad Palamon above the rest appears,
In sable garments, dewed with gushing tears;
His auburn locks on either shoulder flowed,
Which to the funeral of his friend he vowed:
But Emily, as chief, was next his side,
A virgin-widow, and a mourning bride.°
And, that the princely obsequies might be
Performed according to his high degree,
The steed that bore him living to the fight,   930
Was trapped with polished steel, all shining bright,
And covered with the achievements of the knight.
The riders rode abreast; and one his shield,
His lance of cornel-wood another held;°
The third his bow; and, glorious to behold,
The costly quiver, all of burnished gold.
The noblest of the Grecians next appear,
And weeping, on their shoulders bore the bier;
With sober pace they marched, and often stayed,
And through the master-street the corpse conveyed.°   940
The houses to their tops with black were spread,
And e'en the pavements were with mourning hid.
The right side of the pall old Egeus kept,
And on the left the royal Theseus wept;
Each bore a golden bowl, of work divine,
With honey filled, and milk, and mixed with ruddy wine.
Then Palamon, the kinsman of the slain;
And after him appeared the illustrious train.
To grace the pomp, came Emily the bright,
With covered fire, the funeral pile to light.   950
With high devotion was the service made,
And all the rites of pagan honour paid:
So lofty was the pile, a Parthian bow,°

With vigour drawn, must send the shaft below.
The bottom was full twenty fathom broad,°
With crackling straw beneath in due proportion strowed.
The fabric seemed a wood of rising green,
With sulphur and bitumen cast between,
To feed the flames; the trees were unctuous fir,
And mountain-ash, the mother of the spear;   960
The mourner-yew, and builder-oak, were there;
The beech, the swimming alder, and the plane,
Hard box, and linden of a softer grain,
And laurels, which the gods for conquering chiefs ordain.
How they were ranked shall rest untold by me,
With nameless nymphs that lived in every tree;
Nor how the dryads, and the woodland train,
Disherited, ran howling o'er the plain;
Nor how the birds to foreign seats repaired,°
Or beasts that bolted out, and saw the forest bared;   970
Nor how the ground, now cleared, with ghastly fright,
Beheld the sudden sun, a stranger to the light.
    The straw, as first I said, was laid below;
Of chips, and sere-wood, was the second row;
The third of greens, and timber newly felled;
The fourth high stage the fragrant odours held,
And pearls, and precious stones, and rich array;
In midst of which, embalmed, the body lay.
The service sung, the maid, with mourning eyes,
The stubble fired; the smouldering flames arise: 980
This office done, she sank upon the ground;
But what she spoke, recovered from her swound,
I want the wit in moving words to dress;
But, by themselves, the tender sex may guess.
While the devouring fire was burning fast,
Rich jewels in the flame the wealthy cast;
And some their shields, and some their lances threw,
And gave the warrior's ghost a warrior's due.
Full bowls of wine, of honey, milk, and blood,
Were poured upon the pile of burning wood;   990
And hissing flames receive, and, hungry, lick the food.
Then thrice the mounted squadrons ride around
The fire, and Arcite's name they thrice resound.
'Hail and farewell!' they shouted thrice amain,

Thrice facing to the left, and thrice they turned again:
Still, as they turned, they beat their clattering shields;
The women mix their cries, and clamour fills the fields.
The warlike wakes continued all the night,
And funeral-games were played at new-returning light:
Who, naked, wrestled best, besmeared with oil,   1000
Or who, with gauntlets, gave or took the foil,°
I will not tell you, nor would you attend;
But briefly haste to my long story's end.
   I pass the rest. The year was fully mourned,
And Palamon long since to Thebes returned.
When, by the Grecians' general consent,
At Athens Theseus held his parliament;
Among the laws that passed, it was decreed,
That conquered Thebes from bondage should be freed;
Reserving homage to the Athenian throne,   1010
To which the sovereign summoned Palamon.
Unknowing of the cause, he took his way,
Mournful in mind, and still in black array.
   The monarch mounts the throne and placed on high,
Commands into the court the beauteous Emily.
So called, she came; the senate rose, and paid
Becoming reverence to the royal maid.
And first soft whispers through the assembly went;
With silent wonder then they watched the event:
All hushed, the king arose with awful grace,   1020
Deep thought was in his breast, and counsel in his face:
At length he sighed, and, having first prepared
The attentive audience, thus his will declared:
   'The cause and spring of motion, from above,
Hung down on earth, the golden chain of love;
Great was the effect, and high was his intent,
When peace among the jarring seeds he sent:°
Fire, flood, and earth, and air, by this were bound,°
And love, the common link, the new creation crowned.
The chain still holds; for, though the forms decay,   1030
Eternal matter never wears away:
The same first mover certain bounds has placed,
How long those perishable forms shall last;
Nor can they last beyond the time assigned
By that all-seeing, and all-making mind:

Shorten their hours they may; for will is free;
But never pass the appointed destiny.
So men oppressed, when weary of their breath,
Throw off the burden, and suborn their death.
Then, since those forms begin, and have their end,   1040
On some unaltered cause they sure depend:
Parts of the whole are we; but God the whole;
Who gives us life, and animating soul.
For nature cannot from a part derive
That being, which the whole can only give:
He, perfect, stable; but imperfect we,
Subject to change, and different in degree;
Plants, beasts, and man; and, as our organs are,
We more or less of his perfection share.
But, by a long descent, the ethereal fire   1050
Corrupts; and forms, the mortal part, expire.
As he withdraws his virtue, so they pass,
And the same matter makes another mass.
This law the omniscient power was pleased to give,
That every kind should by succession live;
That individuals die, his will ordains;
The propagated species still remains.
The monarch oak, the patriarch of the trees,
Shoots rising up, and spreads by slow degrees;
Three centuries he grows, and three he stays,   1060
Supreme in state, and in three more decays:
So wears the paving pebble in the street,
And towns and towers their fatal periods meet.
So rivers, rapid once, now naked lie,
Forsaken of their springs, and leave their channels dry:°
So man, at first a drop, dilates with heat,
Then formed, the little heart begins to beat;
Secret he feeds, unknowing in the cell;
At length, for hatching ripe, he breaks the shell
And struggles into breath and cries for aid;   1070
Then, helpless, in his mother's lap is laid.
He creeps, he walks, and, issuing into man,
Grudges their life, from whence his own began;
Reckless of laws, affects to rule alone,
Anxious to reign, and restless on the throne;
First vegetive, then feels, and reasons last;

Rich of three souls, and lives all three to waste.
Some thus, but thousands more in flower of age;
For few arrive to run the latter stage.
Sunk in the first, in battle some are slain,   1080
And others whelmed beneath the stormy main.
What makes all this, but Jupiter the king,
At whose command we perish, and we spring?
Then 'tis our best, since thus ordained to die,
To make a virtue of necessity;
Take what he gives, since to rebel is vain;
The bad grows better, which we well sustain;
And could we choose the time, and choose aright,
'Tis best to die, our honour at the height.
When we have done our ancestors no shame,   1090
But served our friends, and well secured our fame;
Then should we wish our happy life to close,
And leave no more for fortune to dispose.
So should we make our death a glad relief
From future shame, from sickness, and from grief;
Enjoying, while we live, the present hour
And dying in our excellence and flower.
Then round our deathbed every friend should run,
And joy us of our conquest early won;
While the malicious world, with envious tears,   1100
Should grudge our happy end, and wish it theirs.
Since then our Arcite is with honour dead,
Why should we mourn, that he so soon is freed,
Or call untimely, what the gods decreed?
With grief as just, a friend may be deplored,
From a foul prison to free air restored.
Ought he to thank his kinsman or his wife,
Could tears recall him into wretched life?
Their sorrow hurts themselves; on him is lost;
And, worse than both, offends his happy ghost.   1110
What then remains, but after past annoy,
To take the good vicissitude of joy;
To thank the gracious gods for what they give,
Possess our souls, and while we live, to live?
Ordain we then two sorrows to combine,
And in one point the extremes of grief to join;
That thence resulting joy may be renewed,

As jarring notes in harmony conclude.
Then I propose that Palamon shall be
In marriage joined with beauteous Emily;   1120
For which already I have gained the assent
Of my free people in full parliament.
Long love to her has borne the faithful knight,
And well deserved, had fortune done him right:
'Tis time to mend her fault, since Emily,
By Arcite's death, from former vows is free;
If you, fair sister, ratify the accord,
And take him for your husband and your lord.
'Tis no dishonour to confer your grace
On one descended from a royal race;   1130
And were he less, yet years of service past,
From grateful souls, exact reward at last.
Pity is heaven's and yours; nor can she find
A throne so soft as in a woman's mind.'
   He said: she blushed; and, as o'erawed by might,
Seemed to give Theseus what she gave the knight.
Then, turning to the Theban, thus he said:
'Small arguments are needful to persuade
Your temper to comply with my command:'°
And, speaking thus, he gave Emilia's hand.   1140
Smiled Venus, to behold her own true knight
Obtain the conquest, though he lost the fight;
And blessed, with nuptial bliss, the sweet laborious night.
Eros and Anteros, on either side,°
One fired the bridegroom, and one warmed the bride;
And long-attending Hymen, from above,
Showered on the bed the whole Idalian grove.
All of a tenor was their after-life,°
No day discoloured with domestic strife;
No jealousy, but mutual truth believed,   1150
Secure repose, and kindness undeceived.
Thus heaven, beyond the compass of his thought,
Sent him the blessing he so dearly bought.
   So may the Queen of Love long duty bless,
And all true lovers find the same success.

To my Honoured Kinsman John Driden, of Chesterton in the County of Huntingdon, Esquire.

How blest is he, who leads a country life,°
Unvexed with anxious cares, and void of strife!
Who studying peace, and shunning civil rage,
Enjoyed his youth, and now enjoys his age:
All who deserve his love, he makes his own,
And to be loved himself, needs only to be known.

    Just, good, and wise, contending neighbours come
From your award, to wait their final doom;
And, foes before, return in friendship home.
Without their cost, you terminate the cause;   10
And save the expense of long litigious laws
Where suits are traversed, and so little won,
That he who conquers, is but last undone.
Such are not your decrees; but so designed,
The sanction leaves a lasting peace behind,
Like your own soul, serene; a pattern of your mind.

    Promoting concord, and composing strife,
Lord of yourself, uncumbered with a wife;
Where, for a year, a month, perhaps a night,
Long penitence succeeds a short delight: 20
Minds are so hardly matched, that e'en the first,
Though paired by heaven, in paradise were cursed.
For man and woman, though in one they grow,
Yet, first or last, return again to two.
He to God's image, she to his was made;
So, farther from the fount, the stream at random strayed.

    How could he stand, when put to double pain,
He must a weaker than himself sustain!
Each might have stood perhaps; but each alone;
Two wrestlers help to pull each other down.   30

    Not that my verse would blemish all the fair;
But yet, if *some* be bad, 'tis wisdom to beware;
And better shun the bait, than struggle in the snare.
Thus have you shunned, and shun the married state,
Trusting as little as you can to fate.

    No porter guards the passage of your door
To admit the wealthy, and exclude the poor:

For God, who gave the riches, gave the heart
To sanctify the whole, by giving part:
Heaven, who foresaw the will, the means has wrought,   40
And to the second son, a blessing brought:
The first-begotten had his father's share;
But you, like Jacob, are Rebecca's heir.°
     So may your stores, and fruitful fields increase;
And ever be you blest, who live to bless.
As Ceres sowed, where'er her chariot flew;
As heaven to deserts rained the bread of dew,
So free to many, to relations most,
You feed with manna your own Israel host.
     With crowds attended of your ancient race,   50
You seek the champian-sports, or sylvan chase:°
With well-breathed beagles, you surround the wood;
E'en then, industrious of the common good
And often have you brought the wily fox
To suffer for the firstlings of the flocks;
Chased e'en amid the folds, and made to bleed,
Like felons, where they did the murderous deed.
This fiery game, your active youth maintained,
Not yet, by years extinguished, though restrained;
You season still with sports your serious hours,   60
For age but tastes of pleasures, youth devours.
The hare, in pastures or in plains is found,
Emblem of human life; who runs the round,
And, after all his wandering ways are done,
His circle fills, and ends where he begun,
Just as the setting meets the rising sun.
     Thus princes ease their cares; but happier he,
Who seeks not pleasure through necessity,
Than such as once on slippery thrones were placed,
And, chasing, sigh to think themselves are chased.   70
     So lived our sires, ere doctors learned to kill,
And multiplied with theirs, the weekly bill:°
The first physicians by debauch were made:
Excess began, and sloth sustains the trade.
Pity the generous kind their cares bestow°
To search forbidden truths (a sin to know)
To which, if human science could attain,
The doom of death, pronounced by God, were vain.

In vain the leech would interpose delay,
Fate fastens first, and vindicates the prey.°   80
What help from art's endeavours can we have?
Gibbons but guesses, nor is sure to save:°
But Maurus sweeps whole parishes, and peoples every grave.°
And no more mercy to mankind will use,
Than when he robbed and murdered Maro's muse.
Wouldst thou be soon dispatched, and perish whole?
Trust Maurus with thy life, and Milbourne with thy soul.°

By chase our long-lived fathers earned their food;
Toil strung the nerves, and purified the blood:
But we, their sons, a pampered race of men,   90
Are dwindled down to threescore years and ten.
Better to hunt in fields, for health unbought,
Than fee the doctor for a nauseous draught.
The wise, for cure, on exercise depend;
God never made his work, for man to mend.

The tree of knowledge, once in Eden placed,
Was easy found, but was forbid the taste;
O, had our grandsire walked without his wife,
He first had sought the better plant of life!
Now both are lost: yet wandering in the dark,   100
Physicians for the tree, have found the bark.
They, labouring for relief of human kind,
With sharpened sight some remedies may find;
The apothecary-train is wholly blind.
From files, a random recipe they take,°
And many deaths of one prescription make.
Garth, generous as his muse, prescribes and gives;°
The shopman sells; and by destruction lives:
Ungrateful tribe! who, like the viper's brood
From medicine issuing, suck their mother's blood!   110
Let these obey; and let the learned prescribe,
That men may die, without a double bribe;
Let them, but under their superiors, kill;
When doctors first have signed the bloody bill:
He scapes the best, who, nature to repair,
Draws physic from the fields, in draughts of vital air.

You hoard not health for your own private use,
But on the public spend the rich produce;
When, often urged, unwilling to be great,

Your country calls you from your loved retreat,   120
And sends to senates, charged with common care,
Which none more shuns, and none can better bear.
Where could they find another formed so fit
To poise with solid sense a sprightly wit?
Were these both wanting, (as they both abound)
Where could so firm integrity be found?

Well-born and wealthy, wanting no support,
You steer betwixt the country and the court;
Nor gratify whate'er the great desire,
Nor grudging give what public needs require.   130
Part must be left, a fund when foes invade;
And part employed to roll the watery trade;
E'en Canaan's happy land, when worn with toil,
Required a sabbath-year to mend the meagre soil.

Good senators (and such are you) so give,
That kings may be supplied, the people thrive.
And he, when want requires, is truly wise,
Who slights not foreign aids nor over-buys,
But on our native strength in time of need relies.
Munster was bought, we boast not the success;°   140
Who fights for gain for greater makes his peace.

Our foes, compelled by need, have peace embraced;
The peace both parties want is like to last;°
Which if secure, securely we may trade;
Or not secure, should never have been made.
Safe in ourselves, while on ourselves we stand,
The sea is ours, and that defends the land.
Be then the naval stores the nation's care,
New ships to build, and battered to repair.

Observe the war, in every annual course;   150
What has been done, was done with British force:
Namur subdued is England's palm alone;°
The rest besieged; but we constrained the town:
We saw the event that followed our success;
France, though pretending arms, pursued the peace,
Obliged by one sole treaty to restore°
What twenty years of war had won before.
Enough for Europe has our Albion fought:
Let us enjoy the peace our blood has bought.
When once the Persian king was put to flight,°   160

The weary Macedons refused to fight:
Themselves their own mortality confessed,
And left the son of Jove to quarrel for the rest.°

Even victors are by victories undone;
Thus Hannibal, with foreign laurels won,
To Carthage was recalled, too late to keep his own.
While sore of battle, while our wounds are green,
Why should we tempt the doubtful die again?
In wars renewed, uncertain of success,
Sure of a share, as umpires of the peace.   170

A patriot both the king and country serves,
Prerogative, and privilege preserves:
Of each our laws the certain limit show;
One must not ebb, nor t'other overthrow:
Betwixt the prince and parliament we stand;
The barriers of the state on either hand:
May neither overflow, for then they drown the land.
When both are full, they feed our blessed abode,
Like those that watered once the paradise of God.

Some overpoise of sway by turns they share;   180
In peace the people, and the prince in war:
Consuls of moderate power in calms were made;
When the Gauls came, one sole dictator swayed.

Patriots in peace assert the people's right,
With noble stubbornness resisting might;
No lawless mandates from the court receive,
Nor lend by force, but in a body give.
Such was your generous grandsire; free to grant°
In parliaments that weighed their prince's want:
But so tenacious of the common cause,   190
As not to lend the king against his laws;
And, in a loathsome dungeon doomed to lie,
In bonds retained his birthright liberty,
And shamed oppression, till it set him free.

O true descendant of a patriot line,
Who, while thou sharest their lustre, lendst them thine,
Vouchsafe this picture of thy soul to see;
'Tis so far good as it resembles thee:
The beauties to the original I owe,
Which when I miss, my own defects I show.   200
Nor think the kindred muses thy disgrace;

A poet is not born in every race.
Two of a house, few ages can afford,
One to perform, another to record.
Praiseworthy actions are by thee embraced;
And 'tis my praise to make thy praises last.
For e'en when death dissolves our human frame,
The soul returns to heaven, from whence it came,
Earth keeps the body, verse preserves the fame.

Meleager and Atalanta

Out of the Eighth Book of
Ovid's *Metamorphoses*

CONNECTION TO THE FORMER STORY

Ovid, having told how Theseus had freed Athens from the tribute of children, which was imposed on them by Minos, king of Crete, by killing the Minotaur, here makes a digression to the story of Meleager and Atalanta, which is one of the most inartificial connections in all the Metamorphoses; for he only says, that Theseus obtained such honour from that combat, that all Greece had recourse to him in their necessities; and, amongst others, Calydon, though the hero of that country, Prince Meleager, was then living.

From him the Calydonians sought relief;
Though valiant Meleagrus was their chief.
The cause, a boar, who ravaged far and near;
Of Cynthia's wrath, the avenging minister.°
For Oeneus with autumnal plenty blessed,
By gifts to heaven his gratitude expressed;
Culled sheafs, to Ceres; to Lyaeus, wine;
To Pan and Pales, offered sheep and kine;
And fat of olives to Minerva's shrine.
Beginning from the rural gods, his hand   10
Was liberal to the powers of high command;
Each deity in every kind was blessed,
Till at Diana's fane the invidious honour ceased.
    Wrath touches e'en the gods: the Queen of Night,
Fired with disdain and jealous of her right,
'Unhonoured though I am, at least,' said she,
'Not unrevenged that impious act shall be.'

Swift as the word, she sped the boar away,
With charge on those devoted fields to prey.
No larger bulls the Egyptian pastures feed,   20
And none so large Sicilian meadows breed:
His eye-balls glare with fire suffused with blood;
His neck shoots up a thick-set thorny wood;
His bristled back a trench impaled appears,
And stands erected, like a field of spears;
Froth fills his chaps, he sends a grunting sound,
And part he churns, and part befoams the ground;
For tusks with Indian elephants he strove,°
And Jove's own thunder from his mouth he drove.
He burns the leaves; the scorching blast invades   30
The tender corn, and shrivels up the blades;
Or, suffering not their yellow beards to rear,
He tramples down the spikes, and intercepts the year.°
In vain the barns expect their promised load,
Nor barns at home, nor ricks are heaped abroad;
In vain the hinds the threshing-floor prepare,
And exercise their flails in empty air.
With olives ever green the ground is strowed,
And grapes ungathered shed their generous blood.
Amid the fold he rages, nor the sheep   40
Their shepherds, nor the grooms their bulls, can keep.
    From fields to walls the frighted rabble run,
Nor think themselves secure within the town;
Till Meleagrus, and his chosen crew,°
Contemn the danger, and the praise pursue.
Fair Leda's twins, (in time to stars decreed,)°
One fought on foot, one curbed the fiery steed;
Then issued forth famed Jason after these,°
Who manned the foremost ship that sailed the seas;
Then Theseus, joined with bold Pirithous, came;°   50
A single concord in a double name:
The Thestian sons, Idas who swiftly ran,°
And Caeneus, once a woman, now a man.°
Lynceus, with eagle's eyes, and lion's heart;
Leucippus, with his never-erring dart;
Acastus, Phileus, Phoenix, Telamon,
Echion, Lelex, and Eurytion,
Achilles' father, and great Phocus' son;

Dryas the fierce, and Hippasus the strong,
With twice-old Iolas, and Nestor then but young;°   60
Laertes active, and Ancaeus bold;
Mopsus the sage, who future things foretold;
And t'other seer, yet by his wife unsold.
A thousand others of immortal fame;
Among the rest, fair Atalanta came,°
Grace of the woods: a diamond buckle bound
Her vest behind, that else had flowed upon the ground,
And showed her buskined legs; her head was bare,
But for her native ornament of hair,
Which in a simple knot was tied above,   70
Sweet negligence, unheeded bait of love!
Her sounding quiver on her shoulder tied,
One hand a dart, and one a bow supplied.
Such was her face, as in a nymph displayed
A fair fierce boy, or in a boy betrayed
The blushing beauties of a modest maid.
The Calydonian chief at once the dame
Beheld, at once his heart received the flame,
With heavens averse. 'O happy youth,' he cried,
'For whom thy fates reserve so fair a bride!'   80
He sighed, and had no leisure more to say;
His honour called his eyes another way,
And forced him to pursue the now neglected prey.
    There stood a forest on a mountain's brow,
Which overlooked the shaded plains below;
No sounding axe presumed those trees to bite,
Coeval with the world, a venerable sight.
The heroes there arrived, some spread around
The toils, some search the footsteps on the ground,°
Some from the chains the faithful dogs unbound.   90
Of action eager, and intent in thought,
The chiefs their honourable danger sought:
A valley stood below; the common drain
Of waters from above, and falling rain;
The bottom was a moist and marshy ground,
Whose edges were with bending osiers crowned;
The knotty bulrush next in order stood,
And all within of reeds a trembling wood.
    From hence the boar was roused, and sprung amain,

Like lightning sudden on the warrior-train;   100
Beats down the trees before him, shakes the ground,
The forest echoes to the crackling sound;
Shout the fierce youth, and clamours ring around.
All stood with their protended spears prepared,
With broad steel heads the brandished weapons glared.
The beast impetuous with his tusks aside
Deals glancing wounds; the fearful dogs divide;
All spend their mouth aloof, but none abide.°
Echion threw the first, but missed his mark,
And stuck his boar-spear on a maple's bark.   110
Then Jason; and his javelin seemed to take,
But failed with over-force, and whizzed above his back.
Mopsus was next; but, ere he threw, addressed
To Phoebus thus: 'O patron, help thy priest!
If I adore and ever have adored
Thy power divine, thy present aid afford,
That I may reach the beast!' The god allowed
His prayer and smiling, gave him what he could:
He reached the savage, but no blood he drew;
Dian unarmed the javelin as it flew.   120
   This chafed the boar, his nostrils' flames expire,
And his red eye-balls roll with living fire.
Whirled from a sling, or from an engine thrown,
Amid the foes so flies a mighty stone,
As flew the beast: the left wing put to flight,
The chiefs o'erborne, he rushes on the right.
Empalamos and Pelagon he laid
In dust and next to death but for their fellows' aid.
Enesimus fared worse, prepared to fly,
The fatal fang drove deep within his thigh,   130
And cut the nerves; the nerves no more sustain
The bulk; the bulk unpropped, falls headlong on the plain.
Nestor had failed the fall of Troy to see,
But leaning on his lance, he vaulted on a tree;
Then, gathering up his feet, looked down with fear,
And thought his monstrous foe was still too near.
Against a stump his tusk the monster grinds,
And in the sharpened edge new vigour finds;
Then, trusting to his arms, young Othrys found,
And ranched his hips with one continued wound.   140

Now Leda's twins, the future stars, appear;
White were their habits, white their horses were;
Conspicuous both, and both in act to throw,
Their trembling lances brandished at the foe:
Nor had they missed; but he to thickets fled,
Concealed from aiming spears, not pervious to the steed.
But Telamon rushed in, and happed to meet
A rising root, that held his fastened feet;
So down he fell, whom, sprawling on the ground,
His brother from the wooden gyves unbound.   150
    Meantime the virgin-huntress was not slow
To expel the shaft from her contracted bow.
Beneath his ear the fastened arrow stood,
And from the wound appeared the trickling blood.
She blushed for joy: but Meleagrus raised
His voice with loud applause, and the fair archer praised.
He was the first to see, and first to show
His friends the marks of the successful blow.
'Nor shall thy valour want the praises due.'
He said; a virtuous envy seized the crew.   160
They shout; the shouting animates their hearts,
And all at once employ their thronging darts;
But out of order thrown, in air they join,
And multitude makes frustrate the design.
With both his hands the proud Ancaeus takes,
And flourishes his double biting axe:
Then forward to his fate, he took a stride
Before the rest, and to his fellows cried,
'Give place, and mark the difference, if you can,
Between a woman-warrior and a man;   170
The boar is doomed; nor, though Diana lend
Her aid, Diana can her beast defend.'
Thus boasted he; then stretched, on tiptoe stood,
Secure to make his empty promise good;
But the more wary beast prevents the blow,
And upward rips the groin of his audacious foe.
Ancaeus falls; his bowels from the wound
Rush out, and clotted blood distains the ground.°
    Pirithous, no small portion of the war,
Pressed on, and shook his lance; to whom from far,   180
Thus Theseus cried: 'O stay, my better part,

My more than mistress; of my heart, the heart!
The strong may fight aloof: Ancaeus tried
His force too near, and by presuming died.'
He said, and, while he spake, his javelin threw;
Hissing in air, the unerring weapon flew;
But on an arm of oak, that stood betwixt
The marksman and the mark, his lance he fixed.

Once more bold Jason threw, but failed to wound
The boar, and slew an undeserving hound;   190
And through the dog the dart was nailed to ground.

Two spears from Meleager's hand were sent,
With equal force, but various in the event;
The first was fixed in earth, the second stood
On the boar's bristled back, and deeply drank his blood.
Now, while the tortured savage turns around,
And flings about his foam, impatient of the wound,
The wound's great author, close at hand, provokes
His rage, and plies him with redoubled strokes;
Wheels as he wheels, and with his pointed dart   200
Explores the nearest passage to his heart.
Quick, and more quick, he spins in giddy gyres,
Then falls, and in much foam his soul expires.
This act with shouts heaven high the friendly band
Applaud, and strain in theirs the victor hand.
Then all approach the slain with vast surprise,
Admire on what a breadth of earth he lies;
And, scarce secure, reach out their spears afar,
And blood their points, to prove their partnership of war.

But he, the conquering chief, his foot impressed   210
On the strong neck of that destructive beast;
And gazing on the nymph with ardent eyes,
'Accept,' said he, 'fair Nonacrine, my prize;°
And, though inferior, suffer me to join
My labours, and my part of praise, with thine.'
At this presents her with the tusky head
And chine, with rising bristles roughly spread.
Glad, she received the gift; and seemed to take
With double pleasure, for the giver's sake.
The rest were seized with sullen discontent,   220
And a deaf murmur through the squadron went:
All envied; but the Thestian brethren showed

The least respect, and thus they vent their spleen aloud:
'Lay down those honoured spoils, nor think to share,
Weak woman as thou art, the prize of war;
Ours is the title, thine a foreign claim,
Since Meleagrus from our lineage came.
Trust not thy beauty; but restore the prize,
Which he, besotted on that face and eyes,
Would rend from us.' At this, inflamed with spite,   230
From her they snatch the gift, from him the giver's right.

    But soon the impatient prince his falchion drew,
And cried, 'Ye robbers of another's due,
Now learn the difference, at your proper cost,
Betwixt true valour and an empty boast.'
At this advanced, and, sudden as the word,
In proud Plexippus' bosom plunged the sword:
Toxeus amazed, and with amazement slow,
Or to revenge, or ward the coming blow,
Stood doubting; and while doubting thus he stood,   240
Received the steel bathed in his brother's blood.

    Pleased with the first, unknown the second news,
Althaea to the temples pays their dues
For her son's conquest; when at length appear
Her grisly brethren stretched upon the bier:
Pale at the sudden sight she changed her cheer,
And with her cheer her robes; but hearing tell
The cause, the manner, and by whom they fell,
'Twas grief no more, or grief and rage were one
Within her soul; at last 'twas rage alone;   250
Which burning upwards, in succession dries
The tears that stood considering in her eyes.

    There lay a log unlighted on the hearth:
When she was labouring in the throes of birth
For the unborn chief, the fatal sisters came,°
And raised it up, and tossed it on the flame;
Then on the rock a scanty measure place
Of vital flax, and turned the wheel apace;
And turning sung, 'To this red brand and thee,
O new-born babe, we give an equal destiny;'   260
So vanished out of view. The frighted dame
Sprung hasty from her bed, and quenched the flame;
The log, in secret locked, she kept with care,

And that, while thus preserved, preserved her heir.
This brand she now produced; and first she strows
The hearth with heaps of chips, and after blows;
Thrice heaved her hand, and heaved, she thrice repressed;
The sister and the mother long contest,
Two doubtful titles in one tender breast;
And now her eyes and cheeks with fury glow,   270
Now pale her cheeks, her eyes with pity flow;
Now louring looks presage approaching storms,
And now prevailing love her face reforms:
Resolved, she doubts again; the tears, she dried
With burning rage, are by new tears supplied;
And as a ship, which winds and waves assail,
Now with the current drives, now with the gale,
Both opposite, and neither long prevail,
She feels a double force; by turns obeys
The imperious tempest, and the impetuous seas: 280
So fares Althaea's mind; she first relents
With pity, of that pity then repents:
Sister and mother long the scales divide,
But the beam nodded on the sister's side.
Sometimes she softly sighed, then roared aloud;
But sighs were stifled in the cries of blood.

    The pious impious wretch at length decreed,
To please her brothers' ghost, her son should bleed;
And when the funeral flames began to rise,
'Receive,' she said, 'a sister's sacrifice;   290
A mother's bowels burn: high in her hand,
Thus while she spoke, she held the fatal brand;
Then thrice before the kindled pile she bowed,
And the three Furies thrice invoked aloud:
'Come, come, revenging sisters, come and view
A sister paying her dead brothers' due;
A crime I punish, and a crime commit;
But blood for blood, and death for death, is fit:
Great crimes must be with greater crimes repaid,
And second funerals on the former laid.   300
Let the whole household in one ruin fall,
And may Diana's curse o'ertake us all.
Shall fate to happy Oeneus still allow
One son, while Thestius stands deprived of two?

Better three lost, than one unpunished go.
Take then, dear ghosts, (while yet, admitted new
In hell, you wait my duty,) take your due;
A costly offering on your tomb is laid,
When with my blood the price of yours is paid.

'Ah! whither am I hurried? Ah! forgive,   310
Ye shades, and let your sister's issue live:
A mother cannot give him death; though he
Deserves it, he deserves it not from me.

'Then shall the unpunished wretch insult the slain,
Triumphant live? not only live, but reign?
While you, thin shades, the sport of winds, are tossed
O'er dreary plains, or tread the burning coast.
I cannot, cannot bear; 'tis past, 'tis done;
Perish this impious, this detested son;
Perish his sire, and perish I withal;   320
And let the house's heir, and the hoped kingdom fall.

'Where is the mother fled, her pious love,
And where the pains with which ten months I strove!
Ah! hadst thou died, my son, in infant years,
Thy little hearse had been bedewed with tears.

'Thou livest by me; to me thy breath resign;
Mine is the merit, the demerit thine.
Thy life by double title I require;
Once given at birth, and once preserved from fire:
One murder pay, or add one murder more,
And me to them who fell by thee restore.

'I would, but cannot: my son's image stands
Before my sight;—and now their angry hands
My brothers hold, and vengeance these exact;
This pleads compassion, and repents the fact.

'He pleads in vain, and I pronounce his doom:
My brothers, though unjustly, shall o'ercome;
But having paid their injured ghosts their due,
My son requires my death, and mine shall his pursue.'

At this, for the last time, she lifts her hand,   340
Averts her eyes, and half-unwilling drops the brand.
The brand, amid the flaming fuel thrown,
Or drew, or seemed to draw, a dying groan;
The fires themselves but faintly licked their prey,
Then loathed their impious food, and would have shrunk away.

Just then the hero cast a doleful cry,
And in those absent flames began to fry;
The blind contagion raged within his veins;
But he with manly patience bore his pains;
He feared not fate, but only grieved to die   350
Without an honest wound, and by a death so dry.
'Happy Ancaeus,' thrice aloud he cried,
'With what becoming fate in arms he died!'
Then called his brothers, sisters, sire, around,
And her to whom his nuptial vows were bound;
Perhaps his mother; a long sigh he drew,
And, his voice failing, took his last adieu;
For, as the flames augment, and as they stay
At their full height, then languish to decay,
They rise and sink by fits; at last they soar   360
In one bright blaze, and then descend no more:
Just so his inward heats at height impair,
Till the last burning breath shoots out the soul in air.
    Now lofty Calydon in ruins lies;
All ages, all degrees, unsluice their eyes;
And heaven and earth resound with murmurs, groans, and
    cries.
Matrons and maidens beat their breasts, and tear
Their habits, and root up their scattered hair.
The wretched father, father now no more,
With sorrow sunk, lies prostrate on the floor;   370
Deforms his hoary locks with dust obscene,
And curses age, and loathes a life prolonged with pain.
By steel her stubborn soul his mother freed,
And punished on herself her impious deed.
    Had I a hundred tongues, a wit so large
As could their hundred offices discharge;
Had Phoebus all his Helicon bestowed,
In all the streams inspiring all the god;
Those tongues, that wit, those streams, that god in vain
Would offer to describe his sisters' pain;   380
They beat their breasts with many a bruising blow,
Till they turned livid, and corrupt the snow.
The corpse they cherish, while the corpse remains,
And exercise and rub with fruitless pains;
And when to funeral flames 'tis borne away,

They kiss the bed on which the body lay;
And when those funeral flames no longer burn,
The dust composed within a pious urn,
E'en in that urn their brother they confess,
And hug it in their arms, and to their bosoms press.   390
  His tomb is raised; then, stretched along the ground,
Those living monuments his tomb surround;
E'en to his name, inscribed, their tears they pay,
Till tears and kisses wear his name away.
  But Cynthia now had all her fury spent,
Not with less ruin, than a race, content;
Excepting Gorge, perished all the seed,
And her whom heaven for Hercules decreed.°
Satiate at last, no longer she pursued
The weeping sisters; but with wings endued,   400
And horny beaks, and sent to flit in air,
Who yearly round the tomb in feathered flocks repair.

Sigismonda and Guiscardo

from Boccaccio

While Norman Tancred in Salerno reigned,
The title of a gracious prince he gained;
Till turned a tyrant in his latter days,
He lost the lustre of his former praise;
And from the bright meridian where he stood,
Descending, dipped his hands in lovers' blood.
  This prince of fortune's favour long possessed,
Yet was with one fair daughter only blessed;
And blessed he might have been with her alone,
But oh! how much more happy, had he none!   10
She was his care, his hope, and his delight,
Most in his thought, and ever in his sight:
Next, nay beyond his life, he held her dear;
She lived by him, and now he lived in her.
For this, when ripe for marriage, he delayed
Her nuptial bonds, and kept her long a maid,
As envy any else should share a part
Of what was his, and claiming all her heart.
At length, as public decency required,

And all his vassals eagerly desired,   20
With mind averse, he rather underwent
His people's will, than gave his own consent.
So was she torn, as from a lover's side,
And made almost in his despite a bride.

   Short were her marriage joys; for in the prime
Of youth, her lord expired before his time:
And to her father's court, in little space
Restored anew, she held a higher place;
More loved, and more exalted into grace.
This princess fresh and young, and fair, and wise,   30
The worshipped idol of her father's eyes,
Did all her sex in every grace exceed,
And had more wit beside than women need.

   Youth, health, and ease, and most an amorous mind,
To second nuptials had her thoughts inclined;
And former joys had left a secret sting behind.
But prodigal in every other grant,
Her sire left unsupplied her only want;
And she, betwixt her modesty and pride,
Her wishes, which she could not help, would hide.   40

   Resolved at last to lose no longer time,
And yet to please herself without a crime,
She cast her eyes around the court, to find
A worthy subject suiting to her mind,
To him in holy nuptials to be tied,
A seeming widow, and a secret bride.
Among the train of courtiers, one she found
With all the gifts of bounteous nature crowned,
Of gentle blood, but one whose niggard fate
Had set him far below her high estate: 50
Guiscard his name was called, of blooming age,°
Now squire to Tancred, and before his page:
To him, the choice of all the shining crowd,
Her heart the noble Sigismonda vowed.

   Yet hitherto she kept her love concealed,
And with close glances every day beheld
The graceful youth; and every day increased
The raging fire that burned within her breast;
Some secret charm did all his acts attend,
And what his fortune wanted, hers could mend: 60

Till, as the fire will force its outward way,
Or, in the prison pent, consume the prey,
So long her earnest eyes on his were set,
At length their twisted rays together met;°
And he, surprised with humble joy, surveyed
One sweet regard, shot by the royal maid.
Not well assured, while doubtful hopes he nursed,
A second glance came gliding like the first;
And he who saw the sharpness of the dart,
Without defence received it in his heart.   70
In public though their passion wanted speech,
Yet mutual looks interpreted for each:
Time, ways, and means of meeting were denied,
But all those wants ingenious love supplied.
The inventive god, who never fails his part,
Inspires the wit, when once he warms the heart.
      When Guiscard next was in the circle seen,
Where Sigismonda held the place of queen,
A hollow cane within her hand she brought,
But in the concave had enclosed a note;   80
With this she seemed to play, and, as in sport,
Tossed to her love, in presence of the court;
'Take it,' she said; 'and when your needs require,
This little brand will serve to light your fire.'
He took it with a bow, and soon divined
The seeming toy was not for nought designed:
But when retired, so long with curious eyes
He viewed the present, that he found the prize.
Much was in little writ; and all conveyed
With cautious care, for fear to be betrayed   90
By some false confident, or favourite maid.
The time, the place, the manner how to meet,
Were all in punctual order plainly writ:
But since a trust must be, she thought it best
To put it out of laymen's power at least,
And for their solemn vows prepared a priest.
      Guiscard (her secret purpose understood)
With joy prepared to meet the coming good;
Nor pains nor danger was resolved to spare,
But use the means appointed by the fair.   100
      Near the proud palace of Salerno stood

A mount of rough ascent, and thick with wood,
Through this a cave was dug with vast expense,
The work it seemed of some suspicious prince,
Who, when abusing power with lawless might,
From public justice would secure his flight.
The passage made by many a winding way,
Reached e'en the room in which the tyrant lay,
Fit for his purpose; on a lower floor
He lodged, whose issue was an iron door,   110
From whence, by stairs descending to the ground,
In the blind grot a safe retreat he found.
Its outlet ended in a brake o'ergrown
With brambles, choked by time, and now unknown.
A rift there was, which from the mountain's height
Conveyed a glimmering and malignant light,
A breathing-place to draw the damps away,
A twilight of an intercepted day.
The tyrant's den, whose use though lost to fame,
Was now the apartment of the royal dame;   120
The cavern only to her father known,
By him was to his darling daughter shown.

  Neglected long she let the secret rest,
Till love recalled it to her labouring breast,
And hinted as the way by heaven designed
The teacher, by the means he taught, to blind.
What will not women do, when need inspires
Their wit, or love their inclination fires!
Though jealousy of state the invention found,
Yet love refined upon the former ground.   130
That way the tyrant had reserved to fly
Pursuing hate, now served to bring two lovers nigh.

  The dame, who long in vain had kept the key,
Bold by desire, explored the secret way;
Now tried the stairs, and wading through the night,
Searched all the deep recess, and issued into light.
All this her letter had so well explained,
The instructed youth might compass what remained:
The cavern-mouth alone was hard to find,
Because the path disused was out of mind: 140
But in what quarter of the copse it lay,
His eye by certain level could survey:

Yet (for the wood perplexed with thorns he knew)°
A frock of leather o'er his limbs he drew:
And thus provided, searched the brake around,
Till the choked entry of the cave he found.

    Thus, all prepared, the promised hour arrived,
So long expected, and so well contrived:
With love to friend, the impatient lover went,°
Fenced from the thorns, and trod the deep descent.   150
The conscious priest, who was suborned before,°
Stood ready posted at the postern door;
The maids in distant rooms were sent to rest,
And nothing wanted but the invited guest.
He came, and knocking thrice, without delay,
The longing lady heard, and turned the key;
At once invaded him with all her charms,
And the first step he made, was in her arms:
The leathern outside, boistrous as it was,
Gave way, and bent beneath her strict embrace: 160
On either side the kisses flew so thick,
That neither he nor she had breath to speak.
The holy man amazed at what he saw,
Made haste to sanctify the bliss by law;
And muttered fast the matrimony o'er,
For fear committed sin should get before.
His work performed, he left the pair alone,
Because he knew he could not go too soon;
His presence odious, when his task was done.
What thoughts he had, beseems not me to say,   170
Though some surmise he went to fast and pray,
And needed both, to drive the tempting thoughts away.

    The foe once gone, they took their full delight;
'Twas restless rage, and tempest all the night:
For greedy love each moment would employ,
And grudged the shortest pauses of their joy.

    Thus were their loves auspiciously begun,
And thus with secret care were carried on.
The stealth itself did appetite restore,
And looked so like a sin, it pleased the more.   180

    The cave was now become a common way,
The wicket often opened, knew the key.

Love rioted secure, and long enjoyed,
Was ever eager, and was never cloyed.

But as extremes are short, of ill and good,
And tides at highest mark regorge the flood;
So fate, that could no more improve their joy,
Took a malicious pleasure to destroy.

Tancred, who fondly loved, and whose delight
Was placed in his fair daughter's daily sight,   190
Of custom, when his state affairs were done,
Would pass his pleasing hours with her alone:
And, as a father's privilege allowed,
Without attendance of the officious crowd.

It happened once, that when in heat of day
He tried to sleep, as was his usual way,
The balmy slumber fled his wakeful eyes,
And forced him, in his own despite, to rise:
Of sleep forsaken, to relieve his care,
He sought the conversation of the fair: 200
But with her train of damsels she was gone,
In shady walks the scorching heat to shun:
He would not violate that sweet recess,
And found besides a welcome heaviness
That seized his eyes; and slumber, which forgot
When called before to come, now came unsought.
From light retired, behind his daughter's bed,
He for approaching sleep composed his head;
A chair was ready, for that use designed,
So quilted, that he lay at ease reclined;   210
The curtains closely drawn, the light to screen,
As if he had contrived to lie unseen:
Thus covered with an artificial night,
Sleep did his office soon, and sealed his sight.

With heaven averse, in this ill-omened hour°
Was Guiscard summoned to the secret bower,
And the fair nymph, with expectation fired,
From her attending damsels was retired:
For, true to love, she measured time so right,
As not to miss one moment of delight.   220
The garden, seated on the level floor,
She left behind, and locking every door,

Thought all secure; but little did she know,
Blind to her fate, she had enclosed her foe.
Attending Guiscard, in his leathern frock,
Stood ready, with his thrice-repeated knock:
Thrice with a doleful sound the jarring grate
Rung deaf, and hollow, and presaged their fate.
The door unlocked, to known delight they haste,
And panting in each other's arms, embraced;   230
Rush to the conscious bed, a mutual freight,
And heedless press it with their wonted weight.

    The sudden bound awaked the sleeping sire,
And showed a sight no parent can desire:
His opening eyes at once with odious view
The love discovered, and the lover knew:
He would have cried; but hoping that he dreamt,
Amazement tied his tongue, and stopped the attempt.
The ensuing moment all the truth declared,
But now he stood collected, and prepared;   240
For malice and revenge had put him on his guard.

    So, like a lion that unheeded lay,
Dissembling sleep, and watchful to betray,
With inward rage he meditates his prey.°
The thoughtless pair, indulging their desires,°
Alternate kindled and then quenched their fires;
Nor thinking in the shades of death they played,
Full of themselves, themselves alone surveyed,
And, too secure, were by themselves betrayed.
Long time dissolved in pleasure thus they lay,   250
Till nature could no more suffice their play;
Then rose the youth, and through the cave again
Returned; the princess mingled with her train.

    Resolved his unripe vengeance to defer,
The royal spy, when now the coast was clear,
Sought not the garden, but retired unseen,
To brood in secret on his gathered spleen,
And methodize revenge: to death he grieved;
And, but he saw the crime, had scarce believed.
The appointment for the ensuing night he heard;   260
And therefore in the cavern had prepared
Two brawny yeomen of his trusty guard.

    Scarce had unwary Guiscard set his foot

Within the farmost entrance of the grot,
When these in secret ambush ready lay,
And rushing on the sudden seized the prey:
Encumbered with his frock, without defence,
An easy prize, they led the prisoner thence,
And, as commanded, brought before the prince.
The gloomy sire, too sensible of wrong°   270
To vent his rage in words, restrained his tongue;
And only said, 'Thus servants are preferred,
And trusted, thus their sovereigns they reward.
Had I not seen, had not these eyes received
So clear a proof, I could not have believed.'
    He paused, and choked the rest. The youth who saw
His forfeit life abandoned to the law,
The judge the accuser, and the offence to him
Who had both power and will to avenge the crime,
No vain defence prepared; but thus replied,   280
'The faults of love by love are justified:
With unresisted might the monarch reigns,
He levels mountains, and he raises plains;
And not regarding difference of degree,
Abased your daughter, and exalted me.'
    This bold return with seeming patience heard,
The prisoner was remitted to the guard.
The sullen tyrant slept not all the night,
But lonely walking by a winking light,
Sobbed, wept, and groaned, and beat his withered breast,   290
But would not violate his daughter's rest;
Who long expecting lay, for bliss prepared,
Listening for noise, and grieved that none she heard;
Oft rose, and oft in vain employed the key,
And oft accused her lover of delay;
And passed the tedious hours in anxious thoughts away.
    The morrow came; and at his usual hour
Old Tancred visited his daughter's bower;
Her cheek (for such his custom was) he kissed,
Then blessed her kneeling, and her maids dismissed.   300
The royal dignity thus far maintained,
Now left in private, he no longer feigned;
But all at once his grief and rage appeared,
And floods of tears ran trickling down his beard.

'O Sigismonda,' he began to say:
Thrice he began, and thrice was forced to stay,
Till words with often trying found their way:
'I thought, O Sigismonda, (but how blind
Are parents' eyes, their children's faults to find!)
Thy virtue, birth, and breeding were above   310
A mean desire, and vulgar sense of love:
Nor less than sight and hearing could convince
So fond a father, and so just a prince,
Of such an unforeseen, and unbelieved offence.
Then what indignant sorrow must I have,
To see thee lie subjected to my slave!
A man so smelling of the people's lee,°
The court received him first for charity;
And since with no degree of honour graced,
But only suffered, where he first was placed: 320
A grovelling insect still; and so designed
By nature's hand, nor born of noble kind:
A thing, by neither man nor woman prized,
And scarcely known enough, to be despised.
To what has heaven reserved my age? Ah why
Should man, when nature calls, not choose to die,
Rather than stretch the span of life, to find
Such ills as fate has wisely cast behind,
For those to feel, whom fond desire to live
Makes covetous of more than life can give!   330
Each has his share of good, and when 'tis gone,
The guest, though hungry, cannot rise too soon.
But I, expecting more, in my own wrong
Protracting life, have lived a day too long.
If yesterday could be recalled again,
E'en now would I conclude my happy reign:
But 'tis too late, my glorious race is run,
And a dark cloud o'ertakes my setting sun.
Hadst thou not loved, or loving saved the shame,
If not the sin, by some illustrious name,   340
This little comfort had relieved my mind,
'Twas frailty, not unusual to thy kind:
But thy low fall beneath the royal blood,
Shows downward appetite to mix with mud:

Thus not the least excuse is left for thee,
Nor the least refuge for unhappy me.
   'For him I have resolved: whom by surprise
I took, and scarce can call it, in disguise;
For such was his attire, as with intent
Of nature, suited to his mean descent: 350
The harder question yet remains behind,
What pains a parent and a price can find
To punish an offence of this degenerate kind.
   'As I have loved, and yet I love thee more
Than ever father loved a child before;
So that indulgence draws me to forgive:
Nature, that gave thee life, would have thee live.
But, as a public parent of the state,
My justice, and thy crime, require thy fate.
Fain would I choose a middle course to steer;   360
Nature's too kind, and justice too severe:
Speak for us both, and to the balance bring
On either side, the father, and the king.
Heaven knows, my heart is bent to favour thee;
Make it but scanty weight, and leave the rest to me.'
   Here stopping with a sigh, he poured a flood
Of tears, to make his last expression good.
   She, who had heard him speak, nor saw alone
The secret conduct of her love was known,
But he was taken who her soul possessed,   370
Felt all the pangs of sorrow in her breast;
And little wanted, but a woman's heart°
With cries, and tears, had testified her smart:
But inborn worth, that fortune can control,
New strung, and stiffer bent her softer soul;
The heroine assumed the woman's place,
Confirmed her mind, and fortified her face:
Why should she beg, or what could she pretend,
When her stern father had condemned her friend!
Her life she might have had; but her despair   380
Of saving his, had put it past her care:
Resolved on fate, she would not lose her breath,
But rather than not die, solicit death.
Fixed on this thought, she, not as women use

Her fault by common frailty would excuse,
But boldly justified her innocence,
And while the fact was owned, denied the offence:
Then with dry eyes, and with an open look,
She met his glance midway, and thus undaunted spoke.

    'Tancred, I neither am disposed to make   390
Request for life nor offered life to take:
Much less deny the deed; but least of all
Beneath pretended justice weakly fall.
My words to sacred truth shall be confined,
My deeds shall show the greatness of my mind.
That I have loved, I own; that still I love,
I call to witness all the powers above:
Yet more I own: to Guiscard's love I give
The small remaining time I have to live;
And if beyond this life desire can be,   400
Not fate itself shall set my passion free.

    'This first avowed; nor folly warped my mind,
Nor the frail texture of the female kind
Betrayed my virtue: for, too well I knew
What honour was, and honour had his due:
Before the holy priest my vows were tied,
So came I not a strumpet, but a bride;
This for my fame: and for the public voice:
Yet more, his merits justified my choice;
Which had they not, the first election thine,   410
That bond dissolved, the next is freely mine:
Or grant I erred, (which yet I must deny)
Had parents power e'en second vows to tie,
Thy little care to mend my widowed nights
Has forced me to recourse of marriage rites,
To fill an empty side, and follow known delights.
What have I done in this, deserving blame?
State laws may alter: nature's are the same;
Those are usurped on helpless woman-kind,
Made without our consent, and wanting power to bind.   420

    'Thou, Tancred, better shouldst have understood,
That as thy father gave thee flesh and blood,
So gavest thou me: not from the quarry hewed,
But of a softer mould, with sense endued;
E'en softer than thy own, of suppler kind,

More exquisite of taste, and more than man refined.
Nor needst thou by thy daughter to be told,
Though now thy sprightly blood with age be cold,
Thou hast been young; and canst remember still,
That when thou hadst the power, thou hadst the will;   430
And from the past experience of thy fires,
Canst tell with what a tide our strong desires
Come rushing on in youth, and what their rage requires.

   'And grant thy youth was exercised in arms,
When love no leisure found for softer charms,
My tender age in luxury was trained,
With idle ease and pageants entertained;
My hours my own, my pleasures unrestrained.
So bred, no wonder if I took the bent
That seemed e'en warrented by thy consent;   440
For, when the father is too fondly kind,
Such seed he sows, such harvest shall he find.
Blame then thyself, as reason's law requires,
(Since nature gave, and thou fomentst my fires;)
If still those appetites continue strong,
Thou mayst consider, I am yet but young:
Consider too, that having been a wife,
I must have taste of a better life,
And am not to be blamed, if I renew,
By lawful means, the joys which then I knew.   450
Where was the crime, if pleasure I procured,
Young, and a woman, and to bliss enured?
That was my case, and this is my defence:
I pleased myself, I shunned incontinence,
And urged by strong desires, indulged my sense.

   'Left to myself, I must avow, I strove
From public shame to screen my secret love,
And well acquainted with thy native pride,
Endeavoured, what I could not help, to hide;
For which, a woman's wit an easy way supplied.   460
How this, so well contrived, so closely laid,
Was known to thee, or by what chance betrayed,
Is not my care: to please thy pride alone,
I could have wished it had been still unknown.

   'Nor took I Guiscard by blind fancy led,
Or hasty choice, as many women wed;

But with deliberate care, and ripened thought,
At leisure first designed, before I wrought:
On him I rested, after long debate,
And not without considering, fixed my fate: 470
His flame was equal, though by mine inspired;
(For so the difference of our birth required:)
Had he been born like me, like me his love
Had first begun what mine was forced to move:
But thus beginning, thus we persevere;
Our passions yet continue what they were,
Nor length of trial makes our joys the less sincere.
    'At this my choice, though not by thine allowed,
(Thy judgment herding with the common crowd)
Thou takest unjust offence; and, led by them   480
Dost less the merit than the man esteem.
Too sharply, Tancred, by thy pride betrayed,
Hast thou against the laws of kind inveighed;
For all the offence is in opinion placed,
Which deems high birth by lowly choice debased:
This thought alone with fury fires thy breast,
(For holy marriage justifies the rest)
That I have sunk the glories of the state,
And mixed my blood with a plebeian mate:
In which I wonder thou shouldst oversee   490
Superior causes, or impute to me
The fault of fortune, or the fates' decree.
Or call it heaven's imperial power alone,
Which moves on springs of justice, though unknown;
Yet this we see, though ordered for the best,
The bad exalted, and the good oppressed;
Permitted laurels grace the lawless brow,
The unworthy raised, the worthy cast below.
    'But leaving that: search we the secret springs,
And backward trace the principles of things;   500
There shall we find, that when the world began,
One common mass composed the mould of man;
One paste of flesh on all degrees bestowed,
And kneaded up alike with moistening blood.
The same almighty power inspired the frame
With kindled life, and formed the souls the same:
The faculties of intellect, and will,

Dispensed with equal hand, disposed with equal skill,
Like liberty indulged with choice of good or ill.
Thus born alike, from virtue first began 510
The difference that distinguished man from man:
He claimed no title from descent of blood,
But that which made him noble made him good:
Warmed with more particles of heavenly flame,
He winged his upward flight, and soared to fame;
The rest remained below, a tribe without a name.
    'This law, though custom now diverts the course,
As nature's institute, is yet in force;
Uncancelled, though disused: and he whose mind
Is virtuous, is alone of noble kind. 520
Though poor in fortune, of celestial race;
And he commits the crime, who calls him base.
    'Now lay the line; and measure all thy court,
By inward virtue, not external port,
And find whom justly to prefer above
The man on whom my judgment placed my love:
So shalt thou see his parts, and person shine;
And thus compared, the rest a base degenerate line.
Nor took I, when I first surveyed thy court,
His valour, or his virtues on report; 530
But trusted what I ought to trust alone,
Relying on thy eyes, and not my own;
Thy praise (and thine was then the public voice)
First recommended Guiscard to my choice:
Directed thus by thee, I looked and found
A man I thought deserving to be crowned;
First by my father pointed to my sight,
Nor less conspicuous by his native light:
His mind, his mien, the features of his face,
Excelling all the rest of human race: 540
These were thy thoughts, and thou couldst judge aright,
Till interest made a jaundice in thy sight.
    'Or should I grant, thou didst not rightly see;
Then thou wert first deceived, and I deceived by thee.
But if thou shalt allege, through pride of mind,
Thy blood with one of base condition joined,
'Tis false; for 'tis not baseness to be poor;
His poverty augments thy crime the more;

Upbraids thy justice with the scant regard
Of worth: whom princes praise, they should reward. 550
Are these the kings entrusted by the crowd
With wealth to be dispensed for common good?
The people sweat not for their king's delight,
To enrich a pimp, or raise a parasite;
Theirs is the toil; and he who well has served
His country, has his country's wealth deserved.

  'E'en mighty monarchs oft are meanly born,
And kings by birth, to lowest rank return;
All subject to the power of giddy chance,
For fortune can depress, or can advance: 560
But true nobility is of the mind,
Not given by chance, and not to chance resigned.

  'For the remaining doubt of thy decree,
What to resolve, and how dispose of me,
Be warned to cast that useless care aside,
My self alone, will for myself provide:
If in thy doting, and decrepit age,
Thy soul, a stranger in the youth to rage,
Begins in cruel deeds to take delight,
Gorge with my blood thy barbarous appetite; 570
For I so little am disposed to pray
For life, I would not cast a wish away.
Such as it is, the offence is all my own;
And what to Guiscard is already done,
Or to be done, is doomed by thy decree,
That, if not executed first by thee,
Shall on my person be performed by me.

  'Away! with women weep, and leave me here,
Fixed, like a man to die, without a tear;°
Or save, or slay us both this present hour, 580
'Tis all that fate has left within thy power.'

  She said; nor did her father fail to find,
In all she spoke, the greatness of her mind;
Yet thought she was not obstinate to die,
Nor deemed the death she promised was so nigh:
Secure in this belief, he left the dame,
Resolved to spare her life, and save her shame;
But that detested object to remove,
To wreak his vengeance, and to cure her love.

Intent on this, a secret order signed,   590
The death of Guiscard to his guards enjoined:
Strangling was chosen, and the night the time,
A mute revenge, and blind as was the crime:
His faithful heart, a bloody sacrifice,
Torn from his breast, to glut the tyrant's eyes,
Closed the severe command: for, (slaves to pay)
What kings decree, the soldier must obey:
Waged against foes; and, when the wars are o'er
Fit only to maintain despotic power:
Dangerous to freedom, and desired alone   600
By kings, who seek an arbitrary throne:
Such were these guards; as ready to have slain
The prince himself, allured with greater gain:
So was the charge performed with better will,
By men enured to blood, and exercised in ill.

Now, though the sullen sire had eased his mind,
The pomp of his revenge was yet behind,
A pomp prepared to grace the present he designed.
A goblet rich with gems, and rough with gold,
Of depth, and breadth, the precious pledge to hold,   610
With cruel care he chose: the hollow part
Enclosed; the lid concealed the lover's heart:
Then of his trusted mischiefs, one he sent,
And bade him with these words the gift present:
'Thy father sends thee this, to cheer thy breast,
And glad thy sight with what thou lovest the best;
As thou hast pleased his eyes, and joyed his mind,
With what he loved the most of human kind.'

Ere this the royal dame, who well had weighed
The consequence of what her sire had said,   620
Fixed on her fate, against the expected hour,
Procured the means to have it in her power:
For this, she had distilled, with early care,
The juice of simples, friendly to despair,
A magazine of death; and thus prepared,°
Secure to die, the fatal message heard:°
Then smiled severe; nor with a troubled look,
Or trembling hand, the funeral present took;
E'en kept her countenance, when the lid removed,
Disclosed the heart unfortunately loved: 630

She needed not be told within whose breast
It lodged; the message had explained the rest.
Or not amazed, or hiding her surprise,
She sternly on the bearer fixed her eyes:
Then thus: 'Tell Tancred on his daughter's part,
The gold, though precious, equals not the heart:
But he did well to give his best; and I,
Who wished a worthier urn, forgive his poverty.'

    At this, she curbed a groan, that else had come,
And pausing, viewed the present in the tomb: 640
Then to the heart adored devoutly glued
Her lips, and raising it, her speech renewed:
'E'en from my day of birth, to this, the bound
Of my unhappy being, I have found
My father's care and tenderness expressed.
But this last act of love excels the rest:
For this so dear a present, bear him back
The best return that I can live to make.'

    The messenger dispatched, again she viewed
The loved remains and, sighing, thus pursued: 650
'Source of my life, and lord of my desires,
In whom I lived, with whom my soul expires;
Poor heart, no more the spring of vital heat,
Cursed be the hands that tore thee from thy seat!
The course is finished, which thy fates decreed,
And thou, from thy corporeal prison freed:
Soon hast thou reached the goal with mended pace,
A world of woes dispatched in little space:
Forced by the worth, thy foe in death become
Thy friend, has lodged thee in a costly tomb;   660
There yet remained thy funeral exequies,
The weeping tribute of thy widow's eyes,
And those, indulgent heaven has found the way
That I, before my death, have leave to pay.
My father e'en in cruelty is kind,
Or heaven has turned the malice of his mind
To better uses that his hate designed;
And made the insult which in his gift appears,
The means to mourn thee with my pious tears;
Which I will pay thee down, before I go,°   670
And save myself the pains to weep below,

If souls can weep; thought once I meant to meet
My fate with face unmoved, and eyes unwet,
Yet since I have thee here in narrow room,
My tears shall set thee first afloat within thy tomb:
Then (as I know thy spirit hovers nigh)
Under thy friendly conduct will I fly
To regions unexplored, secure to share
Thy state; nor hell shall punishment appear,
And heaven is double heaven, if thou art there.'   680
    She said. Her brimful eyes, that ready stood,
And only wanted will to weep a flood,
Released their watery store, and poured amain,
Like clouds low hung, a sober shower of rain;
Mute solemn sorrow, free from female noise,
Such as the majesty of grief destroys:
For, bending o'er the cup, the tears she shed
Seemed by the posture to discharge her head,
O'erfilled before; and oft (her mouth applied
To the cold heart) she kissed at once and cried.   690
Her maids, who stood amazed, nor knew the cause
Of her complaining, nor whose heart it was;
Yet all due measures of her mourning kept,
Did office at the dirge, and by infection wept;°
And oft inquired the occasion of her grief,
(Unanswered but by sighs) and offered vain relief.
At length, her stock of tears already shed,
She wiped her eyes, she raised her drooping head,
And thus pursued: 'O ever faithful heart,
I have performed the ceremonial part,   700
The decencies of grief. It rests behind,
That as our bodies were, our souls be joined:
To thy whate'er abode, my shade convey,
And as an elder ghost, direct the way.'
She said; and bade the vial to be brought,
Where she before had brewed the deadly draught,
First pouring out the medicinable bane,
The heart, her tears had rinsed, she bathed again;
Then down her throat the death securely throws,
And quaffs a long oblivion of her woes.   710
    This done, she mounts the genial bed, and there,°
(Her body first composed with honest care,)

Attends the welcome rest. Her hands yet hold
Close to her heart, the monumental gold;
Nor further word she spoke, but closed her sight,
And quiet, sought the covert of the night.

The damsels, who the while in silence mourned,
Not knowing, nor suspecting death suborned,°
Yet, and their duty was, to Tancred sent,
Who, conscious of the occasion, feared the event.   720
Alarmed, and with presaging heart he came,
And drew the curtains, and exposed the dame
To loathsome light: then with a late relief
Made vain efforts to mitigate her grief.
She, what she could excluding day, her eyes
Kept firmly sealed, and sternly thus replies:

'Tancred, restrain thy tears, unsought by me,
And sorrow, unavailing now to thee:
Did ever man before afflict his mind
To see the effect of what himself designed?   730
Yet, if thou hast remaining in thy heart
Some sense of love, some unextinguished part
Of former kindness, largely once professed,
Let me by that adjure thy hardened breast,
Not to deny thy daughter's last request:
The secret love, which I so long enjoyed,
And still concealed, to gratify thy pride,
Thou hast disjoined; but with my dying breath,
Seek not, I beg thee, to disjoin our death:
Where'er his corpse by thy command is laid,   740
Thither let mine in public be conveyed;
Exposed in open view, and side by side,
Acknowledged as a bridegroom and a bride.'

The prince's anguish hindered his reply;
And she, who felt her fate approaching nigh,
Seized the cold heart, and heaving to her breast,
'Here, precious pledge,' she said, 'securely rest.'
These accents were her last; the creeping death
Benumbed her senses first, then stopped her breath.

Thus she for disobedience justly died;   750
The sire was justly punished for his pride:
The youth, least guilty, suffered for the offence
Of duty violated to his prince;

Who late repenting of his cruel deed,
One common sepulchre for both decreed;
Entombed the wretched pair in royal state,
And on their monument inscribed their fate.

Baucis and Philemon

Out of the Eighth Book of
Ovid's *Metamorphoses*

The author pursuing the deeds of Theseus, relates how he, with his friend Pirithous, were invited by Achelous, the river-god, to stay with him till his waters were abated. Achelous entertains them with a relation of his own love to Perimele, who was changed into an island by Neptune, at his request. Pirithous, being an atheist, derides the legend, and denies the power of the gods to work that miracle. Lelex, another companion of Theseus, to confirm the story of Achelous, relates another metamorphosis of Baucis and Philemon into trees; of which he was partly an eye-witness.

Thus Achelous ends. His audience hear,
With admiration, and admiring, fear
The powers of heaven; except Ixion's son°
Who laughed at all the gods, believed in none.
He shook his impious head, and thus replies,
'These legends are no more than pious lies:
You attribute too much to heavenly sway
To think they give us forms, and take away.'
    The rest, of better minds, their sense declared
Against this doctrine, and with horror heard.   10
Then Lelex rose, an old experienced man,
And thus with sober gravity began:
'Heaven's power is infinite: earth, air, and sea,
The manufacture mass, the making power obey:
By proof to clear your doubt: in Phrygian ground
Two neighbouring trees, with walls encompassed round,
Stand on a moderate rise, with wonder shown,
One a hard oak, a softer linden one:
I saw the place and them, by Pittheus sent
To Phrygian realms, my grandsire's government.   20
Not far from thence is seen a lake, the haunt
Of coots, and of the fishing cormorant.
Here Jove with Hermes came; but in disguise

Of mortal men concealed their deities.
One laid aside his thunder, one his rod;
And many toilsome steps together trod.
For harbour at a thousand doors they knocked,
Not one of all the thousand but was locked.
At last a hospitable house they found,
A homely shed; the roof not far from ground,   30
Was thatched with reeds and straw together bound.
There Baucis and Philemon lived, and there
Had lived long married, and a happy pair:
Now old in love; though little was their store,
Inured to want, their poverty they bore,
Nor aimed at wealth, professing to be poor.°
For master or for servant here to call,
Was all alike, when only two were all.
Command was none, where equal love was paid,
Or rather both commanded, both obeyed.   40
    'From lofty roofs the gods repulsed before,
Now stooping, entered through the little door.
The man (their hearty welcome first expressed)
A common settle drew for either guest,
Inviting each his weary limbs to rest.
But ere they sat, officious Baucis lays
Two cushions stuffed with straw, the seat to raise;
Coarse, but the best she had; then rakes the load
Of ashes from the hearth, and spreads abroad
The living coals, and lest they should expire,   50
With leaves and barks she feeds her infant fire.
It smokes, and then with trembling breath she blows,
Till in a cheerful blaze the flames arose.
With brushwood and with chips she strengthens these,
And adds at last the boughs of rotten trees.
The fire thus formed, she sets the kettle on
(Like burnished gold the little seether shone)°
Next took the coleworts which her husband got
From his own ground (a small well-watered spot)
She stripped the stalks of all their leaves, the best   60
She culled, and then with handy care she dressed.
High o'er the hearth a chine of bacon hung:
Good old Philemon seized it with a prong,
And from the sooty rafter drew it down;

Then cut a slice, but scarce enough for one;
Yet a large portion of a little store,
Which for their sakes alone he wished were more.
This in the pot he plunged without delay,
To tame the flesh and drain the salt away.
The time between, before the fire they sat,   70
And shortened the delay by pleasing chat.
   'A beam there was, on which a beechen pail
Hung by the handle on a driven nail:
This filled with water, gently warmed, they set
Before their guests; in this they bathed their feet,
And after with clean towels dried their sweat:
This done, the host produced the genial bed,°
Sallow the feet, the borders, and the stead,
Which with no costly coverlet they spread,
But course old garments; yet such robes as these   80
They laid alone at feasts, on holidays.°
The good old housewife, tacking up her gown,
The table sets; the invited gods lie down.
The trivet table of a foot was lame,
A blot which prudent Baucis overcame,
Who thrusts beneath the limping leg a shard;
So was the mended board exactly reared:
Then rubbed it o'er with newly gathered mint,
A wholesome herb, that breathed a grateful scent.
Pallas began the feast, where first was seen°   90
The particoloured olive, black and green:
Autumnal cornels next in order served,°
In lees of wine well pickled, and preserved.°
A garden salad was the third supply,
Of endive, radishes and chicory;
Then curds and cream, the flower of country fare,
And new laid eggs, which Baucis' busy care
Turned by a gentle fire, and roasted rare.
All these in earthenware were served to board;
And next in place an earthen pitcher, stored   100
With liquor of the best the cottage could afford.
This was the table's ornament and pride,
With figures wrought: like pages at his side
Stood beechen bowls; and these were shining clean,
Varnished with wax without, and lined within.

By this the boiling kettle had prepared,
And to the table sent the smoking lard;
On which with eager appetite they dine,
A savoury bit, that served to relish wine.
The wine itself was suiting to the rest,   110
Still working in the must, and lately pressed.°
The second course succeeds like that before,
Plums, apples, nuts, and of their wintry store,
Dry figs, and grapes, and wrinkled dates were set
In canisters, to enlarge the little treat.
All these a milk-white honeycomb surround,
Which in the midst the country banquet crowned.
But the kind hosts their entertainment grace
With hearty welcome, and an open face.
In all they did, you might discern with ease,   120
A willing mind, and a desire to please.

    'Meantime the beechen bowls went round, and still
Though often emptied, were observed to fill;
Filled without hands, and of their own accord
Ran without feet, and danced about the board.
Devotion seized the pair, to see the feast
With wine, and of no common grape, increased;
And up they held their hands, and fell to prayer,
Excusing as they could their country fare.

    'One goose they had, ('twas all they could allow)   130
A wakeful sentry, and on duty now,
Whom to the gods for sacrifice they vow:
Her, with malicious zeal, the couple viewed;
She ran for life, and limping they pursued:
Full well the fowl perceived their bad intent,
And would not make her masters' compliment;
But persecuted, to the powers she flies,
And close between the legs of Jove she lies.
He with a gracious ear the suppliant heard,
And saved her life; then what he was declared,   140
And owned the god. "The neighbourhood", said he,°
"Shall justly perish for impiety:
You stand alone exempted; but obey
With speed, and follow where we lead the way;
Leave these accursed; and to the mountain's height
Ascend; nor once look backward in your flight."

'They haste, and what their tardy feet denied,
The trusty staff (their better leg) supplied.
An arrow's flight they wanted to the top,
And there secure, but spent with travel, stop;   150
Then turn their now no more forbidden eyes:
Lost in a lake the floated level lies;°
A watery desert covers all the plains,
Their cot alone, as in an isle, remains.
Wondering with weeping eyes, while they deplore
Their neighbours' fate, and country now no more,
Their little shed, scarce large enough for two,
Seems, from the ground increased, in height and bulk to grow.
A stately temple shoots within the skies;
The crotches of their cot in columns rise;   160
The pavement polished marble they behold,
The gates with sculpture graced, the spires and tiles of gold.
    'Then thus the sire of gods, with look serene,
"Speak thy desire, thou only just of men;
And thou, O woman, only worthy found
To be with such a man in marriage bound."
    'Awhile they whisper; then to Jove addressed,
Philemon thus prefers their joint request.
"We crave to serve before your sacred shrine,
And offer at your altars rites divine;   170
And since not any action of our life
Has been polluted with domestic strife,
We beg one hour of death; that neither she
With widow's tears may live to bury me,
Nor weeping I, with withered arms may bear
My breathless Baucis to the sepulcher."
    'The godheads sign their suit. They run their race°
In the same tenor all the appointed space:
Then, when their hour was come, while they relate
These past adventures at the temple gate,   180
Old Baucis is by old Philemon seen
Sprouting with sudden leaves of sprightly green.
Old Baucis looked where old Philemon stood,
And saw his lengthened arms a sprouting wood.
New roots their fastened feet begin to bind,
Their bodies stiffen in a rising rind:
Then ere the bark above their shoulders grew,

They give and take at once their last adieu.
At once, "Farewell, O faithful spouse," they said;
At once the encroaching rinds their closing lips invade.   190
E'en yet, an ancient Tyanaean shows°
A spreading oak, that near a linden grows;
The neighbourhood confirm the prodigy,
Grave men, not vain of tongue, or like to lie.
I saw myself the garlands on their boughs,
And tablets hung for gifts of granted vows;
And offering fresher up, with pious prayer,
"The good", said I, "are god's peculiar care,
And such as honour heaven, shall heavenly honour share." '

Pygmalion and the Statue

Out of the Tenth Book of
Ovid's *Metamorphoses*

*The Propoetides,° for their impudent behaviour, being turned into stone by Venus,
Pygmalion, prince of Cyprus, detested all women for their sake, and resolved
never to marry. He falls in love with a statue of his own making, which is changed
into a maid, whom he marries. One of his descendants is Cinyras, the father of
Myrrha; the daughter incestuously loves her own father, for which she is changed
into the tree which bears her name. These two stories immediately follow each
other, and are admirably well connected.*

Pygmalion, loathing their lascivious life,
Abhorred all womankind, but most a wife;
So single chose to live, and shunned to wed,
Well pleased to want a consort of his bed.
Yet fearing idleness, the nurse of ill,
In sculpture exercised his happy skill;
And carved in ivory such a maid, so fair,
As nature could not with his art compare,
Were she to work; but in her own defence,
Must take her pattern here, and copy hence.   10
Pleased with his idol, he commends, admires,
Adores; and last, the thing adored desires.
A very virgin in her face was seen,
And, had she moved, a living maid had been;
One would have thought she could have stirred, but strove

With modesty, and was ashamed to move.
Art, hid with art, so well performed the cheat,
It caught the carver with his own deceit.
He knows 'tis madness, yet he must adore,
And still the more he knows it, loves the more;   20
The flesh, or what so seems, he touches oft,
Which feels so smooth, that he believes it soft.
Fired with this thought, at once he strained the breast,°
And on the lips a burning kiss impressed.
'Tis true, the hardened breast resists the gripe,°
And the cold lips return a kiss unripe;
But when, retiring back, he looked again,
To think it ivory was a thought too mean;
So would believe she kissed, and courting more,
Again embraced her naked body o'er;   30
And, straining hard the statue, was afraid
His hands had made a dint, and hurt his maid;
Explored her, limb by limb, and feared to find
So rude a grip had left a livid mark behind.
With flattery now he seeks her mind to move,
And now with gifts, the powerful bribes of love;
He furnishes her closet first; and fills
The crowded shelves with rarities of shells;
Adds orient pearls, which from the conchs he drew.
And all the sparkling stones of various hue;   40
And parrots, imitating human tongue,
And singing-birds in silver cages hung;
And every fragrant flower, and odorous green,
Were sorted well, with lumps of amber laid between;
Rich fashionable robes her person deck;
Pendants her ears, and pearls adorn her neck;
Her tapered fingers too with rings are graced,
And an embroidered zone surrounds her slender waist.
Thus like a queen arrayed, so richly dressed,
Beauteous she showed, but naked showed the best.   50
Then from the floor he raised a royal bed,
With coverings of Sidonian purple spread;°
The solemn rites performed, he calls her bride,
With blandishments invites her to his side,
And as she were with vital sense possessed,
Her head did on a plumy pillow rest.

The feast of Venus came, a solemn day,
To which the Cypriots due devotion pay;
With gilded horns the milk-white heifers led,
Slaughtered before the sacred altars, bled;   60
Pygmalion, offering, first approached the shrine,
And then with prayers implored the powers divine;
'Almighty gods, if all we mortals want,
If all we can require, be yours to grant,
Make this fair statue mine,' he would have said,
But changed his words for shame, and only prayed,
'Give me the likeness of my ivory maid!'
  The golden goddess, present at the prayer,
Well knew he meant the inanimated fair,
And gave the sign of granting his desire;   70
For thrice in cheerful flames ascends the fire.
The youth, returning to his mistress hies,
And impudent in hope, with ardent eyes,
And beating breast, by the dear statue lies.
He kisses her white lips, renews the bliss,
And looks and thinks they redden at the kiss;
He thought them warm before: nor longer stays,
But next his hand on her hard bosom lays;
Hard as it was, beginning to relent,
It seemed the breast beneath his fingers bent;   80
He felt again, his fingers made a print,
'Twas flesh, but flesh so firm, it rose against the dint.
The pleasing task he fails not to renew;
Soft, and more soft at every touch it grew;
Like pliant wax, when chafing hands reduce
The former mass to form, and frame for use.
He would believe, but yet is still in pain,
And tries his argument of sense again,
Presses the pulse, and feels the leaping vein.
Convinced, o'erjoyed, his studied thanks and praise,   90
To her who made the miracle, he pays;
Then lips to lips he joined; now freed from fear,
He found the savour of the kiss sincere.°
At this the wakened image oped her eyes,
And viewed at once the light and lover with surprise.
The goddess, present at the match she made,
So blessed the bed, such fruitfulness conveyed,

That ere ten moons had sharpened either horn,
To crown their bliss, a lovely boy was born;
Paphos his name, who, grown to manhood, walled   100
The city Paphos, from the founder called.

Cinyras and Myrrha

Out of the Tenth Book of
Ovid's *Metamorphoses*

*There needs no connection of this story with the former, for the beginning of this
immediately follows the end of the last. The reader is only to take notice, that
Orpheus, who relates both, was by birth a Thracian, and his country far distant
from Cyprus, where Myrrha was born, and from Arabia, whither she fled. You
will see the reason of this note, soon after the first lines of this fable.*

Nor him alone produced the fruitful queen;
But Cinyras, who like his sire had been
A happy prince, had he not been a sire.
Daughters and fathers, from my song retire!
I sing of horror; and, could I prevail,
You should not hear, or not believe my tale.
Yet if the pleasure of my song be such,
That you will hear, and credit me too much,
Attentive listen to the last event,
And with the sin believe the punishment: 10
Since nature could behold so dire a crime,
I gratulate at least my native clime.
That such a land, which such a monster bore,
So far is distant from our Thracian shore.
Let Araby extol her happy coast,
Her cinnamon and sweet amomum boast;°
Her fragrant flowers, her trees with precious tears.
Her second harvests, and her double years;
How can the land be called so blessed that Myrrha bears?
Not all her odorous tears can cleanse her crime,   20
Her plant alone deforms the happy clime;°
Cupid denies to have inflamed thy heart,
Disowns thy love, and vindicates his dart:
Some fury gave thee those infernal pains,
And shot her venomed vipers in thy veins.

To hate thy sire, had merited a curse;
But such an impious love deserved a worse.
The neighbouring monarchs, by thy beauty led,
Contend in crowds, ambitious of thy bed;°
The world is at thy choice, except but one.   30
Except but him thou canst not choose alone.
She knew it too, the miserable maid,
Ere impious love her better thoughts betrayed,
And thus within her secret soul she said:
'Ah, Myrrha! whither would thy wishes tend?
Ye gods, ye sacred laws, my soul defend
From such a crime as all mankind detest,
And never lodged before in human breast!
But is it sin? Or makes my mind alone
The imagined sin? For nature makes it none.   40
What tyrant then these envious laws began,
Made not for any other beast but man!
The father-bull his daughter may bestride,
The horse may make his mother-mare a bride;
What piety forbids the lusty ram,
Or more salacious goat, to rut their dam?
The hen is free to wed the chick she bore,
And make a husband, whom she hatched before.
All creatures else are of a happier kind,
Whom nor ill-natured laws from pleasure bind,   50
Nor thoughts of sin disturb their peace of mind.
But man a slave of his own making lives;
The fool denies himself what nature gives;
Too busy senates, with an over-care
To make us better than our kind can bear,
Have dashed a spice of envy in the laws,
And, straining up too high, have spoiled the cause.
Yet some wise nations break their cruel chains,
And own no laws, but those which love ordains;
Where happy daughters with their sires are joined,   60
And piety is doubly paid in kind.
O that I had been born in such a clime,
Not here, where 'tis the country makes the crime!
But whither would my impious fancy stray?
Hence hopes, and ye forbidden thoughts, away!
His worth deserves to kindle my desires,

But with the love that daughters bear to sires.
Then had not Cinyras my father been,
What hindered Myrrha's hopes to be his queen?
But the perverseness of my fate is such,   70
That he's not mine, because he's mine too much:
Our kindred-blood debars a better tie;
He might be nearer, were he not so nigh.
Eyes and their objects never must unite,
Some distance is required to help the sight.
Fain would I travel to some foreign shore,
Never to see my native country more,
So might I to myself myself restore;
So might my mind these impious thoughts remove,
And, ceasing to behold, might cease to love.   80
But stay I must, to feed my famished sight,
To talk, to kiss; and more, if more I might:
More, impious maid! What more canst thou design?
To make a monstrous mixture in thy line,
And break all statutes human and divine?
Canst thou be called (to save thy wretched life)
Thy mother's rival, and thy father's wife?
Confound so many sacred names in one,
Thy brother's mother! sister to thy son!
And fear'st thou not to see the infernal bands,°   90
Their heads with snakes, with torches armed their hands,
Full at thy face the avenging brands to bear,
And shake the serpents from their hissing hair?
But thou in time the increasing ill control,
Nor first debauch the body by the soul;
Secure the sacred quiet of thy mind,
And keep the sanctions nature has designed,
Suppose I should attempt, the attempt were vain;
No thoughts like mine his sinless soul profane,
Observant of the right; and O, that he°   100
Could cure my madness, or be mad like me!'
    Thus she; but Cinyras, who daily sees
A crowd of noble suitors at his knees,
Among so many, knew not whom to choose,
Irresolute to grant, or to refuse;
But, having told their names, inquired of her,
Who pleased her best, and whom she would prefer?

The blushing maid stood silent with surprise,
And on her father fixed her ardent eyes,
And, looking, sighed; and, as she sighed, began   110
Round tears to shed that scalded as they ran.
The tender sire, who saw her blush, and cry,
Ascribed it all to maiden modesty;
And dried the falling drops, and, yet more kind,
He stroked her cheeks, and holy kisses joined:
She felt a secret venom fire her blood.
And found more pleasure than a daughter should;
And, asked again, what lover of the crew
She liked the best she answered, 'One like you.'
Mistaking what she meant, her pious will   120
He praised, and bade her so continue still:
The word of 'pious' heard, she blushed with shame
Of secret guilt, and could not bear the name.
    'Twas now the mid of night, when slumbers close
Our eyes, and soothe our cares with soft repose;
But no repose could wretched Myrrha find,
Her body rolling, as she rolled her mind:
Mad with desire, she ruminates her sin,
And wishes all her wishes o'er again:
Now she despairs, and now resolves to try;   130
Would not, and would again, she knows not why;
Stops and returns, makes and retracts the vow;
Fain would begin, but understands not how:
As when a pine is hewn upon the plains,
And the last mortal stroke alone remains,
Labouring in pangs of death, and threatening all,
This way and that she nods, considering where to fall;
So Myrrha's mind, impelled on either side,
Takes every bent, but cannot long abide:
Irresolute on which she should rely,   140
At last, unfixed in all, is only fixed to die.
On that sad thought she rests; resolved on death,
She rises, and prepares to choke her breath;
Then while about the beam her zone she ties,
'Dear Cinyras, farewell,' she softly cries;
'For thee I die, and only wish to be
Not hated, when thou know'st I die for thee:
Pardon the crime in pity to the cause.'

This said, about her neck the noose she draws.
The nurse, who lay without, her faithful guard, 150
Though not the words, the murmurs overheard,
And sighs and hollow sounds; surprised with fright,
She starts, and leaves her bed, and springs a light;
Unlocks the door, and entering out of breath,
The dying saw and instruments of death.
She shrieks, she cuts the zone with trembling haste,
And in her arms her fainting charge embraced;
Next (for she now had leisure for her tears)
She weeping asked, in these her blooming years,
What unforeseen misfortune caused her care, 160
To loathe her life, and languish in despair?
The maid, with downcast eyes, and mute with grief,
For death unfinished, and ill-timed relief,
Stood sullen to her suit: the beldam pressed
The more to know, and bared her withered breast;
Adjured her, by the kindly food she drew
From those dry founts, her secret ill to show.
Sad Myrrha sighed, and turned her eyes aside;
The nurse still urged and would not be denied:
Nor only promised secrecy, but prayed 170
She might have leave to give her offered aid.
'Good will', she said, 'my want of strength supplies,
And diligence shall give what age denies.
If strong desires thy mind to fury move,
With charms and medicines I can cure thy love;
If envious eyes their hurtful rays have cast.
More powerful verse shall free thee from the blast;
If heaven, offended, sends thee this disease,
Offended heaven with prayers we can appease.
What then remains that can these cares procure? 180
Thy house is flourishing, thy fortune sure;°
Thy careful mother yet in health survives.
And, to thy comfort, thy kind father lives.'
    The virgin started at her father's name,
And sighed profoundly, conscious of the shame;
Nor yet the nurse her impious love divined,
But yet surmised, that love disturbed her mind.
Thus thinking, she pursued her point, and laid
And lulled within her lap the mourning maid;

Then softly soothed her thus: 'I guess your grief;   190
You love, my child; your love shall find relief.
My long experienced age shall be your guide;
Rely on that, and lay distrust aside;
No breath of air shall on the secret blow,
Nor shall (what most you fear) your father know.'
Struck once again, as with a thunderclap,
The guilty virgin bounded from her lap,
And threw her body prostrate on the bed,
And, to conceal her blushes, hid her head:
There silent lay, and warned her with her hand   200
To go; but she received not the command;°
Remaining still importunate to know.
Then Myrrha thus: 'Or ask no more, or go;
I prithee go, or, staying, spare my shame;
What thou wouldst hear, is impious e'en to name.'
At this, on high the beldam holds her hands,
And trembling, both with age and terror, stands;
Adjures, and, falling at her feet, intreats,
Soothes her with blandishments, and frights with threats,
To tell the crime intended, or disclose   210
What part of it she knew, if she no further knows;
And last, if conscious to her counsel made,
Confirms anew the promise of her aid.

   Now Myrrha raised her head: but soon, oppressed
With shame, reclined it on her nurse's breast;
Bathed it with tears, and strove to have confessed:
Twice she began, and stopped; again she tried;
The faltering tongue its office still denied;
At last her veil before her face she spread.
And drew a long preluding sigh, and said,   220
'O happy mother, in thy marriage bed!'
Then groaned, and ceased. The good old woman shook,
Stiff were her eyes, and ghastly was her look:
Her hoary hair upright with horror stood.
Made (to her grief) more knowing than she would;
Much she reproached, and many things she said,
To cure the madness of the unhappy maid;
In vain; for Myrrha stood convict of ill;
Her reason vanquished, but unchanged her will;
Perverse of mind, unable to reply,   230

She stood resolved or to possess, or die.
At length the fondness of a nurse prevailed
Against her better sense, and virtue failed:
'Enjoy, my child, since such is thy desire,
Thy love,' she said; she durst not say, 'Thy sire.'
'Live, though unhappy, live on any terms;'
Then with a second oath her faith confirms.

   The solemn feast of Ceres now was near,
When long white linen stoles the matrons wear;
Ranked in procession walk the pious train,   240
Offering first-fruits, and spikes of yellow grain;
For nine long nights the nuptial bed they shun,
And, sanctifying harvest, lie alone.
Mixed with the crowd, the queen forsook her lord,
And Ceres' power with secret rites adored.
The royal couch now vacant for a time,
The crafty crone, officious in her crime,
The cursed occasion took; the king she found
Easy with wine, and deep in pleasures drowned,
Prepared for love; the beldam blew the flame,   250
Confessed the passion, but concealed the name.
Her form she praised; the monarch asked her years,
And she replied, 'The same thy Myrrha bears.'
Wine and commended beauty fired his thought;
Impatient, he commands her to be brought.
Pleased with her charge performed, she hies her home,
And gratulates the nymph, the task was overcome.
Myrrha was joyed the welcome news to hear;
But, clogged with guilt, the joy was insincere.
So various, so discordant is the mind,   260
That in our will, a different will we find.
Ill she presaged, and yet pursued her lust;
For guilty pleasures give a double gust.
'Twas depth of night: Arctophylax had driven°
His lazy wain half round the northern heaven,
When Myrrha hastened to the crime desired.
The moon beheld her first, and first retired;
The stars, amazed, ran backward from the sight,°
And, shrunk within their sockets, lost their light.
Icarius first withdraws his holy flame;°   270
The Virgin sign, in heaven the second name,°

Slides down the belt, and from her station flies,
And night with sable clouds involves the skies.
Bold Myrrha still pursues her black intent;
She stumbled thrice, (an omen of the event;)
Thrice shrieked the funeral owl, yet on she went,
Secure of shame, because secure of sight;°
E'en bashful sins are impudent by night.
Linked hand in hand, the accomplice and the dame,
Their way exploring, to the chamber came;   280
The door was ope, they blindly grope their way,
Where dark in bed the expecting monarch lay:
Thus far her courage held, but here forsakes;
Her faint knees knock at every step she makes.
The nearer to her crime, the more within
She feels remorse, and horror of her sin;
Repents too late her criminal desire,
And wishes, that unknown she could retire.
Her, lingering thus, the nurse, who feared delay
The fatal secret might at length betray,   290
Pulled forward, to complete the work begun,
And said to Cinyras, 'Receive thy own!'
Thus saying, she delivered kind to kind,
Accursed, and their devoted bodies joined.
The sire, unknowing of the crime, admits
His bowels, and profanes the hallowed sheets.°
He found she trembled, but believed she strove
With maiden modesty, against her love:
And sought, with flattering words, vain fancies to remove.
Perhaps he said, 'My daughter, cease thy fears,'   300
Because the title suited with her years;
And, 'Father,' she might whisper him again,
That names might not be wanting to the sin.°
Full of her sire, she left the incestuous bed,°
And carried in her womb the crime she bred.
Another, and another night she came;
For frequent sin had left no sense of shame:
Till Cinyras desired to see her face,
Whose body he had held in close embrace,
And brought a taper; the revealer, light,   310
Exposed both crime, and criminal, to sight.
Grief, rage, amazement, could no speech afford,

But from the sheath he drew the avenging sword;
The guilty fled; the benefit of night,
That favoured first the sin, secured the flight.
Long wandering through the spacious fields, she bent
Her voyage to the Arabian continent;
Then passed the region which Panchaia joined,°
And flying left the palmy plains behind.
Nine times the moon had mewed her horns; at length,° 320
With travel weary, unsupplied with strength,
And with the burden of her womb oppressed,
Sabaean fields afford her needful rest;
There, loathing life, and yet of death afraid,
In anguish of her spirit, thus she prayed:
'Ye powers, if any so propitious are
To accept my penitence, and hear my prayer,
Your judgments, I confess, are justly sent;
Great sins deserve as great a punishment:
Yet, since my life the living will profane, 330
And since my death the happy dead will stain,
A middle state your mercy may bestow,
Betwixt the realms above, and those below;
Some other form to wretched Myrrha give,
Nor let her wholly die, nor wholly live.'
    The prayers of penitents are never vain;
At least, she did her last request obtain,
For, while she spoke, the ground began to rise,
And gathered round her feet, her legs, and thighs;
Her toes in roots descend, and, spreading wide, 340
A firm foundation for the trunk provide;
Her solid bones convert to solid wood,
To pith her marrow, and to sap her blood:
Her arms are boughs, her fingers change their kind,
Her tender skin is hardened into rind.
And now the rising tree her womb invests,
Now, shooting upwards still, invades her breasts,
And shades the neck; when, weary with delay,
She sunk her head within, and met it half the way.
And though with outward shape she lost her sense, 350
With bitter tears she wept her last offence;
And still she weeps, nor sheds her tears in vain;
For still the precious drops her name retain.

Meantime the misbegotten infant grows,
And, ripe for birth, distends with deadly throes
The swelling rind, with unavailing strife,
To leave the wooden womb, and pushes into life.
The mother-tree, as if oppressed with pain,
Writhes here and there, to break the bark, in vain;
And, like a labouring woman, would have prayed,   360
But wants a voice to call Lucina's aid;
The bending bole sends out a hollow sound,
And trickling tears fall thicker on the ground.
The mild Lucina came uncalled, and stood
Beside the struggling boughs, and heard the groaning wood;
Then reached her midwife-hand, to speed the throes,
And spoke the powerful spells that babes to birth disclose.
The bark divides, the living load to free,
And safe delivers the convulsive tree.°
The ready nymphs receive the crying child,   370
And wash him in the tears the parent plant distilled.
They swathed him with their scarfs; beneath him spread
The ground with herbs; with roses raised his head.
The lovely babe was born with every grace;
E'en envy must have praised so fair a face:
Such was his form, as painters, when they show
Their utmost art, on naked loves bestow;
And that their arms no difference might betray,
Give him a bow, or his from Cupid take away.
Time glides along, with undiscovered haste,   380
The future but a length behind the past,
So swift are years; the babe, whom just before
His grandsire got, and whom his sister bore;
The drop, the thing which late the tree enclosed,
And late the yawning bark to life exposed;
A babe, a boy, a beauteous youth appears;
And lovelier than himself at riper years.
Now to the queen of love he gave desires,
And, with her pains, revenged his mother's fires.

The First Book of Homer's Iliad

THE ARGUMENT

Chryses, priest of Apollo, brings presents to the Grecian princes, to ransom his daughter Chryseis, who was prisoner in the fleet. Agamemnon, the general, whose captive and mistress the young lady was, refuses to deliver her, threatens the venerable old man, and dismisses him with contumely. The priest craves vengeance of his god, who sends a plague among the Greeks; which occasions Achilles, their great champion, to summon a council of the chief officers: he encourages Calchas, the high priest and prophet, to tell the reason why the gods were so much incensed against them. Calchas is fearful of provoking Agamemnon, till Achilles engages to protect him: then, emboldened by the hero, he accuses the general as the cause of all, by detaining the fair captive, and refusing the presents offered for her ransom. By this proceeding, Agamemnon is obliged, against his will, to restore Chryseis, with gifts, that he might appease the wrath of Phoebus; but, at the same time, to revenge himself on Achilles, sends to seize his slave Briseis. Achilles, thus affronted, complains to his mother Thetis; and begs her to revenge his injury, not only on the general, but on all the army, by giving victory to the Trojans, till the ungrateful king became sensible of his injustice. At the same time, he retires from the camp into his ships, and withdraws his aid from his countrymen. Thetis prefers° her son's petition to Jupiter, who grants her suit. Juno suspects her errand, and quarrels with her husband for his grant; till Vulcan reconciles his parents with a bowl of nectar, and sends them peaceably to bed.

The wrath of Peleus' son, O muse, resound,°
Whose dire effects the Grecian army found,
And many a hero, king, and hardy knight,
Were sent, in early youth, to shades of night:
Their limbs a prey to dogs and vultures made;
So was the sovereign will of Jove obeyed:
From that ill-omened hour when strife begun,
Betwixt Atrides great, and Thetis' godlike son.
   What power provoked, and for what cause, relate,
Sowed in their breasts the seeds of stern debate:°   10
Jove's and Latona's son his wrath expressed,°
In vengeance of his violated priest,
Against the king of men, who, swollen with pride,°
Refused his presents, and his prayers denied.
For this the god a swift contagion spread
Amid the camp, where heaps on heaps lay dead.
   For venerable Chryses came to buy,

With gold and gifts of price, his daughter's liberty.
Suppliant before the Grecian chiefs he stood,
Awful, and armed with ensigns of his god: 20
Bare was his hoary head; one holy hand
Held forth his laurel crown, and one his sceptre of command.
His suit was common; but above the rest,°
To both the brother-princes thus addressed:°
'Ye sons of Atreus, and ye Grecian powers,
So may the gods, who dwell in heavenly bowers,
Succeed your siege, accord the vows you make,°
And give you Troy's imperial town to take;
So, by their happy conduct, may you come
With conquest back to your sweet native home;   30
As you receive the ransom which I bring,
Respecting Jove, and the far-shooting king,°
And break my daughter's bonds, at my desire,
And glad with her return her grieving sire.'
   With shouts of loud acclaim the Greeks decree
To take the gifts, to set the damsel free.
The king of men alone with fury burned,
And haughty, these opprobrious words returned:
'Hence, holy dotard! and avoid my sight,
Ere evil intercept thy tardy flight;°   40
Nor dare to tread this interdicted strand,°
Lest not that idle sceptre in thy hand,
Nor thy god's crown, my vowed revenge withstand.
Hence, on thy life! the captive maid is mine,
Whom not for price or prayers I will resign;
Mine she shall be, till creeping age and time
Her bloom have withered, and consumed her prime.
Till then my royal bed she shall attend,
And, having first adorned it, late ascend;
This, for the night; by day, the web and loom,   50
And homely household-task, shall be her doom,
Far from thy loved embrace, and her sweet native home.'
He said: the helpless priest replied no more,
But sped his steps along the hoarse-resounding shore.°
Silent he fled; secure at length he stood,
Devoutly cursed his foes, and thus invoked his god:
   'O source of sacred light, attend my prayer,
God with the silver bow, and golden hair,

Whom Chrysa, Cilla, Tenedos obeys,
And whose broad eye their happy soil surveys!   60
If, Smintheus, I have poured before thy shrine°
The blood of oxen, goats, and ruddy wine,
And larded thighs on loaded altars laid,
Hear, and my just revenge propitious aid!
Pierce the proud Greeks, and with thy shafts attest
How much thy power is injured in thy priest.'
   He prayed; and Phoebus, hearing, urged his flight,
With fury kindled, from Olympus' height;
His quiver o'er his ample shoulders threw,
His bow twanged, and his arrows rattled as they flew.   70
Black as a stormy night, he ranged around
The tents, and compassed the devoted ground;
Then with full force his deadly bow he bent,
And feathered fates among the mules and sumpters sent,°
The essay of rage; on faithful dogs the next;
And last, in human hearts his arrows fixed.
The god nine days the Greeks at rovers killed,°
Nine days the camp with funeral fires was filled;
The tenth, Achilles, by the queen's command,°
Who bears heaven's awful sceptre in her hand,   80
A council summoned; for the goddess grieved
Her favoured host should perish unrelieved.
   The kings, assembled, soon their chief enclose;
Then from his seat the goddess-born arose,
And thus undaunted spoke: 'what now remains,
But that once more we tempt the watery plains,
And, wandering homeward, seek our safety hence,
In flight at least, if we can find defence?°
Such woes at once encompass us about,
The plague within the camp, the sword without.   90
Consult, O king, the prophets of the event;
And whence these ills, and what the god's intent,
Let them by dreams explore, for dreams from Jove are sent.
What want of offered victims, what offence
In fact committed could the sun incense,
To deal his deadly shafts? What may remove
His settled hate, and reconcile his love?
That he may look propitious on our toils,
And hungry graves no more be glutted with our spoils.'

Thus to the king of men the hero spoke,   100
Then Calchas the desired occasion took;
Calchas, the sacred seer, who had in view
Things present and the past, and things to come foreknew;
Supreme of augurs, who by Phoebus taught,
The Grecian powers to Troy's destruction brought.
Skilled in the secret causes of their woes,
The reverend priest in graceful act arose,
And thus bespoke Pelides: 'Care of Jove,
Favoured of all the immortal powers above,
Wouldst thou the seeds deep sown of mischief know,   110
And why provoked Apollo bends his bow,
Plight first thy faith, inviolably true,
To save me from those ills that may ensue.

For I shall tell ungrateful truths to those°
Whose boundless powers of life and death dispose;
And sovereigns, ever jealous of their state,
Forgive not those whom once they mark for hate:
E'en though the offence they seemingly digest,
Revenge, like embers raked within their breast,
Bursts forth in flames, whose unresisted power   120
Will seize the unwary wretch and soon devour.
Such, and no less, is he, on whom depends°
The sum of things, and whom my tongue of force offends.°
Secure me then from his foreseen intent,
That what his wrath may doom, thy valour may prevent.'

To this the stern Achilles made reply:
'Be bold, (and on my plighted faith rely,)
To speak what Phoebus has inspired thy soul
For common good, and speak without control.°
His godhead I invoke; by him I swear   130
That while my nostrils draw this vital air,
None shall presume to violate those bands,
Or touch thy person with unhallowed hands;
E'en not the king of men, that all commands.'

At this, resuming heart, the prophet said:
'Nor hecatombs unslain, nor vows unpaid,°
On Greeks accursed this dire contagion bring,
Or call for vengeance from the bowyer king;
But he the tyrant, whom none dares resist,
Affronts the godhead in his injured priest;   140

He keeps the damsel captive in his chain,
And presents are refused, and prayers preferred in vain.°
For this the avenging power employs his darts,
And empties all his quiver in our hearts;
Thus will persist, relentless in his ire,
Till the fair slave be rendered to her sire,
And ransom-free restored to his abode,
With sacrifice to reconcile the god;
Then he, perhaps, atoned by prayer, may cease
His vengeance justly vowed, and give the peace.'   150
    Thus having said, he sat. Thus answered then,
Upstarting from his throne, the king of men,
His breast with fury filled, his eyes with fire,
Which rolling round, he shot in sparkles on the sire:°
'Augur of ill, whose tongue was never found
Without a priestly curse, or boding sound!
For not one blessed event foretold to me
Passed through that mouth, or passed unwillingly;°
And now thou dost with lies the throne invade,°
By practice hardened in thy slandering trade;   160
Obtending heaven, for whate'er ills befall,°
And sputtering under specious names thy gall.
Now Phoebus is provoked, his rites and laws
Are in his priest profaned, and I the cause;
Since I detain a slave, my sovereign prize,
And sacred gold, your idol-god, despise.
I love her well; and well her merits claim,
To stand preferred before my Grecian dame:°
Not Clytemnestra's self in beauty's bloom
More charmed, or better plied the various loom: 170
Mine is the maid, and brought in happy hour,
With every household-grace adorned, to bless my nuptial bower.
Yet shall she be restored, since public good
For private interest ought not to be withstood,
To save the effusion of my people's blood.
But right requires, if I resign my own,
I should not suffer for your sakes alone;°
Alone excluded from the prize I gained,
And by your common suffrage have obtained.
The slave without a ransom shall be sent,   180
It rests for you to make the equivalent.'

To this the fierce Thessalian prince replied:°
'O first in power, but passing all in pride,
Griping, and still tenacious of thy hold,
Would'st thou the Grecian chiefs, though largely souled,°
Should give the prizes they had gained before,
And with their loss thy sacrilege restore?
Whate'er by force of arms the soldier got,
Is each his own, by dividend of lot;
Which to resume were both unjust and base,   190
Not to be borne but by a servile race.
But this we can; if Saturn's son bestows°
The sack of Troy, which he by promise owes,
Then shall the conquering Greeks thy loss restore,
And with large interest make the advantage more.'

To this Atrides answered: 'Though thy boast
Assumes the foremost name of all our host,
Pretend not, mighty man, that what is mine,
Controlled by thee, I tamely should resign.
Shall I release the prize I gained by right,   200
In taken towns, and many a bloody fight,
While thou detain'st Briseis in thy bands,
By priestly glossing on the god's commands?°
Resolve on this, (a short alternative,)
Quit mine, or, in exchange, another give;°
Else I, assure thy soul, by sovereign right
Will seize thy captive in thy own despite;
Or from stout Ajax, or Ulysses, bear
What other prize my fancy shall prefer:
Then softly murmur, or aloud complain,   210
Rage as you please, you shall resist in vain.
But more of this, in proper time and place;
To things of greater moment let us pass.
A ship to sail the sacred seas prepare,
Proud in her trim, and put on board the fair,
With sacrifice and gifts, and all the pomp of prayer.
The crew well chosen, the command shall be
In Ajax; or if other I decree,
In Creta's king, or Ithacus', or, if I please, in thee:°
Most fit thyself to see performed the intent,   220
For which my prisoner from my sight is sent,
(Thanks to thy pious care,) that Phoebus may relent.'

At this Achilles rolled his furious eyes,
Fixed on the king askant, and thus replies:
'O, impudent, regardful of thy own,
Whose thoughts are centered on thyself alone,
Advanced to sovereign sway for better ends
Than thus like abject slaves to treat thy friends!
What Greek is he, that, urged by thy command,
Against the Trojan troops will lift his hand?   230
Not I; nor such enforced respect I owe,
Nor Pergamus I hate, nor Priam is my foe.°
What wrong from Troy remote could I sustain,
To leave my fruitful soil and happy reign,
And plough the surges of the stormy main?
Thee, frontless man, we followed from afar,
Thy instruments of death, and tools of war.
Thine is the triumph; ours the toil alone;
We bear thee on our backs, and mount thee on the throne.
For thee we fall in fight; for thee redress   240
Thy baffled brother,—not the wrongs of Greece.
And now thou threaten'st, with unjust decree,
To punish thy affronting heaven on me;
To seize the prize which I so dearly bought,
By common suffrage given, confirmed by lot.
Mean match to thine; for, still above the rest,
Thy hooked rapacious hands usurp the best.
Though mine are first in fight, to force the prey,
And last sustain the labours of the day.
Nor grudge I thee the much the Grecians give,   250
Nor murmuring take the little I receive;
Yet e'en this little, thou, who wouldst engross°
The whole, insatiate, enviest as thy loss.
Know, then, for Phthia fixed is my return;°
Better at home my ill-paid pains to mourn,
Than from an equal here sustain the public scorn.'
    The king, whose brows with shining gold were bound,
Who saw his throne with sceptered slaves encompassed round,
Thus answered stern: 'Go, at thy pleasure, go;
We need not such a friend, nor fear we such a foe.   260
There will not want to follow me in fight;
Jove will assist, and Jove assert my right:
But thou of all the kings (his care below)

Art least at my command, and most my foe.
Debates, dissensions, uproars are thy joy;
Provoked without offence, and practised to destroy.
Strength is of brutes, and not thy boast alone;
At least 'tis lent from heaven, and not thy own.
Fly then, ill-mannered, to thy native land,
And there thy ant-born Myrmidons command.°   270
But mark this menace; since I must resign
My black-eyed maid, to please the powers divine;
A well-rigged vessel in the port attends,
Manned at my charge, commanded by my friends;
The ship shall waft her to her wished abode,
Full fraught with holy bribes to the far-shooting god.
This thus dispatched, I owe myself the care,
My fame and injured honour to repair;
From thy own tent, proud man, in thy despite,
This hand shall ravish thy pretended right.   280
Briseis shall be mine, and thou shalt see
What odds of awful power I have on thee,
That others at thy cost may learn the difference of degree.'

    At this the impatient hero sourly smiled;
His heart impetuous in his bosom boiled,
And jostled by two tides of equal sway,
Stood for a while suspended in his way.
Betwixt his reason and his rage untamed,
One whispered soft, and one aloud reclaimed;
That only counselled to the safer side,   290
This to the sword his ready hand applied.
Unpunished to support the affront was hard,
Nor easy was the attempt to force the guard;
But soon the thirst of vengeance fired his blood,
Half shone his falchion, and half sheathed it stood.

    In that nice moment, Pallas, from above,
Commissioned by the imperial wife of Jove,
Descended swift; (the white-armed queen was loath
The fight should follow, for she favoured both;)
Just as in act he stood, in clouds enshrined,   300
Her hand she fastened on his hair behind;
Then backward by his yellow curls she drew;
To him, and him alone, confessed in view.°
Tamed by superior force he turned his eyes,

Aghast at first, and stupid with surprise;
But by her sparkling eyes, and ardent look,
The virgin-warrior known, he thus bespoke.
   'Comst thou, Celestial, to behold my wrongs?
Then view the vengeance which to crimes belongs.'
   Thus he. The blue-eyed goddess thus rejoined: 310
'I come to calm thy turbulence of mind,
If reason will resume her sovereign sway,
And, sent by Juno, her commands obey.
Equal she loves you both, and I protect;
Then give thy guardian gods their due respect,
And cease contention; be thy words severe,
Sharp as he merits; but the sword forbear.
An hour unhoped already wings her way,
When he his dire affront shall dearly pay;
When the proud king shall sue, with treble gain,   320
To quit thy loss, and conquer thy disdain.
But thou, secure of my unfailing word,
Compose thy swelling soul, and sheath the sword.'
   The youth thus answered mild: 'Auspicious maid,°
Heaven's will be mine, and your commands obeyed.
The gods are just, and when, subduing sense,
We serve their powers, provide the recompence.'
He said; with surly faith believed her word,
And in the sheath, reluctant, plunged the sword.
Her message done, she mounts the blessed abodes,   330
And mixed among the senate of the gods.
   At her departure his disdain returned;
The fire she fanned with greater fury burned,
Rumbling within till thus it found a vent:
'Dastard and drunkard, mean and insolent!
Tongue-valiant hero, vaunter of thy might,
In threats the foremost, but the lag in fight!
When didst thou thrust amid the mingled preace,°
Content to bide the war aloof in peace?
Arms are the trade of each plebeian soul;   340
'Tis death to fight; but kingly to control.
Lord-like at ease, with arbitrary power,
To peel the chiefs, the people to devour,°
These, traitor, are thy talents; safer far
Than to contend in fields, and toils of war.

Nor couldst thou thus have dared the common hate,
Were not their souls as abject as their state.
But, by this sceptre solemnly I swear,
(Which never more green leaf or growing branch shall bear;
Torn from the tree, and given by Jove to those   350
Who laws dispense, and mighty wrongs oppose,)
That when the Grecians want my wonted aid,
No gift shall bribe it, and no prayer persuade.
When Hector comes, the homicide, to wield
His conquering arms, with corpse to strew the field,
Then shalt thou mourn thy pride, and late confess
My wrong, repented when 'tis past redress.'
He said; and with disdain, in open view,
Against the ground his golden sceptre threw,
Then sat; with boiling rage Atrides burned,   360
And foam betwixt his gnashing grinders churned.

   But from his seat the Pylian prince arose,°
With reasoning mild, their madness to compose;
Words, sweet as honey, from his mouth distilled;
Two centuries already he fulfilled,
And now began the third; unbroken yet,
Once famed for courage, still in council great.

   'What worse', he said, 'can Argos undergo,
What can more gratify the Phrygian foe,°
Than these distempered heats, if both the lights   370
Of Greece their private interest disunites?
Believe a friend, with thrice your years increased,
And let these youthful passions be repressed.
I flourished long before your birth; and then
Lived equal with a race of braver men,
Than these dim eyes shall e'er behold again.
Ceneus and Dryas, and, excelling them,
Great Theseus, and the force of greater Polypheme.
With these I went, a brother of the war,
Their dangers to divide, their fame to share;   380
Nor idle stood with unassisting hands,
When savage beasts, and men's more savage bands,
Their virtuous toil subdued: yet those I swayed,
With powerful speech; I spoke, and they obeyed.
If such as those my counsels could reclaim,

Think not, young warriors, your diminished name
Shall lose of lustre, by subjecting rage
To the cool dictates of experienced age.
Thou, king of men, stretch not thy sovereign sway
Beyond the bounds free subjects can obey;   390
But let Pelides in his prize rejoice,
Achieved in arms, allowed by public voice.
Nor thou, brave champion, with his power contend,
Before whose throne e'en kings their lowered sceptres bend.
The head of action he, and thou the hand,
Matchless thy force, but mightier his command.
Thou first, O king, release the rights of sway;
Power self-restrained, the people best obey.
Sanctions of law from thee derive their source;
Command thyself, whom no commands can force.   400
The son of Thetis, rampire of our host,°
Is worth our care to keep, nor shall my prayers be lost.'
    Thus Nestor said, and ceased. Atrides broke
His silence next, but pondered ere he spoke:
'Wise are thy words, and glad I would obey,
But this proud man affects imperial sway,
Controlling kings, and trampling on our state;
His will is law, and what he wills is fate.
The gods have given him strength; but whence the style
Of lawless power assumed, or licence to revile?'   410
    Achilles cut him short, and thus replied:
'My worth allowed in words is, in effect, denied;
For who but a poltroon, possessed with fear,
Such haughty insolence can tamely bear?
Command thy slaves; my freeborn soul disdains
A tyrant's curb, and restive, breaks the reins.
Take this along, that no dispute shall rise
(Through mine the woman) for my ravished prize;
But, she excepted, as unworthy strife,
Dare not, I charge thee dare not, on thy life,   420
Touch aught of mine beside, by lot my due,
But stand aloof, and think profane to view;
This falchion else, not hitherto withstood,
These hostile fields shall fatten with thy blood.'°
    He said, and rose the first; the council broke,

And all their grave consults dissolved in smoke.
The royal youth retired, on vengeance bent;°
Patroclus followed silent to his tent.°
   Meantime, the king with gifts a vessel stores,
Supplies the banks with twenty chosen oars;   430
And next, to reconcile the shooter god,
Within her hollow sides the sacrifice he stowed;
Chryseis last was set on board, whose hand
Ulysses took, entrusted with command;
They plough the liquid seas, and leave the lessening land.
   Atrides then, his outward zeal to boast,
Bade purify the sin-polluted host.
With perfect hecatombs the god they graced,
Whose offered entrails in the main were cast;
Black bulls and bearded goats on altars lie,   440
And clouds of savoury stench involve the sky.
These pomps the royal hypocrite designed
For show, but harboured vengeance in his mind;
Till holy malice, longing for a vent,
At length discovered his concealed intent.
Talthybius, and Eurybates the just,
Heralds of arms, and ministers of trust,
He called, and thus bespoke: 'Haste hence your way,
And from the goddess-born demand his prey.°
If yielded, bring the captive; if denied,   450
The king (so tell him) shall chastise his pride;
And with armed multitudes in person come
To vindicate his power, and justify his doom.'
   This hard command unwilling they obey,
And o'er the barren shore pursue their way,
Where quartered in their camp the fierce Thessalians lay.
Their sovereign seated on his chair they find,
His pensive cheek upon his hand reclined,
And anxious thoughts revolving in his mind.
With gloomy looks he saw them entering in   460
Without salute; nor durst they first begin,
Fearful of rash offence and death foreseen.
He soon, the cause divining, cleared his brow,
And thus did liberty of speech allow:
   'Interpreters of gods and men, be bold;
Awful your character, and uncontrolled:

Howe'er unpleasing be the news you bring,
I blame not you, but your imperious king.
You come, I know, my captive to demand;
Patroclus, give her to the herald's hand.   470
But you authentic witnesses I bring
Before the gods, and your ungrateful king,
Of this my manifest, that never more°
This hand shall combat on the crooked shore:
No; let the Grecian powers, oppressed in fight,
Unpitied perish in their tyrant's sight.
Blind of the future, and by rage misled,
He pulls his crimes upon his people's head;
Forced from the field in trenches to contend,
And his insulted camp from foes defend.'   480
He said, and soon obeying his intent
Patroclus brought Briseis from her tent,
Then to the entrusted messengers resigned:
She wept, and often cast her eyes behind.
Forced from the man she loved, they led her thence,
Along the shore, a prisoner to their prince.

   Sole on the barren sands the suffering chief
Roared out for anguish, and indulged his grief;
Cast on his kindred seas a stormy look,
And his upbraided mother thus bespoke: 490
   'Unhappy parent of a short-lived son,
Since Jove in pity by thy prayers was won
To grace my small remains of breath with fame,
Why loads he this embittered life with shame,
Suffering his king of men to force my slave,
Whom, well deserved in war, the Grecians gave?'
   Set by old Ocean's side the goddess heard,°
Then from the sacred deep her head she reared;
Rose like a morning mist, and thus begun
To soothe the sorrows of her plaintive son: 500
'Why cries my care, and why conceals his smart?
Let thy afflicted parent share her part.'
   Then, sighing from the bottom of his breast,
To the sea-goddess thus the goddess-born addressed:
'Thou know'st my pain, which telling but recalls;
By force of arms we razed the Theban walls;
The ransacked city, taken by our toils,

We left, and hither brought the golden spoils:
Equal we shared them; but before the rest,
The proud prerogative had seized the best.°   510
Chryseis was the greedy tyrant's prize,
Chryseis, rosy-cheeked, with charming eyes.
Her sire, Apollo's priest, arrived to buy,
With proffered gifts of price, his daughter's liberty.
Suppliant before the Grecian chiefs he stood,
Awful, and armed with ensigns of his god;
Bare was his hoary head; one holy hand
Held forth his laurel-crown, and one his sceptre of command.
His suit was common, but, above the rest,
To both the brother-princes was addressed.   520
With shouts of loud acclaim the Greeks agree
To take the gifts, to set the prisoner free.
Not so the tyrant, who with scorn the priest
Received, and with opprobrious words dismissed.
The good old man, forlorn of human aid,
For vengeance to his heavenly patron prayed:
The godhead gave a favourable ear,
And granted all to him he held so dear;
In an ill hour his piercing shafts he sped,
And heaps on heaps of slaughtered Greeks lay dead,   530
While round the camp he ranged: at length arose
A seer, who well divined, and durst disclose
The source of all our ills: I took the word;
And urged the sacred slave to be restored,
The god appeased: the swelling monarch stormed,°
And then the vengeance vowed he since performed.
The Greeks, 'tis true, their ruin to prevent,
Have to the royal priest his daughter sent;
But from their haughty king his heralds came,
And seized, by his command, my captive dame,   540
By common suffrage given; but thou be won,
If in thy power, to avenge thy injured son!
Ascend the skies, and supplicating move
Thy just complaint to cloud-compelling Jove.
If thou by either word or deed hast wrought
A kind remembrance in his grateful thought,
Urge him by that; for often hast thou said
Thy power was once not useless in his aid,

When he, who high above the highest reigns,
Surprised by traitor gods, was bound in chains;　　550
When Juno, Pallas, with ambition fired,
And his blue brother of the seas conspired,
Thou freed'st the sovereign from unworthy bands,
Thou brought'st Briareus with his hundred hands,°
(So called in heaven, but mortal men below
By his terrestrial name, Aegeon, know;
Twice stronger than his sire, who sat above
Assessor to the throne of thundering Jove.)
The gods, dismayed at his approach, withdrew,
Nor durst their unaccomplished crime pursue.　　560
That action to his grateful mind recall,
Embrace his knees, and at his footstool fall;
That now, if ever, he will aid our foes;
Let Troy's triumphant troops the camp enclose;
Ours, beaten to the shore, the siege forsake,
And what their king deserves, with him partake;
That the proud tyrant, at his proper cost,°
May learn the value of the man he lost.'

　　To whom the mother-goddess thus replied,
Sighed ere she spoke, and while she spoke she cried,　　570
'Ah wretched me! by fates averse decreed
To bring thee forth with pain, with care to breed!
Did envious heaven not otherwise ordain,
Safe in thy hollow ships thou should'st remain,
Nor ever tempt the fatal field again.
But now thy planet sheds his poisonous rays,
And short and full of sorrow are thy days.
For what remains, to heaven I will ascend,
And at the thunderer's throne thy suit commend.°
Till then, secure in ships, abstain from fight;　　580
Indulge thy grief in tears, and vent thy spite.
For yesterday the court of heaven with Jove
Removed; 'tis dead vacation now above.°
Twelve days the gods their solemn revels keep,
And quaff with blameless Ethiops in the deep.
Returned from thence, to heaven my flight I take,
Knock at the brazen gates, and Providence awake;
Embrace his knees, and suppliant to the sire,
Doubt not I will obtain the grant of thy desire.'

She said, and, parting, left him on the place,   590
Swollen with disdain, resenting his disgrace:
Revengeful thoughts revolving in his mind,
He wept for anger, and for love he pined.
     Meantime, with prosperous gales Ulysses brought
The slave, and ship, with sacrifices fraught,
To Chrysa's port; where, entering with the tide,
He dropped his anchors, and his oars he plied,
Furled every sail, and, drawing down the mast,
His vessel moored, and made with hawsers fast.
Descending on the plain, ashore they bring   600
The hecatomb to please the shooter king.
The dame before an altar's holy fire
Ulysses led, and thus bespoke her sire:°
     'Reverenced be thou, and be thy god adored!
The king of men thy daughter has restored,
And sent by me with presents and with prayer.
He recommends him to thy pious care,
That Phoebus at thy suit his wrath may cease,
And give the penitent offenders peace.'
     He said; and gave her to her father's hands,   610
Who glad received her, free from servile bands.
This done, in order they, with sober grace,
Their gifts around the well-built altar place.
Then washed, and took the cakes, while Chryses stood
With hands upheld, and thus invoked his god.
     'God of the silver bow, whose eyes survey
The sacred Cilla! thou, whose awful sway°
Chrysa the blessed, and Tenedos obey!
Now hear, as thou before my prayer hast heard,
Against the Grecians, and their prince, preferred.   620
Once thou hast honoured, honour once again
Thy priest, nor let his second vows be vain;
But from the afflicted host and humbled prince
Avert thy wrath, and cease thy pestilence!'
Apollo heard, and, conquering his disdain,
Unbent his bow, and Greece respired again.
     Now when the solemn rites of prayer were past,
Their salted cakes on crackling flames they cast;
Then, turning back, the sacrifice they sped,
The fatted oxen slew, and flayed the dead;   630

Chopped off their nervous thighs, and next prepared
To involve the lean in cauls, and mend with lard.
Sweetbreads and collops were with skewers pricked°
About the sides, imbibing what they decked.
The priest with holy hands was seen to tine
The cloven wood, and pour the ruddy wine.
The youth approached the fire, and as it burned,
On five sharp broachers ranked, the roast they turned;°
These morsels stayed their stomachs, then the rest
They cut in legs and fillets for the feast;   640
Which drawn and served, their hunger they appease
With savoury meat, and set their minds at ease.

Now when the rage of eating was repelled,
The boys with generous wine the goblets filled;
The first libations to the gods they pour,
And then with songs indulge the genial hour.
Holy debauch! Till day to night they bring,
With hymns and paeans to the bowyer king.
At sunset to their ship they make return,
And snore secure on decks till rosy morn.   650

The skies with dawning day were purpled o'er;
Awaked, with labouring oars they leave the shore;
The power appeased, with wind sufficed the sail,
The bellying canvas strutted with the gale;
The waves indignant roar with surly pride,
And press against the sides, and, beaten off, divide.
They cut the foamy way, with force impelled
Superior, till the Trojan port they held;°
Then, hauling on the strand, their galley moor,
And pitch their tents along the crooked shore.   660

Meantime the goddess-born in secret pined,
Nor visited the camp, nor in the council joined;
But keeping close, his gnawing heart he fed
With hopes of vengeance on the tyrant's head;
And wished for bloody wars and mortal wounds,
And of the Greeks oppressed in fight to hear the dying sounds.

Now when twelve days complete had run their race,
The gods bethought them of the cares belonging to their place.
Jove at their head ascending from the sea,
A shoal of puny powers attend his way.°   670
Then Thetis, not unmindful of her son,

Emerging from the deep to beg her boon,
Pursued their track, and wakened from his rest,
Before the sovereign stood, a morning guest.
Him in the circle, but apart, she found;
The rest at awful distance stood around.
She bowed, and, ere she durst her suit begin,
One hand embraced his knees, one propped his chin;
Then thus: 'If I, celestial sire, in aught
Have served thy will, or gratified thy thought,   680
One glimpse of glory to my issue give,
Graced for the little time he has to live.
Dishonoured by the king of men he stands;
His rightful prize is ravished from his hands.
But thou, O father, in my son's defence,
Assume thy power, assert thy providence.
Let Troy prevail, till Greece the affront has paid
With doubled honours, and redeemed his aid.'
    She ceased; but the considering god was mute,
Till she, resolved to win, renewed her suit,   690
Nor loosed her hold, but forced him to reply:
'Or grant me my petition, or deny;
Jove cannot fear; then tell me to my face
That I, of all the gods, am least in grace.
This I can bear.' The cloud-compeller mourned,
And, sighing first, this answer he returned.
    'Know'st thou what clamours will disturb my reign,
What my stunned ears from Juno must sustain?
In council she gives licence to her tongue,
Loquacious, brawling, ever in the wrong;   700
And now she will my partial power upbraid,
If, alienate from Greece, I give the Trojans aid.
But thou depart, and shun her jealous sight,
The care be mine to do Pelides right.
Go then, and on the faith of Jove rely,
When, nodding to thy suit, he bows the sky.
This ratifies the irrevocable doom;
The sign ordained, that what I will shall come;
The stamp of heaven, and seal of fate,' he said,
And shook the sacred honours of his head: 710
With terror trembled heaven's subsiding hill,
And from his shaken curls ambrosial dews distil.

The goddess goes exulting from his sight,
And seeks the seas profound, and leaves the realms of light.
  He moves into his hall; the powers resort,
Each from his house, to fill the sovereign's court;
Nor waiting summons, nor expecting stood,
But met with reverence, and received the god.
He mounts the throne; and Juno took her place,
But sullen discontent sat louring on her face.   720
With jealous eyes, at distance she had seen,
Whispering with Jove, the silver-footed queen;°
Then, impotent of tongue, her silence broke,
Thus turbulent, in rattling tone, she spoke.
  'Author of ills, and close contriver Jove,
Which of thy dames, what prostitute of love,
Has held thy ear so long, and begged so hard,
For some old service done, some new reward?
Apart you talked, for that's your special care
The consort never must the council share.   730
One gracious word is for a wife too much;
Such is a marriage vow, and Jove's own faith is such.'
  Then thus the sire of gods, and men below:
'What I have hidden, hope not thou to know.
E'en goddesses are women; and no wife
Has power to regulate her husband's life.
Counsel she may; and I will give thy ear
The knowledge first of what is fit to hear.
What I transact with others, or alone,
Beware to learn, nor press too near the throne.'   740
  To whom the goddess, with the charming eyes:
'What hast thou said, O tyrant of the skies!
When did I search the secrets of thy reign,
Though privileged to know, but privileged in vain?
But well thou dost, to hide from common sight
Thy close intrigues, too bad to bear the light.
Nor doubt I, but the silver-footed dame,
Tripping from sea, on such an errand came,
To grace her issue at the Grecians' cost,
And, for one peevish man, destroy a host'.°   750
  To whom the thunderer made this stern reply:
'My household curse! my lawful plague! the spy
Of Jove's designs! his other squinting eye!

Why this vain prying, and for what avail?
Jove will be master still, and Juno fail.
Should thy suspicious thoughts divine aright,
Thou but becom'st more odious to my sight
For this attempt; uneasy life to me,
Still watched and importuned, but worse for thee.
Curb that impetuous tongue, before too late   760
The gods behold, and tremble at thy fate;
Pitying, but daring not, in thy defence,
To lift a hand against omnipotence.'
    This heard, the imperious queen sat mute with fear,
· Nor further durst incense the gloomy thunderer:
Silence was in the court at this rebuke;
Nor could the gods abashed sustain their sovereign's look.
    The limping smith observed the saddened feast,°
And, hopping here and there, himself a jest,
Put in his word, that neither might offend,   770
To Jove obsequious, yet his mother's friend.
'What end in heaven will be of civil war,
If gods of pleasure will for mortals jar?°
Such discord but disturbs our jovial feast;
One grain of bad embitters all the best.
Mother, though wise yourself, my counsel weigh;
'Tis much unsafe my sire to disobey;
Not only you provoke him to your cost,
But mirth is marred, and the good cheer is lost.
Tempt not his heavy hand, for he has power   780
To throw you headlong from his heavenly tower;
But one submissive word, which you let fall,
Will make him in good humour with us all.'
    He said no more, but crowned a bowl unbid,°
The laughing nectar overlooked the lid;
Then put it to her hand, and thus pursued:
'This cursed quarrel be no more renewed:
Be, as becomes a wife, obedient still;
Though grieved, yet subject to her husband's will.
I would not see you beaten; yet afraid   790
Of Jove's superior force, I dare not aid.
Too well I know him, since that hapless hour
When I, and all the gods, employed our power
To break your bonds; me by the heel he drew,

And o'er heaven's battlements with fury threw.
All day I fell; my flight at morn begun,
And ended not but with the setting sun.
Pitched on my head, at length the Lemnian ground
Received my battered skull, the Sinthians healed my wound.'°
　At Vulcan's homely mirth his mother smiled,　　　　800
And, smiling, took the cup the clown had filled.
The reconciler-bowl went round the board,
Which, emptied, the rude skinker still restored.°
Loud fits of laughter seized the guests, to see
The limping god so deft at his new ministry.°
The feast continued till declining light;
They drank, they laughed, they loved, and then 'twas night.
Nor wanted tuneful harp, nor vocal choir;
The muses sung, Apollo touched the lyre.
Drunken at last, and drowsy, they depart　　　　810
Each to his house, adorned with laboured art
Of the lame architect. The thundering god,
E'en he, withdrew to rest, and had his load;°
His swimming head to needful sleep applied,
And Juno lay unheeded by his side.

The Cock and the Fox

OR THE TALE OF THE NUN'S PRIEST, FROM CHAUCER

There lived, as authors tell, in days of yore,
A widow, somewhat old, and very poor;
Deep in a dell her cottage lonely stood,
Well thatched, and under covert of a wood.
　This dowager, on whom my tale I found,
Since last she laid her husband in the ground,
A simple sober life in patience led,
And had but just enough to buy her bread;
But housewifing the little heaven had lent,
She duly paid a groat for quarter rent;°　　　　10
And pinched her belly, with her daughters two,
To bring the year about with much ado.
　The cattle in her homestead were three sows,°
A ewe called Mally, and three brinded cows.

Her parlour-window stuck with herbs around,
Of savoury smell, and rushes strewed the ground.
A maple dresser in her hall she had,
On which full many a slender meal she made:
For no delicious morsel passed her throat;
According to her cloth she cut her coat.   20
No poignant sauce she knew, no costly treat,
Her hunger gave a relish to her meat.
A sparing diet did her health assure;
Or sick, a pepper posset was her cure.°
Before the day was done, her work she sped,
And never went by candle-light to bed.
With exercise she sweat ill humours out;
Her dancing was not hindered by the gout.
Her poverty was glad, her heart content,
Nor knew she what the spleen or vapours meant.   30
Of wine she never tasted through the year,
But white and black was all her homely cheer;
Brown bread and milk, (but first she skimmed her bowls,)
And rashers of singed bacon on the coals;
On holidays an egg, or two at most;
But her ambition never reached to roast.
    A yard she had, with pales enclosed about,
Some high, some low, and a dry ditch without.
Within this homestead lived, without a peer
For crowing loud, the noble Chanticleer;   40
So hight her cock, whose singing did surpass
The merry notes of organs at the mass.
More certain was the crowing of a cock
To number hours, than is an abbey-clock;
And sooner than the matin-bell was rung,
He clapped his wings upon his roost, and sung:
For when degrees fifteen ascended right,
By sure instinct he knew 'twas one at night.
High was his comb, and coral-red withal,
In dents embattled like a castle wall;   50
His bill was raven-black, and shone like jet;
Blue were his legs, and orient were his feet;
White were his nails, like silver to behold,
His body glittering like the burnished gold.
    This gentle cock, for solace of his life,

Six misses had, beside his lawful wife;
Scandal, that spares no king, though ne'er so good,
Says, they were all of his own flesh and blood;
His sisters, both by sire and mother's side,
And sure their likeness showed them near allied.   60
But make the worst, the monarch did no more,
Than all the Ptolemies had done before:°
When incest is for interest of a nation,
'Tis made no sin by holy dispensation.°
Some lines have been maintained by this alone,
Which by their common ugliness are known.
    But passing this as from our tale apart,
Dame Partlet was the sovereign of his heart:
Ardent in love, outrageous in his play,
He feathered her a hundred times a day;°   70
And she, that was not only passing fair,
But was withal discreet, and debonair,
Resolved the passive doctrine to fulfil,
Though loath, and let him work his wicked will:
At board and bed was affable and kind,
According as their marriage-vow did bind,
And as the church's precept had enjoined.
E'en since she was a se'nnight old, they say,
Was chaste and humble to her dying day,
Nor chick nor hen was known to disobey.   80
    By this her husband's heart she did obtain;
What cannot beauty, joined with virtue, gain?
She was his only joy, and he her pride,
She, when he walked, went pecking by his side;
If, spurning up the ground, he sprung a corn,
The tribute in his bill to her was borne.
But oh! what joy it was to hear him sing
In summer, when the day began to spring,
Stretching his neck, and warbling in his throat,
Solus cum sola, then was all his note.°   90
For in the days of yore, the birds of parts°
Were bred to speak, and sing, and learn the liberal arts.
    It happ'd that perching on the parlour-beam,
Amidst his wives, he had a deadly dream,
Just at the dawn; and sighed, and groaned so fast,
As every breath he drew would be his last.

Dame Partlet, ever nearest to his side,
Heard all his piteous moan, and how he cried
For help from gods and men; and sore aghast
She pecked and pulled, and wakened him at last.   100
　　'Dear heart', said she, 'for love of heaven declare
Your pain, and make me partner of your care.
You groan, sir, ever since the morning light,
As something had disturbed your noble sprite.'
　　'And, madam, well I might,' said Chanticleer,
'Never was shrovetide-cock in such a fear.°
E'en still I run all over in a sweat,
My princely senses not recovered yet.
For such a dream I had of dire portent,
That much I fear my body will be shent: 110
It bodes I shall have wars and woeful strife,
Or in a loathsome dungeon end my life.
Know, dame, I dreamt within my troubled breast,
That in our yard I saw a murderous beast,
That on my body would have made arrest.
With waking eyes I ne'er beheld his fellow;
His colour was betwixt a red and yellow:
Tipped was his tail, and both his pricking ears,
With black, and much unlike his other hairs;
The rest, in shape a beagle's whelp throughout,   120
With broader forehead, and a sharper snout:
Deep in his front were sunk his glowing eyes,
That yet, methinks, I see him with surprise.
Reach out your hand, I drop with clammy sweat,
And lay it to my heart, and feel it beat.'
　　'Now fie for shame!' quoth she: 'by heaven above,
Thou hast for ever lost thy lady's love.
No woman can endure a recreant knight;
He must be bold by day, and free by night:
Our sex desires a husband or a friend,   130
Who can our honour and his own defend;
Wise, hardy, secret, liberal of his purse;
A fool is nauseous, but a coward worse:
No bragging coxcomb, yet no baffled knight,
How dar'st thou talk of love, and dar'st not fight?
How dar'st thou tell thy dame thou art affeared?
Hast thou no manly heart, and hast a beard?

'If aught from fearful dreams may be divined,
They signify a cock of dunghill kind.
All dreams, as in old Galen I have read,°   140
Are from repletion and complexion bred;
From rising fumes of indigested food,
And noxious humours that infect the blood:
And sure, my lord, if I can read aright,
These foolish fancies, you have had tonight,
Are certain symptoms (in the canting style)
Of boiling choler, and abounding bile;
This yellow gall, that in your stomach floats,
Engenders all these visionary thoughts.
When choler overflows, then dreams are bred   150
Of flames, and all the family of red;
Red dragons, and red beasts in sleep we view,
For humours are distinguished by their hue.
From hence we dream of wars and warlike things,
And wasps and hornets with their double wings.

'Choler adust congeals our blood with fear,°
Then black bulls toss us, and black devils tear.
In sanguine airy dreams aloft we bound;
With rheums oppressed, we sink in rivers drowned.

'More I could say, but thus conclude my theme,   160
The dominating humour makes the dream.
Cato was in his time accounted wise,
And he condemns them all for empty lies.
Take my advice, and when we fly to ground,
With laxatives preserve your body sound,
And purge the peccant humours that abound.
I should be loath to lay you on a bier;
And though there lives no 'pothecary near,
I dare for once prescribe for your disease,
And save long bills, and a damned doctor's fees.   170

'Two sovereign herbs, which I by practice know,
And both at hand, (for in our yard they grow,)
On peril of my soul shall rid you wholly
Of yellow choler, and of melancholy:
You must both purge and vomit; but obey,
And for the love of heaven make no delay.
Since hot and dry in your complexion join,
Beware the sun when in a vernal sign;

For when he mounts exalted in the Ram,
If then he finds your body in a flame,   180
Replete with choler, I dare lay a groat,
A tertian ague is at least your lot.°
Perhaps a fever (which the gods forefend)
May bring your youth to some untimely end:
And therefore, sir, as you desire to live,
A day or two before your laxative,
Take just three worms, nor over nor above,
Because the gods unequal numbers love.°
These digestives prepare you for your purge;
Of fumetery, centaury, and spurge,°   190
And of ground-ivy add a leaf, or two,
All which within our yard or garden grow.
Eat these, and be, my lord, of better cheer;
Your father's son was never born to fear.'
   'Madam,' quoth he, 'gramercy for your care,
But Cato, whom you quoted, you may spare.
'Tis true, a wise and worthy man he seems,
And (as you say) gave no belief to dreams;
But other men of more authority,
And, by the immortal powers, as wise as he,   200
Maintain, with sounder sense, that dreams forbode;
For Homer plainly says they come from God.°
Nor Cato said it; but some modern fool°
Imposed in Cato's name on boys at school.
   'Believe me, madam, morning dreams foreshow
The events of things, and future weal or woe:
Some truths are not by reason to be tried,
But we have sure experience for our guide.
An ancient author, equal with the best,°
Relates this tale of dreams among the rest: 210
   'Two friends or brothers, with devout intent,
On some far pilgrimage together went.
It happened so, that when the sun was down,
They just arrived by twilight at a town;
That day had been the baiting of a bull,
'Twas at a feast, and every inn so full,
That no void room in chamber, or on ground,
And but one sorry bed was to be found;

And that so little it would hold but one,
Though till this hour they never lay alone.   220
   'So were they forced to part; one stayed behind,
His fellow sought what lodging he could find:
At last he found a stall where oxen stood,
And that he rather chose than lie abroad.
'Twas in a further yard without a door;
But, for his ease, well littered was the floor.
   'His fellow, who the narrow bed had kept,
Was weary, and without a rocker slept:
Supine he snored; but in dead of night,
He dreamt his friend appeared before his sight,   230
Who, with a ghastly look and doleful cry,
Said, "Help me, brother, or this night I die:
Arise, and help, before all help be vain,
Or in an ox's stall I shall be slain."
   'Roused from his rest, he wakened in a start,
Shivering with horror, and with aching heart;
At length to cure himself by reason tries;
'Twas but a dream, and what are dreams but lies?
So thinking changed his side, and closed his eyes.
His dream returns: his friend appears again: 240
"The murderers come, now help, or I am slain:"
'Twas but a vision still, and visions are but vain.
   'He dreamt the third; but now his friend appeared
Pale, naked, pierced with wounds, with blood besmeared:
"Thrice warned, awake," said he; "relief is late,
The deed is done; but thou revenge my fate:
Tardy of aid, unseal thy heavy eyes,
Awake, and with the dawning day arise:
Take to the western gate thy ready way,
For by that passage they my corpse convey: 250
My corpse is in a tumbrel laid, among
The filth and ordure, and enclosed with dung.
That cart arrest, and raise a common cry;
For sacred hunger of my gold I die:"°
Then showed his grisly wounds; and last he drew
A piteous sigh, and took a long adieu.
   'The frightened friend arose by break of day,
And found the stall where late his fellow lay.

Then of his impious host inquiring more,
Was answered that his guest was gone before: 260
"Muttering he went," said he, "by morning light,
And much complained of his ill rest by night."
This raised suspicion in the pilgrim's mind;
Because all hosts are of an evil kind,
And oft to share the spoil with robbers joined.
    'His dream confirmed his thought; with troubled look
Straight to the western gate his way he took;
There, as his dream foretold, a cart he found,
That carried compost forth to dung the ground.
This when the pilgrim saw, he stretched his throat,   270
And cried out "murder" with a yelling note.
"My murdered fellow in this cart lies dead;
Vengeance and justice on the villain's head!
You, magistrates, who sacred laws dispense,
On you I call to punish this offence."
    'The word thus given, within a little space,
The mob came roaring out, and thronged the place.
All in a trice they cast the cart to ground,
And in the dung the murdered body found;
Though breathless, warm, and reeking from the wound.   280
Good heaven, whose darling attribute we find
Is boundless grace, and mercy to mankind,
Abhors the cruel; and the deeds of night°
By wondrous ways reveals in open light:
Murder may pass unpunished for a time,
But tardy justice will o'ertake the crime.
And oft a speedier pain the guilty feels,
The hue and cry of heaven pursues him at the heels,
Fresh from the fact, as in the present case:
The criminals are seized upon the place;   290
Carter and host confronted face to face.
Stiff in denial, as the law appoints,
On engines they distend their tortured joints;
So was confession forced, the offence was known,
And public justice on the offenders done.
    'Here may you see that visions are to dread;
And in the page that follows this, I read
Of two young merchants, whom the hope of gain
Induced in partnership to cross the main;

Waiting till willing winds their sails supplied,   300
Within a trading-town they long abide,
Full fairly situate on a haven's side.
   'One evening it befell, that, looking out,
The wind they long had wished was come about;
Well pleased they went to rest; and if the gale
Till morn continued, both resolved to sail.
But as together in a bed they lay,
The younger had a dream at break of day.
A man, he thought, stood frowning at his side,
Who warned him for his safety to provide,   310
Not put to sea, but safe on shore abide.
"I come, thy genius, to command thy stay;
Trust not the winds, for fatal is the day,
And death unhoped attends the watery way."
   'The vision said, and vanished from his sight.
The dreamer wakened in a mortal fright;
Then pulled his drowsy neighbour, and declared,
What in his slumber he had seen and heard.
His friend smiled scornful, and with proud contempt
Rejects as idle what his fellow dreamt.   320
"Stay, who will stay; for me no fears restrain,
Who follow Mercury, the god of gain;
Let each man do as to his fancy seems,
I wait not, I, till you have better dreams.
Dreams are but interludes, which fancy makes;
When monarch reason sleeps, this mimic wakes;
Compounds a medley of disjointed things,
A mob of cobblers, and a court of kings:
Light fumes are merry, grosser fumes are sad;
Both are the reasonable soul run mad;   330
And many monstrous forms in sleep we see,
That neither were, nor are, nor e'er can be.
Sometimes, forgotten things long cast behind
Rush forward in the brain, and come to mind.
The nurse's legends are for truths received,
And the man dreams but what the boy believed.
   '"Sometimes we but rehearse a former play,
The night restores our actions done by day,
As hounds in sleep will open for their prey.
In short, the farce of dreams is of a piece,   340

Chimeras all; and more absurd, or less.
You, who believe in tales, abide alone;
Whate'er I get this voyage is my own."
    'Thus while he spoke, he heard the shouting crew
That called aboard, and took his last adieu.
The vessel went before a merry gale,
And for quick passage put on every sail;
But when least feared, and e'en in open day,
The mischief overtook her in the way:
Whether she sprung a leak, I cannot find,   350
Or whether she was overset with wind,
Or that some rock below her bottom rent,
But down at once with all the crew she went.
Her fellow-ships from far her loss descried;
But only she was sunk, and all were safe beside.
    'By this example you are taught again,
That dreams and visions are not always vain;
But if, dear Partlet, you are yet in doubt,
Another tale shall make the former out.
    'Kenelm, the son of Kenulph, Mercia's king,   360
Whose holy life the legends loudly sing,
Warned in a dream, his murder did foretell,
From point to point as after it befell:
All circumstances to his nurse he told,
(A wonder from a child of seven years old;)
The dream with horror heard, the good old wife
From treason counselled him to guard his life;
But close to keep the secret in his mind,
For a boy's vision small belief would find.
The pious child, by promise bound, obeyed,   370
Nor was the fatal murder long delayed;
By Quenda slain, he fell before his time,
Made a young martyr by his sister's crime.
The tale is told by venerable Bede,°
Which, at your better leisure, you may read.
    'Macrobius too relates the vision sent°
To the great Scipio, with the famed event;
Objections makes, but after makes replies,
And adds, that dreams are often prophecies.
    'Of Daniel you may read in holy writ,°   380
Who, when the king his vision did forget,

Could word for word the wondrous dream repeat.
Nor less of patriarch Joseph understand,°
Who by a dream enslaved the Egyptian land,
The years of plenty and of dearth foretold,
When, for their bread, their liberty they sold.
Nor must the exalted butler be forgot,°
Nor he whose dream presaged his hanging lot.°

   'And did not Croesus the same death foresee,°
Raised in his vision on a lofty tree?   390
The wife of Hector, in his utmost pride,°
Dreamt of his death the night before he died:
Well was he warned from battle to refrain,
But men to death decreed are warned in vain;
He dared the dream, and by his fatal foe was slain.

   'Much more I know, which I forbear to speak,
For see the ruddy day begins to break:
Let this suffice, that plainly I foresee
My dream was bad, and bodes adversity;
But neither pills nor laxatives I like,   400
They only serve to make a well man sick;
Of these his gain the sharp physician makes,
And often gives a purge, but seldom takes;
They not correct, but poison all the blood,
And ne'er did any but the doctors good.
Their tribe, trade, trinkets, I defy them all,
With every work of 'Pothecaries' Hall.

   'These melancholy matters I forbear;
But let me tell thee, Partlet mine, and swear,
That when I view the beauties of thy face,   410
I fear not death, nor dangers, nor disgrace;
So may my soul have bliss, as when I spy
The scarlet red about thy partridge eye,
While thou art constant to thy own true knight,
While thou art mine, and I am thy delight,
All sorrows at thy presence take their flight.
For true it is, as in principio,°
Mulier est hominis confusio.°
Madam, the meaning of this Latin is,
That woman is to man his sovereign bliss.   420
For when by night I feel your tender side,
Though for the narrow perch I cannot ride,

Yet I have such a solace in my mind,
That all my boding cares are cast behind,
And e'en already I forget my dream.'
He said, and downward flew from off the beam,
For daylight now began apace to spring,
The thrush to whistle, and the lark to sing.
Then crowing, clapped his wings, the appointed call,
To chuck his wives together in the hall.   430
    By this the widow had unbarred the door,
And Chanticleer went strutting out before,
With royal courage, and with heart so light,
As showed he scorned the visions of the night.
Now roaming in the yard, he spurned the ground,
And gave to Partlet the first grain he found.
Then often feathered her with wanton play,
And trod her twenty times ere prime of day;°
And took by turns and gave so much delight,
Her sisters pined with envy at the sight.   440
    He chucked again, when other corns he found,
And scarcely deigned to set a foot to ground;
But swaggered like a lord about his hall,
And his seven wives came running at his call.
    'Twas now the month in which the world began,
(If March beheld the first created man;)
And since the vernal equinox, the sun
In Aries twelve degrees, or more, had run;
When casting up his eyes against the light,
Both month, and day, and hour, he measured right,   450
And told more truly than the Ephemeris;°
For art may err, but nature cannot miss.
    Thus numbering times and seasons in his breast,
His second crowing the third hour confessed.
Then turning, said to Partlet, 'See, my dear,
How lavish nature has adorned the year;
How the pale primrose and blue violet spring,
And birds essay their throats disused to sing:
All these are ours; and I with pleasure see,
Man strutting on two legs, and aping me;°   460
An unfledged creature, of a lumpish frame,
Endued with fewer particles of flame:
Our dame sits cowering o'er a kitchen fire,

I draw fresh air, and nature's works admire;
And e'en this day in more delight abound,
Than, since I was an egg, I ever found.'
    The time shall come, when Chanticleer shall wish
His words unsaid, and hate his boasted bliss;
The crested bird shall by experience know,
Jove made not him his masterpiece below,   470
And learn the latter end of joy is woe.
The vessel of his bliss to dregs is run,
And heaven will have him taste his other tun.
    Ye wise! draw near and hearken to my tale,
Which proves that oft the proud by flattery fall;
The legend is as true I undertake
As Tristram is, and Launcelot of the lake;
Which all our ladies in such reverence hold,
As if in Book of Martyrs it were told.°
    A fox, full-fraught with seeming sanctity,   480
That feared an oath, but, like the devil, would lie;
Who looked like Lent, and had the holy leer,
And durst not sin before he said his prayer;
This pious cheat, that never sucked the blood,
Nor chewed the flesh of lambs, but when he could,
Had passed three summers in the neighbouring wood;
And musing long, whom next to circumvent,
On Chanticleer his wicked fancy bent;
And in his high imagination cast,
By stratagem to gratify his taste.   490
    The plot contrived, before the break of day,
Saint Reynard through the hedge had made his way;
The pale was next, but proudly, with a bound,
He leapt the fence of the forbidden ground;
Yet fearing to be seen, within a bed
Of coleworts he concealed his wily head;
There skulked till afternoon, and watched his time,
(As murderers use,) to perpetrate his crime.
    O hypocrite, ingenious to destroy!
O traitor, worse than Sinon was to Troy!°   500
O vile subverter of the Gallic reign,°
More false than Gano was to Charlemagne!°
O Chanticleer, in an unhappy hour
Didst thou forsake the safety of thy bower;

Better for thee thou hadst believed thy dream,
And not that day descended from the beam!
   But here the doctors eagerly dispute;
Some hold predestination absolute;
Some clerks maintain, that heaven at first foresees,
And in the virtue of foresight decrees.   510
If this be so, then prescience binds the will,
And mortals are not free to good or ill;
For what he first foresaw, he must ordain,
Or its eternal prescience may be vain;
As bad for us as prescience had not been;
For first, or last, he's author of the sin.
And who says that, let the blaspheming man
Say worse e'en of the devil, if he can.
For how can that eternal power be just
To punish man, who sins because he must?   520
Or, how can he reward a virtuous deed,
Which is not done by us, but first decreed?
   I cannot bolt this matter to the bran,°
As Bradwardine and holy Austin can:°
If prescience can determine actions so,
That we must do, because he did foreknow,
Or that foreknowing, yet our choice is free.
Not forced to sin by strict necessity;
This strict necessity they simple call,
Another sort there is conditional.   530
The first so binds the will, that things foreknown°
By spontaneity, not choice, are done.
Thus galley-slaves tug willing at their oar,
Consent to work, in prospect of the shore;
But would not work at all, if not constrained before.
That other does not liberty constrain,
But man may either act, or may refrain.
Heaven made us agents free to good or ill,
And forced it not, though he foresaw the will.
Freedom was first bestowed on human race,   540
And prescience only held the second place.
   If he could make such agents wholly free,
I not dispute; the point's too high for me:
For heaven's unfathomed power what man can sound,
Or put to his omnipotence a bound?

He made us to his image, all agree;
That image is the soul, and that must be,
Or not the maker's image, or be free.
But whether it were better man had been
By nature bound to good, not free to sin,   550
I waive, for fear of splitting on a rock;
The tale I tell is only of a cock;
Who had not run the hazard of his life,
Had he believed his dream, and not his wife:
For women, with a mischief to their kind,
Pervert, with bad advice, our better mind.
A woman's counsel brought us first to woe,
And made her man his paradise forego,
Where at heart's ease he lived; and might have been
As free from sorrow as he was from sin.   560
For what the devil had their sex to do,
That, born to folly, they presumed to know,
And could not see the serpent in the grass?
But I myself presume, and let it pass.

   Silence in times of suffering is the best,
'Tis dangerous to disturb a hornet's nest.
In other authors you may find enough,
But all they say of dames is idle stuff.
Legends of lying wits together bound,
The wife of Bath would throw them to the ground: 570
These are the words of Chanticleer, not mine,
I honour dames, and think their sex divine.

   Now to continue what my tale begun.
Lay madam Partlet basking in the sun,
Breast-high in sand; her sisters, in a row,
Enjoyed the beams above, the warmth below.
The cock, that of his flesh was ever free,
Sung merrier than the mermaid in the sea;
And so befell, that as he cast his eye,
Among the coleworts, on a butterfly,   580
He saw false Reynard where he lay full low;
I need not swear he had no list to crow;
But cried, *Cock, cock*, and gave a sudden start,
As sore dismayed and frighted at his heart.
For birds and beasts, informed by nature, know
Kinds opposite to theirs, and fly their foe.

So Chanticleer, who never saw a fox,
Yet shunned him, as a sailor shuns the rocks.
    But the false loon, who could not work his will
By open force, employed his flattering skill: 590
'I hope, my lord,' said he, 'I not offend;
Are you afraid of me, that am your friend?
I were a beast indeed to do you wrong,
I, who have loved and honoured you so long:
Stay, gentle sir, nor take a false alarm,
For, on my soul, I never meant you harm!
I come no spy, nor as a traitor press,
To learn the secrets of your soft recess:
Far be from Reynard so profane a thought,
But by the sweetness of your voice was brought: 600
For, as I bid my beads, by chance I heard
The song as of an angel in the yard;
A song that would have charmed the infernal gods,
And banished horror from the dark abodes:
Had Orpheus sung it in the nether sphere,
So much the hymn had pleased the tyrant's ear,
The wife had been detained, to keep the husband there.
    'My lord, your sire familiarly I knew,
A peer deserving such a son as you:
He, with your lady-mother, (whom heaven rest!)   610
Has often graced my house, and been my guest:
To view his living features does me good,
For I am your poor neighbour in the wood;
And in my cottage should be proud to see
The worthy heir of my friend's family.
    'But since I speak of singing, let me say,
As with an upright heart I safely may,
That, save yourself, there breathes not on the ground
One like your father for a silver sound.
So sweetly would he wake the winter-day,   620
That matrons to the church mistook their way,
And thought they heard the merry organ play.
And he to raise his voice with artful care,
(What will not beaux attempt to please the fair?)
On tiptoe stood to sing with greater strength,
And stretched his comely neck at all the length:
And while he pained his voice to pierce the skies,

As saints in raptures use, would shut his eyes,
That the sound striving through the narrow throat,
His winking might avail to mend the note.   630
By this, in song, he never had his peer,
From sweet Cecilia down to Chanticleer;
Not Maro's muse, who sung the mighty man,°
Nor Pindar's heavenly lyre, nor Horace when a swan.°
Your ancestors proceed from race divine:
From Brennus and Belinus is your line;°
Who gave to sovereign Rome such loud alarms,
That e'en the priests were not excused from arms.
    'Besides, a famous monk of modern times
Has left of cocks recorded in his rhymes,   640
That of a parish priest the son and heir,
(When sons of priests were from the proverb clear,)
Affronted once a cock of noble kind,
And either lamed his legs, or struck him blind;
For which the clerk his father was disgraced,
And in his benefice another placed.
Now sing, my lord, if not for love of me,
Yet for the sake of sweet saint charity;°
Make hills and dales, and earth and heaven, rejoice,
And emulate your father's angel-voice.'   650
    The cock was pleased to hear him speak so fair,
And proud beside, as solar people are;°
Nor could the treason from the truth descry,
So was he ravished with this flattery:
So much the more, as from a little elf,
He had a high opinion of himself;
Though sickly, slender, and not large of limb,
Concluding all the world was made for him.
    Ye princes, raised by poets to the gods,
And Alexandered up in lying odes,   660
Believe not every flattering knave's report,
There's many a Reynard lurking in the court;
And he shall be received with more regard,
And listened to, than modest truth is heard.
    This Chanticleer, of whom the story sings,
Stood high upon his toes, and clapped his wings;
Then stretched his neck, and winked with both his eyes,
Ambitious, as he sought the Olympic prize.

But while he pained himself to raise his note,
False Reynard rushed, and caught him by the throat.   670
Then on his back he laid the precious load,
And sought his wonted shelter of the wood;
Swiftly he made his way, the mischief done,
Of all unheeded, and pursued by none.

    Alas! what stay is there in human state,
Or who can shun inevitable fate?
The doom was written, the decree was passed,
Ere the foundations of the world were cast!
In Aries though the sun exalted stood,
His patron-planet to procure his good;   680
Yet Saturn was his mortal foe, and he,
In Libra raised, opposed the same degree:
The rays both good and bad, of equal power,
Each thwarting other, made a mingled hour.

    On Friday-morn he dreamt his direful dream,
Cross to the worthy native, in his scheme.
Ah, blissful Venus! goddess of delight!
How couldst thou suffer thy devoted knight,
On thy own day, to fall by foe oppressed,
The wight of all the world who served thee best?   690
Who, true to love, was all for recreation,
And minded not the work of propagation?
Gaufride, who couldst so well in rhyme complain°
The death of Richard with an arrow slain,
Why had not I thy muse, or thou my heart,
To sing this heavy dirge with equal art!
That I like thee on Friday might complain;
For on that day was Coeur de Lion slain.

    Not louder cries, when Ilium was in flames,
Were sent to heaven by woeful Trojan dames,   700
When Pyrrhus tossed on high his burnished blade,
And offered Priam to his father's shade,
Than for the cock the widowed poultry made.
Fair Partlet first, when he was borne from sight,
With sovereign shrieks bewailed her captive knight;
Far louder than the Carthaginian wife,
When Asdrubal her husband lost his life,°
When she beheld the smouldering flames ascend,
And all the Punic glories at an end:

Willing into the fires she plunged her head,   710
With greater ease than others seek their bed.
Not more aghast the matrons of renown,
When tyrant Nero burned the imperial town,
Shrieked for the downfall in a doleful cry,
For which their guiltless lords were doomed to die.
　Now to my story I return again:
The trembling widow, and her daughters twain,
This woeful cackling cry with horror heard,
Of those distracted damsels in the yard;
And starting up, beheld the heavy sight,   720
How Reynard to the forest took his flight,
And cross his back, as in triumphant scorn,
The hope and pillar of the house was borne.
　'The fox, the wicked fox,' was all the cry;
Out from his house ran every neighbour nigh:
The vicar first, and after him the crew,
With forks and staves the felon to pursue.
Ran Coll our dog, and Talbot with the band,°
And Malkin, with her distaff in her hand:
Ran cow and calf, and family of hogs,   730
In panic horror of pursuing dogs;
With many a deadly grunt and doleful squeak,
Poor swine, as if their pretty hearts would break.
The shouts of men, the women in dismay,
With shrieks augment the terror of the day,
The ducks, that heard the proclamation cried,
And feared a persecution might betide,
Full twenty miles from town their voyage take,
Obscure in rushes of the liquid lake.
The geese fly o'er the barn; the bees, in arms,   740
Drive headlong from their waxen cells in swarms.
Jack Straw at London-stone, with all his rout,°
Struck not the city with so loud a shout;
Not when with English hate they did pursue
A Frenchman, or an unbelieving Jew;
Not when the welkin rung with *One and all*,
And echoes bounded back from Fox's hall;
Earth seemed to sink beneath, and heaven above to fall.
With might and main they chased the murderous fox,
With brazen trumpets, and inflated box,   750

To kindle Mars with military sounds,
Nor wanted horns to inspire sagacious hounds.
　　But see how fortune can confound the wise,
And when they least expect it, turn the dice.
The captive-cock, who scarce could draw his breath,
And lay within the very jaws of death;
Yet in this agony his fancy wrought,
And fear supplied him with this happy thought:
'Yours is the prize, victorious prince,' said he,
'The vicar my defeat, and all the village see.   760
Enjoy your friendly fortune while you may,
And bid the churls that envy you the prey
Call back their mongrel curs, and cease their cry:
See fools, the shelter of the wood is nigh,
And Chanticleer in your despite shall die;
He shall be plucked and eaten to the bone.'
　　''Tis well advised, in faith it shall be done;'
This Reynard said: but as the word he spoke,
The prisoner with a spring from prison broke;
Then stretched his feathered fans with all his might,   770
And to the neighbouring maple winged his flight.
　　Whom, when the traitor safe on tree beheld,
He cursed the gods, with shame and sorrow filled:
Shame for his folly; sorrow out of time,
For plotting an unprofitable crime:
Yet, mastering both, the artificer of lies,
Renews the assault, and his last battery tries.
　　'Though I,' said he, 'did ne'er in thought offend,
How justly may my lord suspect his friend?
The appearance is against me, I confess,   780
Who seemingly have put you in distress.
You, if your goodness does not plead my cause,
May think I broke all hospitable laws,
To bear you from your palace-yard by might,
And put your noble person in a fright.
This, since you take it ill, I must repent,
Though heaven can witness, with no bad intent
I practised it, to make you taste your cheer
With double pleasure, first prepared by fear.
So loyal subjects often seize their prince,   790
Forced (for his good) to seeming violence,

Yet mean his sacred person not the least offence.
Descend; so help me, Jove, as you shall find,
That Reynard comes of no dissembling kind.'
    'Nay,' quoth the cock; 'but I beshrew us both,
If I believe a saint upon his oath:
An honest man may take a knave's advice,
But idiots only will be cozened twice:
Once warned is well bewared; no flattering lies
Shall soothe me more to sing with winking eyes,   800
And open mouth, for fear of catching flies.
Who blindfold walks upon a river's brim,
When he should see, has he deserved to swim?
    'Better, Sir Cock, let all contention cease;
Come down,' said Reynard, 'let us treat of peace.'
    'A peace with all my soul,' said Chanticleer;°
'But, with your favour, I will treat it here:
And lest the truce with treason should be mixed,
'Tis my concern to have the tree betwixt.'

THE MORAL

In this plain fable you the effect may see   810
Of negligence, and fond credulity:
And learn besides of flatterers to beware,
Then most pernicious when they speak too fair.
The cock and fox, the fool and knave imply;
The truth is moral, though the tale a lie.
Who spoke in parables, I dare not say;
But sure he knew it was a pleasing way,
Sound sense, by plain example, to convey.
And in a heathen author we may find,
That pleasure with instruction should be joined;   820
So take the corn, and leave the chaff behind.

Theodore and Honoria

From Boccaccio

Of all the cities in Romanian lands,°
The chief and most renowned Ravenna stands;
Adorned in ancient times with arms and arts,
And rich inhabitants with generous hearts.

But Theodore the brave, above the rest
With gifts of fortune and of nature blessed,
The foremost place for wealth and honour held,
And all in feats of chivalry excelled.

    This noble youth to madness loved a dame
Of high degree, Honoria was her name.   10
Fair as the fairest but of haughty mind,
And fiercer than became so soft a kind;
Proud of her birth (for equal she had none);
The rest she scorned, but hated him alone.
His gifts, his constant courtship, nothing gained;
For she, the more he loved, the more disdained.
He lived with all the pomp he could devise,
At tilts and tournaments obtained the prize,°
But found no favour in his lady's eyes.
Relentless as a rock, the lofty maid   20
Turned all to poison that he did or said.
Nor prayers nor tears nor offered vows could move;
The work went backward; and the more he strove
To advance his suit, the farther from her love.

    Wearied at length, and wanting remedy,
He doubted oft and oft resolved to die.
But pride stood ready to prevent the blow,
For who would die to gratify a foe?
His generous mind disdained so mean a fate;
That passed, his next endeavour was to hate,   30
But vainer that relief than all the rest,
The less he hoped with more desire possessed;
Love stood the siege and would not yield his breast.

    Change was the next, but change deceived his care,
He sought a fairer but found none so fair.
He would have worn her out by slow degrees,
As men by fasting starve the untamed disease;
But present love required a present ease.
Looking he feeds alone his famished eyes,
Feeds lingering death, but looking not he dies.   40
Yet still he chose the longest way to fate,
Wasting at once his life, and his estate.
His friends beheld and pitied him in vain,
For what advice can ease a lover's pain!
Absence, the best expedient they could find,

Might save the fortune if not cure the mind:
This means they long proposed but little gained,
Yet after much pursuit at length obtained.
Hard, you may think it was, to give consent,
But struggling with his own desires he went: 50
With large expense, and with a pompous train
Provided, as to visit France or Spain,
Or for some distant voyage o'er the main.
But love had clipped his wings and cut him short,
Confined within the purlieus of his court.
Three miles he went, nor further could retreat;
His travels ended at his country seat.
To Chassis' pleasing plains he took his way,
There pitched his tents and there resolved to stay.

    The spring was in the prime; the neighbouring grove   60
Supplied with birds, the choristers of love;
Music unbought, that ministered delight
To morning-walks, and lulled his cares by night.
There he discharged his friends, but not the expense
Of frequent treats and proud magnificence.
He lived as kings retire, though more at large°
From public business, yet with equal charge;
With house and heart still open to receive;
As well content as love would give him leave.
He would have lived more free, but many a guest,   70
Who could forsake the friend, pursued the feast.

    It happed one morning, as his fancy led,
Before his usual hour he left his bed
To walk within a lonely lawn, that stood°
On every side surrounded by the wood.
Alone he walked to please his pensive mind,
And sought the deepest solitude to find.
'Twas in a grove of spreading pines he strayed;
The winds within the quivering branches played,
And dancing trees a mournful music made.   80
The place itself was suiting to his care,
Uncouth and savage as the cruel fair.
He wandered on unknowing where he went,
Lost in the wood and all on love intent.
The day already half his race had run,
And summoned him to due repast at noon,

But love could feel no hunger but his own.
While listening to the murmuring leaves he stood,
More than a mile immersed within the wood,
At once the wind was laid; the whispering sound   90
Was dumb; a rising earthquake rocked the ground;
With deeper brown the grove was overspread;°
A sudden horror seized his giddy head,
And his ears tinkled, and his colour fled.
Nature was in alarm; some danger nigh
Seemed threatened, though unseen to mortal eye.
Unused to fear, he summoned all his soul
And stood collected in himself and whole;
Not long: for soon a whirlwind rose around,
And from afar he heard a screaming sound   100
As of a dame distressed, who cried for aid
And filled with loud laments the secret shade.
A thicket close beside the grove there stood
With briars and brambles choked and dwarfish wood;
From thence the noise; which now approaching near
With more distinguished notes invades his ear.
He raised his head and saw a beauteous maid,
With hair dishevelled, issuing through the shade;
Stripped of her clothes, and e'en those parts revealed
Which modest nature keeps from sight concealed.   110
Her face, her hands, her naked limbs were torn
With passing through the brakes and prickly thorn;
Two mastiffs, gaunt and grim, her flight pursued
And oft their fastened fangs in blood imbrued;
Oft they came up and pinched her tender side,
'Mercy, O mercy, heaven,' she ran and cried;
When heaven was named they loosed their hold again,
Then sprung she forth, they followed her amain.
Not far behind, a knight of swarthy face,
High on a coal-black steed pursued the chase;   120
With flashing flames his ardent eyes were filled,
And in his hands a naked sword he held.
He cheered the dogs to follow her who fled,
And vowed revenge on her devoted head.
      As Theodore was born of noble kind,
The brutal action roused his manly mind.
Moved with unworthy usage of the maid,

He, though unarmed, resolved to give her aid.
A sapling pine he wrenched from out the ground,
The readiest weapon that his fury found.   130
Thus furnished for offence, he crossed the way
Betwixt the graceless villain and his prey.
The knight came thundering on, but from afar
Thus in imperious tone forbad the war:
'Cease, Theodore, to proffer vain relief,
Nor stop the vengeance of so just a grief;
But give me leave to seize my destined prey,
And let eternal justice take the way.
I but revenge my fate; disdained, betrayed,
And suffering death for this ungrateful maid.'   140
He said, at once dismounting from the steed;
For now the hell-hounds with superior speed
Had reached the dame, and fastening on her side,
The ground with issuing streams of purple dyed.
Stood Theodore surprised in deadly fright,
With chattering teeth and bristling hair upright;
Yet armed with inborn worth, 'Whate'er', said he,
'Thou art, who knowst me better than I thee;
Or prove thy rightful cause, or be defied.'
The spectre, fiercely staring, thus replied: 150
'Know, Theodore, thy ancestry I claim,
And Guido Cavalcanti was my name.
One common sire our fathers did beget,
My name and story some remember yet.
Thee, then a boy, within my arms I laid,
When for my sins I loved this haughty maid;
Not less adored in life, nor served by me,
Than proud Honoria now is loved by thee.
What did I not her stubborn heart to gain?
But all my vows were answered with disdain;   160
She scorned my sorrows and despised my pain.
Long time I dragged my days in fruitless care,
Then loathing life, and plunged in deep despair,
To finish my unhappy life, I fell
On this sharp sword, and now am damned in hell.
Short was her joy; for soon the insulting maid
By heaven's decree in the cold grave was laid,
And as in unrepenting sin she died,

Doomed to the same bad place is punished for her pride;
Because she deemed I well deserved to die,   170
And made a merit of her cruelty.
There, then, we met; both tried and both were cast,
And this irrevocable sentence passed;
That she whom I so long pursued in vain,
Should suffer from my hands a lingering pain.
Renewed to life that she might daily die,
I daily doomed to follow, she to fly;
No more a lover but a mortal foe,
I seek her life (for love is none below).
As often as my dogs with better speed   180
Arrest her flight, is she to death decreed;
Then with this fatal sword on which I died,
I pierce her opened back or tender side,
And tear that hardened heart from out her breast,
Which, with her entrails, makes my hungry hounds a feast.
Nor lies she long but, as her fates ordain,
Springs up to life and, fresh to second pain,
Is saved today, tomorrow to be slain.'
This, versed in death, the infernal knight relates,°
And then for proof fulfilled their common fates;   190
Her heart and bowels through her back he drew,
And fed the hounds that helped him to pursue.
Stern looked the fiend, as frustrate of his will,
Not half sufficed and greedy yet to kill.
And now the soul expiring through the wound,
Had left the body breathless on the ground,
When thus the grisly spectre spoke again:
'Behold the fruit of ill-rewarded pain.
As many months as I sustained her hate,
So many years is she condemned by fate   200
To daily death; and every several place
Conscious of her disdain and my disgrace
Must witness her just punishment; and be
A scene of triumph and revenge to me.
As in this grove I took my last farewell,
As on this very spot of earth I fell,
As Friday saw me die, so she my prey
Becomes e'en here on this revolving day.'
Thus while he spoke, the virgin from the ground

Upstarted fresh, already closed the wound,   210
And unconcerned for all she felt before
Precipitates her flight along the shore.
The hell-hounds, as ungorged with flesh and blood,
Pursue their prey and seek their wonted food;
The fiend remounts his courser, mends his pace,
And all the vision vanished from the place.
   Long stood the noble youth oppressed with awe,
And stupid at the wonderous things he saw
Surpassing common faith, transgressing nature's law.
He would have been asleep and wished to wake,°   220
But dreams, he knew, no long impression make,
Though strong at first. If vision, to what end,
But such as must his future state portend?
His love the damsel, and himself the fiend.
But yet reflecting that it could not be
From heaven, which cannot impious acts decree,
Resolved within himself to shun the snare
Which hell for his destruction did prepare;
And as his better genius should direct
From an ill cause to draw a good effect.   230
Inspired from heaven he homeward took his way,
Nor palled his new design with long delay;
But of his train a trusty servant sent
To call his friends together at his tent.
They came, and usual salutations paid,
With words premeditated thus he said:
'What you have often counselled, to remove
My vain pursuit of unregarded love;
By thrift my sinking fortune to repair,
Though late, yet is at last become my care.   240
My heart shall be my own; my vast expense
Reduced to bounds by timely providence.
This only I require; invite for me
Honoria, with her father's family,
Her friends and mine; the cause I shall display,
On Friday next, for that's the appointed day.'
   Well pleased were all his friends, the task was light;
The father, mother, daughter they invite;
Hardly the dame was drawn to this repast,
But yet resolved, because it was the last.   250

The day was come; the guests invited came,
And, with the rest, the inexorable dame.
A feast prepared with riotous expense,
Much cost, more care, and most magnificence.
The place ordained was in that haunted grove
Where the revenging ghost pursued his love.
The tables in a proud pavilion spread,
With flowers below, and tissue overhead.
The rest in rank; Honoria chief in place,
Was artfully contrived to set her face   260
To front the thicket and behold the chase.
The feast was served; the time so well forecast
That just when the dessert and fruits were placed
The fiend's alarm began; the hollow sound
Sung in the leaves, the forest shook around;
Air blackened; rolled the thunder; groaned the ground.
Nor long before the loud laments arise
Of one distressed, and mastiffs' mingled cries:
And first the dame came rushing through the wood,
And next the famished hounds that sought their food   270
And gripped her flanks and oft essayed their jaws in blood.
Last came the felon on the sable steed,
Armed with his naked sword, and urged his dogs to speed.
She ran and cried; her flight directly bent
(A guest unbidden) to the fatal tent,
The scene of death and place ordained for punishment.
Loud was the noise, aghast was every guest,
The women shrieked, the men forsook the feast;
The hounds at nearer distance hoarsely bayed;
The hunter close pursued the visionary maid;   280
She rent the heaven with loud laments, imploring aid.
The gallants to protect the ladies right,
Their falchions brandished at the grisly sprite;
High on his stirrups, he provoked the fight.
Then on the crowd he cast a furious look,
And withered all their strength before he strook.
'Back on your lives; let be', said he, 'my prey,
And let my vengeance take the destined way.
Vain are your arms, and vainer your defence,
Against the eternal doom of providence.   290
Mine is the ungrateful maid by heaven designed:

Mercy she would not give, nor mercy shall she find.'
At this the former tale again he told
With thundering tone, and dreadful to behold.
Sunk were their hearts with horror of the crime,
Nor needed to be warned a second time
But bore each other back; some knew the face,
And all had heard the much lamented case
Of him who fell for love, and this the fatal place.

  And now the infernal minister advanced,   300
Seized the due victim, and with fury lanced
Her back, and piercing through her inmost heart
Drew backward, as before, the offending part.
The reeking entrails next he tore away,
And to his meagre mastiffs made a prey.
The pale assistants on each other stared°
With gaping mouths for issuing words prepared;
The still-born sounds upon the palate hung,
And died imperfect on the faltering tongue.
The fright was general; but the female band   310
(A helpless train) in more confusion stand;
With horror shuddering, on a heap they run,
Sick at the sight of hateful justice done;
For conscience rung the alarm, and made the case their own.
So spread upon a lake with upward eye
A plump of fowl behold their foe on high;°
They close their trembling troop, and all attend
On whom the sousing eagle will descend.°
But most the proud Honoria feared the event,
And thought to her alone the vision sent.   320
Her guilt presents to her distracted mind
Heaven's justice, Theodore's revengeful kind,
And the same fate to the same sin assigned;
Already sees herself the monster's prey,
And feels her heart and entrails torn away.
'Twas a mute scene of sorrow, mixed with fear.
Still on the table lay the unfinished cheer,
The knight and hungry mastiffs stood around,
The mangled dame lay breathless on the ground;
When on a sudden reinspired with breath,   330
Again she rose, again to suffer death;
Nor stayed the hell-hounds, nor the hunter stayed,

But followed as before the flying maid.
The avenger took from earth the avenging sword
And, mounting light as air, his sable steed he spurred.
The clouds dispelled, the sky resumed her light,
And nature stood recovered of her fright.
    But fear, the last of ills, remained behind,
And horror heavy sat on every mind.
Nor Theodore encouraged more his feast,   340
But sternly looked, as hatching in his breast
Some deep design, which when Honoria viewed,
The fresh impulse her former fright renewed.
She thought herself the trembling dame who fled,
And him the grisly ghost that spurred the infernal steed;
The more dismayed, for when the guests withdrew,
Their courteous host, saluting all the crew,
Regardless passed her o'er, nor graced with kind adieu.
That sting infixed within her haughty mind,
The downfall of her empire she divined;   350
And her proud heart with secret sorrow pined.
Home as they went, the sad discourse renewed
Of the relentless dame to death pursued,
And of the sight obscene so lately viewed.°
None durst arraign the righteous doom she bore,
E'en they who pitied most yet blamed her more.
The parallel they needed not to name,
But in the dead they damned the living dame.
At every little noise she looked behind,
For still the knight was present to her mind;   360
And anxious oft she started on the way
And thought the horseman-ghost came thundering for his prey.
Returned, she took her bed with little rest,
But in short slumbers dreamt the funeral feast;
Awaked, she turned her side, and slept again,
The same black vapours mounted in her brain,
And the same dreams returned with double pain.
    Now forced to wake because afraid to sleep,
Her blood all fevered, with a furious leap
She sprang from bed, distracted in her mind,   370
And feared at every step a twitching sprite behind.
Darkling and desperate with a staggering pace,
Of death afraid and conscious of disgrace;

Fear, pride, remorse, at once her heart assailed,
Pride put remorse to flight but fear prevailed.
Friday, the fatal day, when next it came,
Her soul forethought the fiend would change his game,
And her pursue; or Theodore be slain,
And two ghosts join their packs to hunt her o'er the plain.
   This dreadful image so possessed her mind,   380
That desperate any succour else to find,
She ceased all further hope; and now began
To make reflection on the unhappy man.
Rich, brave, and young, who past expression loved,
Proof to disdain, and not to be removed.
Of all the men respected and admired,
Of all the dames, except herself, desired.
Why not of her? preferred above the rest
By him with knightly deeds and open love professed?
So had another been, where he his vows addressed.   390
This quelled her pride, yet other doubts remained,
That once disdaining she might be disdained.
The fear was just but greater fear prevailed;
Fear of her life by hellish hounds assailed.
He took a louring leave; but who can tell°
What outward hate might inward love conceal?
Her sex's arts she knew, and why not then
Might deep dissembling have a place in men?
Here hope began to dawn; resolved to try,
She fixed on this her utmost remedy;   400
Death was behind but hard it was to die.
'Twas time enough at last on death to call,
The precipice in sight; a shrub was all
That kindly stood betwixt to break the fatal fall.
   One maid she had, beloved above the rest,
Secure of her the secret she confessed;
And now the cheerful light her fears dispelled,
She with no winding turns the truth concealed
But put the woman off and stood revealed;
With faults confessed commissioned her to go,   410
If pity yet had place, and reconcile her foe.
The welcome message made, was soon received;
'Twas what he wished and hoped but scarce believed:
Fate seemed a fair occasion to present,

He knew the sex, and feared she might repent°
Should he delay the moment of consent.
There yet remained to gain her friends (a care
The modesty of maidens well might spare),
But she with such a zeal the cause embraced
(As women where they will are all in haste),   420
The father, mother, and the kin beside,
Were overborne by fury of the tide.
With full consent of all, she changed her state,
Resistless in her love as in her hate.

By her example warned, the rest beware;
More easy, less imperious, were the fair;
And that one hunting which the devil designed
For one fair female, lost him half the kind.

Ceyx and Alcyone

CONNECTION OF THIS FABLE WITH THE FORMER

Ceyx, the son of Lucifer (the morning star), and King of Trachin, in Thessaly, was married to Alcyone, daughter to Aeolus, god of the winds. Both the husband and the wife loved each other with an entire affection. Daedalion, the elder brother of Ceyx, whom he succeeded, having been turned into a falcon by Apollo, and Chione, Daedalion's daughter, slain by Diana; Ceyx prepares a ship to sail to Claros, there to consult the oracle of Apollo, and (as Ovid seems to intimate) to inquire how the anger of the gods might be atoned.

These prodigies afflict the pious prince;
But, more perplexed with those that happened since,
He purposes to seek the Clarian god,
Avoiding Delphos, his more famed abode;
Since Phlegian robbers made unsafe the road.
Yet could not he from her he loved so well,
The fatal voyage, he resolved, conceal;
But when she saw her lord prepared to part,
A deadly cold ran shivering to her heart;
Her faded cheeks are changed to boxen hue,   10
And in her eyes the tears are ever new;
She thrice essayed to speak; her accents hung,
And, faltering, died unfinished on her tongue,
Or vanished into sighs; with long delay

Her voice returned; and found the wonted way.
'Tell me, my lord,' she said, 'what fault unknown
Thy once beloved Alcyone has done?
Whither, ah whither is thy kindness gone!
Can Ceyx then sustain to leave his wife,
And unconcerned forsake the sweets of life?   20
What can thy mind to this long journey move,
Or need'st thou absence to renew thy love?
Yet, if thou goest by land, though grief possess
My soul e'en then, my fears will be the less.
But ah! be warned to shun the watery way,
The face is frightful of the stormy sea.°
For late I saw adrift disjointed planks,
And empty tombs erected on the banks.
Nor let false hopes to trust betray thy mind,
Because my sire in caves constrains the wind,   30
Can with a breath their clamorous rage appease,
They fear his whistle, and forsake the seas:
Not so; for once indulged, they sweep the main,
Deaf to the call, or, hearing, hear in vain;
But bent on mischief, bear the waves before,
And not content with seas insult the shore;
When ocean, air, and earth, at once engage,
And rooted forests fly before their rage;
At once the clashing clouds to battle move,
And lightnings run across the fields above: 40
I know them well, and marked their rude comport,°
While yet a child within my father's court;
In times of tempest they command alone,
And he but sits precarious on the throne;
The more I know, the more my fears augment,
And fears are oft prophetic of the event.
But if not fears, or reasons will prevail,
If fate has fixed thee obstinate to sail,
Go not without thy wife, but let me bear
My part of danger with an equal share,   50
And present what I suffer only fear;°
Then o'er the bounding billows shall we fly,
Secure to live together, or to die.'
    These reasons moved her starlike husband's heart,
But still he held his purpose to depart;

For as he loved her equal to his life,
He would not to the seas expose his wife;
Nor could be wrought his voyage to refrain,
But sought by arguments to soothe her pain:
Nor these availed; at length he lights on one,   60
With which so difficult a cause he won:
'My love, so short an absence cease to fear,
For by my father's holy flame I swear,
Before two moons their orb with light adorn,
If heaven allow me life, I will return.'
    This promise of so short a stay prevails;
He soon equips the ship, supplies the sails,
And gives the word to launch; she trembling views
This pomp of death, and parting tears renews;
Last with a kiss, she took a long farewell,   70
Sighed with a sad presage, and swooning fell.
While Ceyx seeks delays, the lusty crew,
Raised on their banks, their oars in order drew
To their broad breasts,—the ship with fury flew.
    The queen, recovered, rears her humid eyes,
And first her husband on the poop espies,
Shaking his hand at distance on the main;
She took the sign, and shook her hand again.
Still as the ground recedes, contracts her view
With sharpened sight, till she no longer knew   80
The much-loved face; that comfort lost, supplies
With less, and with the galley feeds her eyes;
The galley borne from view by rising gales,
She followed with her sight the flying sails;
When e'en the flying sails were seen no more,
Forsaken of all sight, she left the shore.
Then on her bridal bed her body throws,
And sought in sleep her wearied eyes to close;
Her husband's pillow, and the widowed part
Which once he pressed, renewed the former smart.   90
    And now a breeze from shore began to blow;
The sailors ship their oars, and cease to row;
Then hoist their yards atrip, and all their sails°
Let fall, to court the wind and catch the gales.
By this the vessel half her course had run,
And as much rested till the rising sun;

Both shores were lost to sight, when at the close
Of day, a stiffer gale at east arose;
The sea grew white, the rolling waves from far,
Like heralds, first denounce the watery war.°   100
   This seen, the master soon began to cry,
'Strike, strike the top-sail; let the main sheet fly,°
And furl your sails.' The winds repel the sound,
And in the speaker's mouth the speech is drowned.
Yet of their own accord, as danger taught,
Each in his way, officiously they wrought;
Some stow their oars, or stop the leaky sides;
Another, bolder yet, the yard bestrides,
And folds the sails; a fourth, with labour, laves°
The intruding seas, and waves ejects on waves.   110
   In this confusion while their work they ply,
The winds augment the winter of the sky,
And wage intestine wars; the suffering seas
Are tossed, and mingled as their tyrants please.
The master would command, but in despair
Of safety, stands amazed with stupid care,
Nor what to bid, or what forbid, he knows,
The ungoverned tempest to such fury grows;
Vain is his force, and vainer is his skill,
With such a concourse comes the flood of ill;   120
The cries of men are mixed with rattling shrouds;
Seas dash on seas, and clouds encounter clouds;
At once from east to west, from pole to pole,
The forky lightnings flash, the roaring thunders roll.
   Now waves on waves ascending scale the skies,
And in the fires above, the water fries;
When yellow sands are sifted from below,
The glittering billows give a golden show;
And when the fouler bottom spews the black,
The Stygian dye the tainted waters take;°   130
Then frothy white appear the flatted seas,°
And change their colour, changing their disease.
Like various fits the Trachin vessel finds,
And now sublime she rides upon the winds;°
As from a lofty summit looks from high,
And from the clouds beholds the nether sky;
Now from the depth of hell they lift their sight,

And at a distance see superior light;
The lashing billows make a loud report,
And beat her sides, as battering rams a fort;   140
Or as a lion, bounding in his way,
With force augmented bears against his prey,
Sidelong to seize; or, unappalled with fear,
Springs on the toils, and rushes on the spear;
So seas impelled by winds, with added power,
Assault the sides, and o'er the hatches tower.
    The planks, their pitchy covering washed away,
Now yield; and now a yawning breach display;
The roaring waters with a hostile tide
Rush through the ruins of her gaping side.   150
Meantime, in sheets of rain the sky descends,
And ocean, swelled with waters, upwards tends,
One rising, falling one; the heavens and sea
Meet at their confines, in the middle way;
The sails are drunk with showers, and drop with rain,
Sweet waters mingle with the briny main.
No star appears to lend his friendly light;
Darkness and tempest make a double night;
But flashing fires disclose the deep by turns,
And while the lightnings blaze, the water burns.   160
    Now all the waves their scattered force unite;
And as a soldier, foremost in the fight,
Makes way for others, and, a host alone,
Still presses on, and, urging, gains the town;
So while the invading billows come abreast,
The hero tenth, advanced before the rest,
Sweeps all before him with impetuous sway,
And from the walls descends upon the prey;
Part following enter, part remain without,
With envy hear their fellows' conquering shout,   170
And mount on others' backs, in hope to share
The city, thus become the seat of war.
    A universal cry resounds aloud,
The sailors run in heaps, a helpless crowd;
Art fails, and courage falls, no succour near;
As many waves, as many deaths appear.
One weeps, and yet despairs of late relief;
One cannot weep, his fears congeal his grief;

But, stupid, with dry eyes expects his fate.
One with loud shrieks laments his lost estate,   180
And calls those happy whom their funerals wait.°
This wretch with prayers and vows the gods implores,
And e'en the skies he cannot see, adores.
That other on his friends his thoughts bestows,
His careful father and his faithful spouse.
The covetous worldling in his anxious mind
Thinks only on the wealth he left behind.

　All Ceyx his Alcyone employs,
For her he grieves, yet in her absence joys;
His wife he wishes, and would still be near,   190
Not her with him, but wishes him with her:
Now with last looks he seeks his native shore,
Which fate has destined him to see no more;
He sought, but in the dark tempestuous night
He knew not whither to direct his sight.
So whirl the seas, such darkness blinds the sky,
That the black night receives a deeper dye.

　The giddy ship ran round; the tempest tore
Her mast, and overboard the rudder bore.
One billow mounts; and with a scornful brow,   200
Proud of her conquest gained, insults the waves below;
Nor lighter falls, than if some giant tore
Pindus and Athos, with the freight they bore,°
And tossed on seas; pressed with the ponderous blow,
Down sinks the ship within the abyss below;
Down with the vessel sink into the main
The many, never more to rise again.
Some few on scattered planks with fruitless care
Lay hold, and swim; but, while they swim, despair.

　E'en he who late a sceptre did command,   210
Now grasps a floating fragment in his hand;
And while he struggles on the stormy main,
Invokes his father, and his wife's, in vain:
But yet his consort is his greatest care;
Alcyone he names amidst his prayer;
Names as a charm against the waves and wind,
Most in his mouth, and ever in his mind.
Tired with his toil, all hopes of safety past,
From prayers to wishes he descends at last:

That his dead body, wafted to the sands,   220
Might have its burial from her friendly hands.
As oft as he can catch a gulp of air,
And peep above the seas, he names the fair;
And, e'en when plunged beneath, on her he raves,
Murmuring Alcyone below the waves:
At last a falling billow stops his breath,
Breaks o'er his head, and whelms him underneath.
Bright Lucifer unlike himself appears
That night, his heavenly form obscured with tears;
And since he was forbid to leave the skies,   230
He muffled with a cloud his mournful eyes.
    Meantime Alcyone (his fate unknown)
Computes how many nights he had been gone,
Observes the waning moon with hourly view,
Numbers her age, and wishes for a new;
Against the promised time provides with care,
And hastens in the woof the robes he was to wear;
And for herself employs another loom,
New-dressed to meet her lord returning home,
Flattering her heart with joys that never were to come.   240
She fumed the temples with an odorous flame,°
And oft before the sacred altars came
To pray for him, who was an empty name;
All powers implored, but far above the rest,
To Juno she her pious vows addressed,
Her much-loved lord from perils to protect,
And safe o'er seas his voyage to direct;
Then prayed that she might still possess his heart,
And no pretending rival share a part.
This last petition heard, of all her prayer;   250
The rest, dispersed by winds, were lost in air.
    But she, the goddess of the nuptial bed,
Tired with her vain devotions for the dead,
Resolved the tainted hand should be repelled,°
Which incense offered, and her altar held:
Then Iris thus bespoke, 'Thou faithful maid,°
By whom thy queen's commands are well conveyed,
Haste to the house of Sleep, and bid the god,
Who rules the night by visions with a nod,
Prepare a dream, in figure and in form   260

Resembling him who perished in the storm:
This form before Alcyone present,
To make her certain of the sad event.'
    Endued with robes of various hue she flies,
And flying draws an arch, a segment of the skies;
Then leaves her bending bow, and from the steep
Descends to search the silent house of Sleep.
    Near the Cimmerians, in his dark abode,°
Deep in a cavern, dwells the drowsy god;
Whose gloomy mansion nor the rising sun,   270
Nor setting, visits, nor the lightsome noon;
But lazy vapours round the region fly,
Perpetual twilight, and a doubtful sky;
No crowing cock does there his wings display,
Nor with his horny bill provoke the day;°
Nor watchful dogs, nor the more wakeful geese,
Disturb with nightly noise the sacred peace;
Nor beast of nature, nor the tame, are nigh,
Nor trees with tempests rocked, nor human cry;
But safe repose, without an air of breath,   280
Dwells here, and a dumb quiet next to death.
    An arm of Lethe, with a gentle flow,
Arising upwards from the rock below,
The palace moats, and o'er the pebbles creeps,
And with soft murmurs calls the coming sleeps;
Around its entry nodding poppies grow,
And all cool simples that sweet rest bestow;
Night from the plants their sleepy virtue drains,
And passing sheds it on the silent plains:
No door there was the unguarded house to keep,   290
On creaking hinges turned, to break his sleep.
    But in the gloomy court was raised a bed,
Stuffed with black plumes, and on an ebon stead;
Black was the covering too, where lay the god,
And slept supine, his limbs displayed abroad;
About his head fantastic visions fly,
Which various images of things supply,
And mock their forms; the leaves on trees not more,
Nor bearded ears in fields, nor sands upon the shore.°
    The virgin, entering bright, indulged the day   300
To the brown cave, and brushed the dreams away;

The god, disturbed with this new glare of light
Cast sudden on his face, unsealed his sight,
And raised his tardy head, which sunk again,
And sinking on his bosom, knocked his chin;
At length shook off himself, and asked the dame
(And asking yawned), for what intent she came?

To whom the goddess thus: 'O sacred rest,
Sweet pleasing sleep, of all the powers the best!
O peace of mind, repairer of decay,   310
Whose balm renews the limbs to labours of the day,
Care shuns thy soft approach, and sullen flies away!
Adorn a dream, expressing human form,
The shape of him who suffered in the storm,
And send it flitting to the Trachin court,
The wreck of wretched Ceyx to report:
Before his queen bid the pale spectre stand,
Who begs a vain relief at Juno's hand.'
She said, and scarce awake her eyes could keep,
Unable to support the fumes of sleep;   320
But fled, returning by the way she went,
And swerved along her bow with swift ascent.

The god, uneasy till he slept again,
Resolved at once to rid himself of pain;
And though against his custom, called aloud,
Exciting Morpheus from the sleepy crowd;
Morpheus, of all his numerous train, expressed
The shape of man, and imitated best;
The walk, the words, the gesture could supply,
The habit mimic, and the mien belie;   330
Plays well, but all his action is confined,
Extending not beyond our human kind.
Another birds, and beasts, and dragons apes,
And dreadful images, and monster shapes:
This daemon, Icelos, in heaven's high hall
The gods have named; but men Phobetor call:
A third is Phantasus, whose actions roll
On meaner thoughts, and things devoid of soul;
Earth, fruits, and flowers, he represents in dreams,
And solid rocks unmoved, and running streams.   340
These three to kings and chiefs their scenes display,
The rest before the ignoble commons play:

Of these the chosen Morpheus is dispatched;
Which done, the lazy monarch overwatched,
Down from his propping elbow drops his head,
Dissolved in sleep, and shrinks within his bed.
    Darkling the demon glides, for flight prepared,
So soft that scarce his fanning wings are heard.
To Trachin, swift as thought, the flitting shade
Through air his momentary journey made: 350
Then lays aside the steerage of his wings,
Forsakes his proper form, assumes the king's;
And pale as death, despoiled of his array,
Into the queen's apartment takes his way,
And stands before the bed at dawn of day:
Unmoved his eyes, and wet his beard appears,
And shedding vain, but seeming real tears;
The briny water dropping from his hairs;
Then staring on her, with a ghastly look
And hollow voice, he thus the queen bespoke: 360
    'Knowest thou not me? Not yet, unhappy wife?
Or are my features perished with my life?
Look once again, and for thy husband lost,
Lo! all that's left of him, thy husband's ghost!
Thy vows for my return were all in vain;
The stormy south o'ertook us in the main;
And never shalt thou see thy living lord again.
Bear witness, heaven, I called on thee in death,
And, while I called, a billow stopped my breath.
Think not that flying fame reports my fate;   370
I present, I appear, and my own wreck relate.
Rise, wretched widow, rise, nor undeplored
Permit my ghost to pass the Stygian ford;
But rise, prepared in black to mourn thy perished lord.'
    Thus said the player god; and adding art
Of voice and gesture, so performed his part,
She thought (so like her love the shade appears)
That Ceyx spake the words, and Ceyx shed the tears.
She groaned, her inward soul with grief oppressed,
She sighed, she wept, and sleeping beat her breast: 380
Then stretched her arms to embrace his body bare,
Her clasping arms enclose but empty air:
At this, not yet awake, she cried, 'Oh stay,

One is our fate, and common is our way!'
So dreadful was the dream, so loud she spoke,
That, starting sudden up, the slumber broke;
Then cast her eyes around in hope to view
Her vanished lord, and find the vision true;
For now the maids, who waited her commands,
Ran in with lighted tapers in their hands.   390
Tired with the search, not finding what she seeks,
With cruel blows she pounds her blubbered cheeks;°
Then from her beaten breast the linen tare,
And cut the golden caul that bound her hair.
Her nurse demands the cause; with louder cries
She prosecutes her griefs, and thus replies:
  'No more Alcyone; she suffered death
With her loved lord, when Ceyx lost his breath:
No flattery, no false comfort, give me none,
My shipwrecked Ceyx is for ever gone;   400
I saw, I saw him manifest in view,
His voice, his figure, and his gestures knew:
His lustre lost, and every living grace,
Yet I retained the features of his face:
Though with pale cheeks, wet beard, and dropping hair,
None but my Ceyx could appear so fair;
I would have strained him with a strict embrace,
But through my arms he slipped, and vanished from the place;
There, e'en just there he stood;' and as she spoke,
Where last the spectre was, she cast her look;   410
Fain would she hope, and gazed upon the ground,
If any printed footsteps might be found;
Then sighed, and said, 'This I too well foreknew,
And my prophetic fear presaged too true;
'Twas what I begged, when with a bleeding heart
I took my leave, and suffered thee to part,
Or I to go along, or thou to stay,
Never, ah never to divide our way!
Happier for me, that, all our hours assigned,
Together we had lived, e'en not in death disjoined!   420
So had my Ceyx still been living here,
Or with my Ceyx I had perished there;
Now I die absent, in the vast profound,
And me without myself the seas have drowned:

The storms were not so cruel; should I strive
To lengthen life, and such a grief survive;
But neither will I strive, nor wretched thee
In death forsake, but keep thee company.
If not one common sepulchre contains
Our bodies, or one urn our last remains,   430
Yet Ceyx and Alcyone shall join,
Their names remembered in one common line.'
   No further voice her mighty grief affords,
For sighs come rushing in betwixt her words,
And stopped her tongue; but what her tongue denied,
Soft tears, and groans, and dumb complaints supplied.
   'Twas morning; to the port she takes her way,
And stands upon the margin of the sea;
That place, that very spot of ground she sought,
Or thither by her destiny was brought,   440
Where last he stood; and while she sadly said,
' 'Twas here he left me, lingering here, delayed
His parting kiss, and there his anchors weighed.'
Thus speaking, while her thoughts past actions trace,
And call to mind, admonished by the place,
Sharp at her utmost ken she cast her eyes,
And somewhat floating from afar descries;
It seemed a corpse adrift, to distant sight,
But at a distance who could judge aright?
It wafted nearer yet, and then she knew   450
That what before she but surmised was true;
A corpse it was, but whose it was, unknown,
Yet moved, howe'er, she made the case her own;
Took the bad omen of a shipwrecked man,
As for a stranger wept, and thus began:
   'Poor wretch, on stormy seas to lose thy life,
Unhappy thou, but more thy widowed wife!'
At this she paused; for now the flowing tide
Had brought the body nearer to the side:
The more she looks, the more her fears increase   460
At nearer sight, and she's herself the less:
Now driven ashore, and at her feet it lies;
She knows too much, in knowing whom she sees,—
Her husband's corpse; at this she loudly shrieks,
' 'Tis he, 'tis he,' she cries, and tears her cheeks,

Her hair, her vest; and stooping to the sands,
About his neck she cast her trembling hands.
   'And is it thus, O dearer than my life,
Thus, thus return'st thou to thy longing wife!'
She said, and to the neighbouring mole she strode,°   470
Raised there to break the incursions of the flood;
Headlong from hence to plunge herself she springs,
But shoots along supported on her wings;
A bird new-made about the banks she plies,
Not far from shore, and short excursions tries;
Nor seeks in air her humble flight to raise,
Content to skim the surface of the seas;
Her bill, though slender, sends a creaking noise,
And imitates a lamentable voice;
Now lighting where the bloodless body lies,   480
She with a funeral note renews her cries.
At all her stretch her little wings she spread,
And with her feathered arms embraced the dead;
Then flickering to his pallid lips, she strove
To print a kiss, the last essay of love;
Whether the vital touch revived the dead,
Or that the moving waters raised his head
To meet the kiss, the vulgar doubt alone,
For sure a present miracle was shown.
The gods their shapes to winter birds translate,   490
But both obnoxious to their former fate.°
Their conjugal affection still is tied,
And still the mournful race is multiplied;
They bill, they tread; Alcyone compressed,°
Seven days sits brooding on her floating nest,
A wintry queen: her sire at length is kind,
Calms every storm, and hushes every wind;
Prepares his empire for his daughter's ease,
And for his hatching nephews smooths the seas.°

The Flower and the Leaf; or,
The Lady in the Arbour

A VISION

Now turning from the wintry signs, the sun°
His course exalted through the Ram had run,
And whirling up the skies, his chariot drove
Through Taurus, and the lightsome realms of love,
Where Venus from her orb descends in showers
To glad the ground, and paint the fields with flowers:
When first the tender blades of grass appear,
And buds, that yet the blast of Eurus fear,°
Stand at the door of life, and doubt to clothe the year;
Till gentle heat, and soft repeated rains   10
Make the green blood to dance within their veins:
Then, at their call emboldened, out they come,
And swell the gems, and burst the narrow room;°
Broader and broader yet their blooms display,
Salute the welcome sun, and entertain the day.
Then from their breathing souls the sweets repair
To scent the skies, and purge the unwholesome air:
Joy spreads the heart, and with a general song,
Spring issues out, and leads the jolly months along.
    In that sweet season, as in bed I lay,   20
And sought in sleep to pass the night away,
I turned my weary side, but still in vain,
Though full of youthful health, and void of pain:
Cares I had none, to keep me from my rest,
For love had never entered in my breast;
I wanted nothing fortune could supply,
Nor did she slumber till that hour deny.
I wondered then, but after found it true,
Much joy had dried away the balmy dew:°
Seas would be pools, without the brushing air   30
To curl the waves; and sure some little care
Should weary nature so, to make her want repair.
    When Chanticleer the second watch had sung,°
Scorning the scorner sleep, from bed I sprung;
And dressing, by the moon, in loose array,

Passed out in open air, preventing day,
And sought a goodly grove, as fancy led my way.
Straight as a line in beauteous order stood
Of oaks unshorn a venerable wood;
Fresh was the grass beneath, and every tree,   40
At distance planted in a due degree,
Their branching arms in air with equal space
Stretched to their neighbours with a long embrace;
And the new leaves on every bough were seen,
Some ruddy-coloured, some of lighter green.
The painted birds, companions of the spring,°
Hopping from spray to spray, were heard to sing.
Both eyes and ears received a like delight,
Enchanting music, and a charming sight.
On Philomel I fixed my whole desire,°   50
And listened for the queen of all the choir;
Fain would I hear her heavenly voice to sing,
And wanted yet an omen to the spring.
     Attending long in vain, I took the way,
Which through a path, but scarcely printed, lay;°
In narrow mazes oft it seemed to meet,
And looked as lightly pressed by fairy feet.
Wandering I walked alone, for still methought
To some strange end so strange a path was wrought:
At last it led me where an arbour stood,   60
The sacred receptacle of the wood:°
This place unmarked, though oft I walked the green,
In all my progress I had never seen;
And seized at once with wonder and delight,
Gazed all around me, new to the transporting sight.
'Twas benched with turf, and goodly to be seen,
The thick young grass arose in fresher green:
The mound was newly made, no sight could pass
Betwixt the nice partitions of the grass;
The well-united sods so closely lay,   70
And all around the shades defended it from day.
For sycamores with eglantine were spread,°
A hedge about the sides, a covering over head.
And so the fragrant brier was wove between,
The sycamore and flowers were mixed with green,
That nature seemed to vary the delight,

And satisfied at once the smell and sight.
The master workman of the bower was known
Through fairy-lands, and built for Oberon;
Who twining leaves with such proportion drew,   80
They rose by measure, and by rule they grew;
No mortal tongue can half the beauty tell,
For none but hands divine could work so well.
Both roof and sides were like a parlour made
A soft recess, and a cool summer shade;
The hedge was set so thick, no foreign eye
The persons placed within it could espy;
But all that passed without with ease was seen,
As if nor fence nor tree was placed between.
'Twas bordered with a field; and some was plain   90
With grass, and some was sowed with rising grain,°
That (now the dew with spangles decked the ground)
A sweeter spot of earth was never found.
I looked, and looked, and still with new delight;
Such joy my soul, such pleasures filled my sight;
And the fresh eglantine exhaled a breath,
Whose odours were of power to raise from death.
Nor sullen discontent, nor anxious care.
E'en though brought thither, could inhabit there:
But thence they fled as from their mortal foe;   100
For this sweet place could only pleasure know.
    Thus as I mused, I cast aside my eye,
And saw a medlar-tree was planted nigh.
The spreading branches made a goodly show,
And full of opening blooms was every bough:
A goldfinch there I saw with gaudy pride
Of painted plumes, that hopped from side to side,
Still pecking as she passed; and still she drew
The sweets from every flower, and sucked the dew:
Sufficed at length, she warbled in her throat,   110
And tuned her voice to many a merry note,
But indistinct, and neither sweet nor clear,
Yet such as soothed my soul and pleased my ear.
    Her short performance was no sooner tried,
When she I sought, the nightingale, replied:
So sweet, so shrill, so variously she sung,
That the grove echoed, and the valleys rung;

And I so ravished with her heavenly note,
I stood entranced, and had no room for thought,
But all o'erpowered with ecstasy of bliss,   120
Was in a pleasing dream of paradise;
At length I waked, and looking round the bower,
Searched every tree, and pried on every flower,
If anywhere by chance I might espy
The rural poet of the melody;
For still methought she sung not far away:
At last I found her on a laurel spray.
Close by my side she sat, and fair in sight,
Full in a line, against her opposite;°
Where stood with eglantine the laurel twined,   130
And both their native sweets were well conjoined.

   On the green bank I sat, and listened long;
(Sitting was more convenient for the song:)
Nor till her lay was ended could I move,
But wished to dwell for ever in the grove.
Only methought the time too swiftly passed,
And every note I feared would be the last.
My sight, and smell, and hearing, were employed,
And all three senses in full gust enjoyed.
And what alone did all the rest surpass,   140
The sweet possession of the fairy place;
Single, and conscious to myself alone
Of pleasures to the excluded world unknown;
Pleasures which nowhere else were to be found,
And all Elysium in a spot of ground.

   Thus while I sat intent to see and hear,
And drew perfumes of more than vital air,
All suddenly I heard the approaching sound
Of vocal music, on the enchanted ground:
A host of saints it seemed, so full the choir;   150
As if the blessed above did all conspire
To join their voices, and neglect the lyre.
At length there issued from the grove behind
A fair assembly of the female kind:
A train less fair, as ancient fathers tell,°
Seduced the sons of heaven to rebel.
I pass their forms, and every charming grace;

Less than an angel would their worth debase:
But their attire, like liveries of a kind,
All rich and rare, is fresh within my mind.   160
In velvet white as snow the troop was gowned,
The seams with sparkling emeralds set around:
Their hoods and sleeves the same; and purfled o'er
With diamonds, pearls, and all the shining store
Of Eastern pomp: their long-descending train,
With rubies edged, and sapphires, swept the plain:
High on their heads, with jewels richly set,
Each lady wore a radiant coronet.
Beneath the circles, all the choir was graced°
With chaplets green on their fair foreheads placed;°   170
Of laurel some, of woodbine many more,
And wreaths of agnus castus others bore:°
These last, who with those virgin crowns were dressed,
Appeared in higher honour than the rest.
They danced around; but in the midst was seen
A lady of a more majestic mien;
By stature, and by beauty, marked their sovereign queen.
    She in the midst began with sober grace;
Her servants' eyes were fixed upon her face,
And as she moved or turned, her motions viewed,   180
Her measures kept, and step by step pursued.°
Methought she trod the ground with greater grace,
With more of godhead shining in her face;
And as in beauty she surpassed the choir,
So, nobler than the rest was her attire.
A crown of ruddy gold enclosed her brow,
Plain without pomp, and rich without a show:
A branch of agnus castus in her hand
She bore aloft (her sceptre of command);
Admired, adored by all the circling crowd,   190
For wheresoe'er she turned her face, they bowed:
And as she danced, a roundelay she sung,
In honour of the laurel, ever young:
She raised her voice on high, and sung so clear,
The fawns came scudding from the groves to hear,
And all the bending forest lent an ear.
At every close she made, the attending throng

Replied, and bore the burden of the song:
So just, so small, yet in so sweet a note,
It seemed the music melted in the throat.   200

   Thus dancing on, and singing as they danced,
They to the middle of the mead advanced,
Till round my arbour a new ring they made,
And footed it about the secret shade.
O'erjoyed to see the jolly troop so near,
But somewhat awed, I shook with holy fear;
Yet not so much, but that I noted well
Who did the most in song or dance excel.

   Not long I had observed, when from afar
I heard a sudden symphony of war;   210
The neighing coursers, and the soldiers' cry,
And sounding trumps that seemed to tear the sky:
I saw soon after this, behind the grove
From whence the ladies did in order move,
Come issuing out in arms a warrior train,
That like a deluge poured upon the plain:
On barbed steeds they rode in proud array,
Thick as the college of the bees in May,
When swarming o'er the dusky fields they fly,
New to the flowers, and intercept the sky.   220
So fierce they drove, their coursers were so fleet,
That the turf trembled underneath their feet.

   To tell their costly furniture were long,
The summer's day would end before the song:
To purchase but the tenth of all their store,
Would make the mighty Persian monarch poor.°
Yet what I can, I will: before the rest
The trumpets issued in white mantles dressed;
A numerous troop, and all their heads around
With chaplets green of cerrial-oak were crowned,°   230
And at each trumpet was a banner bound;
Which, waving in the wind, displayed at large
Their master's coat of arms, and knightly charge.°
Broad were the banners, and of snowy hue,
A purer web the silkworm never drew.
The chief about their necks the scutcheons wore,
With orient pearls and jewels powdered o'er:
Broad were their collars too, and every one

Was set about with many a costly stone.
Next these, of kings at arms a goodly train   240
In proud array came prancing o'er the plain:
Their cloaks were cloth of silver mixed with gold,
And garlands green around their temples rolled:
Rich crowns were on their royal scutcheons placed,
With sapphires, diamonds, and with rubies graced:
And as the trumpets their appearance made,
So these in habits were alike arrayed;
But with a pace more sober, and more slow,
And twenty, rank in rank, they rode a-row.
The pursuivants came next, in number more;   250
And like the heralds each his scutcheon bore:
Clad in white velvet all their troop they led,
With each an oaken chaplet on his head.

Nine royal knights in equal rank succeed,
Each warrior mounted on a fiery steed,
In golden armour glorious to behold;
The rivets of their arms were nailed with gold.
Their surcoats of white ermine-fur were made;
With cloth of gold between that cast a glittering shade.
The trappings of their steeds were of the same;   260
The golden fringe e'en set the ground on flame,
And drew a precious trail: a crown divine
Of laurel did about their temples twine.

Three henchmen were for every knight assigned,
All in rich livery clad, and of a kind;
White velvet, but unshorn, for cloaks they wore,
And each within his hand a truncheon bore:
The foremost held a helm of rare device;
A prince's ransom would not pay the price.
The second bore the buckler of his knight,   270
The third of cornel-wood a spear upright,
Headed with piercing steel, and polished bright.
Like to their lords their equipage was seen,
And all their foreheads crowned with garlands green.

And after these came, armed with spear and shield,
A host so great, as covered all the field:
And all their foreheads, like the knights' before,
With laurels ever green were shaded o'er,
Or oak, or other leaves of lasting kind,

Tenacious of the stem, and firm against the wind   280
Some in their hands, beside the lance and shield,
The boughs of woodbine or of hawthorn held,°
Or branches for their mystic emblems took,
Of palm, of laurel, or of cerrial-oak.°
   Thus marching to the trumpet's lofty sound,
Drawn in two lines adverse they wheeled around,
And in the middle meadow took their ground.
Among themselves the tourney they divide,
In equal squadrons ranged on either side,
Then turned their horses' heads, and man to man,   290
And steed to steed opposed, the jousts began.
They lightly set their lances in the rest,
And at the sign, against each other pressed;
They met. I sitting at my ease beheld
The mixed events, and fortunes of the field.
Some broke their spears, some tumbled horse and man,
And round the fields the lightened coursers ran.
An hour and more like tides in equal sway
They rushed, and won by turns, and lost the day:
At length the nine (who still together held)   300
Their fainting foes to shameful flight compelled,
And with resistless force o'erran the field.
   Thus, to their fame, when finished was the fight,
The victors from their lofty steeds alight:
Like them dismounted all the warlike train,
And two by two proceeded o'er the plain;
Till to the fair assembly they advanced,
Who near the secret arbour sung and danced.
   The ladies left their measures at the sight,
To meet the chiefs returning from the fight,   310
And each with open arms embraced her chosen knight.
Amid the plain a spreading laurel stood,
The grace and ornament of all the wood:
That pleasing shade they sought, a soft retreat
From sudden April showers, a shelter from the heat:
Her leafy arms with such extent were spread,
So near the clouds was her aspiring head,
That hosts of birds, that wing the liquid air,
Perched in the boughs, had nightly lodging there.
And flocks of sheep beneath the shade from far   320

Might hear the rattling hail, and wintry war;
From heaven's inclemency here found retreat,
Enjoyed the cool, and shunned the scorching heat:
A hundred knights might there at ease abide,
And every knight a lady by his side:
The trunk itself such odours did bequeath,
That a Moluccan breeze to these was common breath.°
The lords and ladies here, approaching, paid
Their homage, with a low obeisance made,
And seemed to venerate the sacred shade.   330
These rites performed, their pleasures they pursue,
With songs of love, and mix with measures new;
Around the holy tree their dance they frame,
And every champion leads his chosen dame.
  I cast my sight upon the further field,
And a fresh object of delight beheld:
For from the region of the west I heard
New music sound, and a new troop appeared;
Of knights, and ladies mixed a jolly band,
But all on foot they marched, and hand in hand.   340
  The ladies dressed in rich cymars were seen
Of Florence satin, flowered with white and green,
And for a shade betwixt the bloomy gridelin.°
The borders of their petticoats below
Were guarded thick with rubies on a row;
And every damsel wore upon her head
Of flowers a garland blended white and red.
Attired in mantles all the knights were seen,
That gratified the view with cheerful green:
Their chaplets of their ladies' colours were,   350
Composed of white and red, to shade their shining hair.
Before the merry troop the minstrels played;
All their masters' liveries were arrayed,
And clad in green, and on their temples wore
The chaplets white and red their ladies bore.
Their instruments were various in their kind,
Some for the bow, and some for breathing wind;
The sawtry, pipe, and hautboy's noisy band,
And the soft lute trembling beneath the touching hand.
A tuft of daisies on a flowery lay°   360
They saw, and thitherward they bent their way;

To this both knights and dames their homage made,
And due obeisance to the daisy paid.
And then the band of flutes began to play,
To which a lady sung a virelay:
And still at every close she would repeat
The burden of the song, 'The daisy is so sweet,
The daisy is so sweet,' when she began,
The troop of knights and dames continued on.
The concert and the voice so charmed my ear,   370
And soothed my soul, that it was heaven to hear.
   But soon their pleasure passed; at noon of day,
The sun with sultry beams began to play:
Not Sirius shoots a fiercer flame from high,°
When with his poisonous breath he blasts the sky;
Then drooped the fading flowers (their beauty fled)
And closed their sickly eyes, and hung the head,
And, rivelled up with heat, lay dying in their bed.
The ladies gasped, and scarcely could respire
The breath they drew, no longer air but fire;   380
The fainty knights were scorched; and knew not where
To run for shelter, for no shade was near.
And after this the gathering clouds amain
Poured down a storm of rattling hail and rain;
And lightning flashed betwixt: the field, and flowers,
Burnt up before, were buried in the showers.
The ladies and the knights, no shelter nigh,
Bare to the weather and the wintry sky,
Were dropping wet, disconsolate, and wan,
And through their thin array received the rain;   390
While those in white, protected by the tree,
Saw pass the vain assault, and stood from danger free.
But as compassion moved their gentle minds,
When ceased the storm, and silent were the winds,
Displeased at what, not suffering, they had seen,
They went to cheer the faction of the green:
The queen in white array before her band,
Saluting, took her rival by the hand;
So did the knights and dames with courtly grace
And with behaviour sweet their foes embrace.   400
Then thus the queen with laurel on her brow,
'Fair sister, I have suffered in your woe;

Nor shall be wanting aught within my power
For your relief in my refreshing bower.'
That other answered with a lowly look
And soon the gracious invitation took:
For ill at ease both she and all her train
The scorching sun had borne, and beating rain.
Like courtesy was used by all in white,
Each dame a dame received, and every knight a knight.   410
The laurel champions with their swords invade
The neighbouring forests, where the jousts were made,
And sere wood from the rotten hedges took,
And seeds of latent fire from flints provoke:
A cheerful blaze arose, and by the fire
They warmed their frozen feet, and dried their wet attire.
Refreshed with heat, the ladies sought around
For virtuous herbs which, gathered from the ground,
They squeezed the juice, and cooling ointment made,
Which on their sun-burnt cheeks, and their chapped skins, they
        laid;   420
Then sought green salads, which they bade them eat,°
A sovereign remedy for inward heat.
    The Lady of the Leaf ordained a feast,
And made the Lady of the Flower her guest:
When lo, a bower ascended on the plain,
With sudden seats adorned, and large for either train.
This bower was near my pleasant arbour placed,
That I could hear and see whatever passed:
The ladies sat with each a knight between,
Distinguished by their colours, white and green;   430
The vanquished party with the victors joined,
Nor wanted sweet discourse, the banquet of the mind.
Meantime the minstrels played on either side,
Vain of their art, and for the mastery vied:
The sweet contention lasted for an hour,
And reached my secret arbour from the bower.
    The sun was set; and Vesper, to supply°
His absent beams, had lighted up the sky:
When Philomel, officious all the day
To sing the service of the ensuing May,   440
Fled from her laurel shade, and winged her flight
Directly to the queen arrayed in white;

And hopping sat familiar on her hand,
A new musician, and increased the band.
  The goldfinch, who, to shun the scalding heat,
Had changed the medlar for a safer seat,
And hid in bushes scaped the bitter shower,
Now perched upon the Lady of the Flower;
And either songster holding out their throats,
And folding up their wings, renewed their notes;   450
As if all day, preluding to the fight,
They only had rehearsed, to sing by night.
The banquet ended, and the battle done,
They danced by starlight and the friendly moon:
And when they were to part, the laureate queen
Supplied with steeds the lady of the green.
Her and her train conducting on the way,
The moon to follow, and avoid the day.
  This when I saw, inquisitive to know
The secret moral of the mystic show,   460
I started from my shade, in hopes to find
Some nymph to satisfy my longing mind;
And as my fair adventure fell, I found
A lady all in white, with laurel crowned,
Who closed the rear, and softly paced along,
Repeating to herself the former song.
With due respect my body I inclined,
As to some being of superior kind,
And made my court according to the day,
Wishing her queen and her a happy May.   470
'Great thanks, my daughter,' with a gracious bow
She said: and I, who much desired to know
Of whence she was, yet fearful how to break
My mind, adventured humbly thus to speak:
'Madam, might I presume and not offend,
So may the stars and shining moon attend
Your nightly sports, as you vouchsafe to tell,
What nymphs they were who mortal forms excel,
And what the knights who fought in listed fields so well.'
  To this the dame replied: 'Fair daughter, know   480
That what you saw was all a fairy show;
And all those airy shapes you now behold,
Were human bodies once, and clothed with earthly mould:

Our souls, not yet prepared for upper light,
Till doomsday wander in the shades of night;
This only holiday of all the year,
We, privileged, in sunshine may appear;
With songs and dance we celebrate the day,
And with due honours usher in the May.
At other times we reign by night alone,   490
And posting through the skies pursue the moon;
But when the morn arises, none are found,
For cruel Demogorgon walks the round,°
And if he finds a fairy lag in light,
He drives the wretch before, and lashes into night.
  'All courteous are by kind; and ever proud
With friendly offices to help the good.
In every land we have a larger space
Than what is known to you of mortal race;
Where we with green adorn our fairy bowers,   500
And e'en this grove, unseen before, is ours.
Know further; every lady clothed in white,
And, crowned with oak and laurel every knight,
Are servants to the leaf, by liveries known
Of innocence; and I myself am one.
Saw you not her so graceful to behold
In white attire, and crowned with radiant gold?
The sovereign lady of our land is she,
Diana called, the queen of chastity;
And, for the spotless name of maid she bears,   510
That agnus castus in her hand appears,
And all her train, with leafy chaplets crowned,
Were for unblamed virginity renowned;
But those the chief and highest in command
Who bear those holy branches in their hand:
The knights adorned with laurel crowns are they,
Whom death nor danger ever could dismay,
Victorious names, who made the world obey:
Who, while they lived, in deeds of arms excelled,
And after death for deities were held.   520
But those who wear the woodbine on their brow
Were knights of love, who never broke their vow,
Firm to their plighted faith, and ever free
From fears, and fickle chance, and jealousy.

The lords and ladies, who the woodbine bear,
As true as Tristram and Isolda were.'
   'But what are those,' said I, 'the unconquered nine,
Who, crowned with laurel-wreaths, in golden armour shine?
And who the knights in green, and what the train
Of ladies dressed with daisies on the plain?   530
Why both the bands in worship disagree,
And some adore the flower, and some the tree?'
   'Just is your suit, fair daughter,' said the dame:
'Those laurelled chiefs were men of mighty fame;
Nine worthies were they called of different rites,
Three Jews, three Pagans, and three Christian knights.
These, as you see, ride foremost in the field,
As they the foremost rank of honour held,
And all in deeds of chivalry excelled;
Their temples wreathed with leaves that still renew,   540
For deathless laurel is the victor's due.
Who bear the bows were knights in Arthur's reign,
Twelve they, and twelve the peers of Charlemagne;
For bows the strength of brawny arms imply,
Emblems of valour, and of victory.
Behold an order yet of newer date,°
Doubling their number, equal in their state;
Our England's ornament, the crown's defence,
In battle brave, protectors of their prince;
Unchanged by fortune, to their sovereign true,   550
For which their manly legs are bound with blue.
These, of the garter called, of faith unstained,
In fighting fields the laurel have obtained,
And well repaid those honours which they gained.
The laurel wreaths were first by Caesar worn,
And still they Caesar's successors adorn;
One leaf of this is immortality,
And more of worth than all the world can buy.'
   'One doubt remains,' said I; 'the dames in green,
What were their qualities, and who their queen?'   560
'Flora commands,' said she, 'those nymphs and knights,
Who lived in slothful ease and loose delights;
Who never acts of honour durst pursue,
The men inglorious knights, the ladies all untrue
Who, nursed in idleness, and trained in courts,

Passed all their precious hours in plays and sports,
Till death behind came stalking on, unseen,
And withered (like the storm) the freshness of their green.
These, and their mates, enjoy the present hour,
And therefore pay their homage to the flower.   570
But knights in knightly deeds should persevere,
And still continue what at first they were;
Continue, and proceed in honour's fair career.
No room for cowardice, or dull delay;
From good to better they should urge their way.
For this with golden spurs the chiefs are graced,
With pointed rowels armed, to mend their haste;
For this with lasting leaves their brows are bound;
For laurel is the sign of labour crowned,
Which bears the bitter blast, nor shaken falls to ground: 580
From winter winds it suffers no decay,
For ever fresh and fair, and every month is May.
E'en when the vital sap retreats below,
E'en when the hoary head is hid in snow,
The life is in the leaf, and still between
The fits of falling snows appears the streaky green.
Not so the flower, which lasts for little space,
A short-lived good, and an uncertain grace:
This way and that the feeble stem is driven,
Weak to sustain the storms and injuries of heaven.   590
Propped by the spring, it lifts aloft the head,
But of a sickly beauty, soon to shed;
In summer living, and in winter dead.
For things of tender kind, for pleasure made,
Shoot up with swift increase, and sudden are decayed.'
   With humble words, the wisest I could frame,
And proffered service, I repaid the dame;
That of her grace, she gave her maid to know
The secret meaning of this moral show.
And she, to prove what profit I had made   600
Of mystic truth, in fables first conveyed,
Demanded till the next returning May,
Whether the leaf or flower I would obey?
I chose the leaf; she smiled with sober cheer,
And wished me fair adventure for the year,
And gave me charms and sigils, for defence°

Against ill tongues that scandal innocence:
'But I', said she, 'my fellows must pursue,
Already past the plain, and out of view.'
   We parted thus; I homeward sped my way,   610
Bewildered in the wood till dawn of day;
And met the merry crew who danced about the May.
Then late refreshed with sleep, I rose to write
The visionary vigils of the night.
Blush, as thou may'st, my little book, for shame,
Nor hope with homely verse to purchase fame;
For such thy maker chose, and so designed
Thy simple style to suit thy lowly kind.

The Twelfth Book of Ovid's Metamorphoses,

WHOLLY TRANSLATED

Connection to the End of the Eleventh Book

Aesacus, the son of Priam, loving a country life, forsakes the court; living ob-
scurely, he falls in love with a nymph, who, flying from him, was killed by a serpent;
for grief of this, he would have drowned himself; but, by the pity of the gods, is
turned into a cormorant. Priam, not hearing of Aesacus, believes him to be dead,
and raises a tomb to preserve his memory. By this transition, which is one of the
finest in all Ovid, the poet naturally falls into the story of the Trojan War, which
is summed up in the present book; but so very briefly in many places, that Ovid
seems more short than Virgil, contrary to his usual style. Yet the House of Fame,
which is here described, is one of the most beautiful pieces in the whole Metamor-
phoses. The fight of Achilles and Cygnus, and the fray betwixt the Lapithae and
centaurs, yield to no other part of this poet; and particularly the loves and death of
Cyllarus and Hylonome, the male and female centaur, are wonderfully moving.

Priam, to whom the story was unknown,
As dead, deplored his metamorphosed son;
A cenotaph his name and title kept,
And Hector round the tomb, with all his brothers, wept.
This pious office Paris did not share;
Absent alone, and author of the war,
Which, for the Spartan queen, the Grecians drew°
To avenge the rape, and Asia to subdue.
   A thousand ships were manned, to sail the sea;
Nor had their just resentments found delay,   10

Had not the winds and waves opposed their way.
At Aulis, with united powers, they meet,
But there, cross winds or calms detained the fleet.
Now, while they raise an altar on the shore,
And Jove with solemn sacrifice adore,
A boding sign the priests and people see:
A snake of size immense ascends a tree,
And in the leafy summit spied a nest,
Which, o'er her callow young, a sparrow pressed.
Eight were the birds unfledged; their mother flew,   20
And hovered round her care, but still in view;
Till the fierce reptile first devoured the brood,
Then seized the fluttering dam, and drank her blood.
This dire ostent the fearful people view;
Calchas alone, by Phoebus taught, foreknew
What heaven decreed; and, with a smiling glance,
Thus gratulates to Greece her happy chance.°
'O Argives, we shall conquer; Troy is ours,
But long delays shall first afflict our powers;
Nine years of labour the nine birds portend,   30
The tenth shall in the town's destruction end.'
   The serpent, who his maw obscene had filled,
The branches in his curled embraces held;
But as in spires he stood, he turned to stone;
The stony snake retained the figure still his own.
   Yet not for this the windbound navy weighed;
Slack were their sails, and Neptune disobeyed.°
Some thought him loath the town should be destroyed,
Whose building had his hands divine employed;
Not so the seer, who knew, and known foreshowed,   40
The virgin Phoebe, with a virgin's blood,
Must first be reconciled; the common cause
Prevailed; and pity yielding to the laws,
Fair Iphigenia, the devoted maid,
Was, by the weeping priests, in linen robes arrayed.
All mourn her fate, but no relief appeared;
The royal victim bound, the knife already reared;
When that offended power, who caused their woe,
Relenting ceased her wrath, and stopped the coming blow.
A mist before the ministers she cast,   50
And in the virgin's room a hind she placed.

The oblation slain, and Phoebe reconciled,
The storm was hushed, and dimpled ocean smiled;
A favourable gale arose from shore,
Which to the port desired the Grecian galleys bore.
    Full in the midst of this created space,
Betwixt heaven, earth, and skies, there stands a place
Confining on all three, with triple bound;
Whence all things, though remote, are viewed around,
And thither bring their undulating sound;   60
The palace of loud Fame; her seat of power,
Placed on the summit of a lofty tower.
A thousand winding entries, long and wide,
Receive of fresh reports a flowing tide;
A thousand crannies in the walls are made;
Nor gate nor bars exclude the busy trade.
'Tis built of brass, the better to diffuse
The spreading sounds, and multiply the news;
Where echoes in repeated echoes play:
A mart for ever full, and open night and day.   70
Nor silence is within, nor voice express,
But a deaf noise of sounds that never cease;
Confused, and chiding, like the hollow roar
Of tides, receding from the insulted shore.
Or like the broken thunder, heard from far,
When Jove to distance drives the rolling war.
The courts are filled with a tumultuous din
Of crowds, or issuing forth, or entering in;
A thoroughfare of news; where some devise
Things never heard; some mingle truth with lies;   80
The troubled air with empty sounds they beat;
Intent to hear, and eager to repeat.
Error sits brooding there; with added train
Of vain credulity, and joys as vain;
Suspicion, with sedition joined, are near;
And rumours raised, and murmurs mixed, and panic fear.
Fame sits aloft, and sees the subject ground,
And seas about, and skies above, inquiring all around.
    The goddess gives the alarm; and soon is known
The Grecian fleet, descending on the town.   90
Fixed on defence, the Trojans are not slow
To guard their shore from an expected foe.

They meet in fight; by Hector's fatal hand
Protesilaus falls, and bites the strand;
Which with expense of blood the Grecians won,
And proved the strength unknown of Priam's son;
And to their cost the Trojan leaders felt
The Grecian heroes, and what deaths they dealt.
From these first onsets, the Sigaean shore
Was strewed with carcases, and stained with gore.   100
Neptunian Cygnus troops of Greeks had slain;°
Achilles in his car had scoured the plain,
And cleared the Trojan ranks; where'er he fought,
Cygnus, or Hector, through the fields he sought:
Cygnus he found; on him his force essayed;
For Hector was to the tenth year delayed.
His white-maned steeds, that bowed beneath the yoke,
He cheered to courage, with a gentle stroke;
Then urged his fiery chariot on the foe,
And rising shook his lance, in act to throw.°   110
But first he cried, 'O youth, be proud to bear
Thy death, ennobled by Pelides' spear.'
The lance pursued the voice without delay;
Nor did the whizzing weapon miss the way,
But pierced his cuirass, with such fury sent,
And signed his bosom with a purple dint.
At this the seed of Neptune: 'Goddess-born,°
For ornament, not use, these arms are worn;
This helm, and heavy buckler, I can spare,
As only decorations of the war;   120
So Mars is armed, for glory, not for need.
'Tis somewhat more from Neptune to proceed,
Than from a daughter of the sea to spring;
Thy sire is mortal; mine is ocean's king.
Secure of death, I should contemn thy dart,
Though naked, and impassible depart.'
He said, and threw; the trembling weapon passed
Through nine bull-hides, each under other placed
On his broad shield, and stuck within the last.
Achilles wrenched it out; and sent again   130
The hostile gift; the hostile gift was vain.
He tried a third, a tough well-chosen spear;
The inviolable body stood sincere,

Though Cygnus then did no defence provide,
But scornful offered his unshielded side.
  Not otherwise the impatient hero fared,
Than as a bull, encompassed with a guard,
Amid the circus roars; provoked from far
By sight of scarlet, and a sanguine war.
They quit their ground, his bended horns elude,   140
In vain pursuing, and in vain pursued.
  Before to further fight he would advance,
He stood considering, and surveyed his lance.
Doubts if he wielded not a wooden spear
Without a point; he looked, the point was there.
'This is my hand, and this my lance,' he said,
'By which so many thousand foes are dead,
O whither is their usual virtue fled!
I had it once; and the Lyrnessian wall,°
And Tenedos, confessed it in their fall.   150
Thy streams, Caicus, rolled a crimson flood;
And Thebes ran red with her own natives' blood.
Twice Telephus employed this piercing steel,
To wound him first, and afterward to heal.
The vigour of this arm was never vain;
And that my wonted prowess I retain,
Witness these heaps of slaughter on the plain.'
He said, and, doubtful of his former deeds,
To some new trial of his force proceeds.
He chose Menoetes from among the rest;   160
At him he lanced his spear, and pierced his breast;
On the hard earth the Lycian knocked his head,
And lay supine; and forth the spirit fled.
  Then thus the hero: 'Neither can I blame
The hand, or javelin; both are still the same.
The same I will employ against this foe,
And wish but with the same success to throw.'
So spoke the chief, and while he spoke he threw;
The weapon with unerring fury flew,
At his left shoulder aimed; nor entrance found;   170
But back, as from a rock, with swift rebound
Harmless returned; a bloody mark appeared,
Which with false joy the flattered hero cheered.

Wound there was none; the blood that was in view,
The lance before from slain Menoetes drew.

Headlong he leaps from off his lofty car,
And in close fight on foot renews the war;
Raging with high disdain, repeats his blows;
Nor shield nor armour can their force oppose;
Huge cantlets of his buckler strew the ground,   180
And no defence in his bored arms is found.°
But on his flesh no wound or blood is seen;
The sword itself is blunted on the skin.

This vain attempt the chief no longer bears;
But round his hollow temples and his ears,
His buckler beats; the son of Neptune, stunned
With these repeated buffets, quits his ground;
A sickly sweat succeeds, and shades of night;
Inverted nature swims before his sight:
The insulting victor presses on the more,   190
And treads the steps the vanquished trod before,
Nor rest, nor respite gives. A stone there lay
Behind his trembling foe, and stopped his way;
Achilles took the advantage which he found,
O'erturned, and pushed him backward on the ground.
His buckler held him under, while he pressed,
With both his knees above, his panting breast;
Unlaced his helm; about his chin the twist
He tied, and soon the strangled soul dismissed.

With eager haste he went to strip the dead;   200
The vanquished body from his arms was fled.
His sea-god sire, to immortalise his fame,
Had turned it to the bird that bears his name.°
A truce succeeds the labours of this day,
And arms suspended with a long delay.
While Trojan walls are kept with watch and ward,
The Greeks before their trenches mount the guard.
The feast approached; when to the blue-eyed maid,
His vows for Cygnus slain the victor paid,
And a white heifer on her altar laid.   210
The reeking entrails on the fire they threw,
And to the gods the grateful odour flew;
Heaven had its part in sacrifice; the rest

Was broiled and roasted for the future feast.
The chief invited guests were set around;
And hunger first assuaged, the bowls were crowned,
Which in deep draughts their cares and labours drowned.
The mellow harp did not their ears employ,
And mute was all the warlike symphony;
Discourse, the food of souls, was their delight,   220
And pleasing chat prolonged the summer's night.
The subject, deeds of arms and valour shown,
Or on the Trojan side, or on their own.
Of dangers undertaken, fame achieved,
They talked by turns, the talk by turns relieved.
What things but these could fierce Achilles tell,
Or what could fierce Achilles hear so well?
The last great act performed, of Cygnus slain,
Did most the martial audience entertain;
Wondering to find a body, free by fate   230
From steel, and which could e'en that steel rebate.
Amazed, their admiration they renew;
And scarce Pelides could believe it true.
     Then Nestor thus: 'What once this age has known,
In fated Cygnus, and in him alone,
These eyes have seen in Caeneus long before,
Whose body not a thousand swords could bore.
Caeneus in courage and in strength excelled,
And still his Othrys with his fame is filled;°
But what did most his martial deeds adorn,   240
(Though, since, he changed his sex,) a woman born.'
     A novelty so strange and full of fate
His listening audience asked him to relate.
Achilles thus commends their common suit:
'O father, first for prudence in repute,
Tell, with that eloquence so much thy own,
What thou hast heard, or what of Caeneus known;
What was he, whence his change of sex begun,
What trophies, joined in wars with thee, he won?
Who conquered him, and in what fatal strife   250
The youth, without a wound, could lose his life?'
     Neleides then: 'Though tardy age, and time,°
Have shrunk my sinews, and decayed my prime;
Though much I have forgotten of my store,

Yet, not exhausted, I remember more.
Of all that arms achieved, or peace designed,
That action still is fresher in my mind
Than aught beside. If reverend age can give
To faith a sanction, in my third I live.

   'Twas in my second century, I surveyed   260
Young Caenis, then a fair Thessalian maid.
Caenis the bright was born to high command;
A princess, and a native of thy land,
Divine Achilles; every tongue proclaimed
Her beauty, and her eyes all hearts inflamed.
Peleus, thy sire, perhaps had sought her bed,
Among the rest; but he had either led
Thy mother then, or was by promise tied;
But she to him, and all, alike her love denied.

   'It was her fortune once, to take her way   270
Along the sandy margin of the sea;
The power of ocean viewed her as she passed,°
And, loved as soon as seen, by force embraced.
So fame reports. Her virgin treasure seized,
And his new joys the ravisher so pleased,
That thus, transported, to the nymph he cried,
"Ask what thou wilt, no prayer shall be denied."
This also fame relates; the haughty fair,
Who not the rape e'en of a god could bear,
This answer, proud, returned: "To mighty wrongs,   280
A mighty recompense, of right, belongs.
Give me no more to suffer such a shame;
But change the woman for a better name,
One gift for all." She said, and, while she spoke,
A stern, majestic, manly tone she took.
A man she was; and, as the godhead swore,
To Caeneus turned, who Caenis was before.

   'To this the lover adds, without request,
No force of steel should violate his breast.
Glad of the gift, the new-made warrior goes,   290
And arms among the Greeks, and longs for equal foes.

   'Now brave Pirithous, bold Ixion's son,
The love of fair Hippodame had won.
The cloud-begotten race, half men, half beast,°
Invited, came to grace the nuptial feast.

In a cool cave's recess the treat was made,
Whose entrance trees with spreading boughs o'ershade.
They sat: and summoned by the bridegroom, came,
To mix with those, the Lapithaean name:
Nor wanted I; the roofs with joy resound;   300
And "Hymen, Iö Hymen", rung around.
Raised altars shone with holy fires; the bride,
Lovely herself (and lovely by her side
A bevy of bright nymphs, with sober grace,)
Came glittering like a star, and took her place,
Her heavenly form beheld, all wished her joy,
And little wanted, but in vain their wishes all employ.°

'For one, most brutal of the brutal brood,
Or whether wine or beauty fired his blood,
Or both at once, beheld with lustful eyes   310
The bride; at once resolved to make his prize.
Down went the board, and fastening on her hair,
He seized with sudden force the frighted fair.
'Twas Eurytus began; his bestial kind
His crime pursued; and each as pleased his mind,
Or her, whom chance presented, took; the feast
An image of a taken town expressed.

'The cave resounds with female shrieks: we rise,
Mad with revenge, to make a swift reprise:
And Theseus first: "What frenzy has possessed,   320
O Eurytus", he cried, "thy brutal breast,
To wrong Pirithous, and not him alone,
But while I live, two friends conjoined in one?"

'To justify his threat, he thrusts aside
The crowd of centaurs, and redeems the bride.
The monster naught replied; for words were vain,
And deeds could only deeds unjust maintain;
But answers with his hand, and forward pressed,
With blows redoubled, on his face and breast.
An ample goblet stood, of antique mould,   330
And rough with figures of the rising gold;
The hero snatched it up, and tossed in air
Full at the front of the foul ravisher.
He falls, and falling vomits forth a flood
Of wine, and foam, and brains, and mingled blood.
Half roaring, and half neighing through the hall,

"Arms, arms!" the double-formed with fury call,°
To wreak their brother's death. A medley flight
Of bowls and jars, at first supply the fight.
Once instruments of feasts, but now of fate;   340
Wine animates their rage, and arms their hate.

'Bold Amycus from the robbed vestry brings
The chalices of heaven, and holy things
Of precious weight; a sconce, that hung on high,
With tapers filled, to light the sacristy,
Torn from the cord, with his unhallowed hand
He threw amid the Lapithaean band.
On Celadon the ruin fell, and left
His face of feature and of form bereft;
So, when some brawny sacrificer knocks   350
Before an altar led, an offered ox,
His eyeballs, rooted out, are thrown to ground,
His nose dismantled in his mouth is found,
His jaws, cheeks, front, one undistinguished wound.

'This, Pelates, the avenger, could not brook;
But, by the foot, a maple-board he took,
And hurled at Amycus; his chin it bent
Against his chest, and down the centaur sent,
Whom, sputtering bloody teeth, the second blow
Of his drawn sword dispatched to shades below.   360

'Grineus was near; and cast a furious look
On the side-altar, censed with sacred smoke,
And bright with flaming fires; "The gods," he cried,
"Have with their holy trade our hands supplied:
Why use we not their gifts?" Then from the floor
An altar-stone he heaved, with all the load it bore;
Altar and altar's freight together flew,
Where thickest thronged the Lapithaean crew,
And, Broteas and at once Oryus slew.
Oryus' mother, Mycale, was known   370
Down from her sphere to draw the labouring moon.

'Exadius cried: "Unpunished shall not go
This fact, if arms are found against the foe."
He looked about, where on a pine were spread
The votive horns of a stag's branching head:
At Grineus these he throws; so just they fly,
That the sharp antlers stuck in either eye.

Breathless and blind he fell; with blood besmeared,
His eyeballs beaten out hung dangling on his beard.
Fierce Rhoetus from the hearth a burning brand   380
Selects, and whirling waves, till from his hand
The fire took flame; then dashed it from the right,
On fair Charaxus' temples, near the sight:
The whistling pest came on, and pierced the bone,
And caught the yellow hair, that shrivelled while it shone;
Caught, like dry stubble fired, or like sear wood;
Yet from the wound ensued no purple flood,
But looked a bubbling mass of frying blood.
His blazing locks sent forth a crackling sound,
And hissed, like red-hot iron within the smithy drowned.   390
The wounded warrior shook his flaming hair,
Then (what a team of horse could hardly rear,)
He heaves the threshold stone, but could not throw;
The weight itself forbade the threatened blow,
Which dropping from his lifted arms, came down
Full on Cometes' head, and crushed his crown.
Nor Rhoetus then retained his joy; but said,
"So by their fellows may our foes be sped."
Then with redoubled strokes he plies his head:
The burning lever not deludes his pains,   400
But drives the battered skull within the brains.

   'Thus flushed, the conqueror, with force renewed,
Evagrus, Dryas, Corythus, pursued.
First, Corythus, with downy cheeks, he slew;
Whose fall when fierce Evagrus had in view,
He cried, "What palm is from a beardless prey?"
Rhoetus prevents what more he had to say;
And drove within his mouth the fiery death,
Which entered hissing in, and choked his breath.
At Dryas next he flew; but weary chance   410
No longer would the same success advance;
For while he whirled in fiery circles round
The brand, a sharpened stake strong Dryas found,
And in the shoulder's joint inflicts the wound.
The weapon struck; which, roaring out with pain,
He drew; nor longer durst the fight maintain,
But turned his back for fear, and fled amain.
With him fled Orneus, with like dread possessed;

Thaumas and Medon, wounded in the breast,
And Mermeros, in the late race renowned,   420
Now limping ran, and tardy with his wound.
Pholus and Melaneus from fight withdrew,
And Abas maimed, who boars encountering slew;
And augur Astylos, whose art in vain
From fight dissuaded the four-footed train,
Now beat the hoof with Nessus on the plain;
But to his fellow cried, "Be safely slow;
Thy death deferred is due to great Alcides' bow."
   'Meantime strong Dryas urged his chance so well,
That Lycidas, Areos, Imbreus fell;   430
All, one by one, and fighting face to face:
Crenaeus fled, to fall with more disgrace;
For, fearful while he looked behind, he bore,
Betwixt his nose and front, the blow before.
Amid the noise and tumult of the fray,
Snoring and drunk with wine, Aphidas lay.
E'en then the bowl within his hand he kept,
And on a bear's rough hide securely slept.
Him Phorbas with his flying dart transfixed;
"Take thy next draught with Stygian waters mixed,°   440
And sleep thy fill," the insulting victor cried;
Surprised with death unfelt, the centaur died:
The ruddy vomit, as he breathed his soul,
Repassed his throat, and filled his empty bowl.
   'I saw Petraeus' arms employed around
A well-grown oak, to root it from the ground.
This way, and that, he wrenched the fibrous bands;
The trunk was like a sapling in his hands,
And still obeyed the bent: while thus he stood,
Pirithous' dart drove on, and nailed him to the wood.   450
Lycus and Chromis fell, by him oppressed:
Helops and Dictys added to the rest
A nobler palm: Helops, through either ear
Transfixed, received the penetrating spear.
This Dictys saw; and seized with sudden fright,
Leapt headlong from the hill of steepy height,
And crushed an ash beneath, that could not bear his weight.
The shattered tree receives his fall, and strikes,
Within his full-blown paunch, the sharpened spikes.

Strong Aphareus had heaved a mighty stone,   460
The fragment of a rock, and would have thrown;
But Theseus, with a club of hardened oak,
The cubit-bone of the bold centaur broke,°
And left him maimed, nor seconded the stroke;
Then leapt on tall Bienor's back; (who bore
No mortal burden but his own, before,)
Pressed with his knees his sides; the double man,
His speed with spurs increased, unwilling ran.
One hand the hero fastened on his locks;
His other plied him with repeated strokes.   470
The club rung round his ears, and battered brows;
He falls; and, lashing up his heels, his rider throws.
    'The same Herculean arms Nedymnus wound,
And lay by him Lycopes on the ground;
And Hippasus, whose beard his breast invades;
And Ripheus, haunter of the woodland shades;
And Tereus, used with mountain bears to strive,
And from their dens to draw the indignant beasts alive.
    'Demoleon could not bear this hateful sight,
Or the long fortune of the Athenian knight;°   480
But pulled with all his force to disengage
From earth a pine, the product of an age.
The root stuck fast: the broken trunk he sent
At Theseus: Theseus frustrates his intent,
And leaps aside, by Pallas warned, the blow
To shun: (for so he said; and we believed it so.)
Yet not in vain the enormous weight was cast,
Which Crantor's body sundered at the waist:
Thy father's squire, Achilles, and his care;
Whom, conquered in the Dolopeian war,   490
Their king, his present ruin to prevent,°
A pledge of peace implored, to Peleus sent.
    'Thy sire, with grieving eyes, beheld his fate;
And cried, "not long, loved Crantor, shalt thou wait
Thy vowed revenge." At once he said, and threw
His ashen-spear, which quivered as it flew,
With all his force and all his soul applied;
The sharp point entered in the centaur's side:
Both hands, to wrench it out, the monster joined,
And wrenched it out, but left the steel behind.   500

Stuck in his lungs it stood; enraged he rears
His hoofs, and down to ground thy father bears.
Thus trampled under foot, his shield defends
His head; his other hand the lance protends.
E'en while he lay extended on the dust,
He sped the centaur, with one single thrust.
Two more his lance before transfixed from far,
And two his sword had slain in closer war.
To these was added Dorylas; who spread
A bull's two goring horns around his head.   510
With these he pushed; in blood already dyed,
Him, fearless, I approached, and thus defied:
"Now, monster, now, by proof it shall appear,
Whether thy horns are sharper, or my spear."
At this, I threw; for want of other ward,
He lifted up his hand, his front to guard.
His hand it passed, and fixed it to his brow.
Loud shouts of ours attend the lucky blow:
Him Peleus finished, with a second wound,
Which through the navel pierced; he reeled around,   520
And dragged his dangling bowels on the ground;
Trod what he dragged, and what he trod he crushed;
And to his mother earth, with empty belly, rushed.
   'Nor could thy form, O Cyllarus, foreslow°
Thy fate, if form to monsters men allow:
Just bloomed thy beard, thy beard of golden hue;
Thy locks, in golden waves about thy shoulders flew.
Sprightly thy look; thy shapes in every part
So clean, as might instruct the sculptor's art,
As far as man extended; where began   530
The beast, the beast was equal to the man.
Add but a horse's head and neck, and he,
O Castor, was a courser worthy thee.
So was his back proportioned for the seat;
So rose his brawny chest; so swiftly moved his feet.
Coal-black his colour, but like jet it shone;
His legs and flowing tail were white alone.
Beloved by many maidens of his kind,
But fair Hylonome possessed his mind;
Hylonome, for features, and for face,   540
Excelling all the nymphs of double race.

Nor less her blandishments, than beauty move;
At once both loving, and confessing love.
For him she dressed; for him with female care
She combed and set in curls her auburn hair.
Of roses, violets, and lilies mixed,
And sprigs of flowing rosemary betwixt,
She formed the chaplet, that adorned her front;
In waters of the Pegasaean fount,
And in the streams that from the fountain play,   550
She washed her face, and bathed her twice a day.
The scarf of furs, that hung below her side,
Was ermine, or the panther's spotted pride;
Spoils of no common beast. With equal flame
They loved; their sylvan pleasures were the same:
All day they hunted; and when day expired,
Together to some shady cave retired.
Invited to the nuptials both repair;
And side by side they both engage in war.
     'Uncertain from what hand, a flying dart   560
At Cyllarus was sent, which pierced his heart.
The javelin drawn from out the mortal wound,
He faints with staggering steps, and seeks the ground:
The fair within her arms received his fall,
And strove his wandering spirits to recall;
And while her hand the streaming blood opposed,
Joined face to face, his lips with hers she closed.
Stifled with kisses, a sweet death he dies;
She fills the fields with undistinguished cries;
At least her words were in her clamour drowned;   570
For my stunned ears received no vocal sound.
In madness of her grief, she seized the dart
New-drawn, and reeking from her lover's heart;
To her bare bosom the sharp point applied,
And wounded fell; and falling by his side,
Embraced him in her arms, and thus embracing died.
     'E'en still, methinks, I see Phaeocomes;
Strange was his habit, and as odd his dress.
Six lions' hides, with thongs together fast,
His upper part defended to his waist;   580
And where man ended, the continued vest,
Spread on his back, the house and trappings of a beast.°

A stump too heavy for a team to draw,
(It seems a fable, though the fact I saw,)
He threw at Pholon; the descending blow
Divides the skull, and cleaves his head in two.
The brains, from nose and mouth, and either ear,
Came issuing out, as through a colander
The curdled milk; or from the press the whey,
Driven down by weights above, is drained away.   590
    'But him, while stooping down to spoil the slain,
Pierced through the paunch, I tumbled on the plain.
Then Chthonius and Teleboas I slew;
A fork the former armed; a dart his fellow threw:
The javelin wounded me; behold the scar.
Then was my time to seek the Trojan war;
Then I was Hector's match in open field;
But he was then unborn, at least a child;
Now, I am nothing. I forbear to tell
By Periphantes how Pyretus fell,   600
The centaur by the knight; nor will I stay
On Ampyx, or what deaths he dealt that day;
What honour, with a pointless lance, he won,
Stuck in the front of a four-footed man;
What fame young Macareus obtained in fight,
Or dwell on Nessus, now returned from flight.
How prophet Mopsus not alone divined,
Whose valour equalled his foreseeing mind.
    'Already Caeneus, with his conquering hand,
Had slaughtered five, the boldest of their band;   610
Pyracmus, Elymus, Antimachus,
Bromus the brave, and stronger Stiphelus;
Their names I numbered, and remember well,
No trace remaining, by what wounds they fell.
    'Latreus, the bulkiest of the double race,
Whom the spoiled arms of slain Halesus grace,
In years retaining still his youthful might,
Though his black hairs were interspersed with white,
Betwixt the embattled ranks began to prance,
Proud of his helm, and Macedonian lance;   620
And rode the ring around, that either host
Might hear him, while he made this empty boast:
"And from a strumpet shall we suffer shame?

For Caenis still, not Caeneus, is thy name;
And still the native softness of thy kind
Prevails, and leaves the woman in thy mind.
Remember what thou wert; what price was paid
To change thy sex, to make thee not a maid;
And but a man in show; go card and spin,°
And leave the business of the war to men."   630
    'While thus the boaster exercised his pride,
The fatal spear of Caeneus reached his side;
Just in the mixture of the kinds it ran,
Betwixt the nether beast and upper man.
The monster, mad with rage, and stung with smart,
His lance directed at the hero's heart:
It struck; but bounded from his hardened breast,
Like hail from tiles, which the safe house invest;
Nor seemed the stroke with more effect to come,
Than a small pebble falling on a drum.   640
He next his falchion tried in closer fight;
But the keen falchion had no power to bite.
He thrust; the blunted point returned again:
"Since downright blows", he cried, "and thrusts are vain,
I'll prove his side;"—in strong embraces held,
He proved his side; his side the sword repelled;
His hollow belly echoed to the stroke:
Untouched his body, as a solid rock;
Aimed at his neck at last, the blade in shivers broke.
    'The impassive knight stood idle, to deride   650
His rage, and offered oft his naked side;
At length, "Now, monster, in thy turn," he cried,
"Try thou the strength of Caeneus:" at the word
He thrust; and in his shoulder plunged the sword.
Then writhed his hand; and as he drove it down
Deep in his breast, made many wounds in one.
    'The centaurs saw, enraged, the unhoped success,
And rushing on in crowds, together press.
At him, and him alone, their darts they threw;
Repulsed they from his fated body flew.   660
Amazed they stood; till Monychus began,
"O shame, a nation conquered by a man!
A woman-man; yet more a man is he,
Than all our race; and what he was, are we.

Now, what avail our nerves? the united force
Of two the strongest creatures, man and horse?
Nor goddess-born, nor of Ixion's seed
We seem, (a lover built for Juno's bed,)
Mastered by this half man. Whole mountains throw
With woods at once, and bury him below.   670
This only way remains. Nor need we doubt
To choke the soul within, though not to force it out.
Heap weights, instead of wounds:" he chanced to see
Where southern storms had rooted up a tree;
This, raised from earth, against the foe he threw;
The example shown, his fellow brutes pursue.
With forest-loads the warrior they invade;
Othrys and Pelion soon were void of shade,
And spreading groves were naked mountains made.
Pressed with the burden, Caeneus pants for breath,   680
And on his shoulders bears the wooden death.
To heave the intolerable weight he tries;
At length it rose above his mouth and eyes.
Yet still he heaves; and, struggling with despair,
Shakes all aside, and gains a gulp of air;
A short relief, which but prolongs his pain:
He faints by fits, and then respires again.
At last, the burden only nods above,
As when an earthquake stirs the Idaean grove.
Doubtful his death; he suffocated seemed   690
To most; but otherwise our Mopsus deemed,
Who said he saw a yellow bird arise
From out the pile, and cleave the liquid skies.
I saw it too, with golden feathers bright,
Nor e'er before beheld so strange a sight;
Whom Mopsus viewing as it soared around
Our troop and heard the pinions' rattling sound,
"All hail," he cried, "thy country's grace and love;
Once first of men below, now first of birds above!"
Its author to the story gave belief;   700
For us, our courage was increased by grief:
Ashamed to see a single man, pursued
With odds, to sink beneath a multitude,
We pushed the foe, and forced to shameful flight:
Part fell, and part escaped by favour of the night.'

This tale, by Nestor told, did much displease
Tlepolemus, the seed of Hercules;
For often he had heard his father say,
That he himself was present at the fray,
And more than shared the glories of the day.   710
    'Old Chronicle,' he said, 'among the rest,
You might have named Alcides at the least;
Is he not worth your praise?' The Pylian prince
Sighed ere he spoke, then made this proud defence:
'My former woes, in long oblivion drowned,
I would have lost, but you renew the wound;.
Better to pass him o'er, than to relate
The cause I have your mighty sire to hate.
His fame has filled the world, and reached the sky;
Which, oh, I wish with truth I could deny!   720
We praise not Hector, though his name we know
Is great in arms; 'tis hard to praise a foe.
He, your great father, levelled to the ground
Messenia's towers; nor better fortune found
Elis, and Pylas; that, a neighbouring state,
And this, my own; both guiltless of their fate.
To pass the rest, twelve, wanting one, he slew,
My brethren, who their birth from Neleus drew;
All youths of early promise, had they lived;
By him they perished; I alone survived.   730
The rest were easy conquest; but the fate
Of Periclymenos is wondrous to relate.
To him our common grandsire of the main
Had given to change his form, and, changed, resume again.
Varied at pleasure, every shape he tried,
And in all beasts Alcides still defied;
Vanquished on earth, at length he soared above,
Changed to the bird, that bears the bolt of Jove.
The new dissembled eagle, now endued
With beak and pounces, Hercules pursued,   740
And cuffed his manly cheeks, and tore his face,
Then, safe retired, and towered in empty space.
Alcides bore not long his flying foe,
But bending his inevitable bow,
Reached him in air, suspended as he stood,
And in his pinion fixed the feathered wood.

Light was the wound; but in the sinew hung
The point, and his disabled wing unstrung.
He wheeled in air, and stretched his fans in vain;
His fans no longer could his flight sustain;   750
For while one gathered wind, one unsupplied
Hung drooping down, nor poised his other side.
He fell; the shaft, that slightly was impressed,
Now from his heavy fall with weight increased,
Drove through his neck aslant; he spurns the ground,
And the soul issues through the weasand's wound.

   'Now, brave commander of the Rhodian seas,
What praise is due from me to Hercules?
Silence is all the vengeance I decree
For my slain brothers; but 'tis peace with thee.'   760

   Thus with a flowing tongue old Nestor spoke;
Then, to full bowls each other they provoke;
At length, with weariness and wine oppressed,
They rise from table, and withdraw to rest.

   The sire of Cygnus, monarch of the main,
Meantime laments his son in battle slain;
And vows the victor's death, nor vows in vain.
For nine long years the smothered pain he bore;
Achilles was not ripe for fate before;
Then when he saw the promised hour was near,   770
He thus bespoke the god, that guides the year:
'Immortal offspring of my brother Jove,
My brightest nephew, and whom best I love,
Whose hands were joined with mine, to raise the wall
Of tottering Troy, now nodding to her fall;
Dost thou not mourn our power employed in vain,
And the defenders of our city slain?
To pass the rest, could noble Hector lie
Unpitied, dragged around his native Troy?
And yet the murderer lives; himself by far   780
A greater plague, than all the wasteful war.
He lives; the proud Pelides lives, to boast
Our town destroyed, our common labour lost.
O could I meet him! But I wish too late,
To prove my trident is not in his fate.
But let him try (for that's allowed) thy dart,
And pierce his only penetrable part.'

Apollo bows to the superior throne,
And to his uncle's anger adds his own;
Then in a cloud involved, he takes his flight,   790
Where Greeks and Trojans mixed in mortal fight;
And found out Paris, lurking where he stood,
And stained his arrows with plebeian blood.
Phoebus to him alone the god confessed,
Then to the recreant knight he thus addressed:
'Dost thou not blush, to spend thy shafts in vain
On a degenerate and ignoble train?
If fame, or better vengeance, be thy care,
There aim, and with one arrow end the war.'
He said; and showed from far the blazing shield   800
And sword which but Achilles none could wield;
And how he moved a god, and mowed the standing field.
The deity himself directs aright
The envenomed shaft, and wings the fatal flight.
Thus fell the foremost of the Grecian name,
And he, the base adulterer, boasts the fame;
A spectacle to glad the Trojan train,
And please old Priam, after Hector slain.
If by a female hand he had foreseen
He was to die, his wish had rather been   810
The lance and double axe of the fair warrior queen.
And now, the terror of the Trojan field,
The Grecian honour, ornament, and shield,
High on a pile, the unconquered chief is placed;
The god, that armed him first, consumed at last.°
Of all the mighty man, the small remains
A little urn, and scarcely filled, contains,
Yet, great in Homer, still Achilles lives,
And, equal to himself, himself survives.
His buckler owns its former lord, and brings   820
New cause of strife betwixt contending kings;
Who worthiest, after him, his sword to wield,
Or wear his armour, or sustain his shield.
E'en Diomede sat mute, with downcast eyes,
Conscious of wanted worth to win the prize;
Nor Menelaus presumed these arms to claim,
Nor he the king of men, a greater name.
Two rivals only rose; Laertes' son,°

And the vast bulk of Ajax Telamon.
The king, who cherished each with equal love,   830
And from himself all envy would remove,
Left both to be determined by the laws,
And to the Grecian chiefs transferred the cause.

The Speeches of Ajax and Ulysses

From the Thirteenth Book of Ovid's *Metamorphoses*

The chiefs were set, the soldiers crowned the field;
To these the master of the sevenfold shield°
Upstarted fierce; and kindled with disdain,
Eager to speak, unable to contain
His boiling rage, he rolled his eyes around
The shore, and Grecian galleys hauled aground.
Then stretching out his hands, 'O Jove,' he cried,
'Must then our cause before the fleet be tried?
And dares Ulysses for the prize contend
In sight of what he durst not once defend   10
But basely fled, that memorable day,
When I from Hector's hands redeemed the flaming prey?
So much 'tis safer at the noisy bar
With words to flourish, than engage in war.
By different methods we maintain our right,
Nor am I made to talk, nor he to fight.
In bloody fields I labour to be great;
His arms are a smooth tongue and soft deceit.
Nor need I speak my deeds, for those you see;
The sun and day are witnesses for me.   20
Let him who fights unseen relate his own,
And vouch the silent stars, and conscious moon.
Great is the prize demanded, I confess,
But such an abject rival makes it less.
That gift, those honours, he but hoped to gain,
Can leave no room for Ajax to be vain;
Losing he wins, because his name will be
Ennobled by defeat, who durst contend with me.
Were my known valour questioned, yet my blood
Without that plea would make my title good;   30

My sire was Telamon whose arms, employed
With Hercules, these Trojan walls destroyed;
And who before, with Jason, sent from Greece,
In the first ship brought home the golden fleece:
Great Telamon from Aeacus derives
His birth: (the inquisitor of guilty lives
In shades below; where Sisyphus, whose son
This thief is thought, rolls up the restless heavy stone.)°
Just Aeacus the king of gods above
Begot; thus Ajax is the third from Jove.   40
Nor should I seek advantage from my line,
Unless, Achilles, it were mixed with thine:
As next of kin Achilles' arms I claim;
This fellow would ingraft a foreign name
Upon our stock, and the Sisyphian seed
By fraud and theft asserts his father's breed.
Then must I lose these arms, because I came
To fight uncalled, a voluntary name?
Nor shunned the cause, but offered you my aid,
While he long lurking was to war betrayed: 50
Forced to the field he came, but in the rear,
And feigned distraction, to conceal his fear;
Till one more cunning caught him in the snare,°
Ill for himself, and dragged him into war.
Now let a hero's arms a coward vest,
And he who shunned all honours gain the best;
And let me stand excluded from my right,
Robbed of my kinsman's arms who first appeared in fight.
Better for us at home he had remained,
Had it been true the madness which he feigned,   60
Or so believed; the less had been our shame,
The less his counselled crime which brands the Grecian name;
Nor Philocetes had been left enclosed°
In a bare isle to wants and pains exposed;
Where to the rocks, with solitary groans,
His sufferings and our baseness he bemoans,
And wishes (so may heaven his wish fulfil!)
The due reward to him who caused his ill.
Now he with us to Troy's destruction sworn,
Our brother of the war, by whom are borne   70
Alcides' arrows, pent in narrow bounds,

With cold and hunger pinched, and pained with wounds,
To find him food and clothing, must employ
Against the birds the shafts due to the fate of Troy:
Yet still he lives, and lives from treason free,
Because he left Ulysses' company;
Poor Palamede might wish, so void of aid,
Rather to have been left, than so to death betrayed.
The coward bore the man immortal spite,
Who shamed him out of madness into fight; 80
Nor daring otherwise to vent his hate,
Accused him first of treason to the state;
And then for proof produced the golden store
Himself had hidden in his tent before.
Thus of two champions he deprived our host,
By exile one, and one by treason lost.
Thus fights Ulysses, thus his fame extends,
A formidable man but to his friends;
Great, for what greatness is in words and sound;
E'en faithful Nestor less in both is found; 90
But that he might without a rival reign,
He left this faithful Nestor on the plain,
Forsook his friend e'en at his utmost need,
Who, tired, and tardy with his wounded steed,
Cried out for aid, and called him by his name;
But cowardice has neither ears nor shame.
Thus fled the good old man, bereft of aid,
And, for as much as lay in him, betrayed.
That this is not a fable forged by me,
Like one of his, a Ulyssean lie, 100
I vouch e'en Diomede, who though his friend,
Cannot that act excuse, much less defend:
He called him back aloud, and taxed his fear;
And sure enough he heard, but durst not hear.
'The gods with equal eyes on mortals look;
He justly was forsaken, who forsook;
Wanted that succour he refused to lend,
Found every fellow such another friend.
No wonder if he roared that all might hear,
His elocution was increased by fear; 110
I heard, I ran, I found him out of breath,
Pale, trembling, and half-dead with fear of death.

Though he had judged himself by his own laws,
And stood condemned, I helped the common cause:
With my broad buckler hid him from the foe,
(E'en the shield trembled as he lay below,)
And from impending fate the coward freed;
Good heaven forgive me for so bad a deed!
If still he will persist and urge the strife,
First let him give me back his forfeit life;   120
Let him return to that opprobrious field,
Again creep under my protecting shield;
Let him lie wounded, let the foe be near,
And let his quivering heart confess his fear;
There put him in the very jaws of fate,
And let him plead his cause in that estate;
And yet, when snatched from death, when from below
My lifted shield I loosed, and let him go:
Good heavens, how light he rose! with what a bound
He sprung from earth, forgetful of his wound!   130
How fresh, how eager then his feet to ply!
Who had not strength to stand, had speed to fly!
  'Hector came on, and brought the gods along;
Fear seized alike the feeble and the strong;
Each Greek was a Ulysses; such a dread
The approach, and e'en the sound, of Hector bred;
Him, fleshed with slaughter, and with conquest crowned,
I met, and overturned him to the ground.
When after, matchless as he deemed in might,
He challenged all our host to single fight,   140
All eyes were fixed on me; the lots were thrown,
But for your champion I was wished alone.
Your vows were heard; we fought, and neither yield;
Yet I returned unvanquished from the field.
With Jove to friend, the insulting Trojan came.
And menaced us with force, our fleet with flame;
Was it the strength of this tongue-valiant lord,
In that black hour, that saved you from the sword?
Or was my breast exposed alone, to brave
A thousand swords, a thousand ships to save,   150
The hopes of your return? and can you yield,
For a saved fleet, less than a single shield?
Think it no boast, O Grecians, if I deem

These arms want Ajax, more than Ajax them:
Or I with them an equal honour share;
They, honoured to be worn, and I, to wear.
Will he compare my courage with his sleight?
As well he may compare the day with night.
Night is indeed the province of his reign;
Yet all his dark exploits no more contain   160
Than a spy taken, and a sleeper slain,
A priest made prisoner, Pallas made a prey,
But none of all these actions done by day;
Nor aught of these was done, and Diomede away.
If on such petty merits you confer
So vast a prize, let each his portion share;
Make a just dividend; and, if not all,
The greater part to Diomede will fall.
But why for Ithacus such arms as those,
Who naked, and by night, invades his foes?   170
The glittering helm by moonlight will proclaim
The latent robber, and prevent his game;
Nor could he hold his tottering head upright
Beneath that motion, or sustain the weight;
Nor that right arm could toss the beamy lance,
Much less the left that ampler shield advance;
Ponderous with precious weight, and rough with cost
Of the round world in rising gold embossed.
That orb would ill become his hand to wield
And look as for the gold he stole the shield;   180
Which should your error on the wretch bestow,
It would not frighten, but allure the foe.
Why asks he what avails him not in fight,
And would but cumber and retard his flight,
In which his only excellence is placed?
You give him death, that intercept his haste.
Add, that his own is yet a maiden-shield,
Nor the least dint has suffered in the field,
Guiltless of fight; mine, battered, hewed, and bored,
Worn out of service, must forsake his lord.   190
What further need of words our right to scan?
My arguments are deeds, let action speak the man.
Since from a champion's arms the strife arose,
So cast the glorious prize amid the foes;

Then send us to redeem both arms and shield,
And let him wear, who wins them in the field.'
   He said: A murmur from the multitude,
Or somewhat like a stifled shout, ensued;
Till from his seat arose Laertes' son,
Looked down a while, and paused ere he begun;   200
Then to the expecting audience raised his look,
And not without prepared attention spoke;
Soft was his tone, and sober was his face,
Action his words, and words his action grace.
   'If heaven, my lords, had heard our common prayer,
These arms had caused no quarrel for an heir;
Still great Achilles had his own possessed,
And we with great Achilles had been blessed:
But since hard fate, and heaven's severe decree,
Have ravished him away from you and me,'   210
(At this he sighed, and wiped his eyes, and drew,
Or seemed to draw, some drops of kindly dew,)
'Who better can succeed Achilles lost,
Than he who gave Achilles to your host?
This only I request, that neither he°
May gain, by being what he seems to be,
A stupid thing, nor I may lose the prize,
By having sense, which heaven to him denies;
Since, great or small, the talent I enjoyed
Was ever in the common cause employed: 220
Nor let my wit, and wonted eloquence,
Which often has been used in your defence
And in my own, this only time be brought
To bear against myself, and deemed a fault.
Make not a crime, where nature made it none;
For every man may freely use his own.
The deeds of long descended ancestors
Are but by grace of imputation ours,
Theirs in effect; but since he draws his line
From Jove, and seems to plead a right divine,   230
From Jove, like him, I claim my pedigree,
And am descended in the same degree.
My sire, Laertes, was Arcesius' heir,
Arcesius was the son of Jupiter;
No parricide, no banished man, is known

In all my line; let him excuse his own.
Hermes ennobles too my mother's side,
By both my parents to the gods allied.
But not because that on the female part
My blood is better, dare I claim desert, 240
Or that my sire from parricide is free;
But judge by merit betwixt him and me.
The prize be to the best; provided yet
That Ajax for a while his kin forget,
And his great sire, and greater uncle's name,°
To fortify by them his feeble claim.
Be kindred and relation laid aside,
And honour's cause by laws of honour tried.
For, if he plead proximity of blood,
That empty title is with ease withstood. 250
Peleus, the hero's sire, more nigh than he,
And Pyrrhus, his undoubted progeny,
Inherit first these trophies of the field;
To Scyros, or to Phthia, send the shield:
And Teucer has an uncle's right, yet he
Waives his pretensions, nor contends with me.
    'Then, since the cause on pure desert is placed,
Whence shall I take my rise, what reckon last?
I not presume on every act to dwell,
But take these few, in order as they fell. 260
    'Thetis, who knew the fates, applied her care
To keep Achilles in disguise from war;
And till the threatening influence were past,
A woman's habit on the hero cast:
All eyes were cozened by the borrowed vest,
And Ajax (never wiser than the rest)
Found no Pelides there. At length I came
With proffered wares to this pretended dame;
She, not discovered by her mien or voice,
Betrayed her manhood by her manly choice; 270
And, while on female toys her fellows look,
Grasped in her warlike hand, a javelin shook;
Whom by this act revealed, I thus bespoke:
"O goddess born! resist not heaven's decree,
The fall of Ilium is reserved for thee;"
Then seized him, and, produced in open light,

Sent blushing to the field the fatal knight.
Mine then are all his actions of the war;
Great Telephus was conquered by my spear,
And after cured; to me the Thebans owe   280
Lesbos and Tenedos, their overthrow;
Scyros and Cylla; Not on all to dwell,
By me Lyrnessus and strong Chrysa fell;
And, since I sent the man who Hector slew,
To me the noble Hector's death is due.
Those arms I put into his living hand;
Those arms, Pelides dead, I now demand.
    'When Greece was injured in the Spartan prince,
And met at Aulis to revenge the offence,
'Twas a dead calm, or adverse blasts, that reigned,   290
And in the port the windbound fleet detained:
Bad signs were seen, and oracles severe
Were daily thundered in our general's ear,°
That by his daughter's blood we must appease°
Diana's kindled wrath, and free the seas.
Affection, interest, fame, his heart assailed,
But soon the father o'er the king prevailed;
Bold, on himself he took the pious crime,
As angry with the gods as they with him.
No subject could sustain their sovereign's look,   300
Till this hard enterprise I undertook;
I only durst the imperial power control,
And undermined the parent in his soul;
Forced him to exert the king for common good,
And pay our ransom with his daughter's blood.
Never was cause more difficult to plead,
Than where the judge against himself decreed;
Yet this I won by dint of argument.
The wrongs his injured brother underwent;
And his own office, shamed him to consent.   310
    "'Twas harder yet to move the mother's mind,°
And to this heavy task was I designed:
Reasons against her love I knew were vain;
I circumvented whom I could not gain.
Had Ajax been employed, our slackened sails
Had still at Aulis waited happy gales.
    'Arrived at Troy, your choice was fixed on me,

A fearless envoy, fit for a bold embassy.
Secure, I entered through the hostile court,
Glittering with steel, and crowded with resort: 320
There in the midst of arms I plead our cause,
Urge the foul rape, and violated laws;
Accuse the foes as authors of the strife,
Reproach the ravisher, demand the wife.
Priam, Antenor, and the wiser few
I moved; but Paris and his lawless crew
Scarce held their hands, and lifted swords; but stood
In act to quench their impious thirst of blood.
This Menelaus knows; exposed to share
With me the rough preludium of the war.   330
   'Endless it were to tell what I have done,
In arms, or counsel, since the siege begun.
The first encounters past, the foe repelled,
They skulked within the town, we kept the field.
War seemed asleep for nine long years; at length,
Both sides resolved to push, we tried our strength.
Now what did Ajax while our arms took breath,
Versed only in the gross mechanic trade of death?
If you require my deeds, with ambushed arms
I trapped the foe, or tired with false alarms;   340
Secured the ships, drew lines along the plain,°
The fainting cheered, chastised the rebel-train,
Provided forage, our spent arms renewed;
Employed at home, or sent abroad, the common cause pursued.
   'The king, deluded in a dream by Jove,
Despaired to take the town, and ordered to remove.
What subject durst arraign the power supreme,
Producing Jove to justify his dream?
Ajax might wish the soldiers to retain
From shameful flight, but wishes were in vain;   350
As wanting of effect had been his words,
Such as of course his thundering tongue affords.
But did this boaster threaten, did he pray,
Or by his own example urge their stay?
None, none of these, but ran himself away.
I saw him run, and was ashamed to see;
Who plied his feet so fast to get aboard as he?
Then speeding through the place, I made a stand,

And loudly cried, "O base degenerate band,
To leave a town already in your hand!   360
After so long expense of blood for fame,
To bring home nothing but perpetual shame!"
These words, or what I have forgotten since,
For grief inspired me then with eloquence,
Reduced their minds; they leave the crowded port,
And to their late forsaken camp resort.
Dismayed the council met; this man was there,
But mute, and not recovered of his fear.
Thersites taxed the king, and loudly railed,
But his wide opening mouth with blows I sealed.   370
Then, rising, I excite their souls to fame
And kindle sleeping virtue into flame.
From thence, whatever he performed in fight
Is justly mine, who drew him back from flight.
    'Which of the Grecian chiefs consorts with thee?
But Diomede desires my company,
And still communicates his praise with me.
As guided by a god, secure he goes,
Armed with my fellowship, amid the foes;
And sure no little merit I may boast,   380
Whom such a man selects from such an host.
Unforced by lots, I went without affright,
To dare with him the dangers of the night;
On the same errand sent, we met the spy°
Of Hector, double-tongued, and used to lie;
Him I dispatched, but not till, undermined,
I drew him first to tell what treacherous Troy designed.
My task performed, with praise I had retired,
But not content with this, to greater praise aspired.
Invaded Rhesus, and his Thracian crew,   390
And him, and his, in their own strength I slew:
Returned a victor, all my vows complete,
With the king's chariot, in his royal seat.
Refuse me now his arms, whose fiery steeds
Were promised to the spy for his nocturnal deeds;
And let dull Ajax bear away my right,
When all his days outbalance this one night.
    'Nor fought I darkling still; the sun beheld
With slaughtered Lycians when I strewed the field:

You saw, and counted as I passed along, 400
Alaster, Cromius, Ceranos the strong,
Alcander, Prytanis, and Halius,
Noemon, Charopes, and Ennomus,
Choon, Chersidamas, and five beside,
Men of obscure descent, but courage tried;
All these this hand laid breathless on the ground.
Nor want I proofs of many a manly wound;
All honest, all before; believe not me,
Words may deceive, but credit what you see.'

    At this he bared his breast and showed his scars, 410
As of a furrowed field, well ploughed with wars;
'Nor is this part unexercised,' said he;
'That giant bulk of his from wounds is free;
Safe in his shield he fears no foe to try,
And better manages his blood than I.
But this avails me not; our boaster strove
Not with our foes alone, but partial Jove,
To save the fleet. This I confess is true,
Nor will I take from any man his due;
But, thus assuming all, he robs from you. 420
Some part of honour to your share will fall;
He did the best indeed, but did not all.
Patroclus in Achilles' arms, and thought
The chief he seemed, with equal ardour fought;
Preserved the fleet, repelled the raging fire,
And forced the fearful Trojans to retire.

    'But Ajax boasts, that he was only thought
A match for Hector, who the combat sought:
Sure he forgets the king, the chiefs, and me,
All were as eager for the fight as he; 430
He but the ninth, and not by public voice,
Or ours preferred, was only fortune's choice:
They fought; nor can our hero boast the event,
For Hector from the field unwounded went.

    'Why am I forced to name that fatal day,
That snatched the prop and pride of Greece away?
I saw Pelides sink, with pious grief,
And ran in vain, alas! to his relief,
For the brave soul was fled; full of my friend,
I rushed amid the war, his relics to defend; 440

Nor ceased my toil till I redeemed the prey,
And loaded with Achilles marched away.
Those arms, which on these shoulders then I bore,
'Tis just you to these shoulders should restore.
You see I want not nerves, who could sustain
The ponderous ruins of so great a man;
Or if in others equal force you find,
None is endued with a more grateful mind.

   'Did Thetis then, ambitious in her care,
These arms thus laboured for her son prepare,   450
That Ajax after him the heavenly gift should wear?
For that dull soul to stare with stupid eyes
On the learned unintelligible prize?
What are to him the sculptures of the shield,
Heaven's planets, earth, and ocean's watery field?
The Pleiads, Hyads; Less and Greater Bear,
Undipped in seas; Orion's angry star;°
Two differing cities, graved on either hand?
Would he wear arms he cannot understand?

   'Beside, what wise objections he prepares   460
Against my late accession to the wars!
Does not the fool perceive his argument
Is with more force against Achilles bent?
For if dissembling be so great a crime,
The fault is common, and the same in him;
And if he taxes both of long delay,
My guilt is less, who sooner came away.
His pious mother, anxious for his life,°
Detained her son; and me, my pious wife.°
To them the blossoms of our youth were due;   470
Our riper manhood we reserved for you.
But grant me guilty, 'tis not much my care,
When with so great a man my guilt I share;
My wit to war the matchless hero brought,
But by this fool I never had been caught.

   'Nor need I wonder that on me he threw
Such foul aspersions, when he spares not you:
If Palamede unjustly fell by me,
Your honour suffered in the unjust decree.
I but accused, you doomed; and yet he died,   480
Convinced of treason, and was fairly tried.

You heard not he was false; your eyes beheld
The traitor manifest, the bribe revealed.
    'That Philoctetes is on Lemnos left
Wounded, forlorn, of human aid bereft,
Is not my crime, or not my crime alone;
Defend your justice, for the fact's your own.
'Tis true, the advice was mine; that, staying there,
He might his weary limbs with rest repair,
From a long voyage free, and from a longer war.   490
He took the counsel, and he lives at least;
The event declares I counselled for the best;
Though faith is all in ministers of state,
For who can promise to be fortunate?
Now since his arrows are the fate of Troy,
Do not my wit, or weak address, employ;
Send Ajax there, with his persuasive sense,
To mollify the man, and draw him thence:
But Xanthus shall run backward; Ida stand
A leafless mountain; and the Grecian band   500
Shall fight for Troy; if when my counsels fail,
The wit of heavy Ajax can prevail.
    'Hard Philoctetes, exercise thy spleen
Against thy fellows, and the king of men;
Curse my devoted head, above the rest,
And wish in arms to meet me, breast to breast;
Yet I the dangerous task will undertake,
And either die myself, or bring thee back.
    'Nor doubt the same success, as when, before,
The Phrygian prophet to these tents I bore,°   510
Surprised by night, and forced him to declare
In what was placed the fortune of the war;
Heaven's dark decrees and answers to display,
And how to take the town, and where the secret lay.
Yet this I compassed, and from Troy conveyed
The fatal image of their guardian maid.°
That work was mine; for Pallas, though our friend,
Yet while she was in Troy, did Troy defend.
Now what has Ajax done, or what designed?
A noisy nothing, and an empty wind.   520
If he be what he promises in show,
Why was I sent, and why feared he to go?

Our boasting champion thought the task not light
To pass the guards, commit himself to night;
Not only through a hostile town to pass,
But scale, with steep ascent, the sacred place;
With wandering steps to search the citadel,
And from the priests their patroness to steal;
Then through surrounding foes to force my way,
And bear in triumph home the heavenly prey;   530
Which had I not, Ajax in vain had held
Before that monstrous bulk his seven-fold shield,
That night to conquer Troy I might be said,
When Troy was liable to conquest made.
　　Why point'st thou to my partner of the war?
Tydides had indeed a worthy share°
In all my toil and praise; but when thy might
Our ships protected, didst thou singly fight?
All joined, and thou of many wert but one;
I asked no friend, nor had, but him alone;   540
Who, had he not been well assured, that art
And conduct were of war the better part,
And more availed than strength, my valiant friend
Had urged a better right, than Ajax can pretend;
As good at least Eurypylus may claim,
And the more moderate Ajax of the name;°
The Cretan king, and his brave charioteer,°
And Menelaus, bold with sword and spear:
All these had been my rivals in the shield,
And yet all these to my pretensions yield.   550
Thy boisterous hands are then of use, when I
With this directing head those hands apply.
Brawn without brain is thine; my prudent care
Foresees, provides, administers the war:
Thy province is to fight; but when shall be
The time to fight, the king consults with me.
No dram of judgment with thy force is joined,
Thy body is of profit, and my mind.
But how much more the ship her safety owes
To him who steers, than him that only rows;   560
By how much more the captain merits praise
Than he who fights, and fighting, but obeys;
By so much greater is my worth than thine,

Who canst but execute what I design.
What gain'st thou, brutal man, if I confess
Thy strength superior, when thy wit is less?
Mind is the man; I claim my whole desert
From the mind's vigour, and the immortal part.

   'But you, O Grecian chiefs, reward my care,
Be grateful to your watchman of the war;   570
For all my labours in so long a space,
Sure I may plead a title to your grace.
Enter the town; I then unbarred the gates,
When I removed their tutelary fates.
By all our common hopes, if hopes they be,
Which I have now reduced to certainty;
By falling Troy, by yonder tottering towers,
And by their taken gods, which now are ours;
Or if there yet a further task remains
To be performed by prudence or by pains;   580
If yet some desperate action rests behind,
That asks high conduct, and a dauntless mind:
If aught be wanting to the Trojan doom,
Which none but I can manage and o'ercome;
Award those arms I ask by your decree;
Or give to this what you refuse to me.'

   He ceased, and, ceasing, with respect he bowed,
And with his hand at once the fatal statue showed.
Heaven, air, and ocean rung, with loud applause,
And by the general vote he gained his cause.   590
Thus conduct won the prize, when courage failed,
And eloquence o'er brutal force prevailed.

THE DEATH OF AJAX

He who could often, and alone, withstand
The foe, the fire, and Jove's own partial hand,
Now cannot his unmastered grief sustain,
But yields to rage, to madness, and disdain;
Then snatching out his falchion, 'Thou,' said he,
'Art mine; Ulysses lays no claim to thee.
O often tried, and ever trusty sword,
Now do thy last kind office to thy lord!   600
'Tis Ajax who requests thy aid to show

None but himself, himself could overthrow.'
He said, and with so good a will to die,
Did to his breast the fatal point apply,
It found his heart, a way till then unknown,
Where never weapon entered but his own.
No hands could force it thence, so fixed it stood,
Till out it rushed, expelled by streams of spouting blood.
The fruitful blood produced a flower, which grew°
On a green stem, and of a purple hue;   610
Like his, whom unaware Apollo slew.°
Inscribed in both, the letters are the same,°
But those express the grief, and these the name.

The Wife of Bath's Tale

In days of old, when Arthur filled the throne,
Whose acts and fame to foreign lands were blown,
The king of elves, and little fairy queen,
Gambolled on heaths, and danced on every green;
And where the jolly troop had led the round,
The grass unbidden rose, and marked the ground.
Nor darkling did they dance; the silver light
Of Phoebe served to guide their steps aright,
And with their tripping pleased, prolonged the night.
Her beams they followed, where at full she played,   10
Nor longer than she shed her horns they stayed,
From thence with airy flight to foreign lands conveyed.
Above the rest our Britain held they dear;
More solemnly they kept their sabbaths here,
And made more spacious rings, and revelled half the year.
    I speak of ancient times; for now the swain,
Returning late, may pass the woods in vain,
And never hope to see the nightly train;
In vain the dairy now with mints is dressed,
The dairy-maid expects no fairy guest   20
To skim the bowls and after pay the feast.
She sighs and shakes her empty shoes in vain,
No silver penny to reward her pain;
For priests with prayers and other godly gear,
Have made the merry goblins disappear;

And where they played their merry pranks before,
Have sprinkled holy water on the floor;
And friars that through the wealthy regions run,
Thick as the motes that twinkle in the sun,
Resort to farmers rich, and bless their halls   30
And exorcise the beds and cross the walls:
This makes the fairy choirs forsake the place,
When once 'tis hallowed with the rites of grace.
But in the walks where wicked elves have been,
The learning of the parish now is seen;
The midnight parson posting o'er the green,
With gown tucked up to wakes; for Sunday next,
With humming ale encouraging his text;°
Nor wants the holy leer to country-girl betwixt.
From fiends and imps he sets the village free,   40
There haunts not any incubus but he.
The maids and women need no danger fear
To walk by night, and sanctity so near;
For by some haycock or some shady thorn,
He bids his beads both evensong and morn.
    It so befell in this King Arthur's reign,
A lusty knight was pricking o'er the plain;
A bachelor he was, and of the courtly train.
It happened as he rode, a damsel gay,
In russet robes, to market took her way;   50
Soon on the girl he cast an amorous eye,
So straight she walked, and on her pasterns high:
If seeing her behind he liked her pace,
Now turning short, he better liked her face.
He lights in haste and full of youthful fire,
By force accomplished his obscene desire.
This done, away he rode, not unespied,
For swarming at his back the country cried;
And once in view, they never lost the sight,
But seized and pinioned brought to court the knight.   60
    Then courts of kings were held in high renown,
Ere made the common brothels of the town;
There virgins honourable vows received,
But chaste as maids in monasteries lived;
The king himself, to nuptial ties a slave,
No bad example to his poets gave,

And they, not bad but in a vicious age,
Had not, to please the prince, debauched the stage.
  Now what should Arthur do? He loved the knight,
But sovereign monarchs are the source of right: 70
Moved by the damsel's tears and common cry,
He doomed the brutal ravisher to die.
But fair Genevra rose in his defence°
And prayed so hard for mercy from the prince,
That to his queen the king the offender gave,
And left it in her power to kill or save.
This gracious act the ladies all approve,
Who thought it much a man should die for love;
And with their mistress joined in close debate,
(Covering their kindness with dissembled hate,)   80
If not to free him, to prolong his fate.
At last agreed, they called him by consent
Before the queen and female parliament;
And the fair speaker, rising from her chair,°
Did thus the judgment of the house declare:
  'Sir knight, though I have asked thy life, yet still
Thy destiny depends upon my will:
Nor hast thou other surety, than the grace,
Not due to thee from our offended race.
But as our kind is of a softer mould   90
And cannot blood, without a sigh, behold,
I grant thee life; reserving still the power
To take the forfeit when I see my hour;
Unless thy answer to my next demand
Shall set thee free from our avenging hand.
The question whose solution I require.
Is, what the sex of women most desire?
In this dispute thy judges are at strife;
Beware, for on thy wit depends thy life.
Yet (lest surprised, unknowing what to say,   100
Thou damn thyself) we give thee further day;
A year is thine to wander at thy will,
And learn from others, if thou want'st the skill;
But, our proffer not to hold in scorn,°
Good sureties will we have for thy return,
That at the time prefixed thou shalt obey,
And at thy pledge's peril keep thy day.'

Woe was the knight at this severe command.
But well he knew 'twas bootless to withstand.
The terms accepted, as the fair ordain,   110
He put in bail for his return again;
And promised answer at the day assigned,
The best, with heaven's assistance, he could find.
   His leave thus taken, on his way he went
With heavy heart, and full of discontent,
Misdoubting much, and fearful of the event.
'Twas hard the truth of such a point to find,
As was not yet agreed among the kind.
Thus on he went; still anxious more and more,
Asked all he met, and knocked at every door;   120
Inquired of men; but made his chief request
To learn from women what they loved the best.
They answered each according to her mind,
To please herself, not all the female kind.
One was for wealth, another was for place;
Crones, old and ugly, wished a better face.
The widow's wish was oftentimes to wed;
The wanton maids were all for sport abed.
Some said the sex were pleased with handsome lies,
And some gross flattery loved without disguise.   130
Truth is, says one, he seldom fails to win
Who flatters well; for that's our darling sin.
But long attendance, and a duteous mind,
Will work e'en with the wisest of the kind.
One thought the sex's prime felicity
Was from the bonds of wedlock to be free;
Their pleasures, hours, and actions, all their own,
And uncontrolled, to give account to none.
Some wish a husband-fool; but such are cursed,
For fools perverse of husbands are the worst,   140
All women would be counted chaste and wise,
Nor should our spouses see but with our eyes;
For fools will prate; and though they want the wit
To find close faults, yet open blots will hit;
Though better for their ease to hold their tongue,
For womankind was never in the wrong.
So noise ensues, and quarrels last for life;
The wife abhors the fool, the fool the wife.

And some men say, that great delight have we
To be for truth extolled, and secrecy;   150
And constant in one purpose still to dwell,
And not our husbands' counsels to reveal.
But that's a fable; for our sex is frail,
Inventing rather than not tell a tale.
Like leaky sieves no secrets we can hold;
Witness the famous tale that Ovid told.°
   Midas the king, as in his book appears,
By Phoebus was endowed with ass's ears,
Which under his long locks he well concealed;
(As monarchs' vices must not be revealed.)   160
For fear the people have them in the wind,
Who, long ago, were neither dumb nor blind;
Nor apt to think from heaven their title springs,
Since Jove and Mars left off begetting kings.
This Midas knew; and durst communicate
To none but to his wife his ears of state;
One must be trusted, and he thought her fit,
As passing prudent, and a parlous wit.
To this sagacious confessor he went,
And told her what a gift the gods had sent;   170
But told it under matrimonial seal,
With strict injunction never to reveal.
The secret heard, she plighted him her troth,
(And sacred sure is every woman's oath,)
The royal malady should rest unknown,
Both for her husband's honour and her own:
But ne'ertheless she pined with discontent,
The counsel rumbled till it found a vent.
The thing she knew she was obliged to hide;
By interest and by oath the wife was tied,   180
But if she told it not, the woman died.
Loath to betray a husband and a prince,
But she must burst or blab, and no pretence
Of honour tied her tongue from self-defence.
A marshy ground commodiously was near,
Thither she ran, and held her breath for fear,
Lest if a word she spoke of anything,
That word might be the secret of the king.
Thus full of counsel to the fen she went,

Griped all the way, and longing for a vent;   190
Arrived, by pure necessity compelled,
On her majestic marrow-bones she kneeled:°
Then to the water's brink she laid her head,
And as a bittern bumps within a reed,°
'To thee alone, O lake!' she said, 'I tell,
(And, as thy queen, command thee to conceal:)
Beneath his locks, the king my husband wears
A goodly royal pair of ass's ears:
Now I have eased my bosom of the pain,
Till the next longing fit return again.'   200
  Thus through a woman was the secret known;
Tell us, and, in effect, you tell the town.
But to my tale: the knight, with heavy cheer,
Wandering in vain, had now consumed the year;
One day was only left to solve the doubt,
Yet knew no more than when he first set out.
But home he must; and as the award had been,
Yield up his body captive to the queen.
In this despairing state he happed to ride,
As fortune led him, by a forest side;   210
Lonely the vale, and full of horror stood,
Brown with the shade of a religious wood;
When full before him, at the noon of night,
(The moon was up, and shot a gleamy light,)
He saw a choir of ladies in a round,
That featly footing seemed to skim the ground.
Thus dancing hand in hand, so light they were,
He knew not where they trod, on earth or air.
At speed he drove, and came a sudden guest;
In hope, where many women were, at least   220
Some one, by chance, might answer his request.
But faster than his horse the ladies flew,
And in a trice were vanished out of view.
  One only hag remained; but fouler far
Than grandam apes in Indian forests are;
Against a withered oak she leaned her weight.
Propped on her trusty staff, not half upright,
And dropped an awkward curtsy to the knight.
Then said, 'What make you, sir, so late abroad
Without a guide, and this no beaten road?   230

Or want you aught that here you hope to find,
Or travel for some trouble in your mind?
The last I guess; and if I read aright,
Those of our sex are bound to serve a knight.
Perhaps good counsel may your grief assuage,
Then tell your pain, for wisdom is in age.'
  To this the knight: 'Good mother, would you know
The secret cause and spring of all my woe?
My life must with to-morrow's light expire,
Unless I tell what women most desire.   240
Now could you help me at this hard essay,
Or for your inborn goodness, or for pay,
Yours is my life, redeemed by your advice,
Ask what you please, and I will pay the price:
The proudest kerchief of the court shall rest
Well satisfied of what they love the best.'
'Plight me thy faith,' quoth she, 'that what I ask,
Thy danger over, and performed the task,
That shalt thou give for hire of thy demand,
(Here take thy oath, and seal it on my hand,)   250
I warrant thee, on peril of my life,
Thy words shall please both widow, maid, and wife.'
  More words there needed not to move the knight,
To take her offer, and his truth to plight.
With that she spread her mantle on the ground,
And first inquiring whither he was bound,
Bade him not fear, though long and rough the way,
At court he should arrive ere break of day:
His horse should find the way without a guide,
She said: with fury they began to ride,   260
He on the midst, the beldam at his side.
The horse, what devil drove I cannot tell,
But only this, they sped their journey well;
And all the way the crone informed the knight,
How he should answer the demand aright.
  To court they came; the news was quickly spread
Of his returning to redeem his head.
The female senate was assembled soon,
With all the mob of women in the town:
The queen sat lord chief justice of the hall,   270
And bade the crier cite the criminal.

The knight appeared, and silence they proclaim:
Then first the culprit answered to his name;
And after forms of law, was last required
To name the thing that women most desired.
   The offender, taught his lesson by the way,
And by his counsel ordered what to say,
Thus bold began: 'My lady liege,' said he,
'What all your sex desire is *sovereignty*.
The wife affects her husband to command;   280
All must be hers, both money, house, and land:
The maids are mistresses e'en in their name,
And of their servants full dominion claim.
This, at the peril of my head, I say,
A blunt plain truth, the sex aspires to sway,
You to rule all, while we, like slaves, obey.'
   There was not one, or widow, maid, or wife,
But said the knight had well deserved his life.
Even fair Genevra, with a blush, confessed,
The man had found what women love the best.   290
   Up starts the beldam, who was there unseen.
And, reverence made, accosted thus the queen:
'My liege,' said she, 'before the court arise,
May I, poor wretch, find favour in your eyes,
To grant my just request: 'twas I who taught
The knight this answer, and inspired his thought.
None but a woman could a man direct
To tell us women what we most affect.
But first I swore him on his knightly troth,
(And here demand performance of his oath,)   300
To grant the boon that next I should desire;
He gave his faith, and I expect my hire.
My promise is fulfilled: I saved his life,
And claim his debt, to take me for his wife.'
The knight was asked, nor could his oath deny,
But hoped they would not force him to comply.
The women, who would rather wrest the laws,
Than let a sister-plaintiff lose the cause,
(As judges on the bench more gracious are,
And more attent to brothers of the bar,)   310
Cried one and all, the suppliant should have right,
And to the grandam hag adjudged the knight.

In vain he sighed, and oft with tears desired
Some reasonable suit might be required.
But still the crone was constant to her note;
The more he spoke, the more she stretched her throat.
In vain he proffered all his goods, to save
His body, destined to that living grave.
The liquorish hag rejects the pelf with scorn,
And nothing but the man would serve her turn.   320
'Not all the wealth of Eastern kings,' said she,
'Have power to part my plighted love, and me:
And old and ugly as I am, and poor,
Yet never will I break the faith I swore;
For mine thou art by promise, during life,
And I thy loving and obedient wife.'
    'My love! nay rather my damnation thou,'
Said he: 'nor am I bound to keep my vow;
The fiend thy sire has sent thee from below,
Else how couldst thou my secret sorrows know?   330
Avaunt, old witch, for I renounce thy bed:
The queen may take the forfeit of my head,
Ere any of my race so foul a crone shall wed.'
    Both heard, the judge pronounced against the knight;
So was he married in his own despite:
And all day after hid him as an owl,
Not able to sustain a sight so foul.
Perhaps the reader thinks I do him wrong,
To pass the marriage-feast, and nuptial song:
Mirth there was none, the man was *à la mort*,   340
And little courage had to make his court.
To bed they went, the bridegroom and the bride:
Was never such an ill-paired couple tied.
Restless he tossed, and tumbled to and fro,
And rolled, and wriggled further off, for woe.
The good old wife lay smiling by his side,
And caught him in her quivering arms, and cried,
'When you my ravished predecessor saw,
You were not then become this man of straw;
Had you been such, you might have scaped the law.   350
Is this the custom of King Arthur's court?
Are all round-table knights of such a sort?
Remember I am she who saved your life,

Your loving, lawful, and complying wife:
Not thus you swore in your unhappy hour,
Nor I for this return employed my power.
In time of need I was your faithful friend;
Nor did I since, nor ever will offend.
Believe me, my loved lord, 'tis much unkind;
What fury has possessed your altered mind?   360
Thus on my wedding-night—without pretence—
Come turn this way, or tell me my offence.
If not your wife, let reason's rule persuade:
Name but my fault, amends shall soon be made.'
    'Amends! nay that's impossible,' said he,
'What change of age or ugliness can be?
Or could Medea's magic mend thy face,
Thou art descended from so mean a race,
That never knight was matched with such disgrace.
What wonder, madam, if I move my side,   370
When if I turn I turn to such a bridge?'
    'And is this all that troubles you so sore?'
'And what the devil couldst thou wish me more?'
'Ah Benedicite!' replied the crone:
'Then cause of just complaining have you none.
The remedy to this were soon applied,
Would you be like the bridegroom to the bride:
But, for you say a long-descended race,
And wealth and dignity, and power, and place,
Make gentlemen, and that your high degree   380
Is much disparaged to be matched with me,
Know this, my lord, nobility of blood
Is but a glittering and fallacious good:
The nobleman is he, whose noble mind
Is filled with inborn worth, unborrowed from his kind.
The king of heaven was in a manger laid,
And took his earth but from a humble maid:°
Then what can birth, or mortal men, bestow,
Since floods no higher than their fountains flow?
We, who for name and empty honour strive,   390
Our true nobility from his derive.
Your ancestors, who puff your mind with pride,
And vast estates to mighty titles tied,
Did not your honour, but their own, advance;

For virtue comes not by inheritance.
If you tralineate from your father's mind,°
What are you else but of a bastard kind?
Do, as your great progenitors have done,
And by their virtues prove yourself their son.
No father can infuse, or wit, or grace;   400
A mother comes across, and mars the race.
A grandsire or a grandam taints the blood;
And seldom three descents continue good.
Were virtue by descent, a noble name
Could never villanise his father's fame;
But, as the first, the last of all the line,
Would, like the sun, e'en in descending shine.
Take fire, and bear it to the darkest house,
Betwixt King Arthur's court and Caucasus,°
If you depart, the flame shall still remain,   410
And the bright blaze enlighten all the plain;
Nor, till the fuel perish, can decay,
By nature formed on things combustible to prey.
Such is not man, who mixing better seed
With worse, begets a base degenerate breed.
The bad corrupts the good, and leaves behind
No trace of all the great begetter's mind.
The father sinks within his son, we see,
And often rises in the third degree;
If better luck a better mother give,   420
Chance gave us being, and by chance we live.
Such as our atoms were, e'en such are we,
Or call it chance, or strong necessity:
Thus loaded with dead weight, the will is free.
And thus it needs must be; for seed conjoined
Lets into nature's work the imperfect kind;
But fire, the enlivener of the general frame,
Is one, its operation still the same.
Its principle is in itself: while ours
Works, as confederates war, with mingled powers;   430
Or man or woman, whichsoever fails;
And oft the vigour of the worse prevails.
Ether, with sulphur blended, alters hue,
And casts a dusky gleam of Sodom blue.°
Thus in a brute their ancient honour ends,

And the fair mermaid in a fish descends:
The line is gone; no longer duke or earl;
But by himself degraded, turns a churl.
Nobility of blood is but renown
Of thy great fathers, by their virtue known,   440
And a long trail of light, to thee descending down.
If in thy smoke it ends, their glories shine;
But infamy and villeinage are thine.°
Then what I said before is plainly showed,
That true nobility proceeds from God:
Not left us by inheritance, but given
By bounty of our stars, and grace of heaven.
Thus from a captive Servius Tullus rose,°
Whom for his virtues the first Romans chose.
Fabricius from their walls repelled the foe,°   450
Whose noble hands had exercised the plough.
From hence, my lord, and love, I thus conclude,
That, though my homely ancestors were rude,
Mean as I am, yet I may have the grace
To make you father of a generous race.
And noble then am I, when I begin,
In virtue clothed, to cast the rags of sin.
If poverty be my upbraided crime,
And you believe in heaven, there was a time
When he, the great controller of our fate,   460
Deigned to be man, and lived in low estate;
Which he who had the world at his dispose,
If poverty were vice, would never choose.
Philosophers have said, and poets sing
That a glad poverty's an honest thing;
Content is wealth, the riches of the mind,
And happy he who can that treasure find;
But the base miser starves amidst his store,
Broods on his gold, and griping still at more,
Sits sadly pining, and believes he's poor;   470
The ragged beggar, though he wants relief,
Has naught to lose, and sings before the thief.°
Want is a bitter and a hateful good,
Because its virtues are not understood.
Yet many things, impossible to thought,
Have been, by need, to full perfection brought:

The daring of the soul proceeds from thence,
Sharpness of wit, and active diligence;
Prudence at once, and fortitude, it gives,
And if in patience taken, mends our lives;   480
For e'en that indigence, that brings me low
Makes me myself, and him above, to know;
A good which none would challenge, few would choose,
A fair possession, which mankind refuse.
    'If we from wealth to poverty descend,
Want gives to know the flatterer from the friend.
If I am old and ugly, well for you,
No lewd adulterer will my love pursue;
Nor jealousy, the bane of married life,
Shall haunt you for a withered homely wife;   490
For age and ugliness, as all agree,
Are the best guards of female chastity.
    'Yet since I see your mind is worldly bent,
I'll do my best to further your content;
And therefore of two gifts in my dispose,—
Think ere you speak,—I grant you leave to choose:
Would you I should be still deformed and old,
Nauseous to touch, and loathsome to behold;
On this condition, to remain for life
A careful, tender, and obedient wife,   500
In all I can contribute to your ease,
And not in deed, or word, or thought displease?
Or would you rather have me young and fair,
And take the chance that happens to your share?
Temptations are in beauty, and in youth,
And how can you depend upon my truth?
Now weigh the danger with the doubtful bliss,
And thank yourself, if aught should fall amiss.'
    Sore sighed the knight, who this long sermon heard;
At length considering all, his heart he cheered,   510
And thus replied: 'My lady, and my wife,
To your wise conduct I resign my life:
Choose you for me, for well you understand
The future good and ill, on either hand:
But if a humble husband may request,
Provide, and order all things for the best;
Yours be the care to profit, and to please,
And let your subject-servant take his ease.'

'Then thus in peace,' quoth she, 'concludes the strife,
Since I am turned the husband, you the wife: 520
The matrimonial victory is mine,
Which, having fairly gained, I will resign;
Forgive, if I have said or done amiss,
And seal the bargain with a friendly kiss.
I promised you but one content to share,
But now I will become both good and fair.
No nuptial quarrel shall disturb your ease;
The business of my life shall be to please:
And for my beauty, that, as time shall try,
But draw the curtain first, and cast your eye.'   530
    He looked, and saw a creature heavenly fair,
In bloom of youth and of a charming air.
With joy he turned, and seized her ivory arm,
And, like Pygmalion, found the statue warm.°
Small arguments there needed to prevail,
A storm of kisses poured as thick as hail.
    Thus long in mutual bliss they lay embraced,
And their first love continued to the last;
One sunshine was their life, no cloud between,
Nor ever was a kinder couple seen.   540
    And so may all our lives like theirs be led;
Heaven send the maids young husbands fresh in bed:
May widows wed as often as they can,
And ever for the better change their man.
And some devouring plague pursue their lives,
Who will not well be governed by their wives.

Of the Pythagorean Philosophy

From Ovid's *Metamorphoses*, Book Fifteen

The fourteenth book [of Ovid's *Metamorphoses*] *concludes with the death
and deification of Romulus; the fifteenth begins with the election of Numa
to the crown of Rome. On this occasion, Ovid, following the opinion of
some authors, makes Numa the scholar of Pythagoras, and to have begun
his acquaintance with the philosopher at Crotona, a town in Italy; from
thence he makes a digression to the moral and natural philosophy of Pythagoras: on both which our author enlarges; and which are the most learned
and beautiful parts of the whole* Metamorphoses.

A king is sought to guide the growing state,
One able to support the public weight,
And fill the throne where Romulus had sate.
Renown, which oft bespeaks the public voice,
Had recommended Numa to their choice:
A peaceful, pious prince; who not content
To know the Sabine rites his study bent
To cultivate his mind; to learn the laws
Of nature, and explore their hidden cause.
Urged by this care his country he forsook,　　10
And to Crotona thence, his journey took.
Arrived, he first enquired the founder's name
Of this new colony, and whence he came.
Then thus a senior of the place replies,
Well read, and curious of antiquities:
"Tis said Alcides hither took his way
From Spain, and drove along his conquered prey;
Then leaving in the fields his grazing cows,
He sought himself some hospitable house:
Good Croton entertained his godlike guest;　　20
While he repaired his weary limbs with rest.
The hero, thence departing, blessed the place;
"And here," he said, "in time's revolving race
A rising town shall take his name from thee."
Revolving time fulfilled the prophecy:
For Myscelos, the justest man of earth,
Alemon's son, at Argos had his birth.
Him Hercules, armed with his club of oak
O'ershadowed in a dream and thus bespoke,
"Go, leave thy native soil, and make abode　　30
Where Aesaris rolls down his rapid flood;"
He said; and sleep forsook him, and the god.
Trembling he waked, and rose with anxious heart;
His country laws forbade him to depart;
What should he do? 'Twas death to go away,
And the god menaced if he dared to stay:
All day he doubted, and when night came on,
Sleep, and the same forewarning dream began:
Once more the god stood threatening o'er his head,
With added curses if he disobeyed.　　40
Twice warned, he studied flight; but would convey

At once his person, and his wealth away;
Thus while he lingered, his design was heard;
A speedy process formed, and death declared.
Witness there needed none of his offence,
Against himself the wretch was evidence;
Condemned, and destitute of human aid,
To him, for whom he suffered, thus he prayed.
'"O power who has deserved in heaven a throne
Not given, but by the labours made thy own,   50
Pity thy suppliant, and protect his cause,
Whom thou hast made obnoxious to the laws."
'A custom was of old, and still remains,
Which life or death by suffrages ordains;
White stones and black within an urn are cast,
The first absolve, but fate is in the last.
The judges to the common urn bequeath
Their votes, and drop the sable signs of death;
The box receives all black, but poured from thence
The stones came candid forth: the hue of innocence.   60
Thus Aleminodes his safety won,°
Preserved from death by Alcumena's son.°
Then to his kinsman god his vows he pays,
And cuts with prosperous gale the Ionian seas;
He leaves Tarentum favoured by the wind,
And Thurine bays, and Temises behind,
Soft Sybaris, and all the capes that stand
Along the shore, he makes in sight of land;
Still doubling and still coasting till he found
The mouth of Aesaris, and promised ground,   70
Then saw where on the margin of the flood
The tomb that held the bones of Croton stood;
Here, by the god's command, he built and walled
The place predicted, and Crotona called.'
Thus fame from time to time delivers down
The sure tradition of the Italian town.
Here dwelt the man divine whom Samos bore,°
But now self-banished from his native shore,
Because he hated tyrants, nor could bear
The chains which none but servile souls will wear.   80
He, though from heaven remote, to heaven could move,
With strength of mind, and tread the abyss above;

And penetrate with his interior light
Those upper depths which nature hid from sight;
And what he had observed, and learned from thence,
Loved in familiar language to dispense.

   The crowd with silent admiration stand
And heard him, as they heard their god's command;
While he discoursed of heaven's mysterious laws,
The world's original, and nature's cause;   90
And what was god, and why the fleecy snows
In silence fell, and rattling winds arose;
What shook the steadfast earth, and whence begun
The dance of planets round the radiant sun;
If thunder was the voice of angry Jove,
Or clouds with nitre pregnant burst above:
Of these, and things beyond the common reach
He spoke, and charmed his audience with his speech.

   He first the taste of flesh from tables drove,
And argued well, if arguments could move.   100
'O mortals! from your fellows' blood abstain,
Nor taint your bodies with a food profane;
While corn and pulse by nature are bestowed,
And planted orchards bend their willing load;
While laboured gardens wholesome herbs produce,
And teeming vines afford their generous juice;
Nor tardier fruits of cruder kind are lost,
But tamed with fire, or mellowed by the frost;
While kine to pails distended udders bring,
And bees their honey redolent of spring;   110
While earth not only can your needs supply,
But lavish of her store, provides for luxury;
A guiltless feast administers with ease,
And without blood is prodigal to please.
Wild beasts their maws with their slain brethren fill,
And yet not all, for some refuse to kill;
Sheep, goats, and oxen, and the nobler steed
On browse and corn and flowery meadows feed.
Bears, tigers, wolves, the lion's angry brood,
Whom heaven endued with principles of blood,   120
He wisely sundered from the rest, to yell
In forests, and in lonely cave to dwell,
Where stronger beasts oppress the weak by might,
And all in prey, and purple feasts delight.

'O impious use! to nature's laws opposed,
Where bowels are in other bowels closed;
Where fattened by their fellows' fat they thrive;
Maintained by murder, and by death they live.
'Tis then for naught that mother earth provides
The stores of all she shows, and all she hides,   130
If men with fleshy morsels must be fed,
And chew with bloody teeth the breathing bread:
What else is this but to devour our guests,
And barbarously renew cyclopean feasts!°
We, by destroying life, our life sustain,
And gorge the ungodly maw with meats obscene.

  'Not so the golden age, who fed on fruit,
Nor durst with bloody meals their mouths pollute.
Then birds in airy space might safely move,
And timorous hares on heaths securely rove;   140
Nor needed fish the guileful hooks to fear,
For all was peaceful, and that peace sincere.
Whoever was the wretch (and cursed be he)
That envied first our food's simplicity,
The essay of bloody feasts on brutes began,
And after forged the sword to murder man.
Had he the sharpened steel alone employed
On beasts of prey that other beasts destroyed,
Or man invaded with their fangs and paws,
This had been justified by nature's laws,   150
And self-defence; but who did feasts begin
Of flesh, he stretched necessity to sin.
To kill man-killers, man has lawful power,
But not the extended licence to devour.

  'Ill habits gather by unseen degrees,
As brooks make rivers, rivers run to seas.
The sow, with her broad snout for rooting up
The entrusted seed, was judged to spoil the crop,
And intercept the sweating farmer's hope;
The covetous churl of unforgiving kind,   160
The offender to the bloody priest resigned:
Her hunger was no plea; for that she died.
The goat came next in order to be tried;
The goat had cropped the tendrils of the vine;
In vengeance laity and clergy join,
Where one had lost his profit, one his wine.

Here was at least some shadow of offence;
The sheep was sacrificed on no pretence,
But meek and unresisting innocence.
A patient, useful creature, born to bear   170
The warm and woolly fleece, that cloathed her murderer,
And daily to give down the milk she bred,
A tribute for the grass on which she fed.
Living, both food and raiment she supplies,
And is of least advantage when she dies.

   'How did the toiling ox his death deserve,
A downright simple drudge, and born to serve?
O tyrant! with what justice canst thou hope
The promise of the year, a plenteous crop,
When thou destroy'st thy labouring steer, who tilled,   180
And ploughed with pains, thy else ungrateful field?
From his yet reeking neck to draw the yoke,
(That neck with which the surly clods he broke)
And to the hatchet yield thy husband-man,
Who finished autumn and the spring began.

   'Nor this alone! but heaven itself to bribe,
We to the gods our impious acts ascribe;
First recompense with death their creatures' toil,
Then call the blest above to share the spoil:
The fairest victim must the powers appease,   190
(So fatal 'tis sometimes too much to please!)
A purple fillet his broad brows adorns,
With flowery garlands crowned, and gilded horns:
He hears the murderous prayer the priest prefers,
But understands not, 'tis his doom he hears;
Beholds the meal betwixt his temples cast,°
The fruit and product of his labours past;
And in the water views perhaps the knife
Uplifted, to deprive him of his life;
Then, broken up alive, his entrails sees,   200
Torn out for priests to inspect the god's decrees.

   'From whence, O mortal men, this gust of blood
Have you derived, and interdicted food?
Be taught by me this dire delight to shun,
Warned by my precepts, by my practice won;
And when you eat the well deserving beast,
Think, on the labourer of your field you feast!

'Now since the god inspires me to proceed,
Be that whate'er inspiring power obeyed.
For I will sing of mighty mysteries,   210
Of truths concealed before from human eyes,
Dark oracles unveil, and open all the skies.
Pleased as I am to walk along the sphere
Of shining stars, and travel with the year,
To leave the heavy earth, and scale the height
Of Atlas, who supports the heavenly weight;
To look from upper light, and thence survey
Mistaken mortals wandering from the way,
And wanting wisdom, fearful for the state
Of future things, and trembling at their fate!   220
    'Those I would teach; and by right reason bring
To think of death, as but an idle thing.
Why thus affrighted at an empty name,
A dream of darkness, and fictitious flame?
Vain themes of wit, which but in poems pass,
And fables of a world that never was!
What feels the body when the soul expires,
By time corrupted, or consumed by fires?
Nor dies the spirit, but new life repeats
In other forms, and only changes seats.   230
    'E'en I, who these mysterious truths declare,
Was once Euphorbus in the Trojan war;°
My name and lineage I remember well,
And how in flight by Sparta's king I fell.
In Argive Juno's fane I late beheld
My buckler hung on high, and owned my former shield.
    'Then death, so called, is but old matter dressed
In some new figure, and a varied vest;
Thus all things are but altered, nothing dies,
And here and there the unbodied spirit flies,   240
By time, or force, or sickness dispossessed,
And lodges where it lights, in man or beast;
Or hunts without, till ready limbs it find,
And actuates those according to their kind;
From tenement to tenement is tossed;
The soul is still the same, the figure only lost.
And as the softened wax new seals receives,
This face assumes, and that impression leaves;

Now called by one, now by another name,
The form is only changed, the wax is still the same: 250
So death, so called, can but the form deface,
The immortal soul flies out in empty space
To seek her fortune in some other place.

'Then let not piety be put to flight
To please the taste of glutton appetite;
But suffer inmate souls secure to dwell,
Lest from their seats your parents you expel;
With rabid hunger feed upon your kind,
Or from a beast dislodge a brother's mind.

'And since, like Tiphys parting from the shore,°   260
In ample seas I sail, and depths untried before,
This let me further add, that nature knows
No steadfast station, but, or ebbs or flows:
Ever in motion; she destroys her old,
And casts new figures in another mould.
E'en times are in perpetual flux, and run
Like rivers from their fountain rolling on;
For time no more than streams is at a stay:
The flying hour is ever on her way;
And as the fountain still supplies her store,   270
The wave behind impels the wave before,
Thus in successive course the minutes run,
And urge their predecessor minutes on,
Still moving, ever new; for former things
Are set aside, like abdicated kings;
And every moment alters what is done,
And innovates some act till then unknown.

'Darkness we see emerges into light,
And shining suns descend to sable night;
E'en heaven itself receives another dye,   280
When wearied animals in slumbers lie
Of midnight ease; another when the grey
Of morn preludes the splendour of the day.
The disk of Phoebus when he climbs on high,
Appears at first but as a bloodshot eye;
And when his chariot downwards drives to bed,
His ball is with the same suffusion red;
But mounted high in his meridian race
All bright he shines, and with a better face;

For there, pure particles of ether flow   290
Far from the infection of the world below.

   'Nor equal light the unequal moon adorns,
Or in her waxing or her waning horns.
For every day she wanes, her face is less,
But gathering into globe, she fattens at increase.

   'Perceiv'st thou not the process of the year,
How the four seasons in four forms appear,
Resembling human life in every shape they wear?
Spring first, like infancy, shoots out her head,
With milky juice requiring to be fed: 300
Helpless, though fresh, and wanting to be led.
The green stem grows in stature and in size,
But only feeds with hope the farmer's eyes;
Then laughs the childish year, with flowerets crowned,
And lavishly perfumes the fields around,
But no substantial nourishment receives,
Infirm the stalks, unsolid are the leaves.

   'Proceeding onward whence the year began
The summer grows adult, and ripens into man.
This season, as in men, is most replete   310
With kindly moisture, and prolific heat.

   'Autumn succeeds, a sober tepid age,
Not froze with fear, nor boiling into rage;
More than mature, and tending to decay,
When our brown locks repine to mix with odious grey.

   'Last, winter creeps along with tardy pace,
Sour is his front, and furrowed is his face;
His scalp if not dishonoured quite of hair,
The ragged fleece is thin, and thin is worse than bare.

   E'en our own bodies daily change receive,   320
Some part of what was theirs before they leave;
Nor are today what yesterday they were;
Nor the whole same tomorrow will appear.

   'Time was, when we were sowed, and just began,
From some few fruitful drops, the promise of a man;
Then nature's hand (fermented as it was)°
Moulded to shape the soft, coagulated mass;
And when the little man was fully formed,
The breathless embryo with a spirit warmed;
But when the mother's throes begin to come,   330

The creature, pent within the narrow room,
Breaks his blind prison, pushing to repair
His stifled breath, and draw the living air;
Cast on the margin of the world he lies,
A helpless babe, but by instinct he cries.
He next essays to walk, but, downward pressed
On four feet imitates his brother beast:
By slow degrees he gathers from the ground
His legs, and to the rolling chair is bound;
Then walks alone; a horseman now become,   340
He rides a stick, and travels round the room:
In time he vaunts among his youthful peers,
Strong-boned, and strung with nerves, in pride of years,
He runs with mettle his first merry stage,
Maintains the next, abated of his rage,
But manages his strength, and spares his age.
Heavy the third, and stiff, he sinks apace,
And though 'tis downhill all, but creeps along the race.
Now sapless on the verge of death he stands,
Contemplating his former feet, and hands;   350
And Milo-like, his slackened sinews sees,°
And withered arms, once fit to cope with Hercules,
Unable now to shake, much less to tear the trees.
    'So Helen wept when her too faithful glass
Reflected to her eyes the ruins of her face;
Wondering what charms her ravishers could spy,
To force her twice, or e'en but once enjoy!°
    'Thy teeth, devouring time, thine, envious age,
On things below still exercise your rage:
With venomed grinders you corrupt your meat,   360
And then at lingering meals, the morsels eat.
    'Nor those, which elements we call, abide,
Nor to this figure, nor to that are tied:
For this eternal world is said of old
But four prolific principles to hold,
Four different bodies; two to heaven ascend,
And other two down to the centre tend.
Fire first with wings expanded mounts on high,
Pure, void of weight, and dwells in upper sky;
Then air, because unclogged in empty space   370

Flies after fire, and claims the second place;
But weighty water as her nature guides,
Lies on the lap of earth; and mother earth subsides.
  'All things are mixed of these, which all contain,
And into these are all resolved again:
Earth rarefies to dew; expanded more,
The subtle dew in air begins to soar;
Spreads as she flies, and weary of her name
Extenuates still, and changes into flame;
Thus having by degrees perfection won,   380
Restless they soon untwist the web they spun,
And fire begins to lose her radiant hue,
Mixed with gross air, and air descends to dew;
And dew condensing, does her form forego,
And sinks, a heavy lump of earth below.
  'Thus are their figures never at a stand,
But changed by nature's innovating hand;
All things are altered, nothing is destroyed,
The shifted scene, for some new show employed.
  'Then to be born, is to begin to be   390
Some other thing we were not formerly:
And what we call to die, is not to appear,
Or be the thing that formerly we were.
Those very elements which we partake
Alive, when dead some other bodies make:
Translated grow, have sense, or can discourse,
But death on deathless substance has no force.
  'That forms are changed I grant, that nothing can
Continue in the figure it began:
The golden age, to silver was debased;   400
To copper that; our metal came at last.°
  'The face of places and their forms decay,
And that is solid earth, that once was sea;
Seas in their turn retreating from the shore,
Make solid land, what ocean was before;
And far from strands are shells of fishes found,
And rusty achors fixed on mountain ground;
And what were fields before, now washed and worn
By falling floods from high, to valleys turn,
And crumbling still descend to level lands;   410

And lakes, and trembling bogs are barren sands:
And the parched desert floats in streams unknown;
Wondering to drink of waters not her own.
    'Here nature living fountains opes; and there
Seals up the wombs where living fountains were;
Or earthquakes stop their ancient course, and bring
Diverted streams to feed a distant spring.
So Lycus, swallowed up, is seen no more,°
But far from thence knocks out another door.
Thus Erasinus dives; and blind in earth°   420
Runs on, and gropes his way to second birth,
Starts up in Argos meads, and shakes his locks
Around the fields, and fattens all the flocks.
So Mysus by another way is led,
And grown a river now disdains his head;
Forgets his humble birth, his name forsakes,
And the proud title of Caicus takes.
Large Amenane, impure with yellow sands,
Runs rapid often, and as often stands,
And here he threats the drunken fields to drown;   430
And there his dugs deny to give their liquor down.
    'Anigrus once did wholesome draughts afford,
But now his deadly waters are abhorred:
Since, hurt by Hercules, as fame resounds,
The centaurs, in his current washed their wounds.
The streams of Hypanis are sweet no more,
But brackish lose the taste they had before.
Antissa, Pharos, Tyre, in seas were pent,
Once isles, but now increase the continent;
While the Leucadian coast, mainland before,   440
By rushing seas is severed from the shore.
So Zancle to the Italian earth was tied,°
And men once walked where ships at anchor ride.
Till Neptune overlooked the narrow way,
And in disdain poured in the conquering sea.
    'Two cities that adorned the Achaian ground,
Buris and Helice, no more are found,
But whelmed beneath a lake are sunk and drowned;
And boatsmen through the chrystal water show
To wondering passengers the walls below.   450
    'Near Troezen stands a hill, exposed in air

To winter winds, of leafy shadows bare:
This once was level ground; but (strange to tell)
The included vapours, that in caverns dwell,
Labouring with cholic pangs, and close confined,
In vain sought issue for the rumbling wind:
Yet still they heaved for vent, and heaving still
Enlarged the concave, and shot up the hill;
As breath extends a bladder, or the skins
Of goats are blown to enclose the hoarded wines.   460
The mountain yet retains a mountain's face,
And gathered rubbish heals the hollow space.

  'Of many wonders, which I heard or knew,
Retrenching most, I will relate but few.
What, are not springs with qualities opposed,
Endued at seasons, and at seasons lost?
Thrice in a day thine, Ammon, change their form,°
Cold at high noon, at morn and evening warm:
Thine, Athaman, will kindle wood if thrown°
On the piled earth, and in the waning moon.   470
The Thracians have a stream, if any try
The taste, his hardened bowels petrify;
Whate'er it touches it converts to stones,
And makes a marble pavement where it runs.

  'Crathis, and Sybaris her sister flood,
That slide through our Calabrian neighbour wood,
With gold and amber dye the shining hair,
And thither youth resort; for who would not be fair?

  'But stranger virtues yet in streams we find,
Some change not only bodies, but the mind.   480
Who has not heard of Salmacis obscene,
Whose waters into women soften men?
Or Ethiopian lakes which turn the brain
To madness, or in heavy sleep constrain?
Clytorian streams the love of wine expel,
(Such is the virtue of the abstemious well,)
Whether the colder nymph that rules the flood
Extinguishes, and balks the drunken god;
Or that Melampus (so have some assured)
When the mad Proetides with charms he cured;°   490
And powerful herbs, both charms and simples cast
Into the sober spring, where still their virtues last.

'Unlike effects Lyncestis will produce;
Who drinks his waters, though with moderate use,
Reels as with wine, and sees with double sight:
His heels too heavy, and his head too light.
Ladon, once Pheneos, an Arcadian stream,
(Ambiguous in the effects, as in the name)
By day is wholesome beverage; but is thought
By night infected, and a deadly draught.   500
   'Thus running rivers, and the standing lake
Now of these virtues, now of those partake:
Time was (and all things time and fate obey)
When fast Ortygia floated on the sea:
Such were Cyanean isles, when Typhis steered
Betwixt their straits, and their collision feared;
They swam where now they sit; and firmly joined
Secure of rooting up, resist the wind.
Nor Etna vomiting sulphureous fire
Will ever belch; for sulphur will expire,   510
The veins exhausted of the liquid store;
Time was she cast no flames; in time will cast no more.
   'For whether earth's an animal, and air
Imbibes, her lungs with coolness to repair,
And what she sucks remits; she still requires
Inlets for air, and outlets for her fires;
When tortured with convulsive fits she shakes,
That motion chokes the vent till other vent she makes:
Or when the winds in hollow caves are closed,
And subtle spirits find that way opposed,   520
They toss up flints in air; the flints that hide
The seeds of fire, thus tossed in air, collide,
Kindling the sulphur, till the fuel spent,
The cave is cooled, and the fierce winds relent.
Or whether sulphur, catching fire, feeds on
Its unctuous parts, till all the matter gone
The flames no more ascend; for earth supplies
The fat that feeds them; and when earth denies
That food, by length of time consumed, the fire
Famished for want of fuel must expire.   530
   'A race of men there are, as fame has told,
Who shivering suffer hyperborean cold,
Till nine times bathing in Minerva's lake,

Soft feathers, to defend their naked sides, they take.
'Tis said, the Scythian wives (believe who will)
Transform themselves to birds by magic skill;
Smeared over with an oil of wonderous might,
That adds new pinions to their airy flight.
   'But this by sure experiment we know
That living creatures from corruption grow: 540
Hide in a hollow pit a slaughtered steer,
Bees from his putrid bowels will appear;
Who like their parents haunt the fields, and bring
Their honey harvest home, and hope another spring.
The warlike steed is multiplied we find,
To wasps and hornets of the warrior kind.
Cut from a crab his crooked claws and hide
The rest in earth, a scorpion thence will glide
And shoot his sting, his tail in circles tossed
Refers the limbs his backward father lost.   550
And worms, that stretch on leaves their filmy loom,
Crawl from their bags, and butterflies become.
E'en slime begets the frog's loquacious race:
Short of their feet at first, in little space
With arms and legs endued, long leaps they take,
Raised on their hinder part, and swim the lake,
And waves repel: for nature gives their kind
To that intent, a length of legs behind.
   'The cubs of bears, a living lump appear,
When whelped, and no determined figure wear.   560
Their mother licks them into shape, and gives
As much of form, as she herself receives.
   'The grubs from their sexangular abode
Crawl out unfinished, like the maggot's brood,
Trunks without limbs; till time at leisure brings
The thighs they wanted, and their tardy wings.
   'The bird who draws the car of Juno, vain
Of her crowned head, and of her starry train;
And he that bears the artillery of Jove,
The strong-pounced eagle, and the billing dove;   570
And all the feathered kind, who could suppose
(But that from sight, the surest sense he knows)
They from the included yolk, not ambient white, arose.
   'There are who think the marrow of a man,

Which in the spine, while he was living, ran;
When dead, the pith corrupted will become
A snake, and hiss within the hollow tomb.
   'All these receive their birth from other things;
But from himself the phoenix only springs:
Self-born, begotten by the parent flame   580
In which he burned, another and the same;
Who not by corn or herbs his life sustains,
But the sweet essence of amomum drains,
And watches the rich gums Arabia bears,
While yet in tender dew they drop their tears.
He (his five centuries of life fulfilled)
His nest on oaken boughs begins to build,
Or trembling tops of palm, and first he draws
The plan with his broad bill, and crooked claws,
Nature's artificers; on this the pile   590
Is formed, and rises round, then with the spoil
Of cassia, cinnamon, and stems of nard,
(For softness strewed beneath,) his funeral bed is reared:
Funeral and bridal both; and all around
The borders with corruptless myrrh are crowned,
On this incumbent; till etherial flame
First catches, then consumes the costly frame:
Consumes him too, as on the pile he lies;
He lived on odours, and in odours dies.
   'An infant phoenix from the former springs   600
His father's heir, and from his tender wings
Shakes off his parent dust, his method he pursues,
And the same lease of life on the same terms renews.
When grown to manhood he begins his reign,
And with stiff pinions can his flight sustain,
He lightens of its load the tree that bore
His father's royal sepulchre before,
And his own cradle: (this with pious care
Placed on his back) he cuts the buxom air,
Seeks the sun's city, and his sacred church,   610
And decently lays down his burden in the porch.
   'A wonder more amazing would we find?
The hyena shows it, of a double kind,
Varying the sexes in alternate years,
In one begets, and in another bears.

The thin chameleon fed with air, receives
The colour of the thing to which he cleaves.
  'India when conquered, on the conquering god°
For planted vines the sharp-eyed lynx bestowed,
Whose urine shed, before it touches earth,   620
Congeals in air, and gives to gems their birth.
So coral soft, and white in ocean's bed,
Comes hardened up in air, and glows with red.
  'All changing species should my song recite,
Before I ceased, would change the day to night.
Nations and empires flourish and decay,
By turns command, and in their turns obey;
Time softens hardy people, time again
Hardens to war a soft, unwarlike train.
Thus Troy for ten long years her foes withstood,   630
And daily bleeding bore the expense of blood;
Now for thick streets it shows an empty space,
Or only filled with tombs of her own perished race,
Herself becomes the sepulchre of what she was.
  'Mycene, Sparta, Thebes of mighty fame,
Are vanished out of substance into name,
And Dardan Rome that just begins to rise,
On Tiber's banks, in time shall mate the skies;
Widening her bounds, and working on her way;
E'en now she mediates imperial sway: 640
Yet this is change, but she by changing thrives,
Like moons new-born, and in her cradle strives
To fill her infant-horns; an hour shall come
When the round world shall be contained in Rome.
  'For thus old saws foretell, and Helenus
Anchises' drooping son enlivened thus;
When Ilium was now in a sinking state,
And he was doubtful of his future fate.
"O goddess-born, with thy hard fortune strive,°
Troy never can be lost, and thou alive.   650
Thy passage thou shalt free through fire and sword,
And Troy in foreign lands shall be restored.
In happier fields a rising town I see,
Greater than what e'er was, or is, or e'er shall be;
And heaven yet owes the world a race derived from thee.
Sages, and chiefs of other lineage born

The city shall extend, extended shall adorn:
But from Julus he must draw his birth,
By whom thy Rome shall rule the conquered earth:
Whom heaven will lend mankind on earth to reign,   660
And late require the precious pledge again."
This Helenus to great Aeneas told,
Which I retain, e'er since in other mould:
My soul was clothed; and now rejoice to view
My country walls rebuilt, and Troy revived anew,
Raised by the fall: decreed by loss to gain;
Enslaved but to be free, and conquered but to reign.
    "Tis time my hard-mouthed coursers to control,
Apt to run riot, and transgress the goal:
And therefore I conclude; whatever lies   670
In earth, or flits in air, or fills the skies,
All suffer change, and we, that are of soul
And body mixed, are members of the whole.
Then, when our sires, or grandsires, shall forsake
The forms of men, and brutal figures take,
Thus housed, securely let their spirits rest,
Nor violate thy father in the beast.
Thy friend, thy brother, any of thy kin,
Is none of these, yet there's a man within:
O spare to make a Thyestean meal,°   680
To enclose his body, and his soul expel.
    'Ill customs by degrees to habits rise,
Ill habits soon become exalted vice:
What more advance can mortals make in sin
So near perfection, who with blood begin?
Deaf to the calf that lies beneath the knife,
Looks up, and from her butcher begs her life;
Deaf to the harmless kid, that ere he dies
All methods to procure thy mercy tries,
And imitates in vain thy children's cries.   690
Where will he stop, who feeds with household bread,
Then eats the poultry which before he fed?
Let plough thy steers; that when they lose their breath
To nature, not to thee, they may impute their death.
Let goats for food their loaded udders lend,
And sheep from winter-cold thy sides defend;
But neither springes, nets, nor snares employ,

And be no more ingenious to destroy.
Free as in air, let birds on earth remain,
Nor let insidious glue their wings constrain;   700
Nor opening hounds the trembling stag affright,
Nor purple feathers intercept his flight:
Nor hooks concealed in baits for fish prepare,
Nor lines to heave them twinkling up in air.
   'Take not away the life you cannot give;
For all things have an equal right to live.
Kill noxious creatures, where 'tis sin to save;
This only just prerogative we have:
But nourish life with vegetable food,
And shun the sacrilegious taste of blood.'   710
   These precepts by the Samian sage were taught,
Which godlike Numa to the Sabines brought,
And thence transferred to Rome, by gift his own,
A willing people, and an offered throne.
O happy monarch, sent by heaven to bless
A savage nation with soft arts of peace,
To teach religion, rapine to restrain,
Give laws to lust, and sacrifice ordain:
Himself a saint, a goddess was his bride,
And all the muses o'er his acts preside.   720

The Character of a Good Parson

Imitated from Chaucer, and Enlarged°

A parish priest was of the pilgrim train;.
An awful, reverend, and religious man.
His eyes diffused a venerable grace,
And charity itself was in his face.
Rich was his soul, though his attire was poor,
(As God had clothed his own ambassador;)
For such on earth his blessed redeemer bore.
Of sixty years he seemed, and well might last
To sixty more, but that he lived too fast;
Refined himself to soul, to curb the sense,   10
And made almost a sin of abstinence.

Yet had his aspect nothing of severe,
But such a face as promised him sincere.
Nothing reserved or sullen was to see,
But sweet regards, and pleasing sanctity;
Mild was his accent, and his action free.
With eloquence innate his tongue was armed,
Though harsh the precept, yet the preacher charmed.
For, letting down the golden chain from high,
He drew his audience upward to the sky;   20
And oft, with holy hymns, he charmed their ears,
(A music more melodious than the spheres,)
For David left him, when he went to rest,
His lyre; and after him he sung the best.
He bore his great commission in his look,
But sweetly tempered awe, and softened all he spoke.
He preached the joys of heaven, and pains of hell,
And warned the sinner with becoming zeal;
But on eternal mercy loved to dwell.
He taught the gospel rather than the law,   30
And forced himself to drive, but loved to draw.
For fear but freezes minds; but love, like heat,
Exhales the soul sublime, to seek her native seat.
    To threats the stubborn sinner oft is hard,
Wrapped in his crimes, against the storm prepared;
But when the milder beams of mercy play,
He melts, and throws his cumbrous cloak away.
    Lightnings and thunder, (heaven's artillery,)
As harbingers before the almighty fly:
Those but proclaim his style, and disappear;   40
The stiller sound succeeds, and God is there.
    The tithes his parish freely paid, he took,
But never sued, or cursed with bell and book;
With patience bearing wrong, but offering none,
Since every man is free to lose his own.
The country churls, according to their kind.
(Who grudge their dues, and love to be behind,)
The less he sought his offerings, pinched the more,
And praised a priest contented to be poor.
    Yet of his little he had some to spare,   50
To feed the famished, and to clothe the bare;
For mortified he was to that degree,

A poorer than himself he would not see.
True priests, he said, and preachers of the word,
Were only stewards of their sovereign Lord;
Nothing was theirs, but all the public store;
Entrusted riches, to relieve the poor;
Who, should they steal, for want of his relief,
He judged himself accomplice with the thief.

Wide was his parish; not contracted close   60
In streets, but here and there a straggling house;
Yet still he was at hand, without request,
To serve the sick, to succour the distressed;
Tempting on foot alone, without affright,
The dangers of a dark tempestuous night.
All this the good old man performed alone,
Nor spared his pains; for curate he had none.
Nor durst he trust another with his care;
Nor rode himself to Paul's, the public fair,
To chaffer for preferment with his gold,   70
Where bishoprics and sinecures are sold;
But duly watched his flock by night and day,
And from the prowling wolf redeemed the prey,
And hungry sent the wily fox away.
The proud he tamed, the penitent he cheered;
Nor to rebuke the rich offender feared.
His preaching much, but more his practice wrought;
(A living sermon of the truths he taught;)
For this by rules severe his life he squared,
That all might see the doctrine which they heard.   80
For priests, he said, are patterns for the rest;
(The gold of heaven, who bear the God impressed;)
But when the precious coin is kept unclean,
The sovereign's image is no longer seen.
If they be foul on whom the people trust,
Well may the baser brass contract a rust.
The prelate, for his holy life he prized,
The worldly pomp of prelacy despised.
His saviour came not with a gaudy show,
Nor was his kingdom of the world below.   90
Patience in want, and poverty of mind,
These marks of church and churchmen he designed,°
And living taught, and dying left behind.

The crown he wore was of the pointed thorn;
In purple he was crucified, not born.
They who contend for place and high degree,
Are not his sons, but those of Zebedee.°

Not but he knew the signs of earthly power
Might well become Saint Peter's successor;
The holy father holds a double reign,   100
The prince may keep his pomp, the fisher must be plain.

Such was the saint, who shone with every grace,
Reflecting, Moses-like, his maker's face.°
God saw his image lively was expressed;
And his own work, as in creation, blessed.

The tempter saw him too with envious eye,
And, as on Job, demanded leave to try.
He took the time when Richard was deposed.°
And high and low with happy Harry closed.°
This prince, though great in arms, the priest withstood: 110
Near though he was, yet not the next of blood.
Had Richard, unconstrained, resigned the throne,
A king can give no more than is his own:
The title stood entailed, had Richard had a son.

Conquest, an odious name, was laid aside;
Where all submitted, none the battle tried.
The senseless plea of right by providence
Was by a flattering priest invented since;°
And lasts no longer than the present sway,
But justifies the next who comes in play.   120

The people's right remains; let those who dare
Dispute their power, when they the judges are.

He joined not in their choice, because he knew
Worse might, and often did, from change ensue.
Much to himself he thought, but little spoke;
And, undeprived, his benefice forsook.

Now, through the land, his cure of souls he stretched,
And like a primitive apostle preached.
Still cheerful; ever constant to his call;
By many followed; loved by most, admired by all.   130
With what he begged, his brethren he relieved,
And gave the charities himself received;
Gave, while he taught; and edified the more,
Because he showed, by proof, 'twas easy to be poor.

He went not, with the crowd, to see a shrine;
But fed us, by the way, with food divine.
  In deference to his virtues, I forbear
To show you what the rest in orders were:
This brillant is so spotless, and so bright,°
He needs no foil, but shines by his own proper light.   140

The Monument of a Fair Maiden Lady

WHO DIED AT BATH AND IS THERE INTERRED

Below this marble monument is laid
All that heaven wants of this celestial maid.
Preserve, O sacred tomb, thy trust consigned,
The mould was made on purpose for the mind:
And she would lose, if, at the latter day,
One atom could be mixed of other clay;
Such were the features of her heavenly face,
Her limbs were formed with such harmonious grace:
So faultless was the frame, as if the whole
Had been an emanation of the soul;   10
Which her own inward symmetry revealed,
And like a picture shone, in glass annealed;
Or like the sun eclipsed, with shaded light;
Too piercing, else, to be sustained by sight.
Each thought was visible that rolled within,
As through a crystal case the figured hours are seen.
And heaven did this transparent veil provide,
Because she had no guilty thought to hide.
All white, a virgin-saint, she sought the skies,°
For marriage, though it sullies not, it dyes.   20
High though her wit, yet humble was her mind;
As if she could not, or she would not find
How much her worth transcended all her kind.
Yet she had learned so much of heaven below,
That when arrived, she scarce had more to know;
But only to refresh the former hint,
And read her maker in a fairer print.
So pious, as she had no time to spare
For human thoughts, but was confined to prayer.

Yet in such charities she passed the day,   30
'Twas wondrous how she found an hour to pray.
A soul so calm, it knew not ebbs or flows,
Which passion could but curl, not discompose.
A female softness, with a manly mind;
A daughter duteous, and a sister kind;
In sickness patient, and in death resigned.

Cymon and Iphigenia

From Boccaccio

Poeta loquitur°

Old as I am, for ladies' love unfit,
The power of beauty I remember yet,
Which once inflamed my soul, and still inspires my wit.
If love be folly, the severe divine°
Has felt that folly, though he censures mine;
Pollutes the pleasures of a chaste embrace,
Acts what I write, and propagates in grace
With riotous excess, a priestly race.
Suppose him free, and that I forge the offence,
He showed the way, perverting first my sense: 10
In malice witty, and with venom fraught,
He makes me speak the things I never thought.
Compute the gains of his ungoverned zeal;
Ill suits his cloth the praise of railing well.°
The world will think that what we loosely write,
Though now arraigned, he read with some delight;
Because he seems to chew the cud again,
When his broad comment makes the text too plain:
And teaches more in one explaining page,
Than all the double meanings of the stage.   20
    What needs he paraphrase on what we mean?
We were at worst but wanton; he's obscene.
I, nor my fellows, nor myself excuse;
But love's the subject of the comic muse:
Nor can we write without it, nor would you
A tale of only dry instruction view;

Nor love is always of a vicious kind,
But oft to virtuous acts inflames the mind.
Awakes the sleepy vigour of the soul,
And, brushing o'er, adds motion to the pool. 30
Love, studious how to please, improves our parts
With polished manners, and adorns with arts.
Love first invented verse, and formed the rhyme,
The motion measured, harmonized the chime;
To liberal acts enlarged the narrow-souled,
Softened the fierce, and made the coward bold;
The world when waste, he peopled with increase,
And warring nations reconciled in peace.
Ormonde, the first, and all the fair may find,°
In this one legend to their fame designed, 40
When beauty fires the blood, how love exalts the mind.

In that sweet isle, where Venus keeps her court,
And every grace and all the loves resort;
Where either sex is formed of softer earth,
And takes the bent of pleasure from their birth;
There lived a Cyprian lord, above the rest,
Wise, wealthy, with a numerous issue blessed.
    But as no gift of fortune is sincere,
Was only wanting in a worthy heir:
His eldest born a goodly youth to view 50
Excelled the rest in shape, and outward show;
Fair, tall, his limbs with due proportion joined,
But of a heavy, dull, degenerate mind.
His soul belied the features of his face;
Beauty was there, but beauty in disgrace.
A clownish mien, a voice with rustic sound,
And stupid eyes, that ever loved the ground.
He looked like nature's error; as the mind
And body were not of a piece designed,
But made for two, and by mistake in one were joined. 60
    The ruling rod, the father's forming care,
Were exercised in vain, on wit's despair;
The more informed the less he understood,
And deeper sunk by floundering in the mud.
Now scorned of all and grown the public shame,
The people from Galesus changed his name,

And Cymon called, which signifies a brute;
So well his name did with his nature suit.
  His father, when he found his labour lost,
And care employed, that answered not the cost,   70
Chose an ungrateful object to remove,
And loathed to see what nature made him love;
So to his country farm the fool confined:
Rude work well suited with a rustic mind.
Thus to the wilds the sturdy Cymon went,
A squire among the swains, and pleased with banishment.
His corn and cattle were his only care,
And his supreme delight a country fair.
  It happened on a summer's holiday,
That to the greenwood shade he took his way;   80
For Cymon shunned the church, and used not much to pray.
His quarter-staff, which he could ne'er forsake,
Hung half before, and half behind his back.
He trudged along unknowing what he sought,
And whistled as he went, for want of thought.
  By chance conducted, or by thirst constrained,
The deep recesses of the grove he gained;
Where in a plain, defended by the wood,
Crept through the matted grass a crystal flood,
By which an alabaster fountain stood: 90
And on the margin of the fount was laid,
Attended by her slaves, a sleeping maid.
Like Dian, and her nymphs, when tired with sport,
To rest by cool Eurotas they resort.
The dame herself the goddess well expressed,°
Not more distinguished by her purple vest,
Than by the charming features of her face,
And e'en in slumber a superior grace:
Her comely limbs composed with decent care,
Her body shaded with a slight cymar;   100
Her bosom to the view was only bare:
Where two beginning paps were scarcely spied,
For yet their places were but signified:
The fanning wind upon her bosom blows,
To meet the fanning wind the bosom rose;
The fanning wind and purling streams continue her repose.
  The fool of nature stood with stupid eyes

And gaping mouth, that testified surprise,
Fixed on her face, nor could remove his sight
New as he was to love, and novice in delight: 110
Long mute he stood, and leaning on his staff,
His wonder witnessed with an idiot laugh;
Then would have spoke, but by his glimmering sense
First found his want of words, and feared offence:
Doubted for what he was he should be known,
By his clown accent, and his country tone.

    Through the rude chaos thus the running light
Shot the first ray that pierced the native night:
Then day and darkness in the mass were mixed,
Till gathered in a globe, the beams were fixed: 120
Last shone the sun who radiant in his sphere
Illumined heaven and earth, and rolled around the year.
So reason in this brutal soul began:
Love made him first suspect he was a man;
Love made him doubt his broad barbarian sound,
By love his want of words, and wit he found:
That sense of want prepared the future way
To knowledge, and disclosed the promise of a day.

    What not his father's care, nor tutor's art
Could plant with pains in his unpolished heart,   130
The best instructor, love at once inspired,
As barren grounds to fruitfulness are fired;
Love taught him shame, and shame with love at strife
Soon taught the sweet civilities of life.
His gross material soul at once could find
Somewhat in her excelling all her kind;
Exciting a desire till then unknown,
Somewhat unfound, or found in her alone.
This made the first impression in his mind,
Above, but just above, the brutal kind.   140
For beasts can like, but not distinguish too,
Nor their own liking by reflection know;
Nor why they like or this or t'other face,
Or judge of this or that peculiar grace;
But love in gross, and stupidly admire;°
As flies, allured by light, approach the fire.
Thus our man-beast, advancing by degrees,
First likes the whole, then separates what he sees;

On several parts a several praise bestows,
The ruby lips, the well-proportioned nose,   150
The snowy skin, the raven-glossy hair,
The dimpled cheek, the forehead rising fair,
And e'en in sleep itself a smiling air.
From thence his eyes descending viewed the rest,
Her plump round arms, white hands, and heaving breast.
Long on the last he dwelt, though every part
A pointed arrow sped to pierce his heart.
        Thus in a trice a judge of beauty grown,
(A judge erected from a country clown,)
He longed to see her eyes in slumber hid,   160
And wished his own could pierce within the lid.
He would have waked her, but restrained his thought,
And love new-born the first good manners taught.
An awful fear his ardent wish withstood,
Nor durst disturb the goddess of the wood;
For such she seemed by her celestial face,
Excelling all the rest of human race;
And things divine, by common sense he knew,
Must be devoutly seen at distant view:
So checking his desire, with trembling heart   170
Gazing he stood, nor would nor could depart;
Fixed as a pilgrim wildered in his way,°
Who dares not stir by night, for fear to stray;
But stands with awful eyes to watch the dawn of day.
        At length awaking, Iphigene the fair
(So was the beauty called who caused his care)
Unclosed her eyes, and double day revealed,
While those of all her slaves in sleep were sealed.
        The slavering cudden, propped upon his staff,°
Stood ready gaping with a grinning laugh   180
To welcome her awake, nor durst begin
To speak, but wisely kept the fool within.
Then she: 'What make you, Cymon, here alone?'
(For Cymon's name was round the country known,
Because descended of a noble race,
And for a soul ill sorted with his face.)
        But still the sot stood silent with surprise,
With fixed regard on her new opened eyes,
And in his breast received the envenomed dart,

A tickling pain that pleased amid the smart.   190
But conscious of her form, with quick distrust
She saw his sparkling eyes, and feared his brutal lust.
This to prevent, she waked her sleepy crew,
And rising hasty took a short adieu.

Then Cymon first his rustic voice essayed,
With proffered service to the parting maid
To see her safe; his hand she long denied,
But took at length, ashamed of such a guide.
So Cymon led her home, and leaving there,
No more would to his country clowns repair,   200
But sought his father's house with better mind,
Refusing in the farm to be confined.

The father wondered at the son's return,
And knew not whether to rejoice or mourn;
But doubtfully received, expecting still
To learn the secret causes of his altered will.
Nor was he long delayed: the first request
He made, was like his brothers to be dressed,
And, as his birth required, above the rest.

With ease his suit was granted by his sire,   210
Distinguishing his heir by rich attire:
His body thus adorned, he next designed
With liberal arts to cultivate his mind;
He sought a tutor of his own accord
And studied lessons he before abhorred.

Thus the man-child advanced, and learned so fast
That in short time his equals he surpassed:
His brutal manners from his breast exiled,
His mien he fashioned, and his tongue he filed;
In every exercise of all admired,   220
He seemed, nor only seemed, but was inspired:
Inspired by love, whose business is to please;
He rode, he fenced, he moved with graceful ease,
More famed for sense, for courtly carriage more,
Than for his brutal folly known before.

What then of altered Cymon shall we say,
But that the fire which choked in ashes lay,
A load too heavy for his soul to move,
Was upward blown below, and brushed away by love?
Love made an active progress through his mind,   230

The dusky parts he cleared, the gross refined,
The drowsy waked; and, as he went, impressed
The maker's image on the human beast.
Thus was the man amended by desire,
And though he loved perhaps with too much fire,
His father all his faults with reason scanned,
And liked an error of the better hand;
Excused the excess of passion in his mind,
By flames too fierce, perhaps too much refined:
So Cymon, since his sire indulged his will,   240
Impetuous loved, and would be Cymon still;
Galesus he disowned, and chose to bear
The name of fool, confirmed and bishoped by the fair.

　　To Cipseus by his friends his suit he moved,
Cipseus the father of the fair he loved;
But he was pre-engaged by former ties,
While Cymon was endeavouring to be wise;
And Iphigene, obliged by former vows,
Had given her faith to wed a foreign spouse:
Her sire and she to Rhodian Pasimond,   250
Though both repenting, were by promise bound,
Nor could retract; and thus, as fate decreed,
Though better loved, he spoke too late to speed.

　　The doom was past; the ship already sent
Did all his tardy diligence prevent;
Sighed to her self the fair unhappy maid,
While stormy Cymon thus in secret said:
'The time is come for Iphigene to find
The miracle she wrought upon my mind;
Her charms have made me man, her ravished love   260
In rank shall place me with the blessed above.
For mine by love, by force she shall be mine,
Or death, if force should fail, shall finish my design.'

　　Resolved he said; and rigged with speedy care
A vessel strong, and well equipped for war.
The secret ship with chosen friends he stored,
And bent to die or conquer, went aboard.
Ambushed he lay behind the Cyprian shore,°
Waiting the sail that all his wishes bore;
Nor long expected, for the following tide   270
Sent out the hostile ship and beauteous bride.

To Rhodes the rival bark directly steered,
When Cymon sudden at her back appeared,
And stopped her flight: then standing on his prow,
In haughty terms he thus defied the foe:
'Or strike your sails at summons, or prepare°
To prove the last extremities of war.'
Thus warned, the Rhodians for the fight provide;
Already were the vessels side by side,
These obstinate to save, and those to seize the bride.   280
But Cymon soon his crooked grapples cast,
Which with tenacious hold his foes embraced,
And armed with sword and shield amid the press he passed.
Fierce was the fight, but hastening to his prey,
By force the furious lover freed his way;
Himself alone dispersed the Rhodian crew,
The weak disdained, the valiant overthrew;
Cheap conquest for his following friends remained,
He reaped the field, and they but only gleaned.

His victory confessed, the foes retreat,   290
And cast their weapons at the victor's feet.
Whom thus he cheered: 'O Rhodian youth, I fought
For love alone, nor other booty sought;
Your lives are safe; your vessel I resign,
Yours be your own, restoring what is mine;
In Iphigene I claim my rightful due,
Robbed by my rival, and detained by you:
Your Pasimond a lawless bargain drove,
The parent could not sell the daughter's love;
Or if he could, my love disdains the laws,   300
And like a king by conquest gains his cause;
Where arms take place, all other pleas are vain;°
Love taught me force, and force shall love maintain.
You, what by strength you could not keep, release,
And at an easy ransom buy your peace.'

Fear on the conquered side soon signed the accord,
And Iphigene to Cymon was restored.
While to his arms the blushing bride he took,
To seeming sadness she composed her look;
As if by force subjected to his will,   310
Though pleased, dissembling, and a woman still.
And, for she wept, he wiped her falling tears,

And prayed her to dismiss her empty fears;
'For yours I am,' he said, 'and have deserved
Your love much better, whom so long I served,
Than he to whom your formal father tied°
Your vows, and sold a slave, not sent a bride.'
Thus while he spoke, he seized the willing prey,
As Paris bore the Spartan spouse away.°
Faintly she screamed, and e'en her eyes confessed   320
She rather would be thought, than was, distressed.
    Who now exults but Cymon in his mind?
Vain hopes and empty joys of human kind,
Proud of the present, to the future blind!
Secure of fate, while Cymon ploughs the sea,
And steers to Candy with his conquered prey,°
Scarce the third glass of measured hours was run,°
When like a fiery meteor sunk the sun,
The promise of a storm; the shifting gales
Forsake by fits and fill the flagging sails;   330
Hoarse murmurs of the main from far were heard,
And night came on, not by degrees prepared,
But all at once; at once the winds arise,
The thunders roll, the forky lightning flies.
In vain the master issues out commands,
In vain the trembling sailors ply their hands;
The tempest unforeseen prevents their care,°
And from the first they labour in despair.
The giddy ship betwixt the winds and tides,
Forced back and forwards, in a circle rides,   340
Stunned with the different blows; then shoots amain,
Till counterbuffed she stops, and sleeps again.
Not more aghast the proud archangel fell,
Plunged from the height of heaven to deepest hell,
Than stood the lover of his love possessed,
Now cursed the more, the more he had been blessed;
More anxious for her danger than his own,
Death he defies, but would be lost alone.
    Sad Iphigene to womanish complaints
Adds pious prayers, and wearies all the saints;   350
E'en if she could, her love she would repent,
But since she cannot, dreads the punishment:
Her forfeit faith and Pasimond betrayed

Are ever present, and her crime upbraid.
She blames herself, nor blames her lover less;
Augments her anger as her fears increase;
From her own back the burden would remove,
And lays the load on his ungoverned love,
Which interposing durst, in heaven's despite,
Invade and violate another's right: 360
The powers incensed awhile deferred his pain,
And made him master of his vows in vain:
But soon they punished his presumptuous pride;
That for his daring enterprise she died,
Who rather not resisted than complied.

   Then, impotent of mind, with altered sense,°
She hugged the offender, and forgave the offence,
Sex to the last. Meantime with sails declined°
The wandering vessel drove before the wind,
Tossed and retossed, aloft, and then alow;°   370
Nor port they seek, nor certain course they know,
But every moment wait the coming blow.
Thus blindly driven, by breaking day they viewed
The land before them, and their fears renewed.
The land was welcome, but the tempest bore
The threatened ship against a rocky shore.

   A winding bay was near; to this they bent,°
And just escaped; their force already spent.
Secure from storms, and panting from the sea,
The land unknown at leisure they survey;   380
And saw (but soon their sickly sight withdrew)
The rising towers of Rhodes at distant view;
And cursed the hostile shore of Pasimond,
Saved from the seas, and shipwrecked on the ground.

   The frighted sailors tried their strength in vain
To turn the stern, and tempt the stormy main;
But the stiff wind withstood the labouring oar,
And forced them forward on the fatal shore!
The crooked keel now bites the Rhodian strand,
And the ship moored constrains the crew to land.   390
Yet still they might be safe, because unknown;
But as ill fortune seldom comes alone,
The vessel they dismissed was driven before,
Already sheltered on their native shore;

Known each, they know, but each with change of cheer;
The vanquished side exults; the victors fear;
Not them but theirs, make prisoners ere they fight,
Despairing conquest, and deprived of flight.

 The country rings around with loud alarms,°
And raw in fields the rude militia swarms: 400
Mouths without hands, maintained at vast expense,
In peace a charge, in war a weak defence;
Stout once a month they march, a blustering band,
And ever, but in times of need, at hand;
This was the morn when, issuing on the guard,
Drawn up in rank and file they stood prepared
Of seeming arms to make a short essay,°
Then hasten to be drunk, the business of the day.

 The cowards would have fled, but that they knew
Themselves so many, and their foes so few;   410
But crowding on, the last the first impel,
Till overborne with weight the Cyprians fell.
Cymon enslaved, who first the war begun,
And Iphigene once more is lost and won.

 Deep in a dungeon was the captive cast,
Deprived of day, and held in fetters fast;
His life was only spared at their request,
Whom taken he so nobly had released:
But Iphigenia was the ladies' care,
Each in their turn addressed to treat the fair;   420
While Pasimond and his the nuptial feast prepare.

 Her secret soul to Cymon was inclined,
But she must suffer what her fates assigned;
So passive is the church of womankind.
What worse to Cymon could his fortune deal,
Rolled to the lowest spoke of all her wheel?
It rested to dismiss the downward weight,°
Or raise him upward to his former height;
The latter pleased; and love (concerned the most)
Prepared the amends for what by love he lost.   430

 The sire of Pasimond had left a son,
Though younger, yet for courage early known,
Ormisda called, to whom, by promise tied,
A Rhodian beauty was the destined bride;
Cassandra was her name, above the rest

Renowned for birth, with fortune amply blessed.
Lysimachus, who ruled the Rhodian state,
Was then by choice their annual magistrate:
He loved Cassandra too with equal fire,
But fortune had not favoured his desire;   440
Crossed by her friends, by her not disapproved,
Nor yet preferred, or like Ormisda loved:
So stood the affair; some little hope remained,
That, should his rival chance to lose, he gained.

 Meantime young Pasimond his marriage pressed,
Ordained the nuptial day, prepared the feast;
And frugally resolved (the charge to shun,
Which would be double should he wed alone,)
To join his brother's bridal with his own.

 Lysimachus, oppressed with mortal grief,   450
Received the news, and studied quick relief:
The fatal day approached; if force were used,
The magistrate his public trust abused;
To justice liable, as law required,
For when his office ceased, his power expired:
While power remained, the means were in his hand
By force to seize, and then forsake the land:
Betwixt extremes he knew not how to move,
A slave to fame, but more a slave to love:
Restraining others, yet himself not free,   460
Made impotent by power, debased by dignity,
Both sides he weighed: but after much debate,
The man prevailed above the magistrate.

 Love never fails to master what he finds,
But works a different way in different minds,
The fool enlightens, and the wise he blinds.
This youth proposing to possess and scape,
Began in murder, to conclude in rape:°
Unpraised by me, though heaven sometime may bless
An impious act with undeserved success: 470
The great, it seems, are privileged alone,
To punish all injustice but their own.
But here I stop, not daring to proceed,
Yet blush to flatter an unrighteous deed;
For crimes are but permitted, not decreed.

 Resolved on force, his wit the praetor bent°

To find the means that might secure the event;
Nor long he laboured, for his lucky thought
In captive Cymon found the friend he sought.
The example pleased: the cause and crime the same,   480
An injured lover and a ravished dame.
How much he durst he knew by what he dared,
The less he had to lose, the less he cared
To manage loathsome life when love was the reward.

    This pondered well, and fixed on his intent,
In depth of night he for the prisoner sent;
In secret sent, the public view to shun,
Then with a sober smile he thus begun:
'The powers above, who bounteously bestow
Their gifts and graces on mankind below,   490
Yet prove our merit first, nor blindly give
To such as are not worthy to receive:
For valour and for virtue they provide
Their due reward, but first they must be tried:
These fruitful seeds within your mind they sowed:
'Twas yours to improve the talent they bestowed;
They gave you to be born of noble kind,
They gave you love to lighten up your mind
And purge the grosser parts; they gave you care
To please, and courage to deserve the fair.   500
    'Thus far they tried you, and by proof they found
The grain entrusted in a grateful ground:
But still the great experiment remained,
They suffered you to lose the prize you gained,
That you might learn the gift was theirs alone,
And, when restored, to them the blessing own.
Restored it soon will be; the means prepared,
The difficulty smoothed, the danger shared:
Be but yourself, the care to me resign,
Then Iphigene is yours, Cassandra mine.   510
Your rival Pasimond pursues your life,
Impatient to revenge his ravished wife,
But yet not his; tomorrow is behind.
And love our fortunes in one band has joined:
Two brothers are our foes, Ormisda mine
As much declared as Pasimond is thine;

Tomorrow must their common vows be tied;
With love to friend, and fortune for our guide,
Let both resolve to die, or each redeem a bride.
    'Right I have none, nor hast thou much to plead; 520
'Tis force, when done, must justify the deed:
Our task performed, we next prepare for flight;
And let the losers talk in vain of right:
We with the fair will sail before the wind;
If they are grieved, I leave the laws behind.
Speak thy resolves: if now thy courage droop,
Despair in prison and abandon hope;
But if thou darst in arms thy love regain,
(For liberty without thy love were vain:)
Then second my design to seize the prey, 530
Or lead to second rape, for well thou knowest the way.'
    Said Cymon, overjoyed: 'Do thou propose
The means to fight, and only show the foes:
For from the first, when love had fired my mind
Resolved, I left the care of life behind.'
    To this the bold Lysimachus replied,
'Let heaven be neuter and the sword decide:°
The spousals are prepared, already play
The minstrels, and provoke the tardy day:
By this the brides are waked, their grooms are dressed:° 540
All Rhodes is summoned to the nuptial feast,
All but myself, the sole unbidden guest.
Unbidden though I am, I will be there,
And, joined by thee, intend to joy the fair.
    'Now hear the rest; when day resigns the light,
And cheerful torches gild the jolly night,
Be ready at my call; my chosen few
With arms administered shall aid thy crew.
Then entering unexpected will we seize
Our destined prey, from men dissolved in ease, 550
By wine disabled, unprepared for fight,
And hastening to the seas, suborn our flight:
The seas are ours, for I command the fort,
A ship well manned expects us in the port:
If they, or if their friends, the prize contest,
Death shall attend the man who dares resist.'

It pleased; the prisoner to his hold retired,
His troop with equal emulation fired,
All fixed to fight, and all their wonted work required.°
   The sun arose; the streets were thronged around,   560
The palace opened, and the posts were crowned.
The double bridegroom at the door attends
The expected spouse, and entertains the friends:
They meet, they lead to church, the priests invoke
The powers, and feed the flames with fragrant smoke.
This done, they feast, and at the close of night
By kindled torches vary their delight,
These lead the lively dance, and those the brimming bowls invite.
   Now, at the appointed place and hour assigned,
With souls resolved the ravishers were joined: 570
Three bands are formed; the first is sent before
To favour the retreat and guard the shore;
The second at the palace gate is placed,
And up the lofty stairs ascend the last:
A peaceful troop they seem with shining vests,
But coats of mail beneath secure their breasts.
   Dauntless they enter, Cymon at their head,
And find the feast renewed, the table spread:
Sweet voices, mixed with instrumental sounds,
Ascend the vaulted roof, the vaulted roof rebounds.   580
When, like the harpies, rushing through the hall°
The sudden troop appears, the tables fall,
Their smoking load is on the pavement thrown;
Each ravisher prepares to seize his own:
The brides, invaded with a rude embrace,
Shriek out for aid, confusion fills the place.
Quick to redeem the prey their plighted lords
Advance, the palace gleams with shining swords.
   But late is all defence, and succour vain;
The rape is made, the ravishers remain: 590
Two sturdy slaves were only sent before
To bear the purchased prize in safety to the shore.
The troop retires, the lovers close the rear,
With forward faces not confessing fear:
Backward they move, but scorn their pace to mend;
Then seek the stairs, and with slow haste descend.

Fierce Pasimond, their passage to prevent,
Thrust full on Cymon's back in his descent,
The blade returned unbathed, and to the handle bent.
Stout Cymon soon remounts, and cleft in two   600
His rival's head with one descending blow:
And as the next in rank Ormisda stood,
He turned the point; the sword enured to blood
Bored his unguarded breast, which poured a purple flood.

    With vowed revenge the gathering crowd pursues,
The ravishers turn head, the fight renews;
The hall is heaped with corpse; the sprinkled gore
Besmears the walls, and floats the marble floor.°
Dispersed at length the drunken squadron flies,
The victors to their vessel bear the prize,   610
And hear behind loud groans, and lamentable cries.

    The crew with merry shouts their anchors weigh,
Then ply their oars, and brush the buxom sea,
While troops of gathered Rhodians crowd the quay.
What should the people do when left alone?
The governor and government are gone;
The public wealth to foreign parts conveyed;
Some troops disbanded, and the rest unpaid.
Rhodes is the sovereign of the sea no more;
Their ships unrigged, and spent their naval store;   620
They neither could defend nor can pursue,
But grind their teeth, and cast a helpless view:
In vain with darts a distant war they try,
Short, and more short, the missive weapons fly.
Meanwhile the ravishers their crimes enjoy,
And flying sails and sweeping oars employ:
The cliffs of Rhodes in little space are lost;
Jove's isle they seek, nor Jove denies his coast.°

    In safety landed on the Candian shore,
With generous wines their spirits they restore;   630
There Cymon with his Rhodian friend resides,
Both court and wed at once the willing brides.
A war ensues, the Cretans own their cause,
Stiff to defend their hospitable laws:
Both parties lose by turns, and neither wins,
Till peace, propounded by a truce, begins.

The kindred of the slain forgive the deed,
But a short exile must for show precede:
The term expired, from Candia they remove,
And happy each at home enjoys his love.   640

The Secular Masque

Enter Janus

JANUS.   Chronos, Chronos, mend thy pace,
      A hundred times the rolling sun
      Around the radiant belt has run°
      In his revolving race.
      Behold, behold, the goal in sight,
      Spread thy fans, and wing thy flight.°

Enter CHRONOS, *with a scythe in his hand, and a great globe
on his back, which he sets down at his entrance*

CHRONOS.   Weary, weary of my weight,
      Let me, let me drop my freight,
        And leave the world behind.
          I could not bear   10
          Another year
        The load of humankind.

Enter Momus *laughing*°

MOMUS.   Ha! ha! ha! ha! ha! ha! well hast thou done,
        To lay down thy pack,
        And lighten thy back,
      The world was a fool, e'er since it begun,
      And since neither Janus, nor Chronos, nor I,
        Can hinder the crimes,
        Or mend the bad times,
      'Tis better to laugh than to cry.   20

Chorus of all three
'Tis better to laugh than to cry.

JANUS.   Since Momus comes to laugh below,
      Old Time, begin the show,
      That he may see, in every scene,
      What changes in this age have been.
CHRONOS.   Then, goddess of the silver bow, begin.
          [*Horns, or hunting music, within.*

Enter Diana

DIANA.   With horns and with hounds, I waken the day,
         And hie to the woodland walks away;
         I tuck up my robe, and am buskined soon,
         And tie to my forehead a waxing moon,   30
         I course the fleet stag, unkennel the fox,
         And chase the wild goats o'er summits of rocks,
         With shouting and hooting we pierce through the
             sky,
         And Echo turns hunter, and doubles the cry.

                 Chorus of all

         With shouting and hooting we pierce through the
             sky,
         And Echo turns hunter, and doubles the cry.
JANUS.   Then our age was in its prime:
CHRONOS. Free from rage,
DIANA.   And free from crime.
MOMUS.   A very merry, dancing, drinking,
         Laughing, quaffing, and unthinking time.   40

                 Chorus of all

         Then our age was in its prime,
         Free from rage, and free from crime;
         A very merry, dancing, drinking,
         Laughing, quaffing, and unthinking time.

         Dance of Diana's attendants.   Enter Mars

MARS.   Inspire the vocal brass, inspire;°
         The world is past its infant age:
             Arms and honour,
             Arms and honour,
         Set the martial mind on fire,
         And kindle manly rage.   50
         Mars has looked the sky to red;°
         And peace, the lazy good, is fled.
         Plenty, peace, and pleasure fly;
             The sprightly green,
         In woodland walks, no more is seen;
         The sprightly green has drunk the Tyrian dye.°

                 Chorus of all.
                 Plenty, peace, etc.

MARS.   Sound the trumpet, beat the drum;
          Through all the world around,
          Sound a reveille, sound, sound,   60
          The warrior god is come.

Chorus of all.
Sound the trumpet, etc.

MOMUS.   Thy sword within the scabbard keep,
           And let mankind agree;
          Better the world were fast asleep,
           Than kept awake by thee.
          The fools are only thinner,
           With all our cost and care;
          But neither side a winner,
           For things are as they were.   70

Chorus of all.
The fools are only, etc.

Enter Venus

VENUS.   Calms appear, when storms are past;
          Love will have his hour at last:
          Nature is my kindly care;
          Mars destroys, and I repair;
          Take me, take me, while you may,
          Venus comes not every day.

Chorus of all.
Take her, take her, etc.

CHRONOS. The world was then so light,
          I scarcely felt the weight;   80
          Joy ruled the day, and love the night.
          But since the queen of pleasure left the ground,
           I faint, I lag,
           And feebly drag
          The ponderous orb around.

MOMUS.   All, all of a piece throughout;
          Thy chase had a beast in view;
                         [Pointing to Diana
          Thy wars brought nothing about;
                             [To Mars
          Thy lovers were all untrue.
                             [To Venus

JANUS.   'Tis well an old age is out,   90
CHRONOS. And time to begin a new.

Chorus of all.

All, all of a piece throughout;
Thy chase had a beast in view;
Thy wars brought nothing about;
Thy lovers were all untrue.
'Tis well an old age is out,
And time to begin a new.

Dance of huntsmen, nymphs, warriors and lovers

A Song

I

Fair, sweet, and young, receive a prize
Reserved for your victorious eyes:
From crowds, whom at your feet you see,
O pity, and distinguish me!
As I from thousand beauties more
Distinguish you, and only you adore.

II

Your face for conquest was designed,
Your every motion charms my mind:
Angels, when you your silence break,
Forget their hymns, to hear you speak;   10
But when at once they hear and view,
Are loath to mount, and long to stay with you.

III

No graces can your form improve,
But all are lost, unless you love;
While that sweet passion you disdain,
Your veil and beauty are in vain.
In pity then prevent my fate,
For after dying all reprieve's too late.

Aesacus Transformed into a Cormorant

These some old man sees wanton in the air,
And praises the unhappy constant pair.
Then to his friend the long-necked Cormorant shows,°
The former tale reviving others' woes:
'That sable bird,' he cries, 'which cuts the flood
With slender legs, was once of royal blood;
His ancestors from mighty Tros proceed,°
The brave Laomedon and Ganymede,°
Whose beauty tempted Jove to steal the boy,
And Priam, hapless prince! who fell with Troy.   10
Himself was Hector's brother, and, had fate
But given this hopeful youth a longer date,
Perhaps had rivalled warlike Hector's worth,
Though on the mother's side of meaner birth;
Fair Alexiroë, a country maid,
Bore Aesacus by stealth in Ida's shade.
He fled the noisy town and pompous court,
Loved the lone hills and simple rural sport,
And seldom to the city would resort.
Yet he no rustic clownishness professed,   20
Nor was soft love a stranger to his breast;
The youth had long the nymph Hesperie wooed,
Oft through the thicket or the mead pursued.
Her haply on her father's bank he spied,
While fearless she her silver tresses dried;
Away she fled; not stags with half such speed,
Before the prowling wolf, scud o'er the mead;
Not ducks, when they the safer flood forsake,
Pursued by hawks, so swift regain the lake.
As fast he followed in the hot career;   30
Desire the lover winged, the virgin fear.
A snake unseen now pierced her heedless foot,
Quick through the veins the venomed juices shoot;
She fell, and scaped by death his fierce pursuit.
Her lifeless body, frighted, he embraced,
And cried, "Not this I dreaded, but thy haste;
O had my love been less, or less thy fear!
The victory thus bought is far too dear.

Accursed snake! yet I more cursed than he!
He gave the wound; the cause was given by me.   40
Yet none shall say that unrevenged you died."
He spoke; then climbed a cliff's o'erhanging side,
And, resolute, leaped on the foaming tide.
Tethys received him gently on the wave;°
The death he sought, denied, and feathers gave.
Debarred the surest remedy of grief,
And forced to live, he cursed the unasked relief;
Then on his airy pinions upward flies,
And at a second fall successless tries,
The downy plume a quick descent denies.   50
Enraged, he often dives beneath the wave,
And there in vain expects to find a grave.
His ceaseless sorrow for the unhappy maid
Meagred his look, and on his spirits preyed.
Still near the sounding deep he lives; his name°
From frequent diving and emerging came.'

NOTES

I have used the following abbreviations (the place of publication is London unless otherwise specified):

C   *The Works of John Dryden*, ed. E. N. Hooker *et al.* (Berkeley and Los Angeles, 1956–   ). In progress ('The California edition').

D   Dryden.

Elliot   *Virgil: 'The Georgics' with John Dryden's translation*, ed. Alistair Elliot (Ashington, Northumberland, 1981).

Hopkins and   *The Beauties of Dryden*, by David Hopkins and Tom Mason
Mason   (Bristol 1982).

J   *A Dictionary of the English Language*, by Samuel Johnson (1755).

Kinsley   *The Poems of John Dryden*, ed. by James Kinsley (4 vols., Oxford, 1958).

Noyes   *The Poetical Works of Dryden*, ed. by George R. Noyes (2nd edn., Boston, 1950).

Scott   *The Works of John Dryden*, ed. Walter Scott, revised by George Saintsbury (18 vols., London and Edinburgh, 1882–93).

Spingarn   *Critical Essays of the Seventeenth Century*, ed. J. E. Spingarn (3 vols., Oxford, 1908–9).

Watson   *'Of Dramatic Poesy' and Other Critical Essays*, ed. George Watson (2 vols., 1962).

 1 *To John Hoddesdon on his Divine Epigrams*. First published as a commendatory poem in John Hoddesdon's *Sion and Parnassus or Epigrams on Several Texts of the Old and New Testament* (1650) where it has the title 'To his friend the Author on his Divine Epigrams'. Hoddesdon was probably at Westminster with Dryden.

l. 16. *look the sun ... i'the face*: alluding to the popular story of an eagle's testing its young. If they looked at the sun without blinking, they would be cherished. Pliny, *Natural History* 10. 3.

l. 20. *Helicon*: a mountain in Boeotia where Hesiod was visited by the muses.

Heroic Stanzas. First published in *Three Poems upon the Death of his late Highness Oliver Lord Protector of England* ..., 1659. Cromwell died on 3 September 1658. 'Heroic' in the title refers to the stanza form (abab⁵), popularized by Davenant's *Gondibert* (1651), and used again in *Annus Mirabilis*.

l. 1. *now*: after Cromwell's funeral was celebrated on 23 November 1658.

l. 4. *eagle*. The Romans released an eagle, whose upward flight figured the ascent of the soul of a dead emperor to join the gods.

2 l. 9. *liberal*: generous. Also a pun: the 'liberal arts' were those 'becoming a gentleman' (J).

l. 32. *Pompey*. A Roman general; his fortunes began to decline at 45, the age at which Cromwell's began to rise.   l. 33. *He*: Cromwell.

3 l. 50. *bold Greek*: Alexander the Great.

ll. 54–5. *Till ... conquests*: 'till the island might by new maps be shown thick of conquests'.

l. 57. *weights*: palms were supposed to thrive when weighted.

l. 63. *Bologna's walls*. At the siege in 1512, it was said, a chapel in the walls dedicated to the Blessed Virgin was blown up only to land exactly in place.

l. 65. *rescued Ireland*. Cromwell reconquered Ireland in 1649 and 1650.

ll. 66–7. *treacherous Scotland . . . blessed that fate*. Cromwell conquered Scotland and united it with England in 1654.

4 l. 77. *Feretrian Jove*: the god to whom the Romans consecrated captured arms. Livy, I. 10.

5 ll. 105–8. *When . . . pay*. 'The stars balk at, and reluctantly obey, Heaven's decree to create "such heroic virtue" because it drains them of their store of good influences; just as the commons balk at passing an extraordinary tax levy' (C).

l. 113. *He made us freemen*. Cromwell's influence in Europe was very great.   l. 120. *Alexander*: Alexander VII, pope from 1655 to 1667.

l. 121. *boldly crossed the line*. Cromwell attacked San Domingo and Jamaica. No gold was found, except what the English stole from the Spanish.

6 l. 136. *vestal*. The virgin Tarpeia betrayed Rome to the Sabines and was crushed under their shields. Livy, I. 11.

l. 138. *giant-prince*. A great whale came up the Thames on 3 June 1658. Evelyn, *Diary*.

To . . . Sir Robert Howard. First published as a commendatory poem prefixed to Howard's *Poems*, 1660, which contained Howard's translation of the fourth book of Virgil's *Aeneid*, and Statius' *Achilleis*, referred to in ll. 55–86. Howard's *Poems* appeared soon after Charles's entry into London.

   Sir Robert Howard (1626–98) was to become Dryden's brother-in-law in 1663.

l. 16. *Samson's riddle*: Judges 14: 14–18.

7 l. 26. *curious net*. 'Rete Mirabile [wonderful net]' (D). The reference is to

the theory of Galen that there is a marvellous network in the human brain.   l. 58. *Maro's*: Virgil's. See headnote above.

8 l. 59. *Elissa's*: Dido's.

l. 77. *perspective*: mirrors arranged to make a distorted picture offer true proportions.

l. 83. *curious notes*: 'annotations on Statius' (D).

ll. 93–5. *To Charles . . . With Monck*. Howard's first poem was a panegyric on Charles; his last on George Monck (1608–70), instrumental in the arrangements for Charles's restoration.

l. 96. *Rufus'*. Dryden's note quotes Pliny, Letter 6. 10: 'Here lies Rufus, who once, having routed Vindex, claimed the sovereignty not for himself but for his country.'

9 *Astraea Redux*. First published in 1660. The title means 'Justice restored'. Charles II landed at Dover on 25 May 1660, and entered London on his birthday, 29 May.

motto. *Iam redit . . . regna*: from Virgil's 'Messianic' fourth eclogue, l. 6: 'now the maiden [Astraea] returns, and Saturn reigns again'.

ll. 9–12. *The ambitious Swede . . . bequeathed*. Charles X of Sweden invaded Poland in 1655. In 1656 he went to war with Denmark, abandoning the Polish campaign. In 1658 the Danes sued for peace. Charles died in February 1660, to be succeeded by a minor ('now guideless kingdom', l. 12).

ll. 13–18. *And heaven . . . lily's side*. In 1659 France and Spain ceased their long fight, and the next year Louis XIV married the Infanta Maria Theresa.

10 l. 35. *purple . . . scarlet*: of bishops and lords.

l. 36. *sanguine dye to elephants*. 'Apparently an inversion of the common notion that "they that govern elephants never appear before them in white; and the masters of bulls keep from them all garments of blood and scarlet . . ."' (Kinsley, quoting Jeremy Taylor).

l. 37. *Typhoeus*: otherwise Typhon, a many-headed monster; see Ovid, *Metamorphoses* 5. 325 ff.

l. 45. *Cyclops*: a one-eyed giant. At the time the word was associated with puritans and the Commonwealth.

11 l. 67. *Otho*. He helped Galba to become emperor, and then Galba adopted Piso (l. 70) as his heir; Otho had Galba murdered, to be defeated in his turn by Vitellius.

l. 74. *Worcester*: where the parliamentary army defeated Charles II, 3 September 1651.

l. 79. *David*: 2 Samuel 15–21.   l. 98. *grandsire*: Henri IV of France.

l. 101. *covenanting League's*. A comparison of the Covenanters of 1643 with the Catholic League of 1576 in France.

864   NOTES TO PAGES 12–20

12 l. 108. *epochés*. D intends the (English) plural of the Greek ἐποχή, cessation, check.

l. 145. *Booth's*. Sir George Booth's royalist rebellion at Chester was
crushed in August 1659.

13 l. 151. *Monck*: General George Monck orchestrated Charles's return.

l. 182. *Legion*: Mark 5: 9.

twice before. Cromwell ejected the rump of the long parliament in 1653;
reconstituted, it was dissolved again by Lambert in 1659.

14 l. 188. *forbidden bowls*: of wine.

l. 201. *Sforza*: arch-intriguer Ludovico Sforza (1451–1508).

l. 205. *helots*: slaves. According to Plutarch, the Spartans got them drunk,
and then displayed them before their children as a caution.

l. 216. *swarms of English*. They hurried to join the king in Holland.

l. 217. *Batavia*: Holland.

l. 219. *Scheveline's*: Scheveningen, near The Hague, where Charles
embarked for England.

15 l. 230. *Naseby*: a ship named after the parliamentary victory in 1645. She
was renamed the *Royal Charles*.

l. 235. *Gloucester's*: Henry, Duke of Gloucester, third son of Charles I.

l. 246. *Amphitrite*: a sea nymph, wife of Poseidon.

l. 249. *fasces*: emblems of authority.

ll. 262–5. *the almighty . . . name*. Exodus 23: 20–3; 24: 5–7.

16 l. 268. *rigid letter*. The 'hard' *g* of *leges* yields to 'laws'; or possibly just a
reference to 'the letter of the law'.

l. 288. *That star*. On the day Charles was born (29 May 1630), men
seemed to see a bright constellation at midday.

l. 292. *whiter*: more fortunate; a Latinism.

17 *To His Sacred Majesty, a Panegyric on His Coronation*. First published in
1661. Charles's coronation was on St George's day, 23 April 1661.

l. 1. *deluge . . . drowned*: Genesis 8.

l. 18. *guilty months*. 'Had Charles been crowned earlier than 25 March,
the first day of 1661, two months of his exile (25 March–25 May 1660)
would have been included in his coronation year' (Kinsley).

18 l. 23. *As heaven*: Exodus 16: 13–15.

20 l. 104. *Caesar's heart that rose above the waves*. Caesar was crossing the
river Anio in disguise. A storm arose and the master ordered his sailors to
turn back. 'Caesar, perceiving this, disclosed himself . . . and said "Come,
man, be bold and fear naught; thou carryest Caesar and Caesar's fortune
in thy boat" ' (Plutarch, *Life of Caesar* 38).

l. 127. *Two kingdoms*: Spain and Portugal. In May 1661 Charles announced his intention of marrying the Portuguese princess, Catherine of Braganza.

l. 129. *Royal Oak*. Four triumphal arches were erected for the coronation: on one was depicted the 'Royal Oak', in which Charles hid from the parliamentary forces after the battle of Worcester.

21 *To ... Dr Charleton*. First published as a commendatory poem prefaced to Charleton's *Chorea Gigantum* (1663), which was an answer to Inigo Jones's theory that Stonehenge was built by the Romans as a temple to the god Coelus; Charleton held it to be the work of the Danes.

Walter Charleton (1619–1707) was a physician and scientist.

l. 3. *the Stagirite*: Aristotle.

l. 25. *Gilbert*. William Gilbert (1540–1603), physician to Elizabeth and to James I, made discoveries in magnetism.

l. 27. *Boyle*: Robert Boyle (1627–91), chemist.

l. 28. *great brother*: Roger Boyle (1621–79), Earl of Orrery, politician and dramatist.

l. 31. *Harvey's*: William Harvey (1578–1657) discovered the circulation of the blood.

l. 32. *Ent*: Sir George Ent (1604–89) defended Harvey's theory.

22 l. 54. *when ... fled*: after the defeat at Worcester, 1651.

Prologue to The Rival Ladies. First published in Dryden's *The Rival Ladies* (1664).

23 *Annus Mirabilis: The Year of Wonders, 1666*. Completed in November 1666, published early in 1667. The title-page of the first edition continues: *A Historical Poem: containing the progress and various successes of our naval war with Holland, under the conduct of his highness Prince Rupert, and his grace the duke of Albemarle, and describing the fire of London.*

Dryden's dedication (not reprinted here) 'To the Metropolis of Great Britain, the most renowned and late flourishing City of London, in its representatives the Lord Mayor and Court of Aldermen, the Sheriffs and Common Council' ends 'your sufferings are at an end; and that one part of my poem has not been more a history of your destruction, than the other a prophecy of your restoration.'

The verse form, a quatrain with alternate rhymes, had been used before by Sir William Davenant in his heroic poem *Gondibert* (1651) to which Dryden refers in the 'account of the ensuing poem'.

An account of the ensuing poem, in a letter to the honourable Sir Robert Howard. On Howard, Dryden's brother-in-law, see 'To my Honoured Friend' (above, pp. 6–9).

24 *a play*: probably *Secret Love* (1668).

royal admiral: James, Duke of York, brother of and successor to Charles II.

generals: Prince Rupert and George Monck, Duke of Albemarle.

single Iliad: a single book of the *Iliad*.

Lucan: first-century-AD Roman author of *Pharsalia*, a historical epic on the civil wars.

Silius Italicus: first-century-AD Roman author of historical epic *Punica*.

25 *Alaric*: by Georges de Scudéry (1654).   *Pucelle*: by Jean Chapelain (1656).

translation of Homer, by Chapman. Chapman's translation of the *Iliad* (1598–1611) is in lines of seven feet, his *Odyssey* (1615) in lines of five.

Gondibert: by William Davenant. See Spingarn, 2. 19.

descriptas . . . salutor: Horace, *Ars poetica* 86–7.

27 *Dido*: in *Aeneid* 1 and 4.   *Ovid*: see *Metamorphoses* 10.

28 *totamque . . . miscet*: *Aeneid* 6. 726–7.   *lumenque . . . auro*: ibid. 1. 590–3.

materiam . . . opus: *Metamorphoses* 2. 5.   *dixeris . . . novum*: *Ars poetica* 47–8.

29 *et nova . . . detorta*: ibid. 52–3.   *Juvenal*: *Satires* 8. 3.

Virgil: *Aeneid* 6. 847.   *humi serpere*: *Ars poetica* 28.

Horace: ibid. 19.

30 l. 18. *God's people passed*: Exodus 14: 21–2.   l. 29. *Moses*: Exodus 17: 11–13.

31 *Pliny*: *Epistles* 7. 28.

32 *Annus Mirabilis*. The events of the 'year of wonders' with which Dryden deals can be briefly sketched. Essentially this 'historical' poem is about the naval battles in the summer of 1666 between the English and the Dutch, and the Great Fire of London in the autumn of 1666.

The Second Dutch War was declared in March 1665. Before this, England and Holland had been commercial rivals (this had resulted in an earlier war in 1652–4). In 1664 a fleet under Sir Robert Holmes took Dutch possessions in West Africa. In December 1664 Sir Thomas Allin attacked a Dutch fleet from Smyrna off Cadiz.

ll. 73–92: James, Duke of York, engaged the Dutch off Lowestoft on 3 June 1665.

ll. 93–152: The English fleet attempted to attack the Dutch at Bergen in Norway on 3 August 1665, but the plan miscarried. 'Dryden deals discreetly with a shameful episode' (Kinsley).

ll. 153–72: Louis XIV declared war on Holland's side in January 1666, but did not take much active part thereafter.

ll. 213–564: Albemarle engaged the Dutch on 1–4 June 1666.

ll. 565–600: The battered fleet was repaired quickly.

ll. 665–740: The English fleet sailed to meet the Dutch on 19 July 1666. The fleets met on 25 July.

ll. 813–32: On 8 August 1666, Sir Robert Holmes was sent to attack a Dutch fleet off Vlie island.

ll. 857–1044: The fire of London began in Pudding Lane 1–2 September 1666. It burned for four days.

l. 8. *so base a coast*: a pun on *pays bas*, 'low country'.

l. 10. *In eastern quarries*. 'Precious stones at first are dew, condensed and hardened by the warmth of the sun, or subterranean fires' (D).

l. 11. *Idumaean*. Idumaea is in southern Palestine.

l. 14. *Each waxing*. 'According to their opinion who think that great heap of waters under the line is depressed into tides by the moon towards the poles' (D).

33 l. 20. *our second Punic war*. Cromwell fought the first Dutch war in 1653–4.   l. 29. *Iberian*: 'the Spaniard' (D).

l. 32. *babe of Spain*. Philip IV of Spain died in 1665 leaving his throne to his infant son.   l. 40. *usurpers*: like Cromwell.

34 l. 59. *Proteus*. In his note D quotes incorrectly from Virgil's *Georgics* 4. 387–9: 'Under the waves, Proteus, of sea-green hue, feeds his monstrous sea-herd and mighty seals.'

l. 80. *Lawson*. Sir John Lawson, vice-admiral, was fatally wounded at the battle of Lowestoft.

35 l. 85. *Their chief*: Opdam, 'the admiral of Holland' (D).

l. 95. *southern climates*: 'Guinea' (D).

l. 97. *castors*: beavers, which secrete a substance used in making scent.

36 l. 137. *shipwreck everywhere*. In his note D cites a recollection of Petronius' *Satyricon* 115: 'if you work it out properly, you can be shipwrecked anywhere'.

37 l. 145. *Munster's prelate*: Bernhard von Galen, bishop of Munster, England's ally. He made peace with Holland in April 1666.

l. 146. *German faith*. D's note cites Tacitus, *Annals* 13. 54: 'no men excel the Germans in arms or loyalty'.

38 l. 181. *doubled charge*. Parliament voted two and a half million pounds towards the Dutch war in November 1664, and a further one and a quarter million in October 1665.

l. 197. *Scipio*: Scipio Africanus (236–184 BC), Roman general who defeated Hannibal at Carthage in 202 BC.

l. 199. *fasces of the main*. Herodotus tells (4. 3–4) how the Scythians shook horsewhips instead of spears against a slave uprising.

l. 204. *future people*. D's note cites Pliny, *Panegyricus ad Trajanum* 26: 'swarms of children and a people unborn'.

39 l. 214. *famed commander*: De Ruyter.

l. 220. *bloody crosses*: the crosses of St George.

868 NOTES TO PAGES 39–47

l. 223. *The Elean plains*: 'where the Olympic games were celebrated' (D).

l. 228. *lands . . . strove*. D's note refers to Virgil's *Aeneid* 8. 691 f., which he later translated:

> It seems as if the Cyclades again
> Were rooted up, and jostled in the main
> Or floating mountains floating mountains meet
> Such is the fierce encounter of the fleet.

l. 235. *such port the elephant bears*: Pliny, *Natural History* 8. 29.

40 ll. 241–4. *Our . . . leaves*. 'The duke lost his breeches "to his skin"' (Kinsley).

l. 252. *the godlike fathers saw*: Livy, 5. 41.　　l. 253. *Patroclus' body lay*: *Iliad* 17.

l. 267. Berkeley: Sir William Berkeley (1639–1666). His ship, the *Swiftsure*, was cut off and taken.

l. 268. *Creusa*: Aeneas' blameless wife, who died at Troy; see *Aeneid* 2. 736–40.

41 l. 292. *His face spoke hope*. D's note cites Virgil, *Aeneid* 1. 209: 'he feigns hope with his look and holds his grief pressed deep in his heart'.

42 l. 326. *chase-guns*: cannon mounted in bows or stern of ship.

l. 330. *Cacus*: a fire-breathing monster, son of Vulcan; see Virgil, *Aeneid* 8. 225–61.

43 l. 344. *flies at check*: pursues some smaller game.

l. 367. *Thus Israel*: Exodus 13: 21–22.

44 l. 371. *Xenophon*: Greek historian who led the surviving Greeks back after their invasion of Persia.

l. 374. *he that touched the Ark*: Uzza: 1 Chronicles 13: 9–10.

l. 384. *And slowly moves*. D's note cites Virgil's *Aeneid* 9. 797–8: 'he retraces his steps unhurriedly'.

l. 391. *weary waves*. D's note cites Statius' *Silvae* 5. 4. 5–6, roughly translated in ll. 391–2.

l. 396. *succeeding day*. 'The third of June, famous for two former victories' (D); they were in 1653 and 1665.

45 ll. 417–20. *For . . . arose*. This stanza was felt to be blasphemous, and Dryden substituted:

> For now brave Rupert from afar appears,
> 　Whose waving streamers the glad general knows;
> With full spread sails his eager navy steers,
> 　And every ship in swift proportion grows.

47 l. 472. *when . . . still*: Joshua 10: 12–14.

l. 491. *So glides*. D's note cites Virgil, *Georgics* 3. 423–4, roughly translated in ll. 491–2.   l. 494. *throwing*: throwing dice.

49 l. 536. *o'ercome*. D's note cites Horace, *Odes* 4. 4. 51–2, translated in ll. 535–6.   l. 545. *fiends*. Mark 3: 11.

l. 553. *unripe*: still growing. Metals were supposed to 'grow'.

50 l. 573. *grows warm*. 'Fervet opus: [the work grows warm: *Georgics* 4. 169] the same similitude in Virgil' (D).

l. 582. *oakum*: loose fibre drawn from old rope.

l. 590. *cerecloth*: wrap in waxed cloth.

51 l. 595. *big-corned*: big-grained.

l. 601. *London*. The *London* blew up in 1665, and was replaced by the *Loyal London* in 1666.

52 l. 629. *Saturn*. Saturn, dethroned by his son Jupiter, fled by sea to Italy, where he brought arts and laws, *Aeneid* 8. 319 f.

l. 639. *Beyond the year*. D's note cites Virgil, *Aeneid* 6. 796: 'beyond the course of the years and the sun'.

l. 649. *Instructed*: 'By a more exact measure of longitudes' (D).

53 l. 653. *our globe's last verge*: 'The horizon, where ... the terrestrial and celestial spheres meet' (Kinsley).

l. 655. *our rolling neighbours*: the planets.

l. 656. *Royal Society*: founded in 1660. D was an early member.

l. 682. *Straits*: of Gibraltar.

l. 683. *working*: restlessly moving.

54 l. 685. *Allin*: Sir Thomas Allin (1612–85), always loyal to the royalist cause. He attacked the Dutch convoy from Smyrna in the Straits of Gibraltar.   l. 687. *Holmes*: Sir Robert Holmes (1622–92).

l. 692. *tempting fruits*. 'When the Roman senators admired some figs which Cato the Censor dropped in the Senate, "he told them that the country where that fine fruit grew was but three days' sail from Rome"' (Kinsley, citing Plutarch's *Lives*).

l. 693. *Spragge*: Sir Edward Spragge (d. 1673).

l. 695. *Harman*: Sir John Harman (d. 1673). During the four days' battle he saved the *Henry*, twice fired by Dutch fireships.

l. 697. *Holles*: Sir Frescheville Holles (1641–72). His left arm was shot off in the battle of 2 June.

55 l. 728. *gross*: main body.   l. 731. *blind*: concealed.

l. 736. *trident*. D's note cites Virgil, *Aeneid* 1. 145–6: 'he lifts them himself with his trident, and opens the huge quicksands, and calms the sea'.

l. 742. *battles*: ships in battle order.

56 l. 760. *And ... so*. D's note cites Virgil, *Aeneid* 5. 231: 'they are able because they seem to be able'.

l. 762. *Caesar's*: Charles II's.   l. 773. *famous leader*: De Ruyter.

l. 775. *Varro*: Roman commander at defeat of Cannae.

57 l. 788. *patron saint*: 'St James, on whose day [25 July] this victory was gained' (D).

l. 790 *Philip's manes*. 'Philip the second of Spain, against whom the Hollanders rebelling, were aided by Queen Elizabeth' (D).

l. 803. *Bourbon foe*: Louis XIV, imagined 'disowned' by his grandfather Henri IV.

58 l. 825. *vexed*: (1) twisted; (2) tormented.   l. 829 *rummage*: empty.

l. 832. *Transitum*: transition.

59 l. 847. *Great as the world's*. D's note cites Ovid, *Metamorphoses* 1. 257–8: 'when the sea, when the land, and the royal palace of heaven will catch fire and burn'.

60 l. 881. *like crafty*. D's note cites Terence, *Heauton Timorumenos* 366–7: 'she ruled the man's desire artfully, to inflame his passion by frustrating it'.

61 l. 914. *buckets to the hallowed choir*. Fire buckets were kept in churches.

l. 922. *And lightened*. D's note cites Virgil, *Aeneid* 2. 312: 'the wide Sigaean straits reflect the fire'.

l. 926. *Simois*. Hephaestus attacked the river Simois with flames for trying to drown Achilles (*Iliad* 21. 305 ff.).

62 l. 944. *Lombard bankers*. The bankers in Lombard Street, near the Exchange.

64 l. 1006. *permitted*: abandoned.

66 l. 1066. *spotted deaths*: the plague of 1665.

l. 1069. *frequent funerals*: numerous corpses.

l. 1093. *orphans*: of Christ's Hospital.

67 l. 1099. *poet's song*: alluding to Edmund Waller's 'Upon his majesty's repairing of Paul's'.

l. 1100. *poet's songs*. Amphion built the walls at Thebes by playing the harp.

68 ll. 1157–60 *Not ... went*: Ezra 1–3.

69 l. 1165. *trines*. A trine is the relative position of two planets 120° apart.

l. 1169. *chemic*: alchemic, turning to gold.

l. 1171. *which gives the Indies name*: 'Mexico' (D).

l. 1177. *august*: 'Augusta, the old name of London' (D).

70 l. 1193. *Tagus*: a river of Spain.   l. 1210. *left behind*: still to come.

An Essay of Dramatic Poesy. First published in 1668, and revised in 1684: the revisions, a few of which are adopted here, are largely grammatical. In the early editions the title is *Of Dramatic Poesy: An Essay.* I have preferred the form given in the running-title.

Buckhurst. Charles Sackville (1643–1706), Lord Buckhurst, later Earl of Dorset. The 'Eugenius' of the ensuing dialogue, though he was present at the battle on 3 June 1665.

plague. During the Great Plague of 1665, Dryden stayed with his father-in-law in Wiltshire.

71 *Pompey: Pompey the Great* (1664), from Pierre Corneille's *La Mort de Pompée* (1644). Buckhurst presumably translated the fourth act.

Spurina. See Valerius Maximus, *De verecundia* 4. 5.

pars ... cubilia: Horace, *Epodes* 16. 37–8.

72 *Le jeune ... passage*: I. de Benserade, *Ballet royale de la nuit* (1653).

As nature . . . sing: Sir William Davenant, 'Poem to the King's most Sacred Majesty' (1663), slightly misquoted.   *Homer: Iliad* 16. 124 ff.

to defend mine own: 'To the Right Honourable Roger, Earl of Orrery' (1664).   *first made public*: in the dedication to *The Rival Ladies* (1664).

73 *one of his dialogues*: Cicero's *De legibus.*

Cicero: Cicero, *Ad Atticum* 12. 40; Plutarch, *Life of Julius Caesar* 54.

sine ... ira: *Annals* 1. 1 (adapted).

74 *memorable day*: 3 June 1665. The English won a victory over the Dutch off Lowestoft.   *late war*: the 'second Dutch war', 1665–7.

highness: James, Duke of York.

the park: St James's park, a fashionable gathering place.

Eugenius ... Neander. The speakers in the debate are traditionally associated with Dryden and three of his friends. Eugenius ('well-born') represents Buckhurst, to whom the work is dedicated; Crites ('censorious') seems to be a portrait of Sir Robert Howard (1626–98), Dryden's brother-in-law; Lisideius (the name is a Latinized anagram) is Sir Charles Sedley or Sidley (1639–1701), poet and wit; and Neander ('new man') is Dryden himself.

75 *Quem ... scriberet*: Cicero, *Pro archia poeta* 10. 25.

two poets: possibly Richard Wild (1609–79), and Richard Flecknoe (d. 1678), who had written poems on the battle.

76 *catachresis*: the improper use of a word. Dryden associates the fault particularly with John Cleveland (1612–58), a late metaphysical poet.

leveller: the Levellers were a sect of democrats thrown up by the English civil war.   *pauper ... pauper*: *Epigrams* 8. 19.

77 *Withers*: George Wither (1588–1667), prolific puritan poet.

Change-time. The time during which business was transacted.

candles' ends. Bidding at auctions stopped when the candles burned out.

Qui Bavium . . . : Virgil, *Eclogues* 3. 90.

nam . . . *contemnimus*: quotation untraced.

pace . . . *perdidistis*: *Satyricon* 2.   *indignor* . . . *nuper*: *Epistles* 2. 1. 76–7.

78 *si* . . . *annus*: ibid. 2. 1. 34–5.

Sir John Suckling: (1609–42), cavalier poet.

Mr Waller: Edmund Waller (1606–87), cavalier poet.

Sir John Denham: (1615–69), cavalier poet.

Mr Cowley: Abraham Cowley (1618–67), late metaphysical poet.

79 *only a genere et fine.* The definition has the class (*genere*) and the purpose (*fine*), but not the differentiation which would exclude other literary forms.

80 *school*: medieval philosophers and theologians ('schoolmen').

Aeschylus, Euripides, Sophocles, Lycophron. The first three are the main surviving ancient Greek dramatists; Lycophron was an Alexandrian poet whose plays have not survived.

Alit . . . *accendit. Historia Romana* 1. 17.

81 περὶ τῆς Ποιητικῆς: '*On the Art of Poetry*' (the *Poetics*). Only the sections on tragedy are extant.

extracted: by Ludovico Castelvetro (1505–71) in the commentary to his Italian translation of the *Poetics* (*On the Art of Poetry*). Castelvetro was followed by Pierre Corneille's *Discours des trois unités* (1660), which Dryden follows.

82 *Discoveries: Timber or Discoveries* (1641).

83 *Corneille: Discours des trois unités* (1660).

Menander: see Horace, *Epistles* 2. 1. 57–9; and Aulus Gellius, *Noctes Atticae* 2. 23.

half-Menander: see Suetonius, *Vita Terentii*. Caius Caesar said Terence was only half as great a poet as Menander.

Varius . . . *Paterculus*: Horace, *Odes* 1. 6; *Satires* 1. 9. 23, 1. 10. 44; *Ars poetica* 55; Martial, *Epigrams* 8. 18. 7. The reference in Velleius is undiscovered.

Macrobius: in *Conviviorum saturnaliorum libri septem* 4.

84 *audita* . . . *credimus: Historia Romana* 2. 92.

85 *Aristotle.* Dryden's divisions come not from Aristotle, but from the *Poetices* of J. C. Scaliger (1561).

λύσις: literally, 'untying', like *dénouement*.

neu . . . *actu*: misquoted from *Ars poetica* 189.

86　*late writer*: Sir Robert Howard, in the preface to *Four New Plays* (1665).

　　Jonson: ed. C. H. Herford and P. and E. Simpson, 8. 641.

　　good cheap: easily.　　*Juno ... opem*: Terence, *Andria* 3. 1. 15.

87　*precept of the stage*. This precept derives not from the French poets, but from Castelvetro.　　*Scaliger: Poetices* 6. 3.

88　*French poet*: Pierre Corneille, *Troisième discours*.

89　*Medea*: in Euripides' play.

　　sock and buskin: standing for comedy and tragedy.

　　tandem ... triduum: Terence, *Eunuch* 2. 1. 17–18.

90　*sed ... stolide*: *Ars poetica* 270–2.　　*multa ... loquendi*: ibid. 70–2.

　　mistaque ... acantho: *Eclogues* 4. 20.　　*mirantur ... carinas*: *Aeneid* 8. 91–3.

　　si ... caeli: *Metamorphoses* 1. 175–6.　　*et ... pompas*: ibid. 1. 561.

91　*Had ... home*: Cleveland, 'The Rebel Scot', 63–4.

　　Si ... dixisset: Juvenal, *Satires* 10. 123–4.

　　For ... destroys: Cleveland, 'To Prince Rupert', 39–40; 'white powder' is arsenic.　　*omne ... vincit*: Ovid, *Tristia* 2. 381.

　　Myrrha ... Caunus ... Biblis: *Metamorphoses* 10. 312 ff; 9. 453 ff.

92　*anima ... ψυκή*: Juvenal, *Satires* 6. 195.

93　*sum ... notus*: *Aeneid* 1. 378–9.　　*fanfaron*: braggart.

　　si ... aevum: *Satires* 1. 10. 68.　　*quos ... sacravit*: Horace, *Epistles* 2. 1. 49.

94　*leaving the world*: Beaumont died in 1616, Fletcher in 1625, and Jonson in 1637.

95　*Red Bull*: a rowdy playhouse in St John's Street, Clerkenwell.

　　atque ... poscunt: Horace, *Epistles* 2. 1. 185–6.

　　admiration: not a requirement of Aristotle. Dryden may have got the idea from Sidney's *Apology for Poetry*.

　　ex ... sequar: *Ars poetica* 240.　　*atque ... imum*: ibid. 151–2.

96　*old age*: see Justinus, *Universal History* 1. 8; Xenophon, *Cyropaedia* 8. 7.

　　perspective: telescope.　　*quodcumque ... odi*: Horace, *Ars poetica* 188.

　　τὰ ἔτυμα ... ἐτύμοισιν ὁμοῖα: Hesiod, *Theogony* 27; Homer, *Odyssey* 19. 203.

　　Rollo. The anonymous play *The Bloody Brother* was retitled *The Tragedy of Rollo* in 1640.　　*Herodian*: books 3 and 4.

97　*oleo*: hotchpotch.　　*Golias*: a comic figure of medieval literature.

　　Sejanus: 2. 1.　　*Catiline*: 3. 3; 2. 1.

ingenious person: perhaps Thomas Sprat (1635–1713), in his *Observations on M. de Sorbier's Voyage into England* (1665).

protatic persons: characters who appear only in the introduction of a play.

interested: concerned.

98 *fight prizes*: engage in prize-fighting.

99 *What the philosophers say*. Perhaps Dryden is drawing on Descartes' *Principia philosophiae* (1644), 2. 37, translated 1664.

Corneille: in *Discours des trois unités*.

100 *segnius ... anguem*: *Ars poetica* 180–7.   *Magnetic Lady*: 3. 2.

Eunuch: 4. 3.   *Sejanus's*: *Sejanus* 5. 10.

King and no King: by Beaumont and Fletcher, 1611.

101 *Scornful Lady*: by Beaumont and Fletcher, 1609–10; the 'conversion' occurs in 5. 1.   *necessary*: inevitable.

great and judicious poets. Sir Robert Howard found rhyme 'unnatural' on the stage in the preface to *Four New Plays* (1665).

an ancient author: Velleius Paterculus, *Historia Romana* 1. 17.

102 *Liar*: *Le Menteur* (1643); translated as *The Mistaken Beauty or The Liar* (1661).   *He tells you himself*: in *Discours du poème dramatique*.

younger Corneille: Thomas Corneille (1625–1709), brother of Pierre.

Quinault: Phillipe Quinault (1635–88). His plays inspired some of Dryden's.   *Richelieu*: he died in 1642.

103 *Adventures*: *Adventures of Five Hours* (1663), by Sir Samuel Tuke.

bait upon a journey: stop for food.

104 *primum mobile*: the outermost, ninth sphere of the old Ptolemaic astronomy; it gives motion to the eight inner spheres.

Crites: by a slip, Dryden wrote 'Eugenius'.

co-ordination: the absence of the orderly hierarchical arrangement of things; the opposite of 'subordination'.

Cinna ... Pompey ... Polyeucte: by Pierre Corneille. *Cinna*, 1640; *La Mort de Pompée*, 1644; and *Polyeucte*, 1643.

105 *Maid's Tragedy*: a play by Beaumont and Fletcher, 1619.

The Alchemist: by Ben Jonson, 1610.

The Silent Woman: by Ben Jonson, 1609.

The Fox: *Volpone* by Ben Jonson, 1607.

107 *judgment*: in the prologue to *Every Man in his Humour*, ll. 8–16.

108 *one of their newest plays*: *L'Amour à la mode*, by Thomas Corneille, 1651.

109 *all*: an exaggeration.   *examen*: critical analysis.

110 *quantum ... cupressi*: Virgil, *Eclogues* 1. 25.

Mr Hales of Eton: John Hales (1584–1656).   *verses*: *Epigrams* 55.

111 *Philaster*: first produced about 1609.

112 *The Silent Woman*: *Epicoene*, acted 1609.

113 τὸ γελοῖον: Aristotle, *Poetics* 5.   *Socrates ... spectators*: in Aristophanes' *Clouds*.

114 *ex ... dicas*: Terence, *Eunuch* 460.

115 *creditur ... minus*: *Epistles* 2. 1. 168–70.

has prevailed himself: French *se prévaloir de* 'to give oneself the advantage of'.

116 *ubi ... maculis*: Horace, *Ars poetica* 351–2.

117 *vivorum ... difficilis*: Velleius Paterculus, *Historia Romana* 2. 36.

118 *etiam ... Laberi*: Macrobius, *Saturnalia* 2. 7.

Aristotle: *Poetics* 4.   *nicking*: corresponding exactly with.

Arcades ... parati: Virgil, *Eclogues* 7. 4–5.

119 *quicquid ... dicere*: Dryden quotes, apparently, from Sidney's *Apology for Poetry*, where the quotation is ascribed to Ovid. It resembles *Tristia* 4. 10. 26.

Seneca. Dryden confuses Marcus Seneca the rhetorician and Lucius Seneca the philosopher.

omnia ... ponto: Ovid, *Metamorphoses* 1. 292.

120 *both to that person*. Editors suspect corruption here. Nichol Smith suggests reading 'both to you and that person'. 'That person' may be Aristotle.   *I ... make*: source unknown.

121 *prevail himself*: see note to p. 115, l. 16 above.

perpetuo ... fluere: Cicero, *De oratore* 6. 21.

122 *The Rival Ladies*: the first of Dryden's plays to be published, 1664.

nations. After 'nations' the second edition (1684) adds 'at least we are able to prove that the eastern people have used it from all antiquity, *viz.*, Daniel's *Defence of Rhyme*'.

Pindaric way. At this time Pindar was a byword for irregularity.

Siege of Rhodes: opera by Sir William Davenant, first produced in 1656.

124 *tentanda ... humo*: Virgil, *Georgics* 3. 8–9.

The Faithful Shepherdess ... Sad Shepherd: by John Fletcher (1610) and Ben Jonson (1641) respectively.

Hopkins and Sternhold's ... Sandys's. The versified sixteenth-century version of the Psalms by the first two was very popular. George Sandys's verse *Paraphrase upon the Psalms* (1636) was vastly superior.

est ... peccat: Horace, *Epistles* 2. 1. 63.

Mustapha: *The Tragedy of Mustapha*, by Roger Boyle, Earl of Orrery (1621–79), performed 1665.

The Indian Queen and Indian Emperor: by Dryden, 1664 and 1665. The first was written in collaboration with Sir Robert Howard.

125 *indignatur … Thyestae: Ars poetica* 90–1.   *effutire … versus*: ibid. 231.

sonnet: a short poem.   *Aristotle*: *Poetics* 26.

126 *quidlibet audendi*: Horace, *Ars poetica* 10.

musas … severiores: Martial, *Epigrams* 9. 11. 17.

127 *water poet's*. The water poet was John Taylor (?1578–1653), waterman on the Thames and doggerel poet.

128 *Delectus … eloquentiae*: Cicero, *Brutus* 72.

reserate … laris: *Hippolytus* 863.

129 *a most acute person*: Sir Robert Howard.

Somerset-Stairs: a landing-place near Somerset House.

130 *Piazze*: in Covent Garden.

Prologues to Secret Love. First published in Dryden's *Secret Love* (1668).

l. 7. *dead colours*: 'the preparatory layers of colour in a painting' (Kinsley).

l. 54. *Throw boldly*: 'a sustained metaphor from gaming. *Sets*, challenges by putting down a stake; *lay*, wager; *nick*, beat with a winning cast; *on tick*, on credit' (Kinsley).

132 *Prologue and Epilogue to Sir Martin Mar-all*. First published in *Sir Martin Mar-all* (1668). The play was first performed in 1667.
Prologue
l. 2. *regalios*: choice entertainments.

l. 6. *statutable ass*: an ass that meets requirements.

l. 10. *woodcocks*: simpletons.

Epilogue
l. 13. *Lilly*: William Lilly (1602–81), astrologer.

133 *Prologue to The Wild Gallant revived*. First published in *The Wild Gallant* (1669). That play was first produced in 1663 and revived in 1667.

l. 8. *Whetstone's Park*: a lane between Holborn and Lincoln's Inn Fields, site of a noted whorehouse.   l. 25. *balked*: ignored.

Prologue and Epilogue to The Tempest. First published in Dryden and Davenant's adaptation of Shakespeare's play, *The Tempest* (1670). The play was first performed in 1667.

Prologue

134 l. 15. *the neighbouring shore*: a reference to the rival King's company in the Theatre Royal, who had recently staged John Fletcher's *The Sea Voyage*.

l. 18. *Enchanted Isle*: the subtitle of *The Tempest*.

Epilogue

135 l. 4. *general rot*. This, and the following two lines, refer to adaptations and imitations of French and Spanish drama.

Prologue and Epilogue to Tyrannic Love: first published in Dryden's *Tyrannic Love* (1670). The play was first performed in 1669.

Epilogue

136 Mrs Ellen: Nell Gwyn (1650–87), actress.

l. 13. *taking*: excitement.

137 l. 30. *St Catharn*: St Catharine, the familiar title of *Tyrannic Love*.

Prologue to The First Part of The Conquest of Granada. First published in Dryden's *The Conquest of Granada* (1672). The first part was acted in 1670.

l. 1. *This jest*: the 'broad-brimmed hat'.   *t'other house's*. The other house was the Duke's.

l. 7. *Nokes*: James Nokes (d. 1696), famous comic actor with the Duke's company.

l. 19. *two*: Nokes, and Edward Angel (d. 1673), another comedian with the Duke's company.   l. 20. *blocks*: blockheads.

Epilogue to The Second Part of The Conquest of Granada. First published in Dryden's *The Conquest of Granada* (1672). The second part was first acted in 1671.

138 l. 6. *Cobb's tankard ... Otter's horse*. Cobb is in Jonson's *Every Man in his Humour* (1601) and Otter in Jonson's *Epicoene* (?1609). Otter calls one of his tankards 'Horse'.

139 l. 16. *grains*: tiny amounts.

Prologue and Epilogue to Aureng-Zebe. First published in Dryden's *Aureng-Zebe* (1676). The play was first performed in 1675.

Prologue

l. 8. *his long-loved mistress, Rhyme*. *Aureng-Zebe* was Dryden's last rhymed heroic play.

140 l. 37. *our neighbours*: the rival Duke's company.

Epilogue

141 l. 21. *silk-weavers*: commercial rivals of the French.

l. 25. *gens barbares*: barbarous people (French).

Epilogue to The Man of Mode. First published in George Etherege's *The Man of Mode* (1676). The play was first performed in 1676.

l. 3. *harlequins*: traditional figures in French and Italian farces.

142 l. 7. *Sir Fopling*: fop in *The Man of Mode*.   l. 9. *cocks*: struts.

l. 12. *graft upon his kind*: add to his natural stock of talents.

l. 16. *knight of the shire*: county member of parliament.

l. 22. *toss*: of the head.   *wallow*: rolling gait.   l. 24. *snake*: wig-tail.   l. 28. *shog*: shake.

MacFlecknoe. First published in a pirated version in 1682, and published in an authorized version in *Miscellany Poems* in 1684. There is strong evidence that it was written in substantially the form given by *1684* in 1676. During this long period (unusual for Dryden) between writing and publication, the poem had a limited circulation in manuscript form. At least 12 manuscript copies are extant, none of them in Dryden's hand. Readings in the manuscripts do not substantially alter the text as given by *1684*, the version given here.

The poem erects a series of ironical parallels between the handing on of the empire of Rome from Aeneas to Ascanius (Julus), and the handing on of the empire of dulness from Flecknoe to Shadwell.

The occasion of *MacFlecknoe* was a prolonged debate on the nature of drama, on plagiarism, and on wit and humour, between Dryden and Thomas Shadwell, a playwright at the rival Duke's House. The debate was sharpened when, in the dedication to *The Virtuoso* (1676), a satire on the new science of the Royal Society, of which Dryden had been a member, Shadwell went over the old ground and included an offensive and gratuitous swipe at Dryden's pension as Laureate and Historiographer Royal.

title. *MacFlecknoe*: son of Flecknoe.

l. 3. *Flecknoe*: Richard Flecknoe (d. 1678?), poetaster and catholic priest. He was not Irish, as is often assumed.   *like Augustus, young*. Octavius Caesar became emperor at the age of 32.

143 l. 15. *Shadwell*. spelt 'Sh—' in *1684*. Proponents for the idea of a faecal innuendo get no support from the manuscripts. For Thomas Shadwell (?1642–92), playwright, see headnote above.

l. 25. *goodly fabric*: great bulk.

l. 29. *Heywood ... Shirley*. Thomas Heywood (?1570–1641) and James Shirley (1596–1666), dramatists.

l. 32. *sent before but to prepare thy way*: Matthew 3: 3–4.

l. 33. *Norwich drugget*. Drugget is coarse stuff of wool and linen. Shadwell came from Norfolk.

l. 35. *warbling lute*. In 'Flecknoe, an English Priest at Rome' Andrew Marvell tells how Flecknoe tried 'to allure me with his lute' (l. 36).

l. 36. *King John of Portugal*. Flecknoe had boasted of being patronized by the king.

l. 42. *Epsom blankets*. Sir Samuel Hearty in *The Virtuoso* was tossed in blankets. There is also a glance at a recent affray at Epsom, in which Rochester and some friends tossed some fiddlers. Later a man was murdered.   l. 47. *Pissing Alley*: near the Strand.

l. 48. *Ashton Hall*: not satisfactorily explained. Perhaps the town house of Edmund Ashton, a friend of Thomas Shadwell. *1682* reads '*Aston-Hall*'; *1684* '*A— Hall*'.   l. 50. *toast*: waste.

144 l. 53. *St André's*. St André was a French dancing master who choreographed the dances in Shadwell's *Psyche* (1675).

l. 54. *Psyche's*: *Psyche*, an opera by Shadwell (1675).

l. 57. *Singleton*: James Singleton (d. 1686), one of the king's musicians.

l. 59. *Villerius*: a character in Davenant's *The Siege of Rhodes* (1656).

l. 64. *Augusta*. 'The old name of London' (Dryden's note to *Annus Mirabilis* 1177).

l. 67. *Barbican*. A barbican is an outer defence, especially a double tower over a gate.

ll. 72–3. *Where . . . sleep*: parodying Abraham Cowley's *Davideis* (1656) 1. 79–80:

> Where their vast court the mother-waters keep,
> And undisturbed by moons in silence sleep.

ll. 76–7. *Where . . . try*: parodying *Davideis* 1. 75–6:

> Beneath the dens where unfledged tempests lie,
> And infant winds their tender voices try.

l. 78. *Maximins*: Maximin is the ranting hero of Dryden's *Tyrannic Love* (1670).

l. 79. *Fletcher*: John Fletcher (1579–1625), dramatist.   *buskins*: the high boots of Greek tragic actors; 'treads in buskins' means to write tragedy.

l. 80. *Jonson . . . socks*: Jonson as writer of comedy; socks are the light shoes of Greek comic actors.

l. 81. *Simkin*: simpleton.   l. 84. *Panton*: apparently, a character in farce.

l. 87. *Dekker*: Thomas Dekker (1572–1632), city playwright.

ll. 91–2. *Misers . . . Hypocrites*: alluding to early plays of Shadwell: *The Miser* (1672), *The Humourists* (1671), and *The Hypocrite* (lost).

l. 93. *Raymond . . . Bruce*: men of wit in *The Humourists* and *The Virtuoso* (1676) respectively.

145 l. 97. *From . . . Street*: from one end of the town to the other; Bunhill was in the City of London; Watling Street, the Roman road which extends across England.

l. 101. *pies . . . bum*. Unsold unbound printed sheets were used to line baking-dishes, or to furnish privies.

l. 102. *Ogilby*: John Ogilby (1600–76), translator of Virgil and Homer.

l. 104. *Bilked stationers*: booksellers ruined by the failure of their authors.

l. 105. *Herringman*: publisher of both Dryden and Shadwell.

l. 107. *High on a throne of his own labours reared*: alluding to Milton's Satan, 'High on a throne of royal state' (*Paradise Lost* 2. 1).

l. 109. *Rome's other hope*: translating *Aeneid* 12. 168.

ll. 110–11. *His brows . . . face*: parodying *Aeneid* 2. 682–4, which Dryden later rendered (*The Works of Virgil*, *Aeneid* 2. 930–2):

> from young Julus' head
> A lambent flame arose, which gently spread
> Around his brows, and on his temples fed.

l. 122. *Love's Kingdom*: a tragicomedy by Flecknoe (1664).

l. 126. *poppies*: soporifics instead of the expected laurels. Shadwell took opium.   l. 129. *owls*: apparent wisdom masking real stupidity.

l. 131. *Presage . . . took*. Plutarch tells how Romulus won his argument about the site of Rome with Remus by seeing twelve vultures to Remus' six.

146   l. 151. *gentle George*: George Etherege (?1635–91), poet and playwright.

l. 152. *Dorimant . . . Loveit*: characters in Etherege's *The Man of Mode* (1676).

l. 153. *Cully, Cockwood, Fopling*: characters in Etherege's *The Comical Revenge* (1664), *She would if she could* (1668), and *The Man of Mode*.

l. 163. *Sedley*: Sir Charles Sedley (1639–1701), poet. He was said to have helped Shadwell with *Epsom Wells*.

l. 168. *Sir Formal's*: Sir Formal Trifle, a character in Shadwell's *Virtuoso*.

l. 170. *northern dedications*. Shadwell frequently dedicated his plays to the Duke or Duchess of Newcastle.

147   l. 178. *rail at arts*. Shadwell's *The Virtuoso* satirizes the science of the Royal Society.

l. 179. *Nicander's*. Nicander is a character in *Psyche*.

l. 181. *sold he bargains*. To 'sell bargains' is to reply coarsely to a question, as here in the idiom of Sir Samuel Hearty in *The Virtuoso*.

l. 188. *New humours*: parodying the Dedication to *The Virtuoso*, 'four of the humours are entirely new'.

ll. 189–92. *bias . . . will*: parodying Shadwell's Epilogue to *The Humourists* (1671):

> A humour is the bias of the mind
> By which with violence 'tis one way inclined:
> It makes our actions lean on one side still,
> And in all changes that way bends the will.

l. 194. *likeness*: to Jonson.   *tympany*: 'a kind of obstructed flatulence that swells the body like a drum' (J).

l. 195. *tun*: a large wine cask.   l. 196. *kilderkin*: a quarter of a tun.

l. 204. *keen iambics*: sharp satiric verse.

l. 206. *acrostic*: poem whose initial letters spell out a word.

l. 207. *wings display and altars raise*: in shaped poems, such as Herbert's 'Easter Wings' and 'The Altar'.

l. 212. *Bruce ... Longvil*: characters in *The Virtuoso*.

ll. 216–17. *prophet's part ... double portion*: 2 Kings 2: 9–13.

148 *Heads of an Answer to Rymer*. First published by Tonson in his edition of Beaumont and Fletcher (1711). Published from an independent transcription by Samuel Johnson in his *Life of Dryden* (1779). (Johnson's order of paragraphs is 34, 51–3, 35–8, 1–6, 29–33, 39–50, 7–28.) The *Heads* were written on the endpapers of the copy of *The Tragedies of the Last Age* (1677; dated 1678) that Rymer sent to Dryden late in 1677. Dryden expanded the ideas here, but with much less boldness, in 'The Grounds of Criticism in Tragedy' in the preface to *Troilus and Cressida* (1679).

Thomas Rymer (1643–1713) was a lawyer and critic much admired by Dryden, but by few since.

Aristotle places the fable first: *Poetics* 6.

149 *example of Phaedra, cited by Mr Rymer*: Rymer, *Critical Works*, ed. Curt A. Zimansky (New Haven, Conn., 1956), 50–9.

Fletcher: John Fletcher (1579–1625), popular playwright at the time of the Restoration.

those plays which he arraigns: Beaumont and Fletcher's *A King and No King* (1611) and *The Maid's Tragedy* (1610); and *The Bloody Brother* (1619), thought to be by Fletcher and others.

place it upon the actors: Zimansky, p. 19.

150 *nature ... is the same in all places*: Zimansky, p. 19.

with acorns ... or with bread: echoing Rymer (Zimansky, p. 20), but the image was familiar from Jonson's 'Ode to Himself' 11–12.

151 *Rollo*: in *The Bloody Brother*.

poetic justice: apparently the phrase is Rymer's invention (Zimansky, p. 26).

counter-turn: dramatic development.

turned design: a turn is a minor movement in the plot.

152 *Rapin confesses*: *Réflexions sur la poétique d'Aristote* (1674) 2. 10.

The parts of a poem: owing something to *Poetics* 6, and more to Rymer's translation of Rapin's *Réflexions sur la poétique d'Aristote* (1674).

153 *Aristotle: Poetics* 13.

last paragraph save one: paragraph 38, the last but two.

154 *for poetry is an art*: Rapin, *Réflexions* 1. 9.

punishment will be unjust: Aristotle, *Poetics* 13.

third actor: Zimansky, p. 22.

Rapin's words are remarkable: *Réflexions* 1. 26.

Prologue to Oedipus. First published in *Oedipus* (1679), by Dryden and Nathaniel Lee. The play was first performed in 1678–9.

155 l. 25. *at Mons you won.* The Dutch aided by the English beat the French at the battle of Mons, 17 April 1678. Peace between Holland and France was concluded the preceding day, and William of Orange (it was alleged) suppressed the news. 'Dryden may be implying that the English were, though brave, dupes' (David Wykes, *A Preface to Dryden* (1977) 171).

l. 28. *four first councils*: the general councils of Nicaea (325), Constantinople (381), Ephesus (431), and Calcedon (451), accepted by the Church of England as a standard of doctrine equal in authority to the canonical scriptures.

l. 36. *the Woollen Act.* To promote the use of English wool this act decreed that after 1 August 1678, burial shrouds should be of English wool only.

Preface to Ovid's Epistles Translated by Several Hands. Ovid's Epistles was first published in 1680.

Mr Sandys's undertaking: Ovid's Metamorphoses Englished, by George Sandys (1578–1644), published in 1626.

156 *ascribed to him*: Martial, *Epigrams* 11. 20.　　*fulsome*: filthy.

that author's life: by Suetonius.　　*Isis*: *Amores* 2. 13.

Ovid himself complains: *Amores* 3. 12. 7–12.　　*cur ... feci*: *Tristia* 2. 103.

157 *nudam ... Dianam*: *Tristia* 2. 105.　　*The first verses*: *Amores* 3. 12. 13–18.

account of his own life: *Tristia* 4. 10. 43–54.

158 *nescivit ... relinquere*: M. Annaeus Seneca, *Controversiae* 9. 5. 17.

159 *Horace: Ars poetica* 15–16.

Heinsius. Daniel Heinsius' edition appeared in 1652.

vindicated to: asserted ownership.

Art of Love: Ars amatoria 1. 713.　　*quam ... Sabinus*: *Amores* 2. 18. 27–8.

Epistle of Arethusa to Lycotas: Propertius, 4. 3.

160 *Horace's Art of Poetry.* Jonson's translation was first published in 1640.

Virgil's fourth Aeneid: The Passion of Dido (1658), by Edmund Waller and Sidney Godolphin.

to run division on the groundwork: to execute a descant of short notes on a bass passage of long notes.

two odes of Pindar, and one of Horace into English. Pindar's second Olympic and first Nemean odes; Horace, *Odes* 4. 2. Cowley's translations appeared in *Poems* (1656).   *nec . . . interpres*: Ars poetica 133–4.

rendered it: in his translation of the *Ars poetica* (1680).

161 *That . . . fame*: from Denham's complimentary verses to Fanshawe's translation (1647) of Guarini's *Il Pastor Fido* 15–16, 21–4.

    atque . . . ferent: Ovid, *Heroides* 7. 8. The bracketed translation is Dryden's.

    brevis . . . fio: Ars poetica 25–6.   *dic . . . urbes*: Horace, Ars poetica 141–2.

162 ὅς . . . πλάγχθη: Homer, *Odyssey* 1. 1–2.

    practice of one: Sir John Denham. See his preface to *The Destruction of Troy*.

163 *second Aeneid*: *The Destruction of Troy* (1656).

164 *et . . . relinquas*: Horace, Ars poetica 149–50.

    Oenone to Paris: Heroides 5. The paraphrase is by Aphra Behn (1640–89).

165 *Canace to Macareus*. First published in *Ovid's Epistles* (1680). A translation of *Heroides* 11. Dryden wrote a preface to *Ovid's Epistles Translated by Several Hands*, and contributed two *Epistles*, given here, and another, *Helen to Paris*, written in collaboration with the Earl of Mulgrave. Other translators included Samuel Butler, Behn, Otway, Rymer, Settle, and Tate. These *Epistles* were Dryden's first published translations.

169 *Dido to Aeneas*. First published in *Ovid's Epistles* (1680). A translation of *Heroides* 7.

172 l. 139. *Iarbas*: king of the Gaetulians.

173 l. 169. *adjure*: charge under penalty of a curse.

174 l. 205. *dear sister*: Anna.

    Prologue to The Spanish Friar. First published in Dryden's *The Spanish Friar* (1681). The play was first performed in 1680.

    l. 9. *brass money once a year in Spain*. This refers to the chronic instability of Spanish coinage.

    l. 11. *Bromingam*: contemptuous for Birmingham, where coining was rife earlier in the century.

175 l. 21. *notched*: with close-cropped hair.   *sermons*: prentices took notes of sermons for their masters.

l. 26. *mum*: German beer.   l. 39. *Scouring*: roistering on the streets.

l. 48. *Plot*: the popish plot; see headnote to *Absalom and Achitophel* (below).

Epilogue to Mithridates. First published in *A Prologue Spoken at Mithridates* (1681). Dryden had written a different Epilogue for Nathaniel Lee's *Mithridates* when it was first performed and published in 1678. A number of corrections have been made from Narcissus Luttrell's copy, now in the Huntington Library.

l. 3. *the Plot*: the popish plot. See headnote to *Absalom and Achitophel* (below).

176   l. 7. *Salamanca.* See note to *Absalom and Achitophel* 657.

l. 9. *dancing of the rope*: popular entertainment at Bartholomew Fair.

l. 10. *Andrew's*: Andrew is a buffoon.   *clean run off the score*: exhausted his credit with the spectators.

l. 11. *Jacob's*: Jacob Hall, a celebrated rope-dancer.

l. 22. *numps*: a foolish woman; see Ben Jonson, *Bartholomew Fair*, 5. 4. 104.

177   *Absalom and Achitophel* was first published in 1681. The second edition ('Augmented and Revised') adds ll. 180–91 and 957–60, incorporated here.

The application of the biblical story (2 Samuel 13–18) to contemporary political events was common among both supporters and opponents of the king. Here, Dryden applies the story to the Exclusion Crisis of 1678–81. The Whigs, led by the Earl of Shaftesbury, advocated religious toleration for the dissenters and wished to exclude the succession of James, Duke of York, a catholic who, it was feared, would be autocratic. Anti-catholic feeling was further whipped up by the 'popish plot' of the same years, featuring a fictitious Jesuit conspiracy against the king.

At the time the poem was written (at the suggestion of the king, it was said), the exclusionist parliament had been dissolved, and Shaftesbury ('Achitophel') arrested.

To the Reader
Epigraph. si . . . magis: Horace, *Ars poetica* 361–2.

anti-Bromingham: Tory.

commonwealth's men. Dryden suggests that some of the Whigs were working for the restoration of a commonwealth.

178   *Origen*: third-century Church Father.

ense rescindendum: Ovid, *Metamorphoses* 1. 191.

Absalom and Achitophel.

179   l. 7. *Israel's monarch*: Charles II. 1 Samuel 13: 13–14.

l. 11. *Michal*: Catherine of Braganza, wife to Charles II. 2 Samuel 6: 23.

l. 14. *David*: Charles II.

l. 18. *Absalom*: James Scott, Duke of Monmouth (1649–85), bastard son of Charles II. 2 Samuel 14: 25.

l. 34. *Annabel*: Anne, Countess of Buccleuch (1651–1732), Monmouth's wife.

l. 39. *Amnon's murder*. This contemporary reference has not been conclusively identified.

l. 42. *Sion*: London.

180   l. 45. *The Jews*: the English.

l. 57. *Saul*: Oliver Cromwell (1599–1658). 2 Samuel 3–4.

l. 58. *Ishbosheth*: Richard Cromwell (1626–1712).

l. 59. *Hebron*: Scotland. Charles was crowned king by the Scots on 1 January 1651, and by the English on 23 April 1661. 2 Samuel 5: 1–5.

l. 82. *Good Old Cause*: the commonwealth of 1649–53.

181   l. 85. *Jerusalem*: London.   l. 86. *Jebusites*: Roman Catholics. Joshua 15: 63.

l. 88. *the chosen people*: protestants.   l. 94. *deprived*: by the penal laws.

l. 107. *eat and drink*: get their living.   l. 108. *that Plot*: the 'popish plot'.

l. 118. *Egyptian rites*: Roman Catholic rites. 'Egypt' is France.

182   l. 129. *fleece*: income from tithes.   l. 130. *God's anointed*: the king.

l. 150. *Achitophel*: Anthony Ashley Cooper, Earl of Shaftesbury (1621–83).

183   l. 175. *triple bond*: the alliance of England, Holland, and Sweden against France.

l. 188. *Abbethdin*: judge.   l. 200. *to possess*: of possessing.

184   l. 217. *prime*: the beginning of a new cycle.   l. 224. *title*: to the throne.

l. 233. *Their cloudy pillar and their guardian fire*: Exodus 13: 21.

l. 234. *second Moses*: Exodus 14: 15–16.

l. 239. *The young men's vision, and the old men's dream*: Joel 2: 28.

185   l. 264. *Gath*: Brussels. 1 Samuel 27: 1–4.

l. 270. *Jordan's sand*: Dover beach. 2 Samuel 19: 9–15.

l. 273. *prince of angels*: Satan.   l. 281. *Pharaoh's*: Louis XIV's.

186   l. 310. *angel's metal*. 'Angel' was also the name of a coin; 'metal' puns with 'mettle'.

l. 320. *wonders*: as in the 'year of wonders' (*Annus Mirabilis*, above, pp. 32–70).

187   l. 334. *dog-star*: Sirius, thought to cause madness.

l. 353. *His brother*: James, Duke of York.

188   l. 390. *Sanhedrin*: parliament.

191   l. 513. *Solymaean rout:* the London rabble (Solyma is Jerusalem).

     l. 517. *ethnic:* of the gentiles, i.e. the Jebusites or catholics.

     l. 519. *Levites:* dissenting ministers deprived of their benefices in 1662. 2 Chronicles 11: 14–15.   l. 525. *Aaron's race:* priests.

     l. 528. *deepest mouthed:* baying most loudly like dogs.

     l. 529. *saints:* visionaries.

192   l. 544. *Zimri:* George Villiers, Duke of Buckingham (1628–87).

     l. 574. *Balaam:* possibly Theophilus Hastings, Earl of Huntingdon (1650–1701).   *Caleb:* Arthur Capel, Earl of Essex (1632–83). Numbers 13–14.

193   l. 575. *canting Nadab:* William, Lord Howard of Escrick (1626–94), formerly a preacher. Exodus 6: 23.   l. 576. *Paschal Lamb:* Christ.

     l. 581. *Jonas:* Sir William Jones (1631–82), attorney-general.

     l. 585. *Shimei:* Slingsby Bethel (1617–97), one of the two sheriffs of London. 2 Samuel 16: 5–14.

     l. 598. *The sons of Belial:* sons of wickedness. *Paradise Lost* 1. 501–2.

194   l. 617. *Rechabite.* Jeremiah 35: 8.

     l. 632. *Corah:* Titus Oates (1649–1705), chief witness in the popish plot. Numbers 16.   l. 633. *monumental brass:* Numbers 21: 6–9.

     l. 643. *Stephen:* Acts 6: 9–15.

     l. 645. *His tribe were God almighty's gentlemen:* Numbers 16: 8–9.

     l. 649. *Moses' face:* Exodus 34: 29 (the comparison is ironic).

195   l. 676. *Agag's murder:* possibly referring to the catholic Lord Stafford, imprisoned and condemned on Oates's accusation. 1 Samuel 15: 32–3.

     l. 697. *Hybla-drops:* honey from Hybla in Sicily.

196   l. 700. *banished:* Monmouth was sent to Holland in September 1679.

     l. 705. *Tyrus:* Holland.

     l. 710. *Bathsheba's.* Bathsheba is Louise de Kéroualle, Duchess of Portsmouth (1649–1734).

     l. 729. *progress.* Monmouth made a triumphal journey through the west country in the summer of 1680.

     l. 738. *Issachar:* Thomas Thynne (1648–82) of Longleat. Genesis 49: 14–15.

198   l. 786. *flowing ... out:* i.e. the higher the tide, the faster it runs out.

     l. 800. *innovation:* revolution.

     l. 804. *touch our ark:* commit sacrilege.   l. 807. *control:* break.

     l. 817. *Barzillai:* James Butler, Duke of Ormonde (1610–88). 2 Samuel 19: 31–9.

199 l. 830. *more than half a father's name*. Six of Ormonde's ten children were now dead.

l. 831. *eldest hope*: Thomas, Earl of Ossory (1634–80).

l. 842. *Tyrians*: Dutch. Ossory served William of Orange against the French.

l. 851. *starry pole*: the heavens.

200 l. 864. *Zadoc*: William Sandcroft (1617–93), Archbishop of Canterbury. 2 Samuel 8: 17.

l. 866. *Sagan*: the Jewish high priest's deputy. Henry Compton (1632–1713), Bishop of London.

l. 868. *Him of the western dome*: John Dolben (1625–86), Dean of Westminster.

l. 870. *prophets' sons*: the boys at Westminster school.

l. 877. *Adriel*: John Sheffield, Earl of Mulgrave (1648–1721), Dryden's patron.

l. 882. *Jotham*: George Savile (1633–95), Viscount Halifax.

l. 888. *Hushai*: Laurence Hyde (1642–1711), later Earl of Rochester.

l. 899. *Amiel*: Edward Seymour (1633–1708), speaker of the Commons. 1 Chronicles 26: 4–8.

201 l. 910. *the unequal ruler*: Phaeton, who attempted to drive the sun chariot of his father Apollo, lost control, and upset the seasons.

202 l. 944. *The offenders question my forgiving right*. The Commons questioned Charles's right to pardon Danby in 1679, and the sheriffs questioned his right to commute Stafford's sentence in 1680.

l. 955. *Samson*: Judges 15: 25–31.   l. 982. *Esau's hands*: Genesis 27: 22.

203 l. 1008. *hinder parts*: Exodus 33: 23.

204 *From the Second Part of Absalom and Achitophel*. The second part was published late in 1682. It was mostly written by Nahum Tate with some touches by Dryden. These lines (457–509) are indubitably by Dryden.

l. 3. *Og*: Thomas Shadwell; see notes to *MacFlecknoe* (above, p. 878). Shadwell had become a party writer for the Whigs. D was to attack him in this capacity the following year in his *The Vindication of The Duke of Guise* (1683):

> Og may write against the king if he pleases so long as he drinks for him, and his writings will never do the government so much harm as his drinking does it good: for true subjects will not be much perverted by his libels, but the wine duties rise considerably by his claret. He has often called me an atheist in print; I would believe more charitably of him, and that he only goes the broad way because the other is too narrow for him. He may see by this I do not delight to meddle with his

course of life, and his immoralities, though I have a long bead-roll [list] of them. I have hitherto contented myself with the ridiculous part of him, which is enough in all conscience to employ one man; even without the story of his late fall at the Old Devil, where he broke no ribs because the hardness of the stairs could reach no bones; and, for my part, I do not wonder how he came to fall, for I have always known him heavy: the miracle is how he got up again. I have heard of a sea captain as fat as he, who to scape arrests would lay himself flat upon the ground and let the bailiffs carry him to prison, if they could. If a messenger or two, nay, we may put in three or four, should come he has friendly advertisement [advice] how to escape them. But to leave him, who is not worth any further consideration now I have done laughing at him, would every man knew his own talent and that they, who are only born for drinking, would let both poetry and prose alone.

l. 5. *link*: a boy with a torch ('link') of pitch and tar.

205 l. 36. *Doeg*: Elkanah Settle (1648–1724), author of *The Empress of Morocco* (1673).

ll. 50–1. *But of King David's foes . . . Absalom*: 2 Samuel 18: 32.

The Medal was first published early in 1682. A second issue added the two Latin lines at the end. Charles II 'gave Dryden the hint for writing his poem', Joseph Spence tells us.

The Whig Shaftesbury (see headnote to *Absalom and Achitophel*, above) was indicted on a charge of high treason on 24 November 1681. The Whig grand jury returned a verdict of *ignoramus* (lack of competence to judge), and the Whig populace celebrated with bonfires and riots. Early in 1682 a medal was struck to commemorate the events. On one side was a bust of Shaftesbury; on the other a view of London Bridge and the Tower, with the sun breaking through a cloud, and the inscription *Laetamur* (let us rejoice), with the date 24 November 1681.

epigraph. per . . . honorem: Virgil, *Aeneid* 6. 588–9.

Epistle to the Whigs

206 *many a poor Polander*. It was a Tory joke that Shaftesbury wanted to be elected to the Polish throne in 1675.

Bower: George Bower, the engraver of the medal.

No-Protestant Plot: a Whig tract in three parts by Robert Ferguson (1681–2).

Scanderbeg: George Castriota, otherwise Iskander Beg (1404–67), an Albanian who fought for his country's independence from the Turks.

207 *petition in a crowd*. 'In 1661 Parliament passed an act against tumultuous petitioning (13 Car. II, c. 5), making it illegal to obtain more than twenty signatures, or for more than ten persons to present a petition to Parliament' (Kinsley).   *your dead author's*: Andrew Marvell's (1621–78).

The Growth of Popery: An Account of the Growth of Popery and Arbitrary Government in England (1677).

Buchanan: George Buchanan (1506–82), Scotch humanist.

first Covenant: of 1643.

Holy League: the Catholic League which forced Henri IV to join the Roman church (1593).

Davila: Enrico Davila, whose history of the civil wars in France was published in an English translation in 1647.

Beza: Theodore de Bèze (1519–1605), leader of the Genevan church after Calvin.

The Medal

209 l. 3. *Polish*. See note to 'Epistle to the Whigs' (above).

210 l. 14. *shrieval voice*. The sheriffs were Whigs.

l. 15. *Laetamur*: (Latin), let us rejoice.   l. 22. *style*: the engraver's tool.

l. 31. *vermin wriggling in the usurper's ear*: 'earwig … a whisperer; a petty informer' (J).   l. 41. *interlope*: trade illicitly.

211 l. 60. *gears*: harness.

l. 65. *triple hold*: the triple alliance of England, Holland, and Sweden.

l. 67. *Ilium*: Troy; here signifying defeat by treachery.

l. 73. *Samson*: Judges 16.

l. 94. *Pindaric*. Dryden thought that Pindar, Greek lyric poet, typified unruliness.

212 l. 96. *Phocion … Socrates*: an upright general and a philosopher. The crowd condemned them both to death (317 and 399 BC), and then turned against their accusers.

l. 100. *father … son*: Charles I and Charles II.

l. 119. *Jehu*: a tempestuous driver; 2 Kings 9: 20.

l. 129. *their own usurping brave*: Oliver Cromwell.

l. 131. *manna … quails*: Numbers 11.

213 l. 145. *The man who laughed but once*: Marcus Crassus.

l. 167. *emporium*: centre of commerce.

214 l. 178. *Canaanite*. Descendants of Canaan, the son of Ham, the Canaanites inhabited Palestine before the Israelites.

ll. 181–2. *head … hands*: Sir John Moore, Lord Mayor, and the two Whig sheriffs, Samuel Shute and Thomas Pilkington.

l. 202. *the French puritan*: Theodore de Bèze; see note to 'Epistle to the Whigs' (above).

l. 217. *heir … vineyard*: Matthew 21: 33–9.

215 l. 229. *clip his regal rights*. Clipping pieces of silver from a coin was common. Coins clipped 'within the ring' (the circle around the king's head) were not legal tender.   l. 258. *blandishments*: flattery.

216 l. 263. *mercury*: a figure of inconstancy. Mercury was also used in the treatment of syphilis.

l. 268. *renegado priests*: priests expelled from their livings by the Act of Uniformity (1662) or the Test Act (1673).

l. 285. *Bedlam*: Bethlehem hospital in Moorfields; a madhouse.

217 l. 304. *frogs and toads*. See Aesop's fable of the frogs and the stork.

l. 311. *surly:* masterful.

l. 317. *Collatine*: Monmouth. Lucius Tarquinius Collatinus helped expel the Tarquins from Rome, and became the first consul. Then, being a Tarquin, he was expelled in his turn.

ll. 323–4. *pudet ... refelli*. Ovid, *Metamorphoses* 1. 758–9: 'it is shameful that such dishonourable charges can be made against you, and that they cannot be refuted.'

Prologue to the Duchess on her return from Scotland. First published as a separate half-sheet in 1682. During the crisis over the exclusion of the Duke of York from the throne, his brother Charles II kept him out of the way. He was in Scotland for nearly two years until March 1682. He went back to Scotland in May to collect his wife, Mary of Este. They arrived in London on 27 May, amidst rejoicing. Dryden's prologue was written for a special performance of Thomas Otway's *Venice Preserved* on 31 May.

218 l. 16. *Thetis*: a nereid, mother of Achilles.

l. 24. *Joseph's dream*: the dream of Pharaoh, interpreted by Joseph. Genesis 41.   *doom*: (1) judgement; (2) interpretation.

219 *Religio Laici* was first published in November 1682. The title means 'the religion of a layman'.

epigraph. Ornari ... doceri: Manilius, *Astronomica* 3. 39.

Preface
so bold a title. The title challenges comparison with Lord Herbert of Cherbury's *De Religione Laici* (1645), Thomas Browne's *Religio Medici* (1642), and George Mackenzie's *Religio Stoici* (1665).

unhallowed hand upon the ark: 1 Chronicles 13: 10.

sword of Goliath: 1 Samuel 21: 9.   *entitle them to*: father upon them.

220 *St Athanasius*: (?296–373), opposed the Arian heresy.

one only who was accursed: Canaan; Genesis 9: 24–7.

St Paul concludes to be the rule of the heathens: Romans 2: 14–15.

221 *Arius*: (256–336), denied the divinity of Christ and the doctrine of the Trinity.

222 *Socinians*: followers of a Renaissance doctrine which denied the Trinity.

Nicene: the creed formulated by the First Council of Nicene.

223 *Mr Coleman's letters*. Edward Coleman, secretary to the Duke of York,

was executed in 1678 for conspiring, in correspondence with Louis XIV's Jesuit confessor, to re-establish the catholic faith in England.

Mariana ... Simancha: catholic writers of the sixteenth and seventeenth centuries.

Campion ... Parsons. Edmund Campion (1540–81) and Robert Parsons (1546–1610) were English Jesuits. Parsons published under the name of Doleman. *Nebuchadnezzar*: Daniel 4: 28–32.

224 *Bellarmine*: Robert Bellarmine (1542–1621), Jesuit theologian.

villeinage: tenure by service to the feudal lord. *this Plot*: the popish plot.

Cres: Hugh Paulinus Serenus Cressy (?1605–74), English chaplain to Queen Catherine. *this present Pope*: Innocent XI (1611–89).

225 *Tyndale*: William Tyndale (?1490–1536), reformer, and translator of the Bible into English.

Henry the Eighth: Lord Herbert of Cherbury, *The Life and Reign of King Henry the Eighth* (1649).

Hooker. Richard Hooker (?1554–1600) wrote *Of the Laws of Ecclesiastical Polity* (1594–7).

George Cranmer: (1563–1600), friend of Hooker. His letter is included in Izaak Walton's *Life of Hooker*.

Martin Mar-prelate: the assumed name for the author(s) of a series of antiepiscopal tracts (1588–90).

Marvell: Andrew Marvell (1621–78), in his role as pamphleteer.

Billingsgate: a fishmarket, home of abusive invective.

226 *Hacket*: William Hacket, a fanatic who thought he was the Messiah.

Coppinger: Edmund Coppinger, one of Hacket's prophets.

conventiclers: nonconformists.

227 *Simon.* Father Simon's *Histoire critique du vieux testament*, translated in 1682 by Henry Dickinson, challenged the reliability of the Hebrew scriptures.

Religio Laici

228 l. 18. *interfering*: colliding. l. 21. *Stagirite*: Aristotle, born at Stagira in Thrace.

l. 22. *Epicurus*: Athenian philosopher (341–270 BC), who taught that pleasure was the highest good in life.

l. 31. *wiser madmen*: the Stoics, who in contrast to the Epicureans, taught that the highest good was virtue.

229 l. 43. εὕρηκα: I have found it (Archimedes' exclamation).

230 l. 102. *forfeit*: transgression.

231 l. 128. *lustrations*: ceremonial purifications.

232 l. 199. *great apostle*: St Paul; Romans 2: 12–15.

233 l. 211. *rubric-martyrs*: martyrs in the calendar of saints.

l. 213. *The Egyptian bishop*: St Athanasius (?296–373) opposed the Arian heresy.

l. 220. *Arius*: (256–336), denied the divinity of Christ and the doctrine of the Trinity.

l. 241. *Junius and Tremellius*: sixteenth-century Calvinists.

234 l. 269. *brushwood-helps*: insubstantial aids.

l. 283. *cast in the Creed*: with the Creed thrown in.

235 l. 291. *Esdras*: 2 Esdras 14.　　l. 324. *strait gate*: Matthew 7: 14.

236 l. 339. *next*: nearest the source.

l. 346. *Pelagius*: fifth-century monk who denied original sin and believed man was saved by his own exertions.　　*provoke*: appeal.

239 l. 456. *Tom Sternhold's*. Thomas Sternhold (d. 1549) translated the Psalms into English metrical verse.

Ovid's Elegies. Book II. The Nineteenth Elegy. First published in *Miscellany Poems* (1684).
l. 22. *bulks*: stalls projecting from the fronts of shops.

l. 27. *Danae*: daughter of Acrisius king of Argos. It was foretold that her son would kill Acrisius, who therefore kept her locked up in a tower of bronze. But Jove transformed himself into a shower of gold and impregnated her with Perseus, who was to kill Acrisius.

240 l. 31. *Whetstone*: Whetstone Park, a lane between Lincoln's Inn Fields and Holborn, site of a brothel.　　l. 48. *wittol*: a contented cuckold.

l. 50. *pocket up*: submit to an affront without showing resentment.

Prologue to the University of Oxford. First published in *Miscellany Poems* (1684). Written in 1673.

241 l. 14. *Lyceum*: the garden where Aristotle taught.

l. 32. *the Lucretian way*: casually, by chance, as atoms formed the world; see *De rerum natura* 2. 1059–63.

242 l. 40. *Praetorian bands*: the guard of the Roman emperor.

Epilogue to Oxford. First published in *Miscellany Poems* (1684). Written in 1674.

l. 13. *favours past*: previous permissions to perform at Oxford.

l. 17. *Bathurst*: Richard Bathurst (1620–1704), vice-chancellor of Oxford University.

243 *Prologue to the University of Oxford.* First published in *Miscellany Poems* (1684). Tentatively dated 1676 on the basis of a manuscript version in the Bodleian.

l. 17. *grossly*: indiscriminately.

l. 36. *mother-university*: Cambridge.

Prologue Spoken at the Opening of the New House. First published in *Miscellany Poems* (1684). A new Theatre Royal in Drury Lane replaced the old theatre burnt down in 1672. The rival Duke's Theatre built in 1671 had cost £9,000. The King's company could only raise some £4,000 for their theatre.

l. 6. *shining all with gold*: a glance at the gilded proscenium of the Duke's Theatre. l. 9. *druggets*: coarse cloths.

245 l. 46. *ben!*: *bien* (French; 'well').

l. 48. *French machines*: (1) elaborate stage contrivances for opera; (2) crafty political schemes; (3) artificial penises.

l. 53. *tempests*: a glance at Shadwell's operatic version of *The Tempest*, in rehearsal at the Duke's Theatre.

To the Memory of Mr Oldham. First published in *The Remains of Mr John Oldham*, 1684. John Oldham (1653–83) was a poet, satirist, and schoolmaster.

l. 9. *Nisus*. At the funeral of Anchises (*Aeneid* 5) two friends, Nisus and Euryalus run together in a race. Nisus, the older man, slips and falls, but manages to trip the next runner, allowing Euryalus to win. See 'The entire Episode of Nisus and Euryalus' (above, pp. 259–71).

246 l. 14. *numbers*: harmony of verse, or simply versification.

l. 23. *Marcellus*: Augustus' nephew. His early death is celebrated in Virgil's *Aeneid* 6. 860–86, which Dryden echoes in this poem. Cf. his translation (6. 1196–9):

> Observe the crowds that compass him around:
> All gaze and all admire, and raise a shouting sound:
> But hovering mists around his brows are spread;
> And night, with sable shades, involves his head.

Preface to Sylvae. *Sylvae* was published in 1685.

History of the League: Louis Maimbourg, *Histoire de la ligue*. D's translation was published in 1684.

Roscommon's: Wentworth Dillon, Earl of Roscommon (1633–85). His *Essay* is in Spingarn, 2. 297–309. *specious*: plausible.

247 *Ogilbies*: translators like John Ogilby (1600–76), whose Virgil appeared in 1649, his *Iliad* in 1660, and his *Odyssey* in 1665.

248 *late noble painter*: Sir Peter Lely (1618–80).

249 *hand-gallop*: easy gallop. *carpet ground*: smooth ground.

synaloephas: elisions.

Hannibal Caro's in the Italian. The translations of the *Aeneid* by Annibale Caro (1507–68) appeared in 1581.

250 *Tasso tells us in his letters*: not in his letters, but in his *Discorsi dell'Arte Poetica* (1587).

Mezentius and Lausus: Virgil, *Aeneid*, 10. 755–908.

Lord Roscommon justly observes: *Essay on Translated Verse*:

> When in triumphant state the British muse
> True to herself, shall barbarous aid refuse,
> And in the Roman majesty appear,
> Which none know better, and none come so near.

251 *turns*: special points of style or expression.   *breakings*: elisions.

When . . . slain. D quotes from memory: the line actually runs 'When Lausus fell . . .'. This lessens the objection.

four books: the *Georgics*. See above, pp. 463–541.

252 *philosopher of Malmesbury*: Thomas Hobbes (1588–1679).

253 *prosopopeia*: personification.

254 *cashiered*: put away.

Essay on Poetry: by John Sheffield, Lord Mulgrave, published anonymously in 1682.

begins.

> Happy that authour where correct essay
> Repairs so well our old Horatian way.

Much . . . must: 'Of Wit' 45–8.   *fulsome*: offensive.

The scorpion . . . bruised: 'Lucretius: The Fourth Book' 26.

255 *Non ego . . . natura*: *Ars poetica* 351–3.

translator of Lucretius: Thomas Creech (1659–1700); his Lucretius was published in 1682.

Evelyn. John Evelyn (1620–1706), diarist, had written commendatory verses to Creech's translation.

256 *mai . . . bosco*: untraced.

russet: dress of coarse homespun woollen cloth of reddish brown.

numerousness: metrical perfection.

257 *curiosa felicitas*: Petronius, *Satyricon*, 118.

feliciter audere: *Epistles* 2. 1. 166.

present Earl of Rochester: Laurence Hyde (1641–1711).

Cowley: Cowley's 'Pindaric Odes' (*Poems*, 1656).

258 *quod . . . tantum*: Juvenal, *Satires* 7. 56.

fungar . . . secandi: Horace, *Ars poetica* 304–5.

related: D's son Charles. The other contributors remain largely unknown.

NOTES TO PAGES 259–280 895

259 *The entire Episode of Nisus and Euryalus translated from the Fifth and Ninth Books of Virgil's Aeneid.* First published in *Sylvae* (1685). A translation of *Aeneid* 5. 286–361 and 9. 176–449.

Nisus and Euryalus 1
l. 1. *Trojan hero*: Aeneas.

260 l. 33. *well-breathed*: sound of wind.

261 l. 67. *cuts*: passes sharply through.　l. 81. *the prince*: Aeneas.

l. 100. *Neptune's bars*: the gates of a temple dedicated to Neptune.

Nisus and Euryalus 2
264 l. 108. *took the word*: spoke at once.

265 l. 136. *gave*: to Aeneas.

269 l. 302. *perplexing*: tangled.

270 l. 336. *brown*: dark.

l. 337. *poising from his ear*: balancing on a level with his ear.

271 *Translations from Lucretius.* First published in *Sylvae* (1685).

The beginning of the first book: *De rerum natura* 1. 1–40.
272 l. 35. *Memmius*: Lucretius' friend to whom the poem is addressed.

l. 46. *dreadful servant's*: Mars'.

The beginning of the second book: *De rerum natura* 2. 1–61.
273 l. 20. *stints*: limits.　l. 28. *sconces*: bracket candlesticks.

274 l. 58. *wandering errors.* The Latin for 'to wander' is *errare*.

The latter part of the third book. Against the fear of death. De rerum natura 3. 830–1094.
275 l. 4. *Punic arms*: the three wars between Rome and Carthage in the third and second centuries BC.

276 l. 54. *offals*: carrion, waste.

277 l. 99. *brimmers*: goblets.　l. 104. *the god that never thinks*: Bacchus.

279 l. 185. *Tantalus.* He stole the food of the gods.

l. 189. *Tityus.* Two vultures tore at his liver as a punishment for having assaulted Leto.

l. 200. *Sisyphus*: punished by having to roll a stone uphill perpetually.

280 l. 211. *smokes*: rushes.

l. 219. *fifty foolish virgins*: the Danaides, daughters of Danaus. At his command they killed their husbands on their wedding night. They were punished by having to draw water eternally in leaky vessels.

l. 222. *the dog*: Cerberus, guardian of the underworld.

l. 228. *the Tarpeian rock*: from which traitors and murderers were thrown in Rome.

l. 237. *Ancus*: Ancus Marcius, according to tradition the fourth king of Rome.

281   l. 242. *That haughty king*: Xerxes, who bridged the Hellespont.

l. 255. *Democritus*: fifth-century-BC philosopher.

The Fourth Book. Concerning the Nature of Love. De rerum natura 4. 1052-1287.

283   l. 20. *leaky kind*: women.

285   l. 106. *point*: lace.

286   l. 139. *bedlam*: mad.   l. 150. *two-handed*: sturdy.

287   l. 181. *queans*: whores.

289   l. 251. *unsouled*: unanimated.   l. 270. *incrassate*: thicken.

290   *From the Fifth Book. De rerum natura* 5. 222-34.

291   *Daphnis from the twenty-seventh Idyll of Theocritus*. First published in *Sylvae* (1685).

297   *A New Song* ('*Sylvia the fair*'). First published in *Sylvae* (1685).

l. 4. *tousing*: pulling about rudely.

l. 14. *Trimmer*: a temporizer in politics; a 'moderate'.

298   *Song* ('*Go tell Amynta*'). First published in *Sylvae* (1685). This poem was set to music by (among others) Purcell.

299   *The Third Ode of the First Book of Horace*. First published in *Sylvae* (1685).

title. *Roscommon*. Wentworth Dillon, fourth Earl of Roscommon (1637-85).

l. 2. *the twin stars*. The 'Twins', Castor and Pollux, protected sailors.

l. 3. *he*: Aeolus.

l. 6. *Etesian gales*: summer winds from the north-east.

l. 20. *Hyades*: the rainers.

300   l. 55. *unwilling thunder*: Jove punishing despite himself.

The Ninth Ode of the First Book of Horace. First published in *Sylvae* (1685).

301   l. 32. *pointed*: appointed.

302   *The Twenty-ninth Ode of the Third Book of Horace, paraphrased in Pindaric Verse*. First published in *Sylvae* (1685). For 'Pindaric verse' see note to 'Essay of Dramatic Poesy', p. 875 above.

l. 2. *Tuscan sceptre*: Horace's *Ode* is addressed to Maecenas, 'descendant of Etruscan kings'.

l. 21. *give thy soul a loose*: free your soul from restraint.

l. 23. *fit*: short spell.

303 l. 27. *Tyrian loom*: rich purple hangings.

l. 30. *Syrian star*: the dog-star.   l. 37. *sylvans*: countrymen.

ll. 40–3. *new Lord Mayor . . . quiver-bearing foe*. Dryden gives modern examples for Horace's allusions. The quiver-bearing foe 'possibly means the Turks' (C).

305 *The Second Epode of Horace*. First published in *Sylvae* (1685).

306 l. 33. *Priapus*: god of male fertility, and protector of vineyards.

l. 34. *Sylvanus*: wood god.   l. 48. *rear*: rouse.

l. 61. *Sabine matrons*. The Sabine mothers intervened to reconcile their tribe and Rome.

l. 62. *Apulian's bride*. Apulia is a district in southern Italy.

307 l. 78. *heath-poult . . . rarer bird*: young heath-cocks and pheasant.

l. 82. *chards*: artichokes.   l. 96. *Morecraft*: a rich city usurer.

308 *To Sir George Etherege*. First published in the anonymous *History of Adolphus* (1691). Dryden did not authorize publication, and my text is based on that of James Thorpe (*The Poems of Sir George Etherege*, Princeton, 1963), which is based on contemporary manuscripts.

Etherege (?1635–91), playwright and wit, was sent by James II to be envoy to the Diet at Ratisbon in March 1685. Dryden's poem, probably written in the summer of 1686, is the reply to a verse letter Etherege sent to the earl of Middleton (*Letters of Sir George Etherege*, ed. Frederick Bracher (Berkeley, 1974) 272).

ll. 1–4. *chill degree . . . fifty-one*. London is at latitude 51°. Dryden may have assumed that Ratisbon, being colder as Etherege was always complaining, was latitude 53°.

l. 14. *Ad partes infidelium*: (1) to the regions of the infidels; (2) to the regions of the unfaithful ones.

l. 20. *Pinto*: Fernão Mendes Pinto (1509–83), eastern traveller.

l. 27. *the Bear*: (1) an eating house and house of assignation in Drury Lane; (2) the constellation of *Ursa Major*.

l. 30. *Triptolemus*. He was sent to teach men agriculture; Ovid, *Metamorphoses*, 5. 642–61.

l. 37. *French king*: Louis XIV, contemptuous of treaties.

309 l. 45. *rummers*: large drinking glasses.

l. 47. *three . . . hectors*: three German archbishops who, with five secular rulers, made up the college of Electors of the German empire. 'Hectors' are bullies.

l. 65. *defied*: renounced.   l. 73. *St. Aignan*: (?1610–87), reputed author of *Bradamante* (1637).

l. 75. *His Grace of Bucks*: George Villiers (1628–87), Duke of Buckingham, author of *The Rehearsal* (1671). It took, it is said, some ten years to write, and ineptly satirizes Dryden.

310 *To the Memory of Mrs Anne Killigrew*. First published in *Poems by Mrs Anne Killigrew* (1685; dated 1686) and republished in *Examen Poeticum* (1693), some of whose revised readings have been preferred.

Anne Killigrew (1660–85) was maid of honour to James II's wife Mary of Modena. She died of smallpox in 1685. Mrs = mistress. Perhaps thorough-going modernization should read 'Miss' or 'Ms' in the title.

l. 6. *neighbouring star*: planet.

l. 8. *procession fixed and regular*: as of a ('fixed') star.

l. 16. *mortal*: living.

l. 23. *traduction*: a leading across; 'if your mind was not created at birth, but was transmitted from your parents'.

311 l. 33. *Sappho*: Greek lyric poet (*c*.600 BC) from Lesbos.

l. 43. *in trine*: at 120° distant from each other, and thus most benign.

l. 50. *clustering swarm of bees*. Bees settled on the lips of the infant Plato, presaging his eloquence.

312 l. 68. *Arethusian*: pure. Arethusa was changed by Diana into a fountain to save her from the clutches of Alpheus.

l. 82. *Epictetus*: (*c*.55–135), stoic philosopher.

l. 88. *Nine*: nine muses.   l. 93. *painture*: the art of painting.

313 l. 128. *our martial king*: James II.   *strook*: struck.

l. 134. *phoenix-queen*: uniquely beautiful queen.

314 l. 154. *cruel destiny*: smallpox.

l. 162. *Orinda*: Katherine Philips (?1631–64), poet.

l. 165. *her warlike brother*: Henry Killigrew (d. 1712), naval captain.

315 l. 175. *Pleiads*: (1) a cluster of small stars in the constellation Taurus; (2) a sixteenth-century group of French poets, *la Pléiade*, which included Du Bellay and Ronsard.

l. 180. *the valley of Jehosaphat*: Joel 3: 2.

To my Ingenious Friend Henry Higden. First published as a commendatory poem in Higden's *A Modern Essay on the Tenth Satire of Juvenal* (1687). Higden was a lawyer and wit.

l. 2. *pasquins*: pasquinades, squibs.

l. 7. *drolls*: wags.   *bauble*: 'a truncheon with a fool's head and cap upon one end . . . carried by the ancient jester' (Scott).

316 l. 35. *half that cure to you restore*: return you credit for half that cure.

A Song for St Cecilia's Day, 1687. First published in 1687. Written for the concert on 22 November 1687.

l. 2. *universal frame*: the physical universe.

317 l. 14. *compass*: the full range.

l. 15. *diapason*: the combination of all the notes sounding in a harmonious whole. (It does not mean 'all the notes played together'.)

l. 15. *closing full*: ending in a perfect cadence.

l. 17. *Jubal*: 'the father of all such as handle the harp and organ' (Genesis 4: 21).

318 l. 50. *sequacious of*: following. Ovid, *Metamorphoses* 11. 1–2: 'with such songs the bard of Thrace [Orpheus] drew the trees, held beasts enthralled, and constrained stones to follow him'.

l. 63. *untune the sky*: 'when this world and the heavenly bodies are destroyed, the music of the spheres will cease; thus *Music* (the blast of the divine *trumpet*) will untune (make incapable of harmony) the sky' (Noyes).

319 *Prologue to Don Sebastian*. First Published in Dryden's *Don Sebastian* (1690). The play was first performed in 1689.

l. 3. *cast poets*: Dryden had lost his offices of Poet Laureate and Historiographer Royal at the revolution of 1688.

l. 15. *alike to Latian and to Phrygian*: Virgil, *Aeneid* 10. 105–8, which Dryden translated (in part):

> Rutulians, Trojans, are the same to me;
> And both shall draw the lots their fates decree.

320 l. 44. *five pound*. 'At the Revolution an act was passed prohibiting any "Papist or reputed Papist" from possessing a horse worth more than five pounds' (Kinsley).

Prologue to The Mistakes. First published in Joseph Harris's *The Mistakes* (1691). The play was first performed in 1690.

l. 10. *three days*. The proceeds of the third day went to the author.   *as long as Cork*. 'Cork was taken by Marlborough . . . after a five-day siege [in 1690]' (Kinsley).

321 l. 35. *Peace and the butt*: peace and its benefits.

The Lady's Song. Twice published in 1704, in *Poetic Miscellanies*, part 5, and in a slightly different version, in *Miscellaneous Works* by the Duke of Buckingham. My text comes from *Poetic Miscellanies*. *Miscellaneous Works* gives a probable date 1691.

l. 2. *May-lady*: Queen of the May.

l. 7. *Pan . . . Syrinx*. For the myth see 'The First Book of Ovid's

Metamorphoses 951–87 (above, p. 406); but 'Pan' and 'Syrinx' almost
certainly allude to the banished James II and his wife Mary of Este.

322 *To Mr Southerne on his Comedy called The Wives' Excuse.* First published
in *The Wives' Excuse* (1692). The play, by Thomas Southerne (1659–
1746), was first performed in late 1691.

l. 9. *Spanish nymph*. The allusion is unexplained.

l. 15. *Terence*: (c.185–159 BC), Latin comic dramatist.

l. 17. *scene*: play.

l. 18. *Nokes*: James Nokes (d.1696), popular comic actor.

l. 22. *the wife*: Mrs Friendall, the wife in Southerne's play.

l. 28. *Etherege*: Sir George Etherege (?1635–91).

l. 29. *Wycherley*: William Wycherley (1641–1716).

323 *Eleonora: A Panegyric Poem Dedicated to the Memory of the late Countess
of Abingdon.* First published in 1692.

motto. superas ... potuere: Virgil, *Aeneid* 6. 128–31.

title. Earl of Abingdon. James Bertie, an early supporter of William III,
and patron of several poets.

excused the faults of his poetry by his misfortunes. Tristia 1. 39 f.

my disease: perhaps gout.

324 *similitudes*: images, usually similes.   *figures*: non-literal forms of writing.

never seen Mrs Drury: in a letter of 14 April 1612.

325 *Phidias*: Athenian sculptor, born about 490 BC. 'Plutarch, in his Life of
Pericles, tells how Phidias inscribed his name on a statue of Pallas Athena
commissioned by Pericles during the building of the Propylaea' (George
Watson, *John Dryden: 'Of Dramatic Poesy' and other critical essays* (1962)
2. 63).

326 *they*: presumably the men who called in William III in 1688.

sown the dragon's teeth: a reference to Cadmus, who sowed dragon's teeth
and armed men sprang up (Ovid, *Metamorphoses* 3. 104–10).

an elected Speaker of the House: who cannot take part in parliamentary debates.

Eleonora. Eleonora, wife of the Earl of Abingdon, died on 31 May 1691.
Dryden's poem shows a careful study of Donne, and is partly modelled
on Donne's two *Anniversaries*. The Californian editors note 'numerous
comparisons':

  1–11: *1st Anniversary* 43–54
  37: *1st Anniversary* 251–7; *2nd Anniversary* 507–8; *A Valediction For-
    bidding Mourning* 25–6
  40–2: *1st Anniversary* 399–406

46–7: *1st Anniversary* 413–16
75–82: *1st Anniversary* 69–78
90–6: *2nd Anniversary* 417–24
116–25: *2nd Anniversary* 449–67
126–33: *1st Anniversary* 9–10
134–9: *Eclogue 1613* 149–59
143–53: *1st Anniversary* 259–60; *2nd Anniversary* 189–206
154–7: *2nd Anniversary* 127–30
158–9: *2nd Anniversary* 123–6
166–73: *1st Anniversary* 100–6, 179–82
174–5: *1st Anniversary* 227–9; *2nd Anniversary* 306–7; *Canonization* 44–5
191: *2nd Anniversary* 123–6
220–30: *2nd Anniversary* 223–5
257–62: *1st Anniversary* 1–8
270–300: *Funeral Elegy* 78–106
301–4: *Funeral Elegy* 73–5
313–14: *Song* ('Sweetest Love'), 37–40
317–28: *Valediction Forbidding Mourning* 1–4
327–8: *2nd Anniversary* 460–2
330: *Funeral Elegy* 75
341: *2nd Anniversary* 200
342–3: *Obsequies to the Lord Harrington* 5–6
375–6: *1st Anniversary* 473–4
377: *2nd Anniversary* 43, 527–8

327 l. 24. *talents*: Matthew 25: 14–29.   l. 38. *touch her garment*: Mark 5: 24–9.

328 l. 71. *Pharaoh*: Genesis 41.

l. 75. *heaven … shows*. The Californian editors cite Fontenelle's *A Discourse on the Plurality of Worlds* (Dublin, 1687), p. 23: 'The sun … is luminous in himself, by virtue of a nature particular to him, but the planets give no light, but as they are enlightened by him. He sends his light to the moon, she returns it to us.' Dryden draws on Fontenelle frequently in this poem.

l. 80. *latent*: hidden.   l. 91. *high turrets*: Isaiah 2: 12–17.

329 l. 99. *durst*: dared.   l. 114. *orisons*: prayers.

l. 124. *vicissitudes*: alterations. Cf. Milton, *Paradise Lost*, 6. 7–8:

> which makes through heaven
> Grateful vicissitude, like day and night …

l. 130. *offices*: duties.

l. 135. *how can mortal eyes sustain immortal light?*: Exodus 33: 18–23.

330 l. 139. *this watery glass*: this imperfect mirror (Dryden's own poem).

l. 147. *heroic*: either (1) one extraordinary virtue; or (2) hero.

331 l. 197. *Anchises*: *Aeneid* 6. 752–853. Dryden translates in part:

> Thus having said, the father-spirit Anchises leads
> The priestess and his son through swarms of shades,
> And takes a rising ground, from thence to see
> The long procession of his progeny. (6. 1021–4)

l. 201. *Cybele*: mother of Jove (Zeus) and other gods.

l. 207. *the chosen found the pearly grain*: Exodus 16: 11–18.

333 l. 277. *The consul*: alluding to Caninius, who was made consul for one day.

334 l. 330. *livery of the day*. The newly baptized appear in white garments for the services of Whitsunday.

335 l. 343. *pervious*: capable of being passed through.

336 *The sixth satire of Juvenal*. First published in *The Satires of Juvenal and Persius* (1693). Lines 178–81, 340–1, 433–6, 445–6, 449–50, and 464–5 were first published by W. B. Carnochan, 'Some suppressed verses in Dryden's translation of Juvenal VI', *TLS*, 21.1.1972, pp. 73–4. A further 'suppressed' line (290) was added in the fourth volume of the California edition (1974). Some of Dryden's notes are reproduced here.

Argument
Domitian's time. Domitian was Roman emperor AD 81–96.

a loader: a doublet at dice.

337 *play prizes*: engage in a contest.

l. 1. *In Saturn's reign*: 'in the golden age when Saturn reigned' (D).

l. 15. *fat with acorns*. 'Acorns were the bread of mankind, before corn was found' (D).

l. 23. *under Jove*. 'When Jove had driven his father into banishment, the silver age began, according to the poets' (D).

338 l. 28. *uneasy justice*. 'The poet makes justice and charity sisters; and says that they fled to heaven together, and left earth forever' (D).

339 l. 71. *Ceres' feast*: 'when the Roman women were forbidden to bed with their husbands' (D).

l. 84. *Jove and Mars*: 'of whom more fornicating stories are told than any of the other gods' (D).

l. 90. *Secure . . . of*: safe from finding.

340 l. 107. *quail-pipe*: the vocal organs. The Californian editors think there may be an obscene pun. 'Quail-pipe' translates *fibula*, a needle drawn through the foreskin to prevent sex taking place.

l. 118. *wondering Pharos*: 'She fled to Egypt which wondered at the enormity of her crime' (D).

341 l. 163. *The good old sluggard*: Claudius. Juvenal 'tells the famous story of Messalina, wife to the emperor Claudius' (D).

l. 179. *Britannicus*: son of Messalina.

342 l. 208. *Wealth has the privilege*. 'His meaning is that a wife who brings a large dowry may do what she pleases and has all the privileges of a widow' (D).

343 l. 231. *Berenice's ring*: 'a ring of great price, which Herod Agrippa gave to his sister Berenice. He was king of the Jews, but tributary to the Romans' (D).

l. 249. *Cornelia*: 'mother to the Gracchi, of the family of the Cornelii, from whence Scipio the African was descended, who triumphed over Hannibal' (D).

l. 255. *O Paean*: Juvenal 'alludes to the known fable of Niobe in Ovid [*Metamorphoses* 6. 165 ff.]. Amphion was her husband. Paean was Apollo who with his arrows killed her children because she boasted that she was more fruitful than Latona, Apollo's mother' (D).

l. 261. *thirty pigs*. Juvenal 'alludes to the white sow in Virgil [*Aeneid* 3. 390–3], who farrowed thirty pigs' (D).

344 l. 271. *the Grecian cant*. 'Women then learned Greek, as ours speak French' (D).

l. 282. Ζωὴ καὶ ψυχή: life and soul (terms of endearment).

l. 307. *legacy*. 'All the Romans, even the most inferior and most infamous sort of them, had the power of making wills' (D).

346 l. 354. *plastron*: leather-covered breast-plate worn by a fencing-master.

l. 365. *magazine*: jewel-box.   l. 369. *sarcenet*: fine silk.

l. 377. *play a prize*: engage in a contest.

l. 379. *curtain-lecture*: 'a reproof given by a wife to her husband in bed' (J).

347 l. 389. *tiller*: drawer.

l. 424. *eringoes*: the root of sea holly, thought to nourish sexual desire and performance.

348 l. 439. *the goddess named the good*: 'at whose feasts no men were to be present' (D).

l. 449. *flats*: moisture.

l. 450. *rubster*: either one who rubs or something one rubs oneself with.

l. 456. *Nestor*: 'who lived three hundred years' (D).

l. 467. *what singer*. Juvenal 'alludes to the story of P. Clodius who, disguised in the habit of a singing woman, went into the house of Caesar where the feast of the good goddess was celebrated, to find an opportunity with Caesar's wife, Pompeia' (D).

349 l. 504. *navel-string*. It was believed that the navel-string was bound up with prosperity.   l. 510. *Carved*: castrated.

350 l. 522. *Pollio*: 'a famous singing boy' (D).

l. 532. *actors*: 'that such an actor, whom they love, might obtain the prize' (D).

l. 534. *haruspex*: 'he who inspects the entrails of the sacrifice and from thence foretells the successor' (D).

351 l. 586. *Tabors and trumpets*. 'The ancients thought that with such sounds they could bring the moon out of her eclipse' (D).

l. 593. *a mood and figure bride*: 'a woman who had learned logic' (D).

352 l. 601. *Priscian's ... head*: 'a woman grammarian who corrects her husband for speaking false Latin, which is called "breaking Priscian's head"' (D).

l. 622. *A train of these*: 'of she-asses' (D).

353 l. 641. *Sicilian tyrants*: 'are grown to a proverb in Latin for their cruelty' (D).

l. 672. Bellona's priests. 'Bellona's priests were a sort of fortune tellers and their high-priest a eunuch' (D).

354 l. 684. *murrey-coloured vest*: mulberry-coloured vest. D notes: 'a garment was given to the priest which he threw into the river; and that, they thought, bore all the sins of the people which were drowned with it'.

l. 696. *Meroe's burning sand*. Meroe is an island in the Nile.

355 l. 730. *Chaldeans*: 'are thought to have been the first astrologers' (D).

l. 738. *Otho*: 'succeeded Galba in the empire, which was foretold him by an astrologer' (D).

ll. 753-4. *Mars ... Venus*. 'Mars and Saturn are the two unfortunate planets; Jupiter and Venus the two fortunate' (D).

356 l. 761. *twelve houses, and their lords*: 'the twelve astrological divisions of the heavens, and the planets exercising influence in each' (Kinsley).

l. 767. *decumbiture*: horoscope for a bed-ridden person.

l. 769. *Ptolemy*: 'a famous astrologer; an Egyptian' (D).

l. 790. *savin*: shrub used to procure an abortion.

l. 792. *Ethiop's son*. Juvenal's 'meaning is, help her to any kind of slops which may cause her to miscarry; for fear she may be brought to bed of a blackamoor, which thou, being her husband, art bound to father; and that bastard may by law inherit thy estate' (D).

l. 796. *His omen would discolour all the day*. 'The Romans thought it ominous to see a blackamoor in the morning, if he were the first man they met' (D).

357 l. 818. *Caesinia*: 'wife of Caius Caligula, the great tyrant. It is said she gave him a love potion, which flying up in to his head, distracted him, and was the occasion of his committing so many acts of cruelty' (D).

l. 820. *mother's love*. If the hippomanes (a small black membrane on the head of a new-born foal) was removed before the mother could eat it, she refused to nurse her foal.

l. 827. *Agrippina's mushroom*. 'Agrippina was the mother of the tyrant Nero, who poisoned her husband Claudius that Nero might succeed, who was her son, and not Britannicus, who was the son of Claudius by a former wife' (D).

358 l. 848. *Drymon's wife*. 'The widow of Drymon poisoned her sons, that she might succeed to their estate. This was done in the poet's time or just before it' (D).

l. 853. *Medea's legend*. 'Medea, out of revenge to Jason who had forsaken her, killed the children which she had by him' (D).

l. 869. *the Belides*: 'who were fifty sisters, married to fifty young men, who were their cousins-german [first cousins], and killed them all on their wedding night, excepting Hypermnestra, who saved her husband Linus [Lynceus]' (D).

359 *The Tenth Satire of Juvenal*. First published in *The Satires of Juvenal and Persius* (1693).

l. 14. *The brawny fool*. 'Milo of Crotona, who for a trial of his strength going to rend an oak perished in the attempt, for his arms were caught in the trunk of it and he was devoured by wild beasts' (D).

360 l. 49. *lictors*. officers attending a magistrate.

l. 61. *gewgaw*: showy trifle.

361 l. 72. *places*: offices.

l. 93. *Sejanus*: 'was Tiberius' first favourite and while he continued so had the highest marks of honour bestowed on him. Statues and triumphal chariots were everywhere erected to him. But as soon as he fell into disgrace with the emperor, these were all immediately dismounted and the senate and common people insulted over him as meanly as they had fawned on him before' (D).

l. 103. *hanging*: downcast; 'with the quibbling suggestion of "hangable, born to be hanged"' (Kinsley).

362 l. 135. *dipped*: implicated.

363 l. 154. *a narrow isle*: 'the island of Capri which lies about a league out at sea from the Campanian shore was the scene of Tiberius' pleasures in the latter part of his reign. There he lived for some years with diviners, soothsayers, and worse company, and from thence dispatched all his orders to the senate' (D).

l. 173. *greater than the great*: 'Julius Caesar, who got the better of Pompey, that was styled the great' (D).

l. 185. *Tully, or Demosthenes*. 'Demosthenes and Tully [Cicero] both died for their oratory. Demosthenes gave himself poison to avoid being carried

NOTES TO PAGES 363-369

to Antipater, one of Alexander's captains who had made himself master of Athens. Tully was murdered by Mark Antony's order in return for those invectives he made about him' (D).

ll. 190–1. '*Fortune* ... *doom*': 'The Latin of this couplet is a famous verse of Tully's [Cicero's] in which he sets out the happiness of his own consulship, famous for the vanity and the ill poetry of it; for Tully, as he had a good deal of the one, so he had no great share of the other' (D).

364   l. 196. *that Phillippic.* 'The orations of Tully against Mark Antony were styled by him *Philippics* in imitation of Demosthenes, who had given that name to those he made against Philip of Macedonia' (D).

l. 208. *Feretrian Jove*: see note to 'Heroic Stanzas' l. 77 (above, p. 862).

365   l. 257. *huffing*: blustering.

366   l. 277. *the brick-built town*: 'Babylon, where Alexander died' (D).

l. 281. *Athos.* 'Mount Athos made a prodigous promontory in the Aegean sea. [Xerxes] is said to have cut a channel through it and to have sailed round it' (D).

l. 294. *he who points the way*: 'Mercury. ... His statues were anciently placed where roads met, with directions on the fingers of them, pointing out the several ways to travellers' (D).

367   l. 349. *salt*: on heat (of bitches).

368   l. 357. *quondam*: former.   l. 375. *hackney jade*: whore.

l. 388. *Pylian king*: 'Nestor, king of Pylos, who was three hundred years old' (D).

l. 391. *right hand count.* 'The ancients counted by their fingers; their left hands served them till they came up to a hundred; after that they used their right to express all greater numbers' (D).

l. 395. *the midmost sister drew.* 'The fates were three sisters, which had all some peculiar business assigned them by the poets, in relation to the lives of men. The first held the distaff, the second spun the thread, and the third cut it' (D).

369   l. 414. *His last effort.* 'Whilst Troy was sacking by the Greeks, old king Priam is said to have buckled on his armour to oppose them; which he had no sooner done, but he was met by Pyrrhus and slain before the altar of Jupiter in his own palace' (D).

l. 421. *Mithridates*: king of Pontus (120–63 BC), 'after he had disputed the empire of the world for forty years together with the Romans, was at last deprived of life and empire by Pompey the great' (D).

l. 421. *Croesus*': 'in the midst of his prosperity, making his boast to Solon how happy he was, received this answer from the wise man: that no one could pronounce himself happy, till he saw what his end should be. The truth of this Croesus found when he was put in chains by Cyrus and condemned to die' (D).

l. 424. *Marius*: Gaius Marius (157–86 BC), Roman general and consul.

l. 429. *Cimbrian*. The Cimbrians were a German tribe whom Marius defeated in 101 BC.

l. 435. *Pompey*: Roman general, who 'in the midst of his glory, fell into a dangerous fit of sickness at Naples.... He recovered, was beaten at Pharsalia, fled to Ptolemy, king of Egypt, and instead of receiving protection at his court had his head struck off by his order, to please Caesar' (D).

370 l. 440. *Cethegus*: 'was one that conspired with Catiline, and was put to death by the senate' (D).

l. 442. *Sergius*: Sergius Catiline 'died fighting' (D).

l. 453. *Virginia*: 'was killed by her own father to prevent her being exposed to the lust of Appius Claudius who had ill designs upon her' (D).

l. 454. *make*: husband.   l. 479. *springald*: young man.

371 l. 495. *moil*: toil.

l. 503. *modest youth his life*. 'Hippolytus, the son of Theseus, was loved by his mother-in-law, Phaedra, but he not complying with her, she procured his death' (D).

l. 505. *t'other stripling*. 'Bellerophon, the son of king Glaucus, residing some time at the court of Proetus, king of the Argives, the queen, Sthenobaea, fell in love with him; but he refusing her, she turned the accusation upon him, and he narrowly escaped Proetus's vengeance' (D).

l. 513. *Caesar's wife*: 'Messalina, wife to the emperor Claudius, infamous for her lewdness. She set her eyes upon C. Silius, a fine youth, forced him to quit his own wife and marry her with all the formalities of a wedding, whilst Claudius Caesar was sacrificing at Ostia. Upon his return, he put both Silius and her to death' (D).

372 l. 561. *usurps*: occupies.

373 *The First Satire of Persius*. First published in *The Satires of Juvenal and Persius* (1693).

Prologue
ll. 1–2. *cleft Parnassus ... Heliconian stream*. 'Parnassus and Helicon were hills consecrated to the muses and the supposed place of their abode. Parnassus was forked on the top and from Helicon ran a stream, the spring of which was called the Muses' Well' (D).

l. 5. *Pyrene*: 'a fountain in Corinth, consecrated also to the muses' (D).

l. 11. *the shrine*: of Apollo.   l. 13. *pie*: magpie.

The First Satire
374 l. 8. *Labeo's stuff*. Atticus Labeo 'made a very foolish translation of Homer's *Iliad*' (D).

375 l. 35. *They comb*: Persius 'describes a poet preparing himself to rehearse his works in public' (D).   l. 45. *chine*: back.

376 l. 82. *cedar tablets.* 'The Romans wrote on cedar and cypress tablets, in regard to the duration of the wood. Ill verses might justly be afraid of frankincense, for the papers in which they were written were fit for nothing but to wrap it up' (D).

377 l. 98. *citron beds*: 'writings of noblemen, whose bedsteads were of the wood of citron' (D).

l. 111. *Janus-like.* 'He was pictured with two faces, one before and one behind' (D).

l. 115. *Apulian*: from Apulia, in south-west Italy. 'The unusual heat of that region causes great thirst and thus makes tongues long and panting' (C).

l. 127. *riots*: debauches.   l. 130. *pin-feathered*: half-fledged.

l. 134. *trivial*: (1) insignificant; (2) pertaining to the trivium (grammar, rhetoric, and logic).

378 l. 139. *Quintius*: Quintius Cincinnatus, 'a Roman senator, who was called from the plough to be dictator of Rome' (D).

l. 167. *points*: periods.

379 l. 182. *Berecyntian Attis*: Ovid, *Fasti* 4. 221 ff.

ll. 182-5. *'Tis ... Apennine*: 'foolish verse of Nero which the poet repeats' (D).   l. 187. *'Arms ... sing'*: the first line of Virgil's *Aeneid*.

l. 194. *Mimallonian*: Bacchanalian.

ll. 194-200. *'Their ... sound.'* 'Other verses of Nero' (D).

l. 203. *Maenas and Attis*: 'poems on the Maenades, who were priestesses of Bacchus, and of Attis, who made himself a eunuch to attend on the sacrifices of Cybele, called Berecyntia by the poets. She was the mother of the gods' (D).

380 l. 211. *flout*: a mocking speech.

l. 218. *common-shores*: waste lands beside rivers for the deposit of filth.

l. 223. *Lucilius*: early Roman satirist.

l. 240. *Midas.* 'By Midas the poet meant Nero' (D). For Midas, see Ovid, *Metamorphoses* 11. 92 ff.

381 ll. 246-7. *Eupolis ... Cratinus.* 'Eupolis and Cratinus as also Aristophanes mentioned afterward were all Athenian poets, who wrote that sort of comedy which was called the old comedy, where the people were named who were satirized by those authors' (D).

l. 270. *And with his foot.* 'Arithmetic and geometry were taught on floors which were strewn with dust or sand' (D).   l. 275. *drabs*: whores.

382 *The First Book of Ovid's Metamorphoses.* First published in *Examen Poeticum* (1693).

l. 5. *tenor*: continuance.   l. 6. *Deduced*: brought down.

l. 10. *indigested*: disordered.   l. 26. *intestine*: domestic.

383 l. 47. *bounding*: limiting.

384 l. 80. *frozen waggon*: the constellation of the Great Bear (Charles's Wain).

l. 82. *the unwholesome year*: the annual harvest.

385 l. 132. *guiltless of*: unacquainted with.

386 l. 174. *rummaging*: emptying.

387 l. 211. *Lycaon's guilt*. Lycaon was an Arcadian tyrant, impious to Jupiter.

388 l. 248. *nobler parts*: the heart and lungs.

l. 256. *attempted*: assailed.

389 l. 297. *Molossian state*: the state of the Molossi, a people of Epirus.

l. 313. *famished*: wasted.

390 l. 346. *axle-tree*: axle-pin.

391 l. 358. *flaggy*: limp.   l. 368. *Junonian Iris*: Iris the messenger of Juno.

l. 377. *The watery tyrant*: Neptune, god of the sea.

392 l. 426. *bound*: boundary.

393 l. 449. *Tyrian*: purple.

394 l. 489. *our father*: Prometheus.   l. 506. *gradual*: 'an order of steps' (J).

395 l. 532. *diffides*: distrusts.

396 l. 565. *Pharian*: of Pharos.   l. 568. *crusted*: covered with a crust.

l. 574. *temper*: mixture.   l. 586. *Python*: a huge serpent.

397 l. 605. *promiscuous*: undifferentiating.

398 l. 646. *her father*: Peneus.   l. 647. *son*: grandson.

l. 680. *a foe*: as a foe.

399 l. 693. *Delphos*: Delphi.   l. 694. *Patareian*: of Patara.

l. 718. *slipped*: let loose.

400 l. 722. *turn*: 'a hare's change of direction, manœuvred by the dog' (Kinsley).

l. 736. *paternal brook*. Daphne's father was the river-god Peneus.

403 l. 871. *hardly*: 'not softly' (J).

405 l. 921. *his airy messenger*: Mercury (Hermes).

408 l. 1055. *daunted*: overcome with fear.

409 l. 1086. *levée*: rising from bed.

Ovid's Amours. Book I. Elegy 1. First published in Tonson's *Poetical Miscellanies*, part 5 (1704). Possibly written around 1693.

l. 3. *Six feet*. Latin heroic poetry is written in hexameters.

l. 5. *every second verse*. Latin elegiac poetry has a hexameter alternating with a pentameter.

410   l. 34. *unequal verse*: elegiacs. See note to l. 5 above.

Ovid's Amours. Book I. Elegy 4. First published in Tonson's *Poetical Miscellanies*, part 5 (1704).

l. 7. *Hippodamia's*. For Hippodamia see Dryden's translation of *Metamorphoses* 12. 292 ff. (above, p. 773).

412   l. 72. *grubble*: grope.

413   *Ovid's Art of Love. Book I*. First published in *Ovid's Art of Love ... Translated by Several Hands* (1709). Lines 111–51 and 590–635 had been published in Tonson's *Poetical Miscellanies*, part 5 (1704). The translation was written in 1693 (*Letters*, p. 58). The text here is based on *1709*.

l. 13. *Chiron*: unlike other centaurs, he was refined.

l. 29. *Delphian god*: Apollo.

l. 32. *Hesiod*: Greek poet, probably of the eighth century BC.

414   l. 73. *portico*: colonnade.

l. 76. *that other portico*: the portico of the Danaids in the temple of Apollo on the Palatine.

415   l. 86. *hall*: Westminster hall, which featured gossiping women.

l. 99. *chase*: a hunting-ground.

l. 117. *Parian*: from the island of Paros, famed for fine white marble.

416   l. 132. '*The best*': a toast.

417   l. 186. *takes*: captivates.

418   l. 208. *slain Crassi*. Crassus and his son were slain fighting the Parthians at the battle of Carrhae (53 BC).

l. 209. *A youth*: Caius Caesar (Caligula). He was later emperor.

l. 216. *his sire*: Jove.   l. 224. *father*: Germanicus.   l. 239. *averse*: while retreating.

419   l. 243. *thou*: Caligula.   l. 249. *thou*: the reader.

l. 263. *Paphian goddess*: Venus.

420   l. 291. *Baian*: at Baiae, near Naples.

l. 296. *fight a prize*: practise prize-fighting.   l. 299. *myrtle*: sacred to Venus.

l. 319. *Byblis*. She nursed an incestuous passion for her brother Caunus.

l. 321. *Myrrha*. See D's 'Cinyras and Myrrha' (above, pp. 677–86).

421   l. 325. *Ida's*: Mount Ida is in Crete.

l. 330. *The queen*: Pasiphae, wife of Minos, king of Crete.

l. 339. *Minos*: king of Crete.

422 l. 369. *son*: the Minotaur.

l. 370. *Atreus' wife*: Aerope. Her crimes blotted out the sun (ll. 372–3).

l. 374. *Thy daughter*: Scylla.　　l. 380. *Phoenix*: son of Amyntor.

423 l. 443. *dipped*: involved.

424 l. 468. *Allia*: a tributary of the Tiber, near which the Romans were defeated by the Gauls in 390 BC.

425 l. 492. *dribs*: inches.

l. 520. *Cydippe*. Her lover Acontius wrote on an apple 'I swear by Diana to marry Acontius'. Cydippe read it aloud, and was bound by the vow.

427 l. 608. *Mimallonian dames*: women followers of Bacchus.

428 l. 611. *clear*: drunk.

429 l. 665. *Eurytion*: a centaur, made drunk at the feast of the Lapiths.

430 l. 706. *naked Pallas with Jove's queen*: Minerva (Pallas Athene) with Juno. A reference to the judgment of Paris.

431 l. 737. *Phalaris*. Sicilian tyrant who roasted his victims in a bull made by Perillus.　　l. 738. *cow*: for 'bull'.

l. 768. *Phoebe and her sister*. Phoebe and Hilaira, daughters of Leucippus, were raped by Castor and Pollux, sons of Tyndareus.

l. 770. *Deidamia*: mother of Neoptolemus (Pyrrhus) by Achilles.

432 l. 780. *Aeacides*: Achilles, grandson of Aeacus.

433 l. 834. *Orion*: a giant hunter.

434 l. 882. *Bug-words*: words that terrify or alarm.

The Fable of Iphis and Ianthe. First published in *Examen Poeticum* (1693). A translation of *Metamorphoses* 9. 665–796.

l. 1. *of this*: of the preceding story in Ovid.

l. 3. *Gnossian*: Cretan.

435 l. 14. *tits*: girls (used playfully).

l. 37. *Osiris*: Egyptian god of fertility and husband to Isis.

l. 38. *The silent god*: Harpocrates, Egyptian god of silence.

437 l. 109. *daughter of the sun*: Pasiphae.

439 l. 199. *burnish*: increase in breadth.

440 *The Fable of Acis, Polyphemus, and Galatea*. First published in *Examen Poeticum* (1693). A translation of *Metamorphoses* 13. 750–897.

l. 31. *simagres*: grimaces.

443 l. 115. *Phaeacian*: ideal, like the gardens of Alcinous, king of the Phaeacians (*Odyssey* 7).

l. 127. *sweepy*: moving with a sweeping motion.

l. 137. *turtles*: turtle-doves.

446 *A Song to a Fair Young Lady going out of Town in the Spring*. First published in *Examen Poeticum* (1693).

447 *Prologue*. First published in *Examen Poeticum* (1693). Written in 1689. The occasion for this prologue is not known.

l. 15. *new Deserters' Bill*: one of the mutiny acts passed by parliament from 1689.

l. 17. *Mercury*: 'a general term for newspapers' (Kinsley).

l. 25. *the blue*: '"glossy blue" and therefore sound and worth plucking' (Kinsley).

448 *Veni, Creator Spiritus Translated in Paraphrase*. First published in *Examen Poeticum* (1693). A translation of a hymn by an unknown late ninth-century poet.

Title. Veni Creator Spiritus: come creator spirit (Latin).

l. 8. *Paraclete*: the holy spirit. John 14: 16.

l. 14. *sevenfold energy*: 'the energizing force available from the traditional "Seven Gifts of the Holy Spirit", which were wisdom, understanding, counsel, fortitude, knowledge, piety, and fear of the Lord. The list was drawn from Isaiah 11: 2' (C).

449 *Rondelay*. First published in *Examen Poeticum* (1693).

450 *The Last Parting of Hector and Andromache*. First published in *Examen Poeticum* (1693). A translation of *Iliad* 6. 369–502.

l. 19. *the blue-eyed progeny of Jove*: Athene.

451 l. 34. *Eetion's heir*. Eetion, father of Andromache, was king of Thebe in Cilicia.   l. 35. *Hippoplacus*: beneath (*hypo*) the mountain of Placus.

455 l. 194. *restore*: renew.

To my Dear Friend Mr Congreve on his Comedy called The Double Dealer. First published as a commendatory poem in Congreve's *The Double Dealer* (1694). The poem was written in 1693. William Congreve (1670–1729) had been Dryden's friend for some three years.

l. 5. *giant race*: Genesis 6: 4.

l. 7. *Janus*: Roman god who introduced cultivation into Italy (Ovid, *Fasti*, 1. 63).

l. 14. *The second temple*: Ezra 5–6, and Haggai 2: 3.

l. 15. *Vitruvius*: Roman architect in the time of Augustus. As author of the only surviving ancient treatises on architecture, he exerted an enormous influence at the time of the renaissance.

l. 29. *courtship*: courtliness. Etherege was also a diplomat.

l. 30. *manly*: the name of the hero of Wycherley's *The Plain Dealer*.

l. 32. *foiled*: overdone, surpassed.

456 ll. 35–8. *Fabius ... overcome*. 'Had Scipio been as loveable as you the envious Fabius would have rejoiced in his early fame and supported him, even though Fabius had in his own day been unsuccessful against Hannibal' (Kinsley).

ll· 39–40. *Romano ... taught*. Dryden is in error. Giulio Romano (1492–1546) was nine years younger than Raphael (1483–1520) and never his master.

ll. 45–6. *one Edward ... A greater Edward*. Edward II (1284–1327) was deposed and murdered. He was succeeded by his son Edward III (1312–77), 'greater', I suppose, because of his frequent wars against the French.

ll. 47–8. *Tom the second ... Tom the first*. When William of Orange came to the throne in 1688 Dryden lost his offices of Poet Laureate and Historiographer Royal to Thomas Shadwell (see headnote to 'MacFlecknoe' above, p. 878). Next as Historiographer (1692) came Thomas Rymer.

l. 49. *my patron's*. Dryden's patron was Charles Sackville, Earl of Dorset (1638–1706). As Lord Chamberlain, it was his duty to deprive Dryden of his offices.

l. 53. *High on the throne of wit*: *Paradise Lost*, 2. 1: 'High on a throne of royal state.'

l. 55. *Thy first attempt*: Congreve's *The Old Bachelor* (1693).

l. 72. *Be kind to my remains*. Congreve was to edit Dryden's plays in 1717.

457 *To Sir Godfrey Kneller*. First published in *The Annual Miscellany* (1694). Sir Godfrey Kneller (1646–1723) came to England in 1676, and became court painter in 1680. He painted Dryden several times.

l. 9. *pencil*: paint-brush.

l. 22. *Prometheus*. See 'The First Book of Ovid's *Metamorphoses*' 97–112 (above, p. 384). l. 36. *picture*: the art of painting.

458 l. 54. *Bantam's embassy*. Eight ambassadors from the king of Bantam in Java came to England in 1682. l. 57. *sister arts*: poetry and painting.

l. 57. *iron*: deathlike, fast; a Virgilian usage (*Aeneid* 10. 745).

l. 73. *thy gift*: 'Shakespeare's picture drawn by Sir Godfrey Kneller and given to the author' (D's note).

459 l. 78. *Teucer*: Greek archer who took shelter behind the shield of Ajax (Homer, *Iliad* 8. 266–72).

l. 81. *his*: Shakespeare's.

l. 96. *Jacob's race*: Genesis 27.

l. 97. *Apelles' art*. Apelles, a fourth-century-BC Greek painter, was the only man allowed to paint Alexander the Great.

l. 98. *Leo's gold*: Leo X was Raphael's patron.

460 l. 119. *heroes'*: of the painters of the renaissance.

l. 132. *seven cities*: 'a reference to the well-known legend that seven Greek cities claimed to have been Homer's birthplace' (C).

l. 139. *early*. Kneller 'travelled very young into Italy' (D's note).

461 l. 158. *posture sink*: disappear in your judicious placing.

l. 178. *embrown the taint*: darken the colours.

An Ode on the Death of Mr Henry Purcell. First published in 1696. Purcell died on 21 November 1695.

l. 5. *the close of*: the closing in of.   l. 6. *Philomel*: the nightingale.

462 l. 18. *sovereigns'*: Pluto and Prosperpina, 'here thought of as symbolizing disorder' (C).   l. 25. *handed him along*: led him by the hand.

463 *Virgil's Georgics*. The whole poem was first published in *The Works of Virgil* (1697). A second edition followed in 1698. Book III had been published in *The Annual Miscellany* (1694). My text reprints *1697*, incorporating some of the changes (or, in the case of Book III, restorations) of *1698*.

Book I

l. 6. *Maecenas*: Virgil's patron.   l. 15. *thou*: Neptune.

l. 17. *thou*: Aristaeus.   *Caean*: of Ceos, an Aegean island.

l. 23. *fattening oil*: olive oil.

l. 24. *Thou founder of the plough*: Triptolemus, favourite of Demeter, goddess of the cornfield.

464 l. 30. *thou, whose undetermined state*: Augustus, not yet a god.

l. 38. *Caesar*: Augustus.

l. 46. *the Balance*: Libra, a sign of the zodiac like Scorpion and the Maid (l. 48). Virgil suggests Augustus will be 'specially close to Justice' (Elliot).

465 l. 86. *Idume*: biblical Edom.

l. 87. *Pontus*: the Black Sea.   *beaver-stones*: the two small sacs in the groin of the beaver, from which the substance 'castor' is obtained.

l. 89. *Epirus*: in north-west Greece.   *Elean*: of Elis, 'famous for horse-breeding' (Elliot).

l. 93. *Deucalion*. See Dryden's translation of Ovid's *Metamorphoses* 1. 424–556 (above, pp. 392–5).

l. 102. *Arcturus*: a star in the Great Bear.

467 l. 149. *Mysia*: in Asia Minor.

ll. 149–50. *tops of Gargarus*: upper reaches of Mount Ida, in Asia Minor.

l. 179. *Strymonian*: of the Strymon, a river in Greece celebrated for its cranes. l. 183. *The sire of gods and men*: Jove.

468 l. 200. *liquid gold*: honeydew.

l. 210. *Northern Car*: the Great Bear. l. 214. *Drags*: drag-nets.

l. 221. *Dodonian*: of Dodona, in Greece, where oaks were sacred to Zeus.

470 l. 295. *Kids, Dragon*, (1) two stars in the constellation Auriga, bringing storms in spring and autumn; (2) a northern constellation.

l. 297. *Helle's stormy straits*: the Hellespont.

l. 298. *Astraea's balance*: Libra, which the sun enters at the autumn equinox. l. 307. *The Bull*: Taurus, which the sun enters in April.

l. 308. *the Dog*: Sirius, which appears to set at the end of April. Ruaeus, a seventeenth-century editor of Virgil whose edition Dryden used 'explains that the dog is yielding to the constellation *Argo* (the ship) which is "turned away" because it sets stern first. Dryden accepted this too-complicated notion' (Elliot).

l. 310. *Maia*: one of the Pleiades, a constellation which sets in November.

l. 311. *the bright Gnossian diadem*. Ariadne was from Cnossos ('Gnossian') in Crete. Her crown became the Corona Borealis, which, like the Pleiades, sets in November.

471 l. 318. *slow waggoner*. The Waggoner (Boötes) is a constellation near the north star.

l. 320. *Apollo*: in his role as sun-god. l. 346. *furzes*: gorses.

472 l. 374. *the sons of earth*: the Titans. l. 385. *Phoebe's light*: moonlight.

474 l. 460. *Hermes*: the planet Mercury.

l. 466. *mighty Mother's*: Magna Mater, a goddess of Asia Minor. Dryden confuses the Latin, where the reference is to 'great' Ceres.

476 l. 549. *Nisus*: king of Megara, transformed into a sea-bird. See Ovid, *Metamorphoses* 8. 8 ff.

l. 550. *Scylla*: daughter of Nisus, also transformed into a sea-bird.

l. 556. *purple hair*. Scylla stole a purple lock from Nisus, on which the safety of his kingdom depended, in order to gain the love of Minos of Crete.

478 l. 604. *erring*: wandering. l. 614. *firth*: estuary.

479 l. 659. *Emathian*: of Emathia, in Macedonia.

l. 661. *twice*: the battles between Julius Caesar and Pompey at Pharsalus in Thessaly, and the battle of Philippi.

l. 672. *youthful Caesar*: Augustus.

l. 675. *Laomedon*: father of Priam, king of Troy. He was 'cursed' because he refused to pay Apollo and Poseidon for helping him build the walls of Troy. Aeneas, escaping from Troy, took the curse with him to Rome.

Book II

480 l. 4. *Minerva's tree*: the olive.

481 l. 20. *mastful*: full of mast, the fruit of beech, oak, chestnut, and other forest trees.   l. 51. *Ismarus*: a mountain in Thrace.

l. 52. *Taburnus*: mountains in southern Italy.

482 l. 90. *Paphian*: of Paphos.   l. 96. *arbute*: arbutus.

483 l. 120. *Lotes*: lotuses.

l. 124. *Radii*: plural of radius, a variety of olive.   *orchites*: the orchad, another variety of olive.

l. 125. *pausia*: the pausian, another variety of olive.

l. 126. *Alcinous' orchard*. Alcinous was a mythical king of Phaeacia, and father of Nausicaa. For his orchard see Homer, *Odyssey* 7. 112 f.

484 ll. 146–7. *Bumastus ... dugs of cows*: Bumastus is a kind of large grape, whose name is derived from the Greek for 'cow-breasted'.

l. 152. *Eurus*: the south-east wind.   l. 169. *the Seres*: the Chinese.

l. 170. *fleecy forests*: forests of silk.

l. 178. *direful stepdame's*: Medea's. Medea was a sorceress. Stepdame = stepmother.

485 l. 209. *aconite*: wolfsbane.

l. 218. *Larius*: Lake Como.   l. 219. *Benacus*: Garda.

486 l. 225. *where secure the Julian waters glide*. The lakes Lucrinus and Avernus 'had been turned into a large naval port by Agrippa a few years before [the Georgics were written] and was now called Portus Iulius' (Elliot).

l. 236. *greater Scipio's double name*: 'the *two* Scipios, not one Scipio with two names' (Elliot).   l. 237. *Caesar*: Augustus.

l. 241. *Saturnian soil*: Italy, once ruled by Saturn.

l. 246. *Ascraean*: of Ascra, where Hesiod, founder of this tradition of verse, was born.   l. 252. *Palladian plant*: the olive.

487 l. 273. *hapless Mantua*. Land was taken from Mantua (Virgil's native city) to reward Augustus' soldiers.

488 ll. 315–16. *height Of superfice*: level.   l. 333. *rope*: drop stickily.

490 l. 397. *Jove's own tree*: the oak.

491 l. 433. *the white bird*: the stork.

492 l. 484. *when the Dog-star cleaves*: late August.

493 l. 529. *glad with Bacchus*: tipsy.

l. 532. *Saturnian rhymes*: rustic rhymes; also alluding to 'Saturnian metre', a type of early Latin verse.

l. 540. *honest*: handsome; a Latinism.

l. 547. *hazel broach*: spits of hazel-wood.

494 l. 576. *Nor when*: 'Dryden changes his construction at l. 582. "But when" would make a better link with "Yet still"' (Kinsley).

495 l. 612. *Cytorus*: a mountain in Asia Minor.

l. 614. *Narycian*: of Naryx, a Locrian city on the Euboean sea.

Book III

500 l. 1. *Pales*: an Italian rural deity of flocks and herbs.

l. 3. *Amphrysian shepherd*: Apollo, when a shepherd on the banks of the Amphrysus.   *Lycaean*: Of Lycaeaus, a mountain in Arcadia.

l. 7. *Busiris'*: Busiris was a king of Egypt who killed strangers.

l. 8. *Eurystheus*: king of Mycenae and Hercules' taskmaster.

l. 9. *Hylas*: favourite of Hercules.

l. 9. *Latona's ... isle*: Delos. Latona was the mother of Apollo and Artemis (Diana).

l. 10. *Pelops' ivory shoulder*. Pelops was given to the gods to eat. Ceres ate some of his shoulder which was replaced with ivory when he was re-assembled.

ll. 10–11. *toil For fair Hippodame*. Pelops won Hippodame in a chariot race.

l. 19. *Parian stone*: marble from the island of Paros in the Aegean.

l. 20. *Mincius*: a river in Mantua.   l. 23. *Caesar*: Augustus Caesar.

l. 25. *Tyrian*: purple.

501 l. 30. *whirlbat*: 'a certain game . . . among the ancients wherein they whirled leaden plummets at one another' (*OED*, citing Edward Phillips).

l. 39. *interwoven Britons*: conquered Britons displayed woven on the stage curtains.   l. 41. *elephant*: ivory.

l. 42. *Indian*: eastern.   l. 44. *His*: the Nile's.

l. 45. *Niphates*: a mountain in Armenia. Dryden took it for a river.

l. 48. *backward bows*: shot while retreating.

l. 58. *Tros*: the king after whom Troy was named.

l. 59. *he—the god*: Apollo.

502 l. 73. *Cithaeron*: Greek mountain famous for game.

l. 74. *Taygetus*: mountain near Sparta, famous for hunting dogs. *open*: give tongue.

l. 75. *Epidaurus*: town in north-east Greece.

l. 82. *Tithon*: Tithonus, lover of Aurora.

l. 99. *insult*: mounting.

503 l. 143. *god of Thrace*: Mars.

l. 145. *Saturn*. Caught in adultery with the nymph Phillyra, Saturn fled from his wife Rhea disguised as a horse.

504 l. 177. *Erichthonius*: legendary king of Athens.

l. 180. *Lapithae*: legendary tribe of Thessaly.

505 l. 199. *salacious*: tending to provoke lust.

l. 204. *out of case*: out of condition.

l. 207. *the craving kind*: female animals during pregnancy.

l. 224. *The male has done*: 'Here the poet returns to cows' (D).

506 l. 235. *Alburnian*: of Alburnus, a mountain in southern Italy.

l. 238. *Oestros . . . Asylus*: 'gadfly' in Greek and Latin respectively.

l. 242. *Tanagrus*: a tributary of the Silarus, near mount Alburnus.

l. 244. *Io's punishment*. Io, a priestess loved by Jove, was turned into a heifer and plagued by a gadfly.

507 l. 283. *beestings*: the first milk a female gives after the birth of her young.

l. 287. *Pisa's flood*: the river Alpheus, near Pisa in Elis.

510 l. 403. *the youth*: Leander.

l. 414. *Sestian*: of Sestos, where Leander swam to visit Hero.

l. 438. *boring*: sticking their noses out as high as possible.

511 l. 443. *Hippomanes*: a small black membrane on the forehead of a new-born foal, used in making love-potions.

l. 453. *oil*: 'as in "midnight oil"' (Elliot).

l. 476. *new Ram*: Aries, which begins in April.

l. 479. *drunk with Tyrian juice*: dyed purple.

512 l. 487. *camlets*: goatskin coats.   l. 495. *defend*: fend off.

l. 504. *Hesperus*: the evening star, although the context tells that the morning star is needed.

513 l. 541. *Scythian*: of Scythia, now south-west Russia.

l. 543. *Maeotian*: of Maeotis, a sea north of the Black Sea.

l. 544. *Hister*: the Danube.   l. 556. *hostry*: hostelry.

514 l. 585. *barmy*: frothy.

l. 586. *Rhipaean*: of Rhipaea, an imagined range of snowy mountains far to the north.

l. 602. *Did bribe thee*: won your love. But it was Selene, not the chaste Cynthia that Pan seduced.

515 l. 625. *beamy*: antlered.   l. 627. *galbanum*: gum resin.

516 l. 672. *tetter*: skin-eruption.

l. 683. *oint*: anoint.   *mothered oil*: the dregs of oil, boiled down.

517 l. 689. *squills*: sliced and dried lily bulbs, used in medicine.

l. 703. *Gelons*: the Geloni, supposed to have lived in Scythia.

518 l. 742. *Or*: ere.   l. 743. *thriven*: full-grown.

l. 746. *and labours from the chine*: suffers from skin complaints.

l. 759. *ropy*: forming gelatinous threads.

519 l. 808. *phocae*: seals.

520 l. 812. *paddocks*: frogs.

l. 820. *Tisiphone*: one of the furies, bringing plague.

Book IV

521 l. 2. *Aerial honey*. 'The ancients believed that real honey fell like dew from heaven' (Elliot).

l. 19. *Procne*. She was turned into a swallow after killing her son, Itys.

522 l. 57. *Idaean pitch*: pitch from Ida, a mountain of Phrygia in Asia Minor.

523 l. 67. *roast red crabs*. 'Apparently burnt crabs were used as medicine for certain diseases of trees' (Elliot).

l. 87. *milfoil*: common yarrow.   *honeysuckles*: flowers of clover.

524 l. 106. *king's*: queen's. 'Queen'-bees were thought to be males.

l. 115. *shocking*: conflicting.

525 l. 152. *pointed*: appointed.   l. 156. *metheglin*: spiced mead.

l. 167. *god obscene*: Priapus.   l. 168. *lath*: thin strip of wood.

l. 179. *Paestan*: of Paestum, a city in south Italy, famous for roses.

l. 185. *bear's-foot*: hellebore.

526 l. 188. *Corycian*: of Corycus, in Asia Minor.

l. 194. *vervain*: verbena.

l. 212. *by tale*: by number. 'As many blooms had he, he had apples.'

l. 217. *good fellows*: topers.   l. 227. *drive*: conduct.

527 l. 239. *grout*: dregs.

528 l. 306. *Media*: in Asia, south of the Caspian.

529 l. 332. *congenial planets*: planets in conjunction at their birth.

l. 338. *Pleiades*. The Pleiades appear in mid June.

l. 341. *Scorpion*. Scorpio appears in November.

530 l. 371. *shagged*: shaggy.   l. 383. *galbanean*: of galbanum (gum resin).

l. 387. *sodden*: boiled, that is, reduced.

l. 389. *galls*: excrescences caused by insects on oaks.

l. 390. *Cecropian*: from Mount Hymettus in Attica, once ruled by Cecrops.   *centaury*: a healing plant.

l. 392. *Amellus*: the purple Italian star-wort.

531 l. 404. *the Arcadian master*: Aristaeus (l. 451), son of Apollo and Cyrene.

l. 422. *burnished*: 'with the dead skin rubbed off' (Kinsley).

532 l. 446. *gall*: harass.

l. 451. *Tempe*: a valley in north-east Thessaly.

l. 456. *Cyrene*: a nymph, mother of Aristaeus.

533 l. 475. *distaffs*: cleft sticks holding wool.

l. 476. *carded*: teased out.   *Milesian*: of Miletus, a city in Asia Minor.

l. 480. *Lucina's*. Lucina was goddess of childbirth.

ll. 488–9. *theft ... Of Mars*. 'Mars had an affair with Venus, whose husband Vulcan caught them in a net' (Elliot).

l. 505. *paternal stream*: the stream of Peneus, great-grandfather of Aristaeus.

534 l. 534. *fattens*: fertilizes.   l. 536. *pory*: porous.

l. 553. *Vestal fire*: fire of Vesta, goddess of hearth and household.

535 l. 557. *Carpathian*: the Carpathian sea, near Crete.

l. 561. *Pallenian*: near Pallene, in Macedonia.

l. 564. *Nereus*: a sea-god.   l. 577. *southing*: travelling south.

536 l. 636. *part*: depart.

537 l. 657. *For crimes not his, the lover lost his life*. 'The episode of Orpheus and Eurydice begins here, and contains the only machine [supernatural agent] ... in the *Georgics*. I have observed in the epistle before the *Aeneid* that our author seldom employs machines but to adorn his poem and that the action which they seemingly perform is really produced without them' (D).

538 l. 678. *the inexorable king*: Pluto.

l. 686. *Cocytus*: a river in the underworld.

l. 688. *Styx*: a river in the underworld.

539 l. 738. *Strymon's*: of Strymon, a river in Macedonia.

540 l. 761. *Hebrus*: 'the Maritsa which flows through Bulgaria and European Turkey into the northeast Aegean' (Elliot).

l. 776. *Napaean race*: race dwelling in wooded valleys.

l. 787. *poets' king*: Orpheus.

541 l. 809. *Caesar*: Augustus Caesar.

Postscript to the Reader. First published in *The Works of Virgil* (1697).

542 *Ennius*: Quintus Ennius (239–169 BC), the 'father of Roman poetry'.

farthings. A farthing was a quarter of an old penny; an insignificant sum.

Revolution: of 1688, which ousted James II, and deprived Dryden of his offices as Poet Laureate and Historiographer Royal.

Cynthius . . . admonuit: Virgil, *Eclogues* 6. 3–4.

permitted Aeneas to pass freely to Elysium. Cerberus, the three-headed dog that guarded the entrance to the lower world, was drugged by the sibyl to let Aeneas in. Virgil, *Aeneid* 6. 417–25 (in Dryden's version 6. 563–75).

Earls of Derby and Peterborough: William Stanley (1655–1702) and Charles Mordaunt (1658–1735).

difference of interests. Peterborough was an ardent supporter of William III.

543 *last Aeneid*: last book of the *Aeneid*.

William Trumball: (1639–1716), secretary of state.

Extremum . . . Gallo: Virgil, *Eclogues* 10. 1–3.

Gilbert Dolben: (1658–1722), member of parliament.

Delphin. Ruaeus's Delphin edition (Paris, 1675) was Dryden's principal text in his translation of Virgil.

Fabrini: Giovanni Fabrini's translation of and commentary on Virgil (1604–23).

William Bowyer: (1639–1722), Dryden's friend since their days at Trinity College, Cambridge.

Earl of Exeter: John Cecil (1648–1700).

William Walsh: (1663–1708), poet.

Duke of Shrewsbury: Charles Talbot (1660–1718).

544 *good letters*: literature.

'*The Power of Love*': '*Amour omnibus idem* or, The Force of Love in all Creatures' in *Examen Poeticum* (1693); possibly by Lord Lansdowne.

'*Silenus*': Virgil's sixth eclogue, translated in *Miscellany Poems* (1684).

Addison: Joseph Addison (1672–1719), who had translated Virgil's fourth Georgic, on bees, in *The Annual Miscellany* (1694).

Cowley's praise of a country life: Abraham Cowley's (1618–67) imitation of *Georgics* 2. 458–540, in *Several Discourses* (1668).

only one of them. Sir Richard Blackmore (1653–1729), physician and poet.

545 *Alexander's Feast or The Power of Music: an Ode in honour of St Cecilia's Day*. First published in 1697, as a separate poem. It was set to music by Jeremiah Clarke, and in the eighteenth century by Thomas Clayton and by Handel.

title. Alexander the Great, son of Philip of Macedon (l. 2) was fabled to have been fathered by Zeus (Jove: ll. 25 ff.). St Cecilia is the patron saint of music. Her day is 22 November.

l. 9. *Thais*: an Athenian courtesan. *1697* reads 'Lais'.

l. 20. *Timotheus*: Alexander's music master.

l. 28. *belied*: concealed.

l. 29. *Sublime*: high.   *spires*: coils.

l. 30. *Olympia*: Alexander's mother.

546   l. 51. *a purple grace*: wine.

l. 52. *honest*: (1) frank; (2) handsome.   l. 53. *hautboys*: oboes.

547   l. 69. *master*: Timotheus.   l. 70. *ardent*: fiery.

ll. 80–1. *Deserted ... fed*: Darius was killed by his own followers.

l. 97. *Lydian*: soft and plaintive.

549   l. 147. *flambeau*: torch.   l. 162. *vocal frame*: organ.

550   *To Mr Motteux on his Tragedy called Beauty in Distress*. First published
in Motteux's *Beauty in Distress* (1698). Motteux's play had failed earlier
in 1698, and Dryden wrote this poem to help the published version. Peter
Anthony Motteux (1663–1718) was a French Huguenot. He came to
England in 1685 and became a bookseller. He completed Urquhart's
translation of Rabelais into English, edited the *Gentleman's Journal*, and
translated *Don Quixote*.

l. 2. *but the stage*: a glance at Jeremy Collier's *A Short View of the Immorality and Profaneness of the English Stage* (1698), a 300-page pamphlet
much on Dryden's mind in this poem.

l. 19. *Rebellion, worse than witchcraft*: 'Rebellion is as the sin of witchcraft'
(1 Samuel 15: 23).

551   l. 34. *Corneille*: Pierre Corneille (1606–84), French playwright.

l. 35. *tripled unity*: alluding to the 'the three unities', of time, place, and
action. See above, 'An Essay of Dramatic Poesy', p. 81 and note.

l. 39. *Wycherley*: William Wycherley (1641–1716), playwright.

l. 47. *allay*: alloy.

552   *Fables Ancient and Modern*. First published March 1700. It begins with a
fulsome dedication to the Duke of Ormonde, not reprinted here.

Preface
certain nobleman: George Villiers (1628–87) second Duke of Buckingham.
He ridiculed Dryden in *The Rehearsal* (1671) and Dryden portrayed him
as Zimri in *Absalom and Achitophel*.

Sandys. George Sandys (1578–1644) published his translation of Ovid's
Metamorphoses in 1626.

Waller: Edmund Waller (1606–87).

Fairfax. Edward Fairfax (d. 1635) translated Tasso's *Gerusalemme Liber-
ata* as *Godfrey of Bulloigne* (1600).

insinuates: *Fairy Queen* 4. 2. 34.

NOTES TO PAGES 553–560　923

553　*connexion*: *Leviathan* 1. 3.

　　Boccace: Giovanni Boccaccio (1313–75).

　　octave rhyme: *ottava rima*. Boccaccio popularized the form.

　　Rymer: Thomas Rymer (1641–1713), critic and historian, in his *Short View of Tragedy* (1693), ch. 4.

554　*dead-colouring*: the first layer of colour.

　　staved: destroyed.

555　*versus . . . canorae*: Horace, *Ars poetica* 322.

　　remain. The reference in this paragraph is to *A Short View of the Immorality and Profaneness of the English Stage* (1698), in which Jeremy Collier (1650–1726) criticized Dryden's licentiousness.

556　*Dido*: in *Aeneid* 1 and 4.　　*Calypso*: in *Odyssey* 5.

　　Hobbes. Thomas Hobbes published his translation in 1676.

　　impiger . . . acer: Horace, *Ars poetica* 121.

　　quo . . . sequamur: *Aeneid* 5. 709.

557　*Longinus*: *On the Sublime* 12. 4–5.

　　machine: 'supernatural agency in poems' (J). But Agamemnon's dream is before the catalogue of the ships in *Iliad* 2.

　　philology: the study of literature.　　*many*: in fact, none.

　　Arcite. The source for 'The Knight's Tale' ('Palamon and Arcite'), as Dryden was unaware, was Boccaccio's *Teseida*.

　　Chaucer. Chaucer's source for the story of Griselda was Petrarch's Latin version, not Boccaccio's Italian.

　　Lombard author. The source of *Troilus and Cressida* was Boccaccio's *Il Filostrato*.

558　*The Cock and the Fox*: Chaucer's 'Nun's Priest's Tale'.

　　Ennius: Quintus Ennius (239–169 BC), early Roman poet.

　　inopem . . . fecit: *Metamorphoses* 3. 466.

559　*John . . . conceit*: Ben Jonson, *Bartholomew Fair* (1631), 1. 1. 35.

　　Dido: in *Aeneid* 4.

　　One of our late great poets: Abraham Cowley (1618–67).

　　forgive: resist.

560　*Not . . . stand*. This statement is nowhere else recorded.

　　Catullus: not Catullus but Martial, *Epigrams* 3. 44. 4.

　　Tacitus: *De oratoribus* 21.

　　last edition of him: Thomas Speght's edition (1598), reprinted 1687. Speght argued that to a skilful reader Chaucer's verse could scan.

　　heroic: iambic pentameter.　　*dipped*: implicated.

561 *Piers Plowman*: by William Langland (d. 1400). A *Plowman's Tale* was attributed to Chaucer in Speght.

took: detracted.

scandalum magntum: a law term for slandering people in high office.

562 *Drake*. James Drake (1667–1707) published an answer (1699) to Collier's *A Short View of the Immorality and Profaneness of the English Stage* (1698).

prior laesit: Terence, *Eunuchus*, prologue, l. 6.

Porta: Giambattista della Porta (1538–1615), an Italian physiognomist.

564 *But . . . dede*: 'Prologue' 725–42.   *Wincing . . . Bolt*: 'Miller's Tale' 77–8.

Leicester: Philip Sidney (1619–98), third Earl of Leicester.

565 *There . . . Danè*: 'Knight's Tale' 1204–6, misquoted.

Milbourne. Luke Milbourne (1649–1720), who published a translation of Virgil in 1688, criticized Dryden's translation in 1698.

multa . . . loquendi: Horace, *Ars poetica* 70–2.

grandam gold: hoarded wealth.

Scudéry: Madelaine de Scudéry (1607–1701). There is no other record of her translation of Chaucer.

567 *Provençal*. No such version is known to exist.

568 *Dioneo . . . Palemone*. *Decameron* 7. 10 Epilogue.

The Flower and the Leaf. This poem is not now regarded as Chaucer's.

Blackmore: Sir Richard Blackmore (1654–1729), physician and bad poet. He published *Prince Arthur* in 1695 and *King Arthur* in 1697. Blackmore attacked Dryden in the preface to *Prince Arthur*.

Ogilby: John Ogilby (1600–76). His translation of Virgil was published in 1654.

569 *two poems*. See note on Blackmore above.   *Entellus*: *Aeneid* 5. 401.

570 *the zeal . . . up*: Psalms 96. 9.

Senneph: on 11 August 1674. Condé's foe was William III, then Prince of Orange.   *Demetri . . . cathedras*: Horace, *Satires* 1. 10. 90–1.

571 *To the Duchess of Ormonde*
First published in *Fables* (1700). Lady Mary Somerset (1665–1733) married the duke in 1685. Dryden's poem makes an elaborate parallel between the duchess and Chaucer's heroine Emily. It also invokes the *Aeneid* and Virgil's fourth *Eclogue*.

l. 1. *The bard*: Chaucer.

l. 2. *this ancient song*: 'Palamon and Arcite' ('The Knight's Tale'), which follows.

l. 4. *doubtful palm*: i.e. in a contest Virgil's verse would not necessarily be found superior to Chaucer's. Dryden adapts what Juvenal said of Virgil and Homer (*Satires* 11. 180–1): 'the bard of the *Iliad* will be sung [at my feast] and the songs ... of Virgil that contest the palm with his'.

l. 14. *Plantagenet*. The reference may be to Joan, daughter of Edmund of Woodstock, Earl of Kent, second son of Edward I. She married for the third time ('three contending princes') Edward the Black Prince, and was the mother of Richard II. She is believed to have named the Order of the Garter ('the noblest order').

l. 29. *blood*: the Duchess of Ormonde's father was a lineal descendant of Edward III.   *Platonic year*: a cycle imagined by some ancient astronomers, in which the heavenly bodies were supposed to go through all their possible movements and return to their original relative positions (after which, according to some, all events would recur in the same order as before).

572 l. 42. *westward*. In the summer of 1697 the duchess and her daughters paid their first visit to the family seat at Kilkenny, in Ireland, where they were welcomed with enthusiasm.

l. 46. *etesian gale*. Properly, an annual wind. Dryden perhaps misuses the term. 'Gale' means 'breeze'.

l. 48. *Portunus*: Roman protector of harbours.

l. 53. *Hibernia*: Ireland.

l. 54. *pledge*. The duke followed the duchess to Ireland after some months.

l. 56. *father and his grandsire*: Thomas Butler, Earl of Ossory (1634–80), and James Butler, Duke of Ormonde (1610–88) ('Barzillai' in *Absalom and Achitophel*: see ll. 817 ff. and notes (above, pp. 198, 886)).

l. 71. *ark*: Genesis 7–8.

573 l. 99. *abroad*: Genesis 8: 8–11.

l. 117. *four ingredients*: the four temperaments, in the natural philosophy of the Middle Ages, the combination of supposed qualities (*hot* or *cold*, *moist* or *dry*) in a certain proportion, determining one's nature.

574 l. 125. *Vespasian*: Titus Flavius Sabinus Vespasianus, who commanded the siege of Jerusalem but lamented its fate.

l. 130. *the table of my vow*: table meant 'tablet'. Sailors who had escaped shipwreck dedicated a picture of the shipwreck (*tabula votiva*) to Neptune.

l. 131. *Morley's*: Christopher Love Morley, physician.

l. 133. *the Macedon*: Alexander III ('the great') of Macedon. He 'dreamed of a herb which cured [his friend] Ptolomey' (Dryden's note to the second satire of Persius).

l. 158. *Penelope*: wife of Odysseus in the *Odyssey*; she employed her time embroidering and warding off suitors while she waited for her husband's return.

575 l. 162. *Elissa*: Dido, Queen of Carthage. See *Aeneid* 4. 335.

Palamon and Arcite
First published in *Fables* (1700). A rendering of Chaucer's 'The Knight's Tale'.

Book I
l. 7. *Scythia*: the region just north of the Baltic sea.　　*the warrior queen*: Hippolyta, Queen of the Amazons.

576 l. 22. *tilts*: charges on horseback with lance, generally by two armed knights riding on opposite sides of a barrier.　　*tourneys*: martial sports.

l. 56. *sounded*: archaic form of 'swooned'.

577 l. 67. *giddy chance*. In classical and medieval iconography the vicissitudes of life were represented by the goddess Fortuna with her ever-revolving wheel.

l. 76. *Capaneus*: one of the seven heroic kings who besieged Thebes. He was struck dead by lightning as he was scaling the walls, a penalty for impious presumption.

l. 81. *Creon*: brother of Jocasta, the mother of Oedipus. He assumed the rule of Thebes after Oedipus' sons had killed each other.

l. 89. *screaked*: shrilly screamed.

578 l. 107. *sister*: his sister-in-law Emelia.

ll. 109–10. *Where ... car*. 'Where on a silver background the god of war (Mars) was drawn triumphant in his iron chariot.'

l. 114. *sanguine*: blood red.

l. 116. *Minotaur*. Theseus killed the Minotaur (a monster half man, half bull, sent from the sea by Poseidon and Pasiphaë, wife of Minos) in Crete.　　l. 122. *the city*: Thebes.

l. 144. *trophies*: in classical times a structure erected, often on the field of battle, as a memorial of a victory in war, and hung with arms from the enemy.

580 l. 204. *partition*: part.

l. 222. *parterres*: level spaces containing flower beds.

l. 223. *with arms across*: with arms crossed.

581 l. 247. *aspect*. In astrology the aspect of planets was their position relative to one another and to the earth. Saturn (l. 246) was normally a planet of evil influence.

ll. 254–5. *The moment ... faint away*. The first impact of love was typically depicted in terms of Cupid's arrow piercing the lover's eye.

l. 257. *glance*: (1) look; (2) light blow.

l. 258. *Actaeon*. A famous hunter who happened upon Artemis (Diana) bathing naked. She turned him into a stag to be torn to pieces by his own hounds.

l. 261. *the Cyprian queen*: Venus.

582   l. 306. *bad pretence*: what you wrongfully claim.

583   l. 329. *positive*: man-made law (*lex positiva*), contrasted with the law of nature.

l. 342. *Aesop's hounds*. Chaucer does not ascribe the similitude to Aesop. It is much like Aesop's fable of the lion and the bear.

585   l. 404. *love's extremest line*: a translation from Terence (*Eunuchus* 640) *extrema linea amare* 'to love at a distance'. 'But Dryden, with Terence's phrase in mind, adds the image of the sun of beauty, thinking of "line" as the circle of the terrestrial sphere farthest from the path of the sun' (Kinsley).

l. 414. *Fire, water, air, and earth*. The four elements of which everything was composed. These elements were related to the four humours, an imbalance of which would upset man's equilibrium.   *Fates*: three Greek goddesses who were thought to govern man's lot.

l. 427. *guilty of their vows*: guilty in neglecting their vows; a Latinism (*voti reus*).

587   l. 473. *adamant*: 'a stone, imagined by writers, of impenetrable hardness' (J).

l. 495. *threads*: makes his way cleverly through.

l. 498. *Saturn*. The planet Saturn was usually considered to have a malign influence upon the destinies of men. In Book III he contrives Arcite's death.

l. 500. *quartile*. The aspect (see above, note to l. 247) of two heavenly bodies which are 90° distant from each other. When two planets are in a quartile, they are said to be in an adverse aspect to each other and to the earth. Venus and Mars in a quartile would mean disruptions in love.

588   ll. 551–2. *his sire's command ... snaky wand*. Zeus (Jove, 'his sire', l. 551) loved the priestess Io. Hera (Juno) set Argos of the hundred eyes to keep a watch on Io. Sent by Zeus, Hermes lulled Argos to sleep with his magic wand with two serpents coiled about it (*caduceus*), and then killed him (Ovid, *Metamorphoses* 1. 668 ff.).

589   l. 578. *thick resort*: dense crowd.

l. 590. *Philostratus*: 'the man vanquished by love'.

Book II

590   l. 10. *May within the Twins received the sun*: after 21 May, when the sun moves into Gemini ('the Twins').

l. 12. *in causes first*: in their first, or motivating, causes.

591 l. 34. *turn our style*: change to another subject.

l. 51. *against*: towards.

592 l. 83. *her day.* Friday was Venus' day; called after Frigg, her northern equivalent.

l. 95. *Cadmus' doom.* Cadmus founded Thebes. Juno hated the Thebans, because of the affair between Cadmus' daughter Semele and her husband Zeus (Jove). Cadmus was turned into a snake by Zeus (*Metamorphoses* 4. 600).

594 l. 156. *of proof*: of tested quality.

596 l. 232. *the goddess of the silver bow*: Diana.

l. 258. *listed*: converted into lists (places of combat) for tilting.

598 l. 324. *bribes*: influences favourably.

599 l. 364. *proverb*: *amare et sapere vix deo conceditur*, 'to love and to be wise is hardly granted to a god' (Publilius Syrus, *Sententiae* 22).

l. 382. *train*: (1) train of a woman's gown; (2) trail of bait for wild beasts.

600 l. 414. *bars*: palisades of the lists.

601 l. 445. *degrees*: steps.

l. 461. *roses ... myrtle*: plants sacred to Venus.

l. 462. *opposed*: opposite.

602 l. 483. *sigils . . . hours*: occult devices, supposed to have mysterious powers, drawn at times when the appropriate planets were situated most favourably in respect of their influences.

l. 489. *Down-looked*: with a downcast glance.   *cuckoo*: figuring cuckoldry.

l. 491. *carol.* Properly, a carol is a ring dance with the accompaniment of song. Dryden would have understood it in Johnson's sense of 'a song of joy and exultation'.

l. 498. *Idalian mount*: scene of the judgment of Paris in Venus' favour. *Citheron*: Cythera, an island sacred to Venus, associated by Chaucer with Mount Cithaeron, the home of the muses.

l. 502. *Narcissus*: a figure of immature love; Narcissus fell in love with his own reflection in a pool.

l. 503. *Samson ... Solomon*. Both were brought to grief by unwise loves.

l. 505. *Medea's*: Medea was a sorceress who fell in love with Jason. When he left her she killed (among others) her own children in rage. *Circean*: of Circe, the sorceress who changed Odysseus' men into pigs (*Odyssey* 10. 210).

l. 513. *From ocean.* Venus was born by rising out of the sea-foam.

603 l. 519. *turtles.* The turtle-dove was sacred to Venus.

l. 502. *infant Love*: Cupid.   l. 534. *scurf*: a thin layer of turf.

l. 544. *bent*: patch of heath or rough grass.

l. 549. *Blind*: (1) closed off at one end; (2) with no windows to admit light.

604   l. 558. *A tun about*: the size of a barrel around.

l. 589. *strangled*: suffocated. Chaucer, perhaps more plausibly, has his hunter 'strangled with [by] the wild bears'.

l. 590. *overlaid*: smothered.   l. 595. *intercepts*: pins.

605   ll. 602–3. *sword . . . thread*: an allusion to the sword of Damocles, symbol of the insecurity of princes.

l. 604. *ides*: the ides of March (15 March), the day prophesied for the assassination of Julius Caesar, who was Mars-influenced.

l. 614. *geomantic*. Geomancy is a form of divination by mystical figures of dots, once made by casting pinches of earth (*ge*) randomly on a flat surface, later by dotting points on paper.

l. 615. *a warrior and a maid*: 'Rubeus and Puella' (D), two geomantic figures which were sometimes dedicated to Mars.

l. 623. *Callisto*: a nymph of Diana, seduced by Jove. Juno changed her into a bear, and Jove made her a constellation, along with her son Arcas (l. 625).   l. 627. *Actaeon*. See note on l. 258 above.

l. 631. *Peneian Daphne*: a daughter of the river-god Peneus. She was turned into a laurel tree while trying to escape the clutches of Apollo.

l. 634. *the Calydonian beast*: a wild boar, sent by Diana to ravage the land of Oeneus, king of Calydonia. See 'Meleager and Atalanta' (above, pp. 640–50).

l. 635. *Oenides'*: Meleager's. See previous note.

l. 636. *Atalanta's*. See 'Meleager and Atalanta' (below, pp. 640–50).

606   l. 639. *Volscian queen*. Camilla, a foe of Aeneas (*Aeneid* 11. 432). She was mortally wounded by Aruns, who was in turn killed by Diana.

l. 654. *Lucina's*. Lucina was goddess of childbirth; another manifestation of Diana.

Book III

l. 6. *And . . . voice*: and send the pick of warriors by public acclaim.

607   l. 23. *approved*: tested and found good.

608   l. 89. *reclaimed*: trained for bird-hunting.

609   l. 96. *increase of arms*: advancement of feats of arms. (?)

l. 99. *Bacchus*: the son of Zeus (Jove) and Semele. Brought up in India, he rode in triumph to Greece on a panther at the head of his followers, the baccants.

l. 109. *harbinger*: one sent ahead to find lodgings.

l. 120. *Phosphor*: Phosphorus, the morning star.

l. 124. *preventing*: getting up before.

ll. 129–44. *Creator Venus ... repair*: D's addition, an imitation of Lucretius' lines at the beginning of *De rerum natura*. Compare D's translation (p. 271 above).

610 l. 134. *Thy month*: May.

l. 146. *Increase of Jove*: Athene (Minerva), goddess of war.

l. 147. *Adonis*: a beautiful youth loved by Venus; he disregarded her warnings, and to her great sorrow was killed hunting.

l. 168. *fifth orb*. According to Platonic cosmology there were seven hollow mobile harmonious spheres concentric with the earth. The fifth sphere was that of Venus. There is no mention of Venus' sphere at this point in Chaucer.

l. 169. *clue*: the thread of life.   l. 172. *the Sisters*: the Fates.

611 l. 193. *vests*: vestments.   l. 208. *mastless*: without acorns.

l. 212. *Statius*. Publius Papinius Statius (45–96) wrote the *Thebaid*, an epic treatment of the Troy story.

612 l. 217. *Queen of the nether skies*: Proserpina or Diana. See note to l. 232 below.   l. 218. *the gloomy sphere*: the earth, dark at night.

l. 219. *maids*: virgins.

l. 221. *Niobe's devoted issue*: Niobe's doomed children. Artemis (Diana) killed all seven pairs of them. Niobe had boasted that she was superior in fecundity to the titaness Leto (Latona) who had only one pair (Ovid, *Metamorphoses* 6. 165 ff.).

l. 231. *gust*: 'height of sensual enjoyment' (J).

l. 232. *triple shape*. Diana is called *diva triformis* by Horace, alluding to her three natures, as huntress-goddess, moon-goddess, and goddess of the underworld.

613 l. 270. *cornel*: of cornel-wood. The cornel is probably the dogwood.

l. 277. *thunderer's*: Jove's.

l. 284. *Disclaimed*: a term from heraldry, 'declared not to be entitled to bear arms'.

l. 290. *planetary hour*. This refers to the astrological system according to which each day and night is divided into twelve 'hours' (varying in length according to the season). The first hour after sunrise belongs to the planet after which the day is named, and the following hours proceed according to a fixed series.

l. 291. *heptarchy*: government of seven (the planets).

l. 297. *hyperborean*: northern.

614 l. 320. *Vulcan*: Roman god of fire. He caught his wife Venus and Mars making love, enmeshed the lovers in a net, and put them on display (*Metamorphoses* 4. 173–89).

615 l. 344. *arms reversed*: with the coat of arms inverted as a sign of disgrace.

l. 357. *close*: end of the speech.   l. 364. *ruffled*: uneven.

l. 369. *Nor wanted*: there was no lack of.

l. 377. *She . . . plead*. Venus having been the first to grant her supplicant's prayer, was able to plead her case on the grounds of temporal precedence.

l. 379. *his wife*: Juno, notoriously jealous.

616 l. 388. *outridden*: outdone in riding. D's mistranslation of Chaucer's 'atrede' = 'outwit'.

l. 389. *trined*: 120° distant from. A favourable aspect between the planets concerned.

l. 390. *Mars in Capricorn*. The planet Mars in the house of the sign of the Goat, favourable for Mars' aggressive influence.

l. 397. *Wide is my course*: Saturn's orbit took thirty years to complete.

l. 401. *watery sign*. The signs of the zodiac corresponding to the element water are Cancer, Scorpio, and Pisces.

l. 402. *earthy*: the earth signs are Taurus, Virgo, and Capricorn.

l. 405. *wilful death*: suicide.

l. 410. *lion's hateful sign*: the zodiacal sign of Leo, 'hateful' because Leo being one of Mars' signs, Saturn's malevolence would have Mars' aggressiveness added to it.

l. 413. *Unkindly seasons and ungrateful land*: unnatural weather for the season, and barren land.

l. 420. *grandsire's*. Saturn was father of Zeus (Jove), who was in turn father to most of the other gods.

617 l. 422. *complexions*: temperaments; Mars was choleric, Saturn melancholic, and Venus phlegmatic.

l. 426. *Chronos'*: Time's; Chronos was also the Greek name for Saturn.

l. 459. *palfreys*: light saddle-horses.

618 l. 489. *cap-a-pe*: from head to foot.   l. 497. *king at arms*: chief herald.

619 l. 512. *career*: charge.   l. 513. *ash*: spear shafts were made of ashwood.

l. 516. *at mischief*: at a disadvantage.

l. 531. *earl-marshal*: judge of the court of chivalry.

l. 541. *titles*: order of precedence according to status.

620 l. 551. *promiscuous*: mixed and disorderly.

621 l. 589. *adverse*: in opposition.

l. 590. *jostling*: coming into collision.

l. 603. *Hauberks*: coats of mail.   l. 612. *truncheon*: spear shaft.

l. 613. *halting*: limping.   l. 617. *By fits*: at varying intervals.

l. 622. *crupper*: rump of a horse.

622   l. 631. *challenges*: asserts its right to.

    l. 642. *he*: Emetrius.   l. 651. *Unyielded*: not given up.

    l. 658. *tribunal*: seat of judgement.

623   l. 669. *rightful Titan*. The Titans were the pre-Olympian gods, until supplanted by Jove. The reference here is to Chronos (Saturn).

    l. 678. *The blustering fool*: Mars.

    l. 689. *popularly*: to gain the people's favour.

    ll. 699–700. *a flashing fire By Pluto sent*. Chaucer has 'a furie infernal'. Pluto ruled the underworld.

624   l. 711. *the noble parts*: the heart and lungs.

    l. 712. *mortified*: deathly injured.   l. 720. *compelled*: compressed.

    l. 725. *despoiled*: stripped.   l. 727. *Foment*: bathe with warm lotions.

    l. 745. *parts*: sides.

625   l. 749. *impairs*: grows worse.   l. 754. *Corrupts*: putrefies, i.e. becomes infected.

    l. 755. *breathing*: giving air to.   *cupping*: drawing blood.

    l. 772. *against right*: unjustly (Palamon fell in love with her first).

626   l. 817. *impotence of mind*: lack of reason.

    l. 818. *his . . . flame*: Palamon's concurrent passion.

627   l. 838. *seat of life*: the heart.   l. 840. *he*: Arcite.

    l. 847. *demonstrative*: invincibly conclusive.   l. 861. *the sex*: women.

    l. 870. *Hector's death*. Hector was killed by Achilles and his Myrmidons outside the walls of Troy.

628   l. 898. *conscious*: already known. (?) bearing witness. (?)

    l. 908. *Vulcanian food*: fuel.

629   l. 913. *laurel . . . myrtle*. Victors' crowns were made of laurel; myrtle, sacred to Venus, because Arcite was a lover.

    l. 927. *mourning bride*. In *1700* 'Mourning Bride', alluding to Congreve's play of 1697.

    l. 934. *cornel-wood*. See note to l. 270 above.

    l. 940. *master-street*: main street.

    l. 953. *Parthian bow*. The Parthians were skilled in archery.

630   l. 955. *fathom*: about six feet.   l. 969. *foreign seats*: strange nests.

631   l. 1001. *gauntlet*: Dryden uses 'gauntlet' in his Virgil translations to render *cestus*, 'thong of hide reinforced with strips of metal, wound round the hands of Roman boxers'.   *foil*: a wrestling throw that is almost a fall.

    l. 1027. *jarring seeds*: warring atoms.

l. 1028. *Fire ... air*. See note to 1. 414 above.

632 l. 1065. *of*: by.

634 l. 1139. *temper*: inclination.

l. 1144. *Eros and Anteros*: love and reciprocal love.

l. 1148. *tenor*: constant mode.

635 *To my Honoured kinsman John Driden*
First published in *Fables* (1700). Dryden used both spellings of his family
name indifferently. His cousin John Driden (1635–1708) was a justice of
the peace and member of parliament for Huntingdon.

l. 1. *How blest is he, who leads a country life*. The line alludes to Horace's
second epode 'Beatus ille' (for D's translation see above, p. 305) which
gave rise to a tradition of poems on rural retirement.

636 l. 43. *Rebecca's heir*. Jacob, the second son, is helped by his mother
Rebecca to obtain his brother Esau's birthright. See Genesis 27. The
estate of Chesterton descended to John Driden, the second son, through
his mother.

l. 51. *champian-sports*: sports of the open country.

l. 72. *weekly bill*: the weekly announcement of those who had died within
the City of London.   l. 75. *Pity*: it is a pity that.

637 l. 80. *vindicates*: claims its revenge on.   l. 82. *Gibbons*: Dryden's
doctor.

l. 83. *Maurus*: Sir Richard Blackmore (1655–1729), physician and poet.
His ridiculous epic *Prince Arthur* (1695), in which Dryden was libelled,
'murdered Maro's muse', i.e. outraged the memory of Virgil.

l. 85. *Milbourne*: Luke Milbourne had attacked Dryden's translation of
Virgil. See 'Preface to Fables', pp. 568–9 above.

l. 105. *files*: a file is 'a line on which papers are strung to keep them in
order' (J).

l. 107. *Garth*. Sir Samuel Garth (1661–1719), physician and poet, dispensed free medicines to the poor. His *Dispensary* (1699) is a satire on
greedy doctors.

638 l. 140. *Munster*. The bishops of Munster initially supported England
against Holland, in return for bribes. See *Annus Mirabilis* 145–8 and note
(above, pp. 37, 867).

l. 143. *peace*. The treaty of Ryswick established peace between France
and England in 1697.

l. 152. *Namur*: William III took Namur in 1695.

l. 156. *treaty*: of Ryswick.   l. 160. *Persian king*: Darius.

639 l. 163. *son of Jove*. For the belief that Alexander was the son of Jove, see
'Alexander's Feast' 25 ff. (above, p. 545).

l. 188. *grandsire*. Sir Erasmus Driden, also grandfather of John Dryden the poet. Erasmus Driden was imprisoned for resisting an illegal levy by Charles I.

640 *Meleager and Atalanta*
First published in *Fables* (1700). A translation of Ovid's *Metamorphoses* 8. 270–545.

Meleager was the son of Oeneus (l. 5), king of Calydon, and Althaea (l. 243). There are various versions of this story (an earlier version is told in *Iliad* 9). The story as related by Ovid is late, and contains elements of folk-tale (e.g. the fiery brand).

l. 4. *Cynthia's*. Cynthia was a name given to Diana, associated with Artemis, the virgin goddess of hunting. The cause of her displeasure is given at l. 13 below.

641 l. 28. *strove*: 'his tusks were as big as an Indian elephant's'.

l. 33. *intercepts*: stops from accomplishing its purpose.

l. 44. *crew*.. Most are glossed by Dryden in the text. Only the more noteworthy are glossed here.

l. 46. *Leda's twins*: Castor and Pollux. They became the constellation Gemini.

l. 48. *Jason*: son of Aeson, a king of Thessaly. He 'manned' the *Argo* to recover the Golden Fleece.

l. 50. *Theseus … Pirithous*. See 'Palamon and Arcite' 1. 358 ff. (above, p. 584).

l. 52. *Thestian sons*. Sons of Thestius, Plexippus, and Toxeus, brothers of Althaea, the mother of Meleager.

l. 53. *Caeneus*: of Thessaly. He was born a woman. See 'The Twelfth Book of Ovid's *Metamorphoses*' 260–87 (above, p. 773).

642 l. 60. *twice-old Iolas*. 'Iolas was rejuvenated by Hebe at the instigation of his friend Hercules' (Kinsley).

l. 65. *Atalanta*: daughter of Iasus and Clymene. A virgin huntress.

l. 89. *toils*: large piece of cloth bordered with thick ropes, stretched round an enclosure to capture wild beasts or large nets for stags.

643 l. 108. *spend their mouth*: bark, give tongue, on discovering the quarry.

644 l. 178. *distains*: stains.

645 l. 213. *Nonacrine*: Arcadian.

646 l. 255. *fatal sisters*: the Moirai.

650 l. 398. *her whom heaven for Hercules decreed*: Deianira, sister of Meleager.

Sigismonda and Guiscardo
First published in *Fables* (1700). A translation of *Decameron* 4. 1. Dryden keeps close to Boccaccio (except for the addition at ll. 151–72), his main

changes being: (1) Guiscardo's social rank is raised a few notches; (2) the lovers marry before their love is consummated.

651 l. 51. *blooming*: just coming to manhood.

652 l. 64. *rays*: glances. (Before the seventeenth century the eyes were thought to give out light.)

654 l. 143. *perplexed*: entangled. l. 145. *to*: as a.

l. 151. *conscious*: privy to the secret. Lines 151–72 are Dryden's addition. Boccaccio has simply 'After giving each other a rapturous greeting, they made their way into her chamber, where they spent a good part of the day in transports of bliss.'

655 l. 215. *averse*: turned away.

656 l. 244. *meditates*: muses over. l. 245. *thoughtless*: unsuspecting.

657 l. 270. *sensible of*: acutely perceiving.

658 l. 317. *people's lee*: the dregs of the people.

659 l. 372. *And little wanted, but*: there was little wanting but that . . .

664 l. 579. *Fixed*: resolved.

665 l. 625. *magazine*: storehouse. l. 626. *Secure*: certain.

666 l. 670. *down*: immediately.

667 l. 694. *infection*: the 'catching' influence of sympathy.

l. 711. *genial bed*: marriage bed.

668 l. 718. *suborned*: usually 'bribed'. Here probably 'brought to her aid for a sinister reason'.

669 *Baucis and Philemon*
First published in *Fables* (1700). A translation of Ovid's *Metamorphoses* 8. 611–724.

l. 3. *Ixion's son*: Pirithous.

670 l. 36. *professing*: openly declaring themselves.

l. 57. *seether*: a utensil for boiling.

671 l. 77. *genial bed*: a marriage bed. l. 81. *alone*: only.

l. 90. *Pallas*. The olive was sacred to Pallas (Athene).

l. 92. *cornels*: the red fruit of the cornelian cherry tree.

l. 93. *lees of wine*: vinegary dregs of wine.

672 l. 111. *working in the must*: fermenting.

l. 141. *owned*: acknowledge himself.

673 l. 152. *floated level*: flooded surface.

l. 177. *sign their suit*: signal their agreement to what was asked.

674 l. 191. *Tyanaean*: a native of Tyana in Turkey.

Pygmalion and the Statue
First published in *Fables* (1700). A translation of Ovid's *Metamorphoses* 10. 243–97.

headnote: *Propoetides*: young women of Amathus.

675 l. 23. *strained*: clasped tight.   l. 25. *gripe*: grip.

l. 52. *Sidonian*. Sidon was near Tyre, famous for its purple-dyeing industry.

676 l. 93. *kiss sincere*: genuine kiss. There may be a pun: the old etymology of 'sincere' was from *sine cera*, 'without wax'.

677 *Cinyras and Myrrha*
First published in *Fables* (1700). A translation of Ovid's *Metamorphoses* 10. 298–524.

l. 16. *amomum*: a fragrant plant.

l. 21. *Her plant*: Myrrha, a shrub or plant yielding gum-resin, used in scent. See below, ll. 347–69.

678 l. 29. *ambitious of*: aspiring to.

679 l. 90. *the infernal bands*: the Furies, avengers of crime, especially crimes against the ties of kinship.

l. 100. *Observant of*: respectfully attentive to.

681 l. 181. *house*: kindred.

682 l. 201. *received*: heard.

683 l. 264. *Arctophylax*: Boötes.   l. 268. *amazed*: horrified.

l. 270. *Icarius*: Icarus.   l. 271. *Virgin sign*: Erigone.

684 l. 277. *Secure of*: feeling no care from.   l. 296. *His bowels*: his own flesh.

l. 303. *That names might not be wanting to the sin*: 'even the names were not wanting to complete the wickedness'.

l. 304. *Full of her sire*: impregnated by her father.

685 l. 318. *Panchaia*: an island east of Arabia.

l. 320. *mewed*. Properly 'mew' means 'to shed the feathers'; here, more generally, 'changed'.

l. 320. *her horns*: each of the pointed extremities of the moon as she appears in her first and last quarters.

686 l. 369. *convulsive*: shaken with convulsions.

687 *The First Book of Homer's Iliad*
First published in *Fables* (1700).

headnote. *prefers*: presents.

l. 1. *Peleus' son*: Achilles.   l. 10. *debate*: quarrel.

l. 11. *Jove's and Latona's son*: Apollo.   l. 13. *king of men*: Agamemnon.

688 l. 23. *common*: addressed generally.

l. 24. *brother-princes*: Agamemnon and Menelaus.

l. 27. *accord*: agree to.

l. 32. *far-shooting king*: Apollo, associated with archery.

l. 40. *evil*: harm.   l. 41. *interdicted strand*: forbidden shore.

l. 54. *hoarse-resounding*: harsh sounding.

689 l. 61. *Smintheus*. 'Mouse god', a title given to Apollo who had delivered some people from a plague of mice.

l. 74. *feathered fates*: arrows bringing death.   *sumpters*: pack-horses.

l. 77. *at rovers*: at random.   l. 79. *the queen's*: Juno's (Hera's).

l. 88. *defence*: a means of defence.

690 l. 114. *ungrateful*: unpleasing.   l. 122. *he*: Agamemnon.

l. 123. *of force*: unavoidably.   l. 129. *control*: constraint.

l. 136. *hecatombs*: sacrifices of a hundred cattle.

691 l. 142. *preferred*: offered up.   l. 154. *sparkles*: sparks.

l. 158. *or*: or if it did.   l. 159. *invade*: assail.

l. 161. *Obtending*: laying the responsibility on.   l. 168. *dame*: wife.

l. 177. *I should not suffer for your sakes alone*: 'I alone should not . . .'.

692 l. 182. *Thessalian prince*: Achilles, king of Thessaly.

l. 185. *largely souled*: liberal, generous.   l. 192. *Saturn's son*: Jove (Zeus).

l. 203. *glossing*: commenting. (?) reading a different sense into. (?)

l. 205. *Quit*: give recompense for.

l. 219. *Creta's king*: Idomeneus.   *Ithacus'*: Ulysses' (Odysseus').

693 l. 232. *Pergamus*: Troy.   *Priam*: king of Troy.

l. 252. *engross*: seize.   l. 254. *Phthia*: Achilles' home in Thessaly.

694 l. 270. *Myrmidons*: subjects of Peleus and Achilles, created out of ants (*murmekes*).   l. 303. *confessed in view*: acknowledged in sight.

695 l. 324. *Auspicious*: of good omen.

l. 338. *preace*: press.   l. 342. *peel*: rob.

696 l. 362. *Pylian prince*: Nestor, king of Pylos.

l. 369. *Phrygian*: Phrygia was in Asia Minor, allied to Troy.

697 l. 401. *rampire*: old form of 'rampart', a fortified wall.

l. 424. *fatten*: feed.

698 l. 427. *royal youth*: Achilles.   l. 428. *Patroclus*: lover of Achilles.

l. 449. *goddess-born*: Achilles.

699 l. 473. *manifest*: declaration.   l. 497. *goddess*: Thetis.

700 l. 510. *proud prerogative*: the exclusive privilege of the proud man.

l. 535. *swelling*: inflated with pride.

701 l. 554. *Briareus*: a monster who helped Jove (Zeus) in his battle with the Titans.

l. 567. *at his proper*: to his own.   l. 579. *thunderer's*: Jove's (Zeus').

l. 583. *dead vacation*: all activity has ceased.

702 l. 603. *bespoke*: addressed.   l. 617. *Cilla*: Aeolis, in Asia Minor.

703 l. 633. *collops*: small pieces of meat.   l. 638. *broachers*: spits.

l. 658. *held*: were in.   l. 670. *puny*: lesser.

705 l. 722. *silver-footed queen*: Thetis.   l. 750. *peevish*: querulous.

706 l. 768. *limping smith*: Vulcan (Hephaestus).

l. 773. *jar*: clash.   l. 784. *crowned*: filled to overflowing.

707 l. 799. *Sinthians*: inhabitants of Lemnos.

l. 803. *skinker*: one that serves drinks.   l. 805. *deft*: dextrous.

l. 813. *load*: 'as much drink as one can bear' (J).

The Cock and the Fox ... from Chaucer
First published in *Fables* (1700).

l. 10. *groat*: four old pence; about 1½ pence.

l. 13. *cattle*: livestock in general.

708 l. 24. *posset*: a drink made of curdled milk.

709 l. 62. *Ptolemies*: an incestuous dynasty of ancient Egypt.

l. 64. *dispensation*: alluding to 'Henry VIII's marriage with Catherine of Aragon, his brother's widow' (Kinsley).

l. 70. *feathered*: copulated with.

l. 90. *Solus cum sola*. 'The reference is to the monkish Latin proverb "Solus cum sola non cogitabuntur orare pater noster", which could be roughly translated "If a man is alone with a woman, neither of them would be expected to spend their time saying their prayers". *Solus cum sola* is the title of a Pavan for the lute by John Dowland' (Hopkins and Mason).   *all his note*: (1) the entire subject of his song; (2) what he was renowned for.

l. 91. *parts*: (1) accomplishments; (2) part-singing.

710 l. 106. *shrovetide-cock*: 'a cock tied and pelted with stones on Shrove Tuesday' (Kinsley).

711 l. 140. *Galen*: second-century physician, and influential writer on medicine.   l. 156. *Choler adust*: the humour of melancholy.

712 l. 182. *tertian ague*: 'an ague [fever] intermitting but one day, so that there are two fits in three days' (J).

l. 188. *the gods unequal numbers love*: Virgil, *Eclogues* 8. 75.

l. 190. *fumetery, centaury, and spurge*: medicinal plants of uncertain identification.   l. 202. *Homer plainly says*: Iliad 1. 63.

ll. 203–4. *Nor Cato ... school*. 'The distinction is between the Roman Cato and the medieval school-book *Dionysii Catonis disticha de moribus ad filium*, vulgarly ascribed to Cato' (Kinsley).

l. 209. *An ancient author*: Cicero, *De divinatione* 1. 27.

713   l. 254. *sacred*: accursed. An echo of Virgil's *auri sacra fames* (*Aeneid* 3. 57).

714   l. 283. *deeds of night*: evil deeds.

716   l. 374. *Bede*: Dryden's error. Bede lived a century before Kenelm (l. 360).

l. 376. *Macrobius*: Ambrosius Macrobius, fifth-century Roman, who wrote a commentary on Cicero's fictional *Somnium Scipionis* (*Scipio's Dream*), in which Scipio's grandfather appeared in a dream and told him he would conquer Carthage.   l. 380. *Daniel*: Daniel 7.

717   l. 383. *Joseph*: Genesis 37.   l. 387. *butler*: Pharaoh's butler (Genesis 40).

l. 388. *he*: Pharaoh's baker (Genesis 40).

l. 389. *Croesus*: the fabulously rich king of Lydia in Asia Minor.

l. 391. *wife of Hector*: Andromache's dream is a medieval addition to the matter of Troy.   l. 417. *in principio*: 'in the beginning' (Genesis 1. 1).

l. 418. *Mulier ... confusio*: woman is man's undoing.

718   l. 438. *trod*: copulated with.

l. 451. *Ephemeris*: a table showing the predicted positions of a heavenly body for every day during a given period.

l. 460. *man strutting on two legs and aping me*: alluding to Plato's humorous definition of man as 'a two-legged animal without wings' (Diogenes Laertius *Vitae philosophorum* 6. 40).

719   l. 479. *Book of Martyrs*: John Foxe's *Acts and Monuments* (1563).

l. 500. *Sinon*. See *Aeneid* 2. 57–194.

l. 501. *Gallic*: (1) French; (2) a cock's (Latin *gallus* = cock).

l. 502. *Gano*: he betrayed Charlemagne at Roncesvalles.

720   l. 523. *bolt ... bran*: 'I can't make head or tail of it.'

l. 524. *Bradwardine*: Thomas Bradwardine (d. 1349), Archbishop of Canterbury.   *Austin*: St Augustine.

ll. 531–5. *The first ... before*. 'The distinction is between (i) deliberate, rational choice of a course of action, and (ii) "spontaneous" or unreflective action in inevitable circumstances. The slaves do not initially choose whether to row or not; but constrained to work, they pull "willingly" under the natural prompting of a "prospect of the shore"' (Kinsley).

723   l. 633. *Maro's*: Virgil's.   l. 634. *Horace when a swan*: Odes 2. 20.

l. 636. *Brennus*: leader of the Gauls who overran Italy in 390 BC.   *Belinus*: a god of the Gauls.

l. 648. *saint charity*: holy charity.

l. 652. *solar people*: people born under the influence of the sun.

724   l. 693. *Gaufride*: Geoffrey de Vinsauf, who mourned the Friday death of Richard I.

l. 707. *Asdrubal*: Carthaginian leader whom Scipio (note to l. 376 above) overthrew in 146 BC.

725   l. 728. *Talbot with the band*. Chaucer speaks of Talbot and Gerland. Dryden was misled by his corrupt text.

l. 742. *Jack Straw*: a leader of the Peasants' Revolt of 1381.

727   ll. 806–9. '*A peace . . . betwixt*': Dryden's addition.

Theodore and Honoria
First published in *Fables* (1700). A translation of *Decameron* 5. 8.

l. 1. *Romanian*: D's word for Boccaccio's 'di Romagna'.

728   l. 18. *tilts*. See note to 'Palamon and Arcite' 1. 22 (above, p. 926).

729   ll. 66–7 *at large From*: without restraint of.   l. 74. *lawn*: glade.

730   l. 92. *a deeper brown*: it was becoming darker ('brown').

732   l. 189. *versed in*: occupied in.

733   l. 220. *would have been*: wished he were.

735   l. 306. *assistants*: spectators.

l. 316. *plump of fowl*: flock of birds.   l. 318. *sousing*: swooping.

736   l. 354. *obscene*: 'offensive, disgusting' (J).

737   l. 395. *louring*: frowning.

738   l. 415. *repent*: feel sorry for what she had done and change her mind.

Ceyx and Alcyone.
First published in *Fables* (1700). A translation of Ovid's *Metamorphoses* 11. 410–748.

739   l. 26. *face*: surface, appearance.   l. 41. *comport*: behaviour.

l. 51. *present what I suffer only fear*: 'being with you, fear only what I am really suffering' (Kinsley).

740   l. 93. *atrip*: hoisted up.

741   l. 100. *denounce*: announce.   l. 102. *Strike*: lower.

l. 109. *laves*: bails out.   l. 130. *Stygian*: hellish.

l. 131. *flatted*: made calm.   l. 134. *sublime*: elevated.

743   l. 181. *whom their funerals wait*: those dead already.

l. 203. *Pindus . . . Athos*: famous ancient mountains.

l. 241. *fumed*: 'perfumed with odours in the fire' (J).

744 l. 254. *tainted*: because until a dead man had been given proper burial his home and family were regarded as unclean.

l. 256. *Iris*: Greek goddess of the rainbow, and messenger of the gods.

745 l. 268. *Cimmerians*: an ancient people living on the edge of the world.

l. 275. *provoke*: rouse.   l. 299. *bearded ears*: prickly spikes of corn.

748 l. 392. *blubbered*: swollen.

750 l. 470. *mole*: dyke.   l. 491. *obnoxious*: exposed to.

l. 494. *compressed*: pregnant.   l. 499. *nephews*: grandchildren.

751 *The Flower and the Leaf*
First published in *Fables* (1700). A version of an anonymous poem of the late fifteenth century, once thought to have been by Chaucer.

ll. 1–4. *wintry signs … Taurus*: signs of the zodiac. The Ram (Aries) reigns from 21 March, Taurus from 21 April.

l. 8. *Eurus*: the east wind.   l. 13. *gems*: 'the first buds' (J).

l. 29. *the balmy dew*: sleep.   l. 33. *Chanticleer*: a cock.

752 l. 46. *painted*: brilliant of hue; a recollection of *pictae volucres*, 'birds brilliant of hue' (*Georgics* 3. 243).

l. 50. *Philomel*: the nightingale, associated with erotic love.

l. 55. *printed*: 'marked by pressing' (J).

l. 61. *receptacle*: place of shelter.   l. 72. *sycamores*: fig-trees.

753 l. 91. *grass … grain*: these have symbolic values analogous to those of the flower and the leaf respectively. 'Grass' = transience and worldly vanity; 'grain' = virtue, permanence, and the like.

754 l. 129. *her opposite*: the goldfinch.

l. 155. *ancient fathers*. Kinsley cites Burton who cites Lactantius and Justin Martyr.

755 l. 169. *circles*: the coronets.   l. 170. *chaplets*: garlands for the head.

l. 172. *agnus castus*: a shrub believed to be preservative of chastity.

l. 181. *measures*: (1) dance rhythms; (2) qualities of moderation; (3) limits.

756 l. 226. *the mighty Persian monarch*. 'Chaucer' refers to Prester John, legendary king of Asia, who was reputed enormously rich.

l. 230. *cerrial-oak*: the evergreen oak, another symbol of immortality.

l. 233. *charge*: a device borne on an escutcheon.

758 l. 282. *hawthorn*: emblem of fidelity in love.

l. 284. *palm*: emblem of triumph.

759 l. 327. *Moluccan*: from the Spice Islands.

l. 343. *gridelin*: pale purple.   l. 360. *lay*: lea.

760 l. 374. *Sirius*: the dog-star, supposed to breed pestilence.

761 l. 421. *salads*: raw vegetables. Lettuce was an antaphrodisiac.

l. 437. *Vesper*: the evening star.

763 l. 493. *Demogorgon*: a primordial terror.

764 l. 546. *order*: of the Garter.

765 l. 606. *sigils*. See note to 'Palamon and Arcite' 2. 483 (above, p. 928).

Immediately after 'The Flower and the Leaf' in *Fables* Dryden reprinted 'Alexander's Feast' (above, p. 545), which had appeared as a separate poem in 1697.

766 *The Twelfth Book of Ovid's Metamorphoses*
First published in *Fables* (1700).

l. 7. *Spartan queen*: Helen.

767 l. 27. *gratulates to*: congratulates on.

l. 37. *disobeyed*: held back (from aiding the Greeks).

769 l. 101. *Neptunian Cygnus*: Neptune's son Cygnus.

l. 110. *in act to*: just about to.   l. 117. *Goddess-born*: Achilles, son of Thetis.

770 l. 149. *Lyrnessian*: Lyrnessus'.

771 l. 181. *bored arms*: pierced armour.   l. 203. *his name*: swan; Latin *cygnus*.

772 l. 239. *Othrys*: a mountain in Thessaly.

l. 252. *Neleides*: Nestor, son of Neleus.

773 l. 272. *power of ocean*: Neptune.

l. 294. *cloud-begotten race*: the centaurs.

774 l. 307. *little wanted ... employ*: and it had nearly happened that all had wished her joy in vain.

775 l. 337. *double-formed*: centaurs, half-man, half-horse.

777 l. 440. *Stygian*: of Styx.

778 l. 463. *cubit-bone*: elbow-joint.

l. 480. *Athenian knight*: Theseus.   l. 491. *Their king*: Amyntor.

779 l. 524. *form*: beauty.

780 l. 582. *house*: covering attached to saddle.

782 l. 629. *card*: comb wool.

786 l. 815. *The god*: Vulcan.   l. 828. *Laertes' son*: Ulysses.

787 *The Speeches of Ajax and Ulysses*
First published in *Fables* (1700). A translation of *Metamorphoses* 13. 1-398. Ajax and Ulysses contend for the shield of Achilles.

l. 2. *master of the sevenfold shield*: Ajax.

788 l. 38. *This thief*: Ulysses.

l. 53. *one more cunning*: Palamedes. He exposed, and was killed by, Ulysses. See ll. 77, 478 ff. below.

l. 63. *Philoctetes*: son of Poeas; possessor of a deadly bow. He was left wounded on Lemnos, and later persuaded by Ulysses to join the Greek expedition against Troy. See ll. 484 ff. below.

792 l. 215. *he*: Ajax.

793 l. 245. *uncle's name*: Peleus, father of Achilles.

794 l. 293. *general's*: Agamemnon's.   l. 294. *daughter's*: Iphigenia's.

l. 311. *mother's*: Clytemnestra's.

795 l. 341. *lines*: trenches.

796 l. 384. *spy*: Dolon.

798 l. 457. *Undipped in seas*: by Juno's request Neptune was not to allow these constellations to bathe (set) in his waters.

l. 457. *Orion's angry star*: Ovid has 'Orion's gleaming sword'.

l. 468. *mother*: Thetis.   l. 469. *wife*: Penelope.

799 l. 510. *prophet*: Helenus.   l. 516. *maid*: Minerva. Her image is the Palladium.

800 l. 536. *Tydides*: Diomede.

l. 546. *more moderate Ajax*: Ajax son of Oileus; 'Ajax the lesser'.

l. 547. *Cretan king*: Idomeneus.

802 l. 609. *flower*: the hyacinth.

l. 611. *whom unaware Apollo slew*: Hyacinthus (*Metamorphoses* 10. 162).

l. 612. *letters*: Ovid plays on AIAI, a cry of woe, and AIAΣ, Ajax.

The Wife of Bath's Tale
First published in *Fables* (1700). From Chaucer.

803 l. 38. *humming ale*: strong ale.

804 l. 73. *Genevra*: Guinevere, wife of Arthur.   l. 85. *speaker*: spokesperson.

l. 104. *But, our proffer not to hold in scorn*: Kinsley's conjecture for *1700*'s 'But not to hold our proffer in scorn'.

806 l. 156. *Ovid*: *Metamorphoses* 11. 174 ff.

807 l. 192. *marrow-bones*: knees.   l. 194. *bumps*: cries.

811 l. 387. *earth*: earthly form.

812 l. 396. *tralineate*: deviate.

l. 409. *Caucasus*: conventional for any distant place.

l. 434. *Sodom blue*: the blue flame wherewith Sodom perished.

813 l. 443. *villeinage*: serfdom.

l. 448. *Servius Tullus*: Servius Tullius, sixth king of Rome, whose mother was a slave.

l. 450. *Fabricius*: Roman consul cited by Cicero as a typical specimen of Roman virtue.   l. 472. *sings before the thief*: Juvenal 10. 22.

815   l. 534. *Pygmalion*. See 'Pygmalion and the Statue' (above, p. 674).

Of the Pythagorean Philosophy
First published in *Fables* (1700). A translation of Ovid, *Metamorphoses* 15. 1–484.

817   l. 61. *Alemonides*: Alemon's son Myseclos (l. 26).

l. 62. *Alcumena's son*: Hercules.   l. 77. *man divine*: Pythagoras.

819   l. 134. *cyclopean feasts*: *Odyssey* 9 tells how the cyclops Polyphemus eats the followers of Odysseus two by two.

820   l. 196. *meal betwixt his temples cast*: barley was sprinkled as part of the ritual of sacrifice.

821   l. 232. *Euphorbus*: a Dardanian. He wounded Patroclus, and was killed by Menelaus (l. 234).

822   l. 260. *Tiphys*: the helmsman of the Argo; see ll. 505 ff. below.

823   l. 326. *fermented*: 'excited', stirred up.

824   l. 351. *Milo-like*: Milon was a renowned athlete from Croton. He was caught in the trunk of a tree he was trying to split open and was eaten by wolves.

l. 357. *To force her twice*. Helen was abducted by Theseus and, after her marriage, by Paris.

825   l. 401. *our metal*: iron.

826   l. 418. *Lycus*: river of Colchis.   l. 420. *Erasinus*: river in Argolis.

l. 442. *Zancle*: Messana in Sicily.

827   l. 467. *Ammon*: the Egyptian equivalent of Jupiter.

l. 469. *Athaman*. Ovid has 'the Athamanians'.

l. 490. *Proetides*: daughter of Proetus.

831   l. 618. *conquering god*: Bacchus.   l. 649. *goddess-born*: Aeneas, son of Venus.

832   l. 680. *Thyestean meal*. Thyestes' brother served up his sons to him.

833   *The Character of a Good Parson.*
First published in *Fables* (1700). A paraphrase of *The Canterbury Tales* (A), ll. 477–528. It pays tribute to the clergy who lost their benefices by refusing to take the oath of allegiance to William and Mary, who supplanted James II.

title. Enlarged: especially the last forty-seven lines.

835   l. 92. *designed*: set apart.

836 l. 97. *Zebedee*: Mark 10: 35–45.   l. 103. *Moses-like*: Exodus 34: 29–35.

l. 108. *Richard*: Richard II.   l. 109. *Harry*: Henry IV.

l. 118. *flattering priest*: thought to be an allusion to William Sherlock, Dean of St Paul's.

837 l. 139. *brillant*: diamond.

The Monument of a Fair Maiden Lady
First published in *Fables* (1700). The lady was Mary Frampton, who died in 1698, and was buried in Bath Abbey, where these lines are to be seen on her monument.

ll. 19–20. *All white ... dyes*. Lines 19–20 echo Donne; cf. 'A Funeral Elegy' 74–6:

> she soon expired,
> Clothed in her virgin white integrity;
> For marriage, though it do not stain, doth dye.

838 *Cymon and Iphigenia*
First published in *Fables* (1700). Translated from *Decameron* 5. 1.

Poeta loquitur: 'The poet speaks'. A reply to Jeremy Collier. See note to the Preface (above, p. 923).

l. 4. *severe divine*: censuring clergyman.

l. 14. *railing*: 'using insolent or reproachful language' (J).

839 l. 39. *Ormonde*. See above, pp. 571–5, 924–6.

840 l. 95. *expressed*: exactly resembled.

841 l. 145. *in gross*: indiscriminately.

842 l. 172. *wildered*: gone astray.   l. 179. *cudden*: clown.

844 l. 268. *Ambushed*: lying in ambush.

845 l. 276. *strike your sails*: lower your topsail (as a mark of surrender).

l. 302. *take place*: take precedence.

846 l. 316. *formal*: 'solemn, precise' (J).

l. 319. *Paris ... Spartan spouse*: Paris abducted Helen, wife of Menelaus, king of Sparta, thus beginning the Trojan war.

l. 326. *Candy*: Crete.

l. 327. *glass*: sand-glass or hour-glass to measure time. (Usually in half-hours, but here apparently, in hours.)

l. 337. *prevents their care*: anticipates their labour.

847 l. 366. *with altered sense*: with different feelings.

l. 368. *Sex to the last*: a (typical) woman to the last.

l. 370. *aloft ... alow*: high ... low.   l. 377. *winding*: encircling.

848 ll. 399–408. *The country ... the day*: 'Dryden's addition. The mainten-
ance of a standing army was a serious issue between William and the
Commons in 1697' (Kinsley).

l. 407. *seeming arms ... short essay*: briefly to resist with token force.

l. 427. *dismiss*: shake off.

849 l. 468. *rape*: kidnap.   l. 476. *praetor*: magistrate.

851 l. 537. *neuter*: neutral.   l. 540. *this*: this time.

852 l. 559. *fixed*: determined.   *wonted*: accustomed.

l. 581. *harpies*: winged monsters.

853 l. 608. *floats*: floods.   l. 628. *Jove's isle*: Crete.

854 *The Secular Masque.* First published in *The Pilgrim written originally by
Mr Fletcher and now very much altered with several additions* (1700). Dryden contributed a prologue and an epilogue, and this masque. 'Secular'
means belonging to a *saeculum*, an age, or period, and refers to the turn of
the century. Fletcher's play was revived for Dryden's benefit.

l. 3. *the radiant belt:* the zodiac.

l. 6. *fans*: wings.   l. 13. *Momus*: the god of ridicule.

855 l. 45. *Inspire the vocal brass*: sound the warlike trumpet.

l. 51. *looked*: stared.   l. 56. *the Tyrian dye*: blood.

857 *A Song* ('*Fair, sweet, and young*'). First published in *Poetical Miscellanies*
part 5 (1704).

858 *Aesacus Transformed into a Cormorant.* First published in *Ovid's Metamorphoses in Fifteen Books* (1717). A translation of Ovid, *Metamorphoses*,
11. 749–95.

l. 3. *shows*: points out.   l. 7. *Tros*: father of Ganymede.

l. 8. *Laomedon*: father of Priam.

859 l. 44. *Tethys*: sea goddess, wife of Oceanus.   l. 55. *his name*: Mergus, a
diver.

FURTHER READING

(The place of publication is London unless otherwise specified.)

EDITIONS

The Works of John Dryden, ed. Walter Scott (18 vols., 1808). Revised by George Saintsbury (18 vols., London and Edinburgh, 1882–93). The only complete edition, gradually being superseded by the next item.
The Works of John Dryden, ed. E. N. Hooker *et al.* (Berkeley and Los Angeles, 1956–). In progress.
The Poems of John Dryden, ed. James Kinsley (4 vols., Oxford, 1958).
The Poetical Works of John Dryden, ed. George R. Noyes (Boston, 1909; 2nd edn., 1950).
'Of Dramatic Poesy' and other Critical Essays, ed. George Watson (2 vols., 1962).
Dryden: The Dramatic Works, ed. Montague Summers (6 vols., 1931–2).
John Dryden: Four Comedies, and *John Dryden: Four Tragedies*, ed. L. A. Beaurline and Fredson Bowers (Chicago and London, 1967).
The Letters of John Dryden, ed. Charles E. Ward (Chapel Hill, NC, 1942).

LIVES

Samuel Johnson, 'Preface to Dryden', in *Prefaces, Biographical and Critical, to the Works of the English Poets*, vol. 3 (1779). Standard edition in G. B. Hill (ed.), *The Lives of the Poets* (3 vols., Oxford, 1905): widely reprinted.
Walter Scott, 'The Life of John Dryden', in vol. 1 of his edition of the *Works* (above). Perhaps still the best life.
Charles E. Ward, *The Life of John Dryden* (Chapel Hill, NC, 1961).
James A. Winn, *The Life of John Dryden* (New Haven, Conn., forthcoming), promises to become the standard life.

REFERENCE

John M. Aden, *The Critical Opinions of John Dryden: A Dictionary* (Nashville, Tenn., 1963).
James and Helen Kinsley (eds.), *Dryden: The Critical Heritage* (1972). Critical comment to the end of the eighteenth century.
Hugh Macdonald, *John Dryden: A Bibliography of Early Editions and of Drydeniana* (Oxford, 1939).
Samuel Holt Monk, *John Dryden: A List of Critical Studies published from 1895 to 1948* (Minneapolis, 1950).
Guy Montgomery (ed.), *Concordance to the Poetical Works of John Dryden* (Berkeley and Los Angeles, 1957).
James M. Osborn, *John Dryden: Some Biographical Facts and Problems* (New York, 1940: revised edn., Gainsville, Fla., 1965).

H. T. Swedenberg, Jr. (ed.), *Essential Articles for the Study of John Dryden* (Hamden, Conn., 1966).

John A. Zamonski, *An Annotated Bibliography of John Dryden: Texts and Studies 1949–1973* (New York, 1975).

CRITICAL

Louis I. Bredvold, *The Intellectual Milieu of John Dryden: Studies in some Aspects of Seventeenth-Century Thought* (Ann Arbor, Mich., 1934).

Irvin Ehrenpreis and James M. Osborn, *John Dryden II* (Los Angeles, 1978).

T. S. Eliot, *John Dryden: The Poet, the Dramatist, the Critic* (New York, 1932).

William Frost, *Dryden and the Art of Translation* (New Haven, Conn., 1955).

James D. Garrison, *Dryden and the Tradition of Panegyric* (Berkeley and Los Angeles, 1975).

K. G. Hamilton, *John Dryden and the Poetry of Statement* (St Lucia, Qld, 1967).

Phillip Harth, *Contexts of Dryden's Thought* (Chicago, 1968).

Phillip Harth, Alan Fisher, and Ralph Cohen, *New Homage to John Dryden* (Los Angeles, 1983).

Arthur W. Hoffman, *John Dryden's Imagery* (Gainsville, Fla., 1962).

David Hopkins, *John Dryden* (Cambridge, 1986).

Bruce King (ed.), *Dryden's Mind and Art* (Edinburgh, 1969).

George McFadden, *Dryden: The Public Writer 1660–1685* (Princeton, NJ, 1978).

Earl Miner, *Dryden's Poetry* (Bloomington, Ind., 1967).

Earl Miner (ed.), *John Dryden* (1972).

William Myers, *Dryden* (1973).

Edward Pechter, *Dryden's Classical Theory of Literature* (Cambridge, 1975).

Paul Ramsey, *The Art of John Dryden* (Lexington, Ky., 1969).

Alan Roper, *Dryden's Poetic Kingdoms* (1965).

Bernard N. Schilling (ed.), *Dryden: A Collection of Critical Esays* (Englewood Cliffs, NJ, 1963).

Judith Sloman, *Dryden: The Poetics of Translation* (Toronto, 1985).

James Sutherland, *John Dryden: The Poet as Orator* (Glasgow, 1963).

Mark van Doren, *The Poetry of John Dryden* (New York, 1920).

Charles E. Ward and H. T. Swedenberg, *John Dryden* (Los Angeles, 1967).

David Wykes, *A Preface to Dryden* (1977).

Steven N. Zwicker, *Politics and Language in Dryden's Poetry: The Arts of Disguise* (Princeton, NJ, 1984).

GLOSSARY

abate, to lessen

abroad, out of the house, this way and that

accord, agreement

Achates, the faithful companion of Aeneas

achievement, escutcheon granted in memory of some distinguished feat

Acis, son of Faunus and a river-nymph; lover of Galatea

admire, to wonder, to wonder at

adown, down

adulterate, impure

Aeacus, father of Telamon, son of Jove, who made him judge of the underworld

Aeneas, Trojan prince, son of Anchises and Venus

Aeolus, god of winds

afeard, frightened

affect, to desire, to seek

affright, fear; to frighten

Agamemnon, son of Atreus, king of Mycenae, husband of Clytemnestra

Ajax, Greek warrior, son of Telamon

Alcides, Hercules

Alexander (356–323 BC), called 'the great', king of Macedonia, conqueror

alga, seaweed

alienate, to convey title to another

allege, to cite

allow, to approve

alluded, mentioned

Almain, Germany

alow, low

amain, vehemently; immediately

amaze(ment), wonder, consternation

Amazon, warrior-woman

ambient, surrounding

amomum, odoriferous plant

Anchises, father of Aeneas

annoy, vexation

Antenor, Trojan counsellor of Priam

antic, strange, grotesque

Apollo, god of poetry, prophecy, archery

appeach, to accuse

appoint, to equip

Araby, Arabia

Ares, Greek god of war

Argives, Greeks

Argo, Jason's ship

argument, subject-matter

Arion, Greek musician saved from drowning by dolphins

armado, army

armipotent, powerful in arms

arose, arisen

arow, in a row

Artemis, Diana

Ascanius, son of Aeneas

assay, to try

assert, to claim

assist, to be present

assistant, one who is present

Atalanta, virgin huntress

Athene, Greek goddess of war

Atlas, a titan who bore the world on his shoulders

atone, to reconcile; to harmonize

Atrides, Agamemnon, son of Atreus

attend, to wait for; to watch

attent, attentive

Augustus, grandnephew of Julius Caesar; Roman emperor

Aurora, goddess of dawn; the morning star

auspice, patronage

Auster, the south wind
authentic, authoritative, genuine
award, decision, judgement
awful, feeling awe, awe-inspiring

Bacchus, god of wine and revelry
bad, past tense of 'bid'
baffled, disgraced, dishonoured
band, bond
bane, poison
Batavian, Dutchman
bate, to lessen
beak, prow
beamy, large, wide
becoming, properly suiting
beholding, indebted
beldam, old woman
Belgian, Dutch
belie, to counterfeit, to feign
bent, hill, bare field; inclination, bias
beshrew, to wish a curse on
bias, predominant disposition
bid, to count
bid beads, to pray
big, pregnant
bilk, to cheat
blatant, bellowing
bleaky, bleak
bloomy, blooming
boding, auguring
boisterous, strong
bolt, shot
boon, what was asked
Boreas, the north wind
botch, boil
boult, to sift
bowels, offspring
box, wind instrument
boyism, puerility
brave, bully
breathe, to lance
breathed, rested
brew, to mix
bristled, covered with bristles
broke, broken
brown, dark

bubby, breast
buxom, pliant

Calchas, Greek seer
callow, unfledged
candid, white
cant, jargon
cantlet, fragment
caparison, cloth covering over saddle
captive, to make captive
car, chariot
careful, attended with care
cast, to require
Castor, one of the Argonauts who brought back the golden fleece
castor, a beaver
cates, provisions
caul, fold of fat
Cecilia, patron saint of music
cense, to scatter incense
censure, judgement
centaur, descendant of Centaurus, son of Ixion; half man, half horse
Ceres, goddess of harvests, agriculture
chaffer, to haggle
Cham, Ham, son of Noah
champian, of the country
chapman, merchant
charge, accusation
charger, horse; flat dish
Charon, ferryman of souls of dead across Styx
chaw, to chew
cheer, appearance; rich food, food and drink
cherubims, cherubs
cherubin, cherub
Chimera, three-headed monster
choir, company
chose, chosen
chuck, to call together with a clucking noise
circular, perfect, complete
cit, citizen (contemptuous)

clench, pun
clenched, clamped
clew, thread, cord, line
clip, to fly
clog, obstacle
close, secret
Clytemnestra, wife of Agamemnon, whom she killed
cockle, weed in corn
colewort, cabbage
coming, forward
common, general, universal
compass, to encircle
complexion, disposition (mental or physical)
composure, study, composition, reconciliation
conceit, conception, idea
concernment, care
conch, shell
condition, nature, character
confidant, person confided in
confining, bordering
congee, bow
conglobutate, compressed into a globe
connatural, of the same nature
conscious, knowing, privy to a secret
consequent, consequence
considering, reflecting
contain, to keep oneself within
contemn, to disdain, to despise
contended, contended for
content, on content, without examination
conversation, behaviour
converse, talk
convert, to change
convict, convicted, guilty
convince, to prove guilty
cot, dwelling
couch, to lay down
couchee, evening reception
counsel, sagacity
course, of course, to be expected
courser, horse

courtship, courtliness
convent, thicket, shelter
cozen, to deceive, cheat
cozenage, deception
crack, noise, crash
crazy, sickly, flawed
cross, across; to contradict
crotch, fork, support
crowd, rabble
cudden, clown, dolt
cumber, distress, encumbrance
Cupid, child god of love
curious, elaborately wrought
cyclops, a one-eyed giant
cymar, a long loose robe
Cynthia, Diana

dame, wife, woman, girl
dared, frightened, bewildered
darkling, in the dark
dart, spear
dash, to dilute
dauby, sticky
debonair, complaisant
decease, to die
deceive, to frustrate
deck, to adorn
decline, to deviate
deducement, deduction
defy, to repudiate
dell, dale
delude, to evade
deplore, to lament
describe, to survey
designment, design
despite, spite
destined, doomed
detort, to twist
Deucalion, son of Prometheus
devoted, doomed
Diana, daughter of Jupiter and Latona, goddess of childbirth and the moon; a virgin huntress
digestive, digesting, methodizing; aid to digestion
dint, force

Diomede, Diomedes, son of Tydeus; Greek warrior from Argos, friend of Ulysses
dip, to immerse
discover, to make known
disembogue, to flow, to pour out
disheir, to deprive of an heir
disherited, disinherited
dishonest, shameful, hideous
dismission, dismissal
dispose, disposal
distinctly, separately
distraction, madness
doddered, having lost branches through decay
dome, temple
doom, judgment; to decree
doted, doting
doubt, to suspect; to hesitate
doubtful, dubious, uncertain
dryad, wood-nymph
dug, pap, udder
dungeon, tower

earthy, of the earth
economy, arrangement
effect, result
Elysium, the abode of the blessed
embrown, to make brown
empiric, quack
emptiness, vacuum
endlong, straight on
engine, machinery, rack, mechanical device, instrument of war
enterprised, undertaken
enthusiast, fanatic
entranced, unconscious
envy, ill-will
Epicurus, Greek philosopher (341–271 BC) who was believed to advocate the pursuit of pleasure
equal, just, impartial, unruffled
equipage, equestrian trappings
erring, wandering
essay, first effort
estate, condition
etherial, heavenly

event, outcome
evidence, witness
evince, to prove
exact, perfect
exception, objection
exclusive, excluding
exequies, funeral rites
exercise, to try severely; to practise
exert, to bring out, to reveal
expect, to await
expire, to rush forth; to breathe out
explicate, to explain

fabric, building, structure
fact, action, deed, feat, crime
factor, agent; observer
falchion, broadsword
fame, rumour, reputation
fanatic, enthusiast, nonconformist
fane, temple
farmost, furthest off
fatal, fated, fateful
fell, fierce
fix, to determine
flame, passion
flix, fur
flood, river, sea
foin, to lunge, to thrust
forbear, to abstain from, to withhold
forbid, forbidden
forceful, strong
forelay, to waylay
forfeit, forfeited
forfent, to forbid
forgot, forgotten
forslow, to retard
forthright, straightforward
fougue, fury
fowl, bird
fraischeur, freshness
freak, whim
free, courteous, liberal, generous
frequent, crowded
fret, to erode
fright, to frighten
frock, tunic

front, face
frontless, shameless
froze, frozen
fry, to burn; to foam
Furies, snake-haired goddesses in Greek mythology
fustian, bombast

gage, pledge
Galatea, sea-nymph loved by Acis
gale, breeze
galled, rubbed, hurt
gaudry, finery
gaudy, splendid, fine
gem, bud
generous, noble, of good stock; strong, vigorous, high-spirited
genial, natural; what gives cheerfulness; contributing to propagation
genius, tutelary god
gentle, noble
give on, to go violently
glad, to gladden
glebe, land, a field
grabble, to grope after
gramercy, thanks
grant, permission
grateful, agreeable
gratulate, to congratulate
grave, to engrave
graver, engraver
grisly, horrible, nasty
gross, in gross, in a body
gull, to cheat
gust, the height of sensual pleasure
gyves, shackles

halcyon, legendary sea-bird bringing good weather
Hannibal, (247–182 BC) leader of the Carthaginians against Rome
happiness, successful aptitude
happy, propitious
hardly, with difficulty
hardy, courageous
haste, to make haste

hatch, to build
hateful, full of hate
heap, on a heap, into one mass
hear, to obey
heartless, destitute of courage
Hector, Trojan hero, son of Priam and Hecuba
Hecuba, wife of Priam, king of Troy
heir, to inherit
Helen, wife of Menelaus, king of Sparta; she was seduced by Paris
Helenus, Trojan prophet, son of Priam
Helicon, mountain in middle Greece sacred to the muses
Hercules, Heracles, Greek hero, son of Jove
Hermes, Mercury, messenger to the gods
hie, to hasten
hight, is called, was called
hind, farm labourer, peasant
hire, reward
hoary, white
hobby, hawk
hold, to abstain
homely, humble
honest, decent, respectable; glorious
hope, to expect; to wait for
horrid, bristling
Holland, cloth from Holland
humid, moist
humour, fancy, temperament
husband, one who tills the soil
Hydra, a many-headed monster
Hymen, god of marriage, son of Bacchus

Ilium, Troy
imp, to repair wings
impassible, insensible to pain
impassive, not subject to pain
impudent, immodestly presumptuous
inartificial, clumsy
increase, offspring; accumulation, growth; size

incumbent, pressing upon
Inde, India
indignant, angry
inform, to make known to
informing, animating
infuse, to pour
infused, poured in
inly, inwardly
inmate, foreign
innocent, harmless
innovate, to introduce for the first time
insensible, imperceptible
inspire, to breathe into or upon
instop, to fill up
insult, to exult proudly (over), to assail, to triumph over
insulting, exulting
intent, purpose; anxiously diligent
intercept, to stop, to obstruct, to cut off
interlude, play
invidious, envious, malignant
involve, to wrap
Io, daughter of Inachus
Iphigenia, daughter of Agamemnon
Isis, Egyptian goddess whose cult spread to Rome in the early centuries AD
Ixion, king of the Lapiths, and the first parricide; he fathered the centaurs

jambeux, armour for the legs
Janus, god of doors and gates
jar, harsh sound, clash, conflict
jealousy, suspicion
Jove, Jupiter, king of the gods, husband of Juno
joy, to make joyful; to rejoice
judgment, judge, critic
Julus, Ascanius, son of Aeneas
Juno, wife of Jove; goddess of marriage
Jupiter, Jove, Zeus, king of the gods
jupon, close-fitting padded tunic

ken, sight, gaze, view; to see, to know
kern, Irish peasant, Irish foot-soldier
kind, nature; by —, naturally
kindness, infatuation
kindred, kinship
kine, cows
king-at-arms, chief herald
knare, knot in a tree, gnarled protuberance
known, well known

Lacedaemon, Sparta
lade, to load
lag, laggard
lard, bacon
lares, household gods
large, generous, free
laund, clearing
lave, to wash, to bathe
laveer, to tack about
lazar, filthy deformed person, leper
leave, to leave off
lee, sediment of wine
leech, physician
left of, left by
legator, testator
lenitive, sedative
let, to hinder
levée, morning reception
liberal, free, generous
lightsome, gay
like, to please
limbec, alchemical still
limber, pliant
limbo, space
line, the equator
linstock, match-holder to light cannon
liquorish, randy
litter, to cover with straw
lively, like life
loll, to stretch out
loon, scoundrel
lubber, clumsy, idle fellow
lubric, slippery, immoral

Lucina, goddess of childbirth
lug, to pull, to drag
lust, desire
lusty, stout, vigorous
luxury, lust
Lyaeus, Bacchus

Macedon, Macedonian
machine, mechanical device
magazine, storehouse
make, to do
manes, dead ancestors who are worshipped
manifest of, guilty of
mannerly, well-mannered
many, crowd; retinue
marling, small tarred line for winding round ropes
Mars, god of war
martlet, swallow
massy, heavy
mate, equal; to equal
maw, gut, stomach
mechanic, vulgar
medicinal, having the power of healing
mend, to improve upon
mien, look, manner
millenary, of the millennium
Minerva, Athene, goddess of wisdom
mingle, mixture
mint, place of assay
mischief, a bringer of mischief; hurt, harm
miserable, fraught with misery
miss of, to fail to secure
missioner, ambassador, missionary
missive, missile
mistaken, misconceived
morion, helmet
motion, movement
mould, form, shape; material
moulted, affected by moulting
mumble, to chew toothlessly

naked, unprotected

nard, an aromatic plant
nasty, foul, filthy, unclean
naughty, bad, wicked
nereid, Greek sea-nymph
nervous, sinewy
Nestor, king of Pylos, of great age and wisdom
nice, (over-)fastidious, precise
nick, to hit; to censure
noblesse, nobility
noiseful, noisy
nose, to annoy; to detect
notched, marked
note, stigma
now, moment
numbers, metrical feet; versification
nursery, training place for young actors

oaf, simpleton
Oberon, king of the fairies
obligement, obligation
obnoxious, liable to punishment; subject to punishment
obscene, loathsome, horrid
obsequious, obedient, compliant
obstinate, determined
obtend, to allege; to hold out; to present in opposition
occasion, opportunity
o'erinformed, filled to overflowing
offend, to attack
officious, obliging, serviceable
once, some time
or . . . or, either . . . or
orient, bright-coloured; of superior value
ostent, portent
overpoise, superiority of weight
overwatched, weary from lack of sleep
out, to oust
outrageous, violent, excessive, extravagant
owe, to own
owing, indebted

pad, to go out robbing

pain, labour; to exert violently

Pales, Roman goddess of flocks

Pallas, Athene, goddess of war

palm, palm leaf worn as symbol of victory or triumph; victory

Pan, fertility god, god of the country

pard, panther, leopard

pardalis, panther

Paris, son of Priam; he abducted Helen, thus beginning the Trojan war

parlous, formidable, perilous

paronomasia, pun

parts, abilities

Patroclus, Greek warrior, Achilles' lover

pay o'er, to spread over

peccant, injurious

peevish, querulous

Pegasus, a winged horse

Pelides, Achilles, son of Peleus

pencil, paintbrush

pennon, small forked streamer

pent, confined

period, conclusion

Philomel, nightingale

philosophy, natural science

Phoebe, the moon-goddess Diana

Phoebus, Apollo, the sun

physic, to heal

picture, art of painting

pile, troop

pimp, to pander

pious, dutiful

place, office

plagiary, plagiarist

pleasure, to please

plume, pen; to pluck

plump, flock

Pluto, god of the underworld and death

poise, to weigh; to steady

Polyphemus, a cyclops, son of Neptune

pomp, spendour, show, pageant

pompous, splendid

ponderous, heavy

port, demeanour

Portunus, god of harbours

post, to travel, to hurry

pounce, claw, talon

practice, plot, intrigue

practise, to frequent

prelude, to usher in

presage, omen; to predict

pretend, to claim

prevail, to avail

prevent, to go before; to anticipate

Priam, king of Troy

pricking, riding

prime, spring, dawn, early morning

procedure, proceeding

process, trial

proclamation, proscription

prodigious, ominous

prodigy, portent; something monstrous

prolific, generative

Prometheus, a titan, son of Iapetus and father of Deucalion; he made mankind out of clay

prompt, to urge

proof, trial

proponent, propounding

Proserpine, goddess of the underworld

prospective, telescope

protend, to stretch out

protractive, protracting

prove, to try

provident, frugal

Pruce, Prussia

Punic wars, wars between Rome and Carthage, 264–146 BC

punk, whore

purchase, acquisition

purfled, bordered

pursue, to follow

pursuivant, attendant

quail-pipe, throat

quantity, length

quarry, game

queazy, squeamish
quick, flesh

rabbin, rabbi
rack, driving clouds; to move around
rage, fit of madness
rail, to abuse
raillery, abuse; mocking
ranch, to tear
rapt, snatched
rathe, early
raven, to hunger
rebate, to blunt
receipt, prescription
reckless, heedless
reconcile, to appease
recourse, expedient
recreant, cowardly; confessing oneself overcome
reeking, oozing blood
refer, to restore
reflective, reflected
regalio, choice entertainment
regorge, to swallow back
rehearsal, repetition, recitation
rehearse, to relate, to number
relent, to soften
remember, to remind
Remus, twin brother of Romulus
rend, to destroy
renown, to make renowned
repair, resort
repeat, to reseek
repose, to place as a trust
reprise, reprisal
republic, republican
require, to seek again; to demand
resolve, to melt, to dissolve
resort, assembly of people
rest, remainder
resume, to revoke
retire, to draw back; to take away
rheums, watery humours
rivelled, dried, shrivelled
Romulus, legendary founder of Rome
room, place

ropy, sticky
ruin, fall
ruminate, to muse on

salve, to excuse
Sappho, woman poet of Lesbos (sixth century BC)
savourly, with savour
sawtry, psaltery
scan, to examine closely
scandal, to scandalize
scape, escape; to escape
school, scholastic
scour, to race over
scud, to run away
secure of, secure from
Seneca, (c.4 BC–AD 65), Roman philosopher and writer
sennight, week
sentence, maxim
sequacious, following
seraphims, seraphs
sere-wood, dry wood
set, to put down a stake
several, various, different
shade, ghost
share, ploughshare
sheer, to cut, to cut through
shent, destroyed
shore, sewer
show, appearance; to appear
shrieval, sheriff's
sign, sign of the zodiac
simples, medicinal herbs
sincere, unmixed, pure: unhurt
sincerely, purely, without alloy
Sisyphus, son of Aeolus, famous for cunning; putative father of Ulysses
slavering, with running mouth
snip, scrap
sophisticate, artificial
sort, number, collection
sot, fool
sounding, resounding
souse, to swoop down
sovereign, all-powerful

spleen, melancholy; ill-humour

spring, to set going

sprite, spirit

spurn, to kick, to kick up

spurt, to squirt

squander, to disperse

stay, to wait

steep, steep height

steepy, steep

stew, brothel

stickle, to dispute

stickler, umpire

still, always

stint, proportion

stoles, robes

stoop, to swoop

strain, race, stock; to constrain

strait, narrow

strict, tight

stub, stump of a tree

stubborn, hard, rigid

studied, carefully contrived

studious of, concerned for

stum, new wine used for fermenting old

stupid, stupified

style, to describe

Styx, river of the underworld

suage, to assuage

submit, to lower

suborn, to procure secretly

succeed, to make to follow; to prosper

success, outcome

successive, by succession

suit, petition, favour, prayer

suiting, suitable

surcoat, outer coat

surety, pledge

sustain, to endure

swain, attendant, youth, labourer

swound, swoon

sylvan, of the woods

sylvans, wood gods

table, tablet

tale, number, calculation, tally

tally, counterpart

tarnish, to become stained

tawny, yellow

tax, to accuse

Telamon, son of Aeacus and father of Ajax

tell, to count

Tethys, wife of Oceanus

Thebes, Greek city, north-west of Athens

Themis, goddess of justice

theologue, theologian

Thetis, sea-nymph, wife of Peleus and mother of Achilles

thick, quickly following

threat, to threaten

throughly, thoroughly

timbrel, tambourine

timely, in time

tine, to kindle

tire, row of guns

tissue, rich cloth

toil, trap, snare

took, taken

towardly, promising

traduction, transmission

train, procession, retinue

tralineate, to get out of the line

translate, to transplant, to transfer

traverse, to oppose (in law)

treasonous, treasonable

treat, entertainment

trim, adornment

trine, two planets positioned 120° apart: a favourable omen

Triton, Greek merman, son of Neptune

trump, trumpet

try, to test

Tully, Cicero

tumbril, farm cart

tun, large cask

twist, cord

tympany, flatulence; swelling

Ulysses, Greek warrior, king of Ithaca; famous for cunning

uncouth, uncanny
unctuous, oily
uncumbered, unencumbered
undecent, unbecoming
ungodded, having no gods
unhappy, unfortunate
unhoped, unexpected
unknowing, ignorant
unlade, to empty
unrooted, insatiable
unshorn, unclipped
unspell, to disenchant
unteach, to destroy the teaching of
unthrift, prodigal
use, to practise, to be accustomed to

vale, valley
value, good opinion
vapours, hypochondria, hysteria
vare, staff
various, variable, many-coloured
Varro, Roman general defeated by Hannibal at Cannae
vegetive, having the potential of growth
Vesper, the evening star
vest, to clothe
villanize, to make villainous
vindicate, to defend against interference
virelay, poem or song with interlaced lines
virtue, power, force, valour
virtuoso, one skilled in some branch of knowledge

virtuous, curative
volume, coil
vulgar, common people

wain, waggon, carriage
wait, to attend, to accompany
wallow, a rolling walk
want, to be wanted; to be without
wanting, poor
wanton, frolicsome, sportive
watch, constable(s) patrolling streets at night
weal, well-being
weeds, clothes
well-breathed, with good lungs
whilom, formerly
whirlbat, gauntlet
wilder, to bewilder
wilding, crab-apple
wit, intelligence, felicity of thought, propriety of thought
without, outside
withstand, to gainsay, to oppose
witness, evidence
woe, woeful
wreak, to avenge
writhen, twisted

yeaning, just born

zealous, fanatical
Zephyr, the west wind
zone, girdle

INDEX OF TITLES AND FIRST LINES

(Titles are set in italic.)

POETRY

PROSE